# The Random and Fluctuating World

Celebrating Two Decades of
*Fluctuation and Noise Letters*

# The Random and Fluctuating World

## Celebrating Two Decades of *Fluctuation and Noise Letters*

Edited by

**P V E McClintock**

Lancaster University, UK

**L B Kish**

Texas A&M University, USA



World Scientific

NEW JERSEY • LONDON • SINGAPORE • BEIJING • SHANGHAI • HONG KONG • TAIPEI • CHENNAI • TOKYO

*Published by*

World Scientific Publishing Co. Pte. Ltd.
5 Toh Tuck Link, Singapore 596224
*USA office:* 27 Warren Street, Suite 401-402, Hackensack, NJ 07601
*UK office:* 57 Shelton Street, Covent Garden, London WC2H 9HE

**British Library Cataloguing-in-Publication Data**
A catalogue record for this book is available from the British Library.

**THE RANDOM AND FLUCTUATING WORLD**
**Celebrating Two Decades of Fluctuation and Noise Letters**



ISBN 978-981-125-213-6 (hardcover)
ISBN 978-981-125-214-3 (ebook for institutions)
ISBN 978-981-125-215-0 (ebook for individuals)

For any available supplementary material, please visit
https://www.worldscientific.com/worldscibooks/10.1142/12720#t=suppl

Desk Editor: Rhaimie Wahap

Typeset by Stallion Press
Email: enquiries@stallionpress.com

Printed in Singapore

# Acknowledgements

We acknowledge with thanks the multivariate contributions from a huge number of people, both to the journal and to the book. In particular, we are grateful to the authors who wrote almost everything, to the excellent handling editors with which Fluctuation and Noise Letters (FNL) has been blessed from its inception, and to World Scientific Publishing Company (WSPC) and its staff. We are much indebted to Sook Cheng Lim who managed the setting up of the journal, and have greatly appreciated the contributions of successive desk editors including Chelsea Chin, Patrick Tay, Gregory Lee, and particularly Rhaimie Wahap who has now served for almost half the life of the journal and who has assembled the selected papers into the coherent whole that follows. We also remember with gratitude all the authors who have passed on since their papers were published, including especially the late Nico van Kampen, FNL's first Honorary Editor.

# Preface

This book celebrates the first two decades of *Fluctuation and Noise Letters (FNL)*. The journal has its origins in the realisation by Laszlo B. Kish around the turn of the century that, despite the universal importance of fluctuations and noise throughout science, engineering, and human endeavour more generally, there was no specialist journal focused on the topic. With encouragement and help from Sergey Bezrukov who was visiting him at the time, he wrote a formal proposal for such a journal. This was accepted by Dr Phua Kok Khoo, Chairman of the World Scientic Publishing Company (WSPC) in 2000, and WSPC gave birth to *FNL* in March 2001. Laszlo was Editor-in-Chief, with Nico van Kampen as Honorary Editor. Production was from "camera-ready-copy" and all editorial communications were effected via personal e-mail. Laszlo became a second Honorary Editor in 2008 when Peter V. E. McClintock was appointed Editor-in-Chief. Editorial Manager was then introduced for administering the editorial processes and communications, together with copy-editing of all submissions.

Twenty years after its birth, *FNL* still remains the only scientic journal specialising in noise and fluctuations across the whole interdisciplinary area. This breadth makes good sense, because exactly the same problems of analysis and interpretation often crop up quite generally. Thus progress made in one field can also help in another seemingly very different area.

Looking back through the 20 years of *FNL*, we have been struck by the very high quality and significance of many of the papers. Our choices for this celebratory reprint volume have been based mainly on their scientific interest and importance, and we are acutely aware of the many other excellent papers that we have been unable to include. We hope that the collection will prove to be of as much interest to readers as it has been to us in assembling it.


**Peter V. E. McClintock, Editor-in-Chief**
Lancaster University

**Laszlo B. Kish, Honorary Edtitor**
Texas A&M University

September 2021

# Contents

# Introduction

P. V. E. McClintock
*Department of Physics, Lancaster University, Lancaster LA1 4YB, UK*
*p.v.e.mcclintock@lancaster.ac.uk*

L. B. Kish
*Department of Electrical & Computer Engineering, Texas A&M University, TX 77843-3112, USA*
*laszlokish@tamu.edu*

## 1. Fundamentals of Noise

Seeking the physical origin of a noise phenomenon often brings scientists closer to the fundamentals of their research topic. For example, the correct explanation of Brownian motion emerged centuries after Newton's discovery of his three laws of mechanics, and several decades after Maxwell-Boltzman statistical physics, even though they provided the foundations for Einstein's 1905 theory. Perhaps this is why a significant fraction of Physics Nobel Prize lectures contains the words "fluctuations" and/or "noise".

On the other hand, the scientific depth required for a fundamental understanding of the noise in a system often triggers debates between different scientists and schools where some or all of the parties may have partial or incorrect understanding of the reality.

The first two papers by van Kampen [1], and by Mannella and McClintock [2], focus on examples of the latter. In his Foreword to FNL (the historic first paper to be published in the journal) van Kampen discusses, amongst other things, why the Langevin/Focker-Plank stochastic differential equation descriptions can be invalid in certain nonlinear systems, such as a diode. Mannella and McClintock survey the related Ito-Stratonovich debate about which stochastic calculus is "correct" — a debate that seems to re-emerge every generation. They emphasise that both are correct, the Straonovich calculus being applicable to continuous systems, and the Ito calculus to discrete systems like the number of fish in a lake. These two papers describe how science often evolves via debates that lead to a deepening of understanding. Van Kampen's final sentence is prophetic: "My final conclusion is that there is ample material for this journal!"

Starting from the generalized Wiener process, Dubkov and Spagnolo [3] derive a new correlation formula for stochastic functionals. This allows them to obtain the general Kolmogorov equation directly from the Langevin equation with a non-Gaussian white noise source. The well-known Kolmogorov-Feller equation for purely discontinuous Markovian processes, and the fractional Fokker-Planck equation describing fast diffusion in the form of Levy flights, are special cases of this fundamental result.

Dykman and Golding [4] point out that a high-intensity, high-frequency, field can significantly influence escape rates from a potential well, in an unexpected way. In stochastic resonance (SR — see Section 3) systems the modulating field is in the low-frequency limit; thus the new features here enrich the understanding of SR, too. A reader familiar with tape-recorder technology may wonder if the high- frequency, high-amplitude, bias signal superimposed on the acoustic signal driving the magnetic recording head (to improve linearity) has a similar purpose?

During the analysis of measured fluctuations — ranging from biology to economics, — superimposed slow drifts (trends) often pose a problem by influencing the spectra, correlations and information seen at higher frequencies. Grafov *et al* [5] and Lin, *et al* [6] explore advanced techniques for reducing or removing trends: Grafov and his coworkers demonstrate that Chebishev spectra have strongly reduced trend sensitivity; Lin and coworkers introduce a new algorithm that diminishes several of the negative impacts of trends.

There have been many attempts and failed inventions attempting to circumvent the Second Law of Thermodynamics. Sooner or later, of course, physicists identify the error lying at the core of each scheme proposed. The paper of Nieuwenhuizen and Allahverdyan [7] provides a nice example of fundamental criticism of an experimental claim that had stirred up the pot by claiming a Second Law violation.

Everybody knows that no standing wave can exist in a cable below the frequency limit ("wave limit") where the half-wavelength matches the length of the cable. This is the basic design rule of musical instruments, too. And quantum physics and solid state physics also have these same limitations on their states, including the electron states in atomic shells. Yet, researchers working with microwaves often produce kinematic mathematical descriptions of slow processes with waves. But what does physics say, particularly statistical physics? The paper by Chen *et al.* [8] proves that the existence of physical waves below the wave limit would violate the Second Law.

## 2. Noise in Quantum Systems

The interpretation of quantum physical experiments and theory often triggers debates between experts from different schools, and these discussions have impacted the history of quantum science. Most of today's debates are about correlations, coincidence-phenomena, or other statistical properties of the quantum noise itself. Yet there have been disputes — sometimes historically long ones — even within the topic of transport processes manifesting quantum noise phenomena.

Van Kampen's article [9] discusses the proper approach to energy dissipation and damping in quantum systems. The basic formulations of quantum mechanics were Hamiltonian, that is, dissipation-free, which led to open questions and failed approaches later when physicists tried to introduce damping. A related issue is discussed by Alicki [10], who points out that quantum error correction fails for Hamiltonian models.

Klyuev *et al.* [11] discuss the spatial fluctuations of "monopoles" in spin ice, a disordered material with frustrated ferromagnetic interactions that imitate the existence of magnetic monopoles.

Motivated by the need for theoretical tools to study the relationship between the AC spectrum and electron dynamics in nanoelectronic devices, Albareda *et al.* [12] develop formalism using both standard and Bohmian quantum mechanics, obtaining identical results. They point out that, for many-body systems, Bohmian trajectories make numerical calculation of the total current easier and illuminate the interplay between quantum electron-dynamics and displacement currents. Note, orthodox quantum theorist say that these trajectories do not exist and they can have at most a statistical meaning. Perhaps if they were called "Bohmian virtual trajectories", both parties (Bohmian and Orthodox) could agree?

The three debate articles by Cohen and Hrasko [13–15] show that the much-heralded quantum teleportation (of an information bit) is not quite so special as portrayed by the media. The well-known "weirdness" of conditional probabilities is responsible for producing many of its features and even classical physical systems, such as coins in boxes, do so, albeit with less than 100% fidelity. (Note, on the other hand, that practical single-photon quantum measurements typically have even less fidelity than the indicated classical values). As the conclusion of the discussion, it is pointed out that several key features of the quantum teleportation process, aspects that are commonly considered to be non-classical, are still present in Cohen's purely classical version of teleportation. The implication of this is not that quantum teleportation requires only classical transfer of information, but rather that the role of the non-classical component of information transfer within the quantum teleportation process is subtler than is generally assumed.

Green and Das have been arguing [16] that there are flaws in the Landauer-Büttiker formalism of mesoscopic transport. For a long time, they were alone in this, but not any more. Their view has become accepted and recognized by much of the community during the last decade.

And finally in this section, an historical enigma: can the thermal noise of resistors exist at zero temperature? The view of Kish *et al.* [17] is that, whether or not Johnson-Nyquist noise persists at temperatures approaching zero depends on the way in which quantum measurement is used for its detection. Whenever it is visible, it is an uncertainty relation artifact of the instrumentation. Even standard Fermi- Dirac quantum statistics prohibits its fundamental existence.

## 3. Noise in Complex Systems

Noise in complex systems is, in general, hard to deal with because of the double difficulty of the noise and the complexity, sometimes also including dynamical chaos.

Lindner *et al.* [18] study the diffusive motion of an overdamped Brownian particle in a tilted periodic (wasboard-like) potential — which represents a paradigm for flow in ratchets of all kinds, including some of the internal machinery of biological cells. Mapping the continuous dynamics onto a discrete cumulative process they find exact expressions for the diffusion coefficient and the Péclet number (ratio of the advection and diffusive transport rates) characterizing the transport and show that there is a characteristic tilt at which optimized transport occurs for a given noise intensity.

Stochastic resonance (SR), in which the amplitude and/or signal/noise ratio (SNR) of a weak, low-frequency, periodic signal in a noisy nonlinear system can be enhanced by adding extra noise, was still under intense investigation during the early years of FNL. It came to be appreciated that the SNR of a given signal cannot be improved by passing it through a "stochastic resonator", as some had hoped, but that SR can be useful where the signal is initially generated in a nonlinear system of an appropriate kind. Makra *et al.* [19] investigate SR in a Schmitt trigger circuit and in a double well potential, to see what SNR gains SR can provide in these systems, and demonstrate SNR gains much greater than unity for both.

Dating back to 1994, Parrondo's paradox involves a counter-intuitive "survival of the weakest" principle, and refers to circumstances where two losing situations can be combined in order to win. It is exemplified by simple coin tossing games. In his review, Abbott [20] first explains the paradox and then points out the deep connections that exist between the Parrondo effect and a whole host of diverse phenomena that arise in physics and beyond. These include the flashing Brownian ratchet, stochastic resonance, the ratchet and pawl machine, the thermodynamics of games of chance, the Brazil nut paradox, the interplay of redundancy and pleiotropy, longshore drift on beaches, costly signalling in biology, the two-girlfriend paradox, and stock market trading. His two take-home messages are (i) that the process of switching is a nonlinearity, so that switching can radically alter the overall system behavior, and (ii) that the interaction between noise and asymmetry can give rise to directed motion under non-equilibrium conditions, even against a gradient.

Buceta *et al.* [21] build on these ideas by proposing a mechanism whereby a random alternation of two distinct dynamics, each of which by itself leads to a homogeneous state, can generate complex, ordered, patterns and structures. The authors point out the relevance of their study to other situations where pattern formation occurs, such as reaction-diffusion systems.

A novel approach to the treatment of separatrix chaos has enabled Soskin *et al.* [22] to address and solve some long-standing problems in that area. By introducing a new criterion for chaos in the separatrix map, they find analytically the maximum width of the separatrix chaotic layer and describe an example of the dramatic facilitation of the onset of global chaos between separatrices.

Reports of rogue waves on the ocean — fluctuations very much larger than those around them — were traditionally treated sceptically as "mariners' tales" — until relatively recently when the famous New Year Wave was recorded quantitatively with modern instrumentation on the Draupner oil platform (1995). Since then their reality has become well established, and they have been identified in other systems too, including superfluid helium, optics, and Bose-Einstein condensates. Residori *et al.* [23] review experimental results on the statistical properties of giant optical waves in spatially extended systems. They consider different experimental set-ups and discuss the ingredients that are needed for the creation of optical rogue waves.

## 4. Noise in Biological Systems

Without the rare random events that give rise to occasional mutations in DNA there would be no evolution, no life, and certainly no us. Noise and fluctuations are of crucial importance in biology. This is true on all length scales, ranging from chemistry where reaction rates are controlled by thermal activation over energy barriers, through biological cells and their subcellular machinery, interactions between whole systems in an organism, right up to the level of population dynamics. Biological systems are typically characterised by strong internal interactions. So the physicist's usual approach of focusing on a particular subsystem and modelling it in isolation can often be a poor approximation. For the same reason, fluctuations and noise in biology tend to be complicated and the assumption of Gaussian white noise may also be a poor approximation. Conversely, the form of the fluctuations usually carries a lot of information about the underlying biology.

Bezrukov [24] discusses how analysis of the noise in single protein nanocontacts can provides important information about their structure and function, allowing us to study protein dynamics and interactions with different solutes at the single molecule level. In addition to yielding deeper scientific understanding, the noise analysis brings practical consequences by illuminating the molecular mechanisms of human illnesses and the underlying mechanisms of drugs.

Selective conduction by ion channels in cellular membranes, where the favoured ion may be selected over other ions by a ratio of 1000:1, is needed for all forms of life. Tindjong *et al.* [25] describe a nonequilibrium rate model for the stochastic permeation of an ion through an open ion channel, taking explicit account of the relevant electrostatic interactions and leading to transition rates in satisfactory agreement with biological experiments. This general approach subsequently led on to the discovery of ionic Coulomb blockade in biological ion channels and to a full-blooded statistical theory of the permeation process.

Ions are used in the sensing of phage-triggered ion cascades (SEPTIC) for fast bacteral identification, based on fluctuation-enhanced sensing in a fluid medium. The concept was introduced by Kish and Dobozi-King in 2004. Kish *et al.* [26] discuss the sensitivity and detection limits of SEPTIC and show that, under optimal conditions, 200 bacteria is a sufficient number to be detected/identified reliably.

Suki *et al.* [27] consider fluctuations in cardio-pulmonary physiology, relating the scaling behaviour to the structure-function relationship in normal and diseased states of the system. They discuss the application of wavelets and multifractal approaches, avalanche behavior of airway openings and the associated crackling sound, as well as the relationship between the observed scaling properties and the design features of the pulmonary vascular tree. They also show how the network failure of lung tissue leads to emphysema.

There has been significant effort to detect the occurrence of stochastic resonance (see Section 3) in biological systems, and to see how nature may perhaps use SR to improve efficiency. Stocks et al. [28] show that a variant of the phenomenon known as suprathreshold SR can improve speech comprehension by enhancing information transmission along the cochlear nerve. There is evidence from experiments on paddlefish [29] that animals can make use of the increased sensory information provided by SR in their sensory nervous systems, as demonstrated for a vital unlearned natural behavior, namely feeding. By investigating a stochastic generalization of the Hodgkin–Huxley model, Schmid *et al.* [30] found evidence for SR due to the intrinsic channel noise where there were sufficiently large clusters of ion channels whose internal noise strength was below its optimal value.

Many, perhaps most, biological processes are oscillatory in character. The resultant fluctuations are also oscillatory, but with the special feature that they are in general non-periodic. This is attributable to the strong interactions within biological systems — so that any given oscillatory process is subject both to the influences of its changing environment and to the effects of other local processes that may also be oscillatory, usually at different frequencies. Even where a biological oscillation is very strong, e.g. the heartbeat measured in arterial blood pressure, its frequency and amplitude are time-varying, and require special analysis methods. The interactions between different oscillations can result in their synchronisation, not necessarily on a 1:1 basis but sometimes e.g. 1:4 in cardio-pulminary synchronisation (4 heartbeats per breath). Synchronisation can be maintained even if the frequencies are varying. Rosenblum *et al.* [31] review the application of synchronization theory to the analysis of multivariate biological signals. They address the problem of phase estimation from measured data and the detection and quantification of weak inter-oscillator interaction, as well as quantification of the direction of coupling, and they discuss the advantages and limitations of the approach.

The phase coherence between two signals is a more subtle quantity than synchronisation. It takes a value between zero and unity and quantifies the tendency for the phase difference between the two signals at a particular frequency to remain constant. A phase coherence of unity would imply synchronisation; at the other extreme, where two signals are unrelated, their phase difference changes continuously in time and so their phase coherence is near zero. Bernjak *et al.* [32] establish the wavelet phase coherence (allowing for time variability) between fluctuations in skin blood flow and in oxygen saturation. They find clear evidence for coherence over a range of frequencies, thus yielding the first detailed insight into the dynamics of blood oxygenation.

## 5. Noise in Materials, Circuitry, Devices and Sensing

Electronic noise appears in materials, circuitry and devices where its role is not always negative. Noise can even have practical applications, such as in sensing. A well-known example is thermal noise thermometry. Obviously, the topic of electronic noise is extremely wide, so that we can sample only a small fraction of the published papers, even within FNL. Our choice here is designed to give the flavor of research within a selected set of sub-topics.

In the first paper that we consider, Sundqvist *et al.* [33] discuss the physical aspects of Chua's seminal memristor theory of 1971, which claims to introduce the "missing" passive circuit element (additional to resistors, inductors and capacitors). The paper [33] proves that Chua's mathematical model is unphysical because its thermodynamics is undefined due to its lack of thermal noise. In other words, Chua's equations cannot address the question of whether the device is passive. They are therefore unable to define the "missing" passive circuit element. The topic has triggered several debates during the last 50 years and, to date, this paper seems to have concluded the discussion.

Barkhausen noise during the magnetization/demagnetization of ferromagnetic materials is due to jumps in domain walls and avalanches of these phenomena measured by a pickup coil. Mills *et al.* [34] use a modified version of the Haar transform to observe the markers of time asymmetry. The method works even with overlapping avalanches, without need of an ability to detect discrete pulses.

Electronic shot noise is a current noise that arises when charge carriers are crossing a potential barrier. It is the simplest quantum noise because its spectrum is proportional to the value of elementary charge $q$ carrying the current. In classical, Maxwellian, continuum charge transport this noise would not exist. In a typical case the current noise spectrum is $S_i = 2FIq$, where $I$ is the DC current value and the Fano factor $F = 1$ represents the situation where the electrons are crossing the potential barrier independently (Poisson statistics). The present frontier of research lies in situations where the Fano factor differs from 1. For example, when the charge carriers form groups when crossing (correlated crossing), such as in Josephson junctions, $F$ is greater than 1. In the opposite situation — anticorrelated crossing — $F$ can be less than 1. However, there are also other situations of anomalous Fano factor. Marconcini *et al.* [35] discuss aspects of diffusive transport (where there is no shot noise). They consider the thickness of the potential barrier in relation to the Debye length in the context of Fano factor reduction. Gokce *et al.* [36] describe an interesting experimental study with magnetic tunnel junctions and their arrays. One interesting finding is that serial junctions show a voltage-noise-based Fano factor that scales inversely with the number of junctions.

The history of $1/f$ (flicker) noise research is full of debates. This is also true of device noise research. Vandamme [37] discusses the origin of $1/f$ noise in MOSFETs. The famous (but incorrect) McWhorter model from 1957 was probably the first to assume that the number of free electrons was the core fluctuating quantity (due to interface trapping). The alternative explanation is based on mobility fluctuations, which is not at all an obvious matter. Vandamme's paper is an effort to clarify the situation.

Rumyantsev *et al.* [38] study and compare different types of heterostructure field effect transistors (FETs). The results indicate that, in these devices, the core fluctuating quantity is the number of free electrons (due to trapping).

Danneville *et al.* [39] offer an explanation of how the 1/*f* noise, which dominates at low frequencies, can become an issue in the microwave frequency range. The low-frequency components are transferred up to the high-frequency regime by nonlinear mixing. It is the nonlinear features of FETs that are the vehicles making this happen.

The electrical breakdown of PN junctions with breakdown voltage beyond 7V are dominated by the avalanche effect, which is very noisy, and it surprisingly shows strong 1/*f* noise, too. Marinov and Deen [40] offer a complex theoretical model utilizing microplasma noise, diffusion and threshold phenomena to explain the low-frequency dynamics.

The next two papers in this section deal with electrical circuitry for low-noise applications: amplifiers and cold resistor circuitries. The design of low-noise amplifiers for the low-frequency range is challenging for a variety of reasons, such as 1/*f* noise, drift and ageing. Cold resistor circuitry imitates a resistor with strongly reduced noise temperature, but there have also been contradicting theories. The paper by Ciofi *et al.* [41] presents the state-of-the-art of low-frequency, low-noise, amplifier design. Song and Kish [42] deduce a systematic theory of cold resistor circuitry with a useful equation for design purposes.

The following two papers utilize noise measurements for sensing. Fluctuation-Enhanced Sensing (FES) uses the noise spectral pattern of chemical sensors to extract chemical information. Smulko *et al.* [43] explore new, higher order, statistical tools, including bispectra, to enrich the extracted information. However, these tools work only if non-Gaussian components are present in the noise. In the other paper, Smulko [44] explores noise measurements for corrosion detection: a tool that can potentially be useful to avoid catastrophes due to corroded building elements.

Finally, the last paper in this section is about circuitry on a much larger scale: power line networks. Crucitti *et al.* [45] analyze practical the power line networks in France, Italy and Spain and identify the "hot-spots" where damage would make the largest impact on the network. They also propose improvements. (We note here: it is lucky that terrorists do not read academic noise papers.)

## 6. Noise, Computation and Energy Dissipation

Logic operations involve switching events and each such event dissipates energy. Every laptop-user who has experienced an overheated processor knows that energy dissipation in computing is a potentially serious limitation of computing performance. Google's mean computation power reached 12.7 terawatt-hours in 2019, up from 2.86 terawatt-hours in 2011.

The related scientific research has two main frontiers: Finding the fundamental lower limit of energy dissipation in logic operations; and inventing special-purpose computing schemes (such as quantum computing) that require progressively less dissipation to carry out the corresponding classical computation.

The history of the fundamental lowest limit of energy dissipation is rich in controversies and mistakes and famous physicists had also contributed such as Leo Szilard and Leon Brillouin. The former came up with the famous $k_BT\ln(2)$ lower limit, which is nowadays mistakenly called the "Landauer limit". Szilard presented his result as the minimum energy dissipation to generate 1 bit of information. In his related papers, Brillouin first confirmed the Szilard result but later refined it to the correct $k_BT\ln(1/\varepsilon)$ value, where $\varepsilon$ is the bit error probability (BEP) due to thermal noise, and the result holds in the small error limit. According to information theory, $\varepsilon = 0.5$ represents the zero information case and, even though 0.5 is not a small error probability, this fact also indicates that the Szilard result is only a lower limit but not the ultimate one, the "highest lower limit". Recently it was shown by Kish that the Brillouin limit holds only within the autocorrelation time of the thermal excitation and that, for longer observations, the formula has a logarithmic correction by the observation time. Finally, it is worthwhile to note that the energy dissipation vs. information entropy reduction is a flawed approach because in accordance with Renyi's comments on information entropy, deterministic systems, such as a digital computer, do not change the information entropy. Thus there is no justification to use the above formulas for information entropy. Correctly, the proper wording should be "number of switching operations". Many leading schools and papers in top journals fell into this trap. Most notable among these is the Landauer principle, which claims energy dissipation for information erasure.

The first paper [46] in this section illustrates the typical kind of debate that arises about the energy dissipation limits. It points out that a paper in a leading magazine neglected the energy dissipation in the control circuitry of a cantilever-based logic gate. This is the reason for the published dissipation value being unphysically low.

The papers that follow deal with various issues in noise-based logic, a classical physical logic scheme that provides exponential performance improvement for special-purpose computational tasks, similarly to quantum computers. An $N$-bit "quantum-computer-mimicking" noise-based logic (QCM-NBL) represents an exponentially large $2N$-dimensional Hilbert space where the vectors are all the $N$-bit long binary numbers.

Wen and Kish [47] show why noise as an information carrier has both physical and complexity advantages compared to deterministic signals in QCM-NBL.

Peper and Kish [48] demonstrate that the binary version of noise-based logic is universal.

The paper by Stacho [49] introduces an algorithmic method for decoding the product signal strings that carry the *N*-bit binary numbers in QCM-NBL.

Kish and Daugherity [50] introduce new types of logic gate construction for QCM-NBL. As an example, they realize the CNOT gates and their non-commutative chain operations, which had been a tough call.

## 7. Noise in Finance

One has only to open a newspaper to appreciate that almost all financial quantities, e.g. currency exchange rates or share prices, are noisy and subject to continuing fluctuations. Sometimes these fluctuations have an obvious cause, like wars or pandemics, but more often they seem at first sight to be spontaneous and intrinsic to the international financial system in its current form. To try to understand the fluctuations and model their effects, it is natural to apply the methods developed and used successfully for analysing fluctuations in scientific and engineering contexts.

By applying methods drawn from statistical physics to 16 stocks traded on the London Stock Exchange, Lillo and Farmer demonstrate [51] the crucially important role played by fluctuations in financial liquidity in determining large price changes. Their findings not only illuminate the mechanisms giving rise to "fat tails" (i.e. greater than Gaussian) in the probability distribution and long memory processes, but also offer some understanding of what determines financial risks and clues about how to reduce them.

Han *et al.* [52] investigate the correlation structure of the Chinese stock market around the time of the Great Crash in 2008. They apply random matrix theory to show that Chinese stocks had stronger average correlation and partial correlation in 2008 than in 2007, that the average partial correlation was significantly weaker than the average correlation in each period and, accordingly, that the largest eigenvalue of the correlation matrix is greater than that of the partial correlation matrix in each period. They point out that the calculated eigenvalues could in principle be used to construct a measure of systemic risk and that the eigenportfolio of the largest eigenvalue could provide the basis of a new stock market index.

## 8. Noise and Security in Communications

The heart of any symmetric key-based secure communication system is the secret key that is used by Alice and Bob (at the communicating stations) to encrypt and decrypt the messages. The system cannot be more secure than its key. Alice and Bob must possess the same key (their shared secret), and this key must be refreshed frequently by secure key exchange/distribution to reduce the chance of statistical attacks.

Most commercial secure communication systems utilize the exchange of random numbers obtained by proper algorithms. Processing these numbers will eventually result in a common secret key on Alice's and Bob's sides. Eve (the eavesdropper) can record these communicated random numbers, so that she has all the information needed to calculate the key. However the protocol utilizes such problems as the solution for Eve being mathematically hard (requiring an exponentially long time and/or hardware complexity) while, for Alice and Bob (who know also the input of their algorithm and their hardware), it is much simpler (polynomial time/hardware complexity). Because Eve has all the information required for extracting the key, these schemes offer zero security at the information entropy level. Yet, with the conditions that Eve's computational power is limited, and that there is no efficient algorithm (requiring polynomial time/hardware complexity) immediately available to her, the system is conditionally secure. These schemes offer no future proof of their security, and even their security in the present is based on unproven assumptions. However, they are cheap and easy to use.

Real security: i.e. unconditional (information theoretic) security is at the frontier of security research. In this case, neither infinite calculating power, nor infinite time for processing the information, can compromise secrecy. But these schemes do have their own issues: bit errors, information leak towards Eve, a much higher price, and less versatility than conditionally secure schemes. Unconditionally secure key exchange requires specific hardware which itself offers attack opportunities for Eve because building elements are never ideal. Thus a thorough security proof is needed and, even then, future surprises can still happen with device issues that had not been foreseen and considered earlier. A good example of an unconditionally secure key exchange is the well-known quantum key distribution (QKD) scheme with all the historical attacks and full cracks against their former commercial versions.

In this short selection of papers on unconditional security, that by Kim [53] proposes a new key exchange method utilizing phase and intensity fluctuations of the communicated signals. Its ultimate security and possible ways of attack are unknown, yet it is an interesting and original scheme.

The rest of the papers deal with various aspects of the Kirchhoff-law-Johnson-noise (KLJN) key exchange protocol, which is based on hardware mimicking the thermal noise of resistors. Mingesz *et al.* [54] prove the earlier conjecture that only Gaussian noises guarantee perfect security. The next paper, by Kish [55], reports the surprising finding that the KLJN scheme is naturally immune against the famous man-in-the-middle attack. The last paper by Cao *et al.* [56] explores how the KLJN system can offer unconditional security for autonomous vehicle systems.

## References

1. N. G. van Kampen, "Foreword: The theory of noise", *Fluct. Noise Lett.* 1 (2001) 3–7.
2. R. Mannella and P. V. E. McClintock, "Ito^ versus Stratonovich: 30 years later", *Fluct. Noise Lett.* 11 (2012) 1240010.
3. A. Dubkov and B. Spagnolo, "Generalized Wiener process and Kolmogorov's equation for diffusion induced by non-Gaussian noise source", *Fluct. Noise Lett.* 5 (2005) L267– L274.
4. M. I. Dykman and B. Golding, "Controlling activated processes", *Fluct. Noise Lett.* 1 (2001) C1–C6.
5. B. M. Grafov, A. L. Kluev, T. B. Kabanova and A. D. Davydov, "Chebyshev spectra, resistance to trend of random noise", *Fluct. Noise Lett.* 17 (2018) 1850028.
6. A. Lin, P. Shang and H. Ma, "The orthogonal V–system detrended fluctuation analysis", *Fluct. Noise Lett.* 10 (2011) 189–206.
7. T. M. Nieuwenhuizen and A. E. Allahverdyan, "Comment on "Experimental demonstration of violations of the Second Law of Thermodynamics for small systems and short time scales"", *Fluct. Noise Lett.* 5 (2005) C23–C24.
8. H. Chen, L. B. Kish, C. Granqvist and G. Schmera, "Do electromagnetic waves exist in a short cable at low frequencies? What does physics say?", *Fluct. Noise Lett.* 13 (2014) 1450016.
9. N. G. van Kampen, "Damping and noise in quantum mechanics", *Fluct. Noise Lett.* 1 (2001) C7–C13.
10. R. Alicki, "Quantum error correction fails for Hamiltonian models", *Fluct. Noise Lett.* 6 (2006) C23–C28.
11. A. V. Klyuev, M. I. Ryzhkin and A. V. Yakimov, "Statistics of fluctuations of magnetic monopole concentration in spin ice", *Fluct. Noise Lett.* 16 (2017) 1750035.
12. G. Albareda, F. L. Traversa, A. Benali and X. Oriols, "Computation of quantum electrical currents through the Ramo–Shockley–Pellegrini theorem with trajectories", *Fluct. Noise Lett.* 11 (2012) 1242008.
13. O. Cohen, "Classical teleportation of classical states", *Fluct. Noise Lett.* 6 (2006) C1–C8.
14. P. Hrasko', "On Cohen's classical teleportation of classical states", *Fluct. Noise Lett.* 6 (2006) C29–C30.
15. O. Cohen, "Reply to: 'On Cohen's classical teleportation of classical states'", *Fluct. Noise Lett.* 6 (2006) C31–C32.
16. F. Green and M. P. Das, "Noise and transport in mesoscopics: Physics beyond the Landauer-Büttiker formalism", *Fluct. Noise Lett.* 5 (2005) C1–C14.
17. L. B. Kish, G. B. Niklasson and C. Granqvist, "Zero thermal noise in resistors at zero temperature", *Fluct. Noise Lett.* 15 (2016) 1640001.

18. B. Lindner, M. Kostur and L. Schimansky-Geier, "Optimal diffusive transport in a tilted periodic potential", *Fluct. Noise Lett.* 1 (2001) R25–R39.
19. P. Makra, Z. Gingl and L. B. Kish, "Signal-to-noise ratio gain in non-dynamical and dynamical bistable stochastic resonators", *Fluct. Noise Lett.* 2 (2002) L147–L155.
20. D. Abbott, "Asymmetry and disorder: A decade of Parrondo's paradox", *Fluct. Noise Lett.* 9 (2010) 129–156.
21. J. Buceta, K. Lindenberg and J. M. R. Parrondo, "Spatial patterns induced by random switching", *Fluct. Noise Lett.* 2 (2002) L21–L29.
22. S. M. Soskin, R. Mannella, O. M. Yevtushenko, I. A. Khovanov and P. V. E. McClintock, "A new approach to the treatment of separatrix chaos", *Fluct. Noise Lett.* 11 (2012) 1240002.
23. S. Residori, U. Bortolozzo, A. Montina, F. Lenzini and F. T. Arecchi, "Rogue waves in spatially extended optical systems", *Fluct. Noise Lett.* 11 (2012) 1240014.
24. S. M. Bezrukov, "Noise analysis in studies of protein dynamics and molecular transport", *Fluct. Noise Lett.* 4 (2004) L23–L31.
25. R. Tindjong, I. K. Kaufman, P. V. E. McClintock, D. G. Luchinsky and R. S. Eisenberg, "Nonequilibrium rate theory for conduction in open ion channels", *Fluct. Noise Lett.* 11 (2012) 1240016.
26. L. B. Kish, M. Cheng, J. U. Kim, S. Seo, M. D. King, R. Young and A. Der, "Estimation of detection limits of the phage-invasion based identification of bacteria", *Fluct. Noise Lett.* 5 (2005) L105–L108.
27. B. Suki, A. M. Alencar, U. R. S. Frey, P. C. Ivanov, S. V. Buldyrev, A. Majumbdar, H. E. Stanley, C. A. Dawso, G. S. Krenz and M. Mishima, "Fluctuations, noise and scaling in the cardio-pulmonary system", *Fluct. Noise Lett.* 3 (2003) R1–R25.
28. N. G. Stocks, D. Allingham and R. P. Morse, "The application of suprathreshold stochastic resonance to cochlear implant coding", *Fluct. Noise Lett.* 2 (2002) L169– L181.
29. D. F. Russell, A. Tucker, B. A. Wettring, A. B. Neiman and L. Wilkens, "Noise effects on the electrosense-mediated feeding behavior of small paddlefish", *Fluct. Noise Lett.* 1 (2001) L71–L86.
30. G. Schmid, I. Goychuk, P. Hänggi and P. Jung, "Stochastic resonance and optimal clustering for assemblies of ion channels", *Fluct. Noise Lett.* 4 (2004) L33–L42.
31. M. G. Rosenblum, A. S. Pikovsky and J. Kurths, "Synchronization approach to analysis of biological systems", *Fluct. Noise Lett.* 4 (2004) L53–L62.
32. A. Bernjak, A. Stefanovska, P. V. E. McClintock, P. J. Owen-Lynch and P. B. M. Clarkson, "Coherence between fluctuations in blood flow and oxygen saturation", *Fluct. Noise Lett.* 11 (2012) 1240013.
33. K. M. Sundqvist, D. K. Ferry and L. B. Kish, "Memristor equations: Incomplete physics and undefined passivity/activity", *Fluct. Noise Lett.* 16 (2017) 1771001.
34. C. Mills, M. B. Weissman and G. Durin, "Use of the one-and-a-half-spectrum to determine Barkhausen noise asymmetry", *Fluct. Noise* Lett. 7 (2007) L13–L18.
35. P. Marconcini, M. Macucci, D. Logoteta and M. Totaro, "Is the regime with shot noise suppression by a factor 1/3 achievable in semiconductor devices with mesoscopic dimensions?", *Fluct. Noise Lett.* 11 (2012) 1240012.

36. A. Gokce, R. Stearrett, E. R. Nowak and C. Nordman, "Shot noise suppression in individual and series arrays of magnetic tunnel junctions", *Fluct. Noise Lett.* 10 (2011) 381–394.
37. L. K. J. Vandamme, "On the origin of 1/*f* noise in MOSFETs", *Fluct. Noise Lett.* 7 (2007) L321–L339.
38. S. L. Rumyantsev, N. Pala, M. S. Shur, M. E. Levinshtein, P. A. Ivanov, M. A. Khan, G. Simin and J. Yang, "Low-frequency noise in AlGaN/GaN heterostructure field effect transistors and 7 metal oxide semiconductor heterostructure field effect transistors", *Fluct. Noise Lett.* 1 (2001) L221–L226.
39. F. Danneville, B. Tamen, A. Cappy, J. Juraver, O. Llopis and J. Graffeuil, "Low frequency noise conversion in FETs under nonlinear operation", *Fluct. Noise Lett.* 1 (2001) L189–L195.
40. O. Marinov and M. J. Deen, "Physical model for low frequency noise in avalanche breakdown of PN junctions", *Fluct. Noise Lett.* 4 (2004) L287–L296.
41. C. Ciofi, G. Giusi and G. S. B. Neri, "Dedicated instrumentation for high sensitivity, low frequency noise measurement systems", *Fluct. Noise Lett.* 4 (2004) L385–L402.
42. J. Song and L. B. Kish, "On the theory and design of cold resistors", *Fluct. Noise Lett.* 20 (2021) 2150001.
43. J. Smulko, C. Granqvist and L. B. Kish., "On the statistical analysis of noise in chemical sensors and its application for sensing", *Fluct. Noise Lett.* 1 (2001) L147– L153.
44. J. Smulko, "Methods of electrochemical noise analysis for investigation of corrosion processes", *Fluct. Noise Lett.* 6 (2006) R1–R9.
45. P. Crucitti, V. Latora and M. Marchiori, "Locating critical lines in high-voltage electrical power grids", *Fluct. Noise Lett.* 5 (2005) L201–L208.
46. L. B. Kish, "Comments on "sub-KBT micro-electromechanical irreversible logic gate", *Fluct. Noise Lett.* 15 (2016) 1620001.
47. H. Wen and L. B. Kish, "Noise-based logic: Why noise? A comparative study of the necessity of randomness out of orthogonality", *Fluct. Noise Lett.* 11 (2012) 1250021.
48. F. Peper and L. B. Kish, "Instantaneous, non-squeezed, noise-based logic", *Fluct. Noise Lett.* 10 (2011) 231–237.
49. L. L. Stachó, "Fast measurement of hyperspace vectors in noise-based logic", *Fluct. Noise Lett.* 11 (2012) 1250001.
50. L. B. Kish and W. C. Daugherity, "Noise-based logic gates by operations on the reference system", *Fluct. Noise Lett.* 17 (2018) 1850033.
51. F. Lillo and J. D. Farmer, "The key role of liquidity fluctuations in determining large price changes", *Fluct. Noise Lett.* 5 (2005) L209–L216.
52. R. Han, W. Xie, X. Xiong, W. Zhang and W. Zhou, "Market correlation structure changes around the great crash: A random matrix theory analysis of the Chinese stock market", *Fluct. Noise Lett.* 16 (2017) 1750018.
53. A. Kim, "Protocols for distributing key strings based on phase and intensity fluction", *Fluct. Noise Lett.* 8 (2008) L333–L339.
54. R. Mingesz, G. Vadai and Z. Gingl, "What kind of noise guarantees security for the Kirchhoff-law–Johnson-noise key exchange?", *Fluct. Noise Lett.* 13 (2014) 1450021.

55. L. B. Kish, "Protection against the man-in-the-middle-attack for the Kirchhoff-loop-Johnson(-like)-noise cipher and expansion by voltage-based security", *Fluct. Noise Lett.* 6 (2006) L57–L63.
56. X. Cao, Y. Saez, G. Pesti and L. B. Kish, "On KLJN-based secure key distribution in vehicular communication networks", *Fluct. Noise Lett.* 14 (2015) 1550008.

# 1

# Fundamentals of Noise

# FOREWORD: THE THEORY OF NOISE

N.G. VAN KAMPEN
***Honorary Editor of Fluctuation and Noise Letters***
*Institute for Theoretical Physics, Utrecht University, P.O.B. 80000, 3508 TA Utrecht, The Netherlands*

The appearance of a new journal on fluctuations indicates that there must be reasons for a special interest in the subject. In fact, there are several reasons. One of them is that the existence of fluctuations is universal. A second reason is that fluctuations appear as perturbing noise in many systems, in particular in electronic systems with their high amplification factor. The third reason is that fluctuations also serve a useful purpose: they provide information about the physics of a system. A case in point is how Brownian motion provided evidence for the existence of molecules at a time when that was still doubted by people like Ernst Mach. Specifically it led to Einstein's determination of Avogadro's number [1]. Another example is the cosmological information culled from the existence of the universal background fluctuations of the electromagnetic field.

What is the basic cause of fluctuations? The natural world is composed of particles – atoms, molecules, electrons, and suchlike. It may happen that their mutual interaction causes them to coalesce and form a solid, or they may fly around, occasionally colliding, and form a gas. Other possibilities are plasmas, liquids, and glasses. The amazing thing therefore is *not* that they give rise to irregular phenomena. The amazing thing is that a collective behavior emerges, which is regular and can be described by general laws, namely the ones we know from our traditional macroscopic physics, such as heat conduction, Ohm's law, and Navier-Stokes. Yet these laws are not completely exact, because on a small scale the particulate nature of matter makes itself felt by the appearance of small deviations. These deviations we call fluctuations or noise. The reason why nevertheless the laws for the collective behavior are so very precise is that the individual particles are extremely small and numerous.

In fact the existence of a macroscopic physics, with deterministic laws, would be regarded as a miracle if it were not true that we ourselves are macroscopic and are confronted with them every day. They are not as exact as Newton's laws of mechanics; in particular the Second Law of Thermodynamics is limited by the existence of the fluctuating deviations. Fortunately the deviations are small and may be treated as corrections to our familiar macroscopic physics. In the few cases where this is not so, where no clear-cut separation of the macroscopic laws and the deviations is available (as in the cases of turbulence and of granular systems), physics is reduced to phenomenology.

The deviations are so small that they were not noticed until they manifested themselves in the shape of Brownian motion. The botanist Robert Brown, although not the first to observe them, was the first to draw attention to them. He showed that it was a general phenomenon and had to have a physical origin. Yet the true nature became clear only many years later, after it was realized that the observed motion was not the actual movement in space of the observed particle, but rather the net displacement resulting from a succession of numerous individually invisible steps. This gave rise to two different ways of describing them theoretically.

Einstein, with his inimitable intuition, created on the basis of general principles, a theory for the motion of a heavy particle in a fluid of light molecules, which turned out to provide an admirably suitable description of the observed Brownian motion [2]. At the same time Smoluchowski arrived at an equivalent description on the basis of a detailed study of the underlying physics [3]. These two different approaches (or should I say different mentalities?), can still be traced in much of the research on fluctuations that has developed since then: those who argue from general principles and those who delve into the microscopic physics.

If one looks carefully at these theories it appears that the "heavy particle" must be heavy compared to individual molecules, but not so heavy that its fluctuations are invisible and that the particle behaves like a macroscopic body obeying the old deterministic laws. This balance between microscopic and macroscopic aspects permeates the whole theory of fluctuations. To emphasize this fact I introduced some time ago the word "mesoscopic", but it has since been used in a different connotation.

Langevin gave a description that is mathematically equivalent to Einstein's, but which agreed better with the physicist's picture of what is actually going on [4]. He said (in modern language) that the Brownian particle with velocity $v$ is subject to a force from the surrounding medium consisting of two parts,

$$m\,dv/dt = -fv + \xi(t). \tag{1}$$

First there is a systematic part $-fv$ proportional to the velocity and in opposite direction, as one expects from a friction force. The second part is a rapidly varying function $\xi(t)$, which is irregular and whose detailed dependence on time cannot be determined. Yet some of its overall features (for instance the average over a small time interval) do obey simple laws. To display these features one replaces $\xi(t)$ with a *stochastic function*, that is, an ensemble of functions with a prescribed probability distribution among them. This probability distribution is fully determined by the three basic assumptions (in modern notation)

$$\xi(t) \text{ is Gaussian}, \qquad \langle\xi(t)\rangle = 0, \qquad \langle\xi(t_1)\,\xi(t_2)\rangle = \Gamma\delta(t_1 - t_2). \tag{2}$$

As a consequence we talk henceforth about an ensemble of Brownian particles. When they are in a stationary situation the fluctuations have the tendency to make them wander away from the equilibrium but the damping pushes them back; the result of this tug-of-war is the equilibrium distribution of $v$ – which must be the known Maxwell distribution. This provides a relation between the constant $\Gamma$ and the friction coefficient $f$, called the fluctuation-dissipation theorem.

Equation (1) is an equation for the velocity of the Brownian particle, but what is actually observed is the position $x(t)$. The connection between both has been established by

Ornstein and Uhlenbeck [5], who integrated (1), using (2), to get for the average position, when the initial position $x(0)$ is given,

$$\langle x(t)\rangle = x(0), \qquad \langle\{x(t) - x(0)\}^2\rangle = 2\,\Gamma t\,. \tag{3}$$

This procedure is applicable to many systems. Its strong point is that it enables one to fully determine the noise in a system, as soon as the macroscopic damped equation is known. Thus the way is opened for studying the effect of noise in a large variety of different systems, in particular electronic circuits. An enormous literature and many useful results have emerged from this. In fact there is a tendency to think that the study of noise coincides with this approach, and the term "stochastic differential equation" is often used as synonymous to this Langevin equation. The mathematicians write it a bit differently and then call "Itô equation".

These are elementary facts that can be found in the literature [6]. My reason for spelling them out here is that I want to emphasize the shortcomings of this procedure. The first one is that its very generality implies that the observation of noise cannot yield much new information about the internal structure of the system.

The second shortcoming is that the scheme only works reliably for systems obeying *linear* equations, such as Ohm's law. As soon as there are nonlinear terms, as in a non-Ohmic resistor, one cannot just add a term $\xi(t)$, because it turns out to be impossible to adjust $\Gamma$ in such a way that the stationary solution coincides with the equilibrium distribution (except in some trivial cases) [7]. This feature has given rise to the notorious Itô-Stratonovich dilemma. The third shortcoming of the convenient Langevin scheme is that it cannot be extended to quantum mechanics. Attempts have been made to simply generalize (1) so as to turn it into an equation for operators, but that did not work [8]. The first mentality has reached its limitation.

Our conclusion is that the cavalier way of introducing fluctuations by simply adding a Langevin term may lead to correct results in the case of a classical linear system, but not in nonlinear ones, let alone in quantum systems. Rather *it is necessary to go back to the actual cause of the fluctuations and take into account the mechanism that causes them.* It may be added parenthetically that it is possible to write a Fokker-Planck equation which is mathematically equivalent to (1) with (2), but it is subject to the same difficulties. A similar equation had been obtained by Rayleigh, but it failed to gain the attention it deserved [9].

In practice one does not use a realistic description of the actual mechanism, but one takes a simple model instead, hoping that the general features of the noise are independent of the precise form of the mechanism. The procedure is therefore as follows. To the system under consideration, S, a second system is coupled, which is large and is regarded as a reservoir or bath B. The total system, T = S + B, is isolated from the rest of the world and may therefore be described by means of a Hamiltonian $H_{\mathrm{T}}$ and the corresponding Schrödinger equation. If this equation can be solved one knows the expectation values of all operators referring to T, in particular also those that represent quantities referring to S alone. In one case this procedure can actually be carried out, namely when S is a harmonic oscillator, linearly coupled to a bath B consisting itself of harmonic oscillators as well [10].

In all other cases one has to resort to perturbation theory. The idea is that

$$H_{\mathrm{T}} = H_{\mathrm{S}} + H_{\mathrm{B}} + \lambda H_{\mathrm{I}} \quad , \tag{4}$$

where the coupling constant $\lambda$ serves as expansion parameter. An essential point is the choice of initial condition. Traditionally one supposes that initially the bath B is in thermal equilibrium, while the system S may have any initial state uncorrelated to B. (Incidentally, note that, rather than to talk in terms of a wave function $\psi$, we have to work with the density matrix $\rho$ in order to be able to describe the equilibrium.) In zeroth order S and B are uncoupled , so that B remains in thermal equilibrium while S evolves by itself. To first order in $\lambda$ the thermal equilibrium of B does not affect S, but the presence of S does perturb the equilibrium of B. In second order this deviation from equilibrium acts back on S and that is what we are interested in. No-one has looked at higher orders.

It is not hard to provide a formal solution of the second order evolution equation of the total system T. From it follows as a special consequence the evolution of S. However this is not the aim; *the aim is to obtain a differential equation for the evolution of S by itself*, which includes the damping and fluctuations. Unfortunately this is not possible, because in the course of their combined evolution correlations build up between S and B, that is, their states become entangled. It is therefore necessary to resort to a drastic assumption and proceed as follows.

First compute the evolution of S to second order in the way described above, but the result is utilized only during a short time interval $\Delta t$. Next utilize the same result again for the next time step $\Delta t$ as if the same initial condition were again valid. That is, as if B starts again in equilibrium and uncorrelated with S (although the change of S during the first time step is taken into account). Thus the correlation built up between them during the previous $\Delta t$ is ignored. This is the *repeated randomness assumption*, the quantum version if the "molecular chaos" assumption or "Stosszahlansatz".

In this way the sequence of time steps amounts to a differential equation describing how the density matrix of the system S under consideration evolves. The approximations appear acceptable, provided $\Delta t$ is chosen suitably. Its role is to distinguish between molecular motion on a microscopic time scale and the macroscopically observable evolution. Yet there is a serious objection to this procedure.

The reason why it cannot be accepted is the following. Any density matrix must by its very definition be hermitian, have unit trace, and be positive definite. If it obeys a differential equation the equation must be such that these features are maintained in time. That requirement imposes certain conditions on the equation, formulated by Kossakowski and by Lindblad [11]. *The equation obtained by the above procedure does not obey the condition with respect to the positive definiteness.* As a consequence it may lead to negative probabilities. The reason is that there cannot be a differential equation for the evolution of the density matrix of S. In fact, our derivation does not take into account the tiny time interval that elapses between the moment when S acts on B to modify its equilibrium, and the moment when this modification acts back on S. This time lapse has the effect that a differential equation for the density matrix of S becomes illusory; in other words the evolution of $\rho$ is not a semi-group.

This problem can be overcome in the following way. Rather than trying to establish the behavior of S after starting unrealistically in an initial state in which S and B are uncorrelated, we take the *total* system T to be in thermal equilibrium, which necessarily implies correlation between S and B. For any observable $F$ one may then ask its autocorrelation function

$$\langle F(t_1)\, F(t_2)\rangle^{eq}, \tag{5}$$

which is a function only of the time difference $t_2 - t_1 = \tau$. In principle this contains all information which we previously tried to cull from the evolution equation for $\rho$.

This approach is worked out in detail in a separate article, necessarily using again perturbation theory [12]. Yet, it should be said that even this is still a special case inasmuch as there is still a bath B consisting of harmonic oscillators. The quantity (5) describes of course the fluctuations in entirely general systems, even if they are not subdivided into an S and B. It goes without saying that in general the actual calculation is beyond the power of theoreticians, even if the physics of the system is known. That is why much experimental work is needed. The challenge is even greater in cases where the physical nature is unknown and the mechanism that is responsible for the noise has to be deduced from the observed fluctuations. A case in point is the elusive $1/f$ noise. Thus both theoretical and experimental research is needed, provided it is done in a critical investigative spirit.

So far I have not mentioned applications to complex biological systems and in particular population problems. These problems are fundamentally different from physics as in their case there is no underlying microscopic world, let alone a quantum mechanical one. The shape of the macroscopic equations and the fluctuations must therefore be obtained (or guessed) directly from the phenomenology of the system one studies.

Another difference is that the elementary units are much larger and less numerous than the molecules that physics deals with, so that the separation between the macroscopic laws and the noise is less pronounced. For these reasons one must be careful in trying to transfer the ideas and concepts of physics. In particular there is a strong temptation to talk about entropy and the increase of randomness without the proper foundation.

My final conclusion is that there is ample material for this journal!

## References


[1] For the early history see S.G. Brush, *The Kind of Motion We Call Heat, Vol. 2*, North-Holland, Amsterdam (1976) ch. 15.

[2] A. Einstein, *Annalen der Physik [4]* **17** (1905) 549; *Investigations on the Theory of Brownian Movement* (A.D. Cowper transl.), Methuen, London (1926) and Dover, New York (1956).

[3] M. von Smoluchowski, *Annalen der Physik [4]* **21** (1906) 756.

[4] P. Langevin, *Comptes Rendus (Paris)* **146** (1908) 530.

[5] G.E. Uhlenbeck & L.S. Ornstein, *Phys. Rev.* **34** (1930) 823.

[6] For instance: N.G. van Kampen, *Stochastic Processes in Physics and Chemistry*, North-Holland, Amsterdam (1981,1992); U. Weiss, *Quantum Dissipative Systems*, World Scientific, Singapore (1993).

[7] D.K.C. MacDonald, *Phil. Mag.* **45** (1954) 63; *Phys. Rev.* **108** (1957) 541.

[8] R. Benguria & M. Kac, *Phys. Rev. Letters* **46** (1980) 1.

[9] Lord Rayleigh, *Phil. Mag.* **32** (1891) 424.

[10] P. Ullersma, *Physica* **32** (1966) 27,56,74,90. Nowadays usually called the Caldeira-Leggett model.

[11] A. Kossakowski, *Bull. Acad. Pol. Sci., Série Math. Astr. Phys.* **20** (1971) 1021; G. Lindblad, *Commun. Mathem. Phys.* **40** (1975) 147 ; See also: R. Alicki & K. Lendi, *Quantum Dynamical Semigroups and Applications*, Springer, Berlin (1987).

[12] N.G. van Kampen, the second issue of FNL.

# ITÔ VERSUS STRATONOVICH: 30 YEARS LATER

RICCARDO MANNELLA
*Dipartimento di Fisica, Università di Pisa*
*Largo B Pontecorvo 3, I-56127 Pisa, Italy*
*mannella@df.unipi.it*

PETER V. E. McCLINTOCK
*Department of Physics, Lancaster University*
*Lancaster LA1 4YB, UK*
*p.v.e.mcclintock@lancaster.ac.uk*

Received 15 September 2011
Accepted 29 December 2011
Published 31 March 2012

Communicated by Aneta Stefanovska

The Itô versus Stratonovich controversy, about the "correct" calculus to use for integration of Langevin equations, was settled to general satisfaction some 30 years ago. Recently, however, it has started to re-emerge, following the advent of new experimental techniques. We briefly review the historical background and discuss critically some of the most recent contributions. We show that some of the new findings are not well based.

*Keywords*: Itô stochastic calculus; Stratonovich stochastic calculus; anticipating stochastic differential equations; Brownian motion; Langevin equations; nanoscale forces.

## 1. Introduction

In the macroscopic world, physical systems are usually nonlinear and subject to noise (random fluctuations). The nonlinearity introduces subtleties into how noise influences a system, and seemingly vexing conundrums arise where the noise is quasi-white and enters multiplicatively in one of the parameters of a model equation. In the presence of multiplicative noise, it turns out that a choice must be made as to which is the appropriate stochastic calculus to be used: this choice appears to be somehow arbitrary, which sparked a widespread debate, usually referred to as the Itô versus Stratonovich controversy. It attracted considerable attention in the physics community for almost a decade. The controversy was eventually settled to general satisfaction but, as so often happens in such cases, a few years later it has started to re-emerge. One of the reasons has been the introduction of new experimental techniques that allow thermodynamic properties to be probed on the nanometre scale. In this paper we briefly review the controversy and the basic mathematics that underlay it, and recall the main conclusions reached in the earlier debate. We then consider some of the most recent papers, and point out where some of their conclusions appear not to be well founded.

## 2. Background and Historical Notes

The Itô versus Stratonovich controversy took place in the physics literature (mainly) from the late 1970s to the early 1980s, in the burgeoning field of nonlinear stochastic physics. To appreciate the reasons behind the controversy, we need first to recall briefly the basic ideas behind stochastic processes.

### 2.1. *Stochastic calculus*

The root of the controversy lay with the counter-intuitive nature of stochastic calculus in the presence of nonlinearities, when considered from a physicist's point of view. Here we will provide intuitive arguments; a more formal derivation can be found in [1]. We start from the stochastic differential equation (SDE)

$$dx = f(x)dt + g(x)dW \tag{1}$$

representing the increment of an observable $x$ (for example, the position of a Brownian particle). Here $dW$ is the increment of a Wiener process $W(t)$ defined in probabilistic terms (we assume $W(0) = 0$ for compactness of notation, without loss of generality) via

$$P(W(t)) = \frac{1}{\sqrt{2\pi t}} e^{-\frac{W(t)^2}{2t}}.$$

Amongst others quantities, we need to deal with those of form $\int g(x)dW$. A possible approach to the integration of Eq. (1) is through a Taylor expansion where, in the simplest nontrivial case ($f(x) = 0$, $g(x) \neq 0$), we can write

$$x(t) - x(0) = \int_0^t g(x)dW \approx \int_0^t dW[g(x(0)) + g'(x(0))(x(s) - x(0))]. \tag{2}$$

To lowest order, $x(t) - x(0) = g(x(0))W(t)$. If $g'(x) \neq 0$, at the next order an integral of the form $\int W(t)dW$ appears. To compute it, the standard approach is through a discretization, in the mean square limit,

$$\int_0^t WdW = \text{m.s.} \lim_{n\to\infty} \sum_{i=1}^{n} W(t_i^*)[W(t_i) - W(t_{i-1})] \tag{3}$$

using a suitable partition $0 = t_0 < t_i < t_n = t$, and where $t_{i-1} \leq t_i^* \leq t_i$: in what follows, we take $t_i^* = t_{i-1} + \alpha(t_i - t_{i-1})$, with $0 \leq \alpha \leq 1$. The mean square limit is defined as

$$\text{m.s.} \lim_{n\to\infty} X_n = X \leftrightarrow \lim_{n\to\infty} \left\langle (X_n - X)^2 \right\rangle = 0, \tag{4}$$

where the average $\langle \cdots \rangle$ is taken over the realizations of the Wiener process. In the evaluation of the mean square limit appearing in Eq. (3), we need to estimate terms like $\langle W(t')W(s')\rangle$. Assuming that $t' < s'$, we have that $\langle W(t')W(s')\rangle = \langle W(t')[W(s')-W(t')]\rangle + \langle W(t')W(t')\rangle$. Recalling that the increment $[W(s')-W(t')]$ is independent of $W(t')$, it follows that $\langle W(t')W(s')\rangle = \langle W(t')W(t')\rangle = t'$ and in general $\langle W(t')W(s')\rangle = \min(t', s')$. We have that

$$\text{m.s.} \lim_{n\to\infty} \left\langle \sum_{i=1}^{n} W(t_i^*)[W(t_i) - W(t_{i-1})] \right\rangle = \sum t_{i-1} + \alpha(t_i - t_{i-1}) - t_{i-1} = \alpha t.$$

Note that the value of the integral *depends on the point* within the interval $[t_i, t_{i-1}]$ where the process $W(t)$ is evaluated: in other words, it depends on $\alpha$. Although in principle any value of $\alpha$ in the range $[0, 1]$ is possible, in the literature only two values are commonly found: $\alpha = 0$ (Itô calculus [2]) and $\alpha = 1/2$ (Stratonovich calculus [3]). If $W(t)$ were a smooth function, clearly $\langle \int W dW\rangle = \langle \frac{1}{2}W(t)^2\rangle = \frac{1}{2}t$, which means that the "standard" result (holding for Riemann–Stieltjes integrals) is recovered by setting $\alpha = 1/2$.

From Eq. (1), it is possible (see for instance [4]) to write the Fokker–Planck equation driving the probability distribution $P(x, t)$

$$\frac{\partial P(x,t)}{\partial t} = \frac{\partial}{\partial x}\left\{-f(x) - \alpha g(x)g'(x) + \frac{1}{2}\frac{\partial}{\partial x}g^2(x)\right\} P(x,t) \tag{5}$$

which obviously depends on $\alpha$ and where $g'(x) = \partial g(x)/\partial x$; the ensuing equilibrium distribution will depend on $\alpha$ too, and on the functions $f(x)$ and $g(x)$.

In a nutshell, the controversy centered on what is the "correct" choice of $\alpha$ for the description of natural phenomena. We note that it is possible to have sets of *different* $f(x)$ and $\alpha$ which lead algebraically to the *same* $f(x) + \alpha g(x)g'(x)$: this implies that there could be systems characterized by different $f(x)$ and $\alpha$ but which have the same equilibrium distribution.

Rewriting $\int g(x)dW = \sum g(x(t_i^*)[dW(t_i) - dW(t_{i-1})]$ we note that $x(t)$ is given by the solution of the SDE (1). Intuitively, this means that, assuming we integrated the SDE up to $x(t_{i-1})$, we get the value of $x(t_i)$ using, in general, values of $x(t)$ for times in the range $[t_{i-1}, t_i]$ even though these are not yet known. This is not a difficulty for continuous functions because, in the limit $t_{i-1} \to t_i$, $x(t)$ is well behaved $(x(t_i) - x(t_{i-1}) \propto t_i - t_{i-1})$. But it poses a problem for stochastic processes $(W(t_i) - W(t_{i-1}) \propto \sqrt{t_i - t_{i-1}})$. Itô calculus elegantly solves this problem, by evaluating $g(x)$ at time $t_{i-1}$, where it is known. Hence Itô prescription is termed *nonanticipating*, whereas all other prescriptions are called *anticipating*.

### 2.2. *History*

It is impossible to mention here all of the papers that tackled the controversy: we review briefly those that we feel were particularly helpful in shaping the growing understanding on the part of the nonlinear and stochastic physics community.

Perhaps one of the earliest papers to question what is the applicable stochastic calculus in nature was [5], where Stratonovich calculus was used in models of population growth. In [6], following some theoretical works on the correspondence between stochastic calculus and ordinary calculus [7, 8], coupled stochastic differential equations were simulated on a hybrid computer. It was shown that any calculus could be achieved in practice through the tuning of parameters. In [9] it was argued that in theoretical biology Itô calculus should be preferred. In two companion papers [10, 11], nonadditive stochastic processes were considered, and the Itô–Stratonovich controversy discussed, looking at the rôle of noise correlation in some physical and chemical systems. In [12] a different approach was followed: starting from a stochastic system with inertia, via contraction an SDE of the form of Eq. (1) was obtained, and in the process it turned out that the "correct" stochastic calculus was Stratonovich. A similar approach to derive the "correct" calculus via contraction from a system with inertia was used in [13]. In [14] a criterion was presented which purposely allowed selection of the correct stochastic calculus. From a theoretical point of view, [15] eventually settled the argument. The consensus that had emerged was that —

- The parameter $\alpha$ is part of the model: it must be chosen on physical grounds and it cannot be inferred through algebraic manipulations.
- In an experiment, a probability distribution is measured: knowledge of this distribution function is not enough to infer $\alpha$, but additional information is needed, e.g., knowledge of $f(x)$). Equation (5) is more fundamental than Eq. (1).
- In many real cases, Eq. (1) is an effective (mesoscopic) equation. One should be careful in using some known microscopic force as the term $f(x)$: in principle, in the passage from the microscopic to the mesoscopic level, the "deterministic" microscopic force might not coincide with the mesoscopic force which, inserted in Eq. (1), reproduces the observed dynamics.
- In a typical, continuous, real physical system, we expect Stratonovich calculus to apply; whereas in a system which is intrinsically discontinuous, e.g., in the stock exchange or in the evolution of biological populations, we expect Itô calculus to apply.

Analogue simulations [16, 17] confirmed that continuous physical systems indeed obey Stratonovich calculus: the same SDE's numerically integrated on a digital computer enforcing Itô calculus clearly reproduced the dynamics theoretically expected of Itô calculus. In [18] it was reported that the equilibrium distribution of a stochastic system driven by two weakly autocorrelated additive and multiplicative noises behaved more Itô (Stratonovich) like when the additive noise was faster (slower) than the multiplicative noise (the physical interpretation being that the faster the additive noise, the less continuous the system would appear on the timescale of the multiplicative noise).

## 3. Itô Versus Stratonovich Again

Following a few years which saw little further interest in the problem, since the beginning of the nineties a few papers started to appear in the literature which again focused on the stochastic calculus realized in nature. In [19] the rôle and use of an anticipating SDE is discussed. The different integral calculi in quantum mechanical SDEs are discussed in [20, 21]. The stochastic calculus to use in bacterial interactions is discussed in [22]. Different calculi in stochastic partial differential equations are discussed in [23]. Mixed stochastic calculi in systems with different time scales are discussed in [24], whereas [25] tackles a similar problem, i.e., the appropriate calculus to use for an overdamped system obtained as a contraction of the dynamics of an underdamped system in the presence of correlated noise. In [26] the problem of the appropriate stochastic calculus in different coordinate systems is discussed. The different calculi are considered in [27], with application to magnetic systems. Itô and Stratonovich calculi have also been studied recently with applications in oceanography [28], population growth [29], and optimal filtering [30]. A recent paper relates the Itô versus Stratonovich problem to thermodynamics [31].

### 3.1. *A recent case study: anticipating SDE's*

A number of theoretical papers [32–34] have also appeared advocating the possibility of $\alpha = 1$. As counter-intuitive as this may seem, some experiments [35, 36] have nonetheless claimed to have found empirical evidence that there could indeed be physical systems where $\alpha = 1$. We now focus on a discussion of [35, 36], noting that [35] has been the subject of a comment [37] and a reply [38]: we will show that some of the arguments in [35] are not well founded.

The work of [35] reports experiments on a colloidal particle near a wall in the presence of a gravitational field, electrostatic repulsion from the wall and random scattering from the solvent, the latter being modeled as space-dependent noise. In [35] two different approaches are suggested to derive from the experiments the force acting on the colloidal particle: one approach is based on drift measurements; the other uses the equilibrium distribution (in space) of the colloidal particle. The central result of [35] is a striking difference between these two forces. From this discrepancy the authors of [35] infer the stochastic calculus realized in the system.

Let us first recast our SDE (1) in the form used in [35]:

$$dx = f(x)dt + g(x)dW = \frac{F(x)}{\gamma(x)}dt + \sqrt{2D_\perp(x)}dW, \tag{6}$$

where $F(x)$ in the SDE, following [35], is *assumed* to be equal to the deterministic force, and $\gamma(x)$ is some damping, assumed to be very large, so that an overdamped SDE can be considered; $D_\perp(x) = \frac{k_B T}{\gamma(x)}[= g^2(x)/2]$. A drift measurement [35] looks at the distance $\Delta x$ travelled during a short time $\Delta t$. From (5) we can relate $\Delta x/\Delta t$

to our model quantities

$$\langle \Delta x/\Delta t\rangle \equiv \overline{v}_d = \left\langle \frac{dx}{dt}\right\rangle = f(x) + \alpha g(x)g'(x) = \frac{F(x)}{\gamma(x)} + \alpha\frac{dD_\perp(x)}{dx}. \quad (7)$$

Note that $\overline{v}_d$ is *not* proportional to $F(x)$, i.e., to the deterministic force: this means that, unlike the case when the diffusion $g(x)$ is constant, a drift measurement cannot, in general, be used to infer the microscopic deterministic force, contrary to the assumption of [35] that $F(x) = \gamma(x)\overline{v}_d$.

In [35] the force is also computed from the equilibrium distribution $P(x)$, which is assumed to exist (it is in fact measured in the experiments). This force is defined as $F_e(x) = -dU(x)/dx$ where $U(x) = -k_BT\ln(P(x))$. From (5) we obtain

$$U(x) = -k_BT\int\frac{f(x)+(\alpha-1)g(x)g'(x)}{g^2(x)/2}dx,$$

$$\frac{F_e(x)}{\gamma(x)} = -\frac{1}{\gamma(x)}\frac{dU(x)}{dx} = f(x) + (\alpha-1)g(x)g'(x), \quad (8)$$

where we used $g^2(x) = 2k_BT/\gamma(x) = 2D_\perp(x)$. Equations (7) and (8) differ by $-g(x)g'(x) = -\frac{dD_\perp(x)}{dx}$, which is *exactly* the experimental discrepancy reported in [35] between the two "forces"; this difference is independent of $\alpha$, i.e., independent of the stochastic calculus used to describe the physical system. However, the fact that the two "forces" differed by $-g'(x)g(x)$ was construed by the authors of [35] as evidence that $\alpha = 1$.

There is another reason why an experiment like the one described in [35, 36] cannot be used to infer the value of $\alpha$, even if the force $f(x)$ were known [39]. Going back to Eq. (1), a possible discretization algorithm for numerical integration of $x(t)$ at first order in the integration time-step $h$ is[a] [40]

$$x(h) = x(0) + W(h)g(x(0)) + hf(x(0)) + \alpha g'(x(0))g(x(0))h. \quad (9)$$

The evolution of $x(t)$ depends *explicitly* on the stochastic calculus, hence, in principle we could infer $\alpha$. But the experiments of [35, 36] deal with a system where the overdamped limit has been taken: the SDE where this limit has not yet been taken (we assume the mass of the particle to be unity for compactness) reads:

$$dv = [F(x) - \gamma(x)v]dt + \sqrt{2k_BT\gamma(x)}dW,$$
$$dx = vdt. \quad (10)$$

The relationship between diffusion and the term multiplying $v$ on the r.h.s. follows from the existence of an equilibrium distribution, i.e., detailed balance. To first

[a]The expression for $x(h)$ which follows from the evaluation of the stochastic integrals is given by

$$x(h) = x(0) + W(h)g(x(0)) + hf(x(0)) + \frac{1}{2}g'(x(0))g(x(0))[W(h)^2 + (2\alpha-1)h]$$

which coincides with Eq. (9) at order $h$, owing to the fact that $\langle W(h)^2\rangle = h$.

order in $h$, Eq. (10) is integrated as

$$v(h) = v(0) + W(h)\sqrt{2k_BT\gamma(x(0))} + h[F(x(0)) - \gamma(x(0))v(0)],$$
$$x(h) = x(0) + hv(0), \tag{11}$$

which does *not* depend on $\alpha$, contrary to the scheme of Eq. (9)! Hence, it is not possible to guess the "correct" $\alpha$ from observations of the model given by Eq. (10). The quantity $\alpha$ appears when we go from Eq. (10) to Eq. (1), taking some limit. In doing so, however, one finds that the assumption that $F(x)$ in Eq. (1) equals $F(x)$ in Eq. (10) is *wrong* [39]. The correct procedure to obtain the model of Eq. (1) is: (a) write the Fokker–Planck equation corresponding to Eq. (10); (b) adiabatically eliminate $v$ from the Fokker–Planck equation, which yields [13]

$$\frac{\partial P(x,t)}{\partial t} = \frac{\partial}{\partial x}\frac{1}{\gamma(x)}\left(-F(x) + k_BT\frac{\partial}{\partial x}\right)P(x,t); \tag{12}$$

and (c) write the SDE corresponding to Eq. (12). It is at this latter stage that $\alpha$ appears, as a choice that we must make, and it determines the correct relationship between the $F(x)$'s in Eqs. (1) and (10). In [35, 36], however, their equality was assumed [39], so that, in effect, a choice of stochastic calculus had already been made. Equation (12) yields a force measured from the probability distribution, equal to $F(x)$ regardless of the stochastic calculus. Inspection of Eq. (8) shows that the stochastic calculus implicitly assumed in writing Eq. (1) was $\alpha = 1$. It is a legitimate choice, but should not be taken as a "proof" that $\alpha = 1$ when inferring the "correct" (in reality, "picked at a previous step of the derivation") $\alpha$.

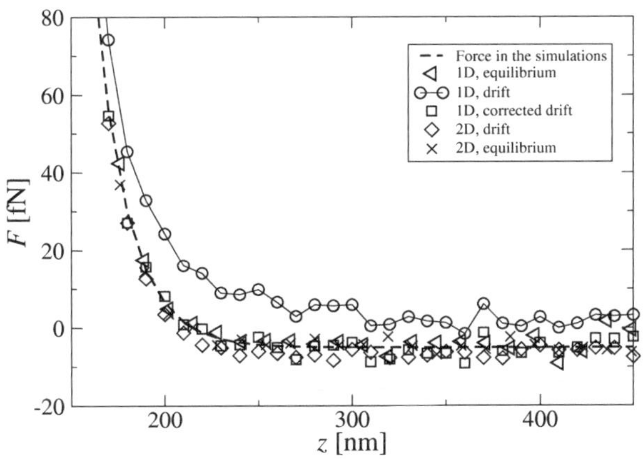


Fig. 1. Forces computed from a simulation of (10) and of the SDE obtained from contraction of (10) using Stratonovich calculus (see [39] for more details). To make contact with the experiment, we took $F(x) = Be^{-kx} + C$ with $B = 770\,\text{pN}$, $C = -5\,\text{fN}$, $k = (18\,\text{nm})^{-1}$, and $D_\perp(x) = k_BT/\gamma(x)$, $\gamma(x) = 6\pi\eta R\frac{x+a}{x}$, with $x$ in nm, $a = 700\,\text{nm}$, $2R = 1.31\,\mu\text{m}$, $T = 300\,\text{K}$, $\eta = 8.5\times10^{-3}\,\text{Pa\,s}$, mass $m = 6.3\times10^{-16}\,\text{kg}$.

Figure 1 summarizes these arguments: integrating Eq. (10) [41, 42], the forces obtained from drift and equilibrium distribution measurements coincide; and they coincide, both with the deterministic force, and with the force obtained from equilibrium distribution measurements by integration of the correct overdamped 1D SDE obtained from contraction of Eq. (10) using Stratonovich calculus. The force from the drift measured in the 1D simulations differs from the corresponding force obtained from equilibrium measurements but, when the correction $g'(x)g(x)$ is applied, we again have coincidence with the deterministic force.

## 4. Conclusion

After being a major topic in nonlinear stochastic physics for an extended period, the Itô versus Stratonovich controversy was finally settled some 30 years ago. The recent resuscitation of the debate has involved theoretical and experimental works whose claim to be able to determine experimentally what is the appropriate stochastic calculus in a system is not soundly based. Although these papers are very interesting, some of the conclusions are incorrect — primarily because the earlier debate seems to have been forgotten. Here we have reviewed this debate and discussed the new findings, showing that the "correct" calculus is still as elusive as ever, and that it can only be inferred from the chosen model. In particular, we discussed how the "force" derived from the drift of a Brownian particle need not necessarily coincide with the "force" obtained from the equilibrium distribution. We showed that the discrepancy reported in [35, 36] has nothing to do with different stochastic calculi as the authors had inferred: it is simply a consequence of having two different definitions of force, neither of which corresponds to the true microscopic force, and which coincide only where the diffusion coefficient is constant. Furthermore, we recalled that some of the simplified models we use may be characterized by quantities that differ from the true microscopic quantities which appear in a full model.

It is evident that stochasticians of all kinds — mathematicians, physicists, engineers and others — need constant reminders that the Itô versus Stratonovich problem was *solved* long ago. The ideas expressed with such clarity and force by van Kampen in the 1980s were amply validated by experiments (e.g., [16, 17]). His classic paper [15] has stood the test of time and is well worth re-reading.

## Acknowledgments

This work is fondly dedicated to the memory of Frank Moss, who died on 4 January 2011. In the 1980s, Frank was one of the main investigators of the Itô versus Stratonovich problem, to which his experiments brought much illumination. We thank J. M. Sancho for drawing [39] to our attention and for providing a copy of the manuscript available prior to publication.

## References

[1] L. Arnold, *Stochastic Differential Equations: Theory and Applications* (Wiley-Interscience, New York, 1974).

[2] K. Itô, Stochastic integral, *Proc. Imp. Acad.* **20** (1944) 519–524.

[3] R. L. Stratonovich, A new representation for stochastic integrals and equations, *SIAM J. Control* **4** (1966) 362–371.

[4] N. G. van Kampen, *Stochastic Processes in Physics and Chemistry* (Elsevier, Amsterdam, 2007).

[5] H. C. Tuckwell, Study of some diffusion models of population growth, *Theor. Popul. Biol.* **5** (1974) 345–357.

[6] J. A. Bullin and A. E. Dukler, Hybrid computer modeling of a stochastic nonlinear dynamic system, *Chem. Engin. Sci.* **30** (1975) 631–635.

[7] E. Wong and M. Zakai, On the relation between ordinary and stochastic differential equations, *Int. J. of Engin. Sci.* **3** (1965) 213–229.

[8] R. E. Mortensen, On conditional expectation of stochastic integrals, *J. Math. Anal. Appl.* **26** (1969) 114.

[9] M. Turelli, Random environments and stochastic calculus, *Theor. Popul. Biol.* **12** (1977) 140–178.

[10] B. J. West, A. R. Bulsara, K. Lindenberg, V. Seshadri and K. E. Shuler, Stochastic processes with non-additive fluctuations. 1. Itô and Stratonovich calculus and the effects of correlations, *Physica A* **97** (1979) 211–233.

[11] A. R. Bulsara, K. Lindenberg, V. Seshadri, K. E. Shuler and B. J. West, Stochastic processes with non-additive fluctuations. 2. Some applications of Itô and Stratonovich calculus, *Physica A* **97** (1979) 234–243.

[12] H. Hasegawa, M. Mabuchi and T. Baba, On the usefulness of stochastic calculus applied to adiabatic elimination in Langevin equations, *Phys. Lett. A* **79** (1980) 273–277.

[13] J. M. Sancho, M. San Miguel and D. Durr, Adiabatic elimination for systems of Brownian particles with non-constant damping coefficients, *J. Stat. Phys.* **28** (1982) 291–305.

[14] T. Morita, On the interpretation of multiplicative white noise, *Phys. Lett. A* **82** (1981) 215–217.

[15] N. G. van Kampen, Itô versus Stratonovich, *J. Stat. Phys.* **24** (1981) 175–187.

[16] J. Smythe, F. Moss and P. V. E. McClintock, Observation of a noise-induced phase-transition with an analog simulator, *Phys. Rev. Lett.* **51** (1983) 1062–1065.

[17] P. V. E. McClintock and F. Moss, Further experimental evidence pertaining to the applicability of the Itô and Stratonovic stochastic calculi to real physical systems, *Phys. Lett. A* **107** (1985) 367–370.

[18] S. Faetti, C. Festa, L. Fronzoni, P. Grigolini and P. Martano, Multiplicative stochatic processes in nonlinear systems — noise-induced transition from the overdamped to the inertial regime, *Phys. Rev. A* **30** (1984) 3252–3263.

[19] T. Masuda, Absolute continuity of distributions of solutions of anticipating stochastic differential equations, *J. Funct. Anal.* **95** (1991) 414–432.

[20] J. Gough, Asymptotic stochastic transformations for nonlinear quantum dynamical systems, *Rep. Math. Phys.* **44** (1999) 313–338.

[21] A. E. Kobryn, T. Hayashi and T. Arimitsu, Quantum stochastic differential equations for boson and fermion systems — method of non-equilibrium thermo field dynamics, *Ann. Phys.* **308** (2003) 395–446.

[22] M. Carletti, On the stability properties of a stochastic model for phage-bacteria interaction in open marine environment, *Math. Biosci.* **175** (2002) 117–131.

[23] O. Carrillo, M. Ibañes, J. García-Ojalvo, J. Casademunt and J. M. Sancho, Intrinsic noise-induced phase transitions: Beyond the noise interpretation, *Phys. Rev. E* **67** (2003) 046110.
[24] D. Givon and R. Kupferman, White noise limits for discrete dynamical systems driven by fast deterministic dynamics, *Physica A* **335** (2004) 385–412.
[25] R. Kupferman, G. A. Pavliotis and A. M. Stuart, Itô versus stratonovich white-noise limits for systems with inertia and colored multiplicative noise, *Phys. Rev. E* **70** (2004) 036120.
[26] E. Martinez, L. Lopez-Diaz, L. Torres and O. Alejos, On the interpretations of Langevin stochastic equation in different coordinate systems, *Physica B* **343** (2004) 252–256.
[27] D. V. Berkov and N. L. Gorn, Stochastic dynamic simulations of fast remagnetization processes: Recent advances and applications, *J. Magn. Magn. Mater.* **290** (2005) 442–448.
[28] R. N. Miller, Topics in data assimilation: Stochastic processes, *Physica D* **230** (2007) 17–26.
[29] C. A. Braumann, Growth and extinction of populations in randomly varying environments, *Comp. Math. Appl.* **56** (2008) 631–644.
[30] A. Amirdjanova and S. Chivoret, New method for optimal nonlinear filtering of noisy observations by multiple stochastic fractional integral expansions, *Comp. Math. Appl.* **52** (2006) 161–178.
[31] I. M. Sokolov, Itô, Stratonovich, Hänggi and all the rest: The thermodynamics of interpretation, *Chem. Phys.* **375** (2010) 359–363.
[32] P. Ao, Potential in stochastic differential equations: Novel construction, *J. Phys. A* **37** (2004) L25–L30.
[33] P. Ao, Laws in Darwinian evolutionary theory, *Phys. Life Rev.* **2** (2005) 117–156.
[34] P. Ao, C. Kwon and H. Qian, On the existence of potential landscape in the evolution of complex systems, *Complexity* **12** (2007) 19–27.
[35] G. Volpe, L. Helden, T. Brettschneider, J. Wehr and C. Bechinger, Influence of noise on force measurements, *Phys. Rev. Lett.* **104** (2010) 170602.
[36] T. Brettschneider, G. Volpe, L. Helden, J. Wehr and C. Bechinger, Force measurement in the presence of Brownian noise: Equilibrium-distribution method versus drift method, *Phys. Rev. E* **83** (2011) 041113.
[37] R. Mannella and P. V. E. McClintock, Comment on 'Influence of noise on force measurements', *Phys. Rev. Lett.* **107** (2011) 078901.
[38] G. Volpe, L. Helden, T. Brettschneider, J. Wehr and C. Bechinger, Volpe *et al.* reply, *Phys. Rev. Lett.* **107** (2011) 078902.
[39] J. M. Sancho, Brownian colloidal particles: Ito, Stratonovich, or a different stochastic interpretation, *Phys. Rev. E*, 062102-1–062102-3.
[40] R. Mannella, Computer experiments in non-linear stochastic physics, *Noise in Nonlinear Dynamical Systems*, eds. F. Moss and P. V. E. McClintock, Vol. 3, Experiments and Simulations (Cambridge University Press, Cambridge, 1988), pp. 189–221.
[41] R. Mannella, Quasisymplectic integrators for stochastic differential equations, *Phys. Rev. E* **69** (2004) 041107-1–041107-8.
[42] R. Mannella, Numerical stochastic integration for quasi-symplectic flows, *SIAM J. Sci. Comp.* **27** (2006) 2121–2136.

# GENERALIZED WIENER PROCESS AND KOLMOGOROV'S EQUATION FOR DIFFUSION INDUCED BY NON-GAUSSIAN NOISE SOURCE

ALEXANDER DUBKOV

*Radiophysics Department, Nizhny Novgorod State University, 23 Gagarin Ave.*
*Nizhny Novgorod, 603950 Russia*
*dubkov@rf.unn.ru*

BERNARDO SPAGNOLO

*INFM and Dipartimento di Fisica e Tecnologie Relative, Group of Interdisciplinary Physics*
*Università di Palermo, Viale delle Science pad. 18, I-90128 Palermo, Italy*
*spagnolo@unipa.it*

Received 23 December 2004
Revised 6 March 2005
Accepted 16 March 2005
Communicated by Werner Ebeling and Bernardo Spagnolo

We show that the increments of generalized Wiener process, useful to describe non-Gaussian white noise sources, have the properties of infinitely divisible random processes. Using functional approach and the new correlation formula for non-Gaussian white noise we derive directly from Langevin equation, with such a random source, the Kolmogorov's equation for Markovian non-Gaussian process. From this equation we obtain the Fokker–Planck equation for nonlinear system driven by white Gaussian noise, the Kolmogorov–Feller equation for discontinuous Markovian processes, and the fractional Fokker–Planck equation for anomalous diffusion. The stationary probability distributions for some simple cases of anomalous diffusion are derived.

*Keywords*: Non-Gaussian white noise; infinitely divisible distribution; Kolmogorov's equation; Wiener process.

## 1. Introduction

Stochastic dynamics is useful to model many biological, chemical, economical and physical systems. The random driving forces have very different origins, in most cases they are Gaussian white or colored noise sources, but often these forces must be considered as non-Gaussian ones, for example, in sensory and biological systems [1]. Moreover, in many physical and biological systems a deviation of real statistics of fluctuations from Gaussian law, leading to anomalous diffusion, is observed [2,3]. A suitable mathematical model to describe the anomalous diffusion is the fractional

Fokker–Planck equation, which can be derived from different theoretical approaches [3–7].

In this paper we show that the increments of generalized Wiener process, representing the integral of non-Gaussian white noise, have the properties of infinitely divisible random processes. This enables us to obtain the characteristic functional of non-Gaussian white noise. Then, by using functional approach to split the correlation between stochastic functionals, we derive the Kolmogorov's equation directly from Langevin equation with non-Gaussian white noise source. From this general equation we can obtain the Fokker–Planck equation for continuous Markovian processes, the Kolmogorov–Feller equation for discontinuous Markovian processes and the fractional Fokker–Planck equation for anomalous diffusion.

## 2. Generalized Wiener Process

It is well known that Wiener process $\eta(t)$ has stationary and independent increments, and its time derivative $\xi(t) = \dot{\eta}(t)$ is Gaussian white noise. The correlation function of the increments $\eta(t+T) - \eta(t)$ depends on time difference $T$ only, and increments of non-overlapping time intervals are statistically independent. Let us consider now a non-Gaussian random process with stationary and independent increments which can be called as *generalized Wiener process*. We can represent the increment $\eta_T \equiv \eta(T) - \eta(0)$ as

$$\eta_T = \sum_{k=1}^{n} [\eta(k\Delta t) - \eta((k-1)\Delta t)], \tag{1}$$

where $\Delta t = T/n$. Since the random variable $\eta_T$ can be divided into the sum of arbitrary number of independent and identically distributed random variables, its probability distribution belongs to the class of infinitely divisible distributions. Hence, we can express the logarithm of characteristic function of the random value $\eta_T$ in the Lévy–Khinchine form [8]

$$\phi_T(u) \equiv \ln \theta_T(u) = \ln \langle e^{iu\eta_T} \rangle = \int_{-\infty}^{\infty} \frac{e^{iux} - 1 - iu \sin x}{x^2} M_T\{dx\}, \tag{2}$$

where $M_T\{dx\}$ is the canonical measure which, according to definition, takes positive values on a finite intervals, and the integrals

$$M_T^+(x) = \int_x^{\infty} \frac{M_T\{dy\}}{y^2}, \qquad M_T^-(-x) = \int_{-\infty}^{-x} \frac{M_T\{dy\}}{y^2}$$

converge for any positive $x$ and $T$. Introducing the non-negative density $\rho_T(x)$ of the canonical measure: $M_T\{dx\} = \rho_T(x)dx$, defined in generalized sense, we can rewrite Eq. (2) as

$$\phi_T(u) = \int_{-\infty}^{\infty} \frac{e^{iux} - 1 - iu \sin x}{x^2} \rho_T(x)\, dx. \tag{3}$$

For two consecutive time intervals $T$ and $S$, we have

$$\theta_{T+S}(u) = \langle e^{iu\eta_{T+S}} \rangle = \left\langle e^{iu(\eta_T+\eta_S)} \right\rangle = \langle e^{iu\eta_T} \rangle \langle e^{iu\eta_S} \rangle = \theta_T(u)\, \theta_S(u),$$

and

$$\phi_{T+S}(u) = \phi_T(u) + \phi_S(u). \tag{4}$$

Because $u$ is an arbitrary parameter, we obtain from Eqs. (3) and (4)

$$\rho_{T+S}(x) = \rho_T(x) + \rho_S(x), \quad \forall x. \tag{5}$$

The differentiable solution of Eq. (5), regarding parameter $T$, is only linear one

$$\rho_T(x) = T\rho(x).$$

So, from Eq. (3) we obtain [9]

$$\phi_T(u) = T\int_{-\infty}^{\infty} \frac{e^{iux} - 1 - iu\sin x}{x^2}\rho(x)\,dx, \tag{6}$$

where the kernel $\rho(x) \geq 0$.

## 3. Functional Correlation Formula for Non-Gaussian White Noise

The time derivative of generalized Wiener process $\xi(t) = \dot{\eta}(t)$ is a stationary random process and has analogy to Gaussian white noise. Now we derive the characteristic functional of this non-Gaussian delta-correlated noise. By definition, we have

$$\begin{aligned}
\Theta_t[u] &= \left\langle \exp\left\{i\int_0^t \xi(\tau)u(\tau)\,d\tau\right\}\right\rangle = \left\langle \exp\left\{i\int_0^t u(\tau)d\eta(\tau)\right\}\right\rangle \\
&= \left\langle \exp\left\{i\lim_{\delta_\tau\to 0}\sum_{k=1}^{n} u(\vartheta_k)[\eta(\tau_k) - \eta(\tau_{k-1})]\right\}\right\rangle \\
&= \lim_{\delta_\tau\to 0}\left\langle \prod_{k=1}^{n}\exp\{iu(\vartheta_k)[\eta(\tau_k) - \eta(\tau_{k-1})]\}\right\rangle = \lim_{\delta_\tau\to 0}\prod_{k=1}^{n}\theta_{\Delta\tau_k}(u(\vartheta_k)),
\end{aligned} \tag{7}$$

where $\vartheta_k$ is some internal point of time interval $(\tau_{k-1}, \tau_k)$, $\delta_\tau = \max_k \Delta\tau_k$, $\Delta\tau_k = \tau_k - \tau_{k-1}$ ($\tau_0 = 0$, $\tau_n = t$), and $\theta_{\Delta\tau_k}$ is the characteristic function of increments. To obtain Eq. (7) we used the statistical independence of increments of generalized Wiener process $\eta(t)$. Further from Eqs. (6) and (7) we obtain

$$\begin{aligned}
\Theta_t[u] &= \lim_{\delta_\tau\to 0}\prod_{k=1}^{n}\exp\left\{\Delta\tau_k\int_{-\infty}^{\infty}\frac{e^{iu(\vartheta_k)x} - 1 - iu(\vartheta_k)\sin x}{x^2}\rho(x)\,dx\right\} \\
&= \exp\left\{\lim_{\delta_\tau\to 0}\sum_{k=1}^{n}\Delta\tau_k\int_{-\infty}^{\infty}\frac{e^{iu(\vartheta_k)x} - 1 - iu(\vartheta_k)\sin x}{x^2}\rho(x)\,dx\right\} \\
&= \exp\left\{\int_0^t d\tau\int_{-\infty}^{\infty}\frac{e^{iu(\tau)x} - 1 - iu(\tau)\sin x}{x^2}\rho(x)\,dx\right\}.
\end{aligned} \tag{8}$$

Now we derive a useful correlation formula for stochastic functionals of non-Gaussian white noise. We use the generalization of Furutsu–Novikov formula [10], to split the correlation between a stochastic functional $R_t[\xi]$ of non-Gaussian random process $\xi(\tau)$, defined on the observation interval $\tau \in (0,t)$, and the process $\xi(t)$ itself

$$\langle \xi(t) R_t[\xi+z]\rangle = \left.\frac{\dot{\Phi}_t[u]}{iu(t)}\right|_{u=\frac{\delta}{i\delta z}} \langle R_t[\xi+z]\rangle . \tag{9}$$

Here $z(t)$ is an arbitrary deterministic function, and $\Phi_t[u] = \ln \Theta_t[u]$. From Eq. (8) we obtain the following expression for variational operator in Eq. (9)

$$\frac{\dot{\Phi}_t[u]}{iu(t)} = \int_{-\infty}^{\infty} \frac{e^{iu(t)x} - 1 - iu(t)\sin x}{iu(t)x^2}\rho(x)\,dx = \int_{-\infty}^{\infty} \frac{\rho(x)}{x^2}\,dx \int_0^x [e^{iu(t)y} - \cos y]dy.$$

Substituting this expression in Eq. (9) we arrive at

$$\langle \xi(t) R_t[\xi+z]\rangle = \int_{-\infty}^{\infty} \frac{\rho(x)}{x^2}\,dx \int_0^x \left[\exp\left\{y\frac{\delta}{\delta z(t)}\right\} - \cos y\right] \langle R_t[\xi+z]\rangle\, dy.$$

By inserting the operator of functional differentiation into the average and by putting $z = 0$, we get finally

$$\langle \xi(t) R_t[\xi]\rangle = \int_{-\infty}^{\infty} \frac{\rho(x)}{x^2}\,dx \int_0^x \left[\left\langle \exp\left\{y\frac{\delta}{\delta \xi(t)}\right\} R_t[\xi]\right\rangle - \langle R_t[\xi]\rangle \cos y\right] dy. \tag{10}$$

## 4. Derivation of Kolmogorov's Equation

Let us consider now the Langevin equation with a non-Gaussian white noise $\xi(t)$

$$\dot{x} = f(x,t) + g(x,t)\,\xi(t) . \tag{11}$$

By differentiating with respect to time the following expression for probability density of the random process $x(t)$

$$W(x,t) = \langle \delta(x - x(t))\rangle , \tag{12}$$

and taking into account Eq. (11), we obtain

$$\frac{\partial W}{\partial t} = -\frac{\partial}{\partial x}[f(x,t)W] - \frac{\partial}{\partial x} g(x,t)\langle \xi(t)\,\delta(x - x(t))\rangle . \tag{13}$$

By using functional differentiation rules and following the same procedure of Ref. [11], from Eq. (11) we get

$$\frac{\delta}{\delta\xi(t)}\delta(x - x(t)) = -\frac{\partial}{\partial x} g(x,t)\,\delta(x - x(t)) . \tag{14}$$

Thus, the operator $\delta/\delta\xi(t)$ of functional differentiation with respect to the function $\delta(x - x(t))$ is equivalent to the ordinary differential operator $-\partial/\partial x\,(g(x,t))$. Using Eq. (10) in Eq. (13) and taking into account Eq. (14), we obtain the Kolmogorov's equation for nonlinear system (11) driven by non-Gaussian white noise

$$\frac{\partial W}{\partial t} = -\frac{\partial[f(x,t)W]}{\partial x} + \int_{-\infty}^{\infty} \frac{\rho(z)}{z^2}\left[\exp\left\{-z\frac{\partial}{\partial x}g(x,t)\right\} - 1 + \sin z\frac{\partial}{\partial x}g(x,t)\right] dzW, \tag{15}$$

which is the main result of our paper. By series expansion of the exponential operator in Eq. (15), we have

$$\frac{\partial W}{\partial t} = -\frac{\partial}{\partial x}\left[f(x,t)\,W\right] + \sum_{n=1}^{\infty}\frac{(-1)^n A_n}{n!}\left[\frac{\partial}{\partial x}\,g(x,t)\right]^n W, \tag{16}$$

where

$$A_1 = \int_{-\infty}^{\infty}\frac{z-\sin z}{z^2}\,\rho(z)\,dz, \qquad A_n = \int_{-\infty}^{\infty} z^{n-2}\rho(z)\,dz, \quad n \geq 2. \tag{17}$$

Equation (16) is valid if all integrals in Eq. (17) converge. It coincides with the result obtained in Ref. [11] for Poissonian white noise $\xi(t)$, if we interpret the coefficients $A_n$ as the moments of random jumps of $x(t)$. Moreover, this equation can be recast in the Kramers–Moyal form

$$\frac{\partial W}{\partial t} = \sum_{n=1}^{\infty}\frac{(-1)^n}{n!}\frac{\partial^n}{\partial x^n}\left[K_n(x,t)\,W\right], \tag{18}$$

where $K_n(x,t)$, $n = 1, 2, \ldots$ are the kinetic coefficients.

## 5. Particular Cases of Non-Gaussian Noise Sources

Following Ref. [8] we analyze different kernel functions $\rho(x)$ to obtain particular cases of Kolmogorov's equation (15), related to different non-Gaussian noise sources. (a) As a first simple case we consider a Gaussian white noise $\xi(t)$. The corresponding kernel function is $\rho(x) = 2D\delta(x)$. Therefore we obtain the ordinary Fokker–Planck equation

$$\frac{\partial W}{\partial t} = -\frac{\partial}{\partial x}\left[f(x,t)\,W\right] + D\frac{\partial}{\partial x}\,g(x,t)\,\frac{\partial}{\partial x}\,g(x,t)\,W. \tag{19}$$

(b) For Poisson kernel $\rho(x) = B\,\delta(x-q)$, Eq. (15) becomes

$$\frac{\partial W}{\partial t} = -\frac{\partial}{\partial x}\left[f(x,t) + \frac{Bg(x,t)(q-\sin q)}{q^2}\right]W + B\sum_{n=2}^{\infty}\frac{(-1)^n q^{n-2}}{n!}\left[\frac{\partial}{\partial x}g(x,t)\right]^n W. \tag{20}$$

(c) When the increment $\eta_T$ of generalized Wiener process has the gamma distribution with exponent $\mu$, the kernel is $\rho(x) = \mu x e^{-ax}$ $(x \geq 0)$. As a result, after integration we find from Eq. (15)

$$\frac{\partial W}{\partial t} = -\frac{\partial}{\partial x}\left[f(x,t) - \mu g(x,t)\,\mathrm{arc\,cot}\,a\right]W - \mu\ln\left[1 + \frac{1}{a}\frac{\partial}{\partial x}\,g(x,t)\right]W. \tag{21}$$

(d) For one-side $\alpha$-stable Lévy distributions the kernel function is $\rho(x) = \rho_0\,x^{1-\alpha}$ $(x \geq 0)$, and we get

$$\frac{\partial W}{\partial t} = -\frac{\partial}{\partial x}\left[f(x,t)\,W\right]$$
$$+\rho_0\int_0^{\infty}\frac{dz}{z^{\alpha+1}}\left[\exp\left\{-z\frac{\partial}{\partial x}\,g(x,t)\right\} - 1 + \frac{\partial}{\partial x}\,g(x,t)\sin z\right]W(x,t).$$

After rearrangements we obtain

$$\frac{\partial W}{\partial t}=-\frac{\partial\left[f(x,t)W\right]}{\partial x}+\frac{\rho_0}{\alpha}\frac{\partial}{\partial x}\,g\left(x,t\right)\int_0^{\infty}\frac{dy}{y^{\alpha}}\left[\cos y-\exp\left\{-y\frac{\partial}{\partial x}\,g\left(x,t\right)\right\}\right]W. \tag{22}$$

This equation is the fractional Fokker–Planck equation, and describes anomalous diffusion in the form of asymmetric Lévy flights.
(e) For additive driving noise $\xi\left(t\right)$, $g\left(x,t\right)=1$ in Eq. (11), and the exponential operator in Eq. (15) reduces to the space shift operator. As a result, we find

$$\frac{\partial W}{\partial t}=-\frac{\partial}{\partial x}\left[f\left(x,t\right)W\right]+\int_{-\infty}^{\infty}\frac{\rho\left(z\right)}{z^2}\left[W\left(x-z,t\right)-W\left(x,t\right)+\sin z\frac{\partial W\left(x,t\right)}{\partial x}\right]dz. \tag{23}$$

Equation (23) is similar to the Kolmogorov–Feller equation for purely discontinuous Markovian processes [4, 12]

$$\frac{\partial W}{\partial t}=\nu\int_{-\infty}^{\infty}Q\left(x-z\right)W\left(z,t\right)dz-\nu W\left(x,t\right), \tag{24}$$

where $Q\left(x\right)$ is probability density of jumps step, and $\nu$ is the constant mean rate of jumps. Putting $f\left(x,t\right)=0$ and comparing Eq. (23) with Eq. (24) we obtain the kernel function for this case

$$\rho\left(x\right)=\nu x^2Q\left(x\right). \tag{25}$$

For non-Gaussian driving force $\xi\left(t\right)$, with symmetric $\alpha$-stable Lévy distribution, the kernel function is $\rho\left(x\right)=D\left|x\right|^{1-\alpha}$. As a result, Eq. (23) takes the following form

$$\frac{\partial W}{\partial t}=-\frac{\partial}{\partial x}\left[f\left(x,t\right)W\right]+D\int_{-\infty}^{\infty}\frac{W\left(z,t\right)-W\left(x,t\right)}{\left|x-z\right|^{\alpha+1}}dz \tag{26}$$

and describes the anomalous diffusion in form of symmetric Lévy flights.
(f) Finally, for multiplicative noise $\xi\left(t\right)$ with $g\left(x,t\right)=x$, the operator in Eq. (15) reduces to a scaling operator, and the generalized Kolmogorov's equation becomes

$$\frac{\partial W}{\partial t}=-\frac{\partial}{\partial x}\left[f\left(x,t\right)-Kx\right]W+\int_{-\infty}^{\infty}\frac{\rho\left(z\right)}{z^2}\left[W\left(e^{-z}x,t\right)-W\left(x,t\right)\right]dz, \tag{27}$$

where

$$K=\int_{-\infty}^{\infty}\frac{\rho\left(z\right)\sin z}{z^2}\,dz.$$

## 6. Simple Examples of Superdiffusion

Now we derive the stationary probability distribution $W_{\infty}(x)$ for some simple cases of anomalous diffusion in a fixed potential $U(x)$ and in overdamped regime (see also [6, 13–15]). The Langevin equation for the particle displacement $x(t)$ reads

$$\dot{x}=-U'\left(x\right)+\xi\left(t\right), \tag{28}$$

where $\xi\left(t\right)$ is a symmetric $\alpha$-stable Lévy process. The corresponding Kolmogorov's equation is Eq. (26), with $f(x,t)=-U'(x)$. The stationary probability distribution

$W_\infty(x)$ can be found from the following equation

$$\frac{d}{dx}\left[U'(x)\,W_\infty\right] + D\int_{-\infty}^{\infty}\frac{W_\infty(z) - W_\infty(x)}{|x-z|^{\alpha+1}}\,dz = 0. \tag{29}$$

(a) For uniform potential well with two reflecting boundaries at $x = \pm L$ we get, from Eq. (29), the uniform stationary probability density

$$W_\infty(x) = \frac{1}{2L}$$

as for ordinary Brownian diffusion.
(b) For linear system with parabolic potential $U(x) = \beta x^2/2$, by Fourier transform of Eq. (29), we obtain the asymptotic characteristic function of the process $x(t)$ as [6]

$$\theta_\infty(k) = \exp\left\{-\frac{2\Gamma(-\alpha)\cos(\pi\alpha/2)}{\alpha\beta}\,|k|^\alpha\right\}. \tag{30}$$

(c) For symmetric potential $U(x) = \gamma|x|$, after Fourier transform of Eq. (29) we find

$$\gamma k G(k) + 2D\,|k|^\alpha\,\Gamma(-\alpha)\,F(k)\cos\frac{\pi\alpha}{2} = 0, \tag{31}$$

where

$$F(k) = \int_0^\infty W_\infty(x)\cos kx\,dx, \qquad G(k) = \int_0^\infty W_\infty(x)\sin kx\,dx.$$

Functions $F(k)$ and $G(k)$ are connected by Hilbert formula

$$G(k) = \frac{1}{\pi}\,v.p.\int_0^\infty\frac{F(q)}{k-q}\,dq, \tag{32}$$

where $v.p.$ means the principal value of integral. After substitution of Eq. (32) in Eq. (31) we obtain finally the following integral equation for the asymptotic characteristic function $\theta_\infty(k) = 2F(k)$

$$v.p.\int_0^\infty\frac{\theta_\infty(q)}{k-q}\,dq + \frac{2\pi D}{\gamma k}\,|k|^\alpha\,\Gamma(-\alpha)\,\theta_\infty(k)\cos\frac{\pi\alpha}{2} = 0\,. \tag{33}$$

(d) For quartic potential $U(x) = \beta x^4/4$, from Eq. (29) we obtain the following differential equation for stationary characteristic function

$$\beta k\,\frac{d^3\theta_\infty(k)}{dk^3} - 2D\,|k|^\alpha\,\Gamma(-\alpha)\,\theta_\infty(k)\cos\frac{\pi\alpha}{2} = 0. \tag{34}$$

Equations (33) and (34) can be solved numerically.

## 7. Conclusions

Starting from the *generalized Wiener process*, whose increments have the probability distribution belonging to the class of infinitely divisible distributions, we obtained the new correlation formula for stochastic functionals. This allows us to derive the Kolmogorov's equation (15) for diffusion, induced by non-Gaussian white noise source, directly from Langevin equation. The Fokker–Planck equation for ordinary Brownian motion, the Kolmogorov–Feller equation for purely discontinuous Markovian processes and the fractional Fokker–Planck equation for anomalous diffusion are obtained as particular cases of the Kolmogorov's equation (15). This equation can be used for determining the stationary probability distributions, probability distributions of first-passage times and other statistical characteristics of nonlinear dynamical systems.

## Acknowledgments

This work has been supported by INTAS Grant 2001-0450, MIUR, INFM-CNR, by Russian Foundation for Basic Research (project 05-02-16405), and by Federal Program "Leading Scientific Schools of Russia" (project 1729.2003.2).

## References

[1] S. M. Bezrukov and I. Vodyanoy, *Noise-induced enhancement of signal transduction across voltage-dependent ion channels*, *Nature* **378** (1995) 362–364.

[2] P. Castiglione, A. Mazzino, P. Muratore-Ginanneschi and A. Vulpiani, *On strong anomalous diffusion*, *Physica D* **134** (1999) 75–93.

[3] R. Metzler and J. Klafter, *The random walk's guide to anomalous diffusion: a fractional dynamics approach*, *Phys. Rep.* **339** (2000) 1–77.

[4] A. I. Saichev and G. M. Zaslavsky, *Fractional kinetic equations: solutions and applications*, *Chaos* **7** (1997) 753–764.

[5] B. J. West, P. Grigolini, R. Metzler and T. F. Nonnenmacher, *Fractional diffusion and Lévy stable processes*, *Phys. Rev. E* **55** (1997) 99–106.

[6] S. Jespersen, R. Metzler and H. C. Fogedby, *Lévy flights in external force fields: Langevin and fractional Fokker–Planck equations and their solutions*, *Phys. Rev. E* **59** (1999) 2736–2745.

[7] V. V. Yanovsky, A. V. Chechkin, D. Schertzer and A. V. Tur, *Lévy anomalous diffusion and fractional Fokker–Planck equation*, *Physica A* **282** (2000) 13–34.

[8] W. Feller, *An Introduction to Probability Theory and its Applications*, Vol. 2, John Wiley & Sons, Inc., New York (1971).

[9] R. L. Stratonovich, *Nonlinear Nonequilibrium Thermodynamics I*, Springer, Berlin (1992).

[10] V. I. Klyatskin, *To statistical theory of light reflection in randomly inhomogeneous medium*, *Sov. Phys. JETP* **38** (1974) 27–34.

[11] P. Hänggi, *Correlation functions and master equations of generalized (non-Markovian) Langevin equations*, *Z. Physik B* **31** (1978) 407–416.

[12] A. Kamińska and T. Srokowski, *Simple jumping process with memory: Transport equation and diffusion*, *Phys. Rev. E* **69** (2004) 062103-1–062103-4.

[13] I. Eliazar and J. Klafter, *Lévy-driven Langevin systems: Targeted stochasticity*, *J. Stat. Phys.* **111** (2003) 739–768.

[14] A. V. Chechkin, V. Yu. Gonchar, J. Klafter, R. Metzler and L. V. Tanatarov, *Lévy flights in a steep potential well*, *J. Stat. Phys.* **115** (2004) 1505–1535.

[15] Y. P. Kalmykov, W. T. Coffey and S. V. Titov, *Bimodal approximation for anomalous diffusion in a potential*, *Phys. Rev. E* **69** (2004) 021105-1–021105-7.

# CONTROLLING ACTIVATED PROCESSES

M. I. DYKMAN and BRAGE GOLDING
*Department of Physics and Astronomy, Michigan State University*
*East Lansing, MI 48824, USA*

Received 13 February 2001
Revised 13 March 2001
Accepted 14 March 2001

The rates of activated processes, such as escape from a metastable state and nucleation, are exponentially sensitive to an externally applied field. We describe how this applies to modulation by high-frequency fields and illustrate it with experimental observations. The results may lead to selective control of diffusion in periodic potentials, novel control mechanisms for crystal growth, and new separation techniques.

*Keywords*: Activated escape, optical trapping, periodic driving, fluctuational paths

## 1. Introduction

Activated processes are responsible for large qualitative changes in broad classes of systems. A well-known example is escape from a potential well, in which fluctuations carry the system over a potential barrier. Activated escape underlies diffusion in crystals, protein folding, and provides a paradigm for activated chemical reactions. Another example is nucleation in phase transitions. It would be advantageous to control the probabilities of activated processes by applying a comparatively weak external force. The idea is that the force need not be solely responsible for driving the system over the barrier; it only must appropriately influence fluctuations.

A familiar phenomenon which has elements of control of activated processes is stochastic resonance (SR) [1]. In SR, an adiabatic modulation of the system parameters by a slowly varying field is usually assumed. The strong effect of the field can be readily understood in this case, if one notices that the probability of a thermally activated process is $W \propto \exp(-R/k_BT)$, where $R$ is the activation energy (the barrier height for escape from a potential well). Even a comparatively small field-induced modulation $\delta R$ of the activation energy, $|\delta R| \ll R$, greatly affects $W$ provided $|\delta R| > k_BT$, with $\ln W$ being linear in the modulation amplitude.

In contrast, one might expect that a high-frequency field would just "heat up" the system by changing its effective temperature. The rate $W$ would then be incremented by a term proportional to the field intensity $I$ rather than the amplitude

$A \propto I^{1/2}$. This is indeed the case in the weak-field limit [2,3]. However, one may ask what happens if the appropriately weighted field amplitude is not small compared to the fluctuation intensity (temperature).

Recent results [4–7] show that, counter-intuitively, for high-frequency driving the change of $\ln W$ is linear in $A$, over a broad range of $A$. The exponential effect of nonadiabatic driving leads to a number of new phenomena not encountered in SR, including resonant (in the field frequency) rate enhancement. This provides the basis for much of the selectivity and flexibility in controlling fluctuations, as we now outline.

## 2. General formulation

One can effectively control activated processes because, although they happen at random, the trajectories of the system in an activated process are close to a specific trajectory. The latter is called the optimal path for the corresponding process [8]. The effect of the driving field accumulates along the optimal path, giving rise to a linear-in-the-field correction to the activation energy $R$. It can be calculated if the fluctuational dynamics of the system is known. Alternatively, response to the driving can be found experimentally, and in fact used to investigate the system dynamics away from the metastable state.

A fairly general case in which the field effect can be evaluated and insight gained into the underlying mechanism is a dynamical system driven by a stationary colored Gaussian noise $f(t)$ with a power spectrum $\Phi(\omega)$ of arbitrary shape [9]. The Langevin equation of motion is of the form:

$$\dot{q} = K(q;t) + f(t), \quad K(q;t) = K_0(q) + F(t), \; F(t+\tau_F) = F(t), \tag{1}$$

where $\tau_F$ is the period of the driving force $F(t)$.

For small characteristic noise intensity $D = \max \Phi(\omega)/2$, the system mainly performs small fluctuations about its periodic metastable state $q_a(t)$. Large fluctuations, like those leading to escape from the basin of attraction to $q_a$, require large bursts of $f(t)$ which would overcome the restoring force $K$. The probability densities of large bursts of $f(t)$ are exponentially small, $\propto \exp[-(2D)^{-1} \int dt\, dt'\, f(t)\hat{\mathcal{F}}(t-t')f(t')]$, and exponentially different depending on the form of $f(t)$ [$\hat{\mathcal{F}}(t)$ is given by the Fourier transform of $2D/\Phi(\omega)$]. Therefore for any state $q_f$ into which the system is brought by the noise at time $t_f$, there exists a realization $f(t) = f_{\mathrm{opt}}(t|q_f, t_f)$ which is exponentially more probable than the others. This optimal realization and the corresponding optimal path of the system $q_{\mathrm{opt}}(t)$ provide the minimum to the functional

$$\mathcal{R}[q(t), f(t)] = \frac{1}{2}\iint dt\, dt'\, f(t)\hat{\mathcal{F}}(t-t')f(t') + \int dt\, \lambda(t)\,[\dot{q} - K(q;t) - f(t)] \tag{2}$$

(the integrals are taken from $-\infty$ to $\infty$). The Lagrange multiplier $\lambda(t)$ relates $f_{\mathrm{opt}}(t)$ and $q_{\mathrm{opt}}(t)$ to each other [cf. Eq. (1); $\lambda(t) = 0$ for $t > t_f$]. The activation rate has the form

$$W = C \exp[-R/D], \; R = \min \mathcal{R}. \tag{3}$$

The exponent $R$ can be obtained for an arbitrary noise spectrum and an arbitrary periodic driving by solving the variational problem (2) numerically, with appropriate boundary conditions [4, 9].

We now turn to the case where the driving force $F(t)$ is comparatively weak, so that the field-induced correction $|\delta R| \ll R$. Nonetheless, $|\delta R|$ may exceed $D$ and thus strongly change the rate $W$ (3). To first order, $\delta R$ can be obtained by evaluating the term $\propto F(t)$ in (2) along the path $q_{\rm opt}^{(0)}(t)$, $f_{\rm opt}^{(0)}(t)$, $\lambda^{(0)}(t)$ calculated for $F = 0$. Special care has to be taken when activated escape and nucleation are analyzed. Here in the absence of driving, the optimal path is an instanton, the optimal fluctuation may occur at any time $t_c$. The field $F(t)$ lifts the time degeneracy of escape paths. It synchronizes optimal escape trajectories, selecting one per period, so as to minimize the activation energy of escape $R$ [4, 5]. The correction $\delta R$ should be evaluated along the appropriate trajectory,

$$\delta R = \min_{t_c} \delta R(t_c), \quad \delta R(t_c) = \int_{-\infty}^{\infty} dt\, \chi(t - t_c) F(t), \quad \chi(t) = -\lambda^{(0)}(t). \tag{4}$$

Eq. (4) provides a closed-form expression for the change of the time-averaged activation rate $\bar{W}$, for an arbitrary spectrum of the driving field $F(t)$. Clearly $\ln \bar{W}$ is linear in $F$, and the corresponding coefficient $\chi$ is therefore called the logarithmic susceptibility (LS) [4]. Because of minimization over $t_c$, the change of $\ln \bar{W}$ is nonanalytic in $F(t)$, which leads to a number of observable consequences. The LS has been evaluated for overdamped and underdamped white-noise driven systems [4]. Extensive numerical and analog simulations of the escape rate in driven systems [6] are in excellent qualitative and quantitative agreement with the theory, including the prefactor [4, 5, 10], over a broad range of field amplitudes.

## 3. Dynamical symmetry breaking in an optical trap

A simple physical system which embodies a number of activated phenomena is a mesoscopic dielectric Brownian particle trapped by a strongly focused laser beam creating an optical gradient trap, i.e. "optical tweezers" [11]. Techniques based on optical tweezers have found broad applications in contactless manipulation of objects such as atoms, colloidal particles, and biological materials. Activated escape can be studied using a dual optical trap generated by two closely spaced parallel light beams. This was used initially to investigate the synchronization of interwell transitions by low-frequency (adiabatic) sinusoidal forcing [12].

Quantitative characterization of activated processes requires that the double-well confining potential of a dual trap $U(\mathbf{r})$ be adequately determined. The corresponding measurement, for a transparent spherical silica particle of diameter $2R = 0.6\ \mu$m optically trapped in water, was reported recently [14].

In the experiment [14], all three coordinates of the particle are determined simultaneously. The double-well potential $U(\mathbf{r})$ is found directly from the measured stationary distribution $\rho(\mathbf{r}) = Z^{-1} \exp[-U(\mathbf{r})/k_B T]$. From the Kramers theory [13], it is possible then to calculate the rates $W_{ij}$ $(i, j = 1, 2)$ of activated transitions between the minima of $U(\mathbf{r})$. For the range of $U(\mathbf{r})$ in which $W_{ij}$ vary by nearly 3 orders of magnitude, the calculated values of $W_{ij}$ are in excellent quantitative agreement with the results of direct measurements. This provides a direct model-free test of the multidimensional Kramers rate theory, with no adjustable parameters.

The double-beam trap can also be used to investigate the effect of ac-modulation on transition rates. An interesting application of this effect is to direct the diffusion of a particle in a spatially periodic potential [15]. For a generic periodic potential, the ac-induced change of the activation barrier differs depending on the direction in which the particle moves (right or left, for example). This makes the probabilities of transitions to the right and to the left exponentially different and results in diffusion in the direction of more frequent transitions.

An effect closely related to directed diffusion, but more amenable to testing using optical trapping, is ac-field induced localization in one of the wells of a symmetric double-well potential. Both effects should occur if the field breaks the spatio-temporal symmetry of the system. The ratio of the period-averaged stationary populations $\bar{w}_1, \bar{w}_2$ of the wells is determined by the ratio of the period-averaged transition rates $\bar{W}_{ij}$,

$$\bar{w}_1/\bar{w}_2 = \bar{W}_{21}/\bar{W}_{12} \propto \exp([\delta R_1 - \delta R_2]/k_B T), \tag{5}$$

where $\delta R_{1,2}$ are the field-induced corrections (4) to the activation energies of escape from wells 1,2.

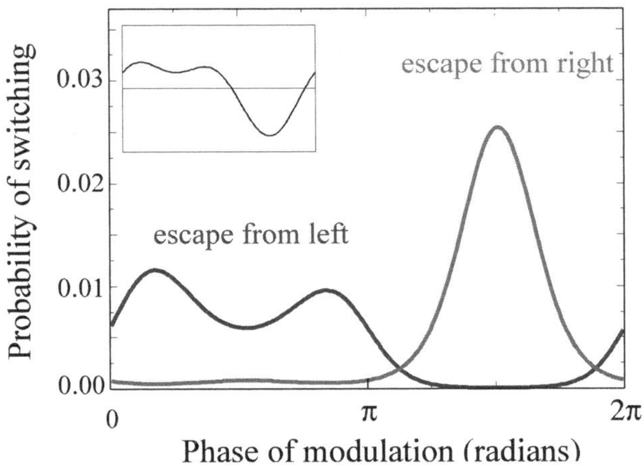


Fig 1. The least-squares fits to the experimentally determined instantaneous time-dependent switching probabilities $W_{ij}(t)$ for a particle in the adiabatically modulated double-beam trap, over a cycle $\omega_F t$ of the modulating waveform. The phase angle between the first and second harmonics is $\phi_{12} = \pi/2$. When the phase angle is incremented by $\pi$, the escape rates from the left and right wells interchange. The inset shows the instantaneous difference between the heights of the potential barriers in the two wells.

The experimental data [7] on effective localization due to the field-induced symmetry breaking are shown in Fig. 1. The experiment is conducted with a symmetric double-well potential and barrier height $\approx 7.5\, k_B T$. The intensity of the laser beams is then modulated so that the well depths are changed by $\delta U_1(t) = -\delta U_2(t) = \text{const} \times [\sin(\omega_F t) + (1/2)\sin(2\omega_F t + \phi_{12})]$. The modulation amplitude is $\approx 2.5\, k_B T$. The frequency $\omega_F/2\pi$ varies between 1 and 100 Hz, which covers the range from adiabatically slow to nonadiabatic modulation. Over this range, field-induced repopulation occurs between the wells for a nonsinusoidal modulation waveform, so that $\bar{w}_1 \neq \bar{w}_2$.

It is the breaking of the spatio-temporal symmetry $t \to t + \pi/\omega_F, \mathbf{r} \to -\mathbf{r}$ that leads to the escape rate from one of the wells being on average much bigger than from the other, as seen from Fig. 1. In turn, this leads to a higher population in one of the wells. Not only is it observed under slow modulation, as evidenced by Fig. 1, but a population difference of 20% is also observed far into the nonadiabatic regime. This is sufficient to create significant directional diffusion, and demonstrates the onset of *dynamical* symmetry breaking. The effect would not arise if nonadiabatic driving led just to "heating" of the particle.

## 4. Conclusions

Investigation of methods to control activated processes is currently at a very exciting stage. The importance of the problem and its relevance to many areas, from condensed-matter physics to biophysics, is becoming increasingly appreciated. The results outlined here show that there is a fairly general approach to controlling fluctuations, and the first experiments on overdamped systems show that such control can indeed be exercised. A broad range of problems remains unexplored. They include such issues as a microscopic theory of driven many-body systems and experimental exploration of underdamped driven systems. We also envision practical applications of these results, starting with the development of new highly selective colloidal separation as well as crystal growth techniques.

## Acknowledgments

We are grateful to Lowell McCann and Vadim Smelyanskiy with whom many of the results discussed above were obtained. This research was supported by the NSF through grants DMR-9971537 and PHY-0071059.

## References

[1] For reviews on stochastic resonance see M.I. Dykman, D.G. Luchinsky, R. Mannella, P.V.E. McClintock, N.D. Stein, and N.G. Stocks, *Stochastic resonance in perspective*, Nuov. Cim. D **17**, pp. 661-83 (1995); L. Gammaitoni, P.H. Hänggi, P. Jung, and F. Marchesoni, *Stochastic resonance*, Rev. Mod. Phys. **70**, pp. 223-88 (1998); R.D. Astumian and F. Moss, *Overview: the constructive role of noise in fluctuation driven transport and stochastic resonance*, Chaos **8**, pp. 533-8 (1998); K. Wiesenfeld and F. Jaramillo, *Minireview of stochastic resonance*, Chaos **8**, pp. 539-48 (1998).

[2] A.I. Larkin and Yu.N. Ovchinnikov, *Resonance reduction of the lifetime of the metastable state of tunnel junctions*, J. Low Temp. Phys. **63**, 317-29 (1986); B.I. Ivlev and V.I. Mel'nikov, *Effect of resonant pumping on activated decay-rates*, Phys. Lett. A **116**, 427-8 (1986); S. Linkwitz and H. Grabert, *Enhancement of the decay rate of a metastable state by an external driving force*, Phys. Rev. B **44** 11901-10 (1991).

[3] M.H. Devoret, D. Esteve, J.M. Martinis, A. Cleland, and J. Clarke, *Resonant activation of a Brownian particle out of a potential well: microwave-enhanced escape from the zero-voltage state of a Josephson junction*, Phys. Rev. B **36**, 58-73 (1987).

[4] M.I. Dykman, H. Rabitz, V.N. Smelyanskiy, and B.E. Vugmeister, *Resonant directed diffusion in nonadiabatically driven systems*, Phys. Rev. Lett. **79**, 1178–81 (1997); V.N. Smelyanskiy, M.I. Dykman, H. Rabitz, and B.E. Vugmeister, *Fluctuations, escape, and nucleation in driven systems: logarithmic susceptibility*, Phys. Rev. Lett. **79**, 3113–16 (1997).

[5] V.N. Smelyanskiy, M.I. Dykman, and B. Golding, *Time oscillations of escape rates in periodically driven systems*, Phys. Rev. Lett. **82**, 3193–3196 (1999).

[6] D.G. Luchinsky, R. Mannella, P.V.E. McClintock, M.I. Dykman, and V.N. Smelyanskiy, *Thermally activated escape of driven systems: the activation energy"*, J. Phys. A **32**, L371-7 (1999); M.I. Dykman, B. Golding, L.I. McCann, V.N. Smelyanskiy, D.G. Luchinsky, R. Mannella, and P.V.E. McClintock, *Activated escape of periodically driven systems*, to be published.

[7] L.I. McCann, M.I. Dykman, and B. Golding, *Controlling diffusion with light*, in preparation.

[8] M. I. Freidlin and A. D. Wentzell, *Random Perturbations of Dynamical Systems*, 2nd ed., Springer, New-York (1998).

[9] M.I. Dykman and B. Golding, *Controlling large fluctuations: theory and experiment*, in Lecture Notes in Physics **557**, eds. J.A. Freund and T. Pöschel, Springer, Berlin (2000), 365–377.

[10] The problem of the prefactor has been also addressed by R.S. Maier and D.L. Stein, *Noise-Activated Escape from a Sloshing Potential Well*, Phys. Rev. Lett. **86** (2001).

[11] A. Ashkin, J.M. Dziedzic, J.E. Bjorkholm, and S. Chu, *Observation of a single-beam gradient force optical trap for dielectric particles*, Optics Lett. **11**, 288–291 (1986).

[12] A. Simon A. and A. Libchaber, *Escape and synchronization of a Brownian particle*, Phys. Rev. Lett. **68**, 3375–3378 (1992)

[13] H. Kramers, *Brownian motion in a field of force and the diffusion model of chemical reactions*, Physica (Utrecht) **7**, 284–304 (1940); R. Landauer and J.A. Swanson, *Frequency factors in the thermally activated process*, Phys. Rev. **121**, 1668–1674 (1961).

[14] L.I. McCann, M.I. Dykman, and B. Golding *Thermally activated transitions in a bistable three-dimensional optical trap*, Nature **402**, 785–787 (1999).

[15] M. Magnasco, *Forced thermal ratchets*, Phys. Rev. Lett. **71**, 1477–80 (1993); for a review see F. Julicher, A. Ajdari, and J. Prost, *Modeling molecular motors*, Rev. Mod. Phys. **69**, 1269-1281, (1997).

# Chebyshev Spectra Resistance to Trend of Random Noise

Boris M. Grafov*, Alexey L. Kluev, Tatyana B. Kabanova
and Alexey D. Davydov

*Frumkin Institute of Physical Chemistry*
*and Electrochemistry, Russian Academy of Sciences*
*Leninskii pr. 31, Moscow 119071, Russia*
**boris.grafov@yandex.ru*

Received 5 February 2018
Accepted 25 April 2018
Published 25 May 2018

Communicated by Igor Khovanov

Spectral analysis of random noise in the space of discrete Chebyshev polynomials is an alternative to spectral Fourier analysis. The importance of Chebyshev spectral approach is associated with the fact that the discrete Chebyshev transformation of the $k$-th order eliminates automatically the polynomial trend of the $(k-1)$ order. Using the method of artificial trend, it was found that, under the real experimental conditions, the intensity of Chebyshev spectral lines with numbers higher than 1 is resistant to a strong trend of random process. This effect is observed when we use both the arithmetic averaging and the median. The Chebyshev spectral approach is a powerful tool for spectral analysis of random time series with a strong trend.

*Keywords*: Random time series; Chebyshev spectroscopy; resistance to trend.

## 1. Introduction

Spectral analysis of a random time series can be performed using any system of orthonormal discrete functions. In practice, the spectral analysis based on the application of discrete Fourier transformation (DFT) is most popular [1, 2]. However, the use of Fourier spectral approach is often complicated by the presence of a trend of random noise [3–10]. Therefore, the Fourier spectral analysis should be preceded by recognizing a trend signal and subtracting it from the measured noise signal using one of the detrending methods. The detrending should be performed with care, because it can significantly distort the Fourier spectrum of the random process under consideration.

We propose the spectral approach [11–16], which is based on the system of orthonormal Chebyshev polynomials of discrete variable [17, 18]. The Chebyshev spectral approach involves no operation of recognizing a trend of random time series.

The elimination of the trend of random process occurs automatically due to the orthogonal properties of discrete Chebyshev polynomials.

It should be added that the Chebyshev orthonormal basis is more appropriate for describing stochastic signals than DFT, which is the orthonormal basis based on a pair of harmonic functions $\{\cos(x)$ and $\sin(x)\}$ more suitable for the expansion of harmonic signals.

In this work, the theory of Chebyshev spectroscopy is briefly discussed and the resistance of Chebyshev spectrum to the trend of the random noise is analyzed. The resistance of Chebyshev high spectral lines to a strong trend of random process is proofed experimentally using the method of introducing an artificial trend in the measurements of corrosion noise.

## 2. Theory of Chebyshev Spectroscopy

One of the most common versions of Fourier spectral approach to the analysis of random time series involves the calculation of the Schuster periodogram [19], by averaging it using the Bartlett method [20]. We propose to do the same in the case of the Chebyshev spectral approach.

Let us denote the sampling frequency by $f_s$. The initial realization $y(t)$ can be divided into $M$ non-overlapping sequential segments $\{y(t, N, f_s, m)\}$, where $t$ is the integer discrete time and $m$ is the segment number. Each segment with number $m$ contains $N$ points. In each segment with number $m$, the discrete time $t$ varies in the range $(0 \leq t \leq N-1)$. The segment number $m$ takes $M$ integer values $(0 \leq m \leq M-1)$. In order to obtain a statistically significant result, the averaging over all $M$ segments is performed. The averaging can be arithmetic and in the median form. The operation of arithmetic averaging is denoted by $\langle \ldots \rangle$. The depth $M$ of statistical averaging depends on the chosen segment length $N$, because the product $N \cdot M$ is equal to the length of initial realization of random time series. The larger the segment length $N$ is, the smaller the statistical averaging depth $M$ will be. In our calculations, a segment commonly contains 4–16 points for the total number of points $M \cdot N = 2^{15} - 2^{20}$.

In each segment with number $m$, the Chebyshev transform $Y_k(N, f_s, m)$ of initial random time series can be calculated:

$$Y_k(N, f_s, m) = \sum_{t=0}^{N-1} P_k(t, N) y(t, N, f_s, m). \tag{1}$$

In Eq. (1), a set $\{P_k(t, N)\}$ is a system of basis functions of $N$-dimensional space in the form of orthonormal Chebyshev polynomials $P_k(t, N)$ of discrete variable $t$ $(0 \leq t \leq N-1)$. The subscript $k$ (the order of the Chebyshev polynomial) takes the integer values from 0 to $N-1$. The right-hand side of Eq. (1) contains the scalar product $(P_k(t, N), y(t, N, f_s, m))$ of basis vector $P_k(t, N)$ by a random vector $y(t, N, f_s, m)$.

The discrete Chebyshev spectrum has a simple interpretation. In accordance with the Plancherel equation $\sigma_y^2 = N^{-1} \sum_{k=0}^{N-1} S_k(N, f_S)$, the intensity of each Chebyshev spectral line $S_k(N, f_S)$ is an estimate of dispersion $\sigma_y^2$ of random process under investigation $y(t)$.

Squaring of the Chebyshev transform (1) leads to an analogue of the Schuster periodogram. The sampling Chebyshev spectrum $S_k(N, f_s)$ is obtained by averaging the squared Chebyshev transform over all $M$ segments:

$$S_k(N, f_s) = \langle [Y_k(N, f_s, m)]^2 \rangle. \tag{2}$$

According to Eq. (2), the intensity $S_k(N, f_s)$ of Chebyshev spectral line with number $k$ is a function of two variables: the segment length $N$ and the sampling frequency $f_s$.

For example, we present the following system of orthonormal Chebyshev polynomials that forms the basis of 4D Chebyshev space [17, 18]:

$$P_0(t, 4) = \frac{1}{2}(1, 1, 1, 1), \tag{3}$$

$$P_1(t, 4) = \frac{1}{2\sqrt{5}}(-3, -1, 1, 3), \tag{4}$$

$$P_2(t, 4) = \frac{1}{2}(1, -1, -1, 1), \tag{5}$$

$$P_3(t, 4) = \frac{1}{2\sqrt{5}}(-1, 3, -3, 1). \tag{6}$$

A direct calculation shows that the mutual scalar product of basis functions (3)–(6) is equal to zero and the scalar product of any basis function (3)–(6) by itself is equal to 1.

## 3. Application of Artificial Trend

Let us add an artificial trend discrete signal $TR(t)$ to the noise discrete signal under investigation $y(t)$. The sum of $y(t)$ discrete signal and $TR(t)$ discrete signal is obtained using a computer, i.e., after analog–digital transformation of $y(t)$ original signal. Then, Eq. (1) for discrete Chebyshev transform takes the following form:

$$Y_k(N, f_s, m) = \sum_{t=0}^{N-1} P_k(t, N)[y(t, N, f_s, m) + TR(t, N, f_s, m)]. \tag{7}$$

An artificial trend signal $TR(t)$ is chosen in the form of one sinusoid period:

$$TR(t) = A \sin\left(2\pi \frac{t}{NM}\right) \quad (0 \le t \le NM - 1). \tag{8}$$

The artificial trend can be controlled by varying amplitude $A$.

If the trend signal $TR(t)$ behaves almost as a linear function of discrete time within any segment $m$, the Chebyshev transform (7) virtually coincides with the

Chebyshev transform (1) for all $k \geq 2$. In this case, the Chebyshev spectrum, starting from the second spectral line, will be resistant to the effect of artificial trend. At the same time, this means that the natural trend has no noticeable effect on the Chebyshev spectral lines with numbers higher than 1. The last statement is valid only if the level of a trend of measured random time series is substantially lower than the level of artificial trend.

## 4. Experimental Procedure

In order to demonstrate the trend resistance of Chebyshev transformation, the electrochemical noise of corroding electrode was measured. Corrosion of metals is a practically important process that has an effect on the service life of all metal structures placed in various environments. The corrosion processes are characterized by the fluctuations (noise) of corrosion potential and/or corrosion current (see, for example, [22]). Measurement and analysis of corrosion noise constitute one of the methods for diagnosis and monitoring of metal parts.

The open-circuit voltage (OCV) between two electrodes of the working couple was measured and digitized using an AD7176-2SBZ evaluation board based on a high-impedance input amplifier and a 24-bit sigma-delta analog-to-digital converter (ADC). The sampling frequency *fs* was 20 Hz; the time $T$ was

$$T = \frac{M \cdot N}{f_s} = \frac{2^{13} \cdot 2^3}{20\,\text{Hz}} = 3276.8\,\text{s}.$$

Prior to each experiment, the noise of the measuring instrument at the short-cut input was recorded under similar experimental conditions.

The end-faces of two identical fragments of Steel 3 wire 1 mm in diameter were used as the working electrodes. The experimental procedure was described in detail in [21]. The electrode couple was placed into the electrochemical cell filled with 3% NaCl solution. A portion of 3% NaCl solution containing VpCI-639 HFB inhibitor (Cortec, USA) was introduced into the cell so that the final dilution factor of inhibitor was 1000. The noise signal was recorded every hour during the entire experiment.

## 5. Experimental Results

Figure 1 shows the noise signal $y(t)$ of measuring instrument, and Fig. 2 shows the original noise signal $y(t)$ and the mixed signal $y(t) + TR(t)$, which contains an artificial trend signal. The level of artificial trend $TR(t)$ is so high that the original noise signal $y(t)$ of instrument looks like a constant value (horizontal straight line). This additional artificial trend with a very large amplitude was added to show the resistance of the method of Chebyshev transformation to even very large trends.

Figure 3 shows the noise of corrosion process. The trend of corrosion noise is clearly seen. Figure 4 gives the mixed noise of corrosion process.

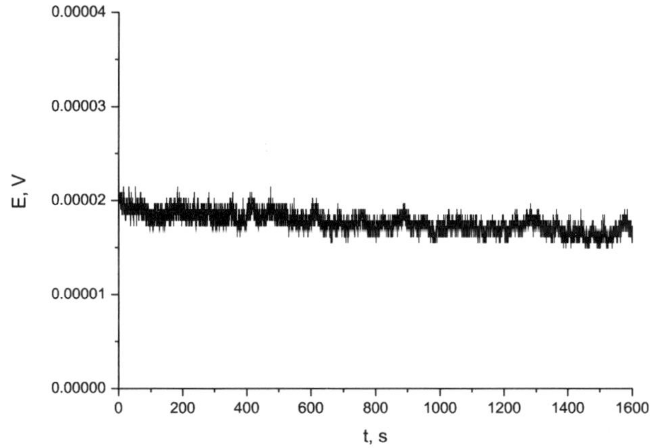


Fig. 1. Noise of measuring instrument.

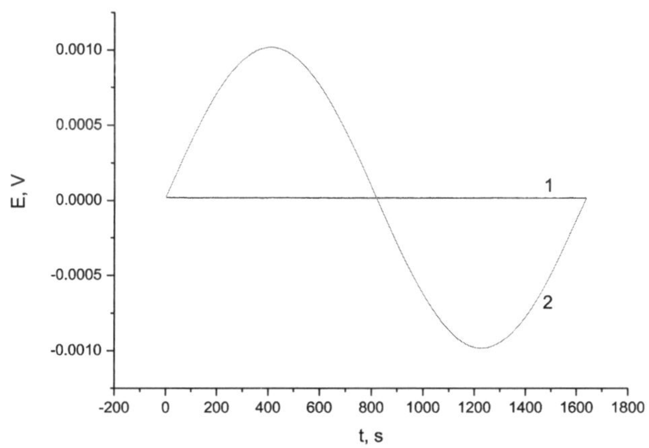


Fig. 2. (1) Original noise of instrument and (2) mixed signal containing artificial trend signal.

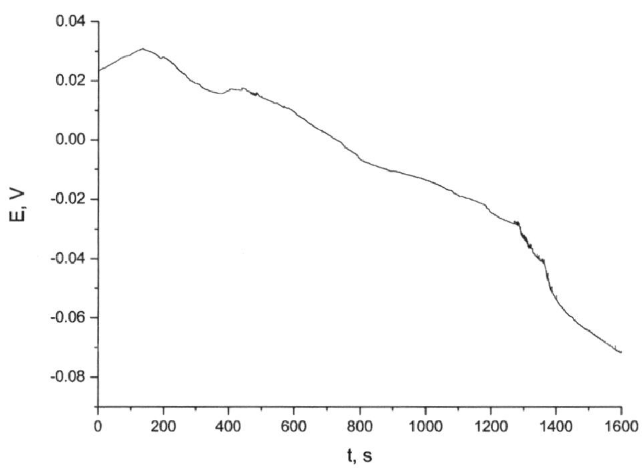


Fig. 3. Time dependence of corrosion noise.

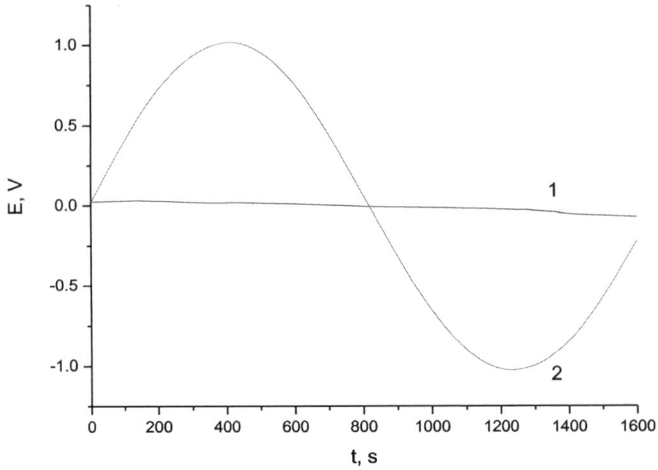


Fig. 4. (1) Original corrosion noise and (2) mixed signal containing artificial trend signal.

It is difficult to recognize the trend of original corrosion noise against the background of artificial trend.

Let us apply the theory of Chebyshev spectra (Eqs. (1)–(8)) at $N = 2^3$, $M = 2^{13}$ and the sampling frequency $f_s = 20\,\text{Hz}$.

## 6. Trend Resistance of Chebyshev Spectra for Measuring Instrument

Figure 5 shows two Chebyshev spectra for measuring instrument. Both spectra are obtained by the method of arithmetic averaging. One spectrum characterizes the

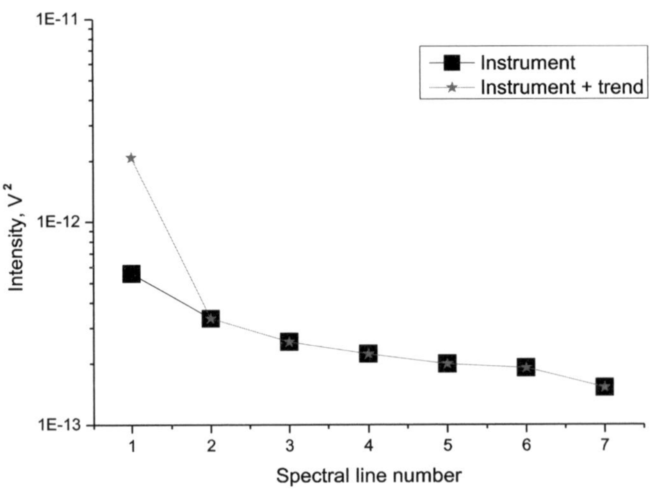


Fig. 5. The effect of artificial trend on Chebyshev spectrum for instrument (arithmetic averaging).

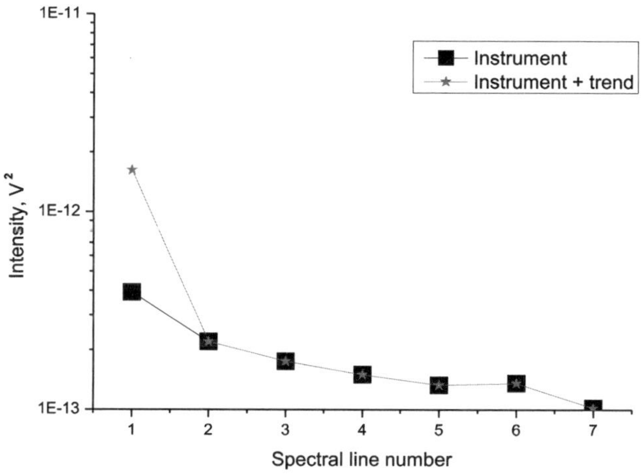


Fig. 6. The effect of artificial trend on Chebyshev spectrum for instrument (method of median).

original noise of instrument and another spectrum characterizes the original noise of instrument with the addition of artificial trend.

Figure 6 is similar to Fig. 5 except for the method of averaging. The data in Fig. 6 were obtained not by the method of arithmetic averaging, but by calculating the median of each Chebyshev spectral line of the noise of measuring instrument [23, 24].

From Figs. 5 and 6, it follows that the Chebyshev spectral lines with numbers 2–7 are resistant to a strong trend signal.

## 7. Trend Resistance of Corrosion Chebyshev Spectra

The curves in Fig. 7 were obtained by the method of arithmetic averaging. Figure 7 compares two spectra of noise corrosion signals, namely, the original noise signal and

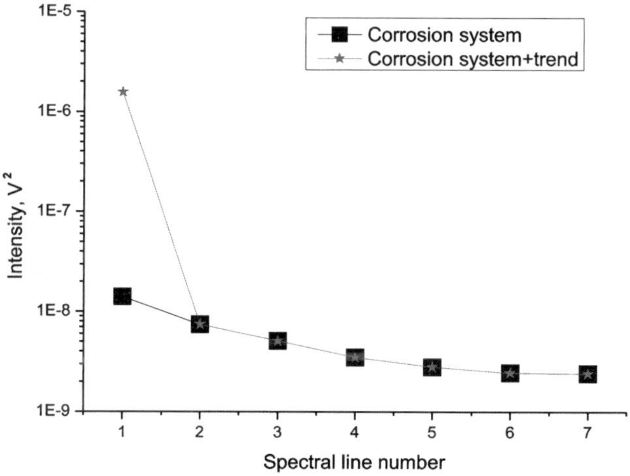


Fig. 7. The effect of artificial trend on corrosion Chebyshev spectrum (arithmetic averaging).

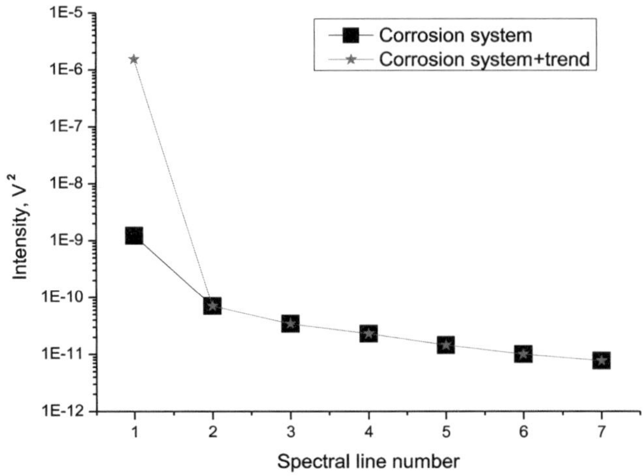


Fig. 8. The effect of artificial trend on corrosion Chebyshev spectrum (method of median).

a mixed noise signal obtained by mixing the original corrosion noise with an artificial trend. Figure 8 is similar to Fig. 7; however, the spectra in Fig. 8 were obtained by the median method. The proximity of Chebyshev spectra (with and without an artificial trend) points to the trend resistance of Chebyshev spectral lines of the corrosion process with numbers from 2 to 7.

## 8. Effect of Trend on Corrosion Discrete Fourier Spectrum

Figure 9 shows the effect of artificial trend on the discrete Fourier corrosion spectrum. The difference between the curves in Fig. 9 by more than an order confirms

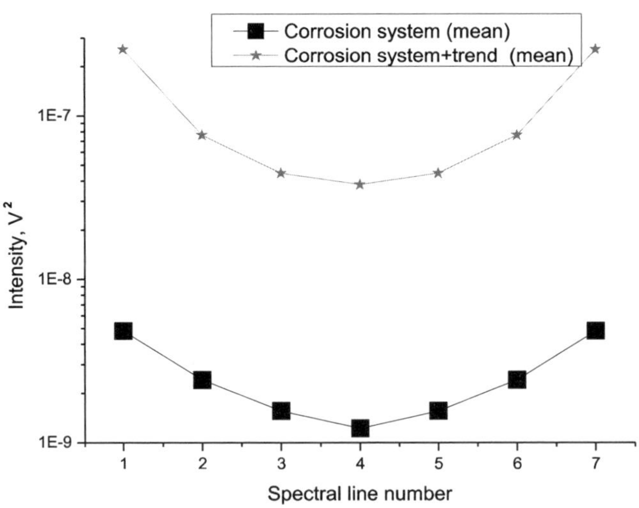


Fig. 9. The effect of artificial trend on corrosion discrete Fourier spectrum (arithmetic averaging).

that discrete Fourier spectroscopy, in contrast to the discrete Chebyshev spectroscopy, requires the detrended noise signal.

## 9. Conclusions

The Chebyshev spectral approach uses the expansion of the realization of random time series in terms of a system of orthonormal Chebyshev polynomials of a discrete variable instead of expansion in terms of Fourier complex exponentials. In this work, two types of statistical averaging are used: the arithmetic mean and the median. In the Chebyshev discrete spectroscopy, a trend is eliminated automatically in the course of calculating Chebyshev spectral line intensity. Figures 5–8 show that all high Chebyshev spectral lines, starting from the second spectral line, exhibit clearly defined resistance to a trend of measured noise.

The discrete Chebyshev spectroscopy is a powerful tool for spectral analysis of random time series with a strong trend.

## References


[1] P. S. Naidu, *Modern Spectrum Analysis of Time Series* (CRC Press, New York, 1996).

[2] S. V. Vaseghi, *Advanced Signal Processing and Digital Noise Reduction* (Wiley, Chichester, 1996).

[3] S. S. Rao, *A Course in Time Series Analysis, Technical Report* (Texas A&M University, College Station, TX, 2017).

[4] P. Mark, B. A. Braen and T. Ihn, Temperature-stabilized differential amplifier for low-noise DC measurements, *Rev. Sci. Instrum.* **88** (2017) 08106.

[5] F. Mansfeld, Z. Sun, C. H. Hsu and A. Nagiub, Concerning trend removal in electrochemical noise measurements, *Corros. Sci.* **43** (2001) 341–352.

[6] U. Bertocci, F. Huet, R. P. Nogueira and P. Rousseau, Trend removal procedures in the analysis of electrochemical noise, *Corrosion* **58** (2002) 337–347.

[7] J. Y. Huang, Y. B. Qiu and X. P. Guo, Comparison of polynomial fitting and wavelet transform to remove trend in electrochemical noise analysis, *Corros. Eng. Sci. Technol.* **45** (2010) 288–294.

[8] S. Y. Arman, N. Reza and P. M. Bijan, Effect of DC trend removal and window functioning methods on correlation between electrochemical noise parameters and EIS data of stainless steel in an inhibited acidic solution, *RSC Adv.* **4** (2014) 39045–39057.

[9] D. H. Xia and B. Yashar, Electrochemical noise: A review of experimental setup, instrumentation and DC removal, *Russ. J. Electrochem.* **51** (2015) 593–601.

[10] L. Lentka and J. Smulko, Analysis of effectiveness and computational complexity of trend removal methods, *Zeszyty Naukowe Wydzialu Elektrotechniki i Automatyki Politechniki Gdańskiej* **51** (2016) 111–114.

[11] M. G. Astaf'ev, L. S. Kanevskii and B. M. Grafov, Analyzing electrochemical noise with Chebyshev's discrete polynomials, *Russ. J. Electrochem.* **43** (2007) 17–24.

[12] B. M. Grafov, Yu. A. Dobrovol'skii, A. D. Davydov, A. E. Ukshe, A. L. Klyuev and E. A. Astaf'ev, Electrochemical noise diagnostics: Analysis of algorithm of orthogonal expansions, *Russ. J. Electrochem.* **51** (2015) 503–507.

[13] A. L. Klyuev, B. M. Grafov, Yu. A. Dobrovol'skii, A. D. Davydov and A. E. Ukshe, Variability of discrete Chebyshev spectra of electrochemical noise, *Russ. J. Electrochem.* **51** (2015) 1180–1185.

[14] A. L. Klyuev, A. D. Davydov, B. M. Grafov, Yu. A. Dobrovolskii, A. E. Ukshe and E. A. Astaf'ev, Electrochemical noise spectroscopy: Method of secondary Chebyshev spectrum, *Russ. J. Electrochem.* **52** (2016) 1001–1005.

[15] B. M. Grafov, Yu. A. Dobrovolskii, A. L. Klyuev, A. E. Ukshe, A. D. Davydov and E. A. Astaf'ev, Median Chebyshev spectroscopy of electrochemical noise, *J Solid State Electrochem.* **21** (2017) 915–918.

[16] B. M. Grafov, A. L. Klyuev, A. D. Davydov and V. P. Lukovtsev, Chebyshev's noise spectroscopy for testing electrochemical systems, *Bulg. Chem. Commun.* **49** (2017) Special Issue C, 102–105.

[17] N. Gogin and M. Hirvensalo, On the generating function of discrete Chebyshev polynomials, *Technical Report No. 819.* Turku Centre for Computer Science, Turku (2007).

[18] A. F. Nikiforov, S. K. Suslov and V. B. Uvarov, *Classical Orthogonal Polynomials of a Discrete Variable* (Springer, Berlin, 1991).

[19] A. Schuster, On the investigation on hidden periodicities with application to a supposed 26-day period of meteorological phenomena, *Terrestrial Magnetism* **3**(1) (1898) 13–41.

[20] M. S. Bartlett, Periodogram analysis and continuous spectra, *Biometrica* **37**(1–2) (1950) 1–16.

[21] B. M. Grafov, A. L. Klyuev and A. D. Davydov, Discrete version of Wiener—Khinchin theorem for Chebyshev's spectrum of electrochemical noise, *J. Solid State Electrochem.* **22**(6) (2018) 1661–1667.

[22] J. R. Kearns, J. R. Scully, P. R. Roberge, D. L. Reichert and J. L. Dawson, *Electrochemical Noise Measurement for Corrosion Applications* (ASTM, West Conshohocken, PA, 1996).

[23] N. Arley and K. R. Buch, *Introduction to the Theory of Probability and Statistics* (Wiley, New York, 1950).

[24] R. R. Wilcox, *Fundamentals of Modern Statistical Methods* (Springer, New York, 2010).

# THE ORTHOGONAL V-SYSTEM DETRENDED FLUCTUATION ANALYSIS

AIJING LIN* and PENGJIAN SHANG

*Department of Mathematics, School of Science*
*Beijing Jiaotong University, Beijing 100044, P. R. China*
**aijinglin128@gmail.com*

HUI MA

*Research Institute of Information Technology*
*Tsinghua University, Beijing 100084, P. R. China*

Received 9 November 2010
Accepted 13 January 2011

Communicated by Zoltan Gingl

The Detrended Fluctuation Analysis (DFA) and its extensions (MF-DFA) have been proposed as robust techniques to determine possible long-range correlations in self-affine signals. However, many studies have reported the susceptibility of DFA to trends which give rise to spurious crossovers and prevent reliable estimations of the scaling exponents. Lately, several modifications of the DFA method have been reported with many different techniques for eliminating the monotonous and periodic trends. In this study, a smoothing algorithm based on the Orthogonal V-system (OVS) is proposed to minimize the effect of power-law trends, periodic trends, assembled trends and piecewise function trends. The effectiveness of the new method is demonstrated on monofractal data and multifractal data corrupted with different trends.

*Keywords*: Orthogonal V-system (OVS); Detrended Fluctuation Analysis (DFA); crossover; trend; smoothing algorithm.

## 1. Introduction

In recent years several studies have provided compelling evidence of power-law processes in nature such as physiological systems [1], DNA sequences [2], internet traffic [3], stock markets [4], traffic system [5–7] and wind-speed records [8, 9]. The characteristics of the correlations in such a process can be investigated by extracting its scaling exponents. While some processes (monofractals) can be quantified by a single scaling exponent, others (multifractals) require a spectrum of exponents to characterize them.

Experimental data is often affected by non-stationarity. These effects have to be well distinguished from the intrinsic fluctuations and correlations of the system in order to find the correct scaling behavior. The Detrended Fluctuation Analysis (DFA) invented by Peng *et al.* [2] has been established as an important tool for the determination of fractal scaling properties and detection of long-range power-law correlations in signals. The method is based on the random walk theory. Its non-detrending predecessors are Hurst's resaled range analysis and fluctuation analysis. The convenience of DFA has led to a broad range of application in very diverse fields where long-range correlations are of interest. The DFA has been successfully applied to fields such as DNA sequences [10–12], heart rate dynamics [13–17], neuron spiking [18], human gait [19], long-time weather records [20–22], cloud structure [23], geology [24], economics time series [25–27], and solid state physics [28].

However, for the reliable characterization of time series, it is essential to distinguish trends from intrinsic fluctuations. Monotonous, periodic or step-like trends are caused by external effects, e.g., by the greenhouse warming and seasonal variations for temperature records. Such trends also prevent reliable detection of the scaling exponents. In addition, recent studies have pointed out the susceptibility of DFA to trends in the given data [29–32]. So it is important to build suitable techniques to accurately capture the scaling exponents. Recently, several modifications of the DFA method have been suggested with diverse techniques for eliminating several types of trends [33–38]. These methods include the Detrended Moving Average (DMA) technique, the Centered Moving Average (CMA), the Continuous DFA (CDFA) technique, the Fourier DFA, a variant of DFA based on empirical mode decomposition (EMD), a variant of DFA based on singular value decomposition (SVD), a variant of DFA based on chaotic singular value decomposition (CSVD), and a variant of DFA based on high-pass filtering.

In the present study, we propose a method based on the Orthogonal V-system (OVS) to minimize the effect of different trends (power-law trends, periodic trends, assembled trends and piecewise function trends) in DFA estimation. The Orthogonal V-system is a complete Orthogonal function system composed by piecewise polynomials with multilevel discontinuity. People have been decomposing many signals with orthogonal functions for many years. The well-known complete orthogonal function systems such as the Fourier trigonometric function system and the orthogonal Legendre polynomial system have been become the essential and primary research tools in frequency spectrum analysis and scientific computing. The Orthogonal V-system is just a generalization of the Haar function system and is a very practical wavelet. The V-system is more practical and flexible in many applications.

Through decomposing of signals, we can get the coefficients of them. They are called frequencies in the frequency domain. The Orthogonal V-system [39] proposed by Hui Ma has been applied in digital image processing. The effectiveness of the proposed technique is demonstrated on both artificial and real world data sets.

The organization of this paper is as follows. The OVS and DFA method are given in Sec. 2. Detailed experimental analysis and the results are presented in Sec. 3. Finally, important conclusions drawn from this study are provided in Sec. 4.

## 2. OVS Method and DFA Technique

### 2.1. *The V-system of degree k*

The V-system [39] is a complete orthogonal function system. The V-system of degree $k$ is defined in the following. If the set of $k+1 (k \geq 0)$ functions $\{f_i(x)\}_1^{k+1}$ defined on the interval $[0, 1]$ satisfies the following three conditions:

(1) $\{f_i(x)\}$ is a $C^{k-i}$ continuous piecewise polynomial of degree $k$ with the knot $x = \frac{1}{2}$, where $C^{-1}$ means $\{f_i(x)\}$ is discontinuous at $x = \frac{1}{2}$;
(2) $\langle f_i(x), f_j(x)\rangle = \delta_{i,j}$, $i, j \in \{1, 2, \ldots, k+1\}$;
(3) $\langle f_i(x), x^j\rangle = 0$, $j \in \{0, 1, \ldots, k\}$

where $\langle .,.\rangle$ is the inner product in $L_2[0, 1]$, then the function set $G(x) = \{f_i(x)\}_1^{k+1}$ is called the function generator of degree $k$. For illustration, we give the function generator $G(x)$ of degree $k$ for $k = 0, 1, 2$ as follows.

The function generator $G(x)$ of degree 0 only contains one function $G(x) = \{f_1(x)\}$ where

$$f_1(x) = \begin{cases} +1, & x \in \left[0, \dfrac{1}{2}\right), \\ -1, & x \in \left(\dfrac{1}{2}, 1\right]. \end{cases} \tag{1}$$

The function generator $G(x)$ of degree 1 contains two functions, that is $G(x) = \{f_1(x), f_2(x)\}$ with

$$f_1(x) = \begin{cases} \sqrt{3}(1-4x), & x \in \left[0, \dfrac{1}{2}\right), \\ \sqrt{3}(4x-3), & x \in \left(\dfrac{1}{2}, 1\right], \end{cases} \tag{2}$$

$$f_2(x) = \begin{cases} 1-6x, & x \in \left[0, \dfrac{1}{2}\right), \\ 5-6x, & x \in \left(\dfrac{1}{2}, 1\right]. \end{cases} \tag{3}$$

The function generator $G(x)$ of degree 2 contains three functions, that is $G(x) = \{f_1(x), f_2(x), f_3(x)\}$ with

$$f_1(x) = \begin{cases} \sqrt{5}(16x^2-10x+1), & x \in \left[0, \dfrac{1}{2}\right), \\ \sqrt{5}(-16(1-x)^2+10(1-x)-1), & x \in \left(\dfrac{1}{2}, 1\right], \end{cases} \tag{4}$$

$$f_2(x) = \begin{cases} \sqrt{3}(30x^2-14x+1), & x \in \left[0, \dfrac{1}{2}\right), \\ \sqrt{3}(30(1-x)^2-14(1-x)+1), & x \in \left(\dfrac{1}{2}, 1\right], \end{cases} \tag{5}$$

$$f_3(x) = \begin{cases} 40x^2 - 16x + 1, & x \in \left[0, \dfrac{1}{2}\right), \\ -40(1-x)^2 - 16(1-x) + 1, & x \in \left(\dfrac{1}{2}, 1\right]. \end{cases} \tag{6}$$

The values of the above function generators at the discontinuous knot $x = \frac{1}{2}$ are the arithmetic averages of the left and the right limits of the function generators at $x = \frac{1}{2}$.

Let the functions $V_{k,1}^1(x), V_{k,1}^2(x), \ldots, V_{k,1}^{k+1}(x)$ be the first $(k+1)$ Legendre polynomials on the interval $[0,1]$ and construct the function generator $G(x) = \{V_{k,2}^i(x)\}_1^{k+1}$ satisfying the three conditions above, then any two functions in the set $G(x)$ are orthogonal each other, and any one function in $G(x)$ is orthogonal with $V_{k,1}^1(x)$, $V_{k,1}^2(x), \ldots, V_{k,1}^{k+1}(x)$. For $n = 3, 4, \ldots, i = 1, 2, \ldots, k+1$ and $j = 1, 2, \ldots, 2^{n-2}$, let

$$V_{k,n}^{i,j}(x) = \begin{cases} \sqrt{2^{n-2}} V_{k,2}^i \left(2^{n-2}\left(x - \frac{j-1}{2^{n-2}}\right)\right), & x \in \left(\dfrac{j-1}{2^{n-2}}, \dfrac{j}{2^{n-2}}\right), \\ 0, & \text{other}. \end{cases} \tag{7}$$

The set of the functions on the interval $[0,1]$

$$\begin{array}{cccc}
V_{k,1}^1(x), & V_{k,1}^2(x), & \ldots, & V_{k,1}^{k+1}(x); \\
V_{k,2}^1(x), & V_{k,2}^2(x), & \ldots, & V_{k,2}^{k+1}(x); \\
& V_{k,3}^{1,1}(x), & V_{k,3}^{1,2}(x), & \\
& V_{k,3}^{2,1}(x), & V_{k,3}^{2,2}(x), & \\
& \cdots & \cdots & \\
& V_{k,3}^{k+1,1}(x), & V_{k,3}^{k+1,2}(x); & \\
V_{k,4}^{1,1}(x), & V_{k,4}^{1,2}(x), & V_{k,4}^{1,3}(x), & V_{k,4}^{1,4}(x), \\
V_{k,4}^{2,1}(x), & V_{k,4}^{2,2}(x), & V_{k,4}^{2,3}(x), & V_{k,4}^{2,4}(x), \\
\cdots & \cdots & \cdots & \cdots \\
V_{k,4}^{k+1,1}(x), & V_{k,4}^{k+1,2}(x), & V_{k,4}^{k+1,3}(x), & V_{k,4}^{k+1,4}(x); \\
\cdots & \cdots & \cdots & \cdots \\
V_{k,n}^{1,1}(x), & V_{k,n}^{1,2}(x), & V_{k,n}^{1,3}(x), & V_{k,n}^{1,2^{n-2}}(x), \\
V_{k,n}^{2,1}(x), & V_{k,n}^{2,2}(x), & V_{k,n}^{2,3}(x), & V_{k,n}^{2,2^{n-2}}(x), \\
\cdots & \cdots & \cdots & \cdots \\
V_{k,n}^{k+1,1}(x), & V_{k,n}^{k+1,2}(x), & V_{k,n}^{k+1,3}(x), & V_{k,n}^{k+1,2^{n-2}}(x); \\
\cdots & \cdots & \cdots & \cdots
\end{array}$$

is called the V-system of degree $k$. The basis functions in the V-system with $k = 0, k = 1, k = 2$ and $k = 3$ are shown in Figs. 1 and 2, respectively.

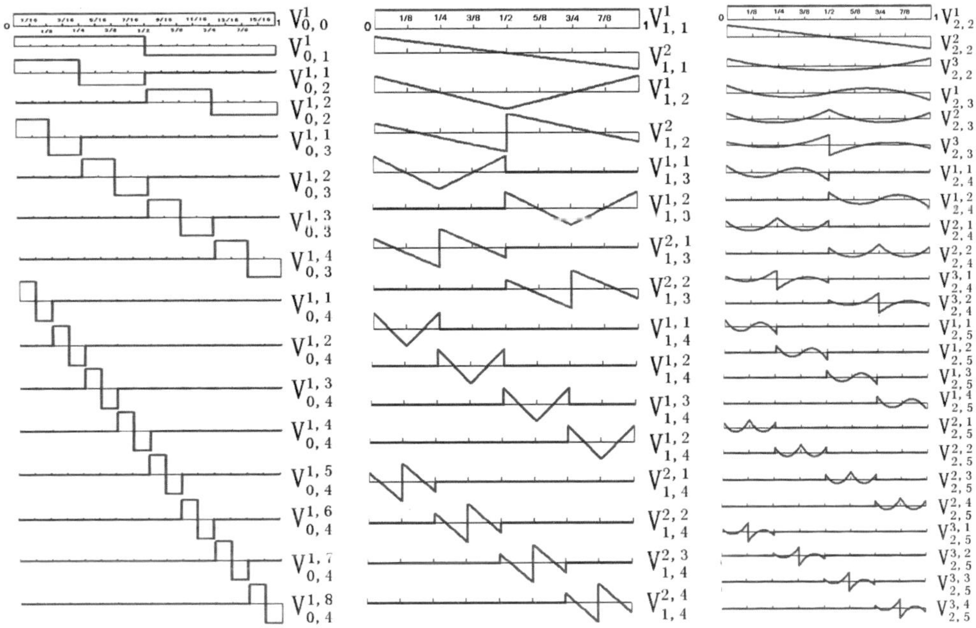

Fig. 1. The basis functions in the V-system. The first 16 basis functions with $k = 0$ (left), the first 16 basis functions with $k = 1$ (middle) and the first 32 basis functions with $k = 2$ (right).

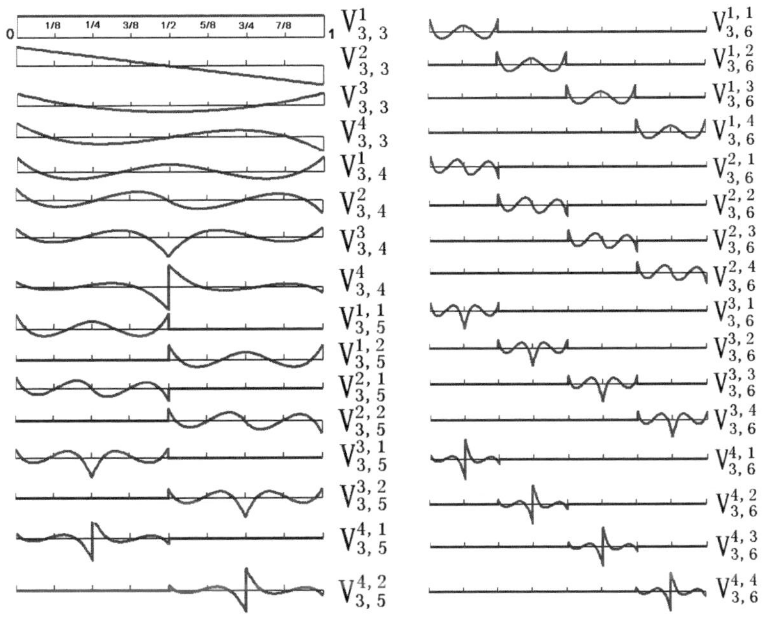

Fig. 2. The 32 basis functions in the V-system with $k = 3$.

It is easy to see that the above V-system of degree $k$ is ordered by class and group, the notation $V_{k,m}^{i,j}$ means that $V_{k,m}^{i,j}$ is the $j$-th function in the $i$-th class of the $m$-th group for $k = 0, 1, 2, \ldots, m = 3, 4, \ldots$, and $j = 1, 2, \ldots, 2^{m-2}$. The second group is the $k+1$ functions from the $(k+2)$-th function to the $(2k+2)$-th function which are all the functions of the function generator of degree $k$. Starting from the $(2k+3)$-th function, we can sequentially perform multi-scale squeezing, shifting and duplicating operations on the function generator composed by these $k+1$ functions to obtain the V-system. Beginning from the third group of the V-system ($n \geq 3$), each group consists of $k+1$ classes and each class has $2^{n-2}$ functions.

### 2.2. *DFA method*

The DFA procedure consists of five steps. Let us suppose that $x_k$ is a series of length $N$, and this series is of compact support, i.e., $x_k = 0$ for an insignificant fraction of the values only.

**Step 1:** Determine the profile

$$Y(i) = \sum_{k=1}^{i} (x_k - \langle x \rangle). \tag{8}$$

**Step 2:** Cut the profile $Y(i)$ into $N_s \equiv [N/s]$ non-overlapping segments of equal length $s$. Since the record length $N$ need not be a multiple of the considered time scale $s$, a short part at the end of the profile will remain in most cases. In order not to disregard this part of the record, the same procedure is repeated starting from the other end of the record. Thus, $2N_s$ segments are obtained altogether.

The subtraction of the mean $\langle x \rangle$ is not compulsory, since it would be eliminated by the later detrending in the third step anyway.

**Step 3:** Calculate the local trend for each segment $\nu$ by a least-square fit of the data. Then we define the detrended time series for segment duration $s$, denoted by $Y_s(i)$, as the difference between the original time series and the fits

$$Y_s(i) = Y(i) - p_\nu(i), \tag{9}$$

where $p_\nu(i)$ is the fitting polynomial in the $\nu$-th segment. Linear, cubic, or higher order polynomials can be used in the fitting procedure (DFA1, DFA3, and higher order DFA). Since the detrending of the time series is done by the subtraction of the polynomial fits from the profile, these methods differ in their capability of eliminating trends in the series. In the $n$-th order DFA, trends of order $n$ in the profile and of order $n-1$ in the original series are eliminated.

We calculate, for each of the $2N_s$ segments, the variance

$$F_s^2(\nu) = \langle Y_s^2(i) \rangle = \frac{1}{s} \sum_{i=1}^{s} Y_s^2[(\nu - 1)s + i] \quad (\nu = 1, 2, \ldots, N_s), \tag{10}$$

$$F_s^2(\nu) = \frac{1}{s} \sum_{i=1}^{s} Y_s^2[N - (\nu - N_s)s + i](\nu = N_s + 1, N_s + 2, \ldots, 2N_s) \tag{11}$$

of the detrended time series $Y_s(i)$ by averaging over all data points $i$ in the $\nu$-th segment.

**Step 4:** Average over all segments to obtain the fluctuation function

$$F(s) = \left[ \frac{1}{2N_s} \sum_{\nu=1}^{2N_s} F_s^2(\nu) \right]^{1/2}. \tag{12}$$

It is apparent that $F(s)$ will increase. Of course, $F(s)$ depends on the DFA order $n$. By construction, $F(s)$ is only defined for $s \geq n+2$.

**Step 5:** Determine the scaling behavior of the fluctuation functions by analyzing log-log plots $F(s)$ versus $s$. If the series $x_k$ are long-range power-law correlated, $F(s)$ increases, for large values of $s$, as a power-law

$$F(s) \sim s^h. \tag{13}$$

The exponent $h$ is called the scaling exponent or correlation exponent, represents the correlation properties of the signal.

The scaling exponent $\tau(q)$ is usually defined via the partition function $Z_q(s)$

$$Z_q(s) = \sum_{v=1}^{2N_s} [Y(vs) - Y((v-1)s)]^q \sim s^{\tau(q)}, \tag{14}$$

where $q$ is a real parameter. Long-range power-law correlated series are monofractal when they are characterized by exponents $\tau(q)$ for different moments $q$ being linearly dependent on $q$; other long-range correlated series are multifractal and are described by $\tau(q)$ that depend nonlinearly on $q$.

## 3. OVS Method for Minimizing the Effect of Trends in DFA

The smoothing technique in the frequency domain can be useful in eliminating the effect of trends while retaining the power-law behavior of the signal. In the present paper, we propose an algorithm based on OVS to minimize the effect of trends. We explain the procedure below to achieve this.

Let the power-law signal be represented by $X(k), k = 1, 2, \ldots, N$ and the trend by $T(k)$, therefore the noise superimposed with the trend is given by $Y(k) = X(k) + T(k)$. We calculate the OVS frequency $F_{OVS}$ using the OVS base function as described in Sec. 2.1. Let the orthogonal base function be presented by $B(k), k = 1, 2, \ldots, N$, where $N$ is the length of the series, the frequency defined by $F_{OVS} = B(k) \cdot Y(k)$. Then we truncate the first few largest $F_{OVS}$ to get the new $F^*_{OVS}$ and inverse transform the series $F^*_{OVS}$ by $X^*(k) = B(k) \cdot F^*_{OVS}$, where $X^*(k)$ is the reconstructed series. In the following, we shall utilize the smoothing filter to minimize the effect of trends on DFA estimation procedures. One example of frequencies calculated by the smoothing filter based on the Orthogonal V-system are presented in Fig. 3. The $F_{OVS}$ differences between the original series and the reconstructed series are very slight, therefore our proposed smoothing filter based on OVS may efficiently minimize the superimposed trend.

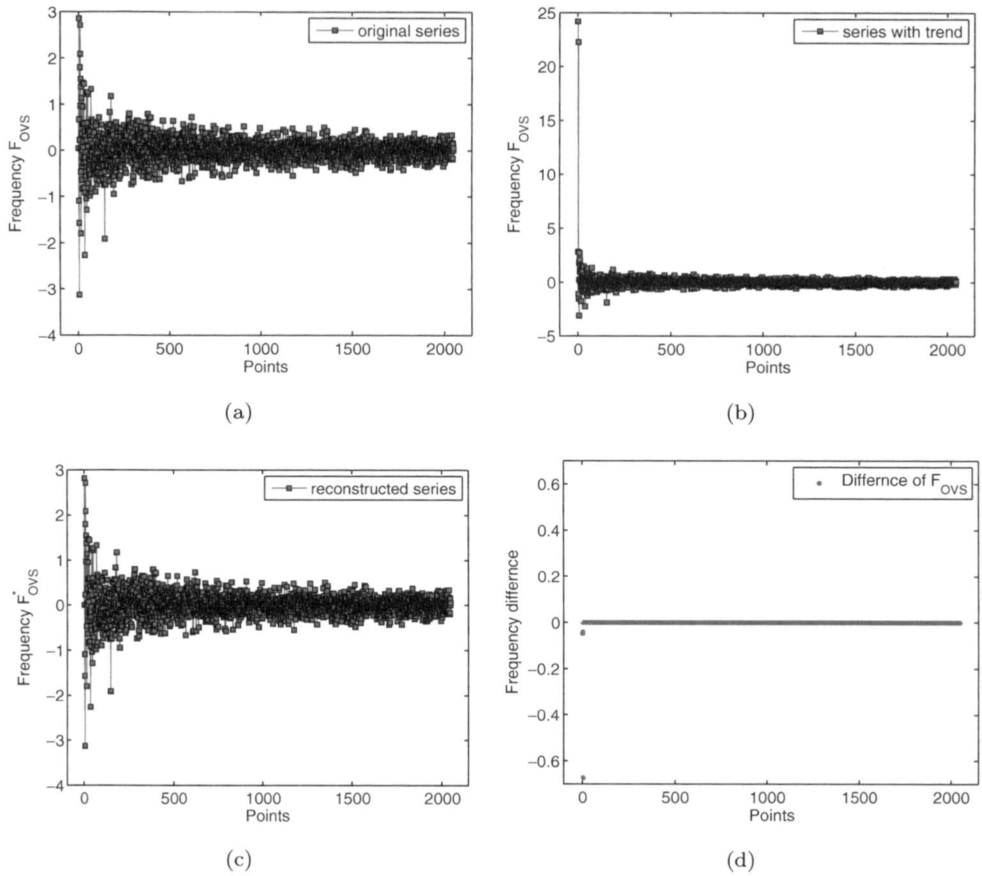


Fig. 3. The frequency calculated by a smoothing filter based on the Orthogonal V-system, (a) $F_{OVS}$ of original series; (b) $F_{OVS}$ of series corrupted with trend; (c) $F^*_{OVS}$ of reconstructed series; (d) the difference between the frequency $F_{OVS}$ of the original series and reconstructed series.

## 3.1. *Artificial data corrupted with trends*

### 3.1.1. *Artificial monofractal data corrupted with trends*

The effectiveness of the proposed technique in minimizing the trends superimposed will be discussed on the artificial monofractal data by using the fractionally autoregressive integrated moving-average process (ARFIMA) [40–42], which consists of $N = 2^{11}$ points. In the following, $X(k), k = 1, 2, \ldots, N$ will be used to denote the trend free time series. The data set with this scaling exponent superimposed with trends, $T(k)$, will be addressed as $Y(k)$. The corresponding filtered data obtained using the proposed algorithm shall be referred to by $X^*(k)$. The power-law, periodic, and also the assembled trend of power-law and periodic trend superimposed on the long-range correlated data ($\alpha = 0.91$) are considered in the following discussion. The original artificial monofractal data is shown in Fig. 4.

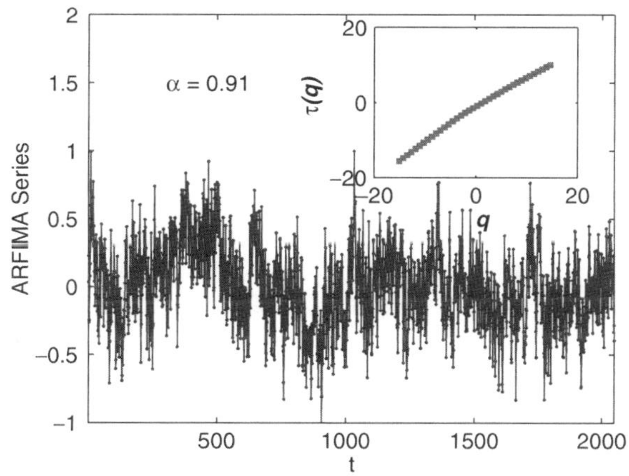


Fig. 4. The original artificial monofractal data by applying ARFIMA ($\alpha = 0.91$).

For the long-range correlated noise corrupted with power law trend $T(k) = [(k - \frac{N}{2})/2^7]^2$ a considerable increase in the magnitude of the scaling exponent is observed, failing to reflect the scaling exponent of the original series $X(k)$, see Fig. 5. The log-log plot of the data superimposed with the power-law trend exhibits a characteristic crossover. It is important to note that the crossover is due to the trend and not the original monofractal data. However, the fluctuation plots obtained after applying the smoothing filter proposed in this paper resembles the straight line (with $\alpha = 0.91 \pm 0.001$) similar to that of the data sets without the power-law trend.

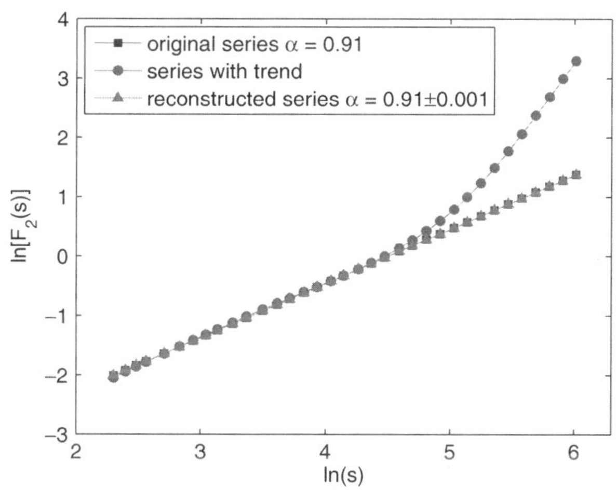


Fig. 5. Comparison of the fluctuation function for monofractal data with scaling $\alpha = 0.91$ (squares), monofractal data corrupted with power trend $T(k) = [(k - \frac{N}{2})/2^7]^2$ (circles), and the reconstructed series using the proposed smoothing filter $\alpha = 0.91 \pm 0.001$ (triangles).

The periodic trends superimposed on long-range correlated noise in this paper are generated as

$$T(k) = A_1 \sin\left(\frac{2\pi f_1 k}{F_1}\right) + A_2 \sin\left(\frac{2\pi f_2 k}{F_2}\right) + A_3 \cos\left(\frac{2\pi f_3 k}{F_3}\right) - A_4 \sin\left(\frac{2\pi f_4 k}{F_4}\right). \tag{15}$$

The periodic trend $T(k) = 0.25\sin(\frac{2\pi k\cdot 7}{2^9}) + 0.5\sin(\frac{2\pi k\cdot 15}{2^9}) - \sin(\frac{2\pi k\cdot 21}{2^9})$ is superimposed on the artificial monofractal data $\alpha = 0.91$. The comparison of the fluctuation function for monofractal data, monofractal data with periodic trend and the reconstructed data using the smoothing filter is shown in Fig. 6. The log-log plot of the artificial monofractal data with periodic trend also exhibit a characteristic crossover, on both the scaling exponents before and behind the crossover change considerably. The distortion introduced by the periodic trend imparts a nonlinear structure to the fluctuation function and prevents the application of linear regression to estimate the scaling exponent. The fluctuation plots obtained after applying the smoothing filter resembles the straight line with $\alpha = 0.91 \pm 0.06$.

To determine the effect of multiple trends, the artificial monofractal data ($\alpha = 0.91$) was corrupted with a combined trend of power-law and periodic trend, where $T(k) = \sin(\frac{2\pi k}{2^5}) + 0.5\cos(\frac{2\pi k}{2^9}) + [(k - \frac{N}{2})/2^9]^4$. The comparison of the fluctuation function for monofractal data, monofractal data with combined trend and the reconstructed data using the smoothing filter is shown in Fig. 7. For the combined trend of the power-law and periodic trend, two obvious crossovers are observed.

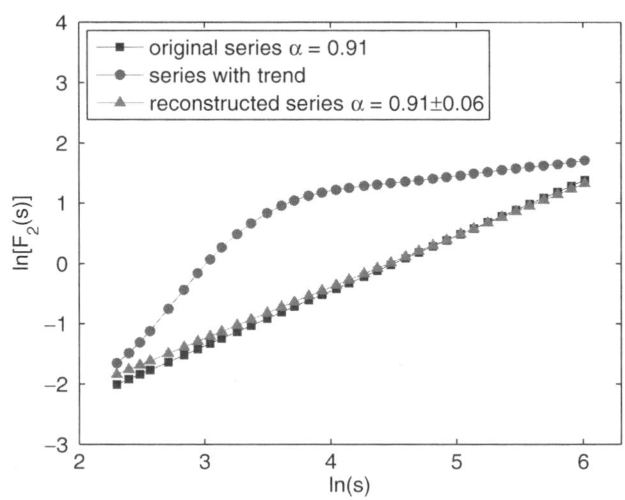


Fig. 6. Comparison of the fluctuation function for monofractal data with scaling $\alpha = 0.91$ (squares), monofractal data corrupted with periodic trend $T(k) = 0.25\sin(\frac{2\pi k\cdot 7}{2^9}) + 0.5\sin(\frac{2\pi k\cdot 15}{2^9}) - \sin(\frac{2\pi k\cdot 21}{2^9})$ (circles), and the reconstructed series using the proposed smoothing filter $\alpha = 0.91 \pm 0.06$ (triangles).

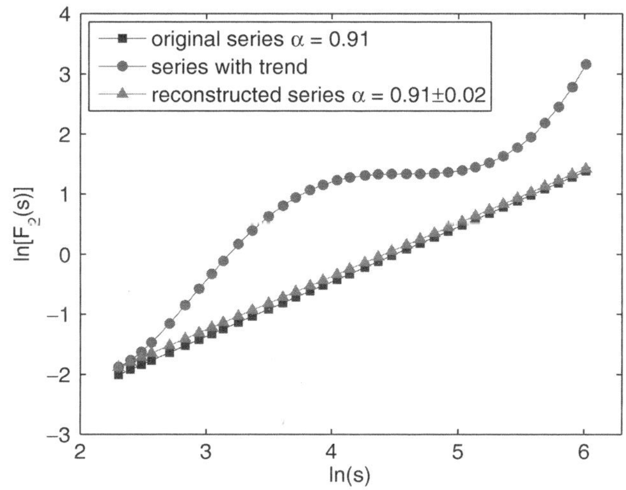


Fig. 7. Comparison of the fluctuation function for monofractal data with scaling $\alpha = 0.91$ (squares), monofractal data corrupted with combined trend $T(k) = \sin(\frac{2\pi k}{2^5}) + 0.5\cos(\frac{2\pi k}{2^9}) + [(k - \frac{N}{2})/2^9]^4$ (circles), and the reconstructed series using the proposed smoothing filter $\alpha = 0.91 \pm 0.02$ (triangles).

It can be observed from Fig. 7 that we retain the original characteristics of the artificial monofractal data and remove the marked change corresponding to the combined trend. These show that our algorithm is robust against the power-law, periodic and the combined trend for artificial monofractal data.

### 3.1.2. *Artificial multifractal data corrupted with trends*

The artificial multifractal data considered is generated by applying the binomial multifractal model [43]. In the binomial multifractal model, a series of $N = 2^{n_{\max}}$ numbers $k$ with $k = 1, 2, \ldots, N$ is defined by

$$X(k) = a^{n(k-1)}(1-a)^{n_{\max}-n(k-1)}, \tag{16}$$

where $0.5 < a < 1$ is a parameter and $n(k)$ is the number of digits equal to 1 in the binary representation of the index $k$, e.g., $n(13) = 3$, since 13 corresponds to binary 1101.

The original artificial multifractal data with the exponent $\alpha = 0.63$ is shown in Fig. 8. Similar to the discussion on the artificial monofractal data, the power-law, periodic and the combined trend are superimposed on the original artificial multifractal data. The effectiveness of the proposed technique in minimizing the power-law trend $T(k) = [(k - \frac{N}{2})/2^7]^2$, periodic trend $T(k) = 0.05\sin(\frac{2\pi k\cdot 3}{2^9}) + 0.05\sin(\frac{2\pi k\cdot 5}{2^{11}}) - 0.05\sin(\frac{2\pi k\cdot 7}{2^{13}})$ and the combined trend $T(k) = 0.01\sin(\frac{2\pi k}{2^9}) + 0.02\cos(\frac{2\pi k}{2^7}) + 0.01[(k - \frac{N}{2})/2^{13}]^3$ are presented in Figs. 9–11, respectively. The proposed method is useful in minimizing the effect of different trends and facilitating the estimation of the scaling exponent.

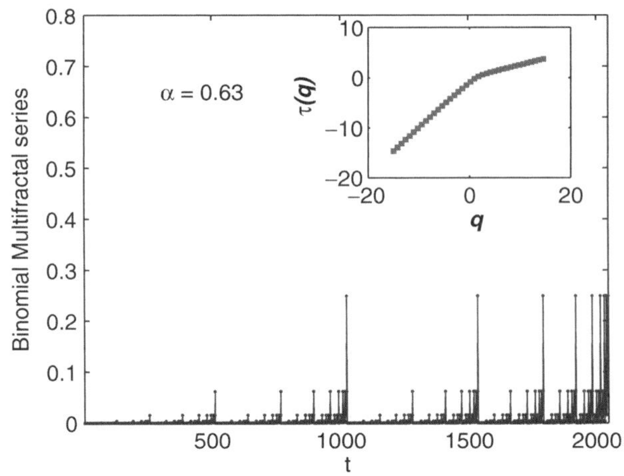


Fig. 8. The original artificial multifractal data by applying binomial multifractal model ($\alpha = 0.63$).

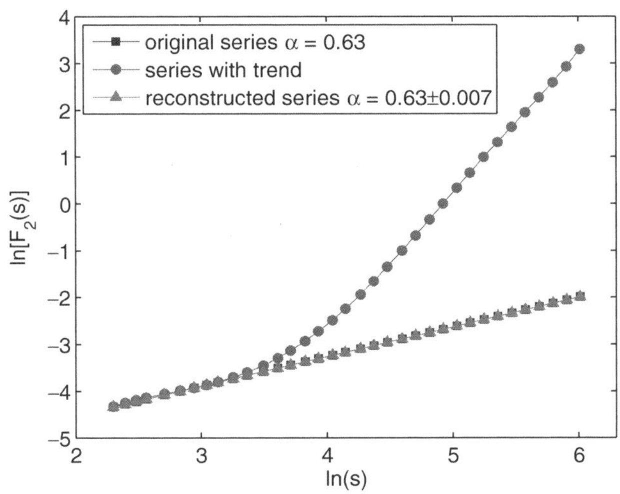


Fig. 9. Comparison of the fluctuation function for multifractal data with scaling $\alpha = 0.63$ (squares), multifractal data corrupted with power-law trend (circles), and the reconstructed series using the proposed smoothing filter $\alpha = 0.63 \pm 0.007$ (triangles).

### 3.2. *Real world data corrupted with trends*

The real world data sets considered are that of stock market (HSI) and traffic flow. The plots of original real world data sets are shown in Fig. 12. The effectiveness of the proposed technique in minimizing the combined trend are presented in Fig. 13.

It is not sufficient to add the same trend on the real world data in all the time intervals, due to the real systems being possibly affected by the staggered

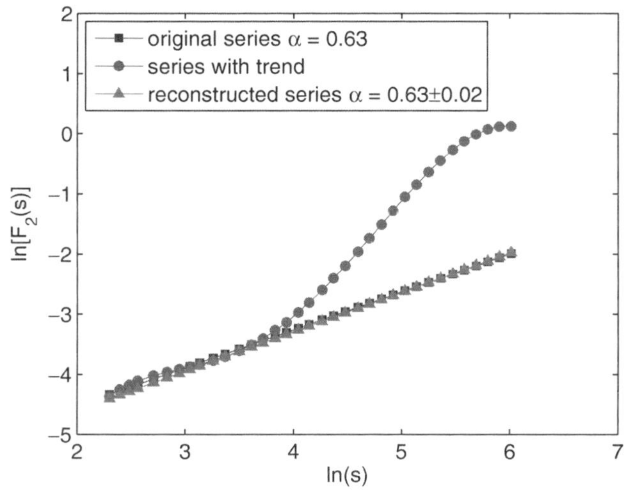


Fig. 10. Comparison of the fluctuation function for multifractal data with scaling $\alpha = 0.63$ (squares), multifractal data corrupted with periodic trend (circles), and the reconstructed series using the proposed smoothing filter $\alpha = 0.63 \pm 0.02$ (triangles).

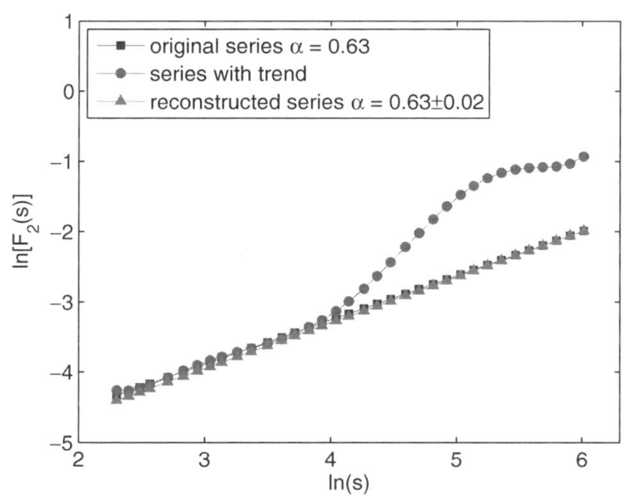


Fig. 11. Comparison of the fluctuation function for multifractal data with scaling $\alpha = 0.63$ (squares), multifractal data corrupted with combined trend (circles), and the reconstructed series using the proposed smoothing filter $\alpha = 0.63 \pm 0.02$ (triangles).

trends, i.e., the trends are different in different time intervals. The effect of piecewise function trends to the DFA method has not been studied before. To study the effectiveness of the proposed technique in minimizing the staggered trend, we add the piecewise function to the real world data. Simply put, a function is a mechanism for taking a number from a person and then giving this person a number back. In a

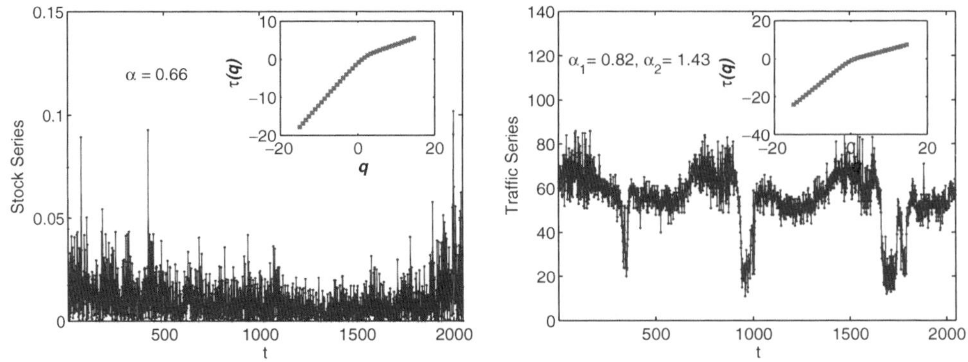


(a) (b)

Fig. 12. The original real world data, (a) HSI ($\alpha = 0.66$); (b) traffic ($\alpha_1 = 0.82, \alpha_2 = 1.43$).

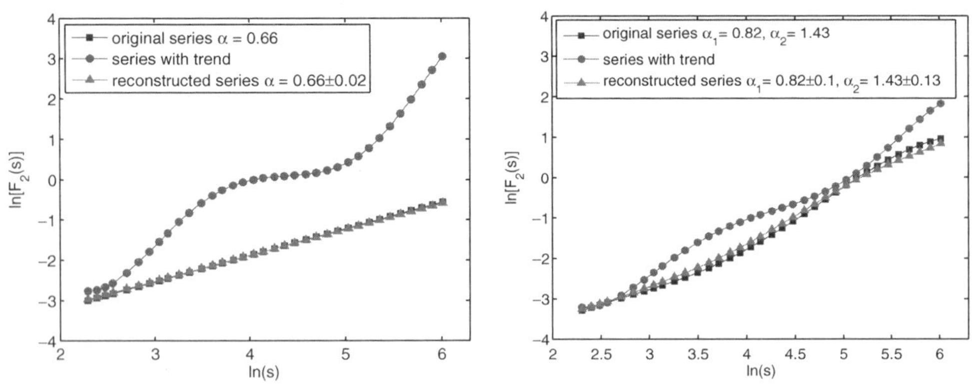


Fig. 13. Comparison of the fluctuation function for real world data HSI ($\alpha = 0.66$) and traffic ($\alpha_1 = 0.82, \alpha_2 = 1.43$).

math class, this is usually a sequence of arithmetic operations. Usually, the function behaves ignorantly, applying the same process no matter what number it is given. A piecewise function behaves more realistically. It looks at the number first, and decides what to do based on the number itself. The example most people see is the absolute value function. This function needs to know whether the number given is positive or negative before deciding what to do with it. If the number is already positive, the absolute value leaves it alone. However if the number is negative, the absolute value function has to change its sign. Figure 14 shows original data, data corrupted with piecewise trend and reconstructed data of HSI and traffic. The piecewise function trends used in this paper are

$$T_1(k) = \begin{cases} 0.05[(k - \frac{N}{2})/2^9]^2, & \text{Sec. 1, 3,} \\ 0.01\sin(\frac{2\pi k \cdot 7}{2^9}) + 0.02\sin(\frac{2\pi k \cdot 15}{2^9}) - 0.1\sin(\frac{2\pi k \cdot 21}{2^9}), & \text{Sec. 2, 4,} \end{cases} \tag{17}$$

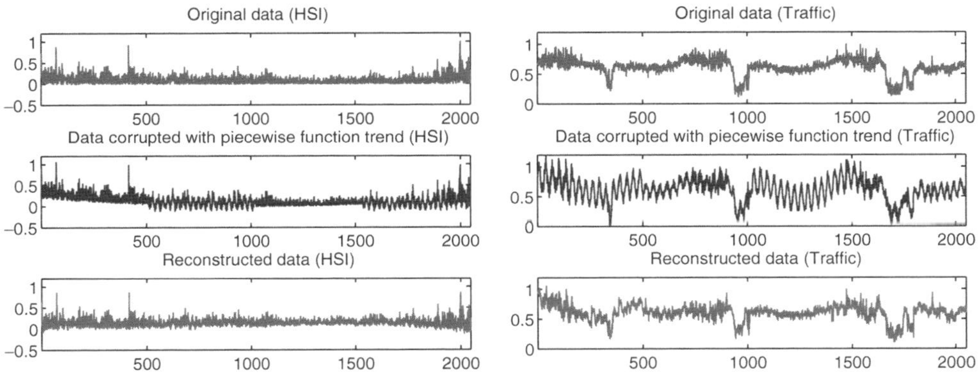


Fig. 14. Original data, data corrupted with piecewise trend and reconstructed data of HSI (left) and traffic (right).

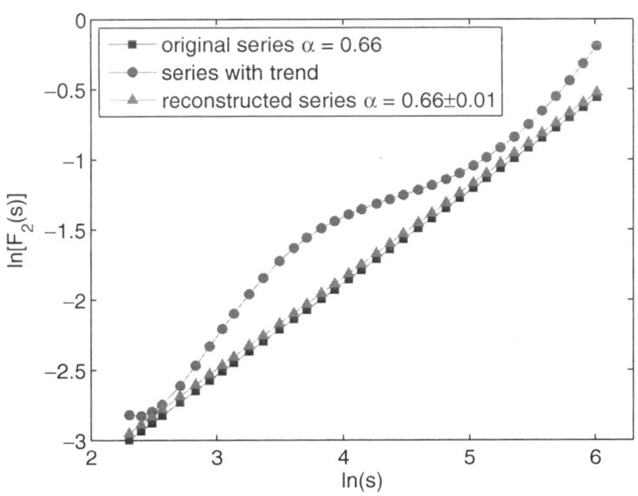


Fig. 15. Comparison of the fluctuation function for HSI data $\alpha = 0.66$ (squares), data corrupted with piecewise function trend (circles), and the reconstructed series using the proposed smoothing filter $\alpha = 0.66 \pm 0.01$ (triangles).

$$T_2(k) = \begin{cases} 0.2\sin(\frac{2\pi k}{2^5}) + 0.1\cos(\frac{2\pi k}{2^9}), & \text{Sec. 1, 3,} \\ 0.1\sin(\frac{2\pi k}{2^5}), & \text{Sec. 2, 4.} \end{cases} \tag{18}$$

Comparison of the fluctuation function for HSI data, data corrupted with piecewise function trend $T_1(k)$, and the reconstructed series using the proposed smoothing filter is demonstrated in Fig. 15. Comparison of the fluctuation function for traffic data, data corrupted with piecewise function trend $T_2(k)$, and the reconstructed series using the proposed smoothing filter is demonstrated in Fig. 16.

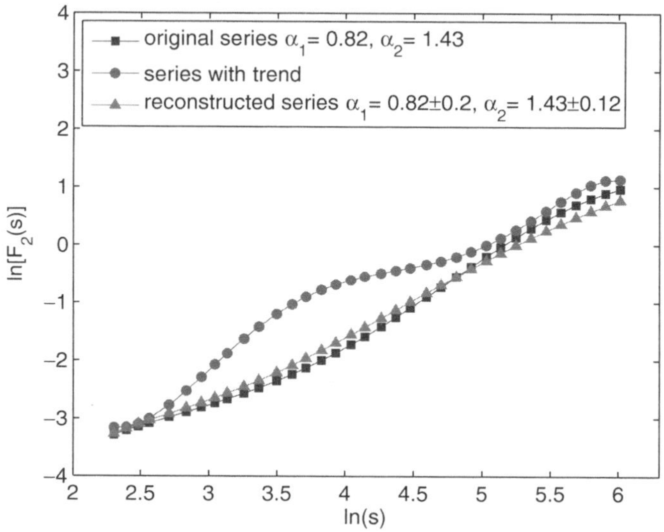


Fig. 16. Comparison of the fluctuation function for traffic data (squares), data corrupted with piecewise function trend (circles), and the reconstructed series using the proposed smoothing filter (triangles).

Both the fluctuation functions of HSI data corrupted with the piecewise function trend $T_1(k)$ and traffic data corrupted with the piecewise function trend $T_2(k)$ exhibit more than two crossovers. The fluctuation plot of HSI data obtained after applying the smoothing filter based on the OVS method resembles the line similar to that of the data sets without the piecewise function trend, see Fig. 15. The fluctuation plot of traffic data almost resembles the line except for very large scales, see Fig. 16. We can come to the conclusion that our OVS-DFA method is robust to the piecewise function trend.

## 4. Conclusions

It is shown that trends can introduce spurious crossovers in the log-log plot of the fluctuation function estimated using DFA. In this paper, the smoothing filter based on OVS, is proposed to minimize the effect of periodic trends. The effectiveness of the smoothing filter on data corrupted with trends are considered. The results show that our proposed method in robust to power-law trends, periodic trends, assembled trends and piecewise function trends. The effect of piecewise function trends to DFA has not been studied before. However, the smoothing filter minimizes the periodic trends — it is an approximation to some extent. This problem required further investigation, both in experimental and theoretical aspects. But we do believe that our proposed method may provide some help to the study of minimizing the trends.

**Acknowledgments**

The financial support from the funds of PhD Innovation of Beijing Jiaotong University (141105522), the National High Technology Research Development Program of China (863 Program) (2007AA11Z212), the China National Science (60772036 & 61071142) and MEDF (20070004002) are gratefully acknowledged.

## References

[1] J. B. Bassingthwaighte, L. S. Liebovitch and B. J. West, *Fractal Physiology* (Oxford University Press, Oxford, 1995).
[2] C. K. Peng, S. V. Buldyrev, S. Havlin, M. Simons, H. E. Stanley and A. L. Goldberger, *Phys. Rev. E* **49** (1994) 1685–1689.
[3] V. J. Levy and R. H. Reidi, *Fractals in Eng.* (1996), pp. 185–202.
[4] P. Gopikrishnan, V. Pleroul, X. Gabaix and H. E. Stanley, *Phys. Rev. E* **62** (2000) 4493–4496.
[5] P. J. Shang, Y. B. Lu and S. Kamae, *Chaos Solitons Fractals* **36** (2008) 82C90.
[6] P. J. Shang, M. Wang and S. Kama, *Comput. Math. Appl.* **54** (2007) 107–116.
[7] P. J. Shang and S. Kama, *Chaos Solitons Fractals* **26** (2005) 997–1007.
[8] R. G. Kavasseri and R. Nagarajan, *IEEE Trans. on Circuits and Syst.* **51** (2004) 2255–2262.
[9] R. G. Kavasseri and R. Nagarajan, *Chaos Solitons Fractals* **24** (2005) 165–173.
[10] S. M. Ossadnik, S. B. Buldyrev, A. L. Goldberger, S. Havlin, R. N. Mantegna, C. K. Peng, M. Simons and H. E. Stanley, *Biophys. J.* **67** (1994) 64–70.
[11] S. V. Buldyrev, A. L. Goldberger, S. Havlin, R. N. Mantegna, M. E. Matsa, C. K. Peng, M. Simons and H. E. Stanley, *Phys. Rev. E* **51** (1995) 5084–5091.
[12] S. V. Buldyrev, N. V. Dokholyan, A. L. Goldberger, S. Havlin, C.-K. Peng, H. E. Stanley and G. M. Viswanathan, *Physica A* **249** (1998) 430–438.
[13] C. K. Peng, S. Havlin, H. E. Stanley and A. L. Goldberger, *Chaos* **5** (1995) 82–87.
[14] P. C. Ivanov, A. Bunde, L. A. N. Amaral, S. Havlin, J. Fritsch-Yelle, R. M. Baevsky, H. E. Stanley and A. L. Goldberger, *Europhys. Lett.* **48** (1999) 594–600.
[15] P. C. Ivanov, L. A. N. Amaral, A. L. Goldberger, S. Havlin, M. G. Rosenblum, H. E. Stanley and Z. Struzik, *Chaos* **11** (2001) 641–652.
[16] Y. Ashkenazy, P. C. Ivanov, S. Havlin, C. K. Peng, A. L. Goldberger and H. E. Stanley, *Phys. Rev. Lett.* **86** (2001) 1900–1903.
[17] A. Bunde, S. Havlin, J. W. Kantelhardt, T. Penzel, J. H. Peter and K. Voigt, *Phys. Rev. Lett.* **85** (2000) 3736–3739.
[18] S. Blesic, S. Milosevic, D. Stratimirovic and M. Ljubisavljevic, *Physica A* **268** (1999) 275–282.
[19] J. M. Hausdorc, S. L. Mitchell, R. Firtion, C. K. Peng, M. E. Cudkowicz, J. Y. Wei and A. L. Goldberger, *J. Appl. Physiol.* **82** (1997) 262–269.
[20] E. Koscielny-Bunde, A. Bunde, S. Havlin, H. E. Roman, Y. Goldreich and H. J. Schellnhuber, *Phys. Rev. Lett.* **81** (1998) 729–732.
[21] K. Ivanova and M. Ausloos, *Physica A* **274** (1999) 349–354.
[22] P. Talkner and R. O. Weber, *Phys. Rev. E* **62** (2000) 150–160.
[23] K. Ivanova, M. Ausloos, E. E. Clothiaux and T. P. Ackerman, *Europhys. Lett.* **52** (2000) 40–46.
[24] B. D. Malamud and D. L. Turcotte, *J. Stat. Plan. Infer.* **80** (1999) 173–196.
[25] R. N. Mantegna and H. E. Stanley, *An Introduction to Econophysics* (Cambridge University Press, Cambridge, 2000).

[26] Y. Liu, P. Gopikrishnan, P. Cizeau, M. Meyer, C. K. Peng and H. E. Stanley, *Phys. Rev. E* **60** (1999) 1390–1400.
[27] N. Vandewalle, M. Ausloos and P. Boveroux, *Physica A* **269** (1999) 170–176.
[28] J. W. Kantelhardt, R. Berkovits, S. Havlin and A. Bunde, *Physica A* **266** (1999) 461–465.
[29] J. W. Kantelhardt, E. Koscielny-Bunde, H. A. H. Rego, S. Havlin and A. Bunde, *Physica A* **295** (2001) 441–454.
[30] K. Hu, P. Ivanov, Z. Chen, P. Carpena and H. E. Stanley, *Phys. Rev. E* **64** (2001) 011114.
[31] Z. Hen, P. Ivanov, K. Hu and H. E. Stanley, *Phys. Rev. E* **65** (2002) 041107.
[32] Z. Chen, *Phys. Rev. E* **71** (2005) 011104.
[33] P. J. Shang, A. J. Lin and L. Liu, *Physica A* **388** (2009) 720–726.
[34] R. Nagarajan and R. G. Kavasseri, *Chaos Solitons Fractals* **26** (2005) 777.
[35] R. Nagarajan and R. G. Kavasseri, *Int. J. Bifur. Chaos* **15** (2005) 1767–1773.
[36] L. Xu, P. C. Ivanov, K. Hu, Z. Chen, A. Carbone and H. E. Stanley, *Phys. Rev. E* **71** (2005) 051101.
[37] A. Bashan, R. Bartsch, J. W. Kantelhardt and S. Havlin, *Physica A* **387** (2008) 5080–5090.
[38] E. Alessio, A. Carbone, G. Castelli and V. Frappietro, *Eur. Phys. J. B* **27** (2002) 197–200.
[39] R. X. Song, H. Ma, T. J. Wang and D. X. Qi, *Commun. Pure Appl. Anal.* **6** (2007) 853–871.
[40] B. Podobnik, D. Horvatic, A. L. Ng, H. E. Stanley and P. C. Ivanov, *Physica A* **387** (2008) 3954–3959.
[41] C. W. J. Granger and R. Joyeux, *J. Time Ser. Anal.* **1** (1980) 15–29.
[42] J. R. M. Hosking, *Biometrika* **68** (1981) 165–176.
[43] J. W. Kantelhardt, S. A. Zschiegner, E. Koscielny-Bunde, S. Havlin, A. Bunde and H. E. Stanley, *Physica A* **316** (2002) 87–114.

# COMMENT ON "EXPERIMENTAL DEMONSTRATION OF VIOLATIONS OF THE SECOND LAW OF THERMODYNAMICS FOR SMALL SYSTEMS AND SHORT TIME SCALES"

THEO M. NIEUWENHUIZEN
*Institute for Theoretical Physics, University of Amsterdam Valckenierstraat 65, 1018 XE Amsterdam, The Netherlands*

ARMEN E. ALLAHVERDYAN
*Institute for Theoretical Physics, University of Amsterdam and Yerevan Physics Institute Valckenierstraat 65, 1018 XE Amsterdam, The Netherlands and Alikhanian Brothers St. 2, Yerevan 375036, Armenia*

Received 1 April 2005
Accepted 6 April 2005
Communicated by Ricardo Mannella

In a recent paper Wang *et al.* [1] report for a colloidal system the entropy production in individual particle trajectories. Some of the trajectories have a negative production, and this is claimed to violate the second law. We stressed that the second law only demands the entropy production *averaged over all trajectories* to be positive, and this is the case for the data of Wang *et al.*

*Keywords*: Second law violation; individual trajectories; colloidal particles; entropy.

## 1. Introduction

Wang *et al.* [1] test the Fluctuation Theorem

$$\frac{\Pr(\Sigma_t = A)}{\Pr(\Sigma_t = -A)} = \exp(A), \tag{1}$$

where $\Sigma_t$ is the entropy production of a given trajectory of a colloidal particle, equal to the ratio of the dispersed energy and $k_B T$. This equality is a strong statement about the second law, a field often described by inequalities. It says that the probability for trajectories with positive entropy production is larger than the one with negative production. As such, the theorem shows that the *average* entropy production is positive. Indeed, a small exercise yields from Eq. (1)

$$\langle \Sigma_t \rangle = \int_{-\infty}^{\infty} \mathrm{d}A\, A\, \Pr(\Sigma_t = A) = \int_0^{\infty} \mathrm{d}A\, A\,(1 - e^{-A})\Pr(\Sigma_t = A) \tag{2}$$

which is obviously positive. The positivity is also clear from the data presented in [1].

The statement made by the second law concerns the *average* entropy production, because the second law is related to the *average* heat transfer between system and bath. Thus, opposite to what the authors claim, their data fully confirm the validity of the second law.

This insight in the meaning of the second law goes back to Maxwell, who invented his so-called Maxwell demon with the motivation *to pick a hole in the second law of thermodynamics*, or, more precisely, *to show that the second law has only a statistical nature* [2,3]. The aim of that exercise clearly was that by proving the non-existence of the demon, the second law in its statistical nature would be proven.

It remains to stress that none of formulations of the second law known to us ever claimed that unaveraged entropy production or unaveraged work must be positive; see e.g. [4–11].

We conclude that the statement "violation of the second law" made by Wang *et al.* is not an adequate term for the observed physics.

## Acknowledgments

This work is part of the research programme of the "Stichting voor Fundamenteel Onderzoek der Materie"(FOM), which is financially supported by the "Nederlandse Organisatie voor Wetenschappelijk Onderzoek (NWO)".

## References

[1] G. M. Wang, E. M. Sevick, E. Mittag, D. J. Searles and D. J. Evans, *Phys. Rev. Lett.* **89** (2002) 050601.

[2] M. J. Klein, *Maxwell, his demon, and the second law of thermodynamics*, *Am. Sci.* **58** (1970) 84–97; reprinted in [3].

[3] H. S. Leff and A. F. Rex, *Maxwell's demon: Entropy, Information, Computing*, Adam Hilger, Bristol (1990); *Maxwell's Demon 2: entropy, classical and quantum information, computing*, eds. H. F. Leff and A. F. Rex, Institute of Physics, Bristol (2003).

[4] L. D. Landau and E. M. Lifshitz, *Statistical Physics, I*, Pergamon Press Oxford (1978).

[5] R. Balian, *From Microphysics to Macrophysics*, Vols. I and II, Springer (1992).

[6] A. Lenard, *J. Stat. Phys.* **19** (1978) 575.

[7] W. Pusz and L. Woronowicz, *Comm. Math. Phys.* **58** (1978) 273.

[8] G. N. Bochkov and Yu. E. Kuzovlev, *Physica A* **106** 443 (1981); *ibid*, 480.

[9] J. Kurchan, cond-mat/0007360.

[10] A. E. Allahverdyan and Th. M. Nieuwenhuizen, *Physica A* **305** (2002) 542.

[11] A. E. Allahverdyan and Th. M. Nieuwenhuizen, *Phys. Rev. E*, to appear; cond-mat/0408699.

# Do Electromagnetic Waves Exist in a Short Cable at Low Frequencies? What Does Physics Say?

Hsien-Pu Chen*, Laszlo B. Kish*, Claes-Göran Granqvist† and Gabor Schmera‡

*Texas A&M University
Department of Electrical and Computer Engineering
College Station, TX 77843-3128, USA

†Department of Engineering Sciences
The Ångström Laboratory, Uppsala University, P. O. Box 534
SE-75121 Uppsala, Sweden

‡Space and Naval Warfare Systems Center
San Diego, CA 92152, USA

Received 27 March 2014
Accepted 7 April 2014
Published 7 May 2014

Communicated by Zoltan Gingl

We refute a physical model, recently proposed by Gunn, Allison and Abbott (GAA) [http://arxiv.org/pdf/1402.2709v2.pdf], to utilize electromagnetic waves for eavesdropping on the Kirchhoff-law–Johnson-noise (KLJN) secure key distribution. Their model, and its theoretical underpinnings, is found to be fundamentally flawed because their assumption of electromagnetic waves violates not only the wave equation but also the second law of thermodynamics, the principle of detailed balance, Boltzmann's energy equipartition theorem, and Planck's formula by implying infinitely strong blackbody radiation. We deduce the correct mathematical model of the GAA scheme, which is based on impedances at the quasi-static limit. Mathematical analysis and simulation results confirm our approach and prove that GAA's experimental interpretation is incorrect too.

*Keywords*: Waves; KLJN secure key exchange; physics in engineering education.

## 1. Introduction

Recently Gunn, Allison and Abbott (GAA) [1] proposed a new, delay-based attack against the Kirchhoff-law–Johnson-noise (KLJN) secure key distribution scheme [2–5]. GAA claim — contradicting earlier statements most recently expounded in work by Kish, Abbott and Granqvist (KAG) [4] — that waves exist in a *finite-size* cable at *arbitrarily low frequencies*.

The theoretical basis of GAA's assertion [1] is the fact that, whereas waveguides usually have a low-frequency cut-off for wave modes vs. the *diameter* of the wave-guide, no such cut-off exists for transversal electromagnetic (TEM) wave modes in the case of *infinitely* long wave-guides. GAA write that, because coaxial cables have TEM wave modes, there is no frequency cut-off of the wave-based component of the electrical transport down to zero frequency. As a consequence of their presumption, GAA use the d'Alembert solution [1]

$$U(t,x) = U_+\left(t - \frac{x}{v}\right) + U_-\left(t + \frac{x}{v}\right) \tag{1}$$

for propagating lossless fluctuations — which may or may not be waves — in a linear medium to model the propagation of voltage in the cable used for key exchange in the KLJN scheme, where $U_+$ and $U_-$ are voltage components of waves propagating to the right and left along the $x$ axis, and $v$ is propagation velocity.

The experimental support of GAA's claim is that they have measured the voltage between the ends of a short coaxial cable at low-frequency sinusoidal voltage drive with an impedance-matched load at the other end and, at first sight, have found that their results confirmed some of the implications of Eq. (1), as further elaborated in Sec. 3 below.

We have analyzed GAA's statements [1] and found most of them invalid. Specifically, our findings and conclusions are the following:

(i) In cables, wave modes with wavelengths greater than twice their length are forbidden states, meaning that such modes do not exist; consequently there are no waves in cables in the frequency range pertinent to the KLJN scheme.
(ii) Instead, time-dependent propagation processes are non-wave type *retarded potentials* in a distributed impedance system; one of the implications of this is that Eq. (1) does not hold.
(iii) GAA's interpretation of their own "wave-verification" experiments [1] is invalid.
(iv) A correct derivation of the problem posed by GAA allows a proof that their attack does not work and an identification of the source of errors in their "attack-demonstration" experiments.
(v) We prove that directional coupler attacks under the KLJN conditions can never offer more information leak than the old wire resistance attacks. It should be noted that directional couplers in KLJN can never separate Alice's and Bob's contribution.
(vi) Finally we are able to formulate a general proof that any future (not yet existing) advanced attack type, that may be able to even *separate* the contribution of Alice and Bob, can never crack the KLJN scheme. Instead such an attack would lead to more advanced KLJN schemes that would not only be faster than earlier versions but would also be immune against these types of attacks.

The results delineated in (i)–(vi) will be derived and presented in three separate publications. The present paper is focused on items (i)–(iii), and moreover points out that the erroneous wave picture at low frequencies, referred to above, seems to be a common misconception in electrical engineering education.

## 2. Refutation of GAA's Theory and Experimental Interpretation

As mentioned above, GAA's attack [1] on the KLJN scheme employs waves and related delays in a cable to extract information. While attempts to utilize time delays in cables for information purposes are to be encouraged, the asserted use of waves, which do not exist, is a fundamental flaw that invalidates GAA's basic considerations, proposed experiments, and interpretation of these experiments.

The wavelengths corresponding to the frequency range of concern in the KLJN scheme are much longer than the physical extent of the cable, and we earlier referred to that range the "no-wave" or "quasi-static" limit [2–5]. As remarked above, GAA argue that TEM wave modes do not exhibit any low-frequency cut-off. It is true that TEM wave modes in a wave-guide do not have a low-frequency cut-off vs. the diameter of the wave-guide, but this argument is irrelevant because wave modes do have a cut-off vs. the length of the cable. This does not imply that the electrical transport itself has a cut-off; it solely means that, when wave modes are forbidden, electrical transport takes place via non-wave phenomena — such as drift and relaxation — which constitute the form of transport in the quasi-static region of electrodynamics.

In this section, we use three separate fields of physics and mathematics to prove, in six different ways, that GAA's assumption of the existence of waves in the short cable within the frequency range pertinent to the KLJN scheme violates the following equations and laws of physics (our proofs are given subsequently):

(a) the wave equation,
(b) essential dynamical behavior of magnetic and electric energy in electromagnetic wave modes,
(c) Boltzmann's energy equipartition theorem of thermodynamics,
(d) the principle of detailed balance of thermodynamics,
(e) the second law of thermodynamics, and
(f) theory and experiments for blackbody radiation.

### 2.1. *Violation of the wave equation*

It was recently pointed out by KAG [4] (including one of the proponents of the GAA model) that the wave equation precludes the existence of waves in the frequency range of concern for the KLNJ scheme. Next we provide more details about this fact and first illustrate the distributed inductance–capacitance–resistance (LCR) model of the cable in the KLJN scheme in Fig. 1.

For the sake of simplicity but without losing generality, we discuss the case of a lossless cable. The main conclusion about the lack of wave modes is general because

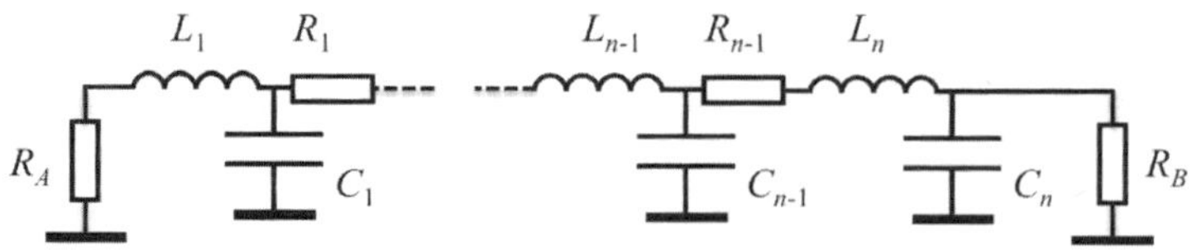


Fig. 1. Outline of the pertinent part of the KLJN scheme with a distributed LCR model of a long and leakage-free cable. When the cable losses can be neglected, one may omit the $R_i$ resistors representing the distributed resistance of the cable. Alice's and Bob's resistors, denoted $R_A$ and $R_B$, respectively, are randomly selected from the set $\{R_L, R_H\}$ with $(R_L \neq R_H)$ at the beginning of each bit-exchange period. These resistors, with associated serial generators (not shown), emulate thermal noise with high noise temperature and strongly limited bandwidth.

the inclusion of damping terms in harmonic differential equations *can never produce new eigen-frequencies*; they can only modify them and their bandwidth.

The wave equations of voltage $U(x,t)$ and current $I(x,t)$ in lossless cables are

$$\frac{\partial^2 U(x,t)}{\partial x^2} = \frac{1}{v_c^2}\frac{\partial^2 U(x,t)}{\partial t^2}, \tag{2}$$

$$\frac{\partial^2 I(x,t)}{\partial x^2} = \frac{1}{v_c^2}\frac{\partial^2 I(x,t)}{\partial t^2}, \tag{3}$$

where the phase propagation velocity of waves in the cable is

$$v_c = \sqrt{\frac{1}{L_u C_u}}. \tag{4}$$

Here $L_u$ and $C_u$ are inductance and capacitance "densities" of the cable (with units of H/m and F/m), i.e., the unit-length (one-meter) cable inductance and capacitance.

The general classical-physical solutions of these equations in infinite ideal cables are superpositions of waves, with arbitrary frequency, propagating in positive and negative directions in accordance with the d'Alembert solution in Eq. (1). However, in a cable with finite length $D$, the frequency-space of solutions is quantized to discrete values so that integer multiples of the half-wavelength fit in the cable. Thus the wavelength $\lambda_{\max}$ of the wave with the lowest frequency $f_{\min}$ can be written as

$$\lambda_{\max} = 2D, \quad f_{\min} = \frac{v_c}{2D}. \tag{5}$$

Frequencies below $f_{\min}$, down to zero frequency, constitute a forbidden band of wave states.

The KLJN key exchanger operation strictly requires for security that its frequency range satisfies the quasi-static condition, i.e.,

$$f \ll f_{\min}. \tag{6}$$

Thus the wave-based scheme and considerations of GAA for eavesdropping violate not only the wave equations in Eqs. (2) and (3), and its d'Alembert solution in

Eq. (1), but also the related other fields of classical and quantum physics of waves, because such non-existent solutions are forbidden states.

We note, in passing, that if the wave-based treatment by GAA [1] were correct, we would not have quantization of atomic electron shells, a forbidden band (energy gap) would not exist in solid state physics, semiconductor devices would not work, and even chemistry would be non-existent or at least very different.

### 2.2. *Violation of the energy dynamics within wave modes*

It is easy to visualize the pertinent physical situation even without the wave equation. While the energy of the wave mode is constant, any electromagnetic wave signifies propagation and oscillatory "bouncing" of energy between magnetic and electric energy forms. To have a physical wave, both aspects are essential, and the propagation must take place via "bouncing" of the energy between its two forms. For example, an $L$–$C$ resonator has its "bouncing" phenomenon between the electric and magnetic energy forms but lacks propagation, and hence this oscillation is not a wave.

Consider now a short cable satisfying relation 6, which is driven by an ac voltage generator $U(t)$ at one end and closed with a resistance $R_w$ given by

$$R_w = \sqrt{\frac{L_u}{C_u}} \tag{7}$$

at the other end. Then the cable has equal amounts of energy in the electric and magnetic forms, as discussed below, and both of these energies are oscillating. However, these energies are not "bouncing" between the electric and magnetic forms but oscillate between these forms and the generator. This fact is obvious from the considerations that follow next.

Quasi-static conditions (relation 6) imply that the time-dependent voltage and current are spatially homogeneous along the wire, i.e.,

$$U(x,t) \simeq U(t), \tag{8}$$

$$I(x,t) \simeq I(t). \tag{9}$$

Thus the electrical and magnetic energies, denoted $E_e(t)$ and $E_m(t)$, respectively, in the cable can be written

$$E_e(t) = \frac{1}{2} C_c U^2(t), \tag{10}$$

$$E_m(t) = \frac{1}{2} L_c I^2(t) = \frac{1}{2} L_c \frac{U^2(t)}{R_w^2}, \tag{11}$$

where $C_c = DC_u$ and $L_c = DL_u$ are the capacitance and inductance of the whole cable, respectively, and the last term of Eq. (11) originates from the well-known fact that a cable of arbitrary length, closed with the wave impedance (resistance)

$R_w$, has an input impedance of exactly $R_w$. Thus

$$\frac{E_e(t)}{E_m(t)} = \frac{C_c}{L_c}R_w^2 = \frac{C_u}{L_u}R_w^2 = 1 \tag{12}$$

at each instant of time. Obviously, no "bouncing" phenomenon takes place between the electric and magnetic forms of energy because the energy is bouncing between the generator and the magnetic and electric fields, not between each other, as would be expected for a wave.

Now suppose that $R_A$ and $R_B$, rather than $R_w$, terminate the cable ends. The only significant change that will occur in the quasi-static limit is that the ratio of electrical and magnetic energies will be different from unity, which further goes against the wave hypothesis.

In conclusion, even the simple picture of wave modes given above proves that waves do not exist in the cable in the quasi-static limit.

## 2.3. *Violation of thermodynamics*

Simple but fundamental thermodynamic considerations also prove that waves cannot exist in the cable in the quasi-static frequency range.

### 2.3.1. *Violation of the energy equipartition theorem, the principle of detailed balance and the second law of thermodynamics*

According to Boltzmann's *energy equipartition theorem* [6] for thermal equilibrium at temperature $T$, each electromagnetic wave mode has a mean thermal energy equal to $kT$, where $k$ is Boltzmann's constant. Half of this mean energy is electrical and the other half is magnetic. For $N$ different wave modes in the system, both the electrical and magnetic fields carry a total energy equal to $NkT/2$.

It is easy to see that the assumption that, in a *hypothetical* wave system, these wave energies are less than the above given values violates not only the *energy equipartition theorem* but also the *principle of detailed balance* [6] and the *second law of thermodynamics*: Coupling this hypothetical wave system to a regular one would hit the *detailed energy balance* of equilibrium between wave modes in the hypothetical and regular systems because it would yield an energy flow toward the hypothetical one. This energy flow could then be utilized for perpetual motion machines of the second kind, i.e., violate the *second law of thermodynamics.*

### 2.3.2. *Violation of Planck's law and experimental facts for blackbody radiation*

Planck [7] deduced his law of thermal radiation, for simplicity, from the properties of a box with black walls, i.e., internal walls with unity absorptivity and emissivity. In thermal equilibrium, thermal radiation within an infinitely large box with walls of arbitrary absorptivity, emissivity and color has a power spectral intensity for each polarization given by

$$I(f) = \frac{4\pi h f^3}{c^2} \frac{1}{e^{hf/kT} - 1}, \tag{13}$$

where $h$ is Planck's constant. The derivation of this formula is based on counting existing wave modes in the frequency range $f > f_{\min}$, where the minimum frequency is obtained by the same frequency quantization as we claim to exist in a cable. GAA's notions (as well as common electrical engineering education) imply that in closing the finite-size cable by the wave impedance (resistance) $R_w$ at its two ends, all of the lower frequencies, at $f < f_{\min}$, will also be available for wave modes as a consequence of the unity absorptivity and emissivity of the impedance match. However, one must realize that a cable closed by the wave impedance (resistance) $R_w$ at its two ends is a one-dimensional realization of Planck's box with perfectly absorbing (black) walls. Allowing the $f < f_{\min}$ frequency range for wave modes results in a non-quantized continuum distribution of wave modes, which yields an infinite number $N$ of wave modes in any finite frequency band. Such a situation in the finite-size cable results in an infinite amount of thermal energy $NkT$ in any finite frequency band within the classical-physical frequency range $f \ll kT/h$. This situation is not a problem in a cable or box with infinite size. However, in a finite-size box, infinite thermal energy in the sum of wave modes would naturally result in infinite intensity of blackbody radiation.

Furthermore, the situation is similar for the KLJN scheme in which the cable ends are terminated by $R_A$ and $R_B$ rather than $R_w$. According to Planck's results, discussed above, the thermal radiation field in the closed box and cable does not depend on the absorptivity and emissivity of the walls — i.e., on the cable termination resistances — since otherwise Planck's Law would be invalid and his formula violate the second law of thermodynamics. Therefore the above argumentation regarding infinite energies and infinite thermal radiation holds for arbitrary termination and wall color.

If there is a loss in the cable and the cable is terminated by $R_A$ and $R_B$, the same situation holds provided $R_A$ and $R_B$, and the cable, have the same (noise) temperature because the system is still in thermal equilibrium.

In conclusion, GAA's claims violate the *Planck formula* and *experimental facts about blackbody radiation.*

### 2.3.3. *Proof that the total energy of the cable is less than the energy required for a single wave mode*

To further support the above conclusions we show that in the quasi-static frequency limit, wherein the KLJN scheme operates, the total energy is not sufficient even to supply a single wave mode, and consequently waves cannot exist without violating several principles and one law of thermodynamics.

As an example, we mathematically analyze how the no-wave situation manifests itself in a lossless cable, which is closed at both ends by resistors equal to the wave impedance $R_w$ at the temperature $T$. These conditions are not necessary but serve to simplify the calculations. Suppose that, in accordance with the KLJN conditions, the Johnson noise of the resistors has an upper cut-off frequency at $f_c$ so that

$$f_c \ll f_{\min} = \frac{v_c}{2D} \tag{14}$$

hold in order to satisfy relation 6. Thus the spatial homogeneity of current and voltage Eqs. (8) and (9) is valid for the wire, and it is straightforward to calculate the electric and magnetic energies of thermal origin as shown next.

According to the Johnson–Nyquist formula [8], the thermal electrical energy in the cable capacitance is

$$\begin{aligned} E_{e,\mathrm{th}} &= \frac{1}{2}C_c\langle U^2(t)\rangle = \frac{1}{2}C_c\int_0^{f_c}\frac{4kTR_w/2}{1+f^2/f_{0C}^2}df \\ &\cong kTC_cR_wf_c = \frac{kT}{2}\frac{f_c}{f_{\min}} \ll \frac{kT}{2} \end{aligned} \tag{15}$$

and the thermal magnetic energy in the cable inductance is

$$\begin{aligned} E_{m,\mathrm{th}} &= \frac{1}{2}L_c\langle I^2(t)\rangle = \frac{1}{2}L_c\int_0^{f_c}\frac{4kT/(2R_w)}{1+f^2/f_{0L}^2}df \\ &\cong kT\frac{L_c}{R_w}f_c = \frac{kT}{2}\frac{f_c}{f_{\min}} \ll \frac{kT}{2}, \end{aligned} \tag{16}$$

where the characteristic frequencies of the Lorentzian spectra in Eqs. (15) and (16) are defined as

$$f_{0C} = \frac{1}{2\pi C_cR_w/2} = \frac{1}{\pi DC_u}\sqrt{\frac{C_u}{L_u}} = \frac{1}{\pi D}\sqrt{\frac{1}{L_uC_u}} = \frac{1}{\pi}\frac{v_c}{D} = \frac{2}{\pi}f_{\min} \tag{17}$$

and

$$f_{0L} = \frac{2R_w}{2\pi L_c} = \frac{1}{\pi L_c}\sqrt{\frac{L_u}{C_u}} = \frac{1}{\pi D}\sqrt{\frac{1}{L_uC_u}} = \frac{1}{\pi}\frac{v_c}{D} = \frac{2}{\pi}f_{\min}. \tag{18}$$

Similar calculations can be carried out for the general case in which the cable ends are not terminated by $R_w$ but with different resistance values $R_A$ and $R_B$. Specifically, the parallel resultant resistance $R_A$ and $R_B$ (instead of $R_w/2$) enters in the left-hand side of Eqs. (15) and (17) and the serial (sum) loop resistance $R_A + R_B$ (instead of $2R_w$) enters in the left-hand side of Eqs. (16) and (18), while the final inequalities shown by Eqs. (15) and (16) remain.

If there is a loss in the cable, and the cable is terminated by $R_A$ and $R_B$, the same situation holds provided $R_A$ and $R_B$, as well as the cable, have the same (noise) temperature, because the system is still in thermal equilibrium. If the cable is cooler then there is an energy flow out of the cable, which further strengthens the inequalities at the right-hand sides of Eqs. (15) and (16).

In conclusion, Eqs. (15) and (16) prove that for a short cable and within the frequency range of interest for the KLJN scheme, the sum of electrical and magnetic energies in all of the hypothetical "wave modes" is much less than the energy needed for a single wave mode in thermal equilibrium.

## 3. Correct Treatment of Cable Delays in the Frequency Range for the KLJN Scheme

### 3.1. *General considerations*

We showed above that wave modes cannot exist in the cable at the KLJN condition $f \ll f_{\text{min}}$. A number of questions then arise naturally, such as (i) what type of system is the cable under these conditions, (ii) what is the nature of the propagating fluctuations caused by Alice's and Bob's noise generators and (iii) are there any other implications of the KLJN condition?

To answer these questions, we first note that the system under consideration is not a waveguide, as implied by GAA [1], but a *distributed impedance network* in the *quasistatic limit*. Second, the propagating fluctuations are not waves but phase-shifted voltages and currents; in the language of physics they are related to *retarded potentials* of non-wave solutions, and in electrical engineering vocabulary they are *spatio-temporal fluctuations in an impedance network*.

The general implications of the KLJN conditions are very pervasive, as elaborated and discussed in Sec. 2 above. The specific consequences for the KLJN scheme are that the mathematical and physical framework used by GAA [1] is invalid and that the same applies to their experimental analysis, as shown in the present paper and as will be further discussed in forthcoming articles.

When the frequency converges towards zero, the impact of the inductance and capacitance of the cable on the cable current and voltage also rapidly diminish. However, the voltage drop over the cable is determined by its serial resistance $R_c$ and inductance $L_c$, because the capacitive shunt currents approach zero. Thus the first-order approximation of the cable impedance is

$$Z_c \cong R_c + j2\pi f L_c. \tag{19}$$

For simplicity, we analyze the situation wherein the cable loss (resistance) is negligible so that

$$Z_c \cong j2\pi f L_c. \tag{20}$$

The corresponding phase delay of the voltage at Bob's end, compared to that of Alice's end, is

$$\varphi_{AB} = -2\pi f L_c / R_B \tag{21}$$

when the voltage is generated by Alice. This phase delay corresponds to a *frequency-independent time delay* according to

$$\tau_{AB} = L_c / R_B, \tag{22}$$

which at first glance seems to suggest that we are dealing with waves and that the d'Alembert equation holds, as stated by GAA. However one must realize that this *time delay depends on the load resistance* $R_B$ at the other end of the cable, which implies that the *time delay and measured phase velocity in the two directions are different* due to the condition $R_A \neq R_B$ during secure bit exchange [2–5], i.e., under circumstances such that GAA's method [1] is supposed to function. To illustrate this dichotomy, we evaluate the phase delay for voltage propagation in the opposite direction, i.e., when the voltage is generated by Bob. Now one finds

$$\varphi_{BA} = 2\pi f L_c / R_A \tag{23}$$

and

$$\tau_{BA} = L_c / R_A. \tag{24}$$

Thus the d'Alembert equation, applied by GAA [1] to prove the existence of waves, cannot be used in the present situation.

### 3.2. *Simulation based on a circuit model for the cable*

To corroborate the theoretical considerations above, we used Linear Technology's LTspice-IV cable simulator software to analyze the experimental situation in GAA's work [1] and confirmed all of their stated results. The simulations proved that a coaxial cable with parameters and conditions similar to those employed by GAA [1] can be modeled with the lumped impedance circuitry shown in Fig. 2, where parts (a) and (b) represent a cable with and without loss, respectively.

Figure 3 shows results of our simulation addressing the experimental data in GAA's article [1]. The conditions are the same as those of GAA [1] and reported in their Fig. 5, but our simulation uses a practical cable model and simple impedance representations (see Fig. 2) which fit the cable data to a high degree of accuracy. The practical lossy cable and the simple impedance model in Fig. 2(a) give identical results, while results for the lossless cable (corresponding to data compensated for loss in GAA's work [1]) are nicely represented by the simple inductance model in Fig. 2(b). Moreover, it is obvious that the cable inductance $L_c$ produces a voltage drop that is the time-derivative of the current, which is determined predominantly by the resistances in the loop. Thus the voltage drop for the lossless cable is the time

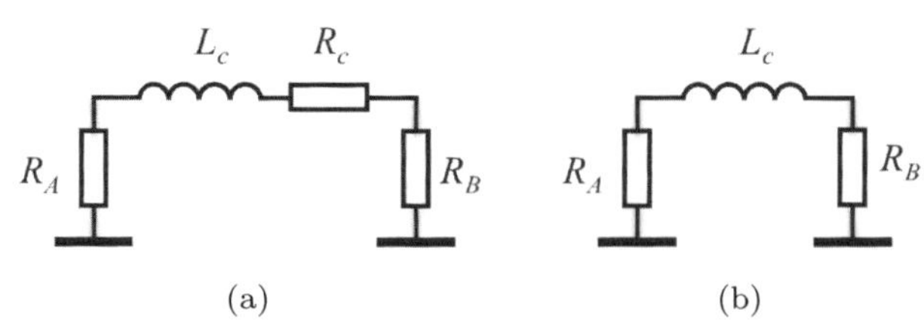


Fig. 2. Lumped impedance-components-based model of a cable at low frequencies for analyzing voltage drop along the cable and phase shift in the limit $f \ll f_{\min}$. Part (a) represents a cable with loss (cable inductance and resistance are designated $L_c$ and $R_c$, respectively), and part (b) represents a lossless cable.

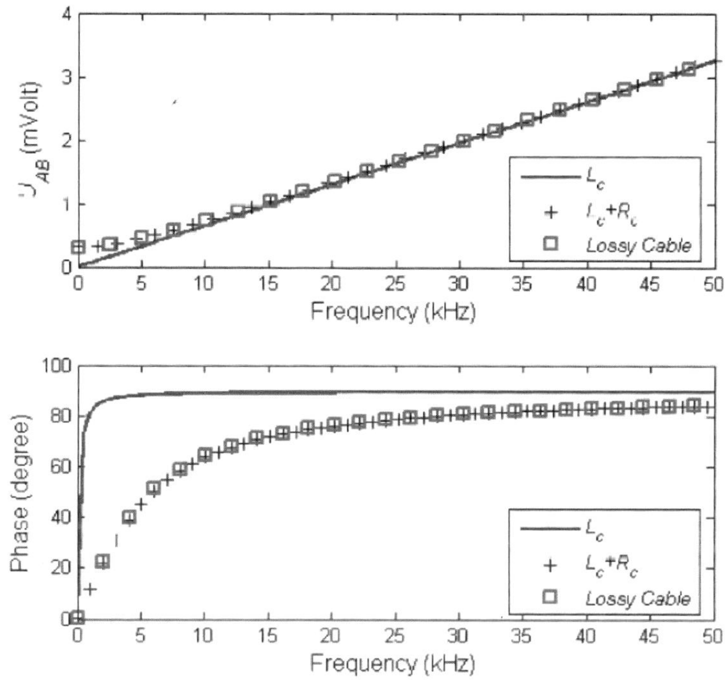


Fig. 3. Comparison of simulated data based on impedance models, using LTspice, with those of a real (lossy) cable. Alice's and Bob's resistors, denoted $R_A$ and $R_B$, have the resistance $R_w$, and Alice drives the cable with a sinusoidal voltage generator (1 V) via Bob's resistor. The cable is characterized by length $D = 1.5\,\text{m}$, $L_c = 1.03\,\mu\text{H}$, $R_c = 0.0315\,\Omega$, and $C_c = 150\,\text{pF}$. Upper panel shows voltage drop $U_{AB}$ over the cable between Alice's and Bob's ends and lower panel shows phase shift of $U_{AB}$ compared to that of the voltage at Alice's end. Squares signify simulations of a lossy cable (model RG58), crosses represent data obtained by use of the lumped-parameters model in Fig. 2(a), and solid line was derived from an inductance model devised to simulate a lossless cable (Fig. 2(b)). These results are in full agreement with the experimental data shown in Fig. 5 of GAA's work [1].

derivative of Alice's generator voltage, and this experimental finding by GAA [1] to "support" the d'Alembert equation is simply an inductor-type voltage response and it has nothing to do with waves.

To subject Eqs. (22) and (24) to a final test, we evaluated the phase delay and corresponding time delay toward Bob when the resistor at Bob's end was varied and the cable was lossless. Data are shown in Table 1 and verify the correctness of our Eq. (22) to an accuracy of five digits. GAA's "propagation velocity" toward Bob is practically independent of frequency but depends on Bob's resistor. During secure bit exchange, the "propagation" times toward Alice and Bob are different. This fact verifies our conclusion that GAA's use of the d'Alembert equation, as the base of their mathematical considerations, is indeed incorrect.

One should observe that, for the cases of 1 and 10 k$\Omega$, GAA's "phase velocity" is greater than the speed of light. This is acceptable and happens often with the phase

Table 1. Simulated equivalent phase velocity calculated from phase shifts between the two ends of the cable vs. driving frequency and load resistance (resistance of termination at the other end). The dependence on the resistance is in violation of the d'Alembert equation, for example, in the KLJN system during secure key exchange where the terminal resistances are different. Cable parameters are given in the caption for Fig. 3.

| | 1 kHz | 5 kHz |
|---|---|---|
| 10 Ω | $3.99998 \times 10^7$ m/s | $4.00018 \times 10^7$ m/s |
| 20 Ω | $7.99996 \times 10^7$ m/s | $8.00038 \times 10^7$ m/s |
| 50 Ω | $1.99999 \times 10^8$ m/s | $2.00007 \times 10^8$ m/s |
| 1 kΩ | $3.99993 \times 10^9$ m/s | $4.00011 \times 10^9$ m/s |
| 10 kΩ | $3.99946 \times 10^{10}$ m/s | $4.00041 \times 10^{10}$ m/s |

velocity of oscillations in a *driven impedance system in the steady-state*; however, it is prohibited in the d'Alembert equation as a consequence of the theory of special relativity [9]. Similar effects can happen in wave-based systems with reflecting boundary conditions in stationary mode after the waves fill the system; however, in wave-based systems the phase velocity would be the same for the left and right directions.

In conclusion, the proper KLJN scheme is a simple impedance circuitry with related phase shifts where the corresponding time shifts are asymmetrical during secure bit exchange.

## 4. Additional Notes

One of the reviewers raised an interesting question: In the light of the results of the present paper, how should one see the situation when a small part of a wave-based system is observed or when that part has different properties? Are there waves in those parts? Such a situation is common in physical systems and the answer is straightforward. We show two examples to elucidate the issue.

(a) First consider that a small part of a large wave-based system is observed, so small that Eq. (6) is satisfied. The answer to the question above is then simple: the part is a non-separable component of the wave system, and the observer views a small spatial fraction of the whole wave field. The energy of the observed wave mode is not the energy within this small part but the integral of related wave energy over the large system. However, if we separate the small system from the large one and terminate its ends, wave solutions would not exist at these frequencies.

(b) Another special example is a quarter-wavelength ($\lambda/4$) cable. If we terminate one end and drive the other end with a generator, the latter end acts as a distributed impedance transformer without wave modes. However if, instead of a generator, the same end is connected to a long cable with the same $R_w$, it will belong to the relevant wave mode that extends over the system and contains a small part of the energy of that wave mode.

## 5. Conclusions

The efforts of GAA [1] to utilize time delays in cables to crack the KLJN scheme represent an interesting and novel approach and, as such, deserve attention. However, it should be mentioned that Liu [10] earlier used a similar technique, but with unphysical conditions for the simulations [11]. As shown in considerable detail above, GAA's efforts can be irrevocably refuted. Further attempts to crack the KLJN scheme are welcome, but it is essential that they be founded on correct physical models.

Finally, we hope that the fact that waves cannot exist below $f_{\mathrm{min}}$ in a short cable, as discussed in Sec. 2, will eventually enrich the physical underpinnings in electrical engineering courses and textbooks.

## Acknowledgments

Discussions with Peter Hrasko, Robert Nevels, Lazar Kish, Krzysztof Michalski, Mehrdad Ehsani, Dave Ferry and Mark Dykman are appreciated. We are grateful to Lachlan Gunn for showing us the results in Ref. 1 and to Derek Abbott for discussions.

## References

[1] L. J. Gunn, A. Allison and D. Abbott, A directional coupler attack against the Kish key distribution system, manuscript http://arxiv.org/abs/1402.2709 (2014) versions 1 and 2.

[2] L. B. Kish and C. G. Granqvist, On the security of the Kirchhoff-law–Johnson-noise (KLJN) communicator, *Quant. Infor. Process.* (2014) (in press), doi:10.1007/s11128-014-0729-7.

[3] R. Mingesz, L. B. Kish, Z. Gingl, C. G. Granqvist, H. Wen, F. Peper, T. Eubanks and G. Schmera, Unconditional security by the laws of classical physics, *Metrol. Measurement Syst.* **XX** (2013) 3–16. Open access: http://www.degruyter.com/view/j/mms.2013.20.issue-1/mms-2013-0001/mms-2013-0001.xml.

[4] L. B. Kish, D. Abbott and C. G. Granqvist, Critical analysis of the Bennett–Riedel attack on secure cryptographic key distributions via the Kirchhoff-law–Johnson-noise scheme, *PLoS ONE* **8** (2013) e81810. Open access: http://www.plosone.org/article/info%3Adoi%2F10.1371%2Fjournal.pone.0081810.

[5] D. Abbott and G. Schmera, Secure communications using the KLJN scheme, *Scholarpedia* **8** (2013) 31157.

[6] L. Boltzmann, *Lectures on Gas Theory* (University of California Press, Berkeley, CA, USA, 1964).

[7] M. Planck and M. Morton, *The Theory of Heat Radiation* (P. Blakiston's Son & Co., Philadelphia, PA, USA, 1914).

[8] H. Nyquist, Thermal agitation of electric charge in conductors, *Phys. Rev.* **32** (1928) 110–113.

[9] A. Einstein, Zur elektrodynamik bewegter Körper, *Annalen der Physik* **17** (1905) 891–921.

[10] P.-L. Liu, A new look at the classical key exchange system based on amplified Johnson noise, *Phys. Lett. A* **373** (2009) 901–904.

[11] L. B. Kish and T. Horvath, Notes on recent approaches concerning the Kirchhoff-law–Johnson-noise-based secure key exchange, *Phys. Lett. A* **373** (2009) 2858–2868.

# 2

# Noise in Quantum Systems

# DAMPING AND NOISE IN QUANTUM MECHANICS

N. G. van KAMPEN

*Institute for Theoretical Physics, University of Utrecht, Leuvenlaan 4, Utrecht, Netherlands*
*kampen@phys.uu.nl*

Received 23 May 2001
Revised 13 June 2001
Accepted 13 June 2001

In a quantum system damping and noise enter through the coupling with an external large bath. The usual treatments computes the density matrix of the system by expanding in the coupling strength to the bath and averaging over the bath. The result, however, is known to be unacceptable as it leads to negative probabilities. It is here proposed to study correlation functions rather than the density matrix itself. Equations for them are derived and compared with the result obtained by the usual method.

*Keywords*: Damping, Noise, Quantum Mechanics.

## 1. Introduction

In the *classical* theory of a Hamiltonian system an easy way of introducing damping consists in supplementing the Hamiltonian equations with a friction term. Moreover fluctuations are introduced by adding a term representing a random force which is postulated to be Gaussian white noise. The strength of the noise is determined by the fluctuation-dissipation theorem, which is the requirement that both terms together give rise to the equilibrium distribution, which is known from equilibrium statistical mechanics.

This easy device is not reliable if the system is nonlinear [1, 2]. In that case it is necessary to take the origin of the damping and noise more seriously. In quantum mechanics the easy device does not even exist. Attempts to generalize the Langevin term to quantum mechanics [3, 4] did not meet with success. It is therefore necessary to introduce explicitly the mechanism that causes the damping and the noise [5]. This is done by attaching to the system under consideration, S an external reservoir or bath, B. The total system T, made up of S and B, is then governed by the Hamiltionian

$$H_T = H_S + H_B + \lambda H_I, \tag{1}$$

where $H_I$ is an interaction term involving a coupling constant $\lambda$.

If the Schrödinger equation for this total system can be solved, the behavior of S itself is obtained by averaging over the variables of B. This can be done in terms of density matrices,

$$\rho_S = \operatorname*{Tr}_{\mathrm{B}} \rho_T. \tag{2}$$

The trace is taken in the Hilbert space of the bath B.

The Schrödinger equation for the total system is

$$\dot{\rho}_T = -i\,[H_T, \rho_T] \equiv \mathcal{L}_T \rho_T. \tag{3}$$

Here $\mathcal{L}_T$ denotes a "superoperator", i.e., an operator acting on the density matrices of the total system. The solution with initial value at $t_0$ is formally

$$\rho_T(t) = e^{(t-t_0)\mathcal{L}_T}\, \rho_T(t_0). \tag{4}$$

Usually one chooses as initial $\rho_T$ the direct product of the equilibrium state $\rho_B^e$ of the bath with an uncorrelated initial state of the system,

$$\rho_T(t_0) = \rho_S(t_0) \otimes \rho_B^e. \tag{5}$$

Then the state of the system at later times is

$$\rho_S\,(t \geq t_0) = \operatorname*{Tr}_{\mathrm{B}} e^{(t-t_0)\mathcal{L}_T}\, \rho_S(t_0) \otimes \rho_B^e. \tag{6}$$

These statements are exact but of no use unless one can actually solve (3), as in the purely harmonic case [6]. For more general cases it is customary to proceed as follows. For small $\Delta t$ one has

$$\rho_S\,(t_0 + \Delta t) - \rho_S(t_0) = \Delta t \operatorname*{Tr}_{\mathrm{B}} \mathcal{L}_T\{\rho_S(t_0) \otimes \rho_B^e\} + \mathcal{O}(\Delta t)^2. \tag{7}$$

It is then tempting to conclude that the system S obeys a differential equation

$$\dot{\rho}_S(t) = \mathcal{D}\rho_S(t), \tag{8}$$

where the superoperator $\mathcal{D}$ acts on the space of density matrices of S. However, this conclusion is unjustified, inasmuch as the special initial condition (5) may hold at $t_0$ but cannot also be true at later times. In fact, the evolution of S and B together builds up correlations between them.

This hurdle is overcome by means of the universal device of physicists: one introduces an assumption (optimistically called 'approximation'). One assumes that it is possible to pick a $\Delta t$ such that these correlations may be ignored; that is, one sets for the purpose of computing subsequent evolution,

$$\begin{aligned} \rho_T\,(t_0 + \Delta t) &\approx \rho_S\,(t + \Delta t) \otimes \rho_B^e \\ &= \left\{ \operatorname*{Tr}_{\mathrm{B}} e^{\Delta t \mathcal{L}_T}\, \rho_S(t_0) \otimes \rho_B^e \right\} \otimes \rho_B^e. \end{aligned} \tag{9}$$

This is the notorious random phase approximation (or Markov assumption or Stosszahlansatz). It is clear that this assumption cannot be true if $\Delta t$ is taken too small.

On the other hand, it must be small enough to permit the omission of the term $\mathcal{O}(\Delta t)^2$ in (7).

More explicitly these two conditions mean the following. The first one implies that any time interval $t - t_0$ may be treated as a sequence of such time steps $\Delta t$ ("repeated randomness assumption"), for which

$$\rho_S\left(t_0 + (n+1)\Delta t\right) = \operatorname*{Tr}_{\mathrm{B}} e^{\Delta t \mathcal{L}_T}\, \rho_S\left(t_0 + n\Delta t\right) \otimes \rho_B^e. \tag{10}$$

The second condition requires that this difference equation can be turned into a differential equation. That is, the change of $\rho_S$ during a time step $\Delta t$ is small in the sense that the averages one wants to compute with it change little. Thus one arrives at the desired differential equation

$$\begin{aligned} \dot{\rho}_S(t) &= \operatorname*{Tr}_{\mathrm{B}} \frac{e^{\Delta t \mathcal{L}_T} - 1}{\Delta t} \rho_S(t) \otimes \rho_B^e \\ &= \left\langle \frac{e^{\Delta t \mathcal{L}_T} - 1}{\Delta t} \right\rangle^e \rho_S(t). \end{aligned} \tag{11}$$

The angular brackets denote the average over the bath in equilibrium. It remains to compute $\langle \exp \Delta t \mathcal{L}_T \rangle^e$ to second order in $\lambda$.

One expects that the resulting formulae for the damping are insensitive to the precise construction of the bath. (If that is not so there cannot be a universal form for damped systems). One may as well choose for B a soluble model, namely a collection of independent harmonic oscillators, whose frequencies lie dense. This choise is universal, although initially more general baths were considered [7]. Also it is literally realized when S is a charged particle and B the electromagnetic field.

As mentioned before the entire system is soluble if S is itself a harmonic oscillator with frequency $\Omega$, and the coupling $H_I$ is bilinear. For this model the scenario can be carried out explicitly and one finds for the damping of the energy $E$ of the oscillator [8]

$$\frac{dE}{dt} = -2\Gamma \left\{ E - \frac{\Omega}{e^{\Omega/kT} - 1} \right\}. \tag{12}$$

Here $T$ is the bath temperature and $\Gamma$ is the strength of the interaction, multiplied with the density of the bath frequencies at the resonance $\Omega$. For $t \to \infty$ the energy tends to the wellknown equilibrium value.

## 2. The problem

Consider the general case, in which S may be any system although B is made up of harmonic oscillators while the interaction is unspecified. It is necessary to resort to a perturbation calculation in powers of the coupling constant $\lambda$. To zeroth order S and B are independent and the motion with initial condition (5) is

$$\rho_T(t) = \rho_S(t) \otimes \rho_B^e = e^{(t-t_0)\mathcal{L}_S}\, \rho_S(t_0) \otimes \rho_B^e. \tag{13}$$

To first order S is affected by the average of $H_I$ but in almost all applications the interaction averages out,

$$\langle H_I \rangle^e = 0. \tag{14}$$

The state of B undergoes a first order modification due to S. This modification affects the equilibrium of B and thereby acts back on S, causing a second order perturbation. This is the effect we are concerned with. Standard perturbation theory yields [8]

$$\dot{\rho}_S(t) = \mathcal{D}\,\rho_S = \left(\mathcal{L}_S + \lambda^2 \mathcal{K}\right)\rho_S, \tag{15}$$

where $\mathcal{K}$ is an operator acting on density matrices of S,

$$\mathcal{K}\rho_S(t) = -\int_0^\infty dt' \operatorname*{Tr}_{\mathrm{B}}\left[H_I, \left[H_I(-t'), \rho_S(t)\rho_B^e\right]\right] \tag{16}$$

$$H_I(-t') = e^{-it'H_0}\, H_I e^{it'H_0}, \;\; H_0 = H_S + H_B. \tag{17}$$

Note that the integration in $\mathcal{K}$ ought to run from $t' = 0$ to $t' = t$. One extends it deliberately to infinity on the ground that the integrand is a correlation function, which virtually vanishes beyond some correlation time $\Delta t$. This assumption assures that (15) does not involve explicitly the initial time and therefore is a true differential equation of the form (8). This is the formal expression of the repeated randomness assumption and has the effect that the evolution of $\rho_S$ becomes a semigroup. This general picture underlies all treatments. Higher orders are considered in quantum electrodynamics, but they are not of our concern. Our concern is that our derivation still involves the initial condition (5), so that the repeated randomness assumption is still needed. Yet there is another reason why (15) is unacceptable.

From the very definition of a density matrix it follows that (8) must be such that at all $t \geq 0$

$$\operatorname*{Tr}_{\mathrm{S}} \rho_S = 1; \quad \rho_S^\dagger = \rho_S; \quad \rho_S \;\; \text{positive definite}. \tag{18}$$

These conditions require according to Kossakowski [9,10] and Lindblad [11,12] that $\mathcal{L}$ be of the general form

$$\mathcal{L}\rho_S = -i\left[G, \rho_S\right] + \frac{1}{2}\sum_\alpha \left(V_\alpha \rho_S V_\alpha^\dagger - \frac{1}{2} V_\alpha^\dagger V_\alpha \rho_S - \frac{1}{2}\rho_S V_\alpha^\dagger V_\alpha\right), \tag{19}$$

with some hermitian operator $G$ and a set of unspecified operators $V_\alpha$. It has turned out that the result (15) of perturbation theory does not obey this requirement but it may happen that the resulting $\rho_S$ ceases to be positive definite at some $t > 0$. This has been verified numerically by Surez et al. [13]; they propose to avoid these negative probabilities by restricting the permissible $\rho_S(0)$, but that does not restore the semigroup.

This shortcoming of perturbation theory cannot be remedied by including higher orders of $\lambda$. The cause of the trouble is the postulated differential equation (8). In fact, $\rho_S$ cannot obey such an equation because the action of S on B and back is not instantaneous. For this reason the repeated randomness assumption requires a minimum value of $\Delta t$ and it is not permissible to make the transition from (10) to (11). To put it differently: the evolution of $\rho_S(t)$ is not a semigroup.

This also implies that the initial value $\rho_S(t_0)$ does not suffice to determine the future $\rho_S(t > t_0)$. In fact, the condition (5) is unrealistic because S and B are in continual interaction with one another. In actual applications, such as Brownian motion, S and B are continually interacting and they are together in equilibrium described by

$$\rho_T^e = Z_\lambda^{-1} e^{-\beta(H_S+H_B+\lambda H_I)} = Z_\lambda^{-1}\, e^{-\beta H_T}, \tag{20}$$

$$Z_\lambda = \operatorname*{Tr}_{\mathrm{T}} e^{-\beta(H_S+H_B+\lambda H_I)} = \operatorname*{Tr}_{\mathrm{T}} e^{-\beta H_T}. \tag{21}$$

The conclusion is that it is no longer possible to compute the decay of a quantity with given initial value. Quantities that remain valid, however, are the correlations between observables at successive times. Let $F$ be an observable of S, and $F\{t\}$ its evolution engendered by the *total* Hamiltonian (to be distinguished from the round brackets in (17), which indicate unperturbed evolution):

$$F\{t\} = e^{itH_T}\, F\, e^{-itH_T}. \tag{22}$$

Its auto-correlation function is $\langle F\{t_1\}F\{t_2\}\rangle_\lambda^e$. Here $\langle\rangle_\lambda^e$ indicates the average over the total equilibrium (20,21). This function depends only on $t_2 - t_1 = \tau$. We shall study the slightly more general correlation between two operators $F, G$ defined by

$$\langle G\, F\{\tau\}\rangle_\lambda^e = Z_\lambda^{-1} \operatorname*{Tr}_{\mathrm{T}} e^{-\beta H_T}\, G\, e^{itH_T}\, F\, e^{-itH_T}. \tag{23}$$

This is the quantity we intend to compute to second order in $\lambda$. Note that $\lambda$ occurs not just in the evolution of $F$, but also in the distribution $\rho_T^e$ of total system, which never entered in the standard treatment (13).

## 3. Computation of the correlation function

To evaluate the exponentials to second order in $\lambda^2$ we utilize the mathematical identity

$$e^{-\beta(H_0+\lambda\, H_I)} = e^{-\beta H_0} \left\{ 1 - \lambda \int_0^\beta H_I(-i\beta')d\beta' + \right.$$

$$\left. +\lambda^2 \int_0^\beta d\beta' \int_0^{\beta'} d\beta''\, H_I(-i\beta')H_I(-i\beta'') + \mathcal{O}(\lambda^3) \right\}, \tag{24}$$

where use has been made of the same abbreviation as in (17). This gives for the equilibrium partition function (21)

$$Z_\lambda = Z_0 - \lambda \int_0^\beta \langle H_I(-i\beta')\rangle_0^e\, d\beta'$$

$$+\lambda^2 \int_0^\beta d\beta' \int_0^{\beta'} d\beta''\, \langle H_I(-i\beta')H_I(-i\beta'')\rangle_0^e. \tag{25}$$

Here $\langle\rangle_0^e$ indicates the average over the unperturbed equilibrium of the total system. The first order in $\lambda$ vanishes on account of (14).

Applying the same methods to the other exponentials in (20) one finally finds after some tedious algebra

$$\begin{aligned}\langle G\,F\{\tau\}\rangle_\lambda^e &= \langle G\,F(\tau)\rangle_0^e\left\{1-\lambda^2\int_0^\beta d\beta'\int_0^{\beta'} d\beta''\,\langle H_I(-i\beta')H_I(-i\beta'')\rangle_0^e\right\}\\ &- \lambda^2\int_0^\tau dt'\int_0^{t'} dt''\,\langle G\,[H_I(t''),[H_I(t'),F(\tau)]]\rangle_0^e\\ &+ \lambda^2\int_0^\beta d\beta'\int_0^{\beta'} d\beta''\,\langle H_I(-i\beta')H_I(-i\beta'')G\,F(\tau)\rangle_0^e\\ &- i\lambda^2\int_0^\beta d\beta'\int_0^\tau dt'\,\langle H(-i\beta')G\,[H_I(t'),F(\tau)]\rangle_0^e\,. \end{aligned}\tag{26}$$

This is the correct result to order $\lambda^2$. The term on the third line may be combined with the one on the first line to give

$$\lambda^2\int_0^\beta d\beta'\int_0^{\beta'} d\beta''\,\langle H_I(-i\beta')H_I(-i\beta'')\,\{G\,F(\tau)-\langle G\,F(t)\rangle_0^e\}\rangle_0^e\,. \tag{27}$$

## 4. Comparison with standard result

According to the standard method, based on (15), one would compute the same correlation function (23) by choosing the initial state of the bath equal to its unperturbed equilibrium

$$\rho_S(0)=\rho_S^e=Z_0^{-1}\exp[-\beta H_S], \tag{28}$$

and computing the time dependence $F\{\tau\}$ by taking the second order of (22). One then gets

$$\langle G\,F\{\tau\}\rangle_0^e=\operatorname*{Tr}_{\mathrm{S}}\rho_0^e\,G\,F\{\tau\}. \tag{29}$$

Rather than evaluating $F\{\tau\}$ explicitly I utilize as a short-cut the fact that the evolution of $F$ is adjoint (in the space density matrices) to that of $\rho_S$,

$$\operatorname*{Tr}_{\mathrm{S}}\rho_S\,F\{\tau\}=\operatorname*{Tr}_{\mathrm{S}}\rho_S(\tau)F. \tag{30}$$

By integrating (15) one has to second order in $\lambda$

$$\rho_S(\tau)=\left[e^{\tau\mathcal{L}_S}+\lambda^2\int_0^\tau e^{(\tau-t')\mathcal{L}_S}\,\mathcal{K}\,e^{t'\mathcal{L}_S}\,dt'\right]\rho_S(0). \tag{31}$$

Accordingly

$$F\{\tau\}=\left[e^{-\tau\mathcal{L}_S}+\lambda^2\int_0^\tau e^{-t'\mathcal{L}_S}\mathcal{K}^\dagger e^{-(\tau-t')\mathcal{L}_S}\,dt'\right]F. \tag{32}$$

Here $\mathcal{K}^\dagger$ is the adjoint of (16). If $\chi$ denotes any trace-class operator in the Hilbert space of S, the action of $\mathcal{K}^\dagger$ on it is found to be

$$\mathcal{K}^\dagger\,\chi=-\int_0^\infty dt''\operatorname*{Tr}_{\mathrm{B}}\rho_B^e\,[H_I(-t''),[H_I,\chi]]\,. \tag{33}$$

Insert (32) into (29) (and remember that round brackets refer to the unperturbed evolution):

$$\begin{aligned}\langle G\,F\{\tau\}\rangle_0^e &= \langle G\,F(\tau)\rangle_0^e - \\ &- \lambda^2 \int_0^\tau dt' \int_0^\infty dt''\,\langle G\,[H_I(t'-t''),[H_I(t'),F(\tau)]]\rangle_0^e. \end{aligned} \quad (34)$$

This is the correlation function according to the standard theory. The second line may trivially be rewritten

$$-\lambda^2 \int_0^\tau dt' \int_{-\infty}^0 dt''\,\langle G\,[H_I(t''),[H_I(t'),F(\tau)]]\rangle_0^e. \quad (35)$$

It appears that there are three differences from (26)

(i) The first line of (26), containing the renormalized partition function $Z_\lambda$, is replaced with the unperturbed $Z_0$.

(ii) All integrals over $\beta$ are absent in (34). It might be said that (34) actually applies to high temperatures, although $\beta$ is still present in the bath distribution $\rho^e$.

(iii) What remains of (26) is its second line but even this one is not reproduced faithfully in (35). The lower integration limit has been replaced with $-\infty$ as a consequence of our extending the integration in (16) with an appeal to the short survival time of the correlation in the integrand. This embodies the repeated randomless assumption, indispensable for obtaining a differential equation for $\rho_S$ of type (8) without memory. The present correct result (26) for the correlation function has no need of this assumption.

## References


[1] D. K. C. MacDonald, *Philos. Mag.* **45** (1954) 63 and 345; *Phys. Rev.* **108** (1957) 541.

[2] N. G. van Kampen, in Fluctuation Phenomena in Solids, ed. R. E. Burgess, Academic Press, New York (1965), pages.

[3] R. Benguria and M. Kac, *Phys. Rev. Letters* **46** (1981), 1.

[4] H. Grabert, U. Weiss and P. Talkner, *Z. Phys. B* **55** (1984), 87.

[5] C. W. Gardiner, *IBM J. Res. Dev.* **32** (1988), 127; *Quantum Noise*, Springer, Berlin (1991).

[6] P. Ullersma, *Physica* **23** (1966) 27, 56, 74, 90.

[7] I. R. Senitzky, *Phys. Rev.* **119** (1960) 670; *Phys. Rev.* **124** (1961) 642.

[8] N. G. van Kampen, *Stochastic Processes in Physics and Chemistry*, North-Holland, Amsterdam (1992).

[9] A. Kossakowski, *Bull. Acad. Pol. Sci., Série Math. Astr. Phys.* **20** (1971) 1021; *Bull. Acad. Pol. Sci., Série Math. Astr. Phys.* **21** (1973) 649.

[10] V. Gorini, A. Kossakowski and E. C. G. Sudarshan, *J. Math. Phys.* **17** (1976) 821.

[11] G. Lindblad, *Commun. Math. Phys.* **40** (1975) 147.

[12] R. Alicki and K. Lendi, *Quantum Dynamical Semigroups and Applications*, Lecture Notes in Physics **286**, Springer, Berlin (1987).

[13] A. Suárez, R. Silbey and I. Oppenheim, *J. Chem. Phys.* **97** (1992) 5101.

# QUANTUM ERROR CORRECTION FAILS FOR HAMILTONIAN MODELS

R. ALICKI
*Institute of Theoretical Physics and Astrophysics, University of Gdańsk, Poland*
*fizra@univ.gda.pl*

Received 13 April 2006
Accepted 30 May 2006
Communicated by Fred Green

It is argued that the existing schemes of fault-tolerant quantum computation, designed for discrete time models and based on quantum error correction, fail for continuous time Hamiltonian models, even with Markovian noise.

*Keywords*: Quantum computer; error correction; Markovian dynamics.

## 1. Introduction

The fundamental challenge to quantum information processing is the devastating influence of decoherence processes due to the interaction of a quantum device with environment. It is quite generally believed, at least in principle, that by suitable active procedures of *quantum error correction* this problem can be solved if the level of noise is lower than a certain threshold. The aim of this note is to show that, unfortunately, the success of existing error correction procedures is due to the discrete-in-time nature of the modeling of quantum evolution. In physical terms, discrete models correspond to invoking infinitely fast, and therefore unphysical, gates. More precisely, we shall argue that the fidelity of the final state of the quantum computer is of the order of

$$\text{fidelity} \simeq \exp(-\lambda^2 t_0 V) \tag{1}$$

where $\lambda^2$ is the decoherence rate for a single qubit, $t_0$ is the duration of a unitary gate and $V$ is the "volume" of an algorithm. As we are going to construct an explicit counterexample, it is enough to discuss the idea of *fault-tolerant quantum information processing* for the simplest model. The task is to protect a bit of classical information for a given, but arbitrarily long, period of time and with arbitrarily high accuracy. The extension of this particular result to general fault-tolerant quantum computation is then straightforward. The framework presented is general enough to cover all the known standard examples of fault-tolerant procedures.

## 2. General Scheme of Error Correction

The mathematical model consists of a $M$-qubit quantum register and a large enough set $A$ of ancillary qubits prepared in a standard state $|\phi_A\rangle = |000...00\rangle$. The state $|0\rangle$ of an ancillary qubit is assumed to be stable with respect to noise. The controlled part of the evolution is governed by a sequence of unitary operations (*quantum gates*) acting on the register and ancillas, and executed in discrete time steps. Noise is modeled by completely positive trace-preserving (error) maps acting on the register and those ancillary qubits which are engaged in the process of error correction at a given time step. Any gate, including the trivial one, is followed by a suitable error map. By a *code* we mean in our specific situation a two-dimensional subspace of the register states spanned by the logical $|0_L\rangle$ and $|1_L\rangle$. We shall use $\|\cdot\|_1, \|\cdot\|_\infty$ and $\|\cdot\|$ for tracial, operator and superoperator norms, respectively.

The dynamics corresponding to $T$ time steps without error corrections is simply given by $(E)^T$ where $E$ is an error map acting on the register. The error map $E$ is composed of elementary error maps $E^{(\alpha)}$ typically corresponding to one- or two-qubit locations and satisfying the estimation

$$\|E^{(\alpha)} - \mathbb{I}\| \leq p \tag{2}$$

where $p$ provides an overall bound on the error rate.

In the next step of construction we add, after any error map at the time step $k$, a sequence of gates producing the unitary map $R_k$ which acts on the register and a subset of ancillas $A_k$ (different for any $k$) and which are designed to recover the input state. Then the evolution is modeled by a sequence of dynamical maps

$$R_T E R_{T-1} E \cdots R_1 E\,. \tag{3}$$

Obviously, the real gates are also accompanied by noise, and should produce completely positive dynamical maps $R'_k$ rather than the unitaries $R_k$. We can always write $R'_k = E'_k R$ and define $E_k = E E'_k$ with $E_0 \equiv E$. The complete evolution in our model is now a composition of maps

$$E'_T R_T E_{T-1} R_{T-1} E_{T-2} \cdots R_1 E_0\,. \tag{4}$$

The error maps $E_k$ and the unitary recovery maps $R_k$ should satisfy the following *error correction property* with the accuracy $\mu$

$$R_k E_{k-1} P_\psi \otimes P_{\phi_{A_k}} \otimes \rho_{k-1} = (1-\mu) P_\psi \otimes \rho_k + \sigma_k\,, \quad \|\sigma_k\|_1 \leq B\mu \tag{5}$$

with a certain fixed $B \geq 1$. Here $\rho_k$ is an unspecified state of ancillas from $\bigcup_{l=1}^k A_l$, $P_\psi = |\psi\rangle\langle\psi|$ is any pure state from the code and $P_{\phi_{A_k}} = |\phi_{A_k}\rangle\langle\phi_{A_k}|$.

One should notice that the condition (5) is more restrictive for $k > 1$ than for $k = 1$, as in the former case more faults have to be corrected and their propagation becomes very important.

We are interested in the "almost final" state of the system (prior to the last error map $E'_T$)

$$\rho_T^{af} = R_T E_{T-1} R_{T-1} E_{T-2} \cdots R_1 E_0 P_\psi \otimes P_{\phi_A}\,. \tag{6}$$

Combining (6) with (5) we obtain a lower bound for the fidelity of the "almost final" state of the register

$$\langle\psi, \mathrm{Tr}_A\big(\rho_T^{af}\big)\psi\rangle \geq (1-\mu)^T - B\big[1-(1-\mu)^T\big]\,. \tag{7}$$

The protection of a single bit is succesful if, for any $T$ and any $\epsilon > 0$, there exists an encoding into a large enough $M$-qubit space, together with recovery maps $R_k$ satisfying the error-correction property (5) with $\mu = \epsilon/(B+1)T$. Moreover, the number of register qubits and ancillas should grow moderately, like poly$[\ln(1/\epsilon)]$. Under these conditions, the fidelity of the register's "almost final" state is bounded from below by $(1-\epsilon)$ and one can show that, using enough redundancy and majority voting, one can extract the classical bit even from this state, when perturbed by the last error map $E'_k$.

The remarkable and highly involved results [1–4] show, by explicit construction of concatenated codes and encoded recovery maps, and for the error maps satisfying reasonable conditions, that the scheme presented above works for general quantum computation. The conditions concern locality, statistical independence and the *threshold condition* ($p < p_c$) imposed on elementary error maps $E^{(\alpha)}$.

## 3. Markovian Physical Model

We shall try to apply this standard scheme, as presented above, to a simple physical model of quantum computation described by continuous, time-irreversible dynamics governed by a Markovian master equation. The strictly Markovian, single-qubit noise acting on the register is explicitly given in terms of the single-qubit semigroup generators

$$L_k\rho = \lambda^2\big(\Phi_k\rho - \rho\big)\ ,\text{where } \Phi_k = \frac{1}{2}\,\mathrm{Tr}_k\,\rho\otimes\mathbf{1}_k\ ,\ \ k = 1,2,\ldots M\,. \tag{8}$$

In this model $\lambda^2$ is a single-qubit decoherence rate and $\lambda^2 t_{clock}$ corresponds to $p$ in the discrete model, where $t_{clock}$ is a time step of the computer's clock.

The total Hamiltonian of the register-and-ancilla system is denoted by $H(t); t \in [0, T\tau]$ and is designed to execute the sequence of unitary-state recovery operations $R_T R_{T-1}\cdots R_1$. Here $\tau > t_{clock}$ is the working period needed to execute the continuous-time analog of the operation $R_k E_{k-1}$. The time-dependent Hamiltonian possesses the following structure:

$$H(t) = \sum_\alpha f_\alpha(t)h^\alpha\,; \tag{9}$$

it consists of well-separated, possibly parallel, pulses $f_\alpha(t)h^\alpha$ such that $f_\alpha(t) \geq 0$ and $\int_{-\infty}^{\infty} f_\alpha(t)dt = 1$, of typical width $t_0 < t_{clock}$. Here $h^\alpha$ are one- or two- qubit Hamiltonians acting on register and ancillas. For the rest of the discussion, the model of noise acting on the ancillas is irrelevant. Taking the limit $f_\alpha(t) \to \delta(t-t_\alpha)$ and using a Markovian model of noise, we obtain the standard discrete-time model (4).

We shall discuss the less demanding error-correction property (5) for $k = 1$, which can be written in terms of fidelity

$$F_1 = \langle\psi, \mathrm{Tr}_A\big(R_1 E P_\psi\otimes P_{\phi_A}\big)\psi\rangle \geq \big(1-(B+1)\mu\big)\,. \tag{10}$$

This condition involves the error map acting on the register in the first time step, combined with the *error-free evolution* of the register and ancillas which is designed to recover the initial state $\psi$. Therefore the continuous-time counterpart to the map $R_1E$ is the propagator $\Lambda$ obtained as a solution of the master equation in the first working period $[0, \tau]$ $(\hbar \equiv 1)$

$$\frac{d}{dt}\rho = -i[H(t), \rho] + \lambda^2 \sum_{k=1}^{M} (\Phi_k \rho - \rho) \,. \tag{11}$$

Introducing the Hamiltonian (super)propagator for $t \geq s$

$$\mathcal{U}(t, s) = \mathbb{T} \exp\Big\{-i \int_s^t [H(u), \cdot] du\Big\}, \; \mathcal{U}(s, t) \equiv \mathcal{U}(t, s)^{-1} \tag{12}$$

we can write the propagator solving (11) in terms of the series expansion

$$\Lambda(t) = \mathcal{U}(t, 0)\Big(e^{-\lambda^2 Mt} \sum_{N=0}^{\infty} (\lambda^{2N} \sum_{k_N,..,k_2,k_1=1}^{M} \int_0^t dt_N \int_0^{t_N} dt_{N-1} \ldots \int_0^{t_2} dt_1$$

$$\times \Phi_{k_N}(t_N)\Phi_{k_{N-1}}(t_{N-1}) \ldots \Phi_{k_1}(t_1)\Big) \tag{13}$$

where

$$\Phi_k(u) = \mathcal{U}(u, 0)^{-1} \Phi_k \mathcal{U}(u, 0) \,. \tag{14}$$

To estimate the fidelity of the protected state after the first working period $[0, \tau]$, it is sufficient to use (13) together with the *recovery condition*

$$\mathcal{U}(\tau, 0) = \mathbb{I} \otimes \mathcal{U}_A(\tau, 0) \,. \tag{15}$$

Denote by $F(t), t \in [0, \tau]$ the following expression

$$F(t) = \Bigg(e^{-\lambda^2 Mt} \sum_{N=0}^{\infty} \lambda^{2N} \sum_{k_N,..,k_2,k_1=1}^{M} \int_0^t dt_N \int_0^{t_N} dt_{N-1} \ldots \int_0^{t_2} dt_1$$

$$\times \mathrm{Tr}\big(P_\psi \otimes \mathbf{1}_A \Phi_{k_N}(t_N)\Phi_{k_{N-1}}(t_{N-1}) \ldots \Phi_{k_1}(t_1) P_\psi \otimes P_{\phi_A}\big)\Bigg) \,. \tag{16}$$

From (16) it follows that the time-dependent error $E(t) = 1 - F(t)$ satisfies the equation

$$\frac{dE}{dt} = -\lambda^2 m E(t) + \lambda^2 m X(t) \tag{17}$$

where

$$X(t) = \Bigg(e^{-\lambda^2 Mt} \sum_{N=0}^{\infty} \lambda^{2N} \sum_{k_N,..,k_2,k_1=1}^{M} \int_0^t dt_N \int_0^{t_N} dt_{N-1} \ldots \int_0^{t_2} dt_1$$

$$\times \Bigg[1 - \Bigg(\frac{1}{M} \sum_{k=1}^{M} \mathrm{Tr}\big([\Phi_k(t) P_\psi \otimes \mathbf{1}_A] \Phi_{k_N}(t_N)\Phi_{k_{N-1}}(t_{N-1}) \ldots \Phi_{k_1}(t_1) P_\psi \otimes P_{\phi_A}\big)\Bigg]\Bigg)$$

$$\geq 0 \,. \tag{18}$$

Using properties of norms we obtain

$$\frac{1}{M}\sum_{k=1}^{M}\mathrm{Tr}\Big(\big[\Phi_k(t)P_\psi\otimes\mathbf{1}_B\big]\Phi_{k_N}(t_N)\Phi_{k_{N-1}}(t_{N-1})\dots\Phi_{k_1}(t_1)P_\psi\otimes P_{\phi_A}\Big)$$
$$\leq \max_k\|\Phi_k(t)P_\psi\otimes\mathbf{1}_A\|_\infty \tag{19}$$

and hence

$$X(t)\geq x(t)=1-\max_k\|\Phi_k(t)P_\psi\otimes\mathbf{1}_A\|_\infty\geq 0\,. \tag{20}$$

Solving equation (17) and using (20) we obtain the lower bound for the error at the end of the working period

$$E(\tau)=\lambda^2M\int_0^\tau e^{-\lambda^2M(\tau-s)}X(s)\,ds\geq\lambda^2M\int_0^\tau e^{-\lambda^2M(\tau-s)}x(s)\,ds\,. \tag{21}$$

Now we have to analyze the time-dependence of the function $x(t); t\in[0,\tau]$. We notice first that due to the condition (15) we have

$$\Phi_k(0)=\Phi_k(\tau)=\Phi_k \tag{22}$$

which gives

$$x(0)=x(\tau)=1-\|\Phi_kP_\psi\|_\infty\geq\frac{1}{2} \tag{23}$$

and the inequality follows from the simple estimation $\|\Phi_kP_\psi\|_\infty\leq 1/2$. The function $x(t)$ is non-negative, continuous and larger then $1/2$ at the ends of the time interval $[0,\tau]$.

In principle one cannot exclude the possibility that, by using a suitable controlling Hamiltonian $H(t)$, we can make $X(t)$ arbitrarily small in the interval $[\delta,\tau-\delta]$ with some arbitrarily small $\delta>0$. But this could be achieved *only* by applying sufficiently fast-acting gates in the physical procedure for "hiding and recovering" of the the initial state $P_\psi$. Any technology applied to implement quantum information processing must inevitably put a finite limit on the speed of gates; this will be given in terms of the typical width of a pulse

$$t_0\geq t_0^{min}\,. \tag{24}$$

Under this assumption, and using (21), we arrive at the basic estimation of the fidelity at the end of the first working period, in terms of the parameter $q\simeq\lambda^2t_0^{min}$

$$F(\tau)\equiv F_1^{cont}\leq 1-\lambda^2Mq\;,\;\text{ for }\lambda^2Mq\ll 1 \tag{25}$$

with saturation setting in for large values of $M$:

$$F_1^{cont}\leq\frac{1}{2}\;,\;\text{ for }\lambda^2Mq\gg 1\,. \tag{26}$$

It follows that the inequality (10) with $\mu=\epsilon/(B+1)T$ cannot be satisfied physically, and the same is true for all working periods. Therefore, successful state protection is not feasible for real quantum systems, at least within existing proposals of quantum error correction.

## 4. Conclusions

Using a tractable continuous-time model of quantum-information processing, with the physically required *limited gate-speed condition* (24), we conclude that, during any working period, the input state's fidelity is reduced by a factor given by (25) and (26). This loss of information cannot be reduced using multi-qubit encodings. It leads to an exponential decrease of the total fidelity with the volume of the algorithm (1). The crucial parameter is the duration of the gate operation, $t_0$, multiplied by the decoherence rate $\lambda^2$. Only in the *unphysical* limit $t_0 \to 0$ one can recover the results of the strictly algorithmic theory of fault-tolerant quantum computation. We also obtain the guiding estimate

$$T_{max} \simeq \frac{1}{\lambda^2 t_0^{min}} \tag{27}$$

for the maximal number of algorithm's steps which can be executed within a given physical implementation. This formula follows from (1) and the fact, that we can tolerate finite error per qubit at the end of quantum computation.

The results we have presented also explain why we do not encounter analogous uncorrectable errors in the classical theory of digital computers. This is because, for the classical digital case, no continuous-in-time reversible operation is possible. Therefore, intrinsically, $t_0 = 0$ in the sense of the physical model we have analyzed above. However, a *continuous* kinetic description of classical computers, which takes into account the physical continuum of bit states lying between the logical $\{0\}$ and $\{1\}$ ones, will lead to results similar to those in quantum theory. This fact may be important for mesoscopic and microscopic realizations of *classical* computers in the future.

## 5. Acknowledgements

The author is extremely grateful to Michał Paweł and Ryszard Horodecki for their countless stimulating and vigorous discussions concerning fault tolerant computation. The exchange of opinions with Daniel Lidar is acknowledged also. This work was supported by the Polish Ministry of Scientific Research under Grant No PBZ-Min-008/P03/03.

## References

[1] D. Aharonov and M. Ben-Or, Fault-tolerant quantum computation with constant error rate, to appear in the *SIAM Journal of Computation*; see *e-preprint* quant-ph/9906129.

[2] E. Knill, R. Laflamme and W. Zurek, Resilient quantum computation, *Science* **279** (1998) 342.

[3] J. Preskill, Fault tolerant quantum computation, *e-preprint* quant-ph/9712048.

[4] M. A. Nielsen and I. L. Chuang, *Quantum Computation and Quantum Information* (Cambridge University Press, Cambridge, UK, 2000).

# Statistics of Fluctuations of Magnetic Monopole Concentration in Spin Ice

Alexey V. Klyuev*,‡, Mikhail I. Ryzhkin† and Arkady V. Yakimov*

*National Research Lobachevsky State University of Nizhni Novgorod
Gagarin Avenue 23, Nizhny Novgorod 603950, Russia

†Institute of Solid State Physics (ISSP) Russian Academy of Sciences
2 Academician Ossipyan str. Chernogolovka, Moscow 142432, Russia
‡klyuev@rf.unn.ru

Received 29 June 2017
Revised 7 September 2017
Published 17 October 2017

Communicated by Igor Goychuk

In this paper, we theoretically investigate the fluctuations in the number of emergent magnetic monopoles in spin ice. The Langevin equation for these fluctuations is deduced. This allows us to calculate the spectrum of relative fluctuations, which can be measured experimentally.

*Keywords*: Magnetic monopoles; spin ice; fluctuations; ice rules; generation–recombination.

## 1. Introduction

Recently a lot of research works have been carried out due to the great interest in the study of spin ice [1]. This interest was caused by the theoretical prediction of unusual magnetic defects in spin ice, which were called as emergent magnetic monopoles [2, 3].

Originally, the term "spin ice" was applied to compounds such as $Ho_2Ti_2O_7$. Magnetic ions in them are located at the vertices of regular tetrahedra linked into a three-dimensional pyrochlore lattice. Due to the strong anisotropy, the magnetic moments can be directed either toward the center of each tetrahedron or away from this center. Note that dual lattice formed by the centers of the tetrahedra is a usual diamond lattice. The ground state of spin system is characterized by the ice rule: two spins of each tetrahedron are directed toward its center and two other ones away from its center (Fig. 1(a)). This corresponds to a distribution of electrical dipole moments, hydrogen bonds in hexagonal ice, solid phase of water in natural conditions [1] that justifies the term "spin ice". Subsequently, this term began to apply to any two-dimensional and three-dimensional systems in which the magnetic moments (spins) are distributed in accordance with the ice rules.

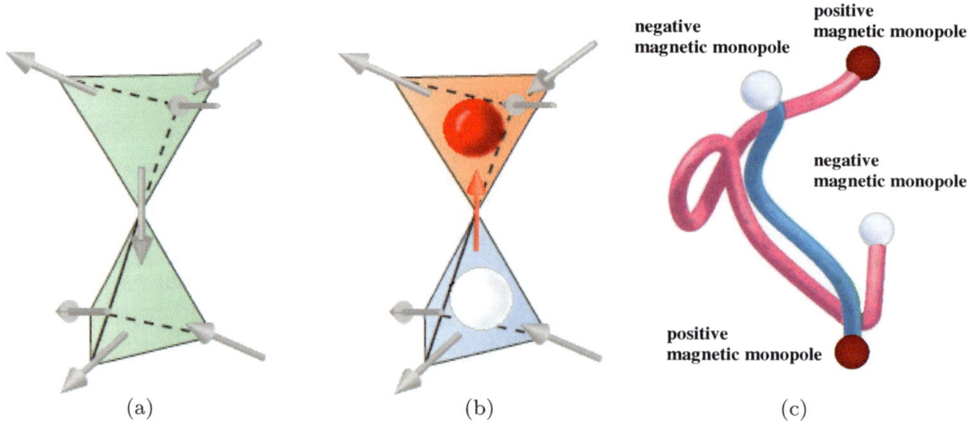


Fig. 1. (a) Two neighboring tetrahedra obeying the ice rule, two spins are directed "in" and two "out". (b) Inverting the shared spin generates a pair of magnetic monopoles. Flips of spins along the way of movement of monopole shown in (c) create a pair of magnetic monopoles bounded by the "spin spaghetti". A movement of monopole along the "spin spaghetti" changes spin configuration on it.

The flipping of an arbitrary spin in ground state configuration, which takes place at finite temperatures, breaks the ice rule in two neighboring tetrahedra. One tetrahedron has three spins directed "in", and one "out", while its neighbor has one "in" and three "out" (Fig. 1(b)). These tetrahedra can be considered as positive and negative emergent magnetic monopoles, respectively. The positive magnetic monopole is a magnetic analog of a positive ion in water ice. Accordingly, the negative magnetic monopole is a magnetic analogue of a negative ionic defect in water ice [2]. As one can see from Fig. 1, the flipping of any of three identical magnetic spins of the defect is equivalent to the defect movement to another site. This movement is not accompanied by the formation of new defect and by the increase of energy. By means of further spin-flips, the magnetic monopoles can move through the lattice and thereby change the spin configuration. Monopoles play the role of classical quasiparticles, in which the movement through the lattice changes the spin configuration on all the distance traveled, resulting in the formation of a "spin spaghetti" (Fig. 1(c)) [3]. Therefore, in comparison with the defects in semiconductors, which may negatively influence on characteristics of devices [4–13], magnetic defects in spin ice [14–20] play the positive role of carriers of magnetic charge.

The purpose of our paper is the investigation of the fluctuations in the number of magnetic monopoles. To implement this, we derive the Langevin equation for these fluctuations by the Einstein–Fokker–Planck (EFP) equation method. This allows us to calculate the spectrum of relative fluctuations in the number of magnetic monopoles. Then we discuss a possible relation of fluctuations in the number of pairs of magnetic monopoles with measurable characteristic of spin ice and an experimental display of obtained results.

## 2. Fluctuations in the Number of Pairs of Magnetic Monopoles

To analyze fluctuations in the number of magnetic monopoles, we consider the processes of generation and recombination of magnetic monopole pairs. It is the simplest stochastic analysis, which shows that these fluctuations can be described by the Langevin equation. Thus, these fluctuations can be considered as generation–recombination noise of quasiparticles (magnetic monopoles). The spectrum of this noise is determined.

To derive the basic equation, we make the following approximations and assumptions:

(a) We assume that monopoles are created and annihilated in pairs. Thus, we introduce the concentration of monopole pairs instead of the concentrations of monopoles.
(b) We neglect by simultaneous generation and recombination of a few pairs, assuming such events are much less probable than generation and recombination of one pair. That is, we assume that only one pair can be created or annihilated over a sufficiently small time $dt$.
(c) We discuss relatively small temperatures at which the concentration of monopoles is low. In this case, we can consider the monopoles as non-interacting quasiparticles.

At the first step, we use standard treatment for generation–recombination noise, see e.g., [21]. We consider probability$P(N,t)$ of the event that $N$ pairs of quasiparticles (magnetic monopoles) are present at time $t$. This probability is changed on value $dP$ during time $dt$ due to the following transitions:

$$\begin{aligned} &N \xrightarrow{g} (N+1); \quad N \xrightarrow{r} (N-1), \\ &(N-1) \xrightarrow{g} N; \quad (N+1) \xrightarrow{r} N. \end{aligned} \tag{1}$$

Here, transitions, which are specified in the first line, reduce the probability, i.e., $dP < 0$. On the contrary, remaining two transitions give $dP > 0$. Transitions, marked by symbol "$g$" above the arrow, are caused by the generation of a pair, respectively, the symbol "$r$" means recombination of the pair.

The rates of generation, $g = g(N)$, and recombination, $r = r(N)$, depend only on number $N$ of the pairs at time $t$. These rates determine the probability $g \cdot dt$ of generation and $r \cdot dt$ of recombination of one pair over an arbitrarily small time $dt$. As a result, the analysis of transitions (1) leads to the following differential–difference equation for probability $P(N,t)$:

$$\begin{aligned} dP = &-P(N,t)\cdot g(N)\cdot dt - P(N,t)\cdot r(N)\cdot dt \\ &+P(N-1,t)\cdot g(N-1)\cdot dt + P(N+1,t)\cdot r(N+1)\cdot dt. \end{aligned} \tag{2}$$

It is seen from this equation that $N = N(t)$ is a random Markovian process because the change in the number of quasiparticles prior to time $(t + dt)$ depends only on the state of the system at time $t$.

In order to derive $P(N,t)$, we use the EFP equation method, see e.g., [22]. This is the second step of our analysis (in difference with [21]). The change of probability

$P(N, t)$, described by Eq. (2), depends on three neighboring points: $(N-1)$, $N$, $(N+1)$. Assuming that $N \gg 1$, we can consider the change $(+1)$ as differential $dN = 1$. Thus, these points are: $(N - dN)$, $N$, $(N + dN)$. In terms of differential calculus, this means that the right part of Eq. (2) contains derivatives of variable $N$ up to second order; derivatives of highest orders are absent. This allows us to expand the last two terms in the right-hand part of Eq. (2) into a power series around $N$. As a result of this, we obtain the EFP equation for $P(N, t)$ as follows:

$$\frac{\partial P}{\partial t} = -\frac{\partial}{\partial N}[K_1(N) \cdot P] + \frac{1}{2}\frac{\partial^2}{\partial N^2}[K_2(N) \cdot P]. \tag{3}$$

Kinetic coefficients in this equation are determined by the generation and recombination rates of quasiparticle pairs:

$$K_1(N) = g(N) - r(N), \quad K_2(N) = g(N) + r(N). \tag{4}$$

Equations (3) and (4) allow one to find all statistical characteristics of fluctuations for the total number of quasiparticle pairs $N(t)$. But here, another treatment of EFP equation is used. Equation (3) corresponds to the following equation for fluctuations in $N(t)$ [23, 24]:

$$\frac{dN}{dt} = K_1(N) + \sqrt{K_2(N)} \cdot \varsigma(t). \tag{5}$$

Here, $\varsigma(t)$ is stationary Gaussian white noise with normalized spectrum $S_\varsigma(f) = 1$ [Hz/Hz]. Note that Eq. (5) coincides in form with the Langevin equation, see e.g. [21]. It is interesting to mention that in [25] the functional method to derive the fractional Fokker–Planck equation for probability distribution from the Langevin equation with Lévy stable noise was proposed.

At this stage, we restrict the analysis by the stationary case. Condition of the stationarity can be represented as follows: $K_1(N_0) = 0$. The average concentration of quasiparticle pairs $N_0 = \langle N(t) \rangle$ is determined in accordance with the first relation in Eq. (4). That is, the rates of generation and recombination of quasiparticles coincide

$$g(N_0) = r(N_0). \tag{6}$$

In further analysis, we consider the relative fluctuations in the number of pairs $\delta N = \delta N(t)$ near the stationary state $N_0$. Thus, the total number of pairs is $N = (1 + \delta N) \cdot N_0$. The variance $\sigma^2_{\delta N}$ of these fluctuations is assumed to be sufficiently small, $\sigma^2_{\delta N} \ll 1$. Linearization of Eq. (5) in vicinity of $N_0$ gives the following equation for the relative fluctuations in the number of pairs:

$$\frac{d}{dt}\delta N = -\frac{1}{\tau_0}\delta N + \sqrt{A_N} \cdot \varsigma(t). \tag{7}$$

Here, $\tau_0$ is the relaxation time, defined via the difference of derivatives from velocities of recombination and generation, taken in point $N_0$ as follows:

$$\tau_0 = \left[\left.\frac{d(r-g)}{dN}\right|_{No}\right]^{-1}. \tag{8}$$

From Eq. (8), one can see that the exact relation for relaxation time depends on exact relations for velocities of recombination and generation. But, as far as the authors know, at the present moment there are no relations for velocities of recombination and generation of pairs of magnetic monopoles in spin ice.

Parameter $A_N$ in Eq. (7) is determined by Eqs. (4) and (6) as follows:

$$A_N = 2g(N_0)/N_0^2. \tag{9}$$

Note, that Eq. (7) coincides in form with the equation for Ornstein–Uhlenbeck process [23]. Thus, we can calculate statistical characteristics of such process. In this work, we focus on the relation between fluctuations in the number of pairs of magnetic monopoles and measurable characteristics of spin ice. For this reason we calculate the spectrum of relative fluctuations in the number of pairs of quasiparticles, which (it will be shown farther) are manifested as measurable fluctuations in the magnetic moment.

Equation (7) allows determining of the spectrum $S_{\delta N}(f)$ of relative fluctuations in the number of pairs as follows:

$$S_{\delta N}(f) = \frac{\tau_0^2 A_N}{1 + (2\pi f \tau_0)^2}. \tag{10}$$

Let us turn to the definition of the parameter $A_N$, described by Eq. (9). To do this, we calculate the variance $\sigma_{\delta N}^2$ of the fluctuations of generation–recombination of pairs of quasiparticles:

$$\sigma_{\delta N}^2 = \int_0^\infty S_{\delta N}(f)df = \frac{1}{4} A_N \tau_0. \tag{11}$$

Taking into account this result, we can find the final expression for the spectrum of relative fluctuations in the number of pairs of quasiparticles:

$$S_{\delta N}(f) = \frac{4\tau_0}{1 + (2\pi f \tau_0)^2} \cdot \sigma_{\delta N}^2. \tag{12}$$

Now, we can investigate the dependence of this spectrum on the magnitude $\tau_0$ (see Fig. 2), assuming the variance of the fluctuations constant, $\sigma_{\delta N}^2 = \text{const}$.

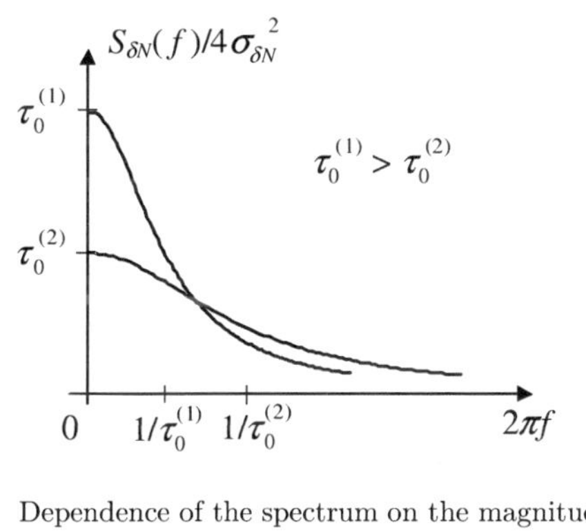


Fig. 2. Dependence of the spectrum on the magnitude $\tau_0$.

While relaxation time $\tau_0$ is increased, the height of the spectrum is increased proportionally to $\tau_0$ and the width is decreased as $\tau_0^{-1}$. Thus, an increase of the lifetime of quasiparticles $\tau_0$ leads to increased low-frequency component of fluctuations of generation–recombination of quasiparticles and to attenuation of high-frequency components.

## 3. Relation of Fluctuations in the Number of Pairs of Magnetic Monopoles with Measurable Characteristic of Spin Ice

Very often in literature, pairs of magnetic monopoles, which are formed by inverting the shared spin (Fig. 1(b)), are called bound. On the other hand, the magnetic monopoles, which move through the lattice (Fig. 1(c)) are called unbound or free. But as we can see from Fig. 1(c), the "unbound" or "free" monopoles are bounded by the "spin spaghetti", which is formed by changing of the spin configuration along the way of monopole.

Following [14], we define number $n_b$ of bound, and unbound (free) $n_u$ pairs, $N_0 = n_b + n_u$ – the total number of pairs, and $\alpha = n_u/N_0$ – the degree of dissociation.

We now assume that the vast majority of pairs of quasiparticles remain in a bound state, i.e., $\alpha \ll 1$, and one pair of bound quasiparticles interacts weakly with another pair. Thus, $N_0 \approx n_b$ – the total number of pairs is approximately constant at a given temperature.

A small change, caused by small perturbation, in concentration of unbound (free) pairs $\Delta n_u$ yields small change $\Delta\alpha$ in degree of dissociation. The relaxation of $\Delta\alpha$ is determined by recombination of unbound pairs. The authors of paper [14], in terms of the theory of magnetic Wien effect in the weak field limit, conclude that fluctuations $\Delta\alpha$ cause proportional fluctuations $\Delta M$ in magnetic moment. In our case of a small change in concentration $n_u$, we have $N = n_b + n_u + \Delta n_u$, or $N = N_0 + \Delta n_u$. That is $N = N_0 \cdot (1 + \Delta n_u/N_0) = N_0 \cdot (1 + \Delta\alpha)$. Thus, $\Delta\alpha = \delta N$ – fluctuations in $N$ through the fluctuations in $\alpha$ (Eq. (8) in [14]) lead to fluctuations in magnetic moment:

$$\Delta M \propto \frac{\mu_0 Q^3}{8\pi kT} \cdot \delta N. \tag{13}$$

Here, $\mu_0$ is the permeability of the vacuum, $Q$ is the elementary magnetic charge, $k$ is the Boltzmann's constant, $T$ is temperature.

For spectra, we can write following equation:

$$S_{\Delta M}(f) \propto \left(\frac{\mu_0 Q^3}{8\pi kT}\right)^2 \cdot S_{\delta N}(f). \tag{14}$$

Here, $S_{\Delta M}(f)$ is the spectrum of absolute fluctuations of magnetic moment. It can be expected that measurement of fluctuations in the magnetic moment may give access to information about fluctuations in the number of pairs of magnetic monopoles in spin ice.

Finally, we note that in a real spin ice a series of relaxation times may exist. As a result, the total spectrum $S_{\delta N}(f)$ in Eq. (14) of relative fluctuations in the number of pairs of monopoles may be a superposition of spectra (12) with different relaxation times.

## 4. Conclusions

We have investigated the fluctuations in the number of pairs of magnetic monopoles. It is shown that these fluctuations can be described by equation of Langevin type. The spectrum of the fluctuations was determined and analyzed. The method of detecting these fluctuations by the measurement of fluctuations in the magnetic moment is suggested.

## Acknowledgments

This research was supported by Grant No. 17-02-00512 of RFBR, by the state task to the educational organizations of the higher education, subordinated to the Ministry of Education and Science of Russia, in the sphere of scientific activity, "Carrying out of research works (fundamental scientific researches, applied scientific researches and experimental development)", code 2183. The research was also supported by the grant (the agreement on August 27, 2013 No. 02.B.49.21.0003 between The Ministry of education and science of the Russian Federation and Lobachevsky State University of Nizhni Novgorod).

## References

[1] M. J. Harris, S. T. Bramwell, D. F. McMorrow, T. Zeiske and K. W. Godfrey, Geometrical frustration in the ferromagnetic pyrochlore $Ho_2Ti_2O_7$, *Phys. Rev. Lett.* **79**(13) (1997) 2554–2557.

[2] I. A. Ryzhkin, Magnetic relaxation in rare-earth oxide pyrochlores, *J. Exp. Theor. Phys.* **101** (2005) 481–486.

[3] C. Castelnovo, R. Moessner and S. L. Sondhi, Magnetic monopoles in spin ice, *Nature* **451** (2008) 42–45.

[4] F. Grüneis, 1/f noise due to the return statistics of mobile point defects in extrinsic semiconductor materials, *Fluct. Noise Lett.* **7**(1) (2007) C1–C18, doi: http://dx.doi.org/10.1142/S0219477507003714.

[5] F. Grüneis, 1/f Noise in extrinsic semiconductor materials interpreted as modulated generation-recombination noise, *Fluct. Noise Lett.* **9**(02) (2010) 229–243, doi: http://dx.doi.org/10.1142/S0219477510000137.

[6] S. M. Bezrukov, L. K. J. Vandamme and L. B. Kish, Special Section on 1/F Noise: A Preface, *Fluct. Noise Lett.* **10**(4) (2011) 417–418, doi: http://dx.doi.org/10.1142/S0219477511000727.

[7] E. I. Shmelev, A. V. Klyuev and A. V. Yakimov, Complexes of spatially multistable defects as the source of 1/f noise in GaAs devices, *Fluct. Noise Lett.* **12**(1) (2013) 1350008-1–1350008-13, doi: 10.1142/S0219477513500089.

[8] A. V. Klyuev, E. I. Shmelev and A. V. Yakimov, Physical origins of 1/f noise in Si delta–doped Schottky diodes, *Fluct. Noise Lett.* **13**(1) (2014) 1450003-1–1450003-16, doi: 10.1142/S0219477514500035.

[9] A. V. Klyuev and A. V. Yakimov, 1/f noise in GaAs nanoscale light-emitting structures, *Physica B* **440** (2014) 145–151, doi: http://dx.doi.org/10.1016/j.physb.2014.01.021.

[10] A. V. Klyuev, A. V. Yakimov and I. S. Zhukova, 1/f noise in *Ti–Au/n*-type *GaAs* Schottky barrier diodes, *Fluct. Noise Lett.* **14**(2) (2015) 1550029-1–1550029-12, doi: 10.1142/S0219477515500297.

[11] B. Spagnolo, D. Valenti, C. Guarcello, A. Carollo, D. Persano Adorno, S. Spezia, N. Pizzolato and B. Di Paola, Noise-induced effects in nonlinear relaxation of condensed matter systems, *Chaos Solitons Fractals, Part B* **81** (2015) 412–424, doi: 10.1016/j.chaos.2015.07.023.

[12] B. Spagnolo, C. Guarcello, L. Magazzù, A. Carollo, D. P. Adorno and D. Valenti, Nonlinear relaxation phenomena in metastable condensed matter systems, *Entropy,* **19**(1) (2017) 20, doi: 10.3390/e19010020.

[13] E. L. Pankratov and B. Spagnolo, Optimization of impurity profile for p-n-junction in heterostructures, *Eur. Phys. J. B* **46** (2005) 15–19, doi: 10.1140/epjb/e2005-00233-1.

[14] S. T. Bramwell, S. R. Giblin, S. Calder, R. Aldus, D. Prabhakaran and T. Fennell, Measurement of the charge and current of magnetic monopoles in spin ice, *Nature* **461** (2009) 956–959, doi: 10.1038/nature08500.

[15] L. D. C. Jaubert and P. C. W. Holdsworth, Signature of magnetic monopole and Dirac string dynamics in spin ice, *Nat. Phys.* **5** (2009) 258–261, doi: 10.1038/NPHYS1227.

[16] S. R. Giblin, S. T. Bramwell, P. C. W. Holdsworth, D. Prabhakaran and I. Terry, Creation and measurements of long-lived magnetic monopole currents in spin ice, *Nat. Phys.* **7** (2011) 252.

[17] M. I. Ryzhkin, I. A. Ryzhkin and S. T. Bramwell, Dynamic susceptibility and dynamic correlations in spin ice, *Europhys. Lett.* **104**(3) (2013) 37005.

[18] C. Paulsen, M. J. Jackson, E. Lhotel, B. Canals, D. Prabhakaran, K. Matsuhira, S. R. Giblin and S. T. Bramwell, Far-from-equilibrium monopole dynamics in spin ice, *Nat. Phys.* **10** (2014) 135–139, doi: 10.1038/NPHYS2847.

[19] A. V. Klyuev, I. A. Ryzhkin and M. I. Ryzhkin, Generalized dielectric permittivity of ice, *J. Exp. Theor. Phys. Lett.* **100**(9) (2015) 604–608.

[20] I. A. Ryzhkin, Thermodynamics of ice: Not obeying the rules, *Nat. Phys.* **12** (2016) 996–997, doi: 10.1038/nphys3853.

[21] A. Van der Ziel, *Noise Sources, Description, Measurement* (Prentice–Hall, Inc., Englewood Cliffs, 1970).

[22] A. V. Yakimov, The Analysis of fluctuations in avalanche transit diode oscillators, *Radio Eng. Electron. Phys.* **21**(9) (1976) 48–50.

[23] C. W. Gardiner, *Handbook of Stochastic Methods*, 2nd edn. (Berlin, Springer, 1985), ISBN 3-540-61634-9.

[24] N. G. van Kampen, *Stochastic Processes in Physics and Chemistry*, 3rd edn. (North Holland, 1981).

[25] A. Dubkov and B. Spagnolo, Langevin approach to Lévy flights in fixed potentials: Exact results for stationary probability distributions, *Acta Phys. Pol. B* **38**(5) (2007) 1745–1758.

# COMPUTATION OF QUANTUM ELECTRICAL CURRENTS THROUGH THE RAMO–SHOCKLEY–PELLEGRINI THEOREM WITH TRAJECTORIES

G. ALBAREDA*, F. L. TRAVERSA, A. BENALI and X. ORIOLS

*Departament d'Enginyeria Electrònica*
*Universitat Autònoma de Barcelona*
*08193 Bellaterra, Spain*
**guillem.albareda@uab.cat*

Received 17 April 2012
Accepted 29 July 2012
Published 5 November 2012

Communicated by Kamal Bardhau

Motivated by a recent approach to solve quantum dynamics with full Coulomb correlations [X. Oriols, *Phys. Rev. Lett.* **98** (2007) 066803], we present here an extension of the Ramo–Shockley–Pellegrini theorem for quantum systems to compute the total (conduction plus displacement) current in terms of quantum (Bohmian) trajectories. By way of test, we derive an extension of the Ramo-Shockley-Pellegrini theorem using standard quantum mechanics and we compare it to our former result. As expected, both formulations give identical results, however we emphasize the numerical viability of computing self-consistently the total current by means of quantum trajectories in front of the difficulties to do it in terms of standard quantum mechanics.

*Keywords*: Electron devices; electron transport; quantum transport; high-frequency regime; displacement current; many-body systems.

## 1. Introduction

The time scale experimentally accessible in electron devices have been, up to now, much larger than the temporal widths of the pulses generated by the electrons crossing a device, mainly because of parasitic RC elements [1]. The computation of the displacement current has been thus mostly disregarded in the recent literature. Nonetheless, due to the importance of the transistor market for high-bandwidth telecommunications and mobile platforms, the influence of these parasitic elements is day by day being reduced, and many emerging devices are envisioned to work close to the Terahertz regime [2–5]. A theoretical approach to provide information on the role of the displacement current in the high-frequency regime of these novel devices would be thus desirable.

The Ramo–Shockley theorem [6,7] represents a powerful tool to understand the relationship between electron dynamics and the behavior of the electrical current at high frequencies. This theorem was named after the work of Shockley [6] and Ramo [7] who were the first to provide an analytical expression for the total (conduction plus displacement) current generated by a moving electron between two parallel plates. Later works generalized the Ramo–Shockley derivation to arbitrary geometries and time-dependent voltages at the electrodes [8–14]. In order to emphasize the noticeable contribution by B. Pellegrini [8, 10, 11], we use here the name Ramo-Shockley-Pellegrini (RSP) theorem to refer to all these works.

The utility of the extension of the RSP theorem for quantum systems, first introduced by Pellegrini [11], is however conditioned by the ability to solve the many-particle time-dependent Schrödinger equation including the Coulomb interaction among electrons. In this regard, motivated by a numerical solvable algorithm on the grounds of Bohmian mechanics to describe many-particle quantum dynamics with full Coulomb correlations [15–17], we generalize here the RSP theorem by using quantum (Bohmian) trajectories. In particular, in order to understand the physical meaning behind the quantum RSP theorem written in terms of trajectories, we derive it from scratch instead of inferring it from Pellegrini's result [11].

A practical demonstration of the usefulness of this generalization and a discussion on how to self-consistently compute quantum displacement currents by means of quantum trajectories can be found in [18, 19, 22].

After this brief introduction, in Sec. 2 we present a simple and compact derivation of the classical RSP theorem. In Sec. 3, we extend it to quantum systems by means of both standard quantum mechanics and quantum (Bohmian) trajectories. We conclude in Sec. 4.

## 2. Classical RSP Theorem

Before introducing the quantum extension of the RSP theorem, let us first present an original demonstration of the classical RSP theorem based on Green's second identity. The final result is identical to that obtained by Pellegrini [8,10,11]. However, our demonstration has the advantage that it will allow a simpler generalization to quantum scenarios in Secs. 3.1 and 3.2.

Let us start considering an arbitrary parallelepiped of volume $\Omega = L_x \cdot L_y \cdot L_z$ limited by a closed surface $S$ defined by of six rectangular surfaces $S = \{S_1, S_2, \ldots, S_6\}$. We restrict $S$ to include the surface where the total current is computed.

In our demonstration we consider two different sets of functions, each one composed by a scalar function plus a vector function defined as the gradient of the former. The first set is the electromagnetic scalar potential $V(\mathbf{r}, t)$ and the vector field $\mathbf{E}(\mathbf{r}, t)$ related through:

$$\mathbf{E}(\mathbf{r}, t) = -\boldsymbol{\nabla} V(\mathbf{r}, t), \tag{1}$$

where we assumed a quasi-static definition of the electric field. According to the uniqueness theorem [20,21], both functions, $V(\mathbf{r}, t)$ and $\mathbf{E}(\mathbf{r}, t)$, are unambiguously defined inside the volume $\Omega$ through the well-known Gauss equation:

$$\nabla(\epsilon(\mathbf{r})\cdot\mathbf{E}(\mathbf{r},t)) = -\nabla(\epsilon(\mathbf{r})\cdot\nabla V(\mathbf{r},t)) = q\sum_{k=1}^{N}\delta(\mathbf{r}-\mathbf{r}_k[t]), \tag{2}$$

together with an appropriate (Dirichlet or Neumann) boundary condition on the whole closed surface $S$. The right-hand side of (2) is just the charge density[a] of electrons, $q$ the electron charge with sign and $\delta(\mathbf{r})$ the 3D delta function. Electrons paths are defined by $\mathbf{r}_k[t]$ with $k = \{1, 2, \ldots, N\}$. Here $N$ includes all electrons inside and outside the volume $\Omega$, so that it does not change with time. For simplicity, along this paper, we will assume a scalar position-dependent permittivity, $\epsilon(\mathbf{r})$.

We define a second set of functions (one set for each particular surface $S_i$) as the scalar functions $\Phi_i(\mathbf{r})$ and its vector functions $\mathbf{F}_i(\mathbf{r})$ through:

$$\mathbf{F}_i(\mathbf{r}) = -\nabla\Phi_i(\mathbf{r}), \tag{3}$$

and fulfilling

$$\nabla(\epsilon(\mathbf{r})\cdot\mathbf{F}_i(\mathbf{r})) = -\nabla(\epsilon(\mathbf{r})\cdot\nabla\Phi_i(\mathbf{r})) = 0. \tag{4}$$

For reasons that will be later fully evident, the following particular Dirichlet boundary conditions on the definition of $\Phi_i(\mathbf{r})$ are considered:

$$\Phi_i(\mathbf{r}) = 1;\ \ \mathbf{r}\in S_i \quad \text{and} \quad \Phi_i(\mathbf{r}) = 0;\ \ \mathbf{r}\in S_{h\neq i}, \tag{5}$$

meaning that $\Phi_i(\mathbf{r}) = 1$ on the surface $S_i$ and zero on the other surfaces.[b] Let us emphasize that, contrarily to the first set of functions, $\Phi_i(\mathbf{r})$ and $\mathbf{F}_i(\mathbf{r})$ have no direct physical meaning. They are just mathematical arbitrary functions.[c] As indicated, the subindex $i$ means that we have six different pairs of functions, one for each surface $S_i$, all of them sharing the same defining equations (3) and (4), but different boundary conditions (5).

Once we have defined the two sets of (well-behaved)[a,b] functions, we introduce them into the following Green's second identity [20, 21] evaluated inside the volume

[a]It can be argued that there are discontinuities in the definition of $V(\mathbf{r})$ at $\mathbf{r} = \mathbf{r}_n[t]$. However, one can apply Green's identities to $V(\mathbf{r})$ by using a limiting approaching process $\mathbf{r} \to \mathbf{r}_n[t]$ which does not consider the "offending" point. See page 41 of Ref. [20].

[b]It can be argued that $\mathbf{F}_i(\mathbf{r})$ does not fulfil expression (3) in the intersection between the surface $S_i$ (with $\Phi_i(\mathbf{r}) = 1$) and the other surfaces $S_{h\neq i}$ (with $\Phi_i(\mathbf{r}) = 0$). However, one can define the functions $\Phi_i(\mathbf{r})$ and $\mathbf{F}_i(\mathbf{r})$ on a new volume $\Omega^*$ equal to the previous volume $\Omega$ except for the elimination of the "offending" points using a limiting approaching. We realize that $I_1^*(t) \to I_1(t)$. See page 41 of Ref. [20] for a similar discussion.

[c]Consistently with expression (3) and Eq. (4), it is argued in the literature that $\Phi_i(\mathbf{r})$ and $\mathbf{F}_i(\mathbf{r})$ are the scalar potential and electric field, respectively, when there is no electron charge inside the volume. However, we will avoid this definition because it can lead to misleading interpretations of the boundary conditions (5) and the units of $\Phi_i(\mathbf{r})$ (without units) and $\mathbf{F}_i(\mathbf{r})$ (the inverse of a distance).

$\Omega$ and on its limiting surface $S$:

$$\int_{\Omega} [V(\mathbf{r},t)\cdot\boldsymbol{\nabla}[\epsilon(\mathbf{r})\boldsymbol{\nabla}\Phi_i(\mathbf{r})] - \Phi_i(\mathbf{r})\cdot\boldsymbol{\nabla}[\epsilon(\mathbf{r})\boldsymbol{\nabla}V(\mathbf{r},t)]\cdot dv$$
$$= \int_S \epsilon(\mathbf{r})[V(\mathbf{r},t)\boldsymbol{\nabla}\Phi_i(\mathbf{r}) - \Phi_i(\mathbf{r})\boldsymbol{\nabla}V(\mathbf{r},t)]\cdot d\mathbf{s}. \tag{6}$$

With (1)–(4), we can rewrite (6) as:

$$\int_{\Omega} q\cdot\Phi_i(\mathbf{r})\cdot\left[\sum_{k=1}^{N}\delta(\mathbf{r}-\mathbf{r}_k[t])\right]\cdot dv$$
$$= -\int_S \epsilon(\mathbf{r})[V(\mathbf{r},t)\cdot\mathbf{F}_i(\mathbf{r}) - \Phi_i(\mathbf{r})\cdot\mathbf{E}(\mathbf{r},t)]\cdot d\mathbf{s}. \tag{7}$$

Finally, evaluating the delta function along the volume $\Omega$ and taking into account the boundary conditions (5) for the last term on the right-hand side of (7), we obtain:

$$\sum_{k=1}^{N} q\cdot\Phi_i(\mathbf{r}_k[t])\Theta(t-t_k^i)\Theta(t_k^o-t)$$
$$= -\int_S \epsilon(\mathbf{r})\cdot\mathbf{F}_i(\mathbf{r})\cdot V(\mathbf{r},t)\cdot d\mathbf{s} + \int_{S_i}\epsilon(\mathbf{r})\cdot\mathbf{E}(\mathbf{r},t)\cdot d\mathbf{s}, \tag{8}$$

where $t_k^i$ and $t_k^o$ are the times when the electron $\mathbf{r}_k[t]$ enters and leaves the volume $\Omega$, respectively. The function $\Theta(t)$ is the Heaviside function (or unit step function). To relate the delta function in (7) and the Heaviside function in (8), we used:

$$\Theta(t-t_k^i)\Theta(t_k^o-t) = \int_{\Omega}\delta(\mathbf{r}-\mathbf{r}_k[t])\cdot dv. \tag{9}$$

An additional index associated to the entering and leaving times would be necessary for those electrons that cross the surface $S$ more than twice, however, in order to simplify the notation, we avoid this last subindex.

The time derivative of the left-hand side of (8) gives:

$$\sum_{k=1}^{N} q\cdot\frac{d}{dt}[\Phi_i(\mathbf{r}_k[t])\Theta(t-t_k^i)\Theta(t_k^o-t)]$$
$$= -\sum_{k=1}^{N} q\cdot\mathbf{F}_i(\mathbf{r}_k[t])\cdot\mathbf{v}_k[t]\Theta(t-t_k^i)\Theta(t_k^o-t)$$
$$+\sum_{k=1}^{N} q\cdot\Phi_i(\mathbf{r}_k[t])\delta(t-t_k^i)\Theta(t_k^o-t) - \sum_{k=1}^{N} q\cdot\Phi_i(\mathbf{r}_k[t])\Theta(t-t_k^i)\delta(t_k^o-t), \tag{10}$$

where we have used $d\Phi_i(\mathbf{r}_k[t])/dt = -\mathbf{F}_i(\mathbf{r}_k[t])\cdot\mathbf{v}_k[t]$ with the electron velocity $\mathbf{v}_k[t] = \frac{d\mathbf{r}_k[t]}{dt}$. According to (9), we can rewrite the first term of the right-hand side of (10) as:

$$\sum_{k=1}^{N} q \cdot \mathbf{F}_i(\mathbf{r}_k[t]) \cdot \mathbf{v}_k(\mathbf{r}_k[t])\Theta\big(t - t_k^i\big)\Theta\big(t_k^o - t\big) = \int_{\Omega} \mathbf{F}_i(\mathbf{r}) \cdot \; \mathbf{J}_c(\mathbf{r}, t) \cdot dv, \quad (11)$$

where we define the current density as:

$$\mathbf{J}_c(\mathbf{r}, t) = \sum_{k=1}^{N} q \cdot \mathbf{v}_k(\mathbf{r}_k[t])\delta(\mathbf{r} - \mathbf{r}_k[t]). \quad (12)$$

The last two terms on the right-hand side of expression (10) can be rewritten as the number of particles crossing the surface $S_i$:

$$q\left[\sum_{\mathbf{r}_k[t_k^i]\in S_i} \delta\big(t - t_k^i\big) - \sum_{\mathbf{r}_k[t_k^o]\in S_i} \delta\big(t_k^o - t\big)\right] = -\int_{S_i} \mathbf{J}_c(\mathbf{r}, t) \cdot d\mathbf{s}, \quad (13)$$

where we have used that $\Phi_i(\mathbf{r}_k[t]) = 0$ for those $k$-electrons entering or leaving the volume $\Omega$ through a surface different from $S_i$, while $\Phi_i(\mathbf{r}_k[t]) = 1$ for those $k$-electrons entering or leaving the device through $S_i$. This is just a consequence of the boundary conditions (5). Putting (13) and (11) together with the time derivative of the right-hand side of (8) and reorganizing terms, we obtain two sets of equivalent expressions to compute the total (conduction plus displacement) current $I_i(t)$ through the surface $S_i$ as:

$$I_i(t) = \Upsilon_i^c(t) + \Upsilon_i^d(t) = \Gamma_i^q(t) + \Gamma_i^e(t), \quad (14)$$

where the standard expressions for the total current $I_i(t)$ are:

$$\Upsilon_i^c(t) = \int_{S_i} \mathbf{J}_c(\mathbf{r}, t) d\mathbf{s}, \quad (15)$$

$$\Upsilon_i^d(t) = \int_{S_i} \epsilon(\mathbf{r}) \cdot \frac{d\mathbf{E}(\mathbf{r}, t)}{dt} d\mathbf{s}, \quad (16)$$

while the RSP expressions are:

$$\Gamma_i^q(t) = -\int_{\Omega} \mathbf{F}_i(\mathbf{r}) \cdot \mathbf{J}_c(\mathbf{r}, t) \cdot dv, \quad (17)$$

$$\Gamma_i^e(t) = \int_{S} \epsilon(\mathbf{r}) \cdot \mathbf{F}_i(\mathbf{r}) \cdot \frac{dV(\mathbf{r}, t)}{dt} \cdot d\mathbf{s}. \quad (18)$$

Expressions (17) and (18) are exactly the ones found by Pellegrini in [8].

Finally, let us note that the terms $\Gamma_i^q(t)$ and $\Gamma_i^e(t)$ cannot be interpreted as the conduction, $\Upsilon_i^c(t)$, and displacement, $\Upsilon_i^d(t)$, currents, respectively. For example, when the electron is not crossing the surface $S_i$, $\Gamma_i^q(t) \neq 0$ while $\Upsilon_i^c(t) = 0$. The term $\Gamma_i^q(t)$ includes itself not only the conduction current, but also part of the displacement current.

## 3. RSP Theorem for Quantum Systems

Hereafter, we present a quantum generalization of the classical theorem presented in Sec. 2. We will consider that the quantum system is described by a many-particle wave-function $\Psi(\mathbf{r}_1, \ldots, \mathbf{r}_N, t)$ solution of the time-dependent many-particle Schrödinger equation for Coulomb-interacting electrons. For simplicity, we assume spinless electrons. Although the quantum RSP theorem is a known result derived by Pellegrini [11], its derivation in Sec. 3.1 will be useful later in order to compare it to our main result based on quantum trajectories in Sec. 3.2.

### 3.1. *Demonstration with standard quantum mechanics*

The scalar potential, $V(\mathbf{r}, t)$ defined in expression (2) has to be interpreted as the value of the scalar potential at the position $\mathbf{r}$ and time $t$ for a distribution of electrons given by $\mathbf{r}_1, \ldots, \mathbf{r}_N$. The probability at each position distribution is described by the (normalized) wave function $|\Psi(\mathbf{r}_1, \ldots, \mathbf{r}_N, t)|^2$. Hereafter, it will be useful to write explicitly the dependence of the scalar potential on all positions, i.e., $V(\mathbf{r}, \mathbf{r}_1, \ldots, \mathbf{r}_N)$. Thus, the mean scalar potential $\overline{V(\mathbf{r}, t)}$ is:

$$\overline{V(\mathbf{r}, t)} = \int_{\Omega_\infty} dv_1 \ldots . \int_{\Omega_\infty} dv_N |\Psi(\mathbf{r}_1, \ldots, \mathbf{r}_N, t)|^2 \cdot V(\mathbf{r}, \mathbf{r}_1, \ldots, \mathbf{r}_N), \tag{19}$$

where $\Omega_\infty$ means an infinite $3D$ space for volume-integration. Identically for the mean value of the electric field:

$$\overline{\mathbf{E}(\mathbf{r}, t)} = \int_{\Omega_\infty} dv_1 \ldots . \int_{\Omega_\infty} dv_N |\Psi(\mathbf{r}_1, \ldots, \mathbf{r}_N, t)|^2 \cdot \mathbf{E}(\mathbf{r}, \mathbf{r}_1, \ldots, \mathbf{r}_N). \tag{20}$$

Similarly, the quantum ensemble value of the charge density, defined with deltas in (7), is now written in terms of the squared modulus of the many-particle wave-function. Therefore, in order to perform the time derivative of the quantum version of the left-hand side of (7), we can use (3) together with the quantum local conservation law:

$$\frac{d|\Psi(\mathbf{r}_1, \ldots, \mathbf{r}_N, t)|^2}{dt} = -\sum_{h=1}^{N} \boldsymbol{\nabla}_{\mathbf{r}_h} \mathbf{J}_{\mathbf{r}_h}(\mathbf{r}_1, \ldots, \mathbf{r}_N, t), \tag{21}$$

where $\mathbf{J}_{\mathbf{r}_h}(\mathbf{r}_1, \ldots, \mathbf{r}_N, t)$ is the standard quantum current density associated to the $\mathbf{r}_h$ particle. We then find:

$$\begin{aligned}
&-q\sum_{k=1}^{N}\sum_{h=1}^{N}\int_{\Omega_\infty} dv_1 .. \int_{\Omega} dv_k .. \int_{\Omega_\infty} dv_N \cdot \Phi_i(\mathbf{r}_k) \cdot \boldsymbol{\nabla}_{\mathbf{r}_h} \mathbf{J}_{\mathbf{r}_h}(\mathbf{r}_1, \ldots, \mathbf{r}_N, t) \\
&= -q\sum_{k=1}^{N}\sum_{h=1}^{N}\int_{\Omega_\infty} dv_1 .. \int_{\Omega} dv_k .. \int_{\Omega_\infty} dv_N \cdot \boldsymbol{\nabla}_{\mathbf{r}_h}[\Phi_i(\mathbf{r}_k) \cdot \mathbf{J}_{\mathbf{r}_h}(\mathbf{r}_1, \ldots, \mathbf{r}_N, t)] \\
&\quad -q\sum_{k=1}^{N}\sum_{h=1}^{N}\int_{\Omega_\infty} dv_1 .. \int_{\Omega} dv_k .. \int_{\Omega_\infty} dv_N \cdot \mathbf{F}_i(\mathbf{r}_k) \cdot \mathbf{J}_{\mathbf{r}_h}(\mathbf{r}_1, \ldots, \mathbf{r}_N, t) \cdot \delta_{kh},
\end{aligned} \tag{22}$$

where $\delta_{kh}$ is the Kronecker delta, which appears because $\boldsymbol{\nabla}_{\mathbf{r}_h}\Phi_i(\mathbf{r}_k)$ is zero when $h \neq k$. Next, we evaluate the first integral of the right-hand side of (22) as:

$$-q\sum_{k=1}^{N}\sum_{h=1}^{N}\int_{\Omega_\infty} dv_1..\int_{\Omega} dv_k..\int_{\Omega_\infty} dv_N\cdot\boldsymbol{\nabla}_{\mathbf{r}_h}(\Phi_i(\mathbf{r}_k)\cdot\mathbf{J}_{\mathbf{r}_h}(\mathbf{r}_1,\ldots,\mathbf{r}_N,t))$$
$$= -q\sum_{k=1}^{N}\int_{\Omega_\infty} dv_1..\int_{S} d\mathbf{s}_k..\int_{\Omega_\infty} dv_N\cdot\Phi_i(\mathbf{r}_k)\cdot\mathbf{J}_{\mathbf{r}_k}(\mathbf{r}_1,\ldots,\mathbf{r}_N,t), \tag{23}$$

where we have used that any integral in the (infinite) volume $\Omega_\infty$ gives zero because it is the same as evaluating the current density on the (infinite) surface $S_\infty$. By using the boundary conditions (5) for $\Phi_i(\mathbf{r}_k)$ on the surface $S$ (meaning $\Phi_i(\mathbf{r}_k) = 0$ except for $S_i$), we can rewrite expression (23) as:

$$-q\sum_{k=1}^{N}\int_{\Omega_\infty} dv_1..\int_{S} d\mathbf{s}_k..\int_{\Omega_\infty} dv_N\cdot\Phi_i(\mathbf{r}_k)\cdot\mathbf{J}_{\mathbf{r}_k}(\mathbf{r}_1,\ldots,\mathbf{r}_N,t) = -\int_{S_i} d\mathbf{s}\cdot\overline{\mathbf{J}_c(\mathbf{r},t)}, \tag{24}$$

where the mean (many-particle) quantum current density $\overline{\mathbf{J}_c(\mathbf{r},t)}$ is defined as:

$$\overline{\mathbf{J}_c(\mathbf{r},t)} = \sum_{k=1}^{N}\left[\int_{\Omega_\infty} dv_1..\int_{\Omega_\infty} d\mathbf{v}_{k-1}\int_{\Omega_\infty} d\mathbf{v}_{k+1}\int_{\Omega_\infty} dv_N\cdot\mathbf{J}_{\mathbf{r}_k}(\mathbf{r}_1,\ldots,\mathbf{r}_N,t)\right]_{\mathbf{r}_k=\mathbf{r}}. \tag{25}$$

Next, we evaluate the second integral of the right-hand side on (22) as:

$$-\sum_{k=1}^{N}\int_{\Omega_\infty} dv_1..\int_{\Omega} d\mathbf{v}_k..\int_{\Omega_\infty} dv_N\cdot\mathbf{F}_i(\mathbf{r}_k)\cdot\mathbf{J}_{\mathbf{r}_k}(\mathbf{r}_1,\ldots,\mathbf{r}_N,t)$$
$$= -\int_{\Omega} dv\cdot\mathbf{F}_i(\mathbf{r})\cdot\overline{\mathbf{J}_c(\mathbf{r},t)}, \tag{26}$$

where again, we use the definition (25). The final quantum version of the RSP theorem can be written as:

$$\overline{I_i(t)} = \overline{\Upsilon_i^c(t)} + \overline{\Upsilon_i^d(t)} = \overline{\Gamma_i^q(t)} + \overline{\Gamma_i^e(t)}, \tag{27}$$

where the standard expressions for the total (conduction plus displacement) current $\overline{I_i(t)}$ are:

$$\overline{\Upsilon_i^c(t)} = \int_{S_i} \overline{\mathbf{J}_c(\mathbf{r},t)}d\mathbf{s}, \tag{28}$$

$$\overline{\Upsilon_i^d(t)} = \int_{S_i} \epsilon(\mathbf{r})\cdot\frac{d\overline{\mathbf{E}(\mathbf{r},t)}}{dt}d\mathbf{s}, \tag{29}$$

and the RSP expressions are:

$$\overline{\Gamma_i^q(t)} = -\int_{\Omega} \mathbf{F}_i(\mathbf{r}) \cdot \overline{\mathbf{J}_c(\mathbf{r},t)} \cdot dv, \tag{30}$$

$$\overline{\Gamma_i^e(t)} = \int_S \epsilon(\mathbf{r}) \cdot \mathbf{F}_i(\mathbf{r}) \cdot \frac{d\overline{V(\mathbf{r},t)}}{dt} \cdot d\mathbf{s}. \tag{31}$$

Let us notice here that expressions (27)–(31) are just the quantum ensemble-average version of expressions (14)–(18).

### 3.2. *Demonstration with Bohmian mechanics*

From a computational point of view, the most relevant property of Bohmian trajectories is the fact that an appropriate ensemble of Bohmian trajectories does exactly reproduce the probability density at any time. Such property is mathematically described by the following identity [22]:

$$|\Psi(\mathbf{r}_1,\ldots,\mathbf{r}_N,t)|^2 = \lim_{M\to\infty} \frac{1}{M} \sum_{j=1}^{M} \prod_{k=1}^{N} \delta(\mathbf{r}_k - \mathbf{r}_k^j[t]), \tag{32}$$

where $\{\mathbf{r}_1^j[t]\ldots\mathbf{r}_N^j[t]\}$ is a many-particle Bohmian trajectory with initial positions $\{\mathbf{r}_1^j[0]\ldots\mathbf{r}_N^j[0]\}$. In order to ensure that expression (32) is accomplished, one has to select $\{\mathbf{r}_1^j[0]\ldots\mathbf{r}_N^j[0]\}$ according to the initial wave-packet probability distribution $|\Psi(\mathbf{r}_1,\ldots,\mathbf{r}_N,0)|^2$. This last condition is called "quantum equilibrium hypothesis" [22]. The super-index $j = 1,\ldots,M$ accounts for the $M \to \infty$ number of different Bohmian trajectories that we have to consider to reproduce the standard quantum statistical results.

According to (32), expression (19) can be written as:

$$\overline{V(\mathbf{r},t)} = \lim_{M\to\infty} \frac{1}{M} \sum_{j=1}^{M} \cdot V\big(\mathbf{r},\mathbf{r}_1^j[t],\ldots,\mathbf{r}_N^j[t]\big), \tag{33}$$

and (20) as:

$$\overline{\mathbf{E}(\mathbf{r},t)} = \lim_{M\to\infty} \frac{1}{M} \sum_{j=1}^{M} \cdot \mathbf{E}\big(\mathbf{r},\mathbf{r}_1^j[t],\ldots,\mathbf{r}_N^j[t]\big). \tag{34}$$

The quantum version of the left-hand side of expression (7) can now be written as:

$$\begin{aligned}\sum_{k=1}^{N} \int_{\Omega_\infty} dv_1 .. \int_{\Omega} dv_k .. \int_{\Omega_\infty} dv_N |\Psi(\mathbf{r}_1,\ldots,\mathbf{r}_N,t)|^2 \cdot q \cdot \Phi_i(\mathbf{r}_k) \\ = \lim_{M\to\infty} \frac{1}{M} \sum_{j=1}^{M} \sum_{k=1}^{N} q \cdot \Phi_i(\mathbf{r}_k^j[t]) \Theta\big(t - t_k^i\big) \Theta\big(t_k^o - t\big),\end{aligned} \tag{35}$$

where we include $\Theta\big(t - t_k^i\big)\Theta\big(t_k^o - t\big)$ to ensure that only those $\mathbf{r}_k[t] \in \Omega$ are considered. To compute the time-derivative of expression (35) we can follow the same steps done in Sec. 2 from expressions (8)–(14) for classical trajectories. Although we have substituted classical trajectories by quantum trajectories, the mathematical procedure is identical. The only difference is that we have to include a sum over all (infinite) possible Bohmian trajectories (the quantum ensemble). Thus, the standard expression of the current is $\overline{I_i(t)} = \overline{\Upsilon_i^c(t)} + \overline{\Upsilon_i^d(t)}$, with:

$$\overline{\Upsilon_i^c(t)} = \lim_{M\to\infty} \frac{1}{M} \sum_{j=1}^{M} \int_{S_i} \mathbf{J}_c\big(\mathbf{r}, r_1^j[t], \ldots, \mathbf{r}_N^j[t]\big) \cdot d\mathbf{s}, \tag{36}$$

$$\overline{\Upsilon_i^d(t)} = \lim_{M\to\infty} \frac{1}{M} \sum_{j=1}^{M} \int_{S_i} \epsilon(\mathbf{r}) \cdot \frac{d\mathbf{E}\big(\mathbf{r}, r_1^j[t], \ldots, \mathbf{r}_N^j[t]\big)}{dt} \cdot d\mathbf{s}, \tag{37}$$

and the RSP expressions $\overline{I_i(t)} = \overline{\Gamma_i^q(t)} + \overline{\Gamma_i^e(t)}$ are:

$$\overline{\Gamma_i^q(t)} = -\lim_{M\to\infty} \frac{1}{M} \sum_{j=1}^{M} \int_{\Omega} \mathbf{F}_i(\mathbf{r}_k^j) \cdot \mathbf{J}_c\big(\mathbf{r}, r_1^j[t], \ldots, \mathbf{r}_N^j[t]\big) \cdot dv, \tag{38}$$

$$\overline{\Gamma_i^e(t)} = \lim_{M\to\infty} \frac{1}{M} \sum_{j=1}^{M} \int_{S} \epsilon(\mathbf{r}) \cdot \mathbf{F}_i(\mathbf{r}) \cdot \frac{dV\big(\mathbf{r}, r_1^j[t], \ldots, \mathbf{r}_N^j[t]\big)}{dt} \cdot d\mathbf{s}. \tag{39}$$

It is quite simple to realize that, as expected, expressions (36)–(39) and (28)–(31) are identical. We have used the Bohmian definition of the quantum current [similar to its classical counterpart (12)]:

$$\mathbf{J}_c\big(\mathbf{r}, \mathbf{r}_1^j[t], \ldots, \mathbf{r}_N^j[t]\big) = \sum_{k=1}^{N} q \cdot \mathbf{v}_k\big(\mathbf{r}_1^j[t], \ldots, \mathbf{r}_N^j[t]\big)\delta\big(\mathbf{r} - \mathbf{r}_k^j[t]\big), \tag{40}$$

where $\mathbf{v}_k\big(\mathbf{r}_1^j[t], \ldots, \mathbf{r}_N^j[t]\big)$ is the Bohmian velocity of the $k$-particle defined as:

$$\mathbf{v}_k\big(\mathbf{r}_1^j[t], \ldots, \mathbf{r}_N^j[t]\big) = \frac{\mathbf{J}_{\mathbf{r}_k}\big(\mathbf{r}_1^j[t], \ldots, \mathbf{r}_N^j[t], t\big)}{\big|\Psi\big(\mathbf{r}_1^j[t], \ldots, \mathbf{r}_N^j[t], t\big)\big|^2}, \tag{41}$$

where $J_{\mathbf{r}_k}(\mathbf{r}_1, \ldots, \mathbf{r}_N, t)$ is the standard quantum current defined in (21). One easily realizes that the mean (many-particle) quantum current density $\overline{\mathbf{J}_c(\mathbf{r}, t)}$ defined in (25) is related to (40) through:

$$\overline{\mathbf{J}_c(\mathbf{r}, t)} = \lim_{M\to\infty} \frac{1}{M} \sum_{j=1}^{M} \mathbf{J}_c\big(\mathbf{r}, r_1^j[t], \ldots, \mathbf{r}_N^j[t]\big). \tag{42}$$

Finally, it is worth introducing a brief reflection on the role of the measuring procedure. Unlike for classical systems, measuring in quantum mechanics means, in general, "collapsing" the wave-function of the system, and thus, special attention must be paid when trying to describe the result of a sequential measurement. In general, expressions (36)–(39) must be understood as the quantum averaged version of expressions (14)–(18). Nonetheless, for some particular cases (such as that of the current measured by a classical ammeter, i.e. an ammeter that obtains information about the current without perturbing the wave function), it can be demonstrated that expressions (14)–(18) have a wide utility [19, 23].

## 4. Conclusion

In order to provide a theoretical tool to study the relation between the AC spectrum and electron dynamics in nanoelectronic devices, we have derived an extension of the RSP theorem [6, 7] for quantum systems using standard and Bohmian quantum mechanics, both, giving identical results. For many-body systems, however, Bohmian trajectories make numerical calculations of the total current easier than with previously existing approaches, and moreover, provides a straightforward way to understand the interplay between quantum electron-dynamics and displacement currents. We believe that this result will provide a renewed interest on the application of the RSP theorem to study the AC behavior of emerging quantum electron devices. The computational viability of this approach to predict transient, AC, and noise behavior of different nanoscale electron devices can be found in [19, 23, 24]. Also, a more practical example elucidating a direct physical intuition behind our theoretical framework has been already published in [25].

## Acknowledgments

This work has been partially supported by the Ministerio de Ciencia e Innovación through the project TEC2009-06986 and by the "Agencia Española de Cooperación Internacional para el Desarrollo" (AECID). G. Albareda would also like to thank the Beatriu de Pinós and the Marie Curie programs for their funding support through the project $2010BP - A00069$.

## References

[1] Y. Taur and T. H. Ning, *Fundamentals of modern VLSI design* (Cambridge University Press, Cambridge, 1998).

[2] M. Asada, S. Suzuki, and N. Kishimoto, Ressonant tunneling diodes for sub-Teraherz and Teraherz oscillators, *Japanese Journal of Applied Physics* **47** (2008) 4375–4384.

[3] N. Orihashi, S. Suzuki, and M. Asada, One THz harmonic oscillation of resonant tunneling diodes, *Appl. Phys. Lett.* **87** (2005) 233501–233503.

[4] P. J. Burke, AC performance of nanoelectronics: Towards a ballistic THz nanotube transistor, *Solid-State Electronics* **48** (2004) 1981–1986.

[5] N. L. Rangel and J. M. Seminario, Graphene Teraherz generators for molecular circuits and systems, *J. Phys. Chem. A* **112** (2008) 13699–13705.

[6] W. Shockley, Currents to conductors induced by a moving point charge, *J. Appl. Phys.* **9** (1938) 635–636.

[7] S. Ramo, Currents induced by electron motion, *Proceedings of the I. R. E.* **27** (1939) 584–585.

[8] B. Pellegrini, Electric charge motion, induced current, energy balance, and noise, *Physical Review B* **34** (1986) 5921–5924.

[9] G. Cavalleri *et al.*, Extension of Ramo's theorem as applied to induced charge in semiconductor detectors, *Nucl. Instrum. and methods.* **92** (1971) 137–140.

[10] B. Pellegrini, Extension of the electrokinematics theorem to the electromagnetic field and quantum mechanics, *Il Nuovo Cimento D* **15** (1993) 855–879.

[11] B. Pellegrini, Elementary applications of quantum-electrokinematics theorem, *Il Nuovo Cimento D* **15** (1993) 881–896.

[12] P. D. Yoder, K. Gärtner, and W. Fichtner, A generalized Ramo–Shockley theorem for classical to quantum transport at arbitrary frequencies. *J. Appl. Phys.* **79** (1996) 1951–1954.

[13] T. W. Berg, A General Reciprocity Theorem for Conducting Media, *Pyisica Scripta* **18** (1978) 375–382.

[14] H. Kim *et al.*, An extended proof of the Ramo-Shockley theorem, *Solid-State Electronics* **34** (1991) 1251–1253.

[15] X. Oriols, Quantum-trajectory approach to time-dependent transport in mesoscopic systems with electron-electron interactions, *Phys. Rev. Lett.* **98** (2007) 066803–066810.

[16] G. Albareda, J. Suné, and X. Oriols, Many-particle Hamiltonian for open systems with full Coulomb interaction: Application to classical and quantum time-dependent simulations of nanoscale electron devices, *Phys. Rev. B* **79** (2009) 075315–075330.

[17] G. Albareda *et al.*, Time-dependent boundary conditions with lead-sample Coulomb correlations: Application to classical and quantum nanoscale electron device simulators, *Phys. Rev. B* **82** (2010) 085301–085321.

[18] A. Alarcón and X. Oriols, Computation of quantum electron transport with local current conservation using quantum trajectories, *J. Stat. Mech.* **2009** (2009) P01051.

[19] F. L. Traversa *et al.*, Time-dependent many-particle simulation for resonant tunneling diodes: Interpretation of an analytical small-signal equivalent circuit, *IEEE Trans. Electr. Dev.* **58** (2010) 2104–2112.

[20] J. D. Jackson, *Classical Electrodynamics* (John Wiley & Sons 1975).

[21] M. Javid and P. M. Brown, *Field Analysis and Electromagnetics* (McGraw-Hill Book Company Inc., New York, 1963).

[22] P. R. Holland, *The Quantum Theory of Motion* (Cambridge University Press, Cambridge, 1993).

[23] Edited by X. Oriols and J. Mompart, *Applied Bohmian Mechanics: From Nanoscale Systems to Cosmology* (Pan Stanford Publishing, 2012).

[24] G. Albareda, F. L. Traversa, and X. Oriols Bohmian formulation of Full Counting Statistics in mesoscopic systems. *Proc. Int. Conf. Noise and Fluct.* (2011) 246–249.

[25] A. Benali *et al.*, Effect of gate-all-around transistor geometry on the high-frequency noise: an analytical discussion. *Fluctuations and Noise Letters* **11** (2012) 1241002–1231013.

# CLASSICAL TELEPORTATION OF CLASSICAL STATES

OLIVER COHEN
*6 Bigpath Farm Cottages, Upham, Hampshire, SO32 1HL, UK*
*oliver@bigglebaggle.com*

Received 24 February 2006
Accepted 27 February 2006
Communicated by Laszlo B. Kish

The standard quantum teleportation scheme is deconstructed, and those aspects of it that appear remarkable and "non-classical" are identified. An alternative teleportation scheme, involving only classical states and classical information, is then formulated, and it is shown that the classical scheme reproduces all of these remarkable aspects, despite the fact that they had seemed non-classical. This leads to a re-examination of quantum teleportation, which suggests that its significance depends on the interpretation of quantum states.

*Keywords*: Teleportation; information; non-classical.

The process of *quantum teleportation* was first described in the seminal paper of Bennett *et al.* [1]. In the thirteen years since its publication, it has had a huge influence on the field of quantum information, triggering a race for the first experimental implementation [2, 3, 4], and generating hundreds of journal citations. Quantum teleportation has also captured the imagination of the general public, with descriptions in the popular science literature that typically evoke images from the TV series Star Trek. More recently, there has been worldwide headline coverage of the successful teleportation of a laser beam [5].

At first sight, the reasons for the huge level of interest and publicity that quantum teleportation has given rise to seem obvious. The instantaneous transfer across an arbitrary spatial distance of a quantum state that can be defined to arbitrary precision, a state that is completely unknown to both sender and receiver, seems startling, remarkable, and, in particular, *non-classical*. Hence it may seem quite natural that teleportation should be considered a key part of the repertoire of quantum information processes.

In the following, our aim is first of all to break down the standard quantum teleportation protocol into a number of steps, and to identify those aspects of the procedure that appear to capture the essence of the non-classicality of the technique. We will then describe a very simple method for teleportation of a type of purely *classical* information, using classical systems. Surprisingly, and notwithstanding what has been said so far, this classical technique is very closely analogous to the standard quantum teleportation protocol. Indeed, it transpires that each of the particular aspects of quantum teleportation that we identified as appearing to display non-classicality is reproduced exactly in the purely classical version of the technique! This leads us to re-examine the significance of quantum teleportation. Our conclusion is that the extent to which quantum teleportation has any special significance depends on our interpretation of quantum states.

We start with a description of quantum teleportation as outlined in [1]. (We also include some comments which are intended to facilitate comparison with the purely classical scheme, a description of which is to follow):

(1) An EPR pair[1] is distributed between Alice and Bob.
(2) Charlie prepares a two-level quantum system in the state $|\psi\rangle$ and gives it to Alice. (Comments: (i) Although the state $|\psi\rangle$ represents a 2-dimensional system, it requires a doubly infinite amount of information to describe it, because it is defined by two continuously variable parameters. (ii) It is impossible to *measure* $|\psi\rangle$, given just a single system in that state.)
(3) Alice carries out a joint measurement on the system that Charlie has just given her and her component of the EPR pair that was originally distributed between her and Bob. This measurement has four possible outcomes. (Comments: (i) As soon as Alice has completed her measurement, she knows that the state of Bob's system will now be the same state $|\psi\rangle$ that Charlie supplied to her, up to a known unitary transformation, although she does now know what $|\psi\rangle$ is. In other words, from Alice's perspective, the teleportation process, up to a known unitary transformation, takes place instantaneously. (ii) After this measurement, Alice has no information whatsoever about the state of either *individual* system in her possession, but only has information about the *joint* state of both systems.)
(4) According to which result is obtained in the measurement described in (3), Alice sends 2 bits of classical information to Bob.
(5) Having received the 2 bits of information from Alice, Bob performs one of four possible operations on his system. (Comment: One of these operations is the identity operation, i.e., it consists of doing nothing.)
(6) Following the operation referred to in (5), the state of Bob's system will be $|\psi\rangle$, i.e., it will be an exact copy of the original state supplied by Charlie to Alice.

At this point it is worth pausing to highlight those aspects of the above process that seem remarkable and non-classical:

(a) The state $|\psi\rangle$ requires a doubly infinite amount of information to define it, yet it is teleported from Alice to Bob with the aid of only two bits of information.
(b) The state $|\psi\rangle$ is teleported faithfully from Alice to Bob without either party at any time having any information about the state whatsoever.
(c) The kernel of the teleportation process, as seen by Alice, takes place instantaneously.

---

[1]An "EPR pair" here refers to a pair of two-dimensional quantum systems in the state $\frac{1}{\sqrt{2}}\left(|01\rangle-|10\rangle\right)$, where the $|0\rangle$ and $|1\rangle$ states would typically be embodied as polarization or spin eigenstates.

(d) All trace of the original state $|\psi\rangle$ at Alice's location is erased in the teleportation process.

(e) If the system that Charlie gives Alice is in a state that is correlated with that of a fourth system, then, at the end of the teleportation process, Bob's system will be correlated with the fourth system in an identical way to that in which Charlie's was initially. This correlation may be quantum in nature (i.e., entanglement), or classical.

We now give an overview of our 100% *classical* teleportation protocol; we will then go on to give a more detailed explanation of the classical protocol. The initial overview is intended to provide a preliminary comparison with the quantum teleportation protocol as described above.

(1) A correlated pair of two-level classical systems is distributed between Alice and Bob. The particular type of correlation, and the method of preparation, will be explained shortly.

(2) Charlie prepares a two-level classical system in the state $|x)$. We will explain what is meant by "the state $|x)$" presently; suffice to say at this stage that the concept of the state $|x)$, and its realization, are purely classical. (Comments: (i) Although the state $|x)$ represents a two-dimensional system, it requires an infinite amount of information to describe it, because it is defined by one continuously variable parameter. (ii) It is impossible to *measure* $|x)$, given just a single system in that state.)

(3) Alice carries out a joint measurement on the system that Charlie has just given her and her component of the correlated pair that was originally distributed between her and Bob. This measurement has two possible outcomes. (Comments: (i) As soon as Alice has completed her measurement, she knows that the state of Bob's system will now be the same state $|x)$ that Charlie supplied to her, up to a known simple transformation, although she does not know what $|x)$ is. In other words, from Alice's perspective, the teleportation process, up to a known simple transformation, takes place instantaneously. (ii) After this measurement, Alice has no information whatsoever about the original state of either *individual* system in her possession; she does, however, have information about the *joint* state of both systems.)

(4) According to which of the two possible outcomes is obtained in the measurement described in (3), Alice sends 1 bit of classical information to Bob.

(5) Having received the bit of classical information from Alice, Bob performs one of two possible operations on his system. (Comment: one of these two possible operations is the identity operation, i.e., it consists of doing nothing.)

(6) Following the operation as described in (5), the state of Bob's system will be $|x)$, i.e., it will be an exact copy of the original state supplied by Charlie to Alice.

Before examining the extent to which the remarkable aspects (a)–(e) of the quantum protocol, as highlighted above, are retained in the classical case, we will explain what we mean by "the state $|x)$", and describe how the classical technique could be implemented.

By "the state $|x)$" we simply mean the following:

A classical system is in one of two possible configurations. One of these possible configurations is the reference configuration. Suppose we do not know which configuration the system is in, but we know that the probability that it is in the reference configuration is x. Then we say that the state of the system is $|x)$.

For example, suppose that we are given an ensemble of N coins, of which Nx are showing heads face up and N(1−x) are showing tails. Suppose that we assign "heads face up" as our reference configuration. If we then select a coin from the ensemble, but do not look to see whether it is showing heads or tails face up, then our description of the coin will be $|x)$.

Note that the notion of a classical state $|x)$, as we have defined it here, is epistemological – it refers to a *state of knowledge* with respect to an observer. However, this is not really very different to the way that many people think about quantum states – that is, that they are a way of encapsulating our knowledge about a system, rather than having any direct ontological interpretation.[2]

We now describe the method we will use for preparing a classical state $|x)$ for the purposes of our classical teleportation protocol. We begin by preparing N ensembles of coins in boxes, where N is large. Each of the N ensembles consists of N coins, each of which is placed inside its own individual box, which is then sealed. The coin is placed in its box showing either heads or tails face up, before the box is sealed. The boxes are symmetrical in the sense that, if a coin is placed in its box showing heads face up, and the box is rotated through 180 degrees, then the system of box plus coin will be indistinguishable from one where the coin was placed in its box showing tails face up and the box was not rotated. We assign "heads face up" (within a box) as our reference configuration.

We can label these N ensembles 1 through N. In ensemble 1, one coin is placed in its box showing heads face up, and (N-1) coins are each placed in their boxes showing tails face up. In ensemble 2, two coins are placed in their boxes showing heads face up, and (N-2) showing tails, and so on, i.e., in ensemble n, n coins are placed in their boxes showing heads face up and (N-n) showing tails.

To prepare a coin in any chosen state $|x)$, we simply select a sealed box from the ensemble numbered Nx. (Regardless of the accuracy to which x is expressed, we can ensure that Nx is a whole number by making N sufficiently large.) If the box is chosen at random from this ensemble, then we know that the probability that the coin inside it is showing heads is x, hence our description of the coin will be $|x)$. (Note that we can convert the state $|x)$ to the state $|1-x)$ by turning the box upside down.)

It is possible to prepare a pair of coins-in-boxes in a correlated state – for example, a state where we know that, with probability y, both coins are showing heads face up within their boxes, and, with probability (1−y), both coins are showing tails face up. We denote such a state $|y)_{HH} + |1-y)_{TT}$. Such a state could be straightforwardly prepared using a method similar to that described above, but this time with Ensemble n containing N pairs of boxes of which n pairs contain coins both of which are showing heads face up within their boxes, and (N-n) pairs contain coins which are both showing tails face up. To prepare the state $|y)_{HH} + |1-y)_{TT}$ we would then select a pair of boxes from Ensemble Ny.

We are now in a position to explain how the classical teleportation scheme can be implemented:

(1) A pair of coins-in-boxes is prepared in the state $|1/2)_{HH} + |1/2)_{TT}$, and the boxes are then distributed to Alice and Bob at their respective locations.

[2]Note also that it is impossible to measure $|x)$ given just a single system in the state $|x)$; in order to determine $|x)$ we would need a whole ensemble in that state. Again, this aspect is closely analogous to the non-measurability of a quantum state $|\psi\rangle$.

The state $|1/2\rangle_{HH} + |1/2\rangle_{TT}$ could be prepared using the ensemble method described above. Alternatively, a simple way of preparing this state using just a single pair of coins-in-boxes (rather than N ensembles of N pairs) is as follows: Two coins are placed in boxes showing heads face up and the boxes are sealed. The boxes are then both rotated through the same random number of 180 degree turns. This rotation could be carried out by a machine coupled to a random number generator, or we could recruit an external protagonist to rotate the two boxes through any number of 180 degree turns they choose and then return them to us. Following the rotation, we will know that both coins are showing the same face within their respective boxes, but we will have no information as to whether that face is heads or tails. Hence our description of the pair will be $|1/2\rangle_{HH} + |1/2\rangle_{TT}$. Once this state has been prepared, the boxes must be distributed to Alice and Bob whilst ensuring that no further rotations take place.

(2) Charlie prepares a single coin-in-a-box in the chosen state $|x\rangle$, using the ensemble method as described earlier. Charlie then passes the box over to Alice.
(3) Alice carries out a joint measurement on the box that was originally distributed to her as part of the correlated pair, and the box that Charlie has just given her, as follows: She arranges that both of these boxes are rotated by the same random number of 180 degree turns. Once again this rotation can be carried out by a machine or by an external protagonist. Following the rotation the boxes are opened (by the machine or the external protagonist) and the coins examined to see if they are showing the same (i.e., both heads or both tails) or different (i.e., one heads and one tails) faces, and the outcome is relayed to Alice as "same" or "different".
(4) Alice sends one bit of information to Bob, according to whether the outcome of the measurement described in (3) is "same" or "different".
(5) If Bob receives the message "same", he leaves his box as it is; if he receives the message "different" he rotates his box by 180 degrees.
(6) Following the operation described in (5), the state of Bob's coin must be $|x\rangle$. This can be seen straightforwardly as follows: The whole sequence of operations just described ensures that the coin inside Bob's box is showing the same face up as the one that was showing in Charlie's box at the time that he (Charlie) originally selected it. When he selects his box, Charlie knows that the probability that the coin inside it is showing heads is x; hence, the probability that Bob's coin is showing heads at the end of the process is x, i.e., the state of Bob's coin is $|x\rangle$.

At no point in the process do Alice or Bob have any knowledge of $|x\rangle$. The state $|x\rangle$, as teleported from Alice to Bob, represents a description of the system *as given by Charlie*. If we compare this to the quantum teleportation protocol we can see that, unless we attach some ontological significance to a quantum state $|\psi\rangle$, then there is an exact correspondence in the sense that, in the quantum case, the teleported state $|\psi\rangle$ represents *Charlie's description of the system*. Indeed, in the quantum case, since Alice and Bob have no information whatsoever about the teleported state, their best final description of the system to which it is teleported is the maximally mixed state.

If, instead of a coin-in-a-box in the state $|x\rangle$, Charlie had given Alice one member of a correlated pair of coins-in-boxes, where the other member of the pair is held by

David, and the pair is initially in the state $|y)_{HH} + |1-y)_{TT}$, then, at the end of the teleportation process, Bob's system will be correlated with David's in an identical way to that in which Charlie's was initially, i.e., Bob's and David's systems will end up in the state $|y)_{HH} + |1-y)_{TT}$. This follows straightforwardly from the fact that the final state of Bob's coin-in-box must be identical to the initial state of the coin-in-box that Charlie gave Alice. There is no remaining residual correlation between Alice's coin-in-box and David's, because the original correlation between these two systems will be washed out by the randomization process that is part of the joint measurement carried out by Alice as part of the protocol.

Now that we have explained how our classical teleportation protocol works, we can re-visit the remarkable and "non-classical" aspects of quantum teleportation that we identified earlier, to see to what extent, if any, these aspects are also present in the classical version. Recall that we earlier highlighted five remarkable features of the quantum protocol as follows:

(a) The state requires a (doubly) infinite amount of information to describe it, yet it is teleported from Alice to Bob with the aid of only two bits of information.
*This also holds in the classical version, the only differences being that an infinite rather than doubly infinite amount of information is required to describe the classical state, and one rather than two bits of information must be sent from Alice to Bob. If we carry out the classical version twice, using two different states, we can match the quantum case, in that the two states will require in total a doubly infinite amount of information to describe them, and Alice will have to send a total of two bits of information to Bob.*

(b) The state is transferred faithfully from Alice to Bob without either party at any time having any information about the state whatsoever.
*This also holds in the classical version.*

(c) The kernel of the teleportation process, as seen by Alice, takes place instantaneously.
*This also holds in the classical version.*

(d) All trace of the original state at Alice's location is erased in the course of the teleportation process.
*This also holds in the classical version.*

(e) If the system that Charlie gives Alice is in a state that is correlated with that of a fourth system, then, at the end of the teleportation process, Bob's system will be correlated with the fourth system in an identical way to that in which Charlie's was initially. This correlation may be quantum in nature (i.e., entanglement), or classical.
*This still holds in the classical version; however, since we cannot have quantum correlations in a classically described system, it is only classical correlations that can be teleported in the classical case.*

Surprisingly, we can see that the "remarkable features" of the quantum teleportation process – the ones that had appeared to capture its uniqueness and *non-classicality* – remain present, in their entirety, in our purely classical version of the process!

It should be pointed out that there is a simpler method by which Alice can transfer the classical probabilistic description $|x)$ from her system to Bob's, without going through the full protocol we have described. Alice can simply open the box that Charlie gives her (without allowing Charlie to see), determine whether the coin is showing heads or tails, and transmit this one bit of information to Bob. Bob then sets his coin in the corresponding configuration. If Charlie is then told that Bob's coin is showing the same face as the one originally showing in his (Charlie's) box when he gave it to Alice, then Charlie's final description of Bob's system will be $|x)$. However, this "cheating" method cannot really be described as teleportation. It does not display (b), (c), (d), or (e) of the features listed above. A closer quantum analogue of this simpler protocol would be a situation where Alice and Bob share a common measurement basis, and Alice transfers an eigenstate of this basis to Bob by measuring it and transmitting the measurement result to Bob so that he can duplicate the state. However, without this shared basis, the cheating method cannot be used to transfer quantum states faithfully.

Some related work should be mentioned. A method for classical teleportation of *quantum* states has been described by Cerf, Gisin, and Massar in [6]. In this scheme Alice and Bob share classical information rather than entanglement. Hardy [7] has described a toy model for teleportation that is local but retains some non-classical features; and Spekkens [8] has described a toy model that can be used to produce analogues of a number of quantum information processes, including teleportation. Spekkens's model is similar to the one described in [7], and likewise is local but retains non-classical aspects. Collins and Popescu have argued [9] that the one-time pad, which can be used to transfer secret bits from Alice to Bob, is analogous to teleportation. (However, the one time pad does not display (a), (b), or (d) of the features listed above.) Even without recourse to a classical model, advocates of the epistemological interpretation of quantum states have argued that, from their standpoint, quantum teleportation is not particularly remarkable [10].

Classical versions of other quantum information processes and results have also been discovered. In particular, a no-cloning theorem for classical probability distributions was presented in [11]; see also the interesting results in [12] and [13]. More recently, Mor [14] has pointed out that, even without any communication between Alice and Bob, a classical protocol can be used by Alice to remotely prepare one of a set of probability distributions at Bob's location (as opposed to the unique distribution implied by the teleportation procedure described here).

What are we to conclude from our analysis? Although quantum states are very different to classical states,[3] it seems that our classical version of teleportation is just as impressive as the original quantum protocol, *if we think of quantum states as representing states of knowledge*. In both cases, the end result of the process is that Charlie's final description of Bob's system is identical to his (Charlie's) original description of his own system. From the epistemological standpoint, it is the status of being the object of Charlie's state of knowledge that is transferred from Alice's system to Bob's. If, on the other hand, we think of a quantum state as having ontological content, representing perhaps a real physical field [15], then our classical version of teleportation is not equivalent to the quantum case, because the latter now involves the transfer of intrinsic physical content from Alice's system to Bob's.

This interpretation-dependence of the significance of quantum teleportation is in contradistinction to other facets of quantum information processing, such as quantum

[3]For example, if a quantum system is in a pure quantum state $|\psi\rangle$, then we can always identify a measurement, the outcome of which, if carried out on the system, can be predicted with certainty. This is not the case for the classical state $|x)$ (which in this sense has more in common with a mixed quantum state such as $x|0\rangle\langle 0| + (1-x)|1\rangle\langle 1|$).

computation and quantum cryptography, the significance of which do not depend on the interpretation of quantum states. This is because the significance of computation and cryptography lies primarily in their practical usefulness, which is independent of any interpretational standpoint. Whilst it is true that quantum teleportation may have limited practical use, for example to enhance the range for quantum cryptography implementations, or as part of a quantum computing implementation [16], our analysis indicates that its fundamental significance as a stand-alone technique cannot be assessed without recourse to the question of interpretation.

### Acknowledgments

I am grateful to Todd Brun and Chris Fuchs for helpful comments on the manuscript.

An earlier version of this paper was posted on the LANL Archive, quant-ph/0310017.

### References

[1] C. H. Bennett *et al.*, *Phys. Rev. Lett.* **70** (1993) 1895–1898.
[2] D. Bouwmeester *et al.*, *Nature* **390** (1997) 575–579.
[3] D Boschi *et al.*, *Phys. Rev. Lett.* **80** (1998) 1121–1124.
[4] A. Furusawa *et al.*, *Science* **282** (1998) 706–709.
[5] W. Bowen *et al.*, *Phys. Rev. A* **67** (2003) 032302–1–032302-4.
[6] N. J. Cerf, N. Gisin and S. Massar, *Phys. Rev. Lett.* **84** (2000) 2521–2524.
[7] L. Hardy, LANL Archive preprint quant-ph/9906123.
[8] R. W. Spekkens, LANL Archive preprint quant-ph/0401052.
[9] D. Collins and S. Popescu, *Phys. Rev. A* **65** (2002) 032321-1–032321-11.
[10] C. A. Fuchs and A. Peres, *Physics Today* **53 (3)** (2000) 70–71.
[11] A. Daffertshofer, A. R. Plastino and A. Plastino, *Phys. Rev. Lett.* **88** (2002) 210601-1–210601-4.
[12] K. A. Kirkpatrick, *Found. Phys. Lett.* **16 (3)** (2003) 199–204.
[13] K. A. Kirkpatrick, *J. Phys. A* **36 (17)** (2003) 4891–4900.
[14] T. Mor, LANL Archive preprint quant-ph/0511172.
[15] See for example D. Bohm and B. J. Hiley, The Undivided Universe: An Ontological Interpretation of Quantum Mechanics, Routledge, London (1993).
[16] D. Gottesman and I. L. Chuang, *Nature* **402** (1999) 390–393.

# ON COHEN'S CLASSICAL TELEPORTATION OF CLASSICAL STATES

P. HRASKÓ
*University of Pécs, Department of Theoretical Physics, Szántó K. J. u. 1/b, Pécs, H- 7633, Hungary*

Received 2 March 2006
Revised 28 April 2006
Accepted 2 May 2006
Communicated by Laszlo Kish

We present additional considerations to elucidate the nature of Cohen's classical teleportation of classical bits.

*Keywords*: Quantum communication; classical communication; totally secure communication; noise.

Recently, Cohen published an interesting study about the possibility of teleportation of classical bits by classical information transfer. He puts the following question:

*Is it true that non-classical transfer of information is a compulsory element of the process of teleportation?*

According to the authors (Bennett, *et al.*) of the original paper [2] the answer is affirmative. Their thesis is:

(*A*) *Precise teleportation of a given quantum state requires (partly) non-classical transfer of information.*

On the other hand, Cohen's thesis [1] is:

(*B*) *Precise teleportation of a given classical state requires only classical transfer of information.*

This latter thesis is logically unrelated to the former one. But Cohen points out the striking similarity of the two protocols, underlying the two theses, and from this similarity he infers that (*B*) disproves (*A*). But this is not a convincing conclusion. In order to disprove (A) one has to prove a third thesis:

(*C*) *For a precise teleportation of a given quantum state the EPR state* $|\Psi_{23}^{(-)}\rangle$ *of Bennett et al.* [2] *is permitted to be replaced by a classical mixture* $\rho_{23}$.

This thesis, however, seems to be incorrect. One may choose e.g.

$$\rho_{23} = \frac{1}{4}\left(I^{(23)} + \eta\vec{\sigma}^{(2)} \otimes \vec{\sigma}^{(3)}\right) \tag{1}$$

$(-1 \le \eta \le 1/3)$ which is the most general spherically symmetric spin density matrix of two spin-1/2 particles. When $\eta = -1$, $\rho_{23}$ represents the pure state $|\Psi_{23}^{(-)}\rangle$, for any other $\eta$ it is a mixture.

If now one repeats the calculation of Bennett *et al.* [2] with $|\Psi_{23}^{(-)}\rangle$ replaced by $\rho_{23}$, one arrives at the conclusion that, except for $\eta = -1$, the given state $|\Phi_1\rangle$ is never transported undisturbed.

Owing to the special choice of $\rho_{23}$, this argument does not strictly disprove (*C*). But our only aim was to point out that it is (*C*) rather than (*B*) which disproves (*A*).

Finally, we note that Cohen [1] stresses himself that his claim is based at least partly on his view that the notions of classical and quantum states are both referred to the state of knowledge of the observer. We do not agree with this view. The coin has already been earlier in that state, in which state it is observed. Hence observation changes only the state of knowledge of the observer. On the contrary, the violation of Bell's inequality shows that it is contradictory to assume, that the particles had been in the observed spin state even before the observation took place.

## References

[1] O. Cohen, Classical teleportation of classical states, *Fluct. Noise Lett.* **6** (2006) C1–C8.
[2] C. H. Bennett *et al.*, *Physical Review Lett.* **70** (1993) 1895–1898.

# REPLY TO: "ON COHEN'S CLASSICAL TELEPORTATION OF CLASSICAL STATES"

OLIVER COHEN
*6 Bigpath Farm Cottages, Upham, Hampshire SO32 1HL, UK*
*oliver@bigglebaggle.com*

Received 2 March 2006
Revised 28 April 2006
Accepted 2 May 2006
Communicated by Laszlo Kish

In response to a recently published protocol for classical teleportation of classical states [1], Hrasko [2] has made some remarks in connection with the impossibility of precise quantum teleportation if a classical mixture is used in place of the EPR state, and with regard to the epistemological interpretation of quantum states. The purpose of this note is to attempt to clarify these issues and their relevance for the argumentation presented in [1].

*Keywords*: Teleportation; classical; epistemological.

Hrasko [2] has pointed out that, for precise teleportation of a given quantum state, an EPR resource, as opposed to a classical mixture, is essential. This is correct and is well known, although it should be pointed out that precise teleportation is possible on a probabilistic basis (i.e. with fidelity = 1 between the state to be teleported and final state of system to which this state is teleported, but with probability of success less than 1 [3]) if the EPR state in the standard teleportation protocol [4] is replaced with a partially entangled state. It is also well known that, if a purely classical resource is used to attempt to teleport a quantum state, this fidelity cannot exceed 2/3 [5].

Hrasko infers, from the necessity of using an EPR state for precise quantum teleportation, that non-classical transfer of information is a compulsory element of the process of teleportation. This seems a reasonable conclusion. However, Hrasko attributes to [1] the opposite conclusion, i.e. that precise teleportation of a given quantum state requires only classical transfer of information. But this attribution is a misrepresentation of [1], which does not in fact make any such claim. Rather, it is pointed out in [1] that several key aspects of the quantum teleportation process, aspects that are *commonly considered* to be non-classical, are still present in the purely classical version of teleportation that is presented in [1]. The implication of this is not that quantum teleportation requires only classical transfer of information, but rather that the role of non-classical transfer of information within the quantum teleportation process is rather more subtle than is generally thought. Indeed, inasmuch as the necessity of using an entangled resource for quantum teleportation, which is already well known as pointed out above, in itself implies non-classical transfer of information, the irreducible nature of the non-classical element of quantum teleportation is self-evident.

In the last paragraph of [2] Hrasko states his objection to the epistemological interpretation of quantum mechanics, according to which quantum states are representative of nothing more than states of knowledge of observers and should be thought of as algorithmic tools for predicting the probabilities for different experimental outcomes. Hrasko writes: "the violation of Bell's inequality shows that it is contradictory to assume that the particles had been in the observed spin state even before the observation took place". In response to this assertion, we should first note that, even if it were true, it would apply only to systems in an entangled state, since entanglement is a prerequisite for violation of a Bell inequality. However, the quoted assertion is in any case somewhat misleading; violation of Bell's inequality is generally understood to rule out the existence of local hidden variables, but there are nonlocal realistic interpretations of quantum mechanics – the best known being due to Bohm [6] – according to which it is perfectly possible to carry out passive measurements that do reveal pre-existing properties of the quantum system. There are also other interpretations of quantum mechanics – see for example [7] – which again allow for passive measurement of pre-existent properties, this time at the expense of modifying the normal classical rules for combining logical assertions.

Moreover, the epistemological view of quantum states makes no assumption about whether or not a system is in a particular state prior to measurement. Rather, it posits that the quantum state itself is not something that the system possesses, but is only a function of the observer, the state of knowledge of whom it encapsulates [8]. According to this view, quantum states are indeed analogous to classical probability states. It was not the intention of [1] to advocate an epistemological interpretation, or indeed any other interpretation, of quantum mechanics, but rather to point out that the significance and "surprise factor" of quantum teleportation is interpretation dependent. If one does adopt the epistemological view of quantum states, then quantum teleportation – notwithstanding the fact that entanglement is required, implying that in some sense non-classical transfer of information is involved – appears to be hardly more remarkable or surprising than classical teleportation.

## References


[1] O. Cohen, Classical teleportation of classical states, *Fluct. Noise Lett.* **6**(2) (2006) C1–C8.

[2] P. Hrasko, On Cohen's classical teleportation of classical states, *Fluct. Noise Lett.* **6**(3) (2006) C29–C30.

[3] P. Agrawal and A. K. Pati, LANL Archive preprint quant-ph/0210004.

[4] C. H. Bennett *et al., Phys. Rev. Lett.* **70** (1993) 1895–1898.

[5] S. Massar and S. Popescu, *Phys. Rev. Lett.* **74** (1995) 1259–1262.

[6] D. Bohm and B. J. Hiley, *The Undivided Universe: An Ontological Interpretation of Quantum Mechanics* (Routledge, London, 1993).

[7] See for example R. B. Griffiths, *Consistent Quantum Theory* (Cambridge University Press, 2003).

[8] See for example C. M. Caves, C. A. Fuchs and R.. Schack, *Phys. Rev. A* (2002) 0222305-1–22305-6.

# NOISE AND TRANSPORT IN MESOSCOPICS: PHYSICS BEYOND THE LANDAUER–BÜTTIKER FORMALISM

FREDERICK GREEN

*School of Physics, The University of New South Wales, NSW 2052, Australia*
*and*
*Department of Theoretical Physics, Institute of Advanced Studies*
*Research School of Physical Sciences and Engineering, The Australian National University*
*ACT 0200, Australia*
*fgreen@phys.unsw.edu.au*

MUKUNDA P. DAS

*Department of Theoretical Physics, Institute of Advanced Studies*
*Research School of Physical Sciences and Engineering, The Australian National University*
*ACT 0200, Australia*
*mukunda.das@anu.edu.au*

Received 13 December 2004
Revised 25 January 2005
Accepted 25 January 2005
Communicated by Robert Vajtai

The standard physical model of contemporary mesoscopic noise and transport consists in a phenomenologically based approach, proposed originally by Landauer and since continued and amplified by Büttiker, Imry and others. Throughout all the years of its gestation and growth, it is surprising that the Landauer–Büttiker approach to mesoscopics has matured with scant attention to the conserving properties lying at its roots: that is, at the level of actual microscopic principles. We systematically apply the sum rules for the electron gas to clarify the issue of conservation within the standard model of mesoscopic conduction. Noise, as observed in quantum point contacts, provides the vital clue.

*Keywords*: Ballistic transport; nonequilibrium noise; quantum kinetics; conservation laws; sum rules; quantum point contacts.

## 1. Introduction

### 1.1. *Background*

If someone were to declare that a successful kinetic theory is one that explains experiment without any need to obey Newton's laws, such an assertion would not be likely to be received unreservedly, even by most philosophers of science. In this paper we ask what it would mean if, in some sense, an analogous proposition were to have been made to a segment of the mesoscopics community, and implicitly accepted. It would be a curious situation indeed; one with intriguing implications for the pursuit of mesoscopic physics and, in particular, for noise theory.

Here, the conservation laws of charge and particle number stand at stage center. The principles of conservation are in every sense as cardinal as Newton's laws. In transport physics, their validity is neither compliant to any quantitative revision nor contingent upon any set of intuitive preconceptions. Conservation is the inalienable heart of every rational model of conduction. There is no exception.

Let us revisit the essentials of transport in mesoscopic conductors. Typically, structures have lengths comparable to, or smaller than, the mean free paths for electron scattering in the bulk. Charge transport can then be regarded as ballistic (collision-free) within the confines of such a metallic channel.[a]

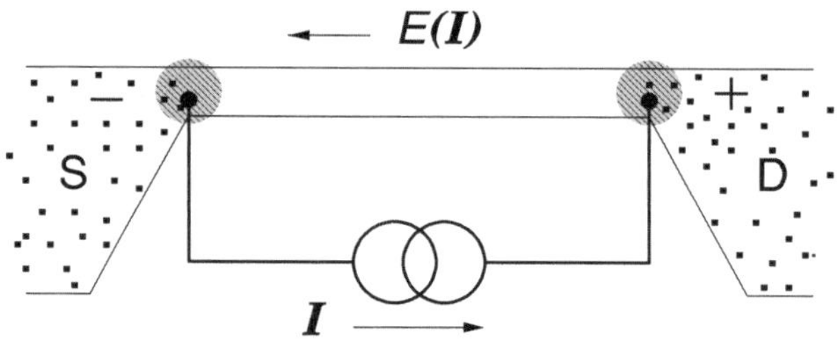


Fig. 1. An ideal, uniform ballistic wire, or quantum point contact. Its diffusive and dissipative leads (S and D, stippled) are permanently neutral and permanently in equilibrium. A paired source and sink of current $I$ at the boundaries drive the transport. Local charge clouds (shaded), induced by the imposed influx and efflux of $I$, set up the dipole voltage $V = E(I)L$ between D and S. The ballistic wire manifests its optimal (Landauer) conductance $I/V$ if, and only if, the mean free paths for elastic and inelastic scattering are equally matched and straddle the whole wire, from its strongly dissipative source terminal to the corresponding drain. See Sec. 3 for a detailed proof.

Figure 1 illustrates this for a so-called "quantum point contact" (QPC). We are dealing with small, potentially quasi-molecular, structures. Consequently they experience a high degree of *openness* to their macroscopic environment. The restricted dimensionality also accentuates the inter-particle *correlations*. It is no secret [1, 2] that these are the very attributes that hold the hardest set of challenges to be met in developing mesoscopic electronics.

Among the challenges, the foremost is how to reconcile *microscopic conservation* with both openness at the device interfaces and strong quantum correlations just about everywhere else. Without a firm anchoring in the conservation laws, nothing prevents an unconstrained model from "creating" or "destroying" charges and currents willy-nilly. Do we need such a model?

Even if seemingly able to explain an experiment, a theory that is nonconserving must be regarded as fundamentally incoherent, for we have learned that Nature does not work in this way. That is the case, anyway, in the realm of condensed

[a]The term "metallic channel" has the specific sense of a device with a band of extended conducting states that can be freely populated and depopulated by controlling the carrier density. Carriers have free access to and from a set of (large) bounding leads which are themselves strongly metallic.

matter. The absolute centrality of conservation in mesoscopic transport needs no further comment.

### 1.2. *Plan of the Paper*

The paper is set up as follows. In the next section we look at the prime relation of mesoscopic transport; that is, the immediate link between finite conductance and finite energy dissipation. This leads into the third section and our first cardinal result: the strictly kinetic derivation of Landauer's quantized conductance formula for an ideal metallic channel [2–4], using only the axioms of microscopic theory and free of any recourse to the standard model's very special assumptions. The message is that a mesoscopic model (one which is faithful to orthodox quantum kinetics) does not need phenomenology for a backbone.

Our result shows that the Landauer–Büttiker (LB) project is already subsumed in a more mature, fundamentally established, and truly first-principles framework: quantum kinetics. This means that the phenomenological assumptions, on which the unique mesoscopic character of LB has been predicated [2], are not foundational.

Later in Sec. 3, then in Sec. 4, we go on to demonstrate how the logic of the conservation laws generates a completely natural microscopic description of a major exemplar of mesoscopic fluctuations: QPC noise. We account for data that went unexplained for nearly a decade. In the process it becomes clearer that the standard Landauer–Büttiker approach to fluctuations does not respect conservation. Because of that, it fails to predict the directly observed consequences of conservation.

Finally we sum up the origin, motivation and cogency of our microscopically conserving analysis. We also retrace the implications for setting up a novel perspective on mesoscopic processes. Interested readers will find an Appendix containing the formal proof that the Landauer–Büttiker noise theory does not satisfy the *perfect-screening* and *compressibility* sum rules. These basic rules govern the dynamics of mesoscopic noise, as experimentally demonstrated for QPCs.

## 2. The Physical Issue

At the outset, we remark that our analysis applies specifically to *metallic* conduction and fluctuations. We examine open mesoscopic structures where the carrier states form a quasi-continuum, are spatially extended, and have considerable overlap. This is territory that the LB approach is supposed to cover extremely well [2]. We are not considering, here, systems in which quite different nonmetallic mechanisms of transport (e.g. hopping and tunneling) involve electronic orbitals that are discrete, highly localized, and with little overlap.

A simple intuitive understanding of mesoscopic transport and noise was established, in the last fifteen years or so, through the insights of Landauer, Büttiker, Imry and many similarly inclined contributors [2–5]. (The sheer proliferation of such works precludes citing all but a few outstanding examples.) In that succinct perspective the physical basis of current flow, mesoscopic and otherwise, is identified as a mismatch of carrier density between two or more metallic reservoirs (terminals) across which the QPC or other low-dimensional conductor is connected.

The hallmark of metallic transport in quantum point contacts, the quantization of conductance as "Landauer steps" in units of $2e^2/h$ ($\approx 0.078$ mS), appears to be adequately explained as coherent transmission of single-electron waves through a perfectly loss-free barrier. However, a picture of quantum-coherent scattering as simple as this cannot address the central theoretical issue of metallic conduction:

*What causes dissipation in a ballistic quantum point contact?*

Such a question is more than academic. In the near future, reliable and effective nano-electronic design will demand models that are credible; credible not just as physical theory but as engineering practice. Therefore a comprehensive theory has to confront the unavoidable presence, and action, of power dissipation.

For conduction, the central issue is plain. Any finite conductance $G$ must *dissipate* electrical energy (inelastically) at the rate $P = IV = GV^2$, where $I = GV$ is the current and $V$ the potential difference across the terminals of the driven channel. This is mandated by the underlying physics of charge conservation in externally driven (open) conduction [6, 7]; as such it cannot be hostage to any transport philosophy insisting on coherent propagation as the exclusive origin of conductance.

Our question has its definitive answer in many-body quantum kinetics [9], free of all "supplementary hand-waving" [2]. The application of well tested microscopic methods leads not only to conductance quantization precisely by the direct inclusion of inelastic energy loss [8–11], but, as we show in Sec. 3 to follow, it also resolves a long-standing experimental enigma [12] in the noise spectrum of a quantum point contact (QPC) [13]. The same developments foreshadow a systematic pathway to truly predictive design of novel structures.

In their dissipative action, inelastic collisions are beyond the reach of transport models that rely on coherent quantum scattering alone to explain the origin of $G$. Coherence implies elasticity, and elastic scattering is always loss-free: it conserves the energy of the scattered particle. The recent review of mesoscopic phenomenology by Agraït and others [1] expresses a similar conclusion on the insufficiency of coherence-based arguments for actual experimental systems.

There is no choice but to explicitly admit the dissipative mechanisms (e.g. phonon emission) by which the energy gained by carriers, transported from source to drain, is returned incoherently to the surroundings. A very detailed microscopic demonstration of this has been given by Sorée *et al.* [8]. Thus, alongside elastic and coherent scattering, inelastic processes are always in place. Together, they fix $G$; yet it is only the energy-dissipating mechanisms that secure *thermodynamic stability* in steady-state conduction. Coherence on its own cannot bring this about.

A serious deficiency underlies purely elastic approaches to transmission. They cannot handle dissipation. We now review the well-defined microscopic resolution of that problem.

## 3. The Physical Solution

The complete microscopic understanding of the ubiquitous power-loss formula $P = GV^2$ [6, 7, 15] is couched in terms of the fluctuation-dissipation theorem (details have long been available in the kinetic literature [15, 16]). It holds for *all* resistive devices at all scales, without exception. The theorem expresses the requirement for thermodynamic stability.[b] With it comes the conclusion that [9–11]

- inelasticity is necessary and sufficient to stabilize current flow at finite conductance;
- ballistic quantum point contacts have finite $G \propto 2e^2/h$; therefore
- processes of energy loss are indispensable to any rational theory of ballistic transport.

To allow for the energy dissipation vital to any microscopic description of ballistic transport, we recall that open-boundary conditions imply the intimate coupling of the QPC channel to its interfaces with the reservoirs. The interface regions must be treated as an integral part of the device model. They are the actual sites for strong scattering effects: *dissipative* many-body events as the current enters and leaves the ballistic channel, and *elastic* one-body events as the carriers interact with background impurities, the potential barriers that confine and funnel the current, and so on. See Fig. 1 for an illustration.

A key idea in our treatment is to subsume the interfaces within the total kinetic description of the ballistic channel. At the same time, strict charge conservation in an *open* device requires the direct supply and removal of charge flux by a generator outside the system of interest [6]. With this canonical precondition, that an open channel must be energized by a strictly *external* driving agent, the current is completely independent of the physics locally governed by the reservoirs.

That fundamental condition sets the quantum kinetic approach totally apart from Landauer-like treatments [2–5]. Instead of the gauge-invariant open-system requirement [6, 7] that any current flow is perforce supplied independently from outside[c] they rely on the intuitive assumption that the current depends, passively, on a notional density "mismatch" between reservoirs. It is nothing other than an *ad hoc* extrapolation, to strongly quantum systems, of purely classical diffusion [4].

### 3.1. *Ballistic Conductance*

One can straightforwardly write the algebra for the conductance in our model system. A uniform, one-dimensional ballistic QPC, of operational length $L$, will be associated with two mean free paths determined by $v_{\rm F}$, the Fermi velocity of the electrons, and a pair of characteristic scattering times $\tau_{\rm el}, \tau_{\rm in}$. Thus

$$\lambda_{\rm el} = v_{\rm F}\tau_{\rm el}; \qquad \lambda_{\rm in} = v_{\rm F}\tau_{\rm in}. \tag{1}$$

Respectively, these are the scattering lengths set by the elastic and inelastic processes active at both interfaces. The device (i.e. the QPC *with* its interfaces) has a

[b] There is an intimate structural correspondence between the dissipation formula $P = IV$, originating from the canonical, many-body Hamiltonian description of a driven conductor [7], and the form of the steady-state solution to the quantum kinetic equation for its carrier distribution. In the latter case, the fluctuation-dissipation theorem follows directly when the limit of a weak driving field is taken. The formalities are in Ref. 11, which also shows the fundamental role played by *electron-hole polarization* in determining the current-current correlator and the conductance.

[c] Transport at fixed current and at fixed voltage are completely equivalent. The electrodynamics of conduction in an open-boundary device driven by an external current generator [6] is isomorphic to that of a closed conducting loop, comprising the "open" device and a voltage source (a battery) [7]. Discussion of the underlying gauge invariance is beyond our scope here. We urge the interested reader to consult, in parallel, both Sols [6] and Magnus & Schoenmaker [7].

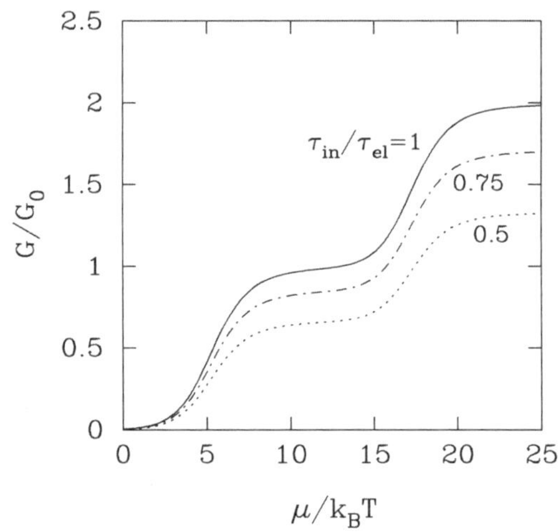


Fig. 2. Conductance quantization in a two-band ballistic point contact, as a function of chemical potential $\mu$, calculated from Eq. (3) within our conserving kinetic theory; see Ref. 9. channels. Broken curves: non-ideal behavior increases with the onset of inelastic phonon emission inside the contact, which preserves the Landauer steps while suppressing their height.

conductive core that is collisionless. It follows that

$$\lambda_{\rm el} = L = \lambda_{\rm in}. \tag{2}$$

Finally, the channel's conductance is given by the familiar formula (deducible straight from Kubo's microscopic prescription [9, 15])

$$G = \frac{ne^2\tau_{\rm tot}}{m^*L} = \frac{2k_{\rm F}}{\pi}\frac{e^2}{m^*L}\left(\frac{\tau_{\rm in}\tau_{\rm el}}{\tau_{\rm el}+\tau_{\rm in}}\right); \tag{3}$$

the effective mass of the carriers is $m^*$. The first factor of the rightmost expression for $G$ has the density $n$ recast in terms of the Fermi momentum $k_{\rm F}$; in the final factor, Matthiessen's rule $\tau_{\rm tot}^{-1} = \tau_{\rm el}^{-1} + \tau_{\rm in}^{-1}$ fixes the total scattering rate.

On applying Equations (1)–(3), the conductance reduces to

$$G = 2\frac{e^2}{\pi\hbar}\frac{\hbar k_{\rm F}}{m^*L}\left(\frac{(L/v_{\rm F})^2}{2L/v_{\rm F}}\right) = \frac{2e^2}{h} \equiv G_0. \tag{4}$$

This is no more, and no less, than the Landauer conductance of a single, one-dimensional, ideal metallic channel.

What is the most important lesson that one can draw from this microscopic result? It is that Occam's Razor can always be used to advantage. Of two contending explanations of a phenomenon, choose the one with fewer hypotheses.

The microscopic interpretation of the Landauer formula *does not need any* of the adventitious assumptions otherwise invoked to explain conductance quantization [2–5]. The tight derivation of Eq. (4) follows from completely standard quantum kinetics for open systems, within which total phase coherence is not even a physical

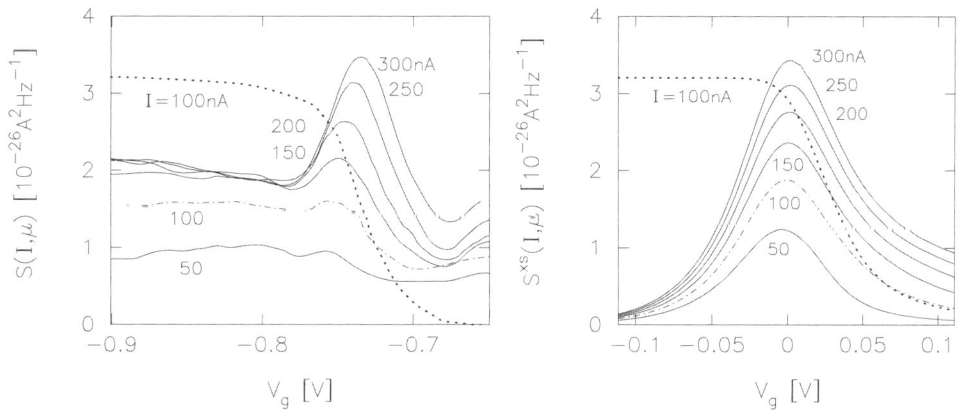


Fig. 3. Nonequilibrium low-frequency current noise of a QPC at constant source-drain current, as a function of gate bias. Left: the data from Reznikov *et al.*, Ref. 12, display a dramatic series of peaks coincident with the energy threshold of the lowest subband, where carriers first populate the states for conduction. Right: calculation from Green *et al.*, Ref. 13. In each case the dotted line at 100nA shows the shot-noise prediction of the Landauer–Büttiker (LB) formula [5], using as its required phenomenological inputs the measured and calculated data for $G$. The LB formula falls well short, qualitatively and quantitatively; compare the dash-dotted lines also at $I = 100\text{nA}$.

possibility let alone a theoretical necessity. Nor are the remaining, singular assumptions of the Landauer–Büttiker approach relevant to actual ballistic transport, any more than coherence is. See Ref. 9 for further discussion.

In Fig. 2 we plot the results of our model for a QPC made up of two one-dimensional conduction bands at energies $5k_BT$ and $17k_BT$, in thermal units at temperature $T$. We use the natural extension of Eq. (4) to cases where one or more channels may be open to conduction [9], depending on $T$ and the size of the chemical potential $\mu$. As the role of inelastic scattering is enhanced ($\tau_{\rm in} < \tau_{\rm el}$) the conductance deviates from the ideally ballistic Landauer limit.

At the core of the quantum-kinetic Landauer formula is the clear and pivotal influence of inelastic energy loss. It is one of the underpinnings of quantum transport. Charge conservation, the complementary underpinning, is guaranteed by the use of microscopically consistent open-boundary conditions at the interfaces [6, 7].

The twin requirements of open-system dissipation and open-system gauge invariance are neither adequately considered nor, as it turns out, respected by any of the phenomenological derivations of Eq. (4) [2–5]. If this is so at the plain level of conduction, such approaches will be far more problematic for current fluctuations, a decisive test of microscopic transport theories. We discuss that very point now.

### 3.2. *Nonequilibrium Noise*

The noise response of a quantum point contact is a fascinating aspect of mesoscopic transport. It is a much more demanding one both experimentally and theoretically.

In 1995, a landmark measurement of nonequilibrium noise was performed by the Weizmann group [12], which yielded a tantalizing result. While conventional models [5] predict no fine structure *whatsoever* in the noise of a QPC driven at constant current levels, the measured data (Fig. 3, left-hand panel) show a prominent and robust set of peaks at threshold, just where the carrier density in the QPC starts to enter the degenerate metallic regime.

We have accounted for the Reznikov *et al.* measurements within a strictly conserving, kinetic formulation of nonequilibrium noise in a quantum point contact [13]. Figure 3 displays, side by side with the experimental data, our computation of excess QPC noise under the same operating conditions.

In contrast to the outcome of popular mesoscopic phenomenology [5] one notes the close affinity between the measurements and the quantum kinetic calculation, as the carrier density is swept across the first conduction-band threshold, where the conductance exhibits its lowest step. At fixed values of source-drain current, the accepted noise account [5] predicts no peaks at all, but rather a featureless monotonic drop in the noise strength as the carrier density passes through threshold.

Remarkable as they are to this day, the Weizmann measurements have remained absolutely unexplained for a decade. Moreover their obvious message, namely that established theories are inadequate to the experiment, was simply ignored by the mesoscopic community. To the contrary, the folklore seemed to spread that Ref. 12 amply confirmed accepted understandings [5].

## 4. The Power of the Conservation Laws

The key to all quantum kinetic descriptions of conductance is the fluctuation-dissipation theorem [16], whose practical implementation is Eq. (3) (where the overall relaxation time $\tau_{\mathrm{tot}}$ encodes all the electron-fluctuation dynamics via the Kubo formula [15]). This universal relation is one of the electron-gas *sum rules* [17]. In this instance, it expresses the conservation of energy, dissipatively transferred from an external source to the thermal surroundings, for any process that involves resistive transport including, notably, that in a ballistic quantum point contact.

A second equally fundamental sum rule concerns the compressibility of an electron fluid in a conductive channel. This sum rule turns out to have an immediate link with the nonequilibrium noise behavior reviewed above. Here we give a brief explanation of that crucial link. For all of the formal details, see Refs. 13, 14, 18 and 19.

Recall that the carriers in a metallic quantum point contact are stabilized by the presence of the large leads, which pin the electron density to fixed values on the outer boundaries of the interfaces (recall too that the interfaces and the channel together define the open system). No matter what the transport processes within the QPC may be, or how extreme, the system's *global neutrality* is guaranteed by the stability of the large and charge-neutral reservoirs.[d]

It follows that the mean total number $N$ of active electrons in the device remains independent of any current that is forced through the channel, for $N$ is always neutralized by the rigid ionic background in its neighborhood as well as the stabilizing

[d] Independently of Sols' Theorem [6] and its consequences, the importance of global neutrality and metallic screening by the reservoirs was driven home long ago, in great microscopic detail, by Fenton [20]. His analysis shows that it is confinement of the electric field within the QPC (via perfect screening), and *not* perfect elastic transmission, that is a crucial prerequisite for conductance quantization.

leads. The presence of the latter means that any and all fringing fields are screened out beyond the device boundaries; hence the global neutrality.

It is readily seen that, if the mean occupation number $N(\mu)$ is independent of any external current as a result of global neutrality (i.e. the perfect-screening sum rule [17]), so is the total mean-square number fluctuation $\Delta N = k_B T \partial N/\partial \mu$ [18]. The carriers' compressibility in the QPC is specified in terms of $N$ and $\Delta N$ by [17]

$$\kappa = \kappa_{cl} \frac{\Delta N}{N} \quad \text{where} \quad \kappa_{cl} \equiv \frac{L}{N k_B T}, \tag{5}$$

which, in consequence, remains strictly unaffected by any transport process.

This a surprising corollary of global neutrality. It asserts that, in an open conductor, the system's equilibrium compressibility completely determines the compressibility of the electrons even away from equilibrium, *regardless* of the strength of the driving field.

The compressibility sum rule expresses the unconditional conservation of carriers in a nonequilibrium conductor. Previously unexamined in mesoscopics, this principle has an immediate importance and applicability. It means that a new, independent and strong criterion has become available to test the consistency of any model for ballistic transport.

How does $\kappa$ determine the noise in a QPC? The strength of the current fluctuations is, in broad physical terms, a convolution of two competing phenomena:

$$S(I,t) \sim \langle I(t)I(0)\rangle \frac{\Delta N}{N} = \langle I(t)I(0)\rangle \frac{\kappa}{\kappa_{cl}}. \tag{6}$$

The first factor represents the self-correlation of the dynamical electron current $I(t)$ evaluated as a trace over the nonequilibrium distribution of excited electrons throughout the device region. The second factor is, evidently, a basic characteristic of the electron gas in the channel. It is independent of $I$, meaning that the invariant compressibility in Eq. (5) dictates the overall scale of the noise spectrum.

Let us analyze the noise curves of Fig. 3 in light of this microscopic result [13,14].

- At large negative bias $V_g$, the channel is depleted. The remnant carriers are classical, so $\Delta N/N \to 1$. The noise is then dominated by strong inelastic processes at high driving fields, represented within the structure $\langle I(t)I(0)\rangle$.

- In the opposite bias limit (right-hand sector of each panel in Fig. 3), the channel is richly populated and thus highly degenerate, with a large Fermi energy $E_F$. Then $\Delta N/N \to k_B T/2E_F \ll 1$. The noise spectrum falls off according to Eq. (6), since the current-correlation factor, now comfortably within the regime of ballistic operation, reaches a set ideal value.

- In the mid-range of $V_g$ there is a point where the carriers' chemical potential matches the energy threshold for populating the first conduction sub-band. Here there is robust competition. As inelastic scattering becomes less effective, the correlation $\langle I(t)I(0)\rangle$ grows while the onset of degeneracy starts to cut down the other factor, the compressibility. Where this interplay is strongest, peaks appear.

The outworking of the compressibility rule is clear: it is, quite directly, the "inexplicable" emergence of the noise peak structures. The striking case of QPC noise gives an insight into the central importance of the conserving sum rules in the physics of metallic transport at meso- and nanoscopic dimensions.

As detailed in the Appendix, the more phenomenological treatments of noise fail to address the explicit action of microscopic conservation in ballistic phenomena, to the point that control over the sum-rule violations does not exist for them. Therefore they cannot offer a rational understanding of the real nature of ballistic conduction, much less its noise.

## 5. Summary

The kinetic analysis of transport provides a detailed, fully microscopic account of conductance and noise. This applies to the specific case of quantum point contacts. We accurately reproduce the entire current response of mesoscopic conductors, in particular conductance quantization. The keys to this newly fruitful picture are *open-system charge conservation* and the physical reality of *dissipative scattering*.

Our unified theory yields a comprehensive understanding of the nonequilibrium fluctuations fundamental to a QPC. We have successfully tested this comprehension by fully explaining the long-standing puzzle posed by the noise measurements of Reznikov *et al.* [12]. The theoretical impact of noise and fluctuation physics is that it carries much more information on the internal dynamics of mesoscopic systems: knowledge that is not accessible through the $I$-$V$ characteristics alone.

The capacity to advance a self-contained microscopic explanation of processes in mesoscopic transport underwrites a matching ability to build up a program for device design that is inherently rational. This goes deeper than assembling, *ad hoc*, any clutch of ideas that happen to fall easily to hand. The reward for more workmanlike efforts is a huge potential to improve the practical engineering of novel generations of devices. These can, and evidently should, be built on a sound and non-speculative knowledge base. Many-body physics is at the heart of the program.

A well grounded mesoscopic theory, at its simplest, ought not to be simplistic. To go so far as to forget or even ignore the conservation laws, just to fit subjective notions of "simplicity", would not be science (though it might pass for good post-modernism). At any rate, it would not be a science for the serious investigation of co-ordinated behavior in matter at small scales. It is the universal principles of conservation that are the gatekeepers to a real mesoscopic understanding.

We have revisited the conserving sum rules in their specific application to small-scale metallic transport. Given its prime role in carrier dynamics, a truly conservative quantum kinetic approach holds opportunities for both traditional many-particle theory applied to mesoscopics, and for expanding noise research. More than ever, the spotlight falls on the *fluctuation* properties of active mesoscopic devices. Noise physics is vital not only for its own sake, but as a robust and trustworthy diagnostic means for electronic concepts to come.

Now is a good time indeed to introduce genuine many-body methods into mesoscopics: a field in no small need of a microscopically cohesive and more far-reaching vision. Some of the tools to make a start on this are freely available [6–11, 13–20].

**Acknowledgments**

Many friends have encouraged us consistently throughout our fluctuation work. First we thank Jagdish Thakur, our collaborator on the QPC noise and compressibility problems. F.G. also recalls the strong and unstinting support of Neil Barrett and Grant Griffiths, whose foresight first empowered him to pursue this line of research.

Our review would be incomplete without recognizing the pioneering contributions of Ed Fenton. Finally we acknowledge the fruitful exchanges we have had with Fernando Sols, Wim Magnus, Wim Schoenmaker and Bart Sorée.

## Appendix A. Sum Rules in the Landauer–Büttiker Theory

In this Appendix we provide, for the record, the demonstration that the Landauer–Büttiker approach violates two crucial sum rules: perfect screening (charge conservation) and compressibility (particle conservation) [17–19].

### A.1. *Perfect Screening Sum Rule*

As we saw in Sec. 4, the overall neutrality of a quantum point contact (indeed, of any bounded conductor) is maintained by the strong screening property of its large conductive leads. Suppose the QPC sustains a carrier density profile $n(x)$. When $n$ is integrated for the total carrier number, the contact's global neutrality implies

$$\int_0^L n(x)dx \equiv N = \text{constant; that is, } \frac{\partial N}{\partial I} = 0 \text{ for all } I. \tag{A.1}$$

Regardless of external current $I$ forced through the structure, and the voltage $V$ measured across it in response, the total of mobile charges dwelling in the QPC (including the active interface regions themselves) cannot depend on $I$ or $V$. The entire structure, though open, stays unconditionally neutral. This is the practical meaning of perfect screening.

Now consider the same situation from the perspective of Landauer and Büttiker [2–5]. In place of the gauge-invariant (and thus microscopically correct) prescription of open, externally driven transport that leads to perfect screening [6,7,19], we posit that the current in a narrow conducting wire is sustained purely by a difference of chemical potentials[e] between an upstream electron reservoir (chemical potential $\mu = \mu_{\rm up}$) and a different downstream one ($\mu = \mu_{\rm dn} = \mu_{\rm up} - eV$).

The density along the conductor would then be some function $n(\mu_{\rm up} - eV(x))$ that took the boundary values $n(\mu_{\rm up})$ and $n(\mu_{\rm dn})$ at the ends of the sample. According to such an account, in the linear-response limit the total number of carriers

[e]One of the central phenomenological assumptions of LB is that there must be a difference of chemical potentials between the leads of a driven device. Otherwise, in LB, the current cannot flow "diffusively" [4]. This phenomenology has absolutely no basis in the quantum kinetic description of mesoscopic transport. For, it would mean that $\mu$, a thermodynamic property which is *always locally invariant* and guarantees the individual stability of each lead's neutral state, is subject to arbitrary change. In reality it is the external driving source, and not any "difference" in $\mu$, that directly sustains the current. Thus in kinetic theory the leads' chemical potentials are left alone to do their proper job: to stabilize the system and maintain its overall neutrality.

in the active structure of length $L$ would *change*, with the field, by

$$N(V) - N(0) = \int_0^L [n(\mu_{\rm up} - eV(x)) - n(\mu_{\rm up})]dx$$
$$\to \int_0^L \frac{dn}{d\mu_{\rm up}}(-eV(x))dx \approx -\Delta N(0)\frac{eV}{2k_{\rm B}T} \neq 0, \qquad \text{(A.2)}$$

a result manifestly counter to the perfect-screening sum rule, Eq. (A.1).

### A.2. *Compressibility Sum Rule*

We remind ourselves that the corollary to perfect screening is the nonequilibrium compressibility sum rule (recall Eq. (5)):

$$\frac{\kappa}{\kappa_{\rm cl}} \equiv \frac{k_{\rm B}T}{N}\frac{\partial N}{\partial \mu} = \text{const.; that is, } \frac{\partial \kappa}{\partial I} = 0 \;\; \text{for all } I. \qquad \text{(A.3)}$$

Note in particular that $\kappa$ is calculable as a strictly *equilibrium*, and not a linear-response, quantity [17]. It cannot depend on any transport parameter.

How does compressibility relate to the so-called Landauer–Büttiker noise formula? The formula emerges from quite a variety of different arguments, which nevertheless all converge to the same final expression (specializing to one subband will not alter our own argument):

$$\mathcal{S}(V) = 4G_0k_{\rm B}T\left[\mathcal{T}^2(v_{\rm F}) + \mathcal{T}(v_{\rm F})\Big(1-\mathcal{T}(v_{\rm F})\Big)\frac{\mu_{\rm up}-\mu_{\rm dn}}{2k_{\rm B}T}\coth\left(\frac{\mu_{\rm up}-\mu_{\rm dn}}{2k_{\rm B}T}\right)\right]. \qquad \text{(A.4)}$$

The coefficient $\mathcal{T}(v_{\rm F})$ for coherent transmission is the probability that an incoming carrier at Fermi velocity $v_{\rm F}$ will propagate coherently from source to drain, right through the subband in the QPC.

The noise formula is the sum of separate contributions [5]. With these same components, one may also reconstruct the LB version of the mean-square number fluctuation. After constructing the would-be "single-carrier" fluctuation,[f] say d$N$ (refer to Martin and Landauer [21] for a clear account of this methodology), we obtain the LB mean-square expectation $\Delta N$ over the quantum point contact[g]:

$$\Delta N \equiv \langle (\mathrm{d}N)^2 - \langle \mathrm{d}N\rangle^2\rangle$$
$$= L\frac{n(\mu_{\rm up})}{2E_{\rm F}}\left[\mathcal{T}^2k_{\rm B}T + \mathcal{T}(1-\mathcal{T})\frac{\mu_{\rm up}-\mu_{\rm dn}}{2}\coth\left(\frac{\mu_{\rm up}-\mu_{\rm dn}}{2k_{\rm B}T}\right)\right], \qquad \text{(A.5)}$$

[f]In a charged Fermi gas the fluctuation that should be averaged for $\Delta N$ is an *electron-hole pair* which is correlated, internally and irreducibly, via conservation of energy, momentum and charge [17]. It *cannot* be expressed as a product of two stochastically independent single-particle terms.
[g]Since in LB the active states are limited to lie within a (thin) shell $k_{\rm B}T$ centered on the Fermi surface, the velocity of all carriers contributing to $S(V)$ always has magnitude $\langle|v|\rangle = v_{\rm F} \gg \langle v\rangle$. Therefore the mean-square LB fluctuation in one dimension simply scales as

$$\Delta N \propto \frac{\langle I^2\rangle - \langle I\rangle^2}{e^2(\langle v^2\rangle - \langle v\rangle^2)} \propto \frac{S(V)}{e^2v_{\rm F}^2}.$$

where $n(\mu_{\rm up})$ is the density in the uniform wire at equilibrium. Note that LB models work only in the degenerate limit of large Fermi energy: $E_{\rm F} \gg k_{\rm B}T$.

Now let the driving field go to zero; the term that depends on $eV = \mu_{\rm up} - \mu_{\rm dn}$ goes to $k_{\rm B}T$. Using $Ln(\mu) = N$, we are left with the ratio

$$\frac{\Delta N}{N} = \mathcal{T}\left[\frac{\Delta N}{N}\right]_{\rm eq}, \text{ where } \left[\frac{\Delta N}{N}\right]_{\rm eq} \equiv \frac{k_{\rm B}T}{2E_{\rm F}} = \frac{\kappa}{\kappa_{\rm cl}}. \tag{A.6}$$

Equation (A.6) very clearly fails to recover the physical compressibility, Eq. (A.3), *even at equilibrium.* It depends manifestly on the transport parameter $\mathcal{T}$.

A moment's reflection shows that Eq. (A.5), the LB form for $\Delta N$ linked immediately to the noise formula, depends explicitly on the voltage. Therefore it visits further violence upon the *invariance* of $\kappa$ under transport.

It follows that the Landauer–Büttiker noise formula is inconsistent with the compressibility sum rule not only at equilibrium, which is fatal, but in every transport situation.

## References

[1] N. Agraït, A. Levy Yeyati and J. M. van Ruitenbeek, *Quantum properties of atomic-sized conductors, Phys. Rep.* **377** (2003) 81–279.

[2] Y. Imry and R. Landauer, *Conductance viewed as transmission, Rev. Mod. Phys.* **71** (1999) S306–312.

[3] Y. Imry, *Introduction to Mesoscopic Physics*, 2nd ed, Oxford University Press, Oxford (2002).

[4] S. Datta, *Electronic Transport in Mesoscopic Systems*, Cambridge University Press, Cambridge, UK (1995).

[5] Ya. M. Blanter and M. Büttiker, *Shot Noise in Mesoscopic Conductors, Phys. Rep.* **336** (2000) 1–166.

[6] F. Sols, *Gauge-Invariant Formulation of Electron Linear Transport, Phys. Rev. Lett.* **67** (1991) 2874–2877.

[7] W. Magnus and W. Schoenmaker, *Quantum Transport in Sub-micron Devices: A Theoretical Introduction*, Springer, Berlin (2002).

[8] B. Sorée, W. Magnus and W. Schoenmaker, *Conductance quantization and dissipation, Phys. Lett. A* **310** (2003) 322–328.

[9] M. P. Das and F. Green, *Landauer formula without Landauer's assumptions, J. Phys.: Condens. Matter* **15** (2003) L687–693.

[10] F. Green and M. P. Das, *High-field noise in metallic diffusive conductors, J. Phys.: Condens. Matter* **12** (2000) 5233–5250.

[11] F. Green and M. P. Das, *Coulomb screening in mesoscopic noise: a kinetic approach, J. Phys.: Condens. Matter* **12** (2000) 5251–5273.

[12] M. Reznikov, M. Heiblum, Hadas Shtrikman and D. Mahalu, *Temporal Correlation of Electrons: Suppression of Shot Noise in a Ballistic Quantum Point Contact, Phys. Rev. Lett.* **75** (1995) 3340–3343.

[13] F. Green, J. S. Thakur and M. P. Das, *Where is the Shot Noise of a Quantum Point Contact?, Phys. Rev. Lett.* **92** (2004) 156804–1–4.

[14] F. Green and M. P. Das, *Classical to Quantum Crossover in High-Field Noise of One-Dimensional Ballistic Wires, Fluctuation and Noise Lett.* **1** (2001) C21–33.

[15] R. Kubo, M. Toda and M. Hashitsume, *Statistical Physics II: Nonequilibrium Statistical Mechanics*, 2nd ed, Springer, Berlin (1991).

[16] N. G. van Kampen, *Stochastic Processes in Physics and Chemistry*, North-Holland, Amsterdam (1981).

[17] D. Pines and P. Nozières, *The Theory of Quantum Liquids*, Benjamin, New York (1966).

[18] J. S. Thakur, F. Green and M. P. Das, *Sum-Rule Constraints for Open Mesoscopic Conductors, Int. J. Mod. Phys. B* **18** (2004) 1479–1488.

[19] F. Green and M. P. Das, *Transport and Noise in Mesoscopic Conductors: a Conservative View*, in *Noise and Fluctuations Control in Electronic Devices*, A. A. Balandin ed, American Scientific Publishers, Stevenson Ranch (2002), pp. 31–48.

[20] E. W. Fenton, *Electric-field conditions for Landauer and Boltzmann-Drude conductance equations, Phys. Rev. B* **46** (1992) 3754–3770; E. W. Fenton, *Electrical and chemical potentials in a quantum-mechanical conductor, Superlattices and Microstructures* **16** (1994) 87–91.

[21] Th. Martin and R. Landauer, *Wavepacket approach to noise in multichannel mesoscopic systems, Phys. Rev. B* **45** (1992) 1742–1755.

# Zero Thermal Noise in Resistors at Zero Temperature

Laszlo B. Kish*,‡, Gunnar A. Niklasson† and Claes-Göran Granqvist†

*Department of Electrical Engineering, Texas A&M University
College Station, TX 77843-3128, USA

†Department of Engineering Sciences
The Ångström Laboratory, Uppsala University
P. O. Box 534, SE-75121 Uppsala, Sweden
‡laszlo.kish@ece.tamu.edu

Received 5 February 2016
Accepted 7 March 2016
Published 13 June 2016

Communicated by Xavier Oriols Pladevall

The bandwidth of transistors in logic devices approaches the quantum limit, where Johnson noise and associated error rates are supposed to be strongly enhanced. However, the related theory — asserting a temperature-independent quantum zero-point (ZP) contribution to Johnson noise, which dominates the quantum regime — is controversial and resolution of the controversy is essential to determine the real error rate and fundamental energy dissipation limits of logic gates in the quantum limit. The Callen–Welton formula (fluctuation–dissipation theorem) of voltage and current noise for a resistance is the sum of Nyquist's classical Johnson noise equation and a quantum ZP term with a power density spectrum proportional to frequency and independent of temperature. The classical Johnson–Nyquist formula vanishes at the approach of zero temperature, but the quantum ZP term still predicts non-zero noise voltage and current. Here, we show that this noise cannot be reconciled with the Fermi–Dirac distribution, which defines the thermodynamics of electrons according to quantum-statistical physics. Consequently, Johnson noise must be nil at zero temperature, and non-zero noise found for certain experimental arrangements may be a measurement artifact, such as the one mentioned in Kleen's uncertainty relation argument.

*Keywords*: Fluctuation–dissipation theorem; Johnson–Nyquist noise; zero-point noise; thermal noise.

## 1. Introduction: Johnson Noise and the Second Law of Thermodynamics

Recently, we have shown [1] that, in the quantum regime $hf > kT$, both the classical Johnson–Nyquist formula [2, 3] and the quantum fluctuation–dissipation theorem

‡Corresponding author.

[4, 5] fail to give correct results when thermal noise of resistors is measured by the force or energy in a shunt capacitor. Here $h$ is Planck's constant, $f$ is frequency, $k$ is Boltzmann's constant and $T$ is temperature. The purpose of the present paper is to prove that the well-known Johnson (–Nyquist) noise [2, 3], which is ubiquitous in resistors and impedances, does not exist at temperatures approaching zero Kelvin ($T \to 0$), which is contrary to the prediction of the fluctuation–dissipation theorem [4, 5]. This answers a long-standing problem, and there is a considerable prior literature related to Johnson noise at $T \to 0$; see, for example, Refs. [6–11].

Johnson noise ensues from spontaneous voltage and current fluctuations due to stochastic motion of charge carriers (electrons) in conducting materials under thermal equilibrium. The second law of thermodynamics requires that the time-average of the instantaneous power flow between two parallel resistors is zero, i.e.,

$$\langle P_{a\Leftrightarrow b}(t,T)\rangle_t = 0, \tag{1}$$

where $t$ is time and $P_{a\Leftrightarrow b}(t,T)$ denotes instantaneous power flow between resistors $R_a$ and $R_b$, as illustrated in Fig. 1. This expression holds for any frequency range.

For a passive impedance $Z(f)$ in thermal equilibrium, Eq. (1) demands that the functional form of the power density spectral functions for noise voltage $u$ and current $i$ is

$$S_u(f,T) = \mathrm{Re}[Z(f)]Q(f,T) = R(f)Q(f,T), \tag{2}$$

$$S_i(f,T) = \mathrm{Re}[Y(f)]Q(f,T) = G(f)Q(f,T), \tag{3}$$

where $Y(f) \equiv 1/Z(f)$ denotes admittance, $R(f)$ and $G(f)$ are the real part of the impedance and admittance, respectively, and $S_u(f)$ and $S_i(f)$ represent power density spectra of Johnson noise voltage and current generators, respectively. Here, we introduced $S_i(f)$ as the Norton-equivalent of $S_u(f)$, expressed by $S_i(f) = S_u(f)/|Z(f)|^2$. $Q(f,T)$ is a universal function of frequency and temperature and is independent of material properties, geometry and mechanism for electrical conduction [2–4].

If we can prove that — in a given system at $T \to 0$ — the function $Q(f,0)$ is zero, then this finding serves as proof that $Q(f,0) = 0$ in *any* system, i.e., that Johnson noise is absent at zero temperature. This proof is given below. We simplify the

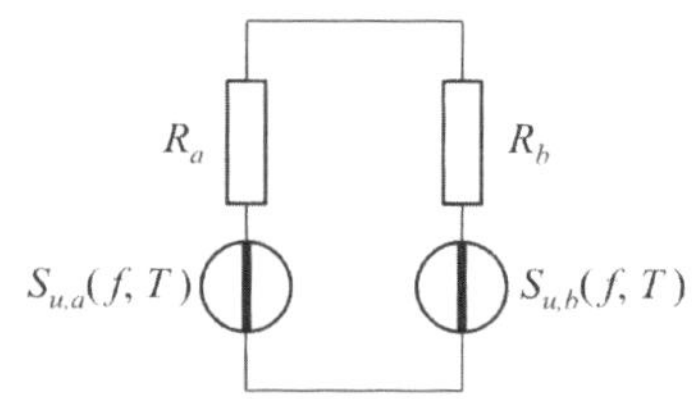


Fig. 1. Circuit for discussing Johnson noise and representing net power flow between resistors $R_a$ and $R_b$. The voltage generators $S_u(f,T)$ produce power spectral functions at frequency $f$ and temperature $T$.

notation in the rest of the paper by considering frequency-independent resistance, but this does not limit the generality of our treatment because all equations and conclusions remain valid if $R$ is substituted by $\mathrm{Re}[Z(f)]$.

## 2. Callen–Welton's Result for Johnson Noise

We now consider the Callen–Welton (C–W) derivation (fluctuation–dissipation theorem) [5] of Johnson voltage noise, which expresses this noise as a sum of the classical Johnson–Nyquist term and a zero-point (ZP) term. C–W's expression can be written

$$S_{u,q}(f,T) = 4Rhf[N(f,T)+0.5] \xrightarrow[f \ll kT/h]{} 4kTR + 2hfR, \tag{4}$$

where $S_{u,q}(f,T)$ is the one-sided power density spectrum for the voltage noise of the resistor. A "Planck number" $N(f,T)$, which was introduced already in Nyquist's work [3], represents the mean number of $hf$ energy quanta in a linear harmonic oscillator and is given by

$$N(f,T) = [\exp(hf/kT) - 1]^{-1}. \tag{5}$$

For the classical physical range, with $f \ll kT/h$, one finds $N(f,T) \approx kT/(hf)$ which leads to the well-known $4kTR$ noise spectrum at low frequencies. Thus, the first term in Eq. (4) is the classical physical (Nyquist) result [3] and the second term is its quantum correction expressed by

$$S_{u,ZP}(f,0) = 2hfR. \tag{6}$$

The classical term disappears at zero temperature, because $N(f,0) = 0$, while the quantum term, known as the ZP noise spectrum, prevails even at $T \to 0$ and is claimed to be a consequence of ZP energy [4, 5].

The Johnson current noise of the resistance follows from the theory of linear operations on the noise and Ohm's law according to

$$S_{i,q}(f,T) = 4Ghf[N(f,T)+0.5] \xrightarrow[f \ll kT/h]{} 4kTG + 2hfG. \tag{7}$$

Similar to the case of voltage noise, the first term in Eq. (7) is the classical physical (Nyquist) result [3] and the second term is its quantum correction

$$S_{i,ZP}(f,0) = 2hfG, \tag{8}$$

which again is asserted to exist even at $T \to 0$ [4, 5].

## 3. Proof that the Quantum ZP Term Does Not Exist in Johnson Noise

We first recall that in order to prove that the ZP term does not exist under general conditions, it is enough to demonstrate that this is so for a special case, and we therefore consider the simplest electron conductor system [12], a metal with non-zero residual resistance at zero temperature and resistance due to randomly located defects. Johnson noise current is studied in a short-circuited resistor positioned along the $x$-axis by measuring the current in the loop. Alternatively, and without changing the essence of our argumentation, a ring-shaped sample could also be used.

The current $I$ is the sum of elementary currents $-e\mathbf{v}(k)$ of occupied single-electron states in $k$-space, i.e.,

$$I = \Sigma_{\{\mathbf{k}\}}(-e\mathbf{v}(\mathbf{k})) \tag{9}$$

where $\{\mathbf{k}\}$ denotes the set of wave-vectors of the occupied states and $\mathbf{v}$ is the velocity of an electron in state $k$. At temperature $T$, the probability of occupation *versus* energy $E$ of the state is given by the Fermi–Dirac distribution $P(E,T)$ according to [12, 13]

$$P(E,T) = \frac{1}{1 + \exp\left(\frac{E-E_F}{kT}\right)}, \tag{10}$$

where $E_{\mathrm{F}}$ denotes Fermi level. At zero temperature and in the absence of an electrical field, the occupation probability satisfies

$$P(E,T=0) = \begin{cases} 1 & \text{for } E \leq E_{\mathrm{F}}, \\ 0 & \text{for } E > E_{\mathrm{F}}, \end{cases} \tag{11}$$

so that all states are occupied below $E_{\mathrm{F}}$ and no states are filled above $E_{\mathrm{F}}$. In $k$-space, this fact is visualized by the Fermi surface, which is the surface given by the set of $k$-vectors corresponding to $E_{\mathrm{F}}$. The symmetry properties of the Fermi surface and Eq. (9) yield that the net current in the material is zero, because electron velocities in states at $k$ and $-k$ are opposite and hence cancel. It is seen in Fig. 2 that for each positive occupied $k$-value in any direction, there is an occupied negative $k$-value with the same absolute magnitude [12]. When an external DC voltage is applied and at zero temperature, the occupied states shift in the positive direction along the $k_x$-axis [12], as also shown in Fig. 2. This means that there will

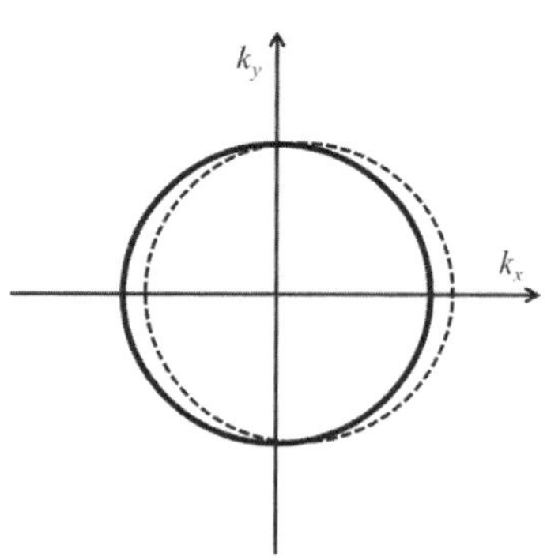


Fig. 2. Fermi surface in two dimensions at zero temperature. Solid and dashed circles represent conditions without and with an external DC current drive, respectively.

be unoccupied states at lower magnitudes of $k$ and lower energies than those of the highest occupied states. Hence, an electron may be scattered from an occupied to an unoccupied state either by collisions with defects or by dissipating energy supplied by the external field, i.e., by emission of a phonon. This argument shows that the residual resistance at $T = 0$ is indeed non-zero, as was assumed above.

However, the situation is different in the absence of an applied voltage. The Fermi sphere is then symmetrical around the origin and no transitions between occupied and unoccupied states are possible in the $T \to 0$ limit. The presence of noise would imply spontaneous violations of the symmetry of the Fermi surface in random directions at different instants of time.

It is obvious from this picture that, in the limit $T \to 0$ and for thermal equilibrium, a non-zero Johnson noise current through the short-circuited resistor would violate the zero-temperature Fermi–Dirac distribution in Eq. (11) and alter the solid curve in Fig. 2. Therefore, the material must be totally "silent" at zero temperature, and no Johnson noise can then occur. Consequently the ZP term of the Johnson noise, expounded by the C–W formula (*cf.* Eqs. (4)–(8)), must vanish at $T \to 0$.

For completeness we now contemplate how the classical term of Johnson noise is generated (Eqs. (5)–(7)). This term is also zero at $T = 0$ according to Eq. (5), but at non-zero temperature and under thermal equilibrium the occupation probability is [12]

$$P(E, T > 0) = \begin{cases} 1 - \varepsilon(E,T) & \text{for } E \leq E_F, \\ \varepsilon(E,T) & \text{for } E > E_F, \end{cases} \tag{12}$$

where the probability $\varepsilon(E,T) > 0$ characterizes the electron states that take part in thermal motion (i.e., the Nyquist noise term), electronic specific heat, etc. [12]. These electrons are the carriers of the classical Johnson noise.

## 4. Comments on Experimental Efforts to Measure ZP Johnson Noise

The ultimate test of ZP Johnson noise is whether it can be documented experimentally. Some relevant work is briefly discussed below, and we argue that experiments have failed to convincingly record ZP noise.

As mentioned above, we have very recently shown by a Gedanken experiment [1] that neither the Johnson–Nyquist formula (the first term of Eq. (4)) nor the C–W formula (Eq. (4)) is valid when measuring Johnson noise by energy/force in a shunting capacitor. We compared this fact with considerations about experimental systems and concluded that, in the quantum limit, it is impossible to propose a Johnson noise formula that identifies a single, measurement-system-independent (external-circuit-independent) noise source in the resistor to account for the measured noise and its effects [1]. We surmise that this fact is in accordance with the principles of quantum physics, *viz.*, that the measurement device interferes with the recorded effect.

Concerning experimental efforts to observe the ZP term of Johnson noise in resistors, we first note that as early as in the beginning of the 1980s, van der Ziel *et al.* [6, 7] were unable to detect ZP Johnson noise through careful direct microwave-based measurements conducted at 1 K and 95 GHz. Almost at the same time, Voss and Webb [8] evaluated shot-noise of Josephson junctions and found that their results could be accurately fitted to a model based on thermal activation over an energy threshold, whereas the experimental data lay 4–10 orders-of-magnitude below the level expected for the zero-point term of Johnson noise. Unfortunately, Voss and Webb neither mentioned nor explored the absence of the latter type of effect.

A famous work by Koch, van Harlingen and Clarke (KvHC) [9] used frequency-selective heterodyne detection applied to Josephson junctions and deserves special attention since it purports to confirm the existence of ZP Johnson noise. However, alternative explanations of KvHC's results can be found in early work by Heffner [10], which formed the basis for a subsequent analysis by Kleen [11]. We consider Heffner's derivation [10] of the noise temperature $T_{ZP}$ for frequency/phase-selective linear amplifiers at $T \to 0$ by use of the uncertainty principle, which yields

$$T_{ZP} = \frac{hf/k}{\ln\left(\frac{2-1/A}{1-1/A}\right)}, \tag{13}$$

where $A$ is the amplification. With a source resistance $R$, the equivalent input noise of the amplifier at zero temperature is

$$S_u = 4kT_{ZP}R = 4R\frac{hf}{\ln\left(\frac{2-1/A}{1-1/A}\right)} = \gamma(A)4Rhf. \tag{14}$$

Clearly $\gamma(A) = 0.5$ leads to the ZP Johnson noise artifact of Eq. (6) above. We evaluated Heffner's formula and found that $A = 1.19$ would give this value of $\gamma$, and that a wide range of amplifications produce similar results for $\gamma$ (specifically $A$ being 1.1, 2, and 10 gave $\gamma(A)$ values of 0.40, 0.91, and 1.34, respectively). Kleen [11] also used Heffner's analysis and argued that the time–energy uncertainty principle could account for KvHC's results. One should note that the large experimental inaccuracies in KvHC's result [9] cannot exclude any of the values of $A$ and $\gamma(A)$ stated above.

## 5. Conclusions

We found that the ZP Johnson noise term in the fluctuation–dissipation theorem, which follows from C–W's well-known analysis [5], is in contradiction with our analysis of a conductor with finite residual resistance in the limit $T \to 0$. Using, correctly, Fermi–Dirac statistics for these charge carriers implies that Johnson noise is absent at zero temperature, and thus the ZP noise term must be non-existent.

We hope these considerations will stimulate further discussions and hopefully experimental work in the field.

## Acknowledgments

We gratefully acknowledge useful discussions with Fred Green, Kyle Sundqvist, Lino Reggiani and Peter Rentzepis.


## References


[1] L. B. Kish, G. A. Niklasson and C. G. Granqvist, Zero point term and quantum effects in the Johnson noise of resistors: A critical appraisal, *J. Stat. Mech. Theory Exp.* arXiv:1504.08229.

[2] J. B. Johnson, Thermal agitation of electricity in conductors, *Nature* **119** (1927) 50–51.

[3] H. Nyquist, Thermal agitation of electric charge in conductors, *Phys. Rev.* **32** (1928) 100–113.

[4] V. L. Ginzburg and L. P. Pitaevskii, Quantum Nyquist formula and the applicability ranges of the Callen–Welton formula, *Usp. Fiz. Nauk* **151** (1987) 333–339; *Sov. Phys. Usp.* **30** (1987) 168–171.

[5] H. B. Callen and T. A. Welton, Irreversibility and generalized noise, *Phys. Rev.* **83** (1951) 34–40.

[6] C. M. Van Vliet, *Equilibrium and Non-Equilibrium Statistical Mechanics* (World Scientific, Singapore, 2008).

[7] A. van der Ziel, Quantum effects at high frequencies and low temperatures, *Proceedings of the 6th International Conference on Noise in Physical Systems*, eds. P. H. E. Meijer, R. D. Mountain and R. J. Soulen (National Bureau of Standards, Washington, DC, USA, 1981), pp. 364–368.

[8] R. F. Voss and R. A. Webb, Pair shot noise and zero-point Johnson noise in Josephson junctions, *Phys. Rev. B* **24** (1981) 7447–7449.

[9] H. Koch, D. J. van Harlingen and J. Clarke, Observation of zero-point fluctuations in a resistively shunted Josephson tunnel junction, *Phys. Rev. Lett.* **47** (1981) 1216—1219.

[10] H. Heffner, The fundamental noise limit of linear amplifiers, *Proc. IRE* **50** (1962) 1604–1608.

[11] W. Kleen, Thermal radiation, thermal noise and zero-point energy, *Solid-State Electron.* **30** (1987) 1303–1304.

[12] L. Solymar and D. Walsh, *Electrical Properties of Materials* (Oxford University Press, Oxford, UK, 2004).

[13] W. Geiner, *Quantum Mechanics: Special Chapters* (Springer, Berlin Heidelberg, Germany, 2008).

# 3

# Noise in Complex Systems

# OPTIMAL DIFFUSIVE TRANSPORT IN A TILTED PERIODIC POTENTIAL

BENJAMIN LINDNER, MARCIN KOSTUR and LUTZ SCHIMANSKY-GEIER
*Institute of Physics, Humboldt-University Berlin*
*Invalidenstr.110 D-10115 Berlin, Germany*

Received 25 January 2001
Revised 14 February 2001
Accepted 15 February 2001

We study the diffusive motion of an overdamped Brownian particle in a tilted periodic potential. Mapping the continuous dynamics onto a discrete cumulative process we find exact expressions for the diffusion coefficient and the Péclet number which characterize the transport. At a sufficiently strong but subcritical bias an optimized transport with respect to the noise strength is observed. These results are confirmed by numerical solution of the Fokker-Planck equation.

*Keywords*: Brownian motion; diffusion; periodic potentials; transport.

## 1. Introduction

There are many examples of stochastic dynamics which can be described by the Brownian motion in a tilted washboard potential ranging from mathematical pendulum with dissipation and noise [1] over superionic conduction [2] to neuronal activity [3, 4]. The qualitative asymptotic behavior of the system is well known. Expressed in terms of real Brownian motion, particles in a tilted washboard subjected to friction and noise will diffuse and drift in the direction of the bias. The determination of mean velocity and effective diffusion coefficient, however, was a challenging task for decades and can be performed in the general damped case only by simulations or by a numerical solution of the Fokker-Planck equation [1].

The *overdamped* limit of this model in turn can be treated analytically to a great extent. Already in 1958, Stratonovich [5] derived a closed expression for the stationary mean velocity and gave also an approximate expression for the effective diffusion coefficient. During the 1970's, an exact expression for the diffusion coefficient could be calculated for the special case of vanishing bias [6]. Another approximation for a *finite* bias has been proposed recently [7].

Stratonovich's approach to the diffusion coefficient holds true for a weak tilting of the potential and small noise intensity. In this case the particle rarely jumps from one potential minimum into the next one to the right or left, respectively, whereby one of the directions is preferred due to the bias. Stratonovich assumed in this regime that the process can be modeled by a biased random walk. Thus, the diffusion coefficient is determined by the rate from the minimum to the left and right potential barriers (maxima of the potential). This approach neglects the relaxation time from barrier to minimum that becomes relevant for stronger tilt. It fails completely for a so called supercritical tilt for which minima and maxima of the potential vanish since in this case the process cannot be described by a Poissonian hopping process anymore. In the case of a very strong tilt, however, the situation is simplified again since the potential shape can be ignored and the diffusion coefficient coincides with that of free diffusion.

The transport of particles is characterized by an average motion in direction of the bias and the counteracting spreading effect due to the presence of noise. While the drift may be desired the dispersion is an unwanted but - especially in case of a subcritical bias - inevitable side effect. Coherent transport refers to the case of large mean velocity at fairly small diffusion. It can be best quantified by the nondimensional Péclet number, i.e., a proper ratio of mean velocity to effective diffusion coefficient. Clearly, the transport is most coherent when this number is maximal.

In this paper we address a situation where the minima and maxima of the effective potential are still present (subcritical tilt) but the tilt is sufficiently strong that the relaxation time scale has to be taken into account. We will restrict ourselves to the overdamped case and derive exact expressions for the diffusion coefficient and the Péclet number. It will be shown that a large relaxation time scale can lead to a coherent (regular) transport of particles when the noise strength is optimally chosen. This is manifested by a maximum of the Péclet number with respect to the noise intensity. We will illustrate the meaning of this maximum in terms of the time dependent probability density of particles.

Furthermore, for specific potential shapes another quality of coherence appears. If a flat potential with small sharp maxima is sufficiently tilted the effective diffusion coefficient itself attains a local minimum as a function of the noise intensity. The surprising effect that increasing noise decreases diffusion is explained by the occurrence of a large ratio of relaxation to escape time for a certain range of noise intensities.

Regarding the calculation of the diffusion coefficient for finite bias, independent work by Reimann et. al. [8] has been done although the approach as well as the physical phenomenon considered in that study differ much from ours.

The paper is organized as follows. In section 2 we introduce the model and the quantities of interest. In section 3 we outline the mapping onto a discrete cumulative process that permits the determination of asymptotic mean and variance of the original process. In sections 4 and 5 we turn to specific potential shapes and the phenomenon of coherent transport. Finally in section 6, we give a summary and discuss the close connection of our findings to *coherence resonance* encountered in excitable systems.

## 2. Model and quantities of interest

Consider the overdamped motion of a Brownian particle in a potential $U(x)$ with period $L$ subject to a tilting force $F$ and a white Gaussian noise

$$\dot{x} = -\frac{d}{dx}(U(x) - Fx) + \sqrt{2D}\xi(t). \tag{1}$$

Here, the friction coefficient $\gamma$ has been omitted realized by rescaling the force term and noise intensity $D$. The force in (1) is often written as the derivative of the *effective* potential $V(x) = U(x) - Fx$. If $U(x)$ is periodic there exists obviously a critical tilt $F_c$ where the corresponding minima and maxima vanish for $V(x)$.
The corresponding probability density for eq. (1) obeys the Fokker-Planck equation

$$\partial_t P(x,t) = \partial_x \left(V'(x) + D\partial_x\right) P(x,t). \tag{2}$$

An ensemble represented by $P(x,t)$ started at the sharp value $x_0 = 0$ moves in direction of the bias and spreads at the same time over several periods of the potential (cf. Fig. 1). For large times, the drift is characterized by the mean stationary velocity

$$v = \langle \dot{x} \rangle = \lim_{t\to\infty} \frac{\langle x(t) - x(0) \rangle}{t} \tag{3}$$

while the dispersion is quantified by the effective diffusion coefficient

$$D_{eff} = \frac{1}{2}\lim_{t\to\infty} \frac{\langle \Delta x(t)^2 \rangle}{t} = \frac{1}{2}\lim_{t\to\infty} \frac{\langle [x(t) - \langle x(t) \rangle]^2 \rangle}{t}. \tag{4}$$

The density itself tends asymptotically to

$$P_{as}(x,t) = P_0(x)\frac{\exp[-(x-vt)^2/4D_{eff}t]}{\sqrt{4\pi D_{eff}t}} \tag{5}$$

where $P_0$ refers to the stationary (necessarily periodic) solution of eq. (2) [1] with the special normalization $\int_0^L dx P_0(x) = L$. Clearly, this function is responsible for the local structure (local minima and maxima) whereas the Gaussian part represents the coarse grained density (thick line in Fig. 1). The asymptotic mean and variance of the true density $P_{as}(x,t)$ coincide with those of the Gaussian factor as we have tacitly assumed in eq. (5). The Gaussian function can be numerically extracted from $P(x,t)$ in different ways (cf. Fig. 1). One may fit $P(x,t)$ to a Gaussian function, filter the density (Fourier transformation, cutting off the high frequencies and perform the back transformation) or carry out a direct coarse graining by $\tilde{P}(\tilde{x},t) = \int_{\tilde{x}-L/2}^{\tilde{x}+L/2} dx P(x,t)/L$ with $\tilde{x} = x_0 + kL, k \in Z$. All three methods yield a satisfying agreement. For the data in Fig. 1 the fitted and filtered functions coincide within line thickness while the histogram obtained from the coarse graining seems to be the discrete counterpart to these functions.
It might be desired that particles started at a sharp value should reach a certain region without much spreading. An optimal transport in this sense is realized at large velocity and small diffusion and will result in a large Péclet number

$$Pe = \frac{vL}{D_{eff}}. \tag{6}$$

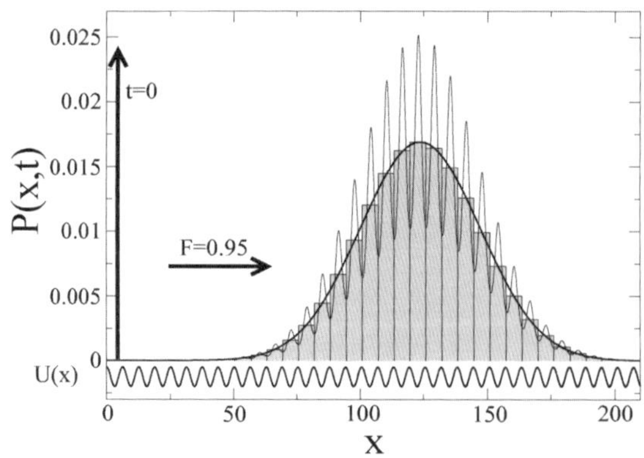


Fig 1. Evolution of the density $P(x,t)$ (thin line) for the potential $U(x) = \cos(x)$, $F = 0.95$ and $D = 2.05$. The density started at $x = 4.4$ (arrow) is calculated by a numerical solution of eq. (2). Additionally shown: the coarse grained density (histogram) and a fit of $P(x,t)$ to a Gaussian function (thick line) which coincides in line thickness with a filtered version of $P(x,t)$ (not shown). The potential $U(x)$ is drawn at the bottom for comparison.

This nondimensional ratio was frequently used in convection problems [9] and also recently applied to the transport problem in ratchet potentials [10].
Beside the trivial case of vanishing potential $U(x) \equiv 0$ (then $v = F$, $D_{eff} = D$), the usual approach to the diffusion problem posed by eq. (4) is to extract the diffusion coefficient from the time dependent solution of the Fokker-Planck equation (2). This works, e.g. in case of piecewise constant potential $U(x)$ [1], however, for a general periodic potential the time dependent solution is not available. Another way is to use a generalized fluctuation-dissipation relation [1]. However, this applies only in case of equilibrium systems, i.e., for vanishing tilt ($F = 0$) and results in an approximate formula for a system with finite bias [7].
Our approach utilizes the fact that the diffusion is entirely determined by the local properties of the system due to the periodicity of the potential. This allows a mapping onto a discrete process yielding the coarse grained density the asymptotic variance of which can be easily calculated.

## 3. Theory

The process described by eq. (1) is physically equivalent to the motion of a particle in an infinite ordered set of segments of the effective potential $V(x)$ as sketched in Fig. 2. The spatial overlap between subsequent segments is given by one period $L$ while the length of a single segment is $2L$. Whenever the particle reaches the left or right boundary it is absorbed and reinjected in the respective neighboring segment. These jump events have no physical meaning since coordinates and potential forces at absorption and injection points coincide. The probability in the $n$-th segment $P_n(x,t)$ obeys a FPE like (2) with absorbing boundary conditions and with an additional source of probability corresponding to the particle injection in the middle

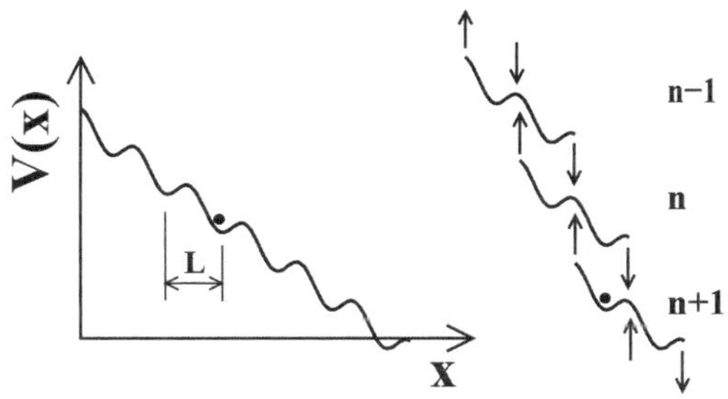


Fig 2. The motion of the particle (circle) in the tilted periodic potential $V(x)$ (left) is equivalent to that one in an infinite number of segments of the potential (right) with length $2L$. The arrows indicate transitions from one segment to the other.

of the segment (see reference [11] for technical details of a similar system of FPE's). The true density $P(x,t)$ from eq. (2) is given by the sum of these densities $P_n(x,t)$.

The seemingly complicated mapping onto the system of segments allows a rough discrete determination of the particle's position in terms of the segment $n_t$. Choosing the boundaries of the segments such that $x = 0$ (starting point) is in the middle of the segment with $n = 0$, we can approximate $x(t) \approx Ln_t$ with an uncertainty $\Delta x = \pm L$ at any time $t$. It is evident that in the asymptotic limit $t \to \infty$

$$\langle x(t)\rangle = L\langle n_t\rangle \ , \ \langle \Delta x(t)^2\rangle = L^2\langle \Delta n_t^2\rangle \tag{7}$$

holds true. The process $n_t$ illustrated in Fig. 3 (r.h.s.) emerges to be a cumulative process with independent increments (see appendix) for which asymptotic mean and variance can be easily found [12]. They are expressed by the first two central moments $\mu$ and $\sigma^2$ of the escape time density for a single segment depicted in Fig. 3 (l.h.s.). From these formulae and by virtue of eqs. (7), (3), (4), and (6) we find

$$v = L\frac{p_+ - p_-}{\mu} \tag{8}$$

$$D_{eff} = \frac{L^2}{2}\left(\frac{1}{\mu} + \frac{(p_+ - p_-)^2}{\mu^3}(\sigma^2 - \mu^2)\right) \tag{9}$$

$$Pe = 2\frac{p_+ - p_-}{1 - (1 - \frac{\sigma^2}{\mu^2})(p_+ - p_-)^2}. \tag{10}$$

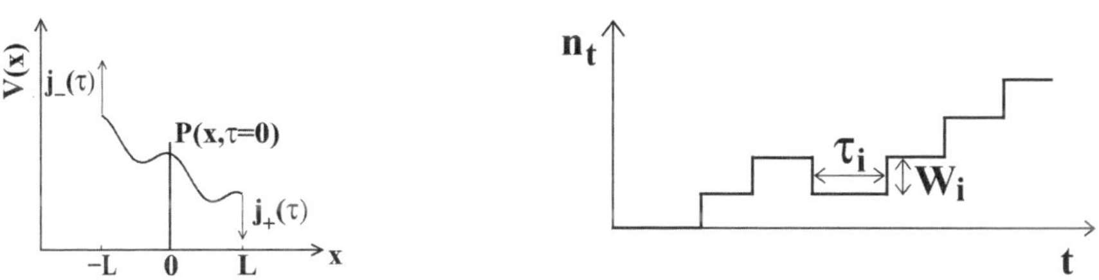


Fig 3. Left: The problem of escape from a single well. The initial condition $P(x, \tau = 0) = \delta(x)$ is indicated by a line. $j_-(\tau), j_+(\tau)$ denote out currents of probability. Right: a sample trajectory of the process $n_t$. The random increments $W_i \in \{-1, 1\}$ and the intervals $\tau_i$ correspond to independent realizations of the escape problem sketched at the l.h.s.

Therein, $p_-$ and $p_+$ denote the total probability to escape from the segment via the left and right boundary, respectively which are given by [13]

$$p_+ = 1 - p_- = \frac{1}{1+e^{-\frac{FL}{D}}}. \tag{11}$$

The first two cumulants of the escape time density $\rho(\tau)$ can be calculated by standard formulae (see e.g. [13]). Assuming without loss of generality that the middle of the segment is at $x = 0$ and the boundaries are at $\pm L$ yields

$$\mu = \frac{1}{D(1+e^{-\frac{FL}{D}})} \int_0^L dx e^{\frac{V(x)}{D}} \int_{x-L}^{x} dy e^{\frac{-V(y)}{D}} \tag{12}$$

$$\sigma^2 = \frac{2}{D(1+e^{-\frac{FL}{D}})} \int_0^L dx e^{\frac{V(x)}{D}} \int_{x-L}^{x} dy e^{\frac{-V(y)}{D}} \left(T_1(y) - \frac{1}{2}\mu\right) \tag{13}$$

where $T_1(x)$ is given in terms of $\Phi(x) = \exp[V(x)/D]$ by

$$T_1(x) = \frac{\left(\int_{-L}^{x} dy\, \Phi(y)\right) \int_x^L dy \int_{-L}^{y} dz\, \frac{\Phi(y)}{\Phi(z)} - \left(\int_x^L dy\, \Phi(y)\right) \int_{-L}^{x} dy \int_{-L}^{y} dz\, \frac{\Phi(y)}{\Phi(z)}}{D \int_{-L}^{L} dy\, \Phi(y)}. \tag{14}$$

We stress that the relations (8),(9) and (10) are exact for an arbitrary shape of the periodic potential. The quadratures (12) and (13) have to be evaluated numerically, in general.

By inserting (12) into eq. (8) and an additional little manipulation of the integrals we recover the classic formula by Stratonovich (see, e.g. [1])

$$v = \frac{LD(1-e^{-\frac{LF}{D}})}{\int_0^L dx\, e^{-\frac{V(x)}{D}} \int_x^{x+L} dy\, e^{\frac{V(y)}{D}}}. \tag{15}$$

For an *unbiased* potential ($F = 0$, i.e., $p_+ = p_-$) the velocity and Péclet number vanish trivially while the diffusion coefficient reduces to $L^2/2\mu$, i.e., to

$$D_{eff}(F=0) = \frac{L^2 D}{\int_0^L dx e^{-U(x)/D} \int_0^L dy e^{U(y)/D}}. \tag{16}$$

Thus, we recover the well known result from [1].

On the other hand, according to Stratonovich's random walk assumption in case of a weak tilt, one may approximate $\sigma = \mu$ and again the second term in eq. (9) vanishes ($D_{eff} = L^2/2\mu$). In this case, the Péclet number approximatively reads

$$Pe(F \ll F_c) \approx 2(p_+ - p_-) = 2\tanh(FL/2D). \tag{17}$$

the absolute value of which is a decreasing function of the noise intensity $D$.
We would like to point out a remarkable property for systems with symmetric potential $U(x) = U(-x)$. In this case, all calculated quantities do not change for the inverse potential $-U(x)$, for instance

$$U(x) = U(-x) \quad \Longrightarrow \quad D_{eff}[U(x)] = D_{eff}[-U(x)] \tag{18}$$

and likewise for $v$ and $Pe$. This can be proven by manipulations of the resulting quadrature formulae for $v$ and $D_{eff}$.
Finally, we remark that the above discretization by segments of length $2L$ is not the only valid choice. Reimann and coworkers [8] have implicitly used a mapping onto a set of infinitely long segments. They obtained the same result for the effective diffusion coefficient like in the present work (the proof is given in appendix Appendix B) and, furthermore, provided an elegant simplification of the quadratures.

## 4. Optimal transport in a cosine potential

Consider the archetypal potential $U(x) = \cos(x)$. The critical tilt at which the minima and maxima of the effective potential vanish is given by $F_c = 1$. For a strong but subcritical tilt $F = 0.95$, velocity and effective diffusion coefficient are increasing functions of the noise intensity (cf. Fig. 4, l.h.s.). The Péclet number, however, attains a maximum at a moderate value of $D$ indicating an optimal transport in this case. These findings are confirmed by results of a numerical solution of the FPE (2).

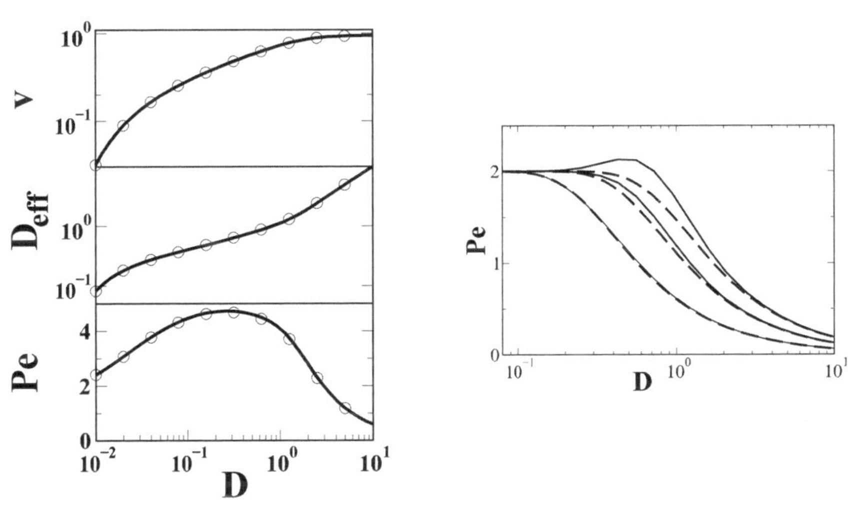


Fig 4. Left: Velocity, diffusion coefficient, and Péclet number according to eqs. (8),(9),(10) versus noise intensity for $F = 0.95$. Circles indicate results from a numerical solution of (2). Right: Péclet number for different tilting forces $F = 0.1, 0.2, 0.3$ (from bottom to top). Theory according to eq. (10) (solid) compared with 'biased-random-walk' approximation eq. (17) (dashed).

At the optimal noise intensity, the motion of the particle is mainly determined by two processes: the noise driven escape from the potential minimum via the right potential barrier followed by a relaxation into the next minimum. The relaxation time depends only weakly on the noise strength and possesses a small variance. The latter fact leads to a certain regularity of the particle motion and accounts for the maximum of the Péclet number.
A weaker maximum of $Pe$ is also observed at smaller tilt. Fig. 4 (r.h.s) shows the Peclet number for $F = 0.1, 0.2, 0.3$ compared to the random walk approximation (17). The latter is sufficient for $F = 0.1$ as well as for small and large noise intensity at arbitrary tilt. It fails, however, to reproduce the nonmonotonous dependence on noise strength that arises numerically for $F > 0.24$.

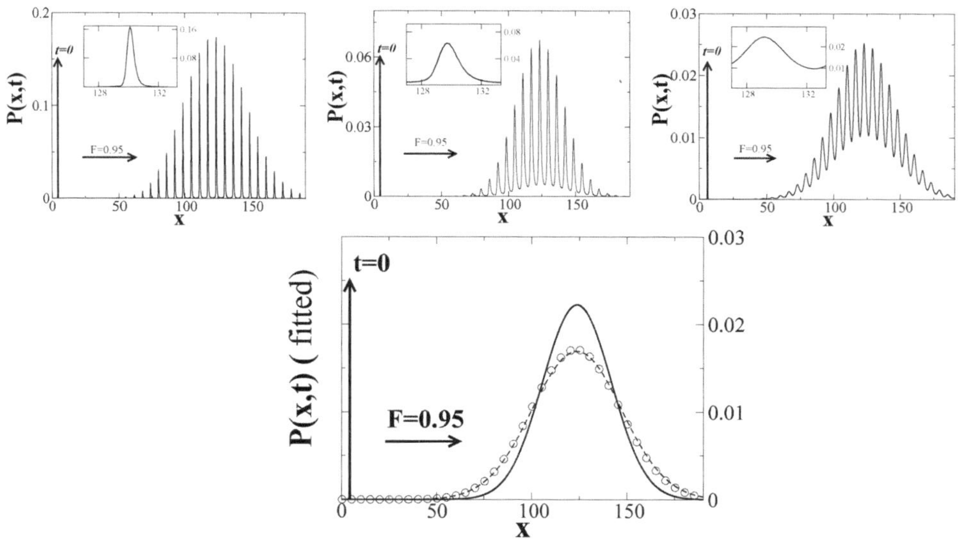


Fig 5. Probability densities versus spatial coordinate for $F = 0.95$ and three different noise levels $D_1 = 0.015$, $D_2 = 0.25$ and $D_3 = 2.05$ at fixed time instances $t \approx 1919, 275$ and 139, respectively. $D_1, D_3$ are chosen such that $Pe(D_1) = Pe(D_3)$ and $D_2$ is the optimal value from Fig. 4. Top: densities obtained by numerical solution of the FPE (2) with $P(x, t = 0) = \delta(x - 4.4)$ (arrow) for $D_1$, $D_2$, $D_3$ (from left to right). The insets show the densities over one period. Bottom: Gaussian fits to these densities, $D_1$ (circles),$D_2$ (solid), $D_3$ (dashed).

The coherent transport for large tilt can be illustrated in the following way. At three different noise intensities $D_1 < D_2 < D_3$ (with $D_2 = D_{max}$ resulting in the maximal Péclet number) an ensemble of particles is started at fixed position $x_0$. The Fokker-Planck equation is integrated until the mean of the respective density reaches a given marker $x_e$. This will, of course, take different times $\Delta t_i$ for the three different noise intensities. For the variances of the densities $\langle \Delta x_i^2 \rangle$ one obtains

$$\langle \Delta x_i^2 \rangle = 2D_{eff\,i}\Delta t_i = 2\frac{Lv_i\Delta t_i}{Pe_i} = 2\frac{L(x_e - x_0)}{Pe_i}\,, \quad i = 1, 2, 3. \tag{19}$$

Thus one may conclude that the density at greatest Péclet number will have the smallest variance. This can be verified numerically for the densities depicted in (Fig. 5, top) and is clearly seen when the densities are fitted to Gaussian functions (Fig. 5, bottom). For a given subcritical tilt, the ordered transport of particles from $x_0$ to $x_e$ is thus optimized at finite noise strength.

## 5. Enhanced transport coherence due to a particular potential shape

The optimal transport discussed in the previous section became apparent by a maximum of the Péclet number with respect to the noise intensity. An even stronger coherence of the particle motion emerges for specific potential shapes which generate a long relaxation period and a quick escape at not too small noise intensity noise. Such shape can be realized, for instance, by the potential

$$U(x) = \frac{\Delta}{\varepsilon} e^{\varepsilon(\cos(x) - 1)} \tag{20}$$

For large $\varepsilon$ the potential is flat apart from periodically occurring small barriers (cf. Fig. 6, r.h.s., top) the height of which can be scaled by the parameter $\Delta$. If $\Delta$ is negative (cf. Fig. 6, r.h.s., top, black dashed line), the potential exhibits small valleys instead of barriers. For the opposite limit $\varepsilon \to 0$ and fixed $\Delta = 1$ in turn, one recovers the cosine potential.

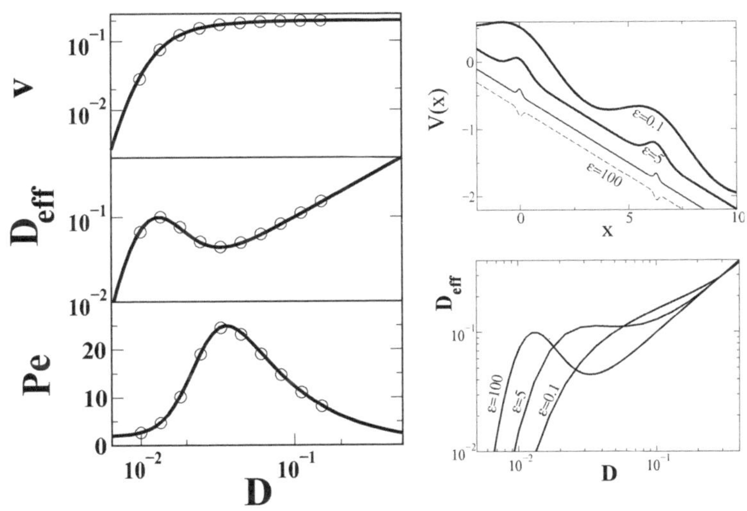


Fig 6. Left: Velocity, diffusion coefficient and Péclet number for the potential eq. (20) according to eqs. (8),(9),(10) versus noise intensity $D$. Circles indicate results from the numerical solution of the FPE (2). Parameters: $\Delta = 10, \varepsilon = 100, F = 0.2$. Right top: The potentials for decreasing values of $\varepsilon = 100, 5, 0.1$ with $\Delta = 10.0, 1.225, 0.282$, respectively. Right bottom: The effective diffusion constant versus noise intensity for the potentials plotted above. The minimum of $D_{eff}$ vanishes at $\varepsilon \approx 5$.

Besides the maximum in the Péclet number we find for large $\varepsilon$ a pronounced minimum in the effective diffusion coefficient at almost the same noise level (see Fig. 6, l.h.s.). This remarkable result (increasing noise decreases diffusion) relies on the

large ratio of relaxation to escape time for the specific potential ($\varepsilon = 100$) at optimal noise intensity. Since the potential tends to a cosine for $\varepsilon \to 0$, we expect the minimum to vanish for decreasing $\varepsilon$. This is shown in Fig. 6 (r.h.s., bottom) for the potentials depicted in Fig. 6 (r.h.s., top). Here, the parameter $\Delta$ was tuned such that the potential barriers to the right and left are the same for all potentials.
Because the potential (20) possesses spatial symmetry we find according to relation (18) the same functions if $\Delta \to -\Delta$, i.e., the same velocity, diffusion coefficient and Péclet number for the potentials drawn as black solid and dashed lines in Fig. 6 (r.h.s., top).
In contrast to the cosine potential, the velocity at the optimal noise intensity is almost saturated ($v \approx F$). Furthermore, the maximal value of $Pe$ exceeds the maximal $Pe$ of a random walk (i.e., the small noise limit with $Pe = 2$) by one order of magnitude. Consequently, a much more pronounced coherence is expected for the same 'experiment' like in the previous section. Indeed, the numerical results in Fig. 7 reveal a larger difference in the dispersions at small, large and optimal noise intensity. In addition, the difference in the times to reach the marker $x_e$ is small for optimal and large noise. Hence, for the special potential the *most coherent diffusive* transport at optimal noise intensity is only slightly slower than the *fastest diffusive* transport at large $D$.

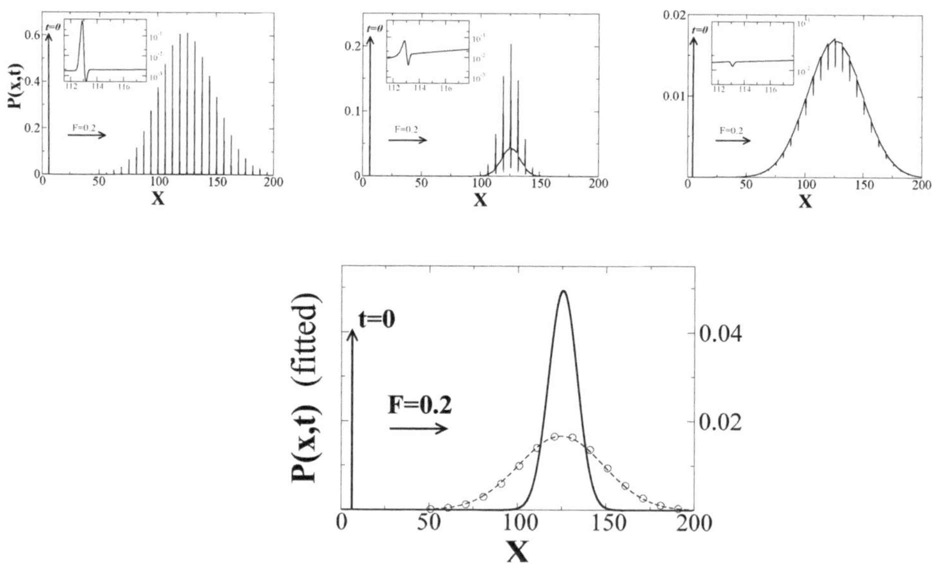


Fig 7. Probability densities versus coordinate for $F = 0.2$ and three different noise levels $D_1 = 0.01$, $D_2 = 0.033$, $D_3 = 0.472$ at fixed time instances $t \approx 4540, 691$ and $625$, respectively. $D_1, D_3$ are chosen such that $Pe(D_1) = Pe(D_3)$ and $D_2$ is the optimal value from Fig. 6. Top: densities obtained by numerical solution of the FPE (2) with $P(x, t = 0) = \delta(x - 6.1)$ (arrow), $D_1, D_2, D_3$ (from left to right). The insets show the logarithm of the densities over one period. Bottom: Gaussian fits to these densities, $D_1$ (dashed),$D_2$ (solid),$D_3$ (circles).

The minimum of the diffusion coefficient trivially implies a minimum of the dispersion for the optimal noise intensity when a fixed time instead of a fixed distance is considered.

## 6. Summary and conclusions

We have derived exact expressions for the diffusion coefficient and the Péclet number of an overdamped Brownian particle in an arbitrary periodic potential with additional tilt. At moderate to strong subcritical bias the Péclet number displays a maximum with respect to the noise intensity indicating a most coherent motion for moderate noise. Especially, for a flat potential with small sharp barriers another effect emerges: the effective diffusion coefficient takes on a local minimum as a function of noise intensity.

The effect of optimal transport is strongly related to the coherence resonance (CR) observed in excitable systems [14] as can be seen in the following. A frequently used measure of this resonance is the coefficient of variation, i.e., the relative standard deviation $R$ of the time interval between subsequent excitations

$$R = \frac{\sqrt{\langle \Delta\tau^2 \rangle}}{\langle \tau \rangle}. \tag{21}$$

As a manifestation of CR, this ratio becomes minimal for a moderate noise intensity, i.e., the excitation sequence is closest to a regular periodic one. In case of the washboard potential it is not obvious how to define a mean time interval since we deal with two different 'excitations': escapes to the left or right side of the segment. If, however, the passage time over one period in direction of the bias [8] is considered we obtain the exact relation (see appendix Appendix B)

$$Pe = \frac{2}{R^2}. \tag{22}$$

It is clearly seen that $Pe$ becomes maximal when $R$ is minimal. A coherence resonance of the above defined passage time hence implies an optimal transport of particles and vice versa.


### Acknowledgements

We acknowledge financial support by DFG: GK 268 (B.L.) and SFB 555 (M.K. and L.S.-G.). M.K. was further supported by KBN Grant No. 2 P03B 160 17 and FNP.


## Appendix A. The discrete process $n_t$

The process $n_t$ depicted in Fig. 3 (r.h.s.) consists of random jumps with random increments $W_i \in \{-1, 1\}$ with probabilities $p_-$ and $p_+$, respectively, at likewise random time instants $t_i$, i.e.,

$$n_t = \sum_{i=1}^{N_t} W_i. \tag{A.1}$$

where $N_t$ denotes the total number of events. Since each realization of an escape event from one segment is independent of all foregoing or subsequent realizations, the escape times $\tau_i = t_i - t_{i-1}$ and increments $W_i$ are independent of all other escape times and increments, respectively. The escape time statistics is given by the problem illustrated in Fig. 3 (l.h.s.), i.e., by the time dependent FPE (2) with initial condition and absorbing boundary conditions

$$P(x, 0) = \delta(x) \qquad P(-L, t) = P(L, t) = 0 \tag{A.2}$$

The absolute values of the outfluxes of probability are related to the conditional escape time densities $\rho_-(\tau)$ and $\rho_+(\tau)$ for left and right jumps, respectively, [13]

$$j_+(t) = p_+\rho_+(\tau) = -D \left.\frac{\partial P(x,t)}{dx}\right|_{x=L} \quad , \quad j_-(t) = p_-\rho_-(\tau) = D \left.\frac{\partial P(x,t)}{dx}\right|_{x=-L} . \tag{A.3}$$

We now show the surprising property

$$\rho_+(\tau) \equiv \rho_-(\tau) \tag{A.4}$$

that permits the conclusion that there exist no correlations between the escape time $\tau$ and the subsequent jump $W$.
The special Laplace transformation $p(x, \lambda) = \exp[(V(x) - V(0))/2D] \int_0^\infty dte^{-\lambda t} P(x, t)$ applied to eq. (2) leads to

$$\hat{L}p = Dp''(x, \lambda) - (\frac{(V'(x))^2}{4D} - \frac{V''(x)}{2} + \lambda)p(x, \lambda) = -\delta(x) \tag{A.5}$$

where the prime denotes the derivative with respect to $x$. The (conventional) Laplace transforms of $j_-(t)$ and $j_+(t)$ in terms of $p(x, \lambda)$ are given by

$$J_+(\lambda) = -De^{\frac{FL}{2D}} \left.\frac{\partial p(x,\lambda)}{dx}\right|_{x=L} \quad , \quad J_-(\lambda) = De^{-\frac{FL}{2D}} \left.\frac{\partial p(x,\lambda)}{dx}\right|_{x=-L} \tag{A.6}$$

It is well known that an equation like (A.5) can be treated by means of homogeneous solutions in $[-L, 0]$ and $[0, L]$ which have to be connected according to

$$\lim_{\varepsilon \to 0}(p(\varepsilon, \lambda) - p(-\varepsilon, \lambda)) = 0 \ , \ \lim_{\varepsilon \to 0}\,(p'(x, \lambda)|_{x=\varepsilon} - p'(x, \lambda)|_{x=-\varepsilon}) = -\frac{1}{D}. \tag{A.7}$$

Provided that $\psi_1(x), \psi_2(x)$ and $\phi_1(x), \phi_2(x)$ (omitting the second argument $\lambda$) are such independent solutions of the homogeneous problem in $[-L, 0]$ and $[0, L]$, respectively

$$\hat{L}\psi_{1,2}(x) = 0 \ , \ -L < x < 0 \qquad \hat{L}\phi_{1,2}(x) = 0 \ , \ 0 < x < L \tag{A.8}$$

then we may assume

$$\phi_1(x) = \psi_1(x - L) \text{ and } \phi_2(x) = \psi_2(x - L) \tag{A.9}$$

because of the periodicity of the force field in (A.5).
The Ansatz

$$p(x, \lambda) = \begin{cases} c_1\psi_1 + c_2\psi_2 \ , & -L < x < 0 \\ d_1\phi_1 + d_2\phi_2 \ , & 0 < x < L \end{cases} \tag{A.10}$$

leads by (A.2,(A.7),(A.9) and (A.6) to the following ratio of currents

$$\frac{J_+}{J_-} = \frac{\psi_1'(0)\psi_2(0) - \psi_1(0)\psi_2'(0)}{\psi_1'(-L)\psi_2(-L) - \psi_1(-L)\psi_2'(-L)} e^{\frac{FL}{D}} \tag{A.11}$$

The expression $\psi_1'(x)\psi_2(x) - \psi_1(x)\psi_2'(x)$ which occurs at different argument in the fraction does not depend on the argument $x$. This can be shown by taking the derivative with respect to $x$ and employing (A.8). Hence, we have for arbitrary $\lambda$

$$\frac{J_+(\lambda)}{J_-(\lambda)} = e^{\frac{FL}{D}} = \frac{p_+}{p_-}. \tag{A.12}$$

*Process with independent increment*

Obviously, the time dependent currents possess the same ratio and the identity (A.4) is proven.
Since all increments and intervals are independent of each other, the process $n_t$ is a so called *cumulative p* whose asymptotic mean and variance are given by [12]

$$\begin{aligned} \langle n_t \rangle &= \langle W \rangle \frac{t}{\mu} \\ \langle \Delta n_t^2 \rangle &= \langle \Delta W^2 \rangle \frac{t}{\mu} + \langle W \rangle^2 \frac{t\sigma^2}{\mu^3}. \end{aligned} \tag{A.13}$$

Here, $\mu$ and $\sigma^2$ are the first two cumulants of the escape time density $\rho(\tau)$. For mean and variance of $W$ one easily verifies

$$\langle W \rangle = p_+ - p_- \ , \ \langle \Delta W^2 \rangle = 4 p_+ p_- . \tag{A.14}$$

Inserting into (A.13) and utilizing eqs. (7), (3), (4), and (6) yield the final results eqs. (8) and (9).

## Appendix B. Equivalence of the result from ref. [8] to formula (9)

In accordance with [8], the velocity and effective diffusion coefficient for $F \geq 0$ are given by

$$v = \frac{L}{\langle T(0 \to L) \rangle} \ , \qquad D_{eff} = \frac{L^2}{2} \frac{\langle \Delta T^2(0 \to L) \rangle}{\langle T(0 \to L) \rangle^3}. \tag{B.15}$$

Here, $\langle T(0 \to L) \rangle$ denotes the mean of the first passage time from 0 to $L$ with absorbing boundary at $-\infty$ and $L$ while $\langle \Delta T^2(0 \to L) \rangle$ is the variance of this time. The formulae (B.15) hold asymptotically for a renewal process corresponding to the discrete mapping depicted in Fig. B.1 (l.h.s.).
Seemingly, the Ansatz in particular at large noise and small bias should fail since the density $p(n)$ is distorted in direction of the bias (obviously, starting an ensemble of particles in the segment with $n = 0$ will not yield any contributions for $n < 0$). Furthermore, it is not obvious which segment number belongs to a certain interval of $x$. However, formula (B.15) emerges to be correct and is equivalent to our results (8) and (9). In order to proof this equivalence we derive and employ a relation between the Laplace transforms of the escape time densities corresponding to finite and infinite segment.

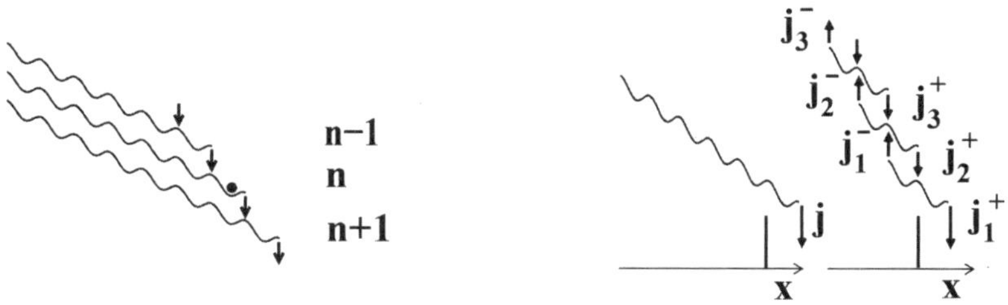


Fig B.1. Left: Renewal process resulting in eqs. (B.15). Right: An ensemble of particles indicated by the thick lines is started at $\tau = 0$, $x = 0$ and absorbed at $x = L$. $j(\tau), j_i^{\pm}(\tau)$ with $i = 1, 2, 3\cdots$ are the absolute values of the respective probability currents. Note that the numbering of the segments is vice versa to that in Fig. 2 for technical reasons.

The renewal event in the above discretization is the escape out of the single *infinite* segment starting in a distance of one period from the absorbing right boundary (Fig. B.1, middle). It can be expressed by the escape via the corresponding right boundary in an *infinite* number of *finite* segments (Fig. B.1, r.h.s.). The escape time density out of the infinite segment is given by the current $j(\tau)$ which is equal to $j_1^+(\tau)$ (cf. Fig. B.1, middle and r.h.s.) since both discrete descriptions apply to the same physical situation. For the currents $j_n^+$ one obtains

$$\begin{aligned} j_1^+(\tau) &= j(\tau) = p_+\rho(\tau) + p_+ \int_0^\tau d\tau'\ j_2^+(\tau')\rho(\tau-\tau') \\ j_n^+(\tau) &= \int_0^\tau d\tau'\ [p_+ j_{n+1}^+(\tau') + p_- j_{n-1}^+(\tau')]\rho(\tau-\tau')\ ,\ n>1 \end{aligned} \tag{B.16}$$

where we have used the relation $j_n^-(\tau) = j_n^+(\tau)p_-/p_+$ derived in Appendix A. Let $J(\lambda), J_n(\lambda)$ and $\varrho(\lambda)$ be the Laplace transforms of $j(\tau), j_n^+(\tau)$ and $\rho(\tau)$, respectively. Then, eqs. (B.16) lead to a one-sided recurrence relation

$$p_+\varrho J_{n+1} - J_n + p_-\varrho J_{n-1} = 0\ ,\ \ n \geq 1\ ,\ \ J_0 \equiv p_+/p_- \tag{B.17}$$

whose solution is given by the periodic continued fraction that can be further simplified [1] to

$$J(\lambda) = J_1(\lambda) = \frac{2p_+\varrho(\lambda)}{1+\sqrt{1-4p_+p_-\varrho(\lambda)^2}}. \tag{B.18}$$

From this relation between the escape time densities the relations between the cumulants $\langle T\rangle, \langle\Delta T^2\rangle$ and $\mu, \sigma^2$ are readily obtained by the first two derivatives of $\ln(J(\lambda))$at $\lambda = 0$ yielding

$$\langle T\rangle = \frac{\mu}{p_+ - p_-}\ ,\ \langle\Delta T^2\rangle = \frac{\sigma^2}{p_+ - p_-} + \frac{4p_+p_-\mu}{(p_+ - p_-)^2}. \tag{B.19}$$

For the diffusion constant $D_{eff} = L^2\langle\Delta T^2\rangle/(2\langle T\rangle^3)$ according to eq. (B.15) one recovers our result eq. (9). Hence, the equivalence of both approaches is established.

In addition, from (B.15) we find immediately eq. (22)

$$Pe = 2\frac{\langle T(0 \to L)\rangle^2}{\langle \Delta T^2(0 \to L)\rangle} = \frac{2}{R^2}. \tag{B.20}$$

## References


[1] H. Risken, *The Fokker-Planck Equation*, Springer, Berlin (1984).

[2] P. Fulde, L. Pietronero, W. R. Schneider, and S. Strässler, *Problem of Brownian Motion in a Periodic Potential, Phys. Rev. Lett.* **35** (1975) 1776-1779.

[3] K. Wiesenfeld, D. Pierson, E. Pantazelou, C. Dames, and F. Moss, *Stochastic resonance on a circle, Phys. Rev. Lett.* **72** (1994) 2125-2129.

[4] Ch. Kurrer and K. Schulten, *Noise-induced synchronous neuronal oscillations, Phys. Rev. E* **51** (1995) 6213-6218.

[5] R. L. Stratonovich, *Oscillator synchronization in the presence of noise*, in *Non-Linear Transformations of Stochastic Processes* ed. by P. I. Kuznetsov, R. L. Stratonovich, and V. I. Tikhonov, Pergamon Press, Oxford, (1965) 269-282.

[6] R. Festa and F. G. d'Agliano, *Diffusion Coeffcient for a Brownian Particle in a Periodic Field of Force: I. Large Friction Limit, Physica* **90 A** (1978) 229-244; D. L. Weaver, *Effective Diffusion Coefficient of a Brownian Particle in a Periodic Potential, Physica* **98 A** (1979) 359-362; R. A. Guyer, *Dynamics of nonlinear systems: The heavy damping limit Phys. Rev. B* **21** (1980) 4484-4499; see also [1]

[7] G. Costantini and F. Marchesoni, *Threshold diffusion in a tilted washboard potential, Europhys. Lett.* **48** (1999) 491-497

[8] P. Reimann, C. Van den Broeck, H. Linke, P. Hänggi, M. Rubi, and A. Perez-Madrid, *Thermal Diffusion in a Tilted Periodic Potential: Enhancement, Universality, and Scaling, submitted to Phys. Rev. E*

[9] L. D. Landau and E. M. Lifschitz, *Lehrbuch der Theoretischen Physik* Vol. 6 (Berlin, Akademie-Verlag, 1971)

[10] J. A. Freund and L. Schimansky-Geier, *Diffusion in discret ratchets, Phys. Rev. E* **60** (1999) 1304-1309.

[11] B. Lindner and L. Schimansky-Geier, *Analytical approach to the stochastic FitzHugh-Nagumo system and coherence resonance, Phys. Rev. E* **60** (1999) 7270-7276.

[12] D. R. Cox, *Renewal Theory*, Methuen, London (1962).

[13] C. W. Gardiner, *Handbook of Stochastic Methods*, Springer, Berlin (1985)

[14] A. Pikovsky and J. Kurths, *Coherence Resonance in a Noise-Driven Excitable System, Phys. Rev. Lett.* **78** (1997) 775-778.

# SIGNAL-TO-NOISE RATIO GAIN IN NON-DYNAMICAL AND DYNAMICAL BISTABLE STOCHASTIC RESONATORS

PETER MAKRA* and ZOLTAN GINGL†
*Department of Experimental Physics, University of Szeged*
*Dóm tér 9., Szeged, H-6720 Hungary*
**phil@neptun.physx.u-szeged.hu*
*†gingl@physx.u-szeged.hu*

LASZLO B. KISH
*Department of Electrical Engineering, Texas A&M University*
*College Station, TX 77843-3128 USA*
*Laszlo.Kish@ee.tamu.edu*

Received 14 July 2002
Revised 16 September 2002
Accepted 18 September 2002

It has recently been reported that in some systems showing stochastic resonance, the signal-to-noise ratio (SNR) at the output can significantly exceed that at the input; in other words, SNR gain is possible. We took two such systems, the non-dynamical Schmitt trigger and the dynamical double well potential, and using numerical and mixed-signal simulation techniques, we examined what SNR gains these systems can provide. In the non-linear response limit, we obtained SNR gains much greater than unity for both systems. In addition to the classical narrow-band SNR definition, we also measured the ratio of the total power of the signal to the power of the noise part, and it showed even better signal improvement. Here we present a brief review of our results, and scrutinise, for both the Schmitt-trigger and the double well potential, the behaviour of the SNR gain by stochastic resonance for different signal amplitudes and duty cycles. We also discuss the mechanism of providing gains greater than unity.

*Keywords:* Stochastic resonance; signal-to-noise ratio gain; Schmitt trigger; double well potential.

## 1. Introduction

Stochastic resonance (SR), whose study has grown to be a significant branch of noise research, is a rather general phenomenon wherein increasing the amount of input noise can optimise signal transfer in a system. To be more exact, we speak of stochastic resonance when the measure reflecting the quality of the transfer (most often the signal-to-noise ratio, or SNR) shows a maximum at non-zero input noise amplitude. Stochastic resonance has attracted considerable attention from various scientific fields, and the term applies to a wide range of systems and processes: the cycle of ice ages, neural models, physiological phenomena such as baroreflex mechanisms, bistable ring lasers, superconducting quantum-interference devices (SQUIDs), electronic circuits, etc (see [1–14] and references therein).

Stochastic resonance means that the output signal gets better with a certain increase in input noise, but it does not usually involve the improvement of the output signal compared to the input. Yet it is a legitimate question whether systems showing SR can possibly function as filters, making the signal passing through them less noisy. A positive answer to this question might have great significance: it might establish the theoretical basis for technical applications and might lead us to a better understanding of several (especially biological) systems.

Early investigations aimed at finding SR-based signal improvement were largely unsuccessful, and the exceptions soon turned out to be artefacts. The search for SNR gain was further discouraged by linear-response theory (LRT) [15], which proved that the output SNR cannot exceed that at the input (SNR gain is not possible) in the linear-response (LR) limit, that is, for signals whose amplitude is small compared to that of the noise. This result, however, did not rule out SNR gain in the non-linear response (NLR) range, so subsequent studies mostly assumed the condition of NLR.

The first theoretical finding that predicted significant SNR improvement involved a level-crossing detector (LCD) driven by a random spike train with additional Gaussian white noise [1]. Here the system produced an SNR gain of 5 orders of magnitude when operated in the strongly non-linear limit. Later, the same set-up also yielded SNR improvement with a deterministic spike train as input [2]. In the studies that followed, significant SNR gains were observed, for example, in saturating threshold device models [3] and neuronal models [4].

In our brief review, we would like to revisit two of our previous results that elaborate on the issue of SNR improvement in bistable systems: one concerning a non-dynamical system, the Schmitt trigger [16], and the other examining a dynamical model, the double-well potential [17]. Our aim is to explore how the gains obtained depend on the duty cycle of the deterministic signal and the non-linearity of the arrangement, and compare the behaviour of the two systems in this respect.

## 2. Signal-to-Noise Ratio Gain by SR in a Schmitt Trigger

The Schmitt trigger is one of the first systems where stochastic resonance was observed. It has two threshold levels and its output is a dichotomous signal with a lower and an upper stage. If the input drops below the lower threshold, the output assumes the lower stage and remains so until the input exceeds the upper threshold, when it switches to the upper stage. In our computer simulations, we considered a model Schmitt trigger with symmetric threshold levels, $-A_T$ and $A_T$. The input consisted of a periodic pulse train $p(t)$ with variable amplitude and duty cycle, plus a Gaussian white noise $w(t)$ with variable amplitude $\sigma$; the output, $x(t)$, was defined in the following way:

$$x(t) = \begin{cases} 1, \text{if } p(t) + w(t) > A_T; \\ -1, \text{if } p(t) + w(t) < -A_T; \\ \textit{previous value}, \text{otherwise}. \end{cases} \quad (1)$$

Figure 1 illustrates the shape of the deterministic input signal $p(t)$: $A$ is the amplitude expressed as a percentage of the threshold value $A_T$, $T$ stands for the period, $\tau$ denotes the width of the pulses and the duty cycle is defined as $2\tau/T$. The noise $w(t)$ we used was band-limited: its spectrum was flat below a cut-off frequency of 50 kHz and was attenuated by 20 dB/decade above this frequency.

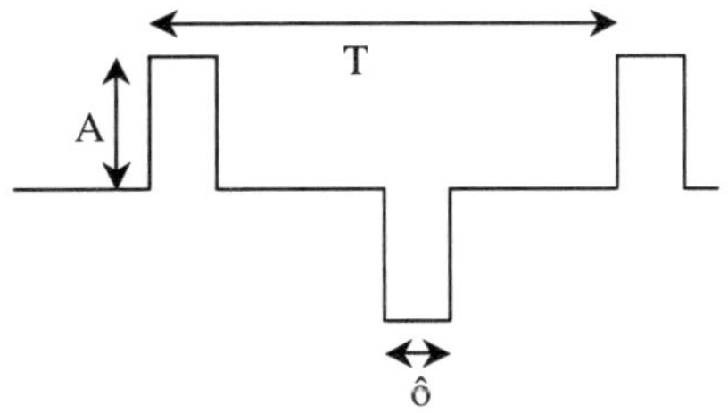


Fig. 1. The shape of the periodic input signal *p(t)*.

To obtain the SNR gain as a function of the input noise amplitude $\sigma$, we recorded 1000 independent samples with lengths of 32768 data points for each $\sigma$ value; the input pulse train was periodic by 1024 data points. Then we calculated the averaged power spectral density (PSD) of these samples and from the PSD we obtained the signal-to-noise ratio. We used two classical definitions for the SNR:

(i) The most common SNR definition, which is the ratio of signal power around the signal frequency $f_0$ and the level of the background noise at the same frequency [18]:

$$SNR := \frac{\lim\limits_{\Delta f \to 0} \int_{f_0-\Delta f}^{f_0+\Delta f} S(f)df}{S_N(f_0)}, \tag{2}$$

(ii) and a more realistic wide-band definition, which is the total power of the periodic signal divided by the total power of the noise (for a detailed discussion, see [17]):

$$SNR_w := \frac{P_S}{P_N} = \frac{\sum\limits_{k=1}^{\infty} \lim\limits_{\Delta f \to 0} \int_{kf_0-\Delta f}^{kf_0+\Delta f} S(f)df}{\int_0^{\infty} S_N(f)df}. \tag{3}$$

In the definitions above, $f_0$ denotes the frequency of the deterministic signal, *S(f)* stands for the PSD of the signal and $S_N(f)$ signifies the PSD of the noise component in the signal. Since the signals we considered were periodic, and the noise had a smooth spectrum, we could easily separate the signal from the noise: at a given harmonic, the PSD of the signal could be obtained as the corresponding data point in the total PSD minus the background noise level, which was interpolated from six neighbouring data points. The SNR gain was simply the ratio of output and input SNRs:

$$G := \frac{SNR_{out}}{SNR_{in}}, \text{ and } G_w := \frac{SNR_{out,w}}{SNR_{in,w}}. \tag{4a-b}$$

We should note here that SNR as a measure to quantify stochastic resonance has attracted widespread criticism. Several objections have emerged, including the argument that SNR as a spectral measure cannot capture the whole probabilistic structure of SR and general information-theoretic distance measures offer a more comprehensive description [19]. For the narrow-band concept, additional shortcomings have been pointed out: it characterises the response of the system only in a narrow frequency interval, and consequently, its applicability is limited to sinusoidal inputs [20]. In spite of these objections, we stuck to SNR, since it is still the most widely used response measure in the literature of SR. We included the wide-band extension so as to avoid some of the problems with SNR.

The results of our simulations are illustrated in Figs. 2 and 3. These figures demonstrate that both kinds of gains can exceed unity, although the wide-band gain remains greater than one over a wider range of input noise amplitude than its narrow-band counterpart. To illustrate the differences between the narrow-band and wide-band SNR, we included Fig. 4, which shows them side by side. Figure 2 shows how narrow-band and wide-band gains depend on the amplitude of the input noise, for three different values of the amplitude of the periodic input signal *p(t)*. Since working in the NLR limit is a prerequisite of producing SNR gain, it does not surprise us that the gains we obtain get higher as the amplitude (thus the non-linearity) increases.

As Fig. 3 reveals, greater duty cycles result in smaller gains. To understand this fact, one should consider that due to the output levels' being fixed, the noise can manifest itself in the output signal only in the form of jitter, that is, switch-time fluctuations. Consequently, $SNR_{out}$ does not depend significantly on the duty cycle if the correlation time of the noise is smaller than the width of the pulses in the periodic signal, and the amplitude of the periodic signal is close to the threshold (these requirements are met in our case). On the other hand, the input SNR is considerably higher for greater pulse width: the power of the deterministic component (the numerators in Eqs. (2) and (3)) increases, while the noise level (the denominators in Eqs. (2) and (3)) remains constant. The SNR gain, therefore, will be smaller if we increase the value of the duty cycle.

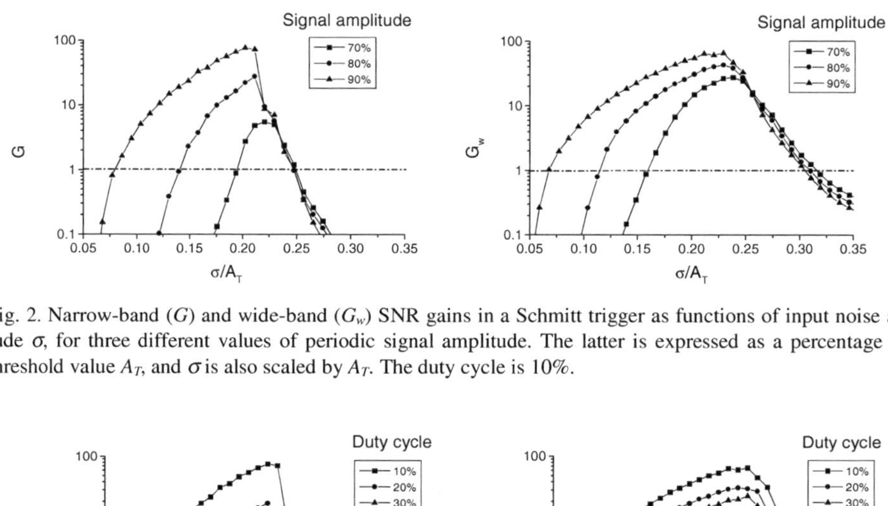


Fig. 2. Narrow-band ($G$) and wide-band ($G_w$) SNR gains in a Schmitt trigger as functions of input noise amplitude $\sigma$, for three different values of periodic signal amplitude. The latter is expressed as a percentage of the threshold value $A_T$, and $\sigma$ is also scaled by $A_T$. The duty cycle is 10%.

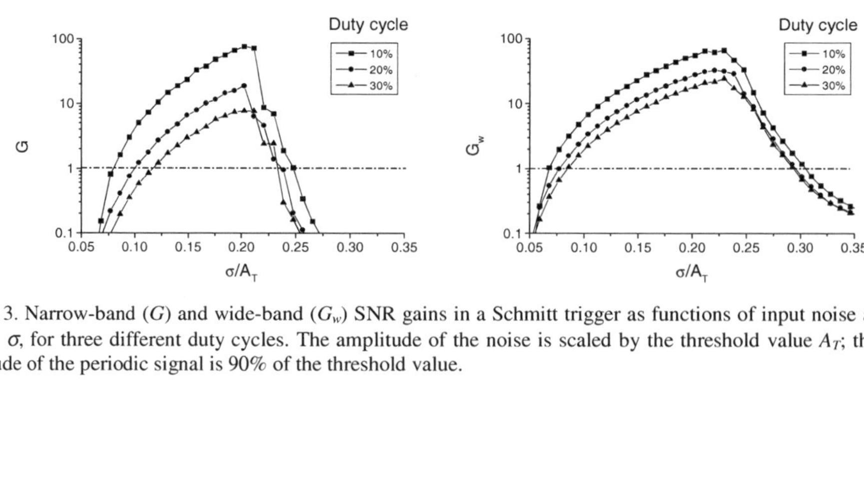


Fig. 3. Narrow-band ($G$) and wide-band ($G_w$) SNR gains in a Schmitt trigger as functions of input noise amplitude $\sigma$, for three different duty cycles. The amplitude of the noise is scaled by the threshold value $A_T$; the amplitude of the periodic signal is 90% of the threshold value.

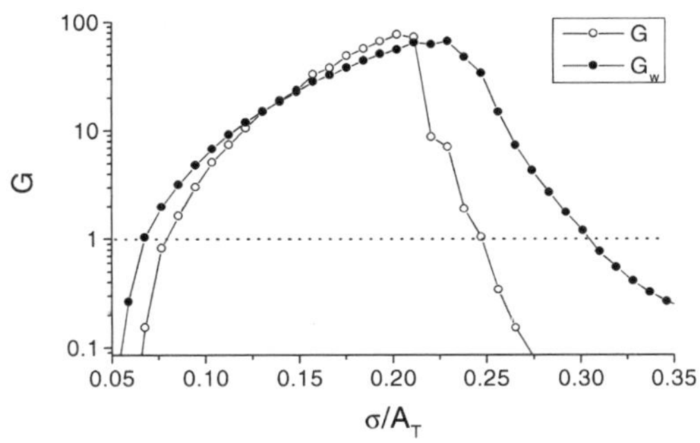


Fig. 4. Narrow-band ($G$) and wide-band ($G_w$) SNR gains compared in the Schmitt trigger. The amplitude of the noise is denoted by $\sigma$; it is scaled by the threshold value $A_T$. The amplitude of the periodic signal is 90% of the threshold value, and the duty cycle is 10%.

## 3. Signal-to-Noise Ratio Gain by SR in a Double-Well Potential

We considered an archetypal SR model: a particle moving in a double-well potential, excited by a periodic pulse train plus a band-limited Gaussian white noise. The over-damped dynamics of this system can be given by the following Langevin equation:

$$\frac{dx}{dt} = x(t) - x^3(t) + p(t) + w(t), \tag{5}$$

where $x(t)$ denotes the position of the particle, $p(t)$ is the periodic signal and $w(t)$ stands for the noise. Since it has recently been shown that if the periodic excitation is sinusoidal, the SNR gain in the double well is almost always below unity even though the system works in the NLR range [20], we used a symmetric periodic pulse train with variable amplitude and duty cycle as periodic input signal $p(t)$ (see Fig. 1). The noise $w(t)$ we applied was a band-limited Gaussian white noise.

We have developed a mixed-signal – both analogue and digital – circuitry for our investigations. We generated the periodic excitation and the noise in the digital domain, then applied D/A converters to convert them into analogue signals. In order to solve Eq. (5) by analogue methods, we transformed it into an integral equation first:

$$x(t) = \int_0^t [x(t') - x^3(t') + p(t') + w(t')]dt'. \tag{6}$$

We applied analogue circuits to realise the mathematical operations required to solve this integral equation, such as integration, multiplication, addition and subtraction. The whole system was driven by a high-performance digital signal processor (DSP) and the data were acquired by high-resolution A/D converters with fast roll-off anti-aliasing filters. We used the same data acquisition equipment (including filters) for input and output signals, guaranteeing the same measurement method and thus avoiding artefacts due to different treatment. A host personal computer running LabVIEW controlled the DSP and performed additional evaluation tasks. The block diagram of the system is shown in Fig. 5.

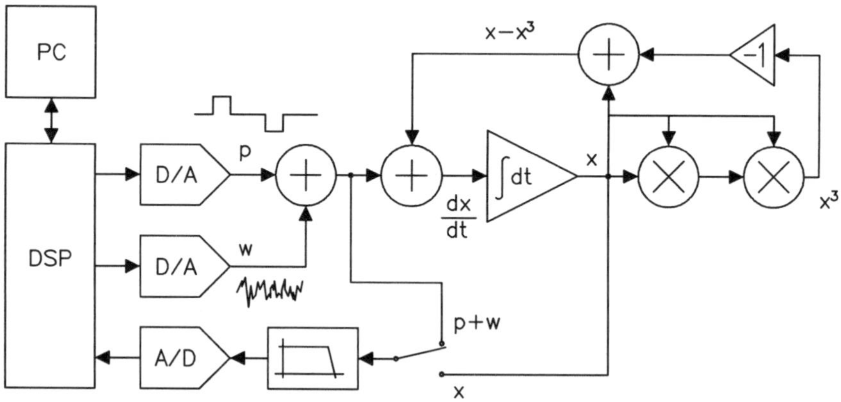


Fig. 5. Block diagram of the mixed-signal simulation system. The periodic signal and the noise are generated digitally and fed into an analogue simulation system. The solution *x(t)* is digitised by an A/D converter and the data are processed by a DSP and the host PC.

It is important to note that the analogue realisation introduces a $1/t_0$ factor in the right side of Eq. (6). Correspondingly, the original Langevin equation (5) is modified in the following way:

$$t_0 \cdot \frac{dx}{dt} = x(t) - x^3(t) + p(t) + w(t). \tag{7}$$

In our case, the value of $t_0$ is $1.2 \cdot 10^{-4}$ s. This implies that the signal frequency and the bandwidth of the noise (to be given in the next paragraph) must be re-scaled if one would like to use the original Langevin equation (5).

We measured the threshold value of the input signal amplitude ($A_T$) — the minimum value at which switching between the two wells can occur without noise — and set input signal amplitudes that were 70, 80 and 90% of this threshold. For every noise amplitude value $\sigma$, we recorded one thousand samples of *x(t)*, each consisting of 2048 data points, and from these records we calculated the averaged power spectral density (PSD) of the signal. Using this averaged PSD, we computed the SNRs according to definitions (2) and (3), then obtained both the narrow-band and the wide-band gains as in (4a-b). The sampling frequency was 8 kHz, the frequency of the periodic input signal was set to 31.25 Hz and the bandwidth of the white noise was 50 kHz (these parameters pertain to Eq. (7) and correspond to $3.75 \cdot 10^{-3}$ Hz and 6 Hz in Eq. (5)).

The results of our simulations can be seen in Figs. 6, 7 and 8. The behaviour reflected in these figures is very similar to that of the Schmitt trigger, though here the obtainable gains are smaller and the choice of the SNR definition makes greater difference: while the wide-band gain can exceed unity for all three signal amplitudes, its narrow-band counterpart always remains smaller than one for signals whose amplitude is 70% of the threshold value. Increasing the amplitude and the duty cycle of the input pulse train has the same kind of effect here as in the case of the Schmitt trigger.

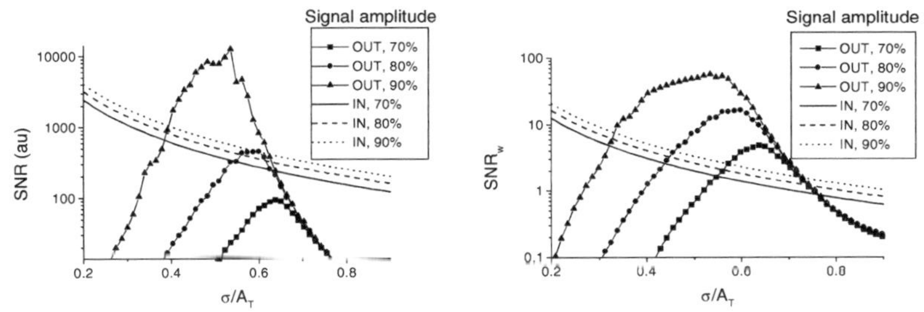


Fig. 6. Input (IN) and output (OUT) SNRs in the double well as functions of the input noise amplitude $\sigma$, both narrow-band (SNR, left panel) and wide-band ($SNR_w$, right panel). The duty cycle is 10%.

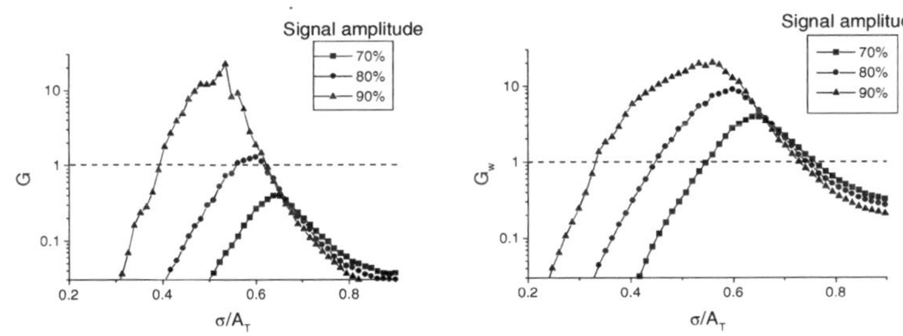


Fig. 7. Narrow-band ($G$) and wide-band ($G_w$) SNR gains in the double well as functions of input noise amplitude $\sigma$, for three different values of periodic signal amplitude. The latter is expressed as a percentage of the threshold value $A_T$, and $\sigma$ is also scaled by $A_T$. The duty cycle is 10%.

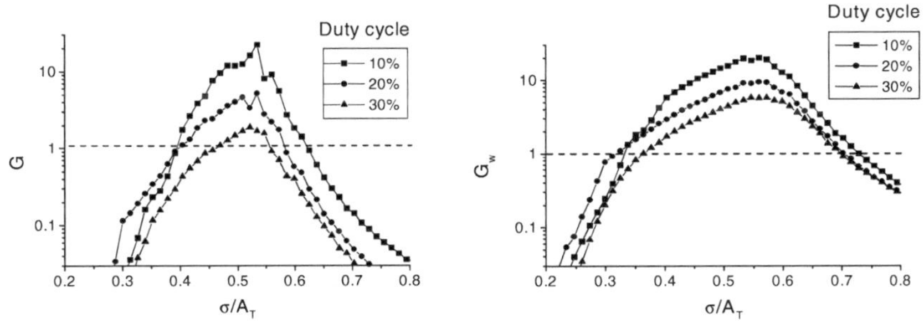


Fig. 8. Narrow-band ($G$) and wide-band ($G_w$) gains as functions of input noise amplitude $\sigma$, for three different duty cycles. The amplitude of the noise is scaled by the threshold value $A_T$; the amplitude of the periodic signal is 90% of the threshold value.

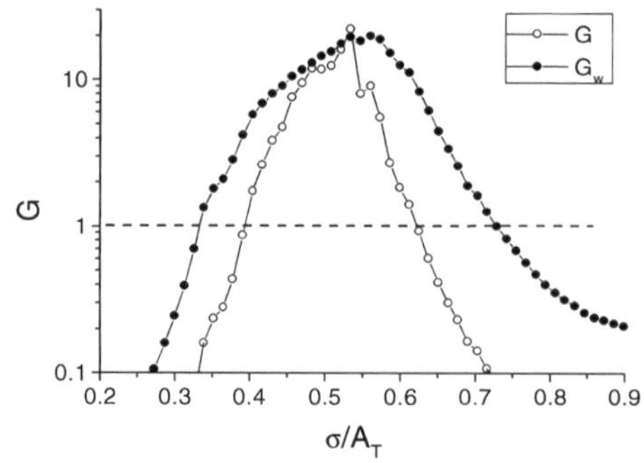


Fig. 9. Narrow-band ($G$) and wide-band ($G_w$) SNR gains compared in the double well. The amplitude of the noise is denoted by $\sigma$; it is scaled by the threshold value $A_T$. The amplitude of the periodic signal is 90% of the threshold value, and the duty cycle is 10%.

## 4. Conclusions

On the basis of our previous studies, we have compared two bistable systems — the non-dynamical Schmitt trigger and the double-well potential, which is dynamical — with regard to the signal-to-noise ratio gain induced by stochastic resonance. We have demonstrated that these systems can provide significant SNR amplification with a periodic pulse train plus Gaussian white noise as input. We have also shown that the dependence of the SNR gain on the amplitude and the duty cycle of the input signal is very similar in these systems: in both cases, greater gains can be obtained by increasing the amplitude and decreasing the duty cycle of the periodic input. This implies a common mechanism in the background, namely, that the input SNR drops considerably if the duty cycle decreases, while the output SNR is less sensitive to the value of the duty cycle, and the sum of these effects may result in SNR amplification. The main difference between the Schmitt trigger and the double well lies in the fact that the non-linear dynamics of the double well introduces more complex signal-transfer effects, which leads to smaller SNR gains and a stronger dependence on the input signal amplitude.

Our findings strengthen the view that SNR gain is significant only for signals wherein the power is concentrated in a narrow, pulse-like segment. These pulse-like signals are not without relevance: the majority of signals encountered in neural systems — which represent one of the main direction for the application of SR-related phenomena — also belong to this class.

We would also like to emphasise that the choice of the SNR definition has considerable significance: in both cases, SNR gain is greater than unity even over a wider range if we apply a much more realistic wide-band definition based on the total power of signal and noise.

## Acknowledgements

We would like to convey our special thanks to the anonymous referee for helpful and constructive comments. Our research has been supported by OTKA (Hungary), under grant T037664.

## References


[1] L. B. Kiss, *Possible breakthrough: significant improvement of signal to noise ratio by stochastic resonance,* in Chaotic, Fractal, and Nonlinear Signal Processing, Proc. Am. Institute Phys., ed. R. Katz, Mystic, Connecticut, USA (1996) 382–396.

[2] K. Loerincz, Z. Gingl and L. B. Kiss, *A stochastic resonator is able to greatly improve signal-to-noise ratio, Phys. Lett. A* **224** (1996) 63–67.

[3] F. Chapeau-Blondeau and X. Godivier, *Theory of stochastic resonance in signal transmission by static non-linear systems, Phys. Rev. E* **55** (1997) 1478–1495.

[4] F. Liu, Y. Yu and W. Wang, *Signal-to-noise ratio gain in neuronal systems, Phys. Rev. E* **63** (2001) 05912.

[5] D. G. Luchinsky and P. V. E. McClintock, *Irreversibility of classical fluctuations studied in analogue electrical circuits, Nature* **389** (1997) 463–466.

[6] S. M. Bezrukov and I. Vodyanoy, *Stochastic resonance in non-dynamical systems without response thresholds, Nature* **385** (1997) 319–321.

[7] S. Kadar, J. Wang and K. Showalter, *Noise-supported traveling waves in sub-excitable media, Nature* **391** (1998) 770–773.

[8] S. M. Bezrukov and I. Vodyanoy, *Noise-induced enhancement of signal-transduction across voltage-dependent ion channels, Nature* **378** (1995) 362–364.

[9] K. Wiesenfeld and F. Moss, *Stochastic resonance and the benefits of noise from ice ages to crayfish and squids, Nature* **373** (1995) 33–36.

[10] J. M. G. Vilar, G. Gomila and J. M. Rubi, *Stochastic resonance in noisy nondynamical systems, Phys. Rev. Lett.* **81** (1998) 14–17.

[11] M. E. Inchiosa, A. R. Bulsara, A. D. Hibbs and B. R. Whitecotton, *Signal enhancement in a nonlinear transfer characteristic, Phys. Rev. Lett.* **80** (1998) 1381–1392.

[12] J. J. Collins, C. C. Chow and T. T. Imhoff, *Aperiodic stochastic resonance in excitable systems, Phys. Rev. E* **52** (1995) 3321–3324.

[13] L. B. Kiss, Z. Gingl, Z. Marton, J. Kertesz, F. Moss, G. Schmera and A. R. Bulsara, *1/f noise in systems showing stochastic resonance, J. Stat. Phys.* **70** (1993) 451–462.

[14] I. Hidaka, D. Nozaki and Y. Yamamoto, *Functional stochastic resonance in the human brain: Noise induced sensitization of baroreflex system, Phys. Rev. Lett.* **85** (2000) 3740-3743.

[15] M. I. Dykman, D. G. Luchinsky, R. Mannella, P. V. E. McClintock, N. D. Stein and N. G. Stocks, *Stochastic resonance in perspective, Il Nuovo Cimento D* **17D** (1995) 661–683.

[16] Z. Gingl, R. Vajtai and L. B. Kiss, *Signal-to-noise ratio gain by stochastic resonance in a bistable system, Chaos, Solitons and Fractals* **11** (2000) 1929–1932.

[17] Z. Gingl, P. Makra and R. Vajtai, *High signal-to-noise ratio gain by stochastic resonance in a double well, Fluctuation and Noise Letters* **1** (2001) L181–L188.

[18] L. Gammaitoni, P. Hänggi, P. Jung and F. Marchesoni, *Stochastic resonance, Rev. Mod. Phys.* **70** (1998) 223–287.

[19] J. W. C. Robinson, D. E. Asraf, A. R. Bulsara and M. E. Inchiosa, *Information-theoretic distance measures and a generalization of stochastic resonance, Phys. Rev. Lett.* **81** (2001) 2850–2853.

[20] P. Hänggi, M. E. Inchiosa, D. Fogliatti and A. R. Bulsara, *Nonlinear stochastic resonance: The saga of anomalous output-input gain, Phys. Rev. E* **62** (2000) 6155–6163.

# ASYMMETRY AND DISORDER: A DECADE OF PARRONDO'S PARADOX

DEREK ABBOTT

*Centre for Biomedical Engineering (CBME)*
*and School of Electrical & Electronic Engineering*
*University of Adelaide, Adelaide, SA 5005, Australia*
*dabbott@eleceng.adelaide.edu.au*

Received 8 June 2009
Accepted 29 June 2009

Communicated by Christian van den Broeck

In 1996, Parrondo's games were first constructed using a simple coin tossing scenario, demonstrating the paradoxical situation where individually losing games combine to win. Parrondo's principle has become paradigmatic for situations where losing strategies or deleterious effects can combine to win. Intriguingly, there are deep connections between the Parrondo effect and a range of physical phenomena, as it turns out that Parrondo's original games are a discrete-time and discrete-space version of a flashing Brownian ratchet. This has been formally established via discretization of the Fokker–Planck equation. Over the past decade, many examples ranging from physics to population genetics have been reported in the literature pointing to the generality of Parrondo's principle. In general terms, the Parrondo effect occurs where there is a nonlinear interaction of random behavior with an asymmetry, and can be mathematically understood in terms of a convex linear combination. Many effects, where randomness plays a constructive role, such as stochastic resonance, volatility pumping, the Brazil nut paradox, etc., can be viewed as being in the class of Parrondian phenomena. We will briefly review the history of Parrondo's paradox, recent developments, and connections to related phenomena.

*Keywords*: Parrondo's paradox; Parrondo's games; Brownian ratchets; game theory; volatility pumping; random mixtures; Allison mixtures.

## 1. Introduction

> *"Can the weaker be the stronger?"*
> Kwai Chang "Grasshopper" Caine
> in "Chains" (Episode 9)
> *Kung Fu*, Season 1, 1972.

Parrondo's paradox is where losing situations combine in order to win, and is exemplified by simple coin tossing games [1] that readily yield to physical and mathematical exploration [2–6]. The Parrondian paradigm is one of "survival of the weakest" and is a counterintuitive nonlinear effect. Parrondo's original games comprise simple coin tossing and are thus not game theory in the von Neumann sense [7] where players make decisions — however, in their original form they can be thought of as game theory in the Blackwell sense [8], and more recently Behrends has extended Parrondo's original games to include player strategy [9, 10] thus ushering them into the von Neumann realm. Consequently, in the following review we will use the term *game-theoretic* in its most inclusive sense — in the new field of quantum game theory, it is interesting to note that the phrase "game theory" is also used broadly. In general, the emerging interest in game theory in the field of physics [11] uses the term in its broadest sense.

This review paper is constructed as follows. Firstly, we take an entertaining look at a number of everyday examples of "losing to win" or where the "weakest is the strongest", to illustrate that the idea is widespread and to motivate the topic. Then we briefly go through the history of Parrondo's games, how they are constructed, how they work, and trace their origins to the flashing ratchet and the Feynman–Smoluchowski ratchet. We review these ratchet devices in order to help the reader develop an understanding of the physical origins of Parrondo's original games. We then review some recent developments in the study of Parrondo effects in a number of diverse fields and also review some interesting closely related phenomena. In particular, we show how *volatility pumping* on the stock market, in its simplest form, can be simulated and point out its similarities as a ratcheting effect.

Finally we conclude with a discussion on the thermodynamics of chance and then exploit thermodynamic analogies to develop a physical picture to explain a new intriguing Parrondo effect: the *Allison mixture* [12]. Here, the Allison mixture is the counterintuitive situation where the random shuffling of random sequences begins to "erase" their randomness. Stated mathematically, two sequences with zero autocovariance can be randomly interleaved resulting in a sequence with nonzero autocovariance.

## 2. The Ubiquity of "Losing in Order to Win"

Is the Parrondian paradigm of losing to win that surprising? After all, many of us are familiar with the concept of a sacrifice in the game of chess. Also in biology, it is known that as a genotype evolves, the fitness landscape is usually not flat but can have a valley, i.e., fitness declines, before the genotype rises to a higher level of fitness (e.g., see [13]). It has been speculated that Winston Churchill deliberately turned a blind eye to the November 14th, 1940, German bombing raid of the city of Coventry [14] — the implication being that Churchill allowed the bombing to proceed to disguise the fact that the German Enigma code had been decrypted at Bletchley Park, in order to save more "important" cities than Coventry. Whilst many historians now believe this to be an urban legend, the anecdote nonetheless illustrates the general notion of sustaining a loss in order to win. In the engineering literature it is known that individually unstable systems can become stable if coupled together [15] and in the physics literature we have the principle that many imperfect devices can be combined to produce a near-perfect device [16]. If one believes that the biogenesis of life occurred in a primordial soup, one is faced with the conundrum that a number of "losing" effects must somehow have cooperated to produce life out of the incipient disorder [17]. Akin to the paradox of biogenesis, is the paradox of measurement: how is that, from the application of "losing" inaccurate stone-age tools, mankind has developed "winning" ways of measuring down to the nanoscale? A similar question can be asked about the nature of knowledge itself: how is that a collection of "losing" assertions, each of which only have a limited probability of being true, can accumulate into a "winning" reliable body of knowledge? This epistemological question, in the context of Parrondo's principle, has been explored by Stjernberg [18].

These and the examples that are about to follow illustrate that the idea of sustaining losses in order to win is ubiquitous and thus prompts us to study the new game theory of "losing to win", motivating the detailed study of Parrondo's paradox in order that we might understand the general principles more deeply.

### 2.1. *The trueling problem*

The *truel* is similar to a traditional duel except three, rather than two, players have a shoot out. The last man standing is the winner. Here we specifically discuss the *sequential* truel where the gunmen take it in turns to shoot. The detailed rules are in the caption of Fig. 1, but essentially the weakest Player A has first shot, then Player B, and so on. If you are the weakest player and you start, what is your best strategy? Should you try to eliminate the strongest out of Player B or Player C? The answer surprisingly is neither! It turns out that your best strategy for survival is in fact to waste your bullet and shoot into the air. For the full analysis see Flitney *et al.* [19] and Amengual *et al.* [20] — however, the general principle is that by sacrificing your turn you leave a greater chance for the more powerful players to fight it out between them. This game is an intriguing example of where "survival of the weakest" relies on making the weakest first move. The rules of the game presently assume unlimited resources — the case of limited rounds of bullets has been considered [21].

The concept of the truel has some bearing on the dynamics of political parties — in large democracies it is interesting to note that invariably there are always two major parties: something akin to "Republican" and "Democrat". All other parties outside the main two tend to be very minor in comparison. Why is this? One possible conjecture is that as soon as a third party starts to become significant it takes more votes away from the politically closest major party than the diametrically opposite party. This becomes self-defeating, as then the diametrical opposition wins! So, it is far more strategic to either stand back and let the two major parties fight it out (rather like the truel), or collude and join with the politically closest party. A famous example of this effect was in the US 2000 election, where Bush won by a

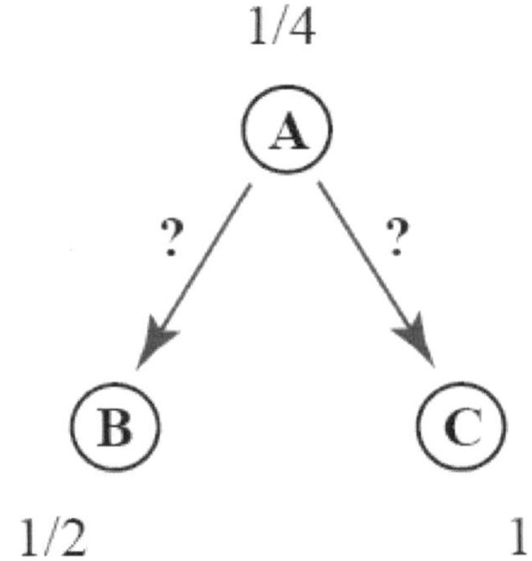


Fig. 1. The trueling problem. This is similar to a wild west dual, except that we now have three instead of two players. The weakest is Player A who only shoots with a success rate of 1/4. Player B is a better marksman who shoots with a success rate of 1/2, and Player C is a gunman who is a guaranteed perfect shot. The rule is that Player A shoots first, then Player B, then C, and so on in sequential order until one man is left standing. Each player must shoot on each turn, and you may assume unlimited resources, i.e., an infinite supply of ammunition. If you are Player A, what is your best starting strategy in order to survive? Should you shoot Player B or C? The solution is surprising, as it turns out that the weakest player can strengthen his chances by making the weakest move.

small margin (a margin so narrow that it may be considered as statistical noise) — however as much as 2.47% of the vote went to a minor party led by Nader. It is often speculated that had Nader not run, then Gore would have risen above the noise level.

Similarly, this type of reasoning has some explicative power for suggesting why in wars there are usually only two major factions: the "enemy" and the "allies". There do not appear to be many major historical cases where there are simultaneously an Enemy 1, an Enemy 2, and an Enemy 3 that fight each other in opposition without collusion. That is because if one is the smaller of the three factions, it is far better to play the truel gambit by stepping back and letting the larger two enemies annihilate each other. It may also be interesting to extend this line of reasoning to explain why a marriage between two partners appears to be more stable than an $n$-partner marriage. Another interesting open question is that in sexual reproduction why are there only two sexes (male and female) and not $n$ distinct sexes that mate either sequentially or simultaneously? — this is in fact a major field of research with rich multidisciplinary activity [22–26].

### 2.2. *The interplay of redundancy and pleiotropy*

The term *pleiotropy* describes an agent that performs multiple tasks [27, 28], while redundancy is when multiple agents perform the same task. This is clearly illustrated in Fig. 2 where we see that pleiotropy can be thought of as the inverse of

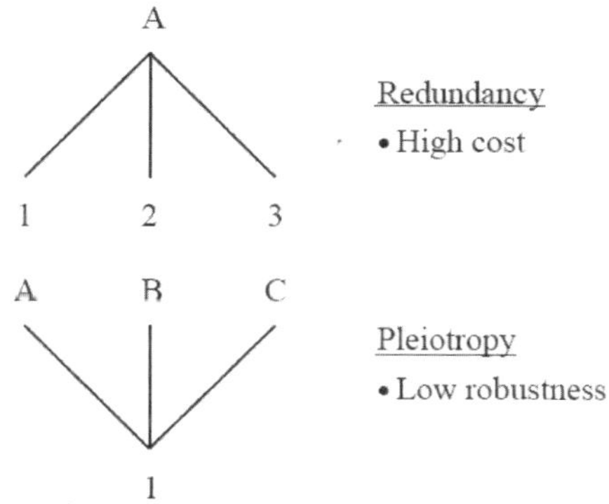


Fig. 2. Redundancy versus pleiotropy. In the top example, we have Agents 1, 2 and 3 all performing one Task A. This redundancy provides robustness at the cost of providing multiple agents. In the bottom example, we have Agent 1, performing three tasks A, B, and C. This pleiotropic situation provides the fulfillment of multiple tasks, but at the expense of low robustness — should Agent 1 fail to function, there would be a catastrophic reduction in output.

redundancy. Pleiotropy and redundancy can be ubiquitously seen in many every day networks, ranging from neural interconnections, telecommunication networks, gene networks, through to client-server based networks made up of server nodes and client nodes.

Figure 2 shows that, individually, pleiotropy and redundancy are rather like "losing games", as redundancy comes at high cost and pleiotropy comes with low robustness. Figure 3 illustrates that a mixture or interplay between pleiotropy and redundancy helps to overcome their individual disadvantages. Biological systems provide important examples of pleiotropy and redundancy [29, 30] — intercellular messenger molecules such as cytokines may act as links between nodes (cells) [31]. A deeper knowledge of how pleiotropy and redundancy operate within the cytokine networks, may improve understanding of how to better manipulate disease states [32–34]. To date, little work has been carried out to explore the trade-offs between pleiotropy and redundancy in an evolutionary computational paradigm — future work in this area may help to explore the general principles

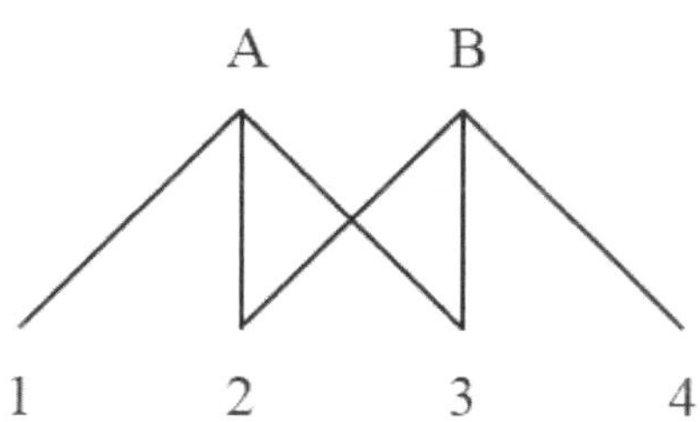


Fig. 3. Mixing redundancy and pleiotropy. Here we see the interplay of pleiotropy and redundancy can overcome their individual disadvantages shown in Fig. 2. We can intuitively see that we have increased robustness at lower average cost per output function.

behind such trade-off in the presence of both limited and unbounded resources. This may enable us to answer a number of fundamental open questions about how real biological, social, and electronic networks are optimally wired.

### 2.3. *Costly signalling*

A large area of research where there is a complex interplay of both losing and winning strategies is that of *costly signalling* [35]. Costly signalling is a term used by evolutionary biologists for the situation whereby an animal advertises its fitness, for example, for procuring a mate. In order to ensure that the signal is "honest" it has been conjectured that it must come at a cost to the animal — otherwise it would be too easy to send out fake signals. The classic example is the fancy plumage of the male peacock. The larger these feathers are the more attractive the male becomes to his entourage of females. However, the feathers come at cost because (a) they make the male easier to spot by a predator, and (b) the feathers are cumbersome when escaping from a predator. Therefore, the conjecture is that the feathers are an honest signal, because they advertise that the male is fit enough to survive despite them. Thus in order to "win" and find the optimal mate, the male plays the losing strategy of becoming vulnerable to predators.

## 3. History of Parrondo's Games

The original Parrondo games [1, 36] were devised in 1996 [37], as a pedagogical analogy of a flashing Brownian ratchet [38]. Since then they have stimulated research in diverse areas from economics [39], through to physical quantum systems [40–42], and population genetics [43–45]. For a review of the early development, see [46].

Part of the original appeal of Parrondo's games is that they clearly illustrated effect of "losing to win" for the first time with a toy model involving simple coin tossing games. Other related phenomena existed prior to Parrondo's games [38, 47–53], but Parrondo was the first to show the effect in a clear game-theoretic form. His work was a landmark discovery because the simple analytical solution to his model enabled many workers to grasp the theory behind the phenomenon of "losing to win". Another significant event was when Parrondo's original games were first shown be formally related to the Fokker–Planck equation [54], then independently confirmed [55], and rigorously systemized [56]. This is significant as it opens up a formal link between thermodynamics and games of chance (see Sec. 5). Parrondo's games were originally inspired by the flashing Brownian ratchet [38, 46], and via the Fokker–Planck equation they are intrinsically related. The flashing ratchet was, in turn, inspired by the Feynman–Smoluchowski ratchet and pawl machine, which we now briefly review below.

### 3.1. *The ratchet and pawl machine*

The ratchet and pawl machine is illustrated in Fig. 4. The idea is that the ratchet wheel is biased to turn in one direction because of the action of a spring loaded

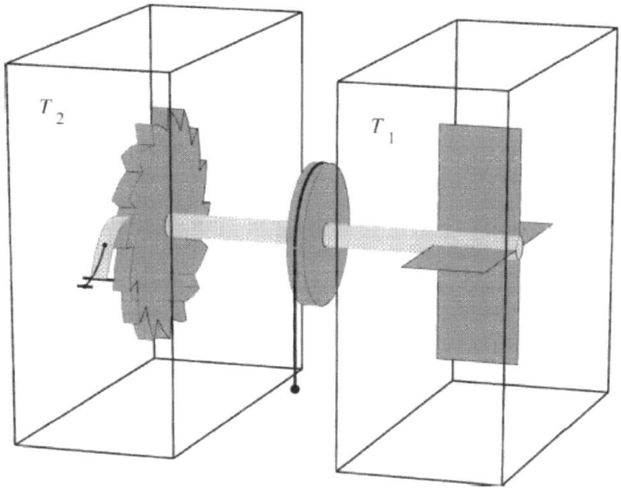


Fig. 4. The Feynman–Smoluchowski Engine (FSE). On the left is a ratchet wheel and a spring loaded pawl at temperature $T_2$. On the right is a vane at temperature $T_1$. The ratchet wheel is connected to the vane via an axle. Now imagine the whole device is small enough so that the bombardment of air molecules causes the vane to rotate. As the whole machine is constrained to rotate in one direction, by the ratchet and pawl, the random motion of the air molecules is in-effect *rectified* to produce directed motion and the rotation of the axle can legally do useful work by lifting a weight via a small pulley. This is allowed provided that $T_1 > T_2$, whence we have net energy coming into the system to maintain this temperature gradient in the first place. Hence there is external energy driving the system to do work. However, when $T_2 = T_1$ there is no longer any net energy into the system, and it becomes therefore impossible to lift the weight. In the case of thermal equilibrium ($T_1 = T_2$) the spring loaded pawl fluctuates to release the wheel to go in the wrong direction. It turns out there is *detailed balance* and thus the weight jiggles up and down, but there is no net displacement on average. After [36].

latch (called the *pawl*). We see in Fig. 4, that this ratchet wheel is connected to a vane via an axle. For generality, we can imagine that the vane is in a box maintained at a temperature $T_1$ and the ratchet is in a box at $T_2$. In 1900, the French Nobel Prize winner, Gabriel Lipmann was the first to do the thought experiment of shrinking this type of apparatus down to the scale of air molecules [57]. Could such a ratchet mechanism be used to rectify the random motion of molecules? Lipmann was the first to ask this question — it was a courageous question at that time considering that the discrete nature of matter had not yet been finally settled. Lipmann also asked if such rectification of random motion would violate the laws of thermodynamics. This caused a flurry of letters to journals, and finally in 1912 Smoluchowski came up with the canonical explanation that we hold to this day [57].

Smoluchowski correctly explained that the machine could legally rectify random motion and do useful work provided $T_1 > T_2$. To maintain such a temperature difference requires external energy. Thus the work output is at the expense of energy in — this principle universally holds for all types of engines and there is no violation of thermodynamics. However, the question is why does the machine stop working when there is no input energy (i.e., when $T_2 = T_1$)? Again, Smoluchowski brilliantly gave the correct answer — he explained that the pawl is also bombarded by air molecules and thus has a certain error rate of releasing the wheel to rotate in the wrong direction. He stated that at thermal equilibrium (i.e., when $T_2 = T_1$) we thus expect the probabilities of the wheel rotating either way to balance out, and therefore the machine can never do any net work.

However, Smoluchowski did not attempt to formally prove that the probabilities satisfied this required *detailed balance* condition. In 1963, Feynman was the first to attempt to do so using Boltzmann statistics showing the probabilities did indeed balance — however, he did not publish the fine details of the calculation [58]. Around 1980, Abbott first attempted to derive Feynman's result from first principles and was unable to do it for 19 years [59]. In 1997, Abbott flew to Madrid to visit Parrondo and show him the problem — at the time a solution was not forthcoming and so Parrondo began discussing ratchets, in general, with Abbott demonstrating his paradoxical games inspired by the flashing ratchet [59]. From that meeting the seminal papers on Parrondo's games were born [1, 36]. Finally, in 1999 the problem of finding Feynman's detailed balance condition was cracked using level crossing statistics rather than Boltzmann statistics [60] — to this day the Boltzmann form remains an unsolved problem.

### 3.2. *"Kitchen sink" examples*

We now briefly sample a few everyday (i.e., "kitchen sink") examples of ratchets to illustrate their generality. In the previous section, the ratchet and pawl machine relied on the asymmetry of the ratchet teeth in order to operate — this is a spatial asymmetry, but ratchet effects are not limited to the spatial variable. Here we will see that an asymmetry in any arbitrary variable can lead to a ratcheting mechanism.

Every child knows that if you randomly jiggle a bowl of sugar, a bag of flour or a backet of sand, the lumps rise to the top — the scientific name for this phenomenon is the *Brazil nut paradox* [53], named after that fact that the large Brazil nuts rise to the top when you shake a bag of mixed nuts. Here, the random shaking of the container drives the large nuts "uphill" against the gravitational gradient and thus this is clearly a Brownian ratchet — but where is the asymmetry? The asymmetry in this case lies in the size distribution of the particles and the fact that gravity is directional.

Another common example is that of *longshore drift* on a beach — here, it is common to find that the sand and shells tend to pile up on one end of the beach. This tends to happen when waves come in at an angle to the beachfront. So for example, if we have a south facing beach, and waves impinging in a north-east direction, then sand and shells will tend to pile up on the east side of the beach. Waves will come in a north-easterly direction, but ebb in a southerly direction, drawing out a ratchet-like profile, and dragging material toward the east. Incoming waves loosen the material, reducing frictional forces, and as the waves ebb away friction increases again. Thus the ratchet asymmetry is in the difference between angle of entry and angle of ebb, as well as difference in frictional forces experienced by the material.

When trading on the stock market, a common injunction is to *buy-low sell-high* in order to ratchet up one's gain. The asymmetry here is in price when we buy and sell, in order to exploit the natural price fluctuations in the market. When paying the restaurant check, at the end of a meal, a client will typically complain if he or she is over charged. However, if the check is accidentally under charged, the client might chose to stay silent. This asymmetry in the transmission of information is used to ratchet up the gain of the client. This is somewhat akin to the previous buy-low sell-high example.

So far we have seen spatial, frictional, informational, and money ratchets — but is a ratchet in the time variable possible? The answer is yes. To illustrate a time ratchet we briefly review the *two-girlfriend paradox*, which is an old chestnut due to Perelman [61] back in the 1930s, although it was later revived in the 1960s by Mosteller [62], and then modernized to be shown to be a ratchet in the early 2000s [51]. The two-girlfriend problem is a brainteaser that goes as follows. Refering to Fig. 5, we are told that Bill arrives at a train station at a random time each day. One train leaves for the east every 10 mins and one train leaves for the west every 10 mins — his strategy is to jump on whichever train arrives first. It turns out on average that he sees Monica nine times more often than Hillary. Why is this so? This seems a little hard to believe given that he arrives at a time random time each day. The answer is that this is a phase (time) ratchet and we must therefore look for an asymmetry in the time variable. In other words, there can be a phase difference between the trains. Imagine a scenario where the eastbound train leaves every 10 mins on the hour, and the westbound train leaves every 10 mins one minute later. If Bill arrives after, say, 10:11 am he will have a nine minute window that

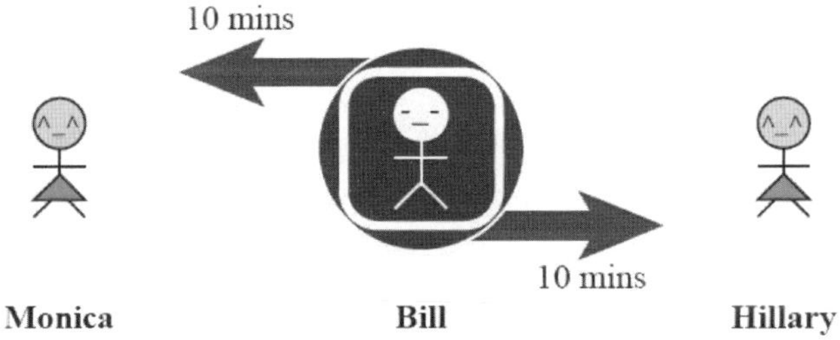


Fig. 5. The 2-girl paradox. The circle represents a city. At the center of the city is a train station, represented by a square. A westbound train leaves every 10 mins and an eastbound train leaves every 10 mins. Bill has two girlfriends, ones lives in the west and the other on the east side of the city. Bill arrives at the station at a random time each day and takes whichever train is there first. It turns out, on average, that Bill ends up visiting the westside girl nine times more than the eastside girl. Why? The solution to this puzzle reveals that the process is in fact a Brownian ratchet, where the asymmetry lies in the phase difference between trains.

Table 1. Everyday examples of Brownian ratchet effects. These examples demonstrate the general ubiquity of Brownian ratchets, where a bias toward a particular direction is a result of the interaction between random behavior and an asymmetry. The traditional focus has been on Brownian ratchets with asymmetry in a spatial variable — the right column shows that other types of variables can yield to asymmetric treatment, leading to directed motion.

| Scenario | Source of randomness | Asymmetry |
|---|---|---|
| Brazil nut paradox | Shaking the container | Particle sizes/field |
| Longshore drift | Waves breaking on the beach | Geometry/friction |
| Restaurant check | Waiter's error rate | Information |
| Buy-low, sell-high | Market fluctuations | Price |
| 2-Girl paradox | Bill's arrival times | Train phase (time) |

captures the eastbound train, but if he arrives after 10:10 am there is a one minute window in which the westbound train will arrive first. Thus if he arrives randomly, he is more likely to end up in the nine minute window, and thus sees Monica nine times more often. Table 1 summarizes the above examples highlighting the different forms of asymmetry we have identified.

### 3.3. *The flashing ratchet*

The principle of the Feynman–Smoluchowski Engine (FSE), in Fig. 4, can be translated from a wheel to a linear mechanism. The flashing ratchet [38] is one example of this — we focus solely on this case as it is the type of Brownian ratchet that inspired Parrondo's games. The operation of the flashing ratchet is explained in the caption of Fig. 6, where we see that particles can be "pumped" uphill against a gradient by flashing the ratchet potential on and off. Energy that we input to toggle the potential on and off does work on the particles to move them uphill. The secret as to why the ratchet works is in the asymmetry of its sawtooth profile. It is this asymmetry that results in $P_{\text{fwd}} > P_{\text{bck}}$, giving rise to the net motion to the right in Fig. 6. Parrondo's genius was in extrapolating from the flashing ratchet to coin tossing games. He visualized going uphill as gaining money, and the random position of a particle as being the accumulated capital. He recognized displacement along the flat potential, $U_{\text{flat}}$, could be simulated by winnings from a simple coin toss, say Game A, and that the gradient could be simulated by bias in the coin. He then recognized that displacement along the sawtooth potential, $U_{\text{saw}}$, could be simulated by winnings from a game composed of two coins, call it Game B. In this case it turns out that two coins are needed as each tooth is composed of two slopes — the longer slope pushes particles in the "winning" direction and the shorter slope pushes them in the "losing" direction. The periodicity of the sawtooth potential is simulated by choosing a selection rule for the coins based on modulo arithmetic. Switching between games A and B simulates the ratchet flashing on and off. In the following subsection, we now examine the construction of the games.

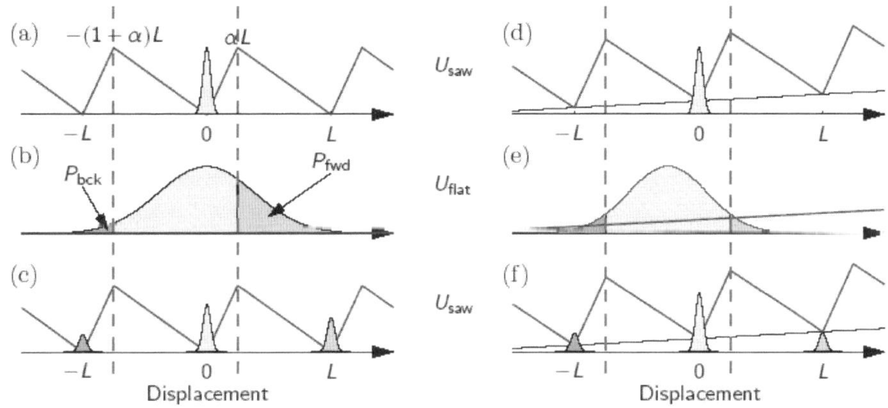


Fig. 6. The flashing ratchet of Adjari and Prost. (a) A ratchet sawtooth potential of pitch $L$. A Gaussian distribution of particles sits inside one of the potential valleys. (b) We now flash off the sawtooth potential off so that it becomes flat. The Gaussian distribution spreads as it is now unconstrained by a potential. Notice for convenience we have exaggerated the size of the Gaussian — in reality the area under the Gaussian is conserved. (c) We flash the ratchet potential back on. A rear tooth captures $P_{bck}$ of the distribution, and a forward tooth captures $P_{fwd}$. A remarkable feature is that it turns out that this ratcheting procedure still operates when working against a gradient, as illustrated in (d)–(f). The flashing ratchet enables the particles to climb "uphill" in a similar fashion to longshore drift on a beach. After [63].

### 3.4. *Parrondo's original games*

The key idea of Parrondo's games is that you can have two or more sets of games that are individually losing — however, if you periodically or randomly switch between the losing strategies, there are conditions under which it is possible to counterintuitively win. The games are constructed as indicated in Fig. 7 to cleverly simulate the action of the flashing ratchet that was expounded in the previous subsection. Game A simulates the flat potential and Game B simulates the sawtooth potential. As we can see, in Fig. 8, Game A and Game B are indeed losing games when played in isolation. Now, when we switch between the two games either periodically or randomly our winnings increase.

It has been shown using Discrete Time Markov Chain (DMTC) analysis [63] that the games are governed by very simple inequalities — Game A is losing provided,

$$\frac{1-p}{p} > 1 \tag{1}$$

and Game B is losing provided,

$$\frac{(1-p_1)(1-p_2)^2}{p_1 p_2^2} > 1 \tag{2}$$

and the random combination of Game A and Game B wins provided,

$$\frac{(1-q_1)(1-q_2)^2}{q_1 q_2^2} < 1, \tag{3}$$

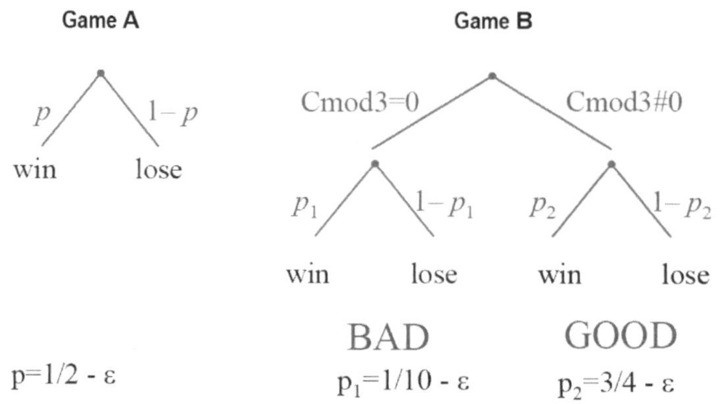


Fig. 7. The construction of Parrondo's original games. Game A is a simple coin toss that simulates the $U_{\text{flat}}$ state of the flashing ratchet. The coin's bias is $\epsilon$, which simulates the gradient of the flashing ratchet. Note that Game A is a losing game. Game B is composed of two coins. The "good" coin is favorable and simulates the ratchet tooth's long slope and the "bad" coin simulates the shorter slope of the ratchet tooth. For simplicity, your capital $C$ goes up or down by \$1 every time you win or lose. You toss the bad coin if your capital is a multiple of three, otherwise you toss the good coin — this modulo arithmetic simulates the periodicity of the ratchet profile. The parameters of Game B are such that it is a losing game overall. When we switch periodically or randomly between the two losing games, surprisingly, we win.

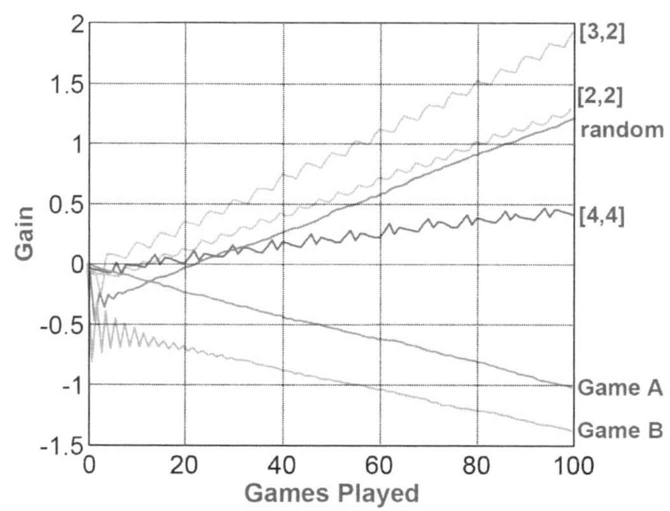


Fig. 8. The output of Parrondo's original games. The graph shows the amount of money gained versus the number of times we play. The parameters in this computer simulation are those listed in Fig. 7, with $\epsilon = 0.005$. As expected, we see that Game A and Game B are individually losing. If we play two rounds of A followed by two rounds of B, indicated as [2, 2], we find that we remarkably win. In fact any periodic combination wins and the graph shows [3,2] and [4,4] as examples. The curve marked "random" is that case where we decide to play A or B on the flip of an unbiased coin. The results are averaged over 50,000 trials to produce smooth curves, however, the trends are still observable for individual trials.

where $p, p_1$, and $p_2$ are defined in Fig. 7 and $q_1 = \gamma p + (1-\gamma)p_1$ and $q_2 = \gamma p + (1-\gamma)p_2$. Here, $\gamma$ is the probability that Game A is selected and $1-\gamma$ is the probability of playing Game B. There are many ways to form a physical picture of why Parrondo's games work as they do — the picture becomes clearer once it is realized that Game A is coupled to Game B via the capital dependent rule. The first physical picture, due to Parrondo, is to simply to view the games as a discrete analogy to the flashing ratchet. An alternative picture is to see that Game B has a state dependence on capital that is forcing it to lose, and that Game A is acting as a source of noise that is breaking up that state dependence — this has been dubbed the Boston Interpretation, as it grew out of discussions at H. E. Stanley's group at Boston University [64]. In fact, it has been shown that as the amount of Game A "noise" is gradually increased, by increasing $\gamma$, we find that the winnings follow a stochastic resonance-like curve [65]. Perhaps the most powerful interpretation is that of Moraal [66], who was the first to show that the games work due to a convex linear combination. This was later, and independently, reported by Costa *et al.* [67], which is recommended as a clearer exposition for the new reader. The realization that a convex linear combination is at the heart of the games is a significant one, as it then more readily links Parrondo's paradox to control and optimization problems. Another key point in understanding why the games work is that the Inequalities 2 and 3 are nonlinear — Parrondo's paradox is essentially a nonlinear phenomenon. The fundamental reason why the governing inequalities are nonlinear is due to the state-dependence in Game B — in mathematical terms this is equivalent to saying that Game B is not a martingale. Mathematical discussion of these various interpretations can be found in [68]. Finally, it should be noted that there are possibly an infinite number of ways to construct different nonlinear games that exhibit Parrondian effects. The open question is to search for the interesting cases that map onto physical and biological systems, and to investigate which display the largest regions of parameter space for the effect to occur. Progress has been made with developing differently constructed Parrondo games [69–72], but so far it is still early days and there is still much to explore in this regard. Another line of enquiry is the investigation of the minimal number of states required for the Parrondo effect to opperate. Based on Parrondo's original games a minimum of three states is required; but by suitable modification two-state processes have been shown to be indeed possible [73, 74] Most of our discussion, thus far, has focussed on Parrondo games with random switching — however, work on finding the optimum periodic sequence that maximizes the payoff has been carried out [75], opening up new interesting lines of enquiry.

## 4. Developments in Parrondo's Paradox and Related Phenomena

From an engineering viewpoint it is known that mixing unstable systems can result in stability, and the connection to Parrondo's paradox has been pointed out [15] — further investigation in this area may be of relevance to optimization problems.

In the area of neural networks it is known that a network can perform better at *network generalization* if noise is added to the training data set [76] — this evokes the idea of losing in order to win and connections to Parrondo effects have yet to be investigated. Parrondo effects in spin systems [66] and quantum game theory [40–42], have been reported. Increasing our understanding of how to control decoherence is one of the motivating factors behind these developments in quantum game theory.

It has been shown [77] that Parrondo's original games can be rather elegantly described in terms Onsager rate equations [78, 79] — indeed the possibility of Parrondo-like effects in chemical kinetics has been explored in a computational chemistry setting [80]. Parrondo effects have also inspired work in the study of negative mobility phenomena [81], reliability theory [82], noise induced synchronization [83], spatial patterns via switching [84], and in controlling chaos [85, 86].

In the area of mathematics, Pinsky and Schuetzow [47] have shown that switching between two transient diffusion processes in random media can form a positive recurrent process — this can be viewed as a continuous-time version of Parrondo's games. It has also been shown that declining random branching processes can be combined to paradoxically increase [50].

The area of biology is still ripe with open questions for the study of Parrondo-like effects. There are many examples in biology of "losing to win" — for example *sickle cell anemia* is deleterious and yet it can protect the host from contracting malaria. Conjectures have been mooted for the application of Parrondo's paradox to biogenesis [17], the dynamics of gene transcription in GCN4 protein [87], and the dynamics of transcription errors in DNA [87]. Parrondo's paradox has been studied in various interesting scenarios involving population genetics [43–45, 88].

In conventional sociobiology the standard dogma is that when chosing an optimal mate we are attracted to beauty, as those features we see in beauty are in fact indicators of a healthy mate — therefore to efficiently propagate our genes we seek attractive mates [89]. If this was really the complete picture, one might expect ugliness to have been selected out by now. The present theory does not seem to account for the common fact that two ugly parents can often produce an attractive child. Perhaps there is a Parrondian payoff in being a "loser" at the dating game, where survival of the weakest can come into play. Arizmendi [90] has recently proposed a Parrondian dating model that fosters survival of the "loser". Satinover and Sornette [91, 92] propose a Parrondian model where they show that short term optimization can turn a positive expected gain into a negative one, thus showing in some circumstances it can pay to be the uncompetitive "loser" who sticks to the minority.

The problem of collective decision making, in the context of Parrondo's paradox, has yielded interesting results [71, 93–97] and opens up fascinating lines of enquiry in the areas of group behavior and voting dynamics.

In terms of the stockmarket, Boman [98, 99] has used a Parrondian game framework as a toy model for studying the dynamics of insider information. A number of Parrondo-like toy models for switching between poor performing investments are well-known. For example, Maslov and Zhang demonstrate a model where switching between volatile assets and non-performing cash reserves produces an increase in the gains [48] in a fashion not too dissimilar from Luenberger's *volatility pumping* method [52]. There are also closely related models such as the *excess growth* model of Fernholz and Shay [100] and Tom Cover's *universal portfolio* [101]. There are open questions yet to be explored in systemizing these effects and drawing out the exact connections with the large body of literature on Parrondo's paradox.

Here, we focus on Luenberger's volatility pumping as it is a simple toy model that rather nicely illustrates the principles of "winning" with poorer stocks in a clear way. Figure 9 simulates two stocks: one stock is stable but is mediocre and in the long run neither wins or loses significantly, the other stock has some growth but is volatile. A very simple toy model of volatility is used here, namely, in that we randomly halve or double the stock value from day to day. Leunberger's method is to then sell both stocks each day and rebuy them implementing *portfolio rebalancing.* The rebalance operation is to take your total cash $\$T$, and buy $\$T/2$ of one stock and $\$T/2$ of the other — thus we maintain a 50:50 portfolio. This operation is repeated each day and remarkably it produces the top curve in Fig. 9 with exponential growth. The simple Matlab code for producing this graph is in Appendix A. Now,

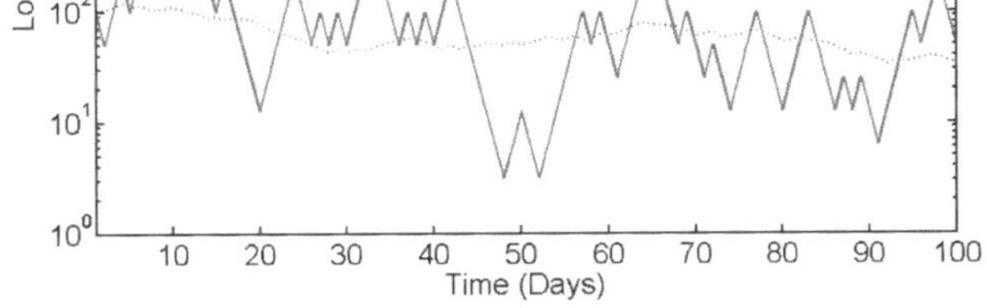


Fig. 9. Volatility pumping a low-risk stock with a high-risk stock. The dotted curve simulates a mediocre low-risk stock that in the long run neither wins nor loses. The solid curve represents a volatile stock that gives a 25% expected return, though is high-risk — a simple toy model of volatility is implemented here, where the stock simply halves or doubles, at random, its previous value at each time-step. The chained curve is found by selling both stocks at the end of each time step, adding the total cash to get $\$T$, then repurchasing them at the beginning of each time-step at a 50:50 split — that is, we purchase $\$T/2$ worth of the high-risk stock and $\$T/2$ worth of the low-risk-stock. This process is called *portfolio rebalancing.* Surprisingly, the chained curve grows exponentially, even though the two stocks individually do not perform as well. Both stocks start at Day 1 priced at $100, and thus the combined portfolio (chained curved) starts at $200. The vertical axis is the return in dollars plotted on a logarithmic scale. The return on the rebalanced portfolio is so large that we would not be able to see the individual curves, without the logarithmic plot.

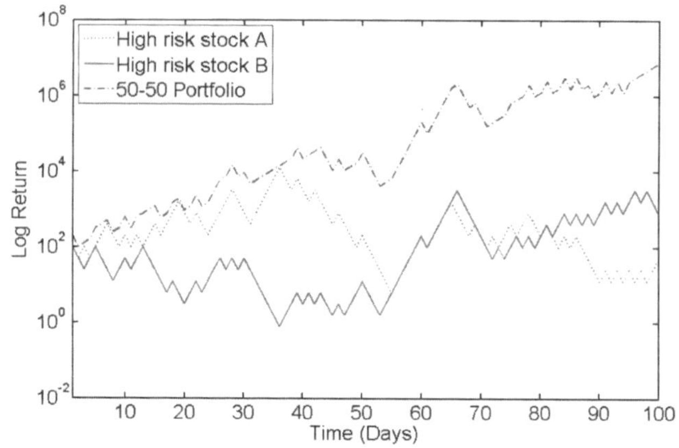


Fig. 10. Volatility pumping a high-risk stock with a high-risk stock. The scheme is identical to that in Fig. 9, with the exception that now both originating stocks are volatile. In this simulation both are generated by random halving and doubling, but are generated independently. Surprisingly, not only does the exponential growth still occur, but the winnings are about a factor of 100 higher than in the previous case.

it should be noted this is a stripped down toy model to illustrate the key idea that rebalancing creates growth. Surprisingly, Fig. 10 shows the process still works even when both stocks are volatile. Of course, in reality one would not buy and sell everyday as transaction costs would be prohibitive. Also the halving and doubling operation is an artificial construct that is cleverly designed so that we never hit rock bottom — in reality, hitting the $y = 0$ axis is always the problem. However, the beauty of any toy model is it enables us to explore the pertinent features of an interesting effect. The open questions are why does volatility pumping work and how should the portfolio rebalancing strategy be optimized for best performance in a real scenario? Whilst both these questions are still being actively debated [102], from the point of view of ratchet science, volatility pumping must clearly be the result of an asymmetry that rectifies fluctuations in the market. This is the principle behind every Brownian ratchet and volatility pumping is no exception. The action of maintaining the 50:50 portfolio split guarantees that we are always buying low and selling high — recalling Sec. 3.2 we argued that this is indeed a ratcheting asymmetry.

## 5. Thermodynamics of Games of Chance

The fact the Parrondo's original games can be exactly derived via discretization of the Fokker–Planck equation [54–56] is of fundamental interest because it can then serve as a useful toy model for investigating the discrete-continuous interface. There is much emerging interest in the so-called discrete-continuous interface due to its importance in optimization and control problems [103]. Furthermore, via adopting a $t \to it$ Wick rotation, it is possible to transform the Fokker–Planck equation into a Schrödinger equation [104] and this opens up interesting directions for future research of both Parrondo's games and the discrete-continuous interface in both classical and quantum regimes.

As Parrondo's games are deeply rooted in the thermodynamics of Brownian ratchets, via the Fokker–Planck equation, this has enabled the use of thermodynamic analogies when understanding the operation of the games. Various authors [36, 87, 105] have suggested simple analogies between Parrondo's games and physical Brownian motors. Amengual *et al.* [106] have taken this a step further and have proposed how to characterize Parrondo's games in terms of a thermodynamic engine efficiency.

This raises the open question of whether a generalized thermodynamic picture can be applied to arbitrary games of chance. To this end, in 2001, Abbott performed the following simple thought experiment [59]. Imagine two players flipping a simple unbiased coin. Player 1 wins $1, given heads, and Player 2 wins $1, given tails. The coin is unbiased, which means that if we take a video of the game, and run the movie backwards we would not be able to tell the difference. Therefore the system displays *time-reversibility*, which is what we expect of physical systems that are in thermal equilibrium. Now, let us imagine we have a biased coin such that, say, Player 1 progressively wins. The situation is no longer time-reversible, because the forward and backward movies are now clearly different — in the forward movie Player 1 appears to win, but in the reverse movie Player 2 appears to win. This time-irreversible situation corresponds to physical systems that are out of equilibrium. Thus the situation of *detailed balance* when the coin is unbiased is analogous to thermodynamic equilibrium, and a biased coin is rather like a system out of equilibrium. Initially, this thought experiment appears trivial and not particularly useful, however, in the next section we put it to good use illustrating a significant link between random manipulation of integer sequences and thermodynamics.

For those readers who are new to the concept that time-reversibility relates to thermal equilibrium, consider the following simple mental picture of Feynman's ratchet and pawl in Fig. 4. The two ends of the system are at different temperatures and the ratchet wheel rotates in one direction. Now if you make the two ends the same temperature, there is no net rotation of the ratchet wheel. Now take a movie of each of these two cases, and run the movie backwards. What do you see?

- Case 1: When the temperatures are different, $T_1 \neq T_2$. Here the backwards movie looks different to the forwards movie. Because the wheel is turning in a particular direction, the backwards movie would have the wheel turning the opposite way. As the movies are different, this is the acid test that tells us the process is irreversible.
- Case 2: When the temperatures are same, $T_1 = T_2$. Here we have thermal equilibrium, and so there is no net energy coming into the system and therefore the wheel does not rotate in a net direction. It just randomly jiggles back and forth, and no net work is done. We run the movie backwards and now we cannot tell

the difference — the backwards movie just looks like random jiggles, as does the forwards movie. Thus this case is reversible.

Thus, in summary, in thermal equilibrium we have reversibility, when out of thermal equilibrium we have irreversibility. This is a well-known principle in thermodynamics, but we have introduced the specific ratchet example as a nice physical picture for visualizing this concept.

## 6. A New Parrondian Effect: The Allison Mixture

A new form of Parrondo's paradox, namely the *Allison mixture*, has recently been reported [12]. Let us imagine two sequences of random numbers; we shall call them Sequence 1 and Sequence 2 — they are random in the sense that they are independent and have zero autocovariance. For simplicity, we can consider these sequences to be random strings of 1s and 0s — however, note that the effect we are about to describe is not limited to binary sequences but is in fact general. If we now randomly scramble these two sequences to generate a third new sequence, naively we would expect this resulting sequence to also have zero autocovariance. It turns out that this is not always the case: counterintuitively the final sequence can have a finite autocovariance $\rho$ even though the $\rho$'s of originating sequences are zero — this is what we call an *Allison mixture.*

Let us now be a little bit more precise about how we actually scramble the two sequences, so we can then write out an analytical expression for $\rho$ to show that it can be in fact nonzero. We start at an arbitrary $n$th position of Sequence 1. We either move to position $n+1$ of Sequence 1 with probability $1-\alpha_1$ or skip to position $n+1$ of Sequence 2 with probability $\alpha_1$. Whenever we find ourselves in Sequence 2, we hop to the next location on Sequence 1 with probability $\alpha_2$ or advance one step within Sequence 2 with probability $1-\alpha_2$. We continue hopping back and forth between the two sequences in this manner and each digit that we sequentially land on is called out to form the new sequence. In this way a third new sequence is generated from the original two sequences by random hopping, using separate transition probabilities $\alpha_1$ and $\alpha_2$ to keep everything perfectly general. For the sake of further generality, let the means of the two originating sequences be $\mu_1$ and $\mu_2$.

It has been shown that the autocovariance $\rho$ for the generated sequence is [12],

$$\rho = \frac{1}{\sigma^2}\frac{\alpha_2}{\alpha_1+\alpha_2}(\mu_1-\mu_2)^2(1-\alpha_1-\alpha_2), \tag{4}$$

where $\sigma^2$ is the variance of the final sequence. The full expression for the variance has been previously reported [12], but is not given here as it is not relevant for the following physical discussion. Our naive expectation is that a random mixture of sequences individually possessing $\rho = 0$ should always result in a final sequence with $\rho = 0$ — however, Eq. (4) reveals that $\rho$ is only zero provided $\mu_1 = \mu_2$ or $\alpha_1 + \alpha_2 = 1$. If we break both these conditions, then we can legally produce a sequence with a nonzero $\rho$. The mathematics dictates to us that this must be the case, but the real question is why? And what is the physical picture and physical basis for what is going on?

The previous section on the thermodynamics of chance — Sec. 5 — contains many of the necessary clues to unravel the physical picture. Firstly, let us address physically why to get $\rho \neq 0$ we must have $\mu_1 \neq \mu_2$ — the implication is that the means of the sequences are analogous to the temperature of a physical process. Loosely speaking, temperature is some measure that is proportional to the average of all the jiggling within a solid object. In the case of the random sequences, $\mu_1$ and $\mu_2$ are the averages of all the jiggling or varying numbers and play the same role as temperature. Thus when $\mu_1 \neq \mu_2$, we have an irreversible situation — the sequences are irreversibly mixed and we therefore get finite autocovariance in the final sequences, because there is information loss. Recall that the originating sequences have $\rho = 0$, and thus have lower compressibility in the Chaitin–Kolmogorov sense and thus contain higher information in the Shannon sense. Thus by subjecting them to an irreversible process we know from thermodynamics that we must *lose* information, and thus further redundancy must have crept into the final sequence leading to $\rho \neq 0$. Now, in the special case, when $\mu_1 = \mu_2$, we have reversible mixing, because this is analogous to thermal equilibrium where $T_1 = T_2$. If the process is reversible then there is no information loss, no extra redundancy is added, and therefore $\rho = 0$.

However, this is only part of the picture, as Eq. (4) also predicts that to obtain a case where $\rho \neq 0$, we must also obey the $\alpha_1 + \alpha_2 \neq 1$ condition. So what is the physical reason why $\alpha_1 + \alpha_2 \neq 1$ is required to get nonzero $\rho$ in the final sequence? To unravel the mystery we draw the state diagrams to illustrate the mechanism. Figure 11 illustrates the case when $\alpha_1 + \alpha_2 \neq 1$ — the caption explains why switching between the two states leads to *memory persistence* that causes correlation in the final sequence (or anticorrelation in the case of antipersistence). Now for the case when $\alpha_1 + \alpha_2 = 1$, Fig. 12 illustrates we get detailed balance between the probability of entering a state and the probability of staying in a state. (Note that staying within a state is also called a *self-transition*). Here, detailed balance implies there is no memory persistence and hence $\rho = 0$. An alternative valid explanation is to use the argument of Sec. 5 that explains why detailed balance implies a reversible process. So essentially we must have $\alpha_1 + \alpha_2 \neq 1$ to ensure irreversibility, which is a necessary condition for obtaining $\rho \neq 0$.

So now we begin to see the connection between an Allison mixture and Parrondian effects that require an asymmetry to interact with random behavior. Figure 12 is the symmetric case where we have detailed balance, and Fig. 11 is the asymmetric case where detailed balance is broken. Symmetry breaking is the essence of all Parrondian and Brownian ratchet phenomena. The $\mu_1 \neq \mu_2$ condition is analogous to the $T_1 \neq T_2$ condition that is required for Feynman's ratchet of Fig. 4 to operate. The $\alpha_1 + \alpha_2 \neq 1$ condition is analogous to the ratcheting mechanism in Fig. 4 as these both are the sources of asymmetry. A number of

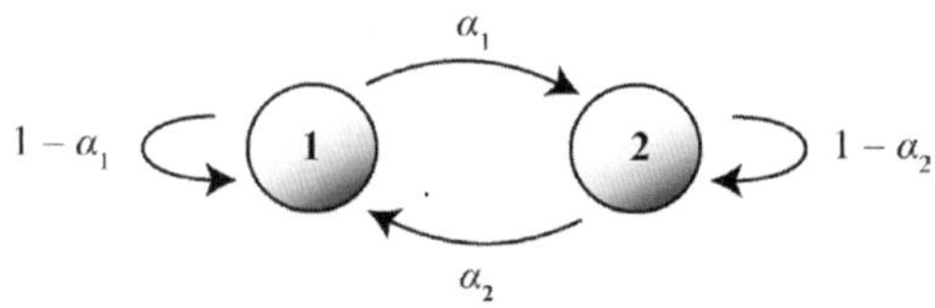


Fig. 11. State diagram for unbalanced switching process, when $\alpha_1 + \alpha_2 \neq 1$, giving rise to persistence. Circle 1 represents the state of landing in random Sequence 1 and Circle 2 represents the state we are in when we land on random Sequence 2. We jump between these two sequences to generate a new sequence. The apparent paradox of this Allison mixture is that the resulting sequence has nonzero autocovariance even when the originating sequences have zero autocovariance. Here, $\alpha_1$ and $\alpha_2$ are the transition probabilities of jumping between the two sequences. Notice, for example, as $\alpha_1 \to 0$ the probability of a self-transition to stay in Sequence 1 is high — so, if we are already in Sequence 1 we are likely to stay there. This can be thought of as a type of "memory" of the system, which causes the new sequence to have nonzero autocovariance. Note: this is *not* a form of memory in the sense that requires storage of a previous state — as there is no clear terminology, in the literature, for our probabilistic type of memory effect, it has been recently dubbed *memory persistence* [59].

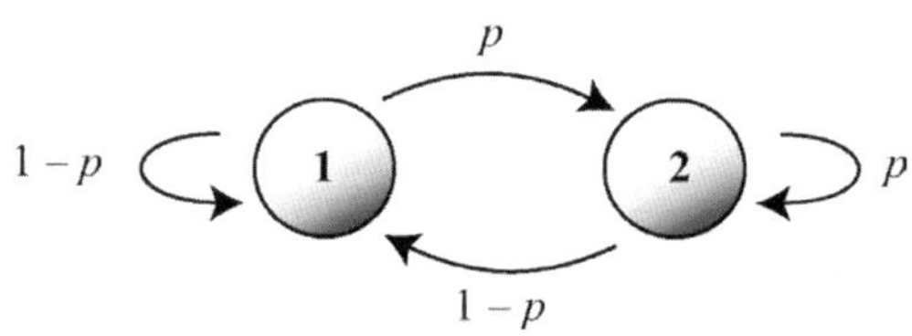


Fig. 12. State diagram for balanced switching process, when $\alpha_1 + \alpha_2 = 1$, resulting in no persistence. For simplicity we have inserted $p = \alpha_1 = 1 - \alpha_2$, which clearly reveals that the probability of entering a state exactly balances the probability of self-transition in that state. This detailed balance implies there is no memory persistence effect. Hence, the new generated sequence also has zero autocovariance.

open questions remain concerning the mathematical features of this switched system that can be thought of as a two-state discrete-time hidden Markov model — for example, Fig. 13 illustrates that as $\alpha_1 \to 0$ and $\alpha_2 \to 0$ the direction of the limits intriguingly affect the final value of $\rho$. Another open question is that of a possible application for Allison mixtures — this remains to be seen, but possible areas of promise might be in encryption and in optimizing file compression. Another open question to ask is if there are any links between Allison mixtures and biological evolution or genetics? Could it be that the redundancy that appears in sequences of non-coding (or "junk") DNA are the result of something along the lines of Allison mixing (i.e., ratcheted random mixing)? In the case of coding DNA, random mutations are a biased process — for example frame shift mutations in DNA are more likely to occur in sequences with runs of a single base and some single base mutations are more probable. This together with the asymmetric process of selection, which again is random but with biases, results in order that is created in a

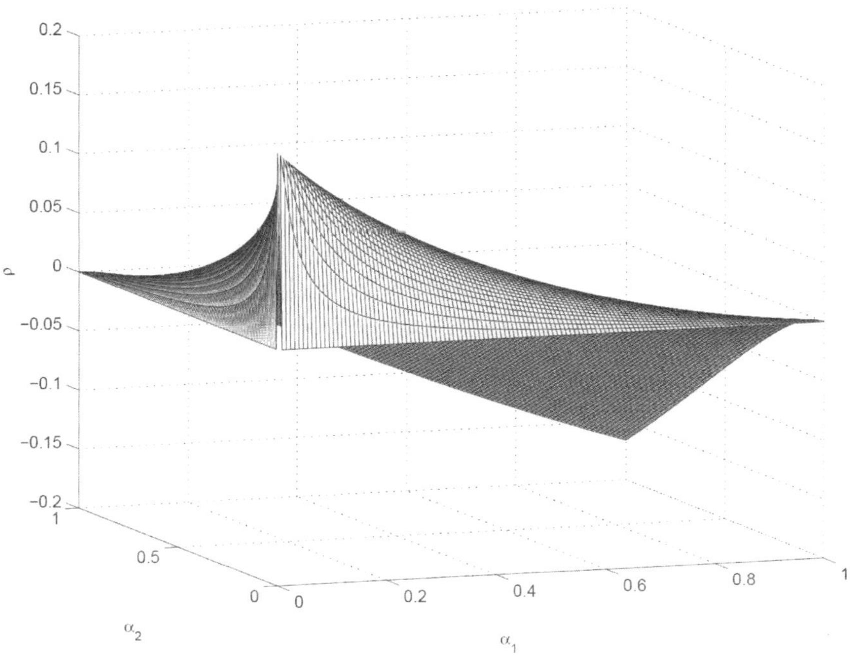


Fig. 13. Plot of the autocovariance $\rho$ of the generated sequence as a function of $\alpha_1$ and $\alpha_2$. In this specific example, the two originating binary sequences have $\mu_1 = 0.2$ and $\mu_2 = 0.6$. Pearce has named the peak of this graph the *pinnacle* [107]. The plot shows that the system displays some mathematically curious features. Surprisingly, $\rho = 0$ does not occur at a unique point in the parameter space. Another intriguing feature is that as $\alpha_1 \to 0$ and $\alpha_2 \to 0$, whether $\rho$ ends up at zero or at Pearce's pinnacle depends on the direction that you approach the limits.

set of DNA sequences. These sequences encapsulate in an ordered way information about the regularities of the organism in its environmental context.

## 7. Conclusion

There are two key take home messages that the study of Parrondo effects reveal: (i) the process of switching is a nonlinearity and therefore switching can radically alter the overall system behavior, and (ii) the interaction between noise and an asymmetry can give rise to directed motion even against a gradient, provided we are out of equilibrium.

Physicists have traditionally sought symmetry in Nature — a new challenge for future research is to now search for asymmetries and observe how they interact with noise or random behavior. On a more philosophical note, we might pose the question "Should we consider noise or randomness as a special form of order?" As well as our discussion on Parrondo effects, there are other examples that point to this: (i) a random walk displays self-similarity, (ii) randomly switched processes can produce fractals, (iii) according to Shannon, noise packs in maximal information, and (iv) as Chaitin points out, even the integers have noisy properties [108]. After all, noise is the most ordered way to avoid redundancy. There are many situations where noise appears to give rise to order and the challenge is to identify the general mathematical principles behind this.

### Acknowledgment

A warm thanks is due to Withawat Withayachumnankul who assisted with the preparation of the diagrams.

### Appendix A

The simple Matlab code for the demonstration of the principle of volatility pumping a high-risk stock together with a low-risk stock, is as follows:

```
   % high-low volatility pumping
clc; clear;
day = 100;
xscale = [1:day];

   % low risk stock initialization
A1(1) = 100;
A2(1) = 100;

   % high risk stock initialization
B1(1) = 100;
B2(1) = 100;

   % random halving and doubling
dh = ceil(2.*rand(day,1));
idx = find(dh==1);
dh(idx) = 0.5;
R = (rand(day,1) - 0.5)./5;

   % portfolio management
for ii=2:day
T(ii-1) = A2(ii-1) + B2(ii-1);
A2(ii) = T(ii-1)/2 + T(ii-1)/2*R(ii-1);
B2(ii) = T(ii-1)/2*dh(ii-1);
A1(ii) = A1(ii-1) + A1(ii-1)*R(ii-1);
B1(ii) = B1(ii-1)*dh(ii-1);
end
T(day) = A2(day) + B2(day);
```

```
  % plot graph
figure;
hx = plot(xscale, A1, xscale, B1, xscale, T);
set(hx(1),'linewidth',2,'linestyle',':');
set(hx(2),'linewidth',2,'linestyle','-');
set(hx(3),'linewidth',2,'linestyle','-.');
set(gca,'linewidth',2,'FontName','Arial',
'FontSize',20,'xlim',[1 100],'yscale','log'); box on;
xlabel('Time (Days)'); ylabel('Log Return');
legend('Low risk stock A','High risk stock B',
'50-50 Portfolio','Location','NorthWest');
```

## References


[1] G. P. Harmer and D. Abbott, Losing strategies can win by Parrondo's paradox, *Nature* **402** (1999) 864.

[2] P. Arena, S. Fazzino, L. Fortuna and P. Maniscalco, Game theory and non-linear dynamics: The Parrondo Paradox case study, *Chaos, Solitons & Fractals* **17** (2003) 545–555.

[3] D. P. Minor, Parrondo's paradox — hope for losers! *The College Mathematics Journal* **34** (2003) 15–20.

[4] E. Behrends, The mathematical background of Parrondo's paradox, *Proc. SPIE Noise in Complex Systems and Stochastic Dynamics II*, Maspalomas, Spain, Ed: Zoltan Gingl, **5471** (2004) 510–517.

[5] J. Miękisz, On two fellows who wanted to make money on fluctuations, *Math. Intel.* **29** (2007) 8–10.

[6] R. A. Epstein, *The Theory of Gambling and Statistical Logic*, 2nd edn. (Academic Press, 2009).

[7] J. von Neumann and O. Morgenstern, *Theory of Games and Economic Behavior* (Princeton University Press, New York, 1954).

[8] D. Blackwell and M. A. Girshick, *Theory of Games and Statistical Decisions* (John Wiley & Sons, New York, 1954).

[9] E. Behrends, Parrondo's paradox: A priori and adaptive strategies, Preprint: A-02-09, www.math.fu-berlin.de (2002).

[10] P. Groeber, On Parrondo's games as generalized by Behrends, *Lect. Notes Cont. Inform. Sci.* **341** (2006) 223–230.

[11] D. Abbott, P. C. W. Davies and C. R. Shalizi, Order from disorder: The role of noise in creative processes: A special issue on game theory and evolutionary processes — overview, *Fluctuation and Noise Letters* **2** (2002) C1–C12.

[12] A. Allison, C. E. M. Pearce and D. Abbott, Finding keywords amongst noise: Automatic text classification without parsing, *Proc. SPIE Noise and Stochastics in Complex Systems and Finance* **6601** (2007) 660113.

[13] N. Beerenwinkel, L. Pachter and B. Sturmfels, Epistasis and shapes of fitness landscapes, arVix:q-bio/0603034v2 (2006).

[14] F. W. Winterbotham, *The Ultra Secret* (Weidenfeld and Nicolson, London, 1974).

[15] A. Allison and D. Abbott, Control systems with stochastic feedback, *Chaos* **11** (2001) 715–724.

[16] D. Challet and N. F. Johnson, Optimal combinations of imperfect objects, *Phys. Rev. Lett.* **89** (2002) 028701.
[17] P. C. W. Davies, Physics and life: The Abdus Salam Memorial Lecture, *6th Trieste Conference on Chemical Evolution* (Kluwer Academic Publishers, 2001), pp. 13–20.
[18] F. Stjernberg, Parrondo's paradox and epistemology — when bad things happen to good cognizers (and conversely), in *Hommage Wlodek. Philosophical Papers Dedicated to Wlodek Rabinowicz*, eds. T. Rønnow-Rasmussen, B. Petersson, J. Josefsson and D. Egonsson (2007). Online: http://www. fil.lu.se/hommageawlodek.
[19] A. P. Flitney and D. Abbott, Quantum two- and three-person duels, *J. Opt. B.* **6** (2004) S860–S866.
[20] P. Amengual and R. Toral, Truels, or survival of the weakest, *Comput. Sci. Eng.* **8** (2006) 88–95.
[21] D. M. Kilgour and S. J. Brams, The truel, *Math. Mag.* **70** (1997) 315–326.
[22] R. A. Beatty, A. McLaren, A. Jost and R. G. Edwards, Genetic basis for the determination of sex, *Phil. Trans. Roy. Soc. Lond. B* **259** (1970) 3–14.
[23] V. Hutson and R. Law, Four steps to two sexes, *Proc. Biol. Sci.* **255** (1993) 43–51.
[24] P. Coker and C. Winter, N-Sex reproduction in dynamic environments, *4th European Conf. on Artificial Life* (MIT Press, 1997).
[25] R. Gorelick, Evolution of dioecy and sex chromosomes via methylation driving Muller's ratchet, *Biol. J. Linn. Soc.* **80** (2003) 353–368.
[26] N. Lane, *Power, Sex, Suicide: Mitochondria and the Meaning of Life* (Oxford University Press, 2005).
[27] S. Coppersmith, R. Black and L. Kadanoff, Analysis of a population genetics model with mutations, selection, and pleiotropy, *J. Stat. Phys.* **97** (1999) 429–457.
[28] M. Morange, Gene function, *C. R. Acad. Science Paris, Série III* **323** (2000) 1147–1153.
[29] Y. Collette, A. Gilles, P. Pontarotti and D. Olive, A co-evolution perspective of the TNFSF and TNFRSF families in the immune system, *Trends Immunol.* **24** (2003) 387–394.
[30] B. Magor and K. Magor, Evolution of effectors and receptors of innate immunity, *Dev. Comp. Immunol.* **25** (2001) 651–682.
[31] L. Coussens and Z. Werb, Inflammation and cancer, *Nature* **420** (2002) 860–867.
[32] D. Mann, Stress-activated cytokines and the heart: From adaptation to maladaptation, *Ann. Rev. Physiol.* **65** (2003) 81–101.
[33] M. Palladino, F. Bahjat, E. Theodorakis and L. Moldawer, Anti-TNF-$\alpha$ therapies: The next generation, *Nat. Rev. Drug Discov.* **2** (2003) 736–746.
[34] M. J. Berryman, W.-L. Khoo, H. Nguyen, E. O'Neil, A. G. Allison and D. Abbott, Exploring trade-offs in pleiotropy and redundancy using evolutionary computing, *Proc. SPIE BioMEMS and Nanotechnology* **5275** (2004) 49–58, arXiv:cs/0404017v1.
[35] J. Maynard Smith and D. Harper, *Animal Signals* (Oxford University Press, 2003).
[36] G. P. Harmer and D. Abbott, Parrondo's paradox, *Stat. Sci.* **14** (1999) 206–213.
[37] J. M. R. Parrondo, How to cheat a bad mathematician, in *EEC HC&M Network on Complexity and Chaos* (#ERBCHRX-CT940546), ISI, Torino, Italy (1996), Unpublished.
[38] A. Adjari and J. Prost, Drift induced by a periodic potential of low symmetry: Pulsed dielectrophoresis, *C. R. Acad. Science Paris, Série II* **315** (1993) 1635–1639.
[39] N. F. Johnson, P. Jeffries and P. M. Hui, *Financial Market Complexity* (Oxford University Press, 2003).

[40] C. F. Lee, N. F. Johnson, F. Rodriguez and L. Quiroga, Quantum coherence, correlated noise and Parrondo games, *Fluctuation and Noise Letters* **2** (2002) L293–L297.

[41] A. P. Flitney and D. Abbott, Quantum Parrondo games, *Phys. A* **314** (2002) 35–42.

[42] D. A. Meyer and H. Blumer, Quantum Parrondo games: Biased and unbiased, *Fluctuation and Noise Letters* **2** (2002) L257–L262.

[43] D. M. Wolf, V. V. Vazirani and A. P. Arkin, Diversity in times of adversity: Probabilistic strategies in microbial survival games, *J. Theor. Biol.* **234** (2005) 227–253.

[44] F. A. Reed, Two-locus epistasis with sexually antagonistic selection: A genetic Parrondo's paradox, *Genetics* **176** (2007) 1923–1929.

[45] N. Masuda and N. Konno, Subcritical behavior in the alternating supercritical Domany–Kinzel dynamics *Eur. Phys. J. B* **40** (2004) 313–319.

[46] G. P. Harmer and D. Abbott, A review of Parrondo's paradox, *Fluctuation and Noise Letters* **2** (2002) R71–R107.

[47] R. Pinsky and M. Scheutzow, Some remarks and examples concerning the transient and recurrence of random diffusions, *Annales de l'Institut Henri Poincaré — Probabilités et Statistiques* **28** (1992) 519–536.

[48] S. Maslov and Y. Zhang, Optimal investment strategy for risky assets, *Int. J. Theor. Appl. Finance* **1** (1998) 377–387.

[49] H. V. Westerhoff, T. Y. Tsong, P. B. Chock, Y. Chen and R. D. Astumian, How enzymes can capture and transmit free energy contained in an oscillating electric field, *Proc. Nat. Acad. Sci.* **83** (1986) 4734–4738.

[50] E. S. Key, Computable examples of the maximal Lyapunov exponent, *Probab. Theor. Relat. Field* **75** (1987) 97–107.

[51] D. Abbott, Overview: Unsolved problems of noise and fluctuations, *Chaos* **11** (2001) 526–538.

[52] D. G. Luenberger, *Investment Science* (Oxford University Press, 1997).

[53] A. Rosato, K. J. Strandburg, F. Prinz and R. H. Swendsen, Why the Brazil nuts are on top: Size segregation of particulate matter by shaking, *Phys. Rev. Lett.* **58** (1987) 1038–1040.

[54] A. Allison and D. Abbott, The physical basis for Parrondo's games, *Fluctuation and Noise Letters* **2** (2002) L327–L341.

[55] R. Toral, P. Amengual and S. Mangioni, Parrondo's games as a discrete ratchet, *Phys. A* **327** (2003) 105–110.

[56] P. Amengual, A. Allison, R. Toral and D. Abbott, Discrete-time ratchets, the Fokker–Planck equation and Parrondo's paradox, *Proc. Royal Society Lond. A* **460** (2004) 2269–2284.

[57] M. von Smoluchowski, Experimentall nachweisbare, der üblichen thermodynamic widersprechende molekularphanomene, *Physikalische Zeitschrift* **13** (1912) 1069–1080.

[58] R. P. Feynman, R. B. Leighton and M. Sands, *The Feynman Lectures on Physics*, Vol. 1 (Addison-Wesley, Reading, MA, 1963), pp. 46.1–46.9.

[59] D. Abbott, Developments in Parrondo's paradox, in *Applications of Nonlinear Dynamics*, eds. V. P. Longhini and A. Palacios (Springer-Verlag, Berlin (2009), pp. 307–321.

[60] D. Abbott, B. R. Davis and J. M. R. Parrondo, The problem of detailed balance for the Feynman–Smoluchowski engine (FSE) and the multiple pawl paradox, *Proc. AIP 2nd Int. Conf. on Unsolved Problems of Noise and Fluctuations* (*UPoN'99*) **511** (2000) 213–218.

[61] Y. I. Perelman, *Zhivaya Matematika* (Nauka, Moscow, 1967), reissue of the 1934 edition.

[62] F. Mosteller, *Fifty Challenging Problems in Probability* (Addison-Wesley, Reading, MA, 1965).

[63] G. P. Harmer, D. Abbott and P. G. Taylor, The paradox of Parrondo's games, *Proc. Roy. Soc. Lond. A* **456** (2000) 247–259.

[64] E. S. Key, M. M. Kłosek and D. Abbott, On Parrondo's paradox: How to construct unfair games by composing fair games, *ANZIAM J.* **47** (2006) 495–511.

[65] A. Allison and D. Abbott, Stochastic resonance in a Brownian ratchet, *Fluctuation and Noise Letters* **1** (2001) L239–L244.

[66] H. Moraal, Counterintuitive behaviour in games based on spin models, *J. Phys. A* **33** (2000) L203–L206.

[67] A. Costa, M. Fackrell and P. G. Taylor, Two issues surrounding Parrondo's paradox, in *Advances in Dynamic Games: Applications to Economics, Finance, Optimization, and Stochastic Control*, eds. A. S. Nowak and K. Szajowski, **7** (2005) 599–609.

[68] S. N. Ethier and J. Lee, Limit theorems for Parrondo's paradox, http://arxiv.org/pdf/0902.2368v1.

[69] J. M. R. Parrondo, G. P. Harmer and D. Abbott, New paradoxical games based on Brownian ratchets, *Phys. Rev. Lett.* **85** (2000) 5226–5229.

[70] R. J. Kay and N. F. Johnson, Winning combinations of history-dependent games, *Phys. Rev. E* **67** (2003) 056128.

[71] R. Toral, Cooperative Parrondo's games, *Fluctuation and Noise Letters* **1** (2001) L7–L12.

[72] R. Artuso, Recycling Parrondo games, *J. Phys. A: Math. Gen.* **39** (2006) 1285–1295.

[73] A. Allison, C. E. M. Pearce and D. Abbott, State-space visualisation and fractal properties of Parrondos games, cond-mat/0205536 (2002).

[74] B. Cleuren and C. Van Den Broeck, Primary Parrondo paradox, *Europhys. Lett.* **67** (2004) 151–157.

[75] L. Dinis, Optimal sequence for Parrondo games, *Phys. Rev. E* **77** (2008) 021124.

[76] C. M. Bishop, *Neural Networks for Pattern Recognition* (Oxford Press, 1996), pp. 346–349.

[77] C. Van den Broeck, P. Reimann, R. Kawai and P. Hänggi, Coupled Brownian motors, in *Lecture Notes in Physics: Statistical Mechanics of Biocomplexity*, eds. D. Reguera, M. Rubi and J. M. G. Vilar (Springer-Verlag, Berlin, Heidelberg, New York, 1999), pp. 93–111.

[78] L. Onsager, Reciprocal relations in irreversible processes I, *Phys. Rev.* **37** (1931) 405–426.

[79] L. Onsager, Reciprocal relations in irreversible processes II, *Phys. Rev.* **38** (1931) 2265.

[80] D. C. Osipovitch, C. Barratt and P. M. Schwartz, Systems chemistry and Parrondo's paradox: Computational models of thermal cycling, *New Journal of Chemistry* (2009), DOI: 10.1039/b900288j.

[81] B. Cleuren and C. Van den Broeck, Random walks with absolute negative mobility, *Phys. Rev. E* **64** (2002) 030101.

[82] A. Di Crescenzo, A Parrondo paradox in reliability theory, *The Mathematical Scientist* **32** (2007), 17–22 arXiv:math/0602308v2.

[83] L. Kocarev and Z. Tasev, Lyanpunov exponents, noise-induced synchronization, and Parrondo's paradox, *Phys. Rev. E* **65** (2002) 046215.

[84] J. Buceta, K. Lindenberg and J. M. R. Parrondo, Pattern formation induced by nonequilibrium global alternation of dynamics, *Phys. Rev. E* **66** (2002) 036216.

[85] J. Almeida, D. Peralta-Salas and M. Romera, Can two chaotic systems give rise to order? *Phys. D* **200** (2005) 124–132.

[86] A. Boyarsky, P. Góra and S. Islam, Randomly chosen chaotic maps can give rise to nearly ordered behavior, *Phys. D* **210** (2005) 284–294.

[87] G. P. Harmer, D. Abbott, P. G. Taylor and J. M. R. Parrondo, Parrondo's games and Brownian ratchets, *Chaos* **11** (2001) 705–714.

[88] D. Atkinson and J. Peijnenburg, Acting rationally with irrational strategies: Applications of the Parrondo effect, in *Reasoning, Rationality, Probability*, eds. M. C. Galavotti, R. Scazzieri and P. Suppes (CSLI Publications, Stanford, 2007).

[89] J. M. Diamond, *Why Sex is Fun?: The Evolution of Human Sexuality* (Harper Collins, 1997).

[90] C. M. Arizmendi, Paradoxical way for losers in a dating game, *Proc. AIP Nonequilibrium Statistical Mechaniucs and Nonliear Physics: XV Conference on Nonequilibrium Statistical Mechanics and Nonlinear Physics*, Mar del Plata, Argentina, 4–8 December, 2006, eds. O. Descalzi, O. A. Rosso and H. A. Larrondo, **913** (2007) 20–25, arXiv:physics/0703189v1.

[91] J. B. Satinover and D. Sornette, "Illusion of control" in time-horizon minority and Parrondo games, *Euro. Phys. J. B* **60** (2007) 369–384.

[92] J. B. Satinover and D. Sornette, Illusion of control in a Brownian game, *Phys. A* **386** (2007) 339–344.

[93] J. M. R. Parrondo and L. Dinis, Brownian motion and gambling: From ratchets to paradoxical games, *Contemp. Phys.* **45** (2004) 147–157.

[94] B. Cleuren and C. Van den Broeck, Optimizing strategies in the primary Parrondo paradox, *Phys. Rev. E* **70** (2004) 067104.

[95] L. Dinis and J. M. R. Parrondo, Inefficiency of voting in Parrondo games, *Phys. A* **343** (2004) 701–711.

[96] L. Dinis and J. M. R. Parrondo, Optimal strategies in collective Parrondo games, *Europhys. Lett.* **63** (2003) 319–325.

[97] J. M. R. Parrondo, L. Dinis, E. García-Toraño and B. Sotillo, Collective decision making and paradoxical games, *Euro. Phys. J. Spec. Top.* **143** (2007) 39–46.

[98] M. Boman, S. J. Johansson and D. Lyback, Parrondo strategies for artificial traders, in *Intelligent Agent Technology: Research and Development*, eds. N. Zhong, J. Liu, S. Ohsuga and J. Bradshaw (World Scientific, publishing, Singapore, 2001), pp. 150–159, arXiv:cs.ce/0204051.

[99] W.-S. Wah-Sui Almberg and M. Boman, An active agent portfolio management algorithm, in *Artif. Intell. Comput. Sci.*, ed. S. Shannon (Nova Science Publishers, Inc., 2005), pp. 123–134.

[100] R. Fernholz and B. Shay, Stochastic portfolio theory and stock market equilibrium, *J. Finance* **37** (1982) 615–624.

[101] T. M. Cover and E. Ordentlich, Universal portfolios with side information, *IEEE Trans. on Inform. Theor.* **42** (1996) 348–363.

[102] M. A. H. Dempster and I. G. Evstigneev, Volatility-induced financial growth, *Quant. Finance* **7** (2007) 151–160.

[103] O. Stein, J. Oldenburg and W. Marquardt, Continuous reformulation of discrete-continuous optimization problems, *Comput. Chem. Eng.* **28** (2004) 1951–1966.

[104] F. Cannata, M. Ioffe, G. Junker and D. Nichnianidze, Intertwining relations of non-stationary Schrödinder operators, *J. Phys. A: Math. Gen.* **32** (1999) 3583–3598.

[105] D. Heath, D. Kinderlehrer and M. Kowalczyk, Discrete and continuous ratchets: From coin toss to molecular motor, *Discrete and Continuous Dynamical Systems — Series B* **2** (2002) 153–167.

[106] P. Amengual, R. Toral, A. Allison and D. Abbott, Efficiency of discrete-time ratchets (2004), arXiv:cond-mat/0410173.

[107] C. E. M. Pearce, A. Allison and D. Abbott, Perturbing singular systems and the correlating of uncorrelated random sequences, *Proc. AIP Int. Conf. on Numerical Analysis and Applied Mathematics*, Corfu, Greece, eds. T. E. Simos, G. Psihoyios and Ch. Tsitouras, **936** (2007) 699.

[108] G. J. Chaitin, *The Unknowable* (Springer-Verlag, 1999).

# SPATIAL PATTERNS INDUCED BY RANDOM SWITCHING

J. BUCETA and KATJA LINDENBERG

*Department of Chemistry and Biochemistry and Institute for Nonlinear Science, University of California, San Diego 9500 Gilman Drive, La Jolla, California 92093-0340, USA klindenberg@ucsd.edu*

J. M. R. PARRONDO

*Departamento de Física Atómica, Molecular y Nuclear, Universidad Complutense de Madrid Ciudad Universitaria s/n, 28040 Madrid, Spain*

Received 26 January 2002
Revised 22 February 2002
Accepted 22 February 2002

We propose a mechanism whereby a random alternation of two dynamics each leading to a different homogeneous state can lead to complex ordered structures. The proposed general formalism, based on the ideas of so-called paradoxical games, is illustrated via numerical simulations of particular examples. The relevance of the present study to other situations that lead to pattern formation, such as reaction-diffusion systems, is noted.

*Keywords*: Random switching; global switching; Parrondo's paradox; pattern formation.

## 1. Introduction

It has recently been established that the outcome of two alternating stochastic dynamics can differ significantly from that of each separate dynamics. The phenomenon has been illustrated with two losing gambling games that turn into a winning game when they are alternated either randomly or periodically [1, 2], and has been extended to other situations such as non-Markovian [3] and cooperative [4] games. This counterintuitive phenomenon, frequently referred to as *Parrondo's paradox*, was originally inspired by Brownian flashing ratchets [5–7], where a Brownian particle exhibits systematic motion in one direction when a periodic potential is alternately switched on and off.

The main message of the paradox is that the alternation of stochastic dynamics can yield new and counterintuitive phenomena. Davies [8] has also pointed out that the paradox provides a simple mechanism to create order (systematic motion or systematic gain) from fluctuations, and that a mechanism of this kind is probably needed for an explanation of the origin of life.

One of the interesting questions prompted by the paradox is whether order and complexity can arise if two dynamics, each of which separately drives the system to a disordered state or to a homogeneous state, are alternated. Here we show that a wide class of spatially extended systems can exhibit this unexpected behavior. These systems are based on the Swift-Hohenberg equation [9], which has been extensively used to model the appearance of spatial patterns in convection, but the mechanism underlying this "paradoxical" behavior can also apply to other spatially extended systems such as reaction-diffusion equations. Moreover, the mechanism presented in this paper helps to understand the role of *global* noise in spatially extended systems. It is known that local (site-dependent) fluctuations may induce pattern formation [10, 11]. On the other hand, global periodic forcing as a method of generating spatial structures has been extensively studied since the seminal work of Faraday [12] and continues to lead to striking new results, such as the recently found *oscillons* or particle-like spatial excitations [13, 14]. However, the use of *global noisy forcing* to obtain spatial patterns is, to our knowledge, uncharted territory both from the theoretical and experimental points of view. The present study sheds light on this issue and gives explicit solutions for a large family of systems.

The paper is organized as follows. In Sec. 2 we state the problem of random alternating dynamics for scalar fields obeying overdamped evolution equations. We also elucidate the general conditions to obtain spatial structures by random switching between dynamics. Section 3 is devoted to the analysis of a particular example. In Sec. 4 we present the main conclusions and point to some ideas for future work.

## 2. General Formalism

Consider the following Langevin dynamics for a scalar field $\varphi$:

$$\dot{\varphi}(\mathbf{r}, t) = -V'(\varphi(\mathbf{r}, t)) + \mathrm{L}\varphi(\mathbf{r}, t) + \xi(\mathbf{r}, t) \,, \tag{1}$$

where $\varphi(\mathbf{r}, t)$ depends on both space, $\mathbf{r}$, and time, $t$. The field could, for example, represent the concentration of a chemical species at a given spatial position and time. The temporal evolution of $\varphi(\mathbf{r}, t)$ is driven by a local force that can be derived from a local potential, $V(\varphi)$, by its coupling with the field at other locations through the operator L, and by thermal fluctuations $\xi(\mathbf{r}, t)$, taken to be Gaussian white noise with zero mean and correlation function

$$\langle \xi(\mathbf{r}, t)\xi(\mathbf{r}', t')\rangle = \sigma^2\delta(\mathbf{r} - \mathbf{r}')\delta(t - t') \,. \tag{2}$$

For a system such as (1), there are two required features for pattern formation: local bistability, that is, the local potential must show two stable equilibrium points, and a morphological instability [15]. A paradigmatic example is the Swift-Hohenberg equation [9], a model that describes Rayleigh-Benard convection near the threshold for the appearance of convective rolls. The Swift-Hohenberg equation is intended as a simplified form of the full Navier-Stokes equation for fluid flow that avoids the intractable complexity of the full equation, and yet is capable of generating some of the patterns observed in convection. A brief qualitative analysis of this model is useful to clarify the features of the systems exhibiting patterns induced by the alternation of dynamics. Equation (1) represents the Swift-Hohenberg model when the local potential and the coupling term are respectively chosen as

$$V(\varphi) = \frac{1}{4}\varphi^4 - \frac{a}{3}\varphi^3 - \frac{b}{2}\varphi^2, \tag{3}$$

$$\mathrm{L} = -(1+\nabla^2)^2. \tag{4}$$

We use this coupling operator throughout this paper. The morphological instabilities leading to pattern formation are identified by determining a uniform solution of the (noiseless) evolution equation (1) and linearizing about this solution. With the local potential (3) we have $V'(0) = 0$ and hence the uniform stationary solution is $\varphi = 0$. Linearization about this field gives the evolution equation

$$\dot{\varphi}(\mathbf{r},t) = -\left[V''(0) + (1+\nabla^2)^2\right]\varphi(\mathbf{r},t) \tag{5}$$

or, performing a spatial Fourier transform (denoted by a hat),

$$\dot{\hat{\varphi}}(\mathbf{k},t) = -\left[V''(0) + (1-|\mathbf{k}|^2)^2\right]\hat{\varphi}(\mathbf{k},t). \tag{6}$$

The least stable modes thus correspond to $|\mathbf{k}| = 1$, and a morphological instability occurs if $V''(0) < 0$. Regarding the local potential, we distinguish different cases according to the values of the parameters $a$ and $b$. If $a$ and $b$ are such that $V(\varphi)$ is monostable, no spatial structures appear. On the other hand, if $a$ and $b$ are such that the local potential is bistable, patterns are obtained. The shape of the patterns depends on the symmetry of $V(\varphi)$: if $V(\varphi) = V(-\varphi)$ the system develops convection rolls, but if Eq. (1) does not possess the symmetry $\varphi \leftrightarrow -\varphi$ then spot-shaped patterns are observed, usually arranged in a hexagonal lattice.

Consider now a random global alternation of the local potential in (1), i.e.,

$$\begin{aligned}\dot{\varphi}(\mathbf{r},t) &= -\Lambda(t)V_1'(\varphi(\mathbf{r},t)) - (1-\Lambda(t))V_2'(\varphi(\mathbf{r},t)) \\ &\quad + \mathrm{L}\varphi(\mathbf{r},t) + \xi(\mathbf{r},t),\end{aligned} \tag{7}$$

where $\Lambda(t)$ is a dichotomous random variable that assumes the values 0 and 1 with equal probability, and with correlation function

$$\frac{\langle\Lambda(t)\Lambda(t')\rangle - \langle\Lambda(t)\rangle\langle\Lambda(t')\rangle}{\langle\Lambda^2(t)\rangle - \langle\Lambda(t)\rangle^2} = \exp(-|t-t'|/\tau), \tag{8}$$

$\tau$ being the correlation time of the switching process. We further require that the local potentials $V_{1,2}(\varphi)$ each have only one stable equilibrium point. Therefore, at a given time, *all* the sites in the system are driven either by $V_1(\varphi)$ or by $V_2(\varphi)$. Notice that each of these dynamics if taken separately can not generate any patterns because both are monostable.

In terms of the potentials

$$V_\pm(\varphi) = \frac{V_1(\varphi) \pm V_2(\varphi)}{2}, \tag{9}$$

and a new dichotomous variable $\mu(t) = \pm 1$, Eq. (7) can be rewritten as

$$\dot{\varphi}(\mathbf{r},t) = -V_+'(\varphi(\mathbf{r},t)) - \mu(t)V_-'(\varphi(\mathbf{r},t)) + \mathrm{L}\varphi(\mathbf{r},t) + \xi(\mathbf{r},t). \tag{10}$$

Note that $\langle\mu(t)\mu(t')\rangle = \exp(-|t-t'|/\tau)$.

The behavior of the system as a function of the switching correlation time is as follows. If $\tau$ is large, the system alternates between two homogeneous states, one associated with $V_1(\varphi)$ and the other with $V_2(\varphi)$. On the other hand, if $\tau \to 0$ but the noise intensity remains finite, the fluctuations are negligible and the noise can simply be approximated by its mean value (only if the intensity of the noise goes to infinity do fluctuations continue to play a role in the vanishing correlation time limit; in this case it is well known [16] that dichotomous noise tends to a white Gaussian noise). The mean value of $\mu(t)$ is zero, and the dynamics in this limit is therefore driven by the effective potential $V_+(\varphi)$. The stability analysis leading to Eq. (6) must now be slightly modified, however, because $V'_+(0)$ is in general not zero, that is, if there is a uniform solution it may not occur at $\varphi = 0$. One must seek a uniform solution $\varphi = \tilde{\varphi}_+$ which satisfies the stationarity condition

$$V'_+(\tilde{\varphi}_+) + \tilde{\varphi}_+ = 0 \tag{11}$$

where the last term arises from the 1 in the coupling operator L. Setting $\varphi = \tilde{\varphi}_+ + \Delta\varphi$, the stability diagnostic equation (6) must now be modified to

$$\Delta\dot{\hat{\varphi}}(\mathbf{k}, t) = -\left[V''_+(\tilde{\varphi}_+) + (1 - |\mathbf{k}|^2)^2\right] \Delta\hat{\varphi}(\mathbf{k}, t)\,, \tag{12}$$

and a morphological instability sets in if $V''_+(\tilde{\varphi}_+) < 0$.

According to the above discussion the potentials $V_{1,2}(x)$ must fulfill the following conditions to produce patterns by random alternation:

- **C1**. $V'_1(\tilde{\varphi}_1) + \tilde{\varphi}_1 = 0$ and $V'_2(\tilde{\varphi}_2) + \tilde{\varphi}_2 = 0$ each have a single stable solution ($V''_1(\tilde{\varphi}_1) > 0$, $V''_2(\tilde{\varphi}_2) > 0$)
- **C2**. $V'_+(\tilde{\varphi}_+) + \tilde{\varphi}_+ = 0$ has at least one unstable solution ($V''_+(\tilde{\varphi}_+) < 0$)

We expect pattern formation when there is sufficiently rapid switching between $V_1$ and $V_2$. Moreover, note that in the limit $\tau \to 0$ the temporal dependence disappears from the local dynamics. Therefore, *stationary* patterns should be obtained as a result of rapid random alternation.

It is easy to demonstrate that conditions (C1–C2) can not be fulfilled if both local potentials are quadratic; we need at least one of the local potentials to be at least quartic. In general, one of the two potentials has to be a non-convex function, i.e., the second derivative can not have the same sign for all $\varphi$. Consequently, nonlinearity is a necessary ingredient for the proposed pattern formation mechanism.

## 3. A Particular Case: Numerical Simulations

Consider the following particular choice of potentials satisfying C1 and C2:

$$\begin{aligned} V_1(\varphi) &= A\left(\frac{\varphi^4}{4} + \frac{\varphi^3}{3} - \frac{\varphi^2}{2} - \varphi\right), \\ V_2(\varphi) &= B\left(\frac{\varphi^4}{4} - \frac{\varphi^3}{3} - \frac{\varphi^2}{2} + \varphi\right), \end{aligned} \tag{13}$$

where $A$ and $B$ are positive constants. For this case

$$V_+(\varphi) = \frac{1}{2}\left(\alpha\frac{\varphi^4}{4} + \beta\frac{\varphi^3}{3} - \alpha\frac{\varphi^2}{2} - \beta\varphi\right), \tag{14}$$

where $\alpha = A + B$ and $\beta = A - B$.Depending on the values of the parameters $A$ and $B$, we can distinguish two cases. If $A = B$, $V_+(\varphi)$ is proportional to the even part of the local potentials. Following the parallels with the Swift-Hohenberg model discussed in the previous section, we expect rolls if the random alternation is sufficiently rapid. On the other hand, if $A \neq B$ the effective local potential $V_+(\varphi)$ is not an even function, but it does have two stable states. In this case we expect spots (cf. Eqs. (14) and (3)).

We perform numerical simulations of Eq. (7) on a $128 \times 128$ lattice using the local potentials as defined in (13). The values of the relevant parameters are $\Delta t = 10^{-3}$, $\Delta x = \Delta y = 0.5$, $L_x = L_y = 64$, $\sigma = 10^{-2}$ and we implement periodic boundary conditions. The most unstable modes for the coupling operator used here are $|\mathbf{k}^*| \simeq 1$ [17]; hence the wavelength of the pattern will be $\lambda = 2\pi/|\mathbf{k}^*| \simeq 2\pi$ and the aspect ratio $L/\lambda \sim 10$. Note that the fluctuations must be sufficiently small not to swamp the potential barrier in $V_+$. In our simulations we take either the initial field to be random according to a Gaussian distribution (in which case the additive fluctuations can actually be omitted entirely), or we start from an arbitrary initial condition (e.g. all points equilibrated with $V_1$), in which case the (small) fluctuations will distribute the field in any case. Note that the role of the additive noise here is different from other systems, where noise terms do have a direct influence on the pattern selection mechanism (for an example in the Swift-Hohenberg context see e.g. [18]).

We focus the numerical simulations on the cases $A = B = 1$ (a) and $A = 1$, $B = 2$ (b). We expect transitions from alternating homogeneous states to rolls in the former and to spots in the latter as we decrease the switching correlation time.

The transition from alternating homogeneous states to pattern formation is clearly observed in the dependence of $\langle S(\mathbf{k})\rangle_\theta$, the spatial structure function averaged over angles, on the random switching correlation time $\tau$. The structure function is defined as usual, $S(\mathbf{k}) = |\hat{\varphi}^2(\mathbf{k}, t)|$, and its angular average as the average over $\theta$ when $\mathbf{k}$ is expressed in polar coordinates. The resulting average is time-independent, that is, the resulting patterns are stationary. Figure 1 shows $\langle S(\mathbf{k})\rangle_\theta$ for case (a) for two different correlation times, $\tau = 1$ and $\tau = 10^{-2}$. The flat spectrum for $\tau = 1$ indicates that the system exhibits no spatial structure; the system is in fact simply alternating between homogeneous states. On the other hand, as evidenced by the peak at the most unstable mode that appears in the structure function for the case $\tau = 10^{-2}$, by sufficiently increasing the flipping rate we obtain an ordered structure, i.e., a pattern. Figure 2 shows the same quantity for case (b). Here again one obtains a pattern by randomly alternating sufficiently quickly between two potentials each of which leads to a homogeneous state.

Figure 3 shows density plots of the field for $\tau = 10^{-2}$ for cases (a) and (b). As expected, (a) leads to a roll-shaped pattern, while case (b) reproduces a spot-shaped pattern. The symmetry of the patterns can be elucidated by analyzing their structure functions $S(\mathbf{k})$. This quantity is plotted in Fig. 4 by means of a density plot on a logarithmic scale, revealing the circular symmetry of the rolls (isotropy), case (a), and the hexagonal symmetry of the spot-like pattern (anisotropy), case (b).

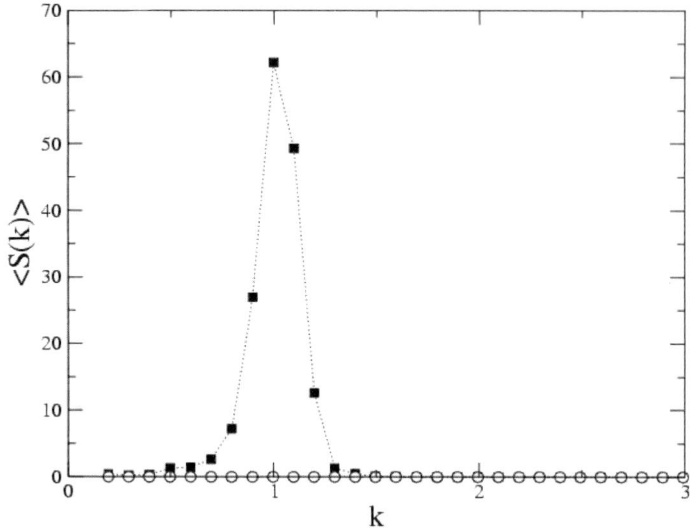


Fig. 1. Structure functions $S(\mathbf{k})$ averaged over angles for case (a) described in the text for two different values of the correlation time of the random switching. Note that no structure appears when alternation between dynamics is slow, as evident from the flat spectrum (circles). Fast switching leads to pattern formation with a well-defined spatial periodicity (squares).

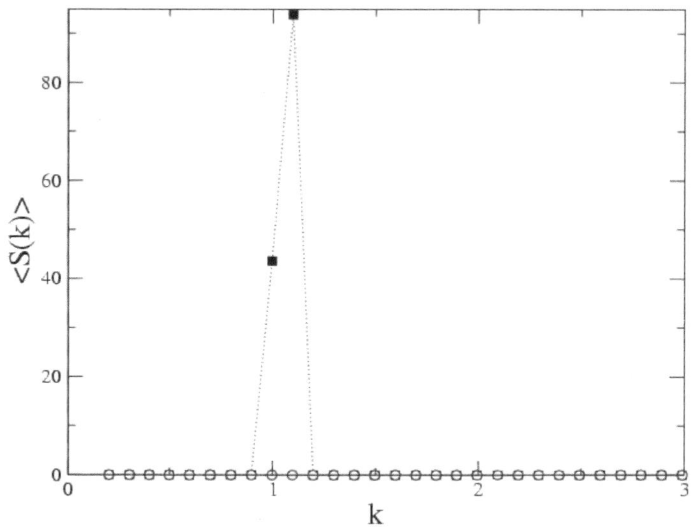


Fig. 2. Angular average of the structure function $S(\mathbf{k})$ vs $\mathbf{k}$ for case (b) described in the text for two different correlation times, $\tau = 1$ (circles) and $\tau = 10^{-2}$ (squares). The appearance of a peak for the mode $k \simeq 1$ upon decreasing the random switching time clearly indicates the emergence of a regular structure.

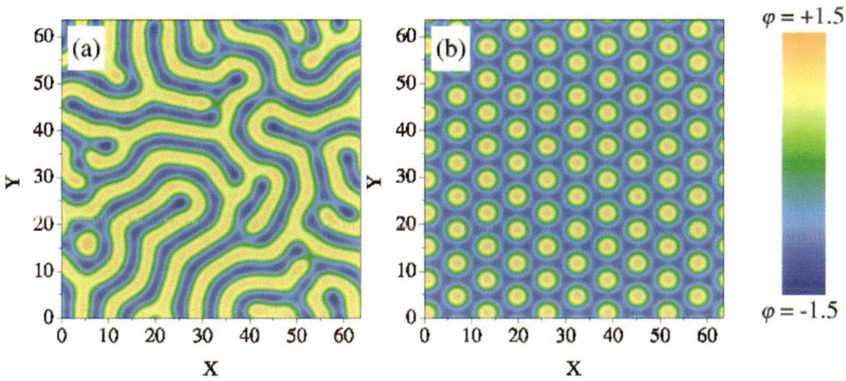


Fig. 3. Density plot of the field for cases (a) and (b) for $\tau = 10^{-2}$. Convection rolls develop in case (a). Hexagonally arranged spot-like patterns appear in case (b). The symmetry of the effective local potential $V_+(\varphi)$ determines the structure.

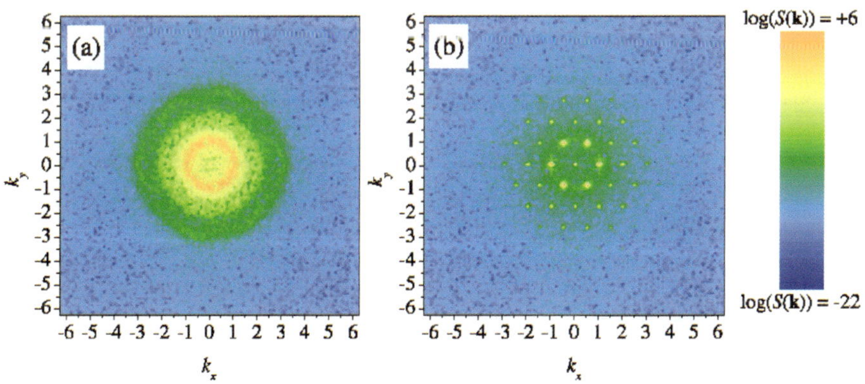


Fig. 4. Density plot of the logarithm of the structure functions associated with the structures of Fig. 3. A clear circular symmetry is obtained for case (a). In case (b) the spectrum indicates a hexagonal pattern.

## 4. Conclusions

We have presented a broad class of spatially extended systems where the alternation of two dynamics induces patterns whereas each separate dynamics drives the system to a homogeneous state. In this way we have partially answered the question of how complexity can arise from the alternation of non-complex dynamics. Moreover, we have demonstrated that global noisy forcing helps to develop ordered spatial structures.

The types of patterns that can be obtained depends on the choice of the local potentials or, more precisely, on their symmetries. For the particular case of a bistable effective potential, convection-rolls are obtained if $V_+(\varphi)$ is an even function, whereas spot-shaped structures appear otherwise. This scenario is in agreement with the known phenomenology of the Swift-Hohenberg model.

If switching between dynamics is periodic rather than random, interesting resonance phenomena between the switching rate and the relaxation rate from non-equilibrium states are observed [14,19]. For example, localized periodic excitations (oscillons) can occur. Multi-phase patterns have attracted a great deal of recent attention in cases where a periodic forcing is applied [20,21]. These patterns raise interesting questions related to the possibility of creating multi-phase patterns in a system that simply alternates between two potentials each of which is associated with a homogeneous state.

It is worth noting that the mechanism we have presented is quite general and does not depend on the particular choice of the local potentials and/or the coupling term: other families of local potentials combined with coupling terms incorporating morphological instabilities can also develop spatial and/or temporal structures. For instance, we expect this mechanism to be relevant in reaction-diffusion systems [22].

## Acknowledgments

We thank R. Kawai for fruitful discussions. This work was supported in part by the National Science Foundation under grant No. PHY-9970699, by DGES-Spain Grant PB-97-0076, by the *New Del Amo* Program, and by MECD-Spain Grant EX2001-02880680.

## References

[1] G. P. Harmer and D. Abbott, *Parrondo's paradox*, *Stat. Sci.* **14** (1999) 206.

[2] G. P. Harmer and D. Abbott, *Game theory — Losing strategies can win by Parrondo's paradox*, *Nature* **402** (1999) 864.

[3] J. M. R. Parrondo, G. P. Harmer and D. Abbott, *New paradoxical games based on Brownian ratchets*, *Phys. Rev. Lett.* **85** (2000) 5226.

[4] R. Toral, *Cooperative Parrondo's games*, *Fluctuations and Noise Letters* **1** (2001) L7.

[5] G. P. Harmer, D. Abbott, P. G. Taylor and J. M. R. Parrondo, *Parrondo's paradoxical games and the discrete Brownian ratchet*, Proc. 2nd Int. Conf. on Unsolved Problems of Noise and Fluctuations, eds. D. Abbott and L. B. Kish, American Institute of Physics, Melville, New York (2000) 189–200.

[6] A. Adjari and J. Prost, *Drift induced by a spatially periodic potential of low symmetry: Pulsed dielectrophoresis*, *C. R. Acad. Sci. Paris II* **315** (1992) 1635.

[7] R. D. Astumian and M. Bier, *Fluctuation driven ratchets: Molecular motors*, *Phys. Rev. Lett.* **72** (1994) 1766.

[8] P. C. W. Davies, *Physics and life*, *Lecture in honour of Abdus Salam*, in The First Steps of Life in the Universe, Proc. Sixth Trieste Conference on Chemical Evolution, eds. J. Chela-Flores, T. Owen and F. Raulin, Trieste, Italy, 18–22 September, 2000, Kluwer, Dordrecht (2001) 13–20.

[9] J. Swift and P. C. Hohenberg, *Hydrodynamic fluctuations at the convective instability*, *Phys. Rev. A* **15** (1977) 319.

[10] J. M. R. Parrondo, C. Van den Broeck, J. Buceta and F. J. de la Rubia, *Noise-induced spatial patterns*, *Physica A* **224** (1996) 153.

[11] A. A. Zaikin and L. Schimansky-Geier, *Spatial patterns induced by additive noise*, *Phys. Rev. E* **58** (1998) 4355.

[12] M. Faraday, *On the forms and states of fluids on vibrating elastic surfaces*, *Phil. Trans. R. Soc. London* **52** (1831) 319.

[13] P. B. Umbanhowar, F. Melo and H. L. Swinney, *Localized excitations in a vertically vibrated granular layer*, *Nature* **382** (1996) 793.

[14] J. Buceta, K. Lindenberg and J. M. R. Parrondo, *Stationary and oscillatory spatial patterns induced by global periodic switching*, *Phys. Rev. Lett.* **88** (2002) 024103.

[15] M. C. Cross and P. C. Hohenberg, *Pattern formation outside of equilibrium*, *Rev. Mod. Phys.* **65** (1993) 851.

[16] W. Horsthemke and R. Lefever, *Noise-Induced Transitions*, Springer-Verlag, Berlin (1984).

[17] See [10] for details on discretization of the Swift-Hohenberg model.

[18] E. Hernández-García, J. Viñals, R. Toral and M. San Miguel, *Fluctuations and pattern selection near an Eckhaus instability*, *Phys. Rev. Lett.* **70** (1993) 3576.

[19] J. Buceta, K. Lindenberg and J. M. R. Parrondo, *Pattern formation induced by nonequilibrium global alternation of dynamics*, submitted for publication.

[20] A. L. Lin, M. Bertram, K. Martínez and H. L. Swinney, *Resonant phase patterns in a reaction-diffusion system*, *Phys. Rev. Lett.* **84** (2000) 4240.

[21] A. L. Lin, A. Hagberg, A. Ardelea, M. Bertram, H. L. Swinney and E. Meron, *Four-phase patterns in forced oscillatory systems*, *Phys. Rev. E* **62** (2000) 3790.

[22] J. Buceta and K. Lindenberg, *Switching-induced Turing instability*, in preparation.

# A NEW APPROACH TO THE TREATMENT OF SEPARATRIX CHAOS

STANISLAV M. SOSKIN

*Institute of Semiconductor Physics*
*03028 Kiev, Ukraine*
*stanislav.soskin@gmail.com*

RICCARDO MANNELLA

*Dipartimento di Fisica*
*Università di Pisa, 56127 Pisa, Italy*
*mannella@df.unipi.it*

OLEG M. YEVTUSHENKO

*Physics Department*
*Ludwig-Maximilians-Universität München*
*D-80333 München, Germany*
*bom@ictp.it*

IGOR A. KHOVANOV

*School of Engineering*
*University of Warwick*
*Coventry CV4 7AL, UK*
*i.khovanov@warwick.ac.uk*

PETER V. E. McCLINTOCK

*Physics Department*
*Lancaster University, Lancaster LA1 4YB, UK*
*p.v.e.mcclintock@lancaster.ac.uk*

Received 18 August 2011
Accepted 22 November 2011
Published 31 March 2012

Communicated by Igor Goychuk

We review an approach to separatrix chaos that has allowed us to solve some significant problems by: (i) finding analytically the maximum width of the chaotic layer, a problem that lay unsolved for 40 years, and showing that the maximum may be much larger than had previously been assumed; (ii) describing the drastic facilitation of the onset of global chaos between neighboring separatrices, a phenomenon discovered eight years ago.

*Keywords*: Chaotic layer; separatrix map; nonlinear resonance; global chaos; superlattice; meandering flow.

## 1. Introduction

In this section, we first introduce the concept of separatrix chaos and demonstrate that it constitutes the seed of any chaos in time-periodically perturbed 1D Hamiltonian systems. We then describe some of the problems that stimulated the development of our new approach to separatrix chaos and, finally, explain why it is difficult to use earlier approaches to solve these and related problems.

### 1.1. *Separatrix chaos*

Hamiltonian dynamics is described by equations of the type:

$$\frac{dq_i}{dt} = \frac{\partial H}{\partial p_i}, \quad \frac{dp_i}{dt} = -\frac{\partial H}{\partial q_i}, \\ i = 1, 2, 3, \ldots \quad H \equiv H(q_1, q_2, \ldots, p_1, p_2, \ldots, t). \tag{1}$$

The simplest case is a one-dimensional (1D) Hamiltonian $H = H(q,p)$, where $H$ is an integral of motion (the energy $E$) and any finite motion is periodic in time:

$$q(t+T) = q(t), \quad p(t+T) = p(t), \quad T \equiv T(E), \quad \omega \equiv \frac{2\pi}{T} \equiv \omega(E). \tag{2}$$

An important notion is that of a *separatrix.* Suppose the phase space of the system possesses a saddle, i.e., an unstable stationary point of the hyperbolic type, which two trajectories approach and from which two trajectories depart, exponentially slowly. These four trajectories form the separatrix. As the energy approaches that of the separatrix $E_s$, the period of oscillation logarithmically diverges:

$$\omega^{-1}(E) \propto T(E) \propto \ln(1/|E - E_s|), \quad E \approx E_s. \tag{3}$$

The separatrix typically separates trajectories of different topology. An archetypal example is a pendulum (Fig. 1): $H = p^2/2 + U(q)$, where $U(q) = -\cos(q)$.

For higher-dimensional Hamiltonians, the motion is typically much more complicated. In this review, we consider the simplest case: D $= \frac{3}{2}$, i.e., a 1D Hamiltonian

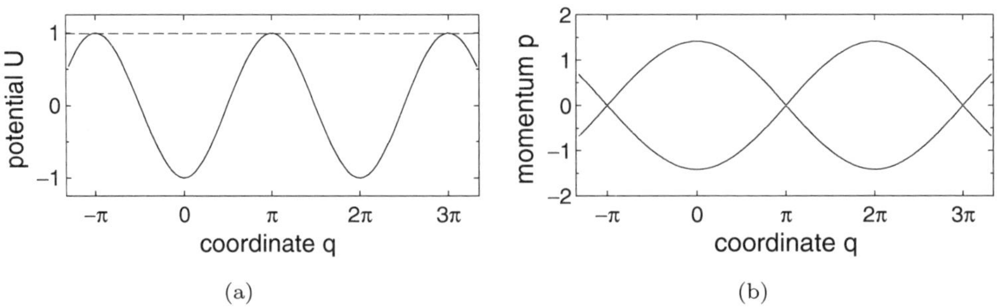


Fig. 1. Pendulum: (a) potential, (b) separatrix in the phase plane.

with a time-dependent perturbation. Moreover, we restrict ourselves to the most characteristic case — a weak time-periodic perturbation.

In the 1950s and 60s, Kolmogorov, Arnold and Moser developed a theory (KAM-theory) which says that the majority of perturbed trajectories remain regular: invariant tori are just slightly deformed rather than destroyed i.e., the perturbation does not modify the topology of a trajectory in a stroboscopic Poincaré section [1].

However, there are two characteristic regions where the KAM theory is inapplicable, so that some of the corresponding trajectories are irregular. The first one relates to the separatrix of the unperturbed system. Due to the divergence of $T(E)$, even a small change of energy caused by the perturbation results in a strong change of the system coordinate after a given interval of time. This causes the separatrix to be replaced by a chaotic layer (Fig. 2), where regions of regular and irregular motion are mixed in a complicated manner. The chaotic layer is a very complicated object: it was studied in hundreds of papers, starting apparently from [2] in 1967.

The second region where the motion may be irregular relates to *resonances* i.e., to regions where

$$E \approx E^{(n)}, \quad n\omega(E^{(n)}) = \omega_f, \quad n = 1, 2, 3, \ldots. \tag{4}$$

The slow dynamics of the deviation of action from that at the resonance energy, and of the related slow angle, are governed by the auxiliary Hamiltonian (sometimes called nonlinear resonance Hamiltonian), which is typically taken as being pendulum-like [3]. The phase plane of this auxiliary Hamiltonian necessarily possesses a separatrix, but the fast oscillations (neglected in a leading-order analysis of the slow dynamics) destroy it, replacing it with a narrow chaotic layer.

Thus, separatrix chaos plays a key role in the onset of any chaos in a periodically perturbed Hamiltonian system.

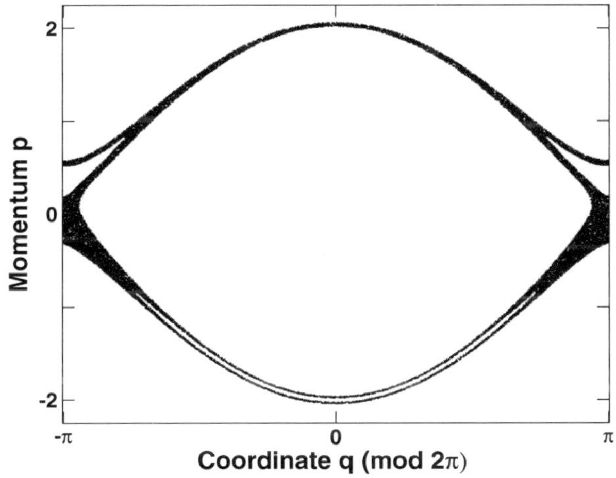


Fig. 2. Stroboscopic Poincare section for the Hamiltonian $H = p^2/2 - \cos(q) + 0.01q\cos(t)$.

### 1.2. *Stimulating problems*

The development of our approach was stimulated in particular by the following open questions concerning separatrix chaos:

(1) How can one describe theoretically the maximal (over the perturbation frequency) width of the separatrix chaotic layer? Neither physicists [4] nor mathematicians [5] had cracked this problem during the 40 years following the pioneering work by Zaslavsky and Filonenko [6].
(2) How can one describe theoretically the involvement of resonant motion in separatrix chaos? This problem became topical with the work of Shevchenko in 1998 [7] although it had been implicit since much earlier [3].

We now describe these problems in more detail. The chaotic layer's energy width depends on the ratio $\omega_f$ between the perturbation frequency and frequency of small oscillations of the unperturbed system. The three characteristic ranges of $\omega_f$ are —

(i) $\omega_f \gg 1$: the width is exponentially small [4, 5].
(ii) $\omega_f \to 0$: in the adiabatic limit, the width is typically[a] small [9, 10], namely of the order of the perturbation amplitude $h$.
(iii) $\omega_f \sim 1$: for the intermediate range qualitative estimates [4, 5] suggested a width of $\sim h$. The range of moderately small frequencies remained almost unexplored (perhaps because it was assumed to be uninteresting), with one exception [7] where it was considered heuristically.

Thus, most theoretical estimates suggested that the maximum should occur at $\omega_f \sim 1$ where it would have a value of order $h$.

However, numerical studies (e.g., [7]) revealed a different picture: typically, many peaks were seen in the range $\omega_f \lesssim 1$, with heights substantially exceeding $h$, suggesting that high peaks actually represent the dominant feature of the dependence of the layer width on the perturbation frequency. The authors of the numerical studies intuitively associated them with nonlinear resonances and an approximate heuristic theory was developed [7] to estimate their positions and maxima. But the mechanism of the involvement of nonlinear resonances in separatrix chaos remained unclear, and a consistent theory of the peaks and a general analysis were lacking.

### 1.3. *Earlier approaches*

Of the many criteria for chaos, two of those most widely used are:

(1) The existence of a positive Lyapunov exponent [11], implying exponential growth in some direction in the separation between initially close trajectories, leading in turn to irregular transport.

[a]In ac-driven spatially periodic systems, the upper energy in the layer diverges as $\omega_f \to 0$ [8]. We do not consider this case. For such models, we consider only the lower energy in the layer.

(2) Destruction of invariant tori (KAM-theory) [4, 5], leading in particular to the replacement in the Poincaré section of regular trajectories (lines i.e., 1D objects) by layers (2D objects) where regions with regular and irregular dynamics are mixed in a complicated manner.

Usually, however, Lyapunov exponents have to be found numerically, it being hard to calculate them analytically [11]; and KAM-theory is difficult to apply in the relevant ranges of parameters, where resonances play a crucial role [12].

## 2. Our Approach to Separatrix Chaos

### 2.1. *Our initial tool: the separatrix map*

The separatrix map was introduced in [6] and rederived rigorously in [13]. As an illustration, consider as an example the ac-driven double-well potential system:

$$H = H_0 - hq\cos(\omega_f t), \quad H_0 = p^2/2 + U(q), \quad U(q) = -q^2 + q^4. \tag{5}$$

For the sake of convenience, we refer to $H_0(q,p) \equiv E$ as the "energy". Differentiating $E$ with respect to time and using the Hamiltonian equations, we obtain $dE/dt = h\cos(\omega_f t)\ dq/dt \propto dq/dt$. Thus, energy changes relatively fast only when the velocity $dq/dt$ is relatively large. If $E$ is close to the barrier value, then the system spends most of its time near the saddle, where the velocity is near zero, i.e., short pulses of significant velocity are separated by long intervals at almost zero velocity. The duration of a long interval is related to the period of the unperturbed system with the energy possessed by the perturbed system during the interval.

Introducing the angle of the perturbation $\varphi \equiv \omega_f t$ and the sign of the velocity $\sigma$, one can map $E$ and $\sigma$ at the beginning of a given pulse, and $\varphi$ in its middle, onto their values in the next pulse: $(E_i, \varphi_i, \sigma_i) \to (E_{i+1}, \varphi_{i+1}, \sigma_{i+1})$. This map is called the *separatrix map* (SM). For our model, it is

$$\begin{aligned} E_{i+1} &= E_i + \sigma_i h\varepsilon \sin(\varphi_i), \\ \varphi_{i+1} &= \varphi_i + \frac{\omega_f \pi (3 - \mathrm{sign}(E_{i+1} - E_s))}{2\omega(E_{i+1})}, \\ \sigma_{i+1} &= \sigma_i \mathrm{sign}(E_s - E_{i+1}), \end{aligned} \tag{6}$$

where the Melnikov integral $\varepsilon \equiv \varepsilon(\omega_f) \sim \omega_f$ [4, 11] for the relevant range of $\omega_f$.

There are both heuristic [4, 11] and rigorous [5] criteria for the onset of chaos but almost all of them miss the relevant range of moderately small $\omega_f$. The one exception [7] appreciated the role of nonlinear resonances, but the resultant calculations led to values of the maxima that were quantitatively incorrect in the asymptotic $h \to 0$ limit. The nature of the connections between nonlinear resonances and separatrix chaos remained unclear as did also the question of the generality of the results.

Our new approach [14–16] has allowed us to solve consistently both these and some related problems.

## 2.2. *Basic elements of the approach*

### 2.2.1. *Decorrelation of angle*

If the system gets onto the separatrix at the $(i+1)$th step i.e., $E_{i+1} = E_s$, then the change in angle at this step, $\varphi_{i+1} - \varphi_i$, diverges due to the divergence of $\omega^{-1}(E_{i+1})$ (see Eq. (3)). This decorrelation of angle leads, in turn, to a decorrelation of the energy since $E_{i+1} - E_i \propto \sin(\varphi_i)$ while $\sin(\varphi_i)$ may take any value between $\pm 1$. Similarly, $\sigma$ is also decorrelated. Obviously, if all variables of the map change randomly at some step, such a trajectory cannot be regular. Thus, *a decorrelation of all variables at some step of an area-preserving map is a sufficient condition for chaos in the map*; in the SM, such a decorrelation is provided by the decorrelation of angle. We hope and surmise that such a condition can be shown to satisfy earlier definitions of chaos, in particular the existence of a positive Lyapunov exponent.

### 2.2.2. *Involvement of a resonant motion into the separatrix chaos*

It might at first appear that a resonant dynamics is impossible near the separatrix because of the divergence of $d\omega/dE \propto 1/|E - E_s|$, given that resonant dynamics requires an approximate constancy of $\omega(E)$. Paradoxically, however, it is possible due to the slowness of the logarithmic divergence: if $x$ is small, then $\ln(1/x)$ remains almost constant even at a rather large variation of $x$.

The energy at some step of the chaotic trajectory takes the value $E_s$: $E_n = E_s$, while $\varphi_n$ and $\sigma_n$ may be arbitrary. Consider changes of energy at the next step which have one and the same sign: let it be positive for example. Then $E_{n+1} - E_s \sim h|\sin(\varphi_n)|$. The properties of the logarithmic function mean that $\omega(E_{n+1}) \simeq \omega(E_s + h)$ for almost all values of $\varphi_n$ (except in the near multiples of $\pi$). Moreover, $\omega(E)$ remains approximately constant for rather large further variation of $E - E_s$ of the same sign. Thus, if $\omega_f \approx \omega(E_s + h)$, a resonant dynamics should be relevant.

By reducing the discrete map equations to the differential equations of motion for the resonance Hamiltonian, we were able to prove the above intuitive ideas explicitly within the relevant range of energies.

### 2.2.3. *Resonant dynamics near the separatrix*

The conventional approximation by which resonant dynamics is described by an auxiliary pendulum-like (Chirikov) Hamiltonian [3, 4, 7, 11] is valid only if $|d\omega/dE|$ is approximately constant within the relevant range of energy. This is not the case near the separatrix, where $|d\omega/dE| \propto 1/|E - E_s|$ varies rapidly. We have developed the first analytic description of the resonant dynamics near the separatrix.

From the quantitative point of view, this part of our approach is of particular importance because resonant dynamics provides the leading term of our asymptotic theory describing the boundaries of the chaotic layers and related quantities in the relevant range of the perturbation frequency.

## 3. Application to the Single-Separatrix Layer

As an illustrative example, we use the paradigmatic case of an ac-driven pendulum $H = p^2/2 - \cos(q) + hq\cos(\omega_f t)$, comparing numerical results with earlier estimates and with our theory. Consider the deviation of the lowest energy in the layer $E_l$ from the separatrix value $E_s$. Figure 3 presents numerical calculations of the deviation $\Delta E^{(-)} \equiv (E_s - E_l)$ normalized by the perturbation amplitude, as a function of frequency, for different values of $h$. One can see a clear growth of tho maximum as $h$ decreases. The inset compares these results with earlier estimates, with which they clearly disagree. In contrast, our asymptotic theory [15] fits the numerical data well, as shown in Fig. 4: the paper also classified all systems in the context of the functional dependence of the width on $h$ and considered characteristic examples. Note also that our theory describes well, not only the width of the layer, but also its boundaries in the Poincaré section.

## 4. Application to Global Chaos Between Separatrices

If an unperturbed system possesses two or more separatrices, then a sufficiently large periodic perturbation gives rise to dynamic (chaotic) transport connecting neighboring separatrices, a phenomenon that is important in many physical applications. Conventionally, the scenario for its onset is as follows. For small perturbation amplitude $h$, the separatrices are replaced by thin chaotic layers. As the amplitude grows, the layers widen and, at some critical value $h_{\rm cr}$, they merge, marking the

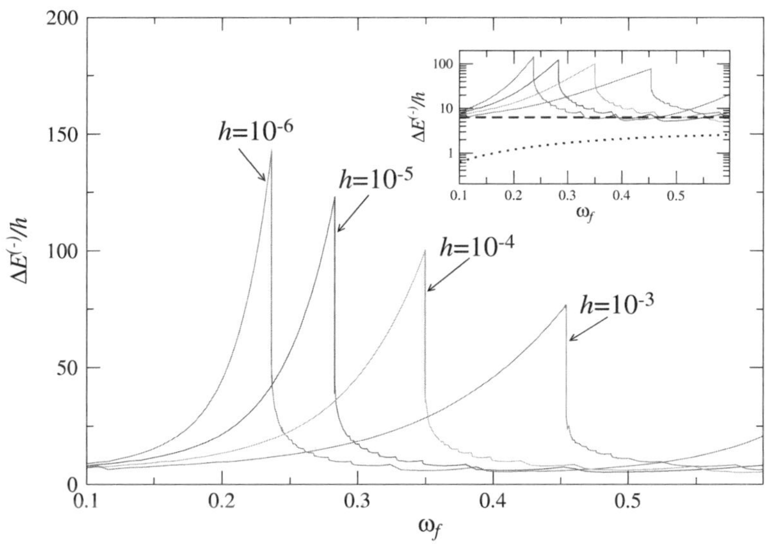


Fig. 3. Deviation of the lowest energy in the layer from the separatrix value, normalized by $h$, as function of frequency, for different $h$. The inset (note the logarithmic ordinate) compares the same results (full curves) with earlier predictions (dashed [4, 10] and dotted [17] lines).

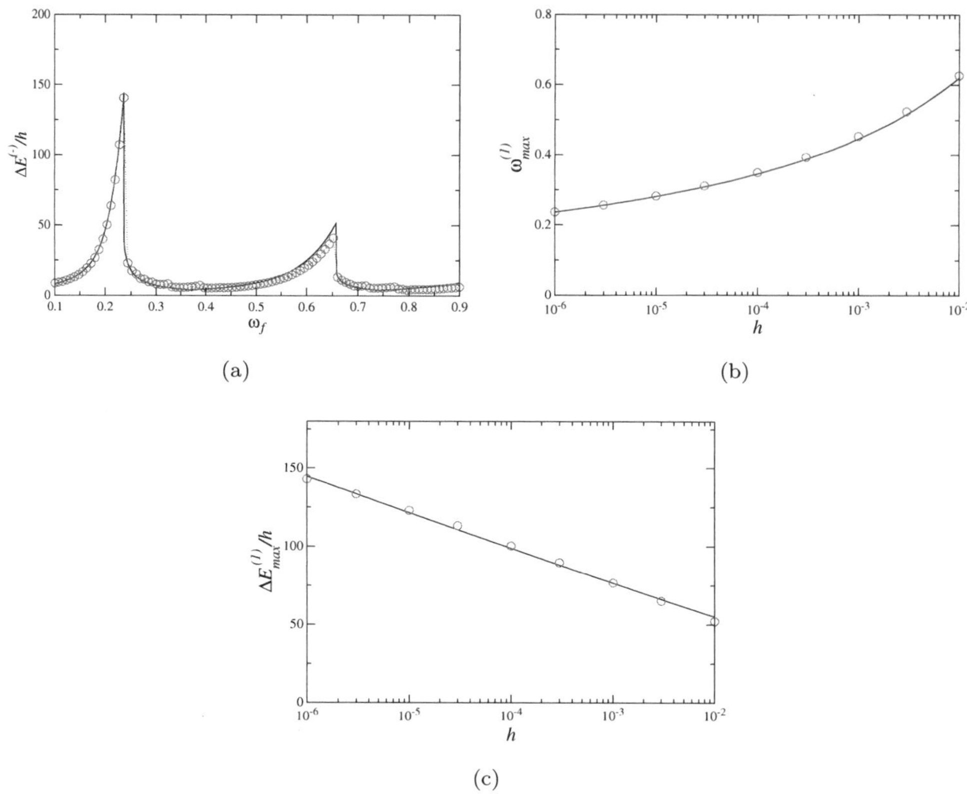


Fig. 4. Comparison of our theory (solid line) and numerical data (circles): (a) $\Delta E^{(-)}/h$ as function of $\omega_f$ for $h = 10^{-6}$; (b) position of the first maximum as function of $h$; (c) value of the first maximum as function of $h$.

onset of global chaos. Typically, $h_{\rm cr}$ is of the order of the difference of energies at the separatrices. However, it was discovered in 2003 [18] that $h_{\rm cr}$ may be much lower if the perturbation frequency $\omega_f$ is close to particular values. We show this using as an example the unperturbed model illustrated in Fig. 5:

$$H_0 = p^2/2 + U(q), \quad U(q) = (\Phi - \sin(q))^2/2, \quad \Phi = 0.2. \tag{7}$$

Figure 6(a) shows a bifurcation diagram of the system with $H = H_0 + hq\sin(\omega_f t)$ in the plane of the perturbation parameters. The shaded area indicates the parameter range for which global chaos exists. Its lower boundary, representing the function $h_{\rm cr}(\omega_f)$, possesses deep minima (spikes). Their existence can be explained as follows. The dependence of the eigenfrequency on energy $\omega(E)$ (Fig. 5(c)) is very flat (in the asymptotic limit $\Phi \to 0$, it approaches a rectangular shape). If $\omega_f$ is close to the local maximum $\omega_m$ of the function $\omega(E)$, then even a weak perturbation remains approximately resonant with eigenoscillations over almost the whole range of energies between the separatrices, so that resonant dynamics may be involved

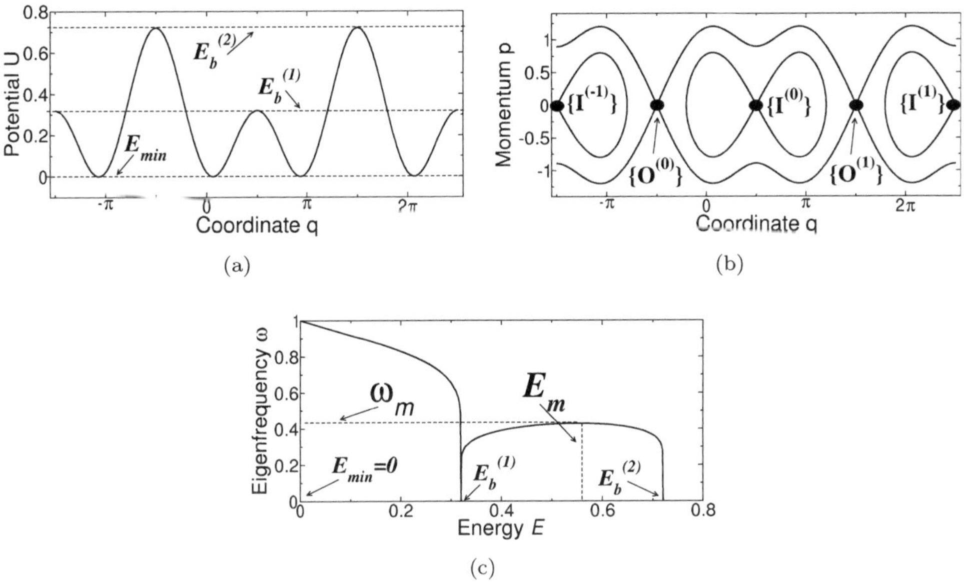


Fig. 5. (a) Potential of Eq. (7). (b) Separatrices in the phase plane. (c) Oscillation frequency as function of energy.

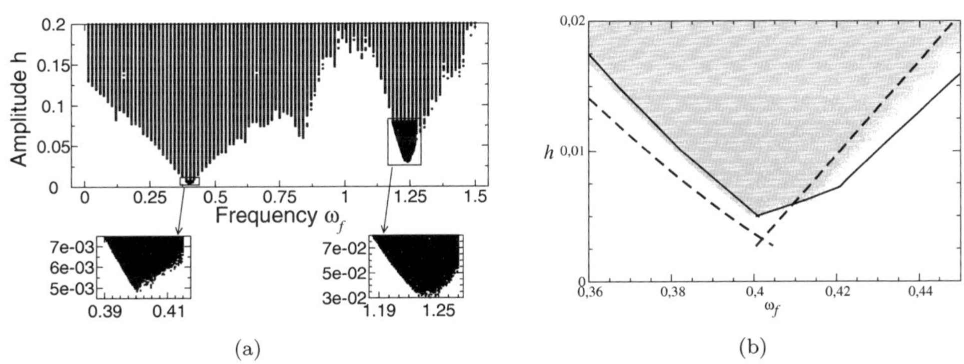


Fig. 6. (a) Bifurcation diagram for $H = H_0 + hq\sin(\omega_f t)$; (b) the first minimum (shaded) and our theory: dashed line — the leading-order theory, solid line — the theory with higher-order corrections.

in the separatrix chaos related to each of them, thus providing transport between them. Using our new approach, we developed a consistent asymptotic theory [14] which nicely describes the spikes over a wide vicinity of their minima (Fig. 6(b)).

## 5. Some Examples of Applications and Relevant Problems

We can summarize the present status of the new method as follows —

I. The approach itself:

- Evaluates the boundaries of separatrix chaotic layers, including their energy width, for a diversity of perturbed systems with separatrices.
- Promises to be useful for the analysis of transport within the region of the separatrix chaos.

II. Results on the thin single-separatrix layer may:

- Be generalized to describe a wide variety of systems with separatrices.
- Provide for a method [15] to facilitate the overlap of resonances of different orders, in turn facilitating the onset of global chaos.

III. Inter-separatrix transport is likely to be important in, for example:

- Spinning pendula [19] and, in some cases, electron gas in magnetic superlattice [20, 21], both being described by the model (7).
- Partial stochastic webs, arising if the resonance is approximate rather than exact [4]; there are many separatrices and, the more separatrices are linked by chaotic transport, the larger the web [22] becomes; it is relevant to quantum electron transport in semiconductor superlattices in magnetic and electric fields [22, 23] when the scattering is very weak.
- Meandering flows [24] e.g., the Gulf Stream [25, 26], where "cross-jet" (inter-separatrix) transport retards transport along the jet.
- Cold atoms in optical lattices [27] can model a variety of potential systems with practically zero dissipation. So, they could model double- or multi-barrier potentials with barriers of different heights.
- Billiards, which may possess more than one separatrix [28].
- Plasma physics [29].

## 6. Conclusion

Our approach to the treatment of separatrix chaos includes three basic elements: a new criterion for chaos in the separatrix map, i.e., decorrelation of angle; involvement of resonant dynamics into separatrix chaos, explained by the matching of the separatrix map and continuous resonant dynamics; explicit analysis of the resonant dynamics near the separatrix. It has enabled us to solve a long-standing problem related to the maximum width of the separatrix chaotic layer and to describe the drastic facilitation of the onset of global chaos between separatrices for a characteristic example. There are many potential applications of these results.

## Acknowledgment

We gratefully acknowledge support by the Royal Society through an International Joint Project, and by Pisa University within its collaboration agreement with the Institute of Semiconductor Physics (Kiev), and by the DFG through SFB TR-12.

## References


[1] V. I. Arnold, *Mathematical Methods of Classical Mechanics* (Springer-Verlag, New York, 1978).

[2] N. N. Filonenko, R. Z. Sagdeev and G. M. Zaslavsky, Destruction of magnetic surfaces by magnetic field irregularities 2, *Nucl. Fusion* **7** (1967) 253–266.

[3] B. V. Chirikov, Universal instability of many-dimensional oscillator systems, *Phys. Rep.* **52** (1979) 263–379.

[4] G. M. Zaslavsky, *Physics of Chaos in Hamiltonian Systems*, 2nd edn (Imperial College Press, London, 2007).

[5] G. N. Piftankin and D. V. Treshchev, Separatrix maps in Hamiltonian systems, *Russ. Math. Surv.* **62** (2007) 219–322.

[6] G. M. Zaslavsky and N. N. Filonenko, Stochastic instability of trapped particles and conditions of application of the quasi-linear approximation, *Sov. Phys.-JETP* **27** (1968) 851–857.

[7] I. I. Shevchenko, Marginal resonances and intermittent behaviour in the motion in the vicinity of a separatrix, *Phys. Scr.* **57** (1998) 185–191.

[8] S. M. Soskin, O. M. Yevtushenko and R. Mannella, Divergence of the chaotic layer width in the adiabatic limit and strong acceleration of the chaotic transport in AC driven periodic systems, *Phys. Rev. Lett.* **95** (2005) 224101.

[9] A. I. Neishtadt, Change in adiabatic invariant at a separatrix, *Sov. J. Plasma Phys.* **12** (1986) 568–573.

[10] Y. Elskens and D. F. Escande, Slowly pulsating separatrices sweep homoclinic tangles where islands must be small — an extension of classical adiabatic theory, *Nonlinearity* **4** (1991) 615–667.

[11] A. J. Lichtenberg and M. A. Lieberman, *Regular and Chaotic Dynamics* (Springer, New York, 1992).

[12] V. Gelfreich, private communication (2009).

[13] V. Rom-Kedar, Transport rates of a class of two-dimensional maps and flows, *Physica D* **43** (1990) 229–268.

[14] S. M. Soskin, R. Mannella and O. M. Yevtushenko, Matching of separatrix map and resonant dynamics, with application to global chaos onset between separatrices, *Phys. Rev. E* **77** (2008) 036221.

[15] S. M. Soskin and R. Mannella, Maximal width of the separatrix chaotic layer, *Phys. Rev. E* **80** (2009) 066212.

[16] S. M. Soskin, R. Mannella, O. M. Yevtushenko, I. A. Khovanov and P. V. E. McClintock, A new approach to the treatment of separatrix chaos and its applications, in *Hamiltonian Chaos Beyond the KAM Theory (dedicated to George Zaslavsky)*, eds. A. Luo and V. Afraimovich (HEP-Springer, Berlin, 2010), pp. 51–141.

[17] G. M. Zaslavsky, R. Z. Sagdeev, D. A. Usikov and A. A. Chernikov, *Weak Chaos and Quasi-Regular Patterns* (Cambridge University Press, PVEMcC, 1991).

[18] S. M. Soskin, O. M. Yevtushenko and R. Mannella, Drastic facilitation of the onset of global chaos, *Phys. Rev. Lett.* **90** (2003) 174101.

[19] A. A. Andronov, A. A. Vitt and S. E. Khaikin, *Theory of Oscillators* (Pergamon, Oxford, 1966).

[20] O. M. Yevtushenko and K. Richter, Effect of an ac electric field on chaotic electronic transport in a magnetic superlattice, *Phys. Rev. B* **57** (1998) 14839–14842.

[21] S. M. Soskin, R. Mannella and P. V. E. McClintock, Zero-dispersion phenomena in oscillatory systems, *Phys. Rep.* **373** (2003) 247–408.

[22] S. M. Soskin, P. V. E. McClintock, T. M. Fromhold, I. A. Khovanov and R. Mannella, Stochastic webs and quantum transport in superlattices: An introductory review, *Contemp. Phys.* **51** (2010) 233–248.

[23] T. M. Fromhold, A. Patanè, S. Bujkiewicz, P. B. Wilkinson, D. Fowler, D. Sherwood, S. P. Stapleton, A. A. Krokhin, L. Eaves, M. Henini, N. S. Sankeshwar and F. W. Sheard, Chaotic electron diffusion through stochastic webs enhances current flow in superlattices, *Nature* **428** (2004) 726–730.

[24] D. del-Castillo-Negrete and P. J. Morrison, Chaotic transport by Rossby waves in shear flow, *Phys. Fluids A* **5** (1993) 948–965.

[25] R. M. Samelson and S. Wiggins, *Lagrangian Transport in Geophysical Jets and Waves* (Springer-Verlag, New York, 2006).

[26] R. M. Samelson, *The Theory of Large-Scale Ocean Circulation* (Cambridge University Press, 2011).

[27] R. Gommers, P. Douglas, S. Bergamini, M. Goonasekera, P. H. Jones and F. Renzoni, Resonant activation in a nonadiabatically driven optical lattice, *Phys. Rev. Lett.* **94** (2005) 143001.

[28] E. D. Leonel and L. A. Bunimovich, Suppressing Fermi acceleration in a driven elliptical billiard, *Phys. Rev. Lett.* **104** (2010) 224101.

[29] M. C. Firpo and D. Constantinescu, Study of the interplay between magnetic shear and resonances using Hamiltonian models for the magnetic field lines, *Phys. Plasmas* **18** (2011) 032506.

# ROGUE WAVES IN SPATIALLY EXTENDED OPTICAL SYSTEMS

S. RESIDORI* and U. BORTOLOZZO†
*INLN, Université de Nice Sophia-Antipolis*
*CNRS, 1361 Route des Lucioles*
*06560 Valbonne, France*
**Stefania.Residori@inln.cnrs.fr*
*†Umberto.Bortolozzo@inln.cnrs.fr*

A. MONTINA
*Perimeter Institute for Theoretical Physics, Waterloo*
*Ontario, Canada N2L 2Y5*
*amontina@perimeterinstitute.ca*

F. LENZINI
*Dipartimento di Fisica, Università di Firenze*
*50125 Firenze, Italy*
*lenzinifrancesco@libero.it*

F. T. ARECCHI
*Dipartimento di Fisica*
*Università di Firenze, and CNR-INO,*
*Largo E. Fermi 6, 50125 Firenze, Italy*
*Tito.Arecchi@inoa.it*

Received 20 September 2011
Revised 29 December 2011
Published 31 March 2012

Communicated by Richard Haley

Giant anomalous events may occur in systems characterized by many waves, with extremely high amplitude waves appearing with a probability much higher than expected for a random dynamics. Such rare, and giant, events are known to occur on the ocean surface and have been named, in that context, "rogue waves". We present here some recent experimental results on the statistical properties of giant optical waves, or "optical rogue waves", observed in spatially extended optical systems. We will present the main statistical features of such extreme events, then, based on both a linear and a nonlinear experiment, we will outline a way enabling us to identify two key ingredients at the origin of optical rogue waves, namely, granularity, that is, a minimal size of the individual light spots, and inhomogeneity, that is, clustering of the light spots into separate domains with different average intensities. In the linear experiment we measure also the statistics of the waiting times between successive rogue waves and we show that it follows a log-Poisson distribution, which is characteristics of event separation observed in cooperative, complex systems and in such different fields as geophysics and biology.

*Keywords*: Rogue waves; non-Gaussian statistics; nonlinear optics.

## 1. Introduction

Rogue waves are extreme and rare events occurring in systems characterized by the presence of many waves. The most known examples are the extreme events that seldom, and unpredictably, appear on the ocean surfaces. The terms "rogue" or "freak" waves have, indeed, been coined in this context, where these extreme events constitute a well-known, and frightening, phenomenon, leading to waves as tall as 20–30 m and representing a threat even for large boats and ocean liners [1, 2]. The origin of rogue waves is still a matter of debate [3] and a large interest has grown in the last years that has brought to the development of different theoretical approaches as well as to setup different kinds of laboratory experiments.

Several models have been developed, most of them relying on the weak nonlinear interaction between many waves and, thus, based on the nonlinear Schroedinger equation [4, 5]. In general, such envelope equation approaches fail, however, to catch the very steep profile that characterizes the extreme events. In this context, numerical simulations play an important role shedding light on the basically involved mechanisms, as the role of large breathers [6] and the Benjamin–Feir instability [7]. Another approach is pointed out in [8], where the emergence of extreme events from wave turbulence is obtained by numerical simulations of the nonlinear Schröedinger equation and is interpreted in view of establishing a link between wave turbulence theory [9] and optical rogue wave phenomena. Note that experimental verification of an optical wave turbulence regime has been reported for a one-dimensional liquid crystal system, where large amplitude solitons are, indeed, observed after the inverse cascade, when modulation instability of the condensate sets in [10]. However, a clear derivation of rogue waves from wave turbulence solitons [11] still remains, up to now, not straightforward to achieve.

From the experimental side we have recently witnessed a growing interest, accompanied by implementations of laboratory scale tests in many different systems. Rogue wave laboratory experiments include implementations in large water tanks, allowing to study directional ocean waves [12], nonlinear light propagation in doped fibers [13], wave turbulence in second sound in superfluid Helium [14], nonlinear optical cavities [15] and microwave systems [16].

The common feature characterizing rogue wave phenomena in such different systems is the observation of large deviations from the Gaussian statistics of the wave amplitude, with long tails of the probability density function (PDF) accounting for the rather frequent emission of giant waves. Despite the specificity of each experiment, other common properties can be identified, as the existence of many uncorrelated "grains" of activity that are inhomogeneously distributed in larger spatial domains. Depending on the system under study and on the nature of the waves considered, grains can be of a different origin, for instance, solitons in nonlinear systems or wave packets in linear propagating waves, and their clustering in spatial domains can occur via different mechanisms, as a temporal delay, a spatial symmetry breaking, a transport phenomenon or a hypercycle type amplification [17].

On the other hand, it has been recently outlined in the microwave experiment that rogue waves can occur even in the absence of nonlinearity [16]. We have, therefore, conjectured that nonlinearity could just have the role of bringing forth granularity, that is, of inducing soliton formation, and to enhance the effect of inhomogeneity, that is, to drive the system through spontaneous self-reconfiguration and space–time dynamics. In order to verify these conjectures we have developed a linear optical experiment and compared the outcomes with those from the nonlinear cavity. The results show that the appearance of rogue waves, identified by a non-Gaussian statistics of the intensity, can be related to the presence, in the system, of a suitable amount of granularity and inhomogeneity, independently if these are induced by a linear or a nonlinear mechanism [18].

Rogue waves appear, therefore, as ubiquitous phenomena, with several different approaches proposed to understand and classify them. The various interpretations often depend on the type of experimental system considered, or on the type of theoretical context from which they emerge. Based on recent developments, we can essentially mention three broad classes of rogue waves: envelope solitons, observed in supercontinuum experiments [19], noise spikes due to nonlinear amplifier response in systems such as Raman fiber amplifiers [20], and modulation instability solitons, such as the Peregrine soliton recently observed in optics [21], in a wave tank [22] and in a plasma experiment [23]. In optics, all these types of abovementioned solutions manifest themselves as intensity spikes in a temporal series of the light intensity. Here the concept of granularity could be introduced as the typical correlation length in the time signal and inhomogeneity could be seen as a consequence of the temporal recursive coupling between the many longitudinal modes characterizing the light propagation dynamics. Thus, it seems possible to draw an analogy between temporal optical rogue waves and rogue waves observed in spatially extended optical systems.

Finally, it is worth mentioning that "deterministic rogue waves" have been claimed in a recent publication [24], where spikes in the intensity time series of a semiconductor laser with injected signal arise as a consequence of deterministic chaos. This is a well-known phenomenon in lasers [25], and it seems doubtful to consider it as a potentially new mechanism for rogue waves, especially when the reported PDF of the intensity do not show clear deviations from a Gaussian behavior. More convincingly, numerical simulations of an unidirectional fiber laser in a nonstationary regime of operation show a probability density function of the intensity with an elevated tail, largely deviating from the exponential, where the high probability of producing extreme pulses seems related to the laser mode-locking dynamics [26].

We review, here, the main results on the statistical properties of giant optical waves in spatially extended systems. The related statistical properties, PDF, have been measured both in nonlinear optical cavities and in the linear propagation experiment. We will see how the comparison between different experimental setups, and their outcomes, enable us to identify granularity and inhomogeneity as the two key ingredients that are at the origin of optical rogue waves. While granularity is a minimal size of the individual light spots, inhomogeneity is due to clustering of the light spots (grains) into separate domains characterized by different average intensities. When performing the PDF of the intensity over all the clusters we obtain a non-Gaussian behavior, with large tails deviating from the exponential.

Finally, we show for the linear system that the statistics of the waiting times between successive rogue waves occurring at the same space position follows a log-Poisson distribution. This behavior is characteristic of temporal separation between large events occurring in cooperative and complex systems, with several examples arising in different fields, as geophysics and biology.

## 2. Rogue Waves in a Nonlinear Optical Cavity

A nonlinear optical cavity is built by using a two-wave mixing process as a gain mechanism. The nonlinear medium is a liquid crystal light-valve, LCLV, a device formed by the assembly of a nematic liquid crystal layer with a photoconductive crystal [27, 28]. An external voltage $V_0$ (typically of 20 V $r.m.s.$ amplitude and with a frequency of 75 Hz) is applied across the cell, whereas the photoconductive substrate transfers it to the liquid crystal layer, within which occurs the wave-mixing.

When two light beams are sent to interfere on the LCLV, a fringe intensity pattern is created on the photoconductor, which, on its turn, generates a refractive index grating through the reorientation of the liquid crystal molecules. The two beams are, therefore, diffracted by the grating and, in undergoing this process, they mix-up, in the sense that part of the photons of one beam are diffracted on the direction of the other and vice versa. If one of the two beams, called the pump, is much more intense than the other, then, the wave-mixing provides a gain mechanism by transferring photons from the pump in the direction of the weak beam.

The nonlinear optical cavity, as illustrated in Fig. 1, is a ring formed by three high-reflectivity dielectric mirrors and a lens of $f = 70\,\text{cm}$ focal length. The total cavity length is $L = 273.3\,\text{cm}$ and the lens is positioned at a distance $L_1 = 88.1\,\text{cm}$ from the entrance plane of the LCLV. The coordinate system is taken such that $z$ is along the cavity axis and $x, y$ are on the transverse plane.

The pump $E_p$, illuminating the LCLV is an enlarged and collimated (10 mm diameter) beam from a solid-state diode pumped laser ($\lambda = 532\,\text{nm}$), linearly polarized in the same direction of the liquid crystal nematic director. Oscillations in the cavity are built from the wave-mixing in the LCLV. The gain is provided by

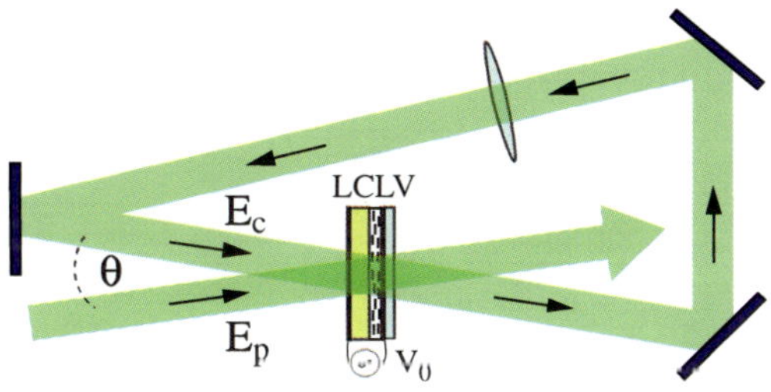


Fig. 1. A schematic illustration of the nonlinear optical cavity. The liquid crystal light-valve, LCLV, provides gain through wave mixing with the pump beam $E_p$, hence, leading to the spontaneous generation of the cavity field $E_c$.

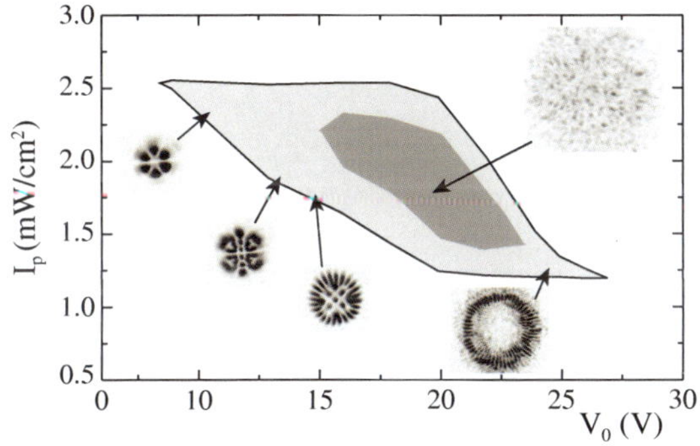


Fig. 2. Experimental phase diagram and instantaneous snapshots taken for parameters in the region indicated by the corresponding arrow. Cavity field oscillations are observed in the whole gray region while large amplitude fluctuations appear in the darker shadowed region (from [29]).

wave-mixing with the pump $E_p$, and the cavity field $E_c$ is spontaneously generated starting from the amplification of noise photons.

Cavity field oscillations occur in a large range of parameter space. A phase space diagram is illustrated in Fig. 2, where the region of oscillations is plotted in the $V_0, I_p$ space, with $I_p = |E_p|^2$ the pump intensity [29]. Gauss–Laguerre modes are observed at the boundaries of the oscillation region, with the order and number of modes increasing as increasing $V_0$. A similar behavior is observed by fixing $V_0$ and then increasing the Fresnel number of the cavity, which is defined as the ratio of the area $a^2$ of the most limiting aperture, the transverse size of the LCLV in our case, to the size $w_0^2 = \lambda L$ of the fundamental Gaussian mode, $F = a^2/\lambda L$. The maximum number of transverse modes hosted in the cavity scales as $F^2$, $F$ being the maximum number of nodes that fits the transverse size along one direction [30]. In the typical experimental conditions $F \sim 100$ and the number of cavity modes is as high as $F^2 \sim 10^4$.

Inside the darker shadowed region of Fig. 2, and at high $F$, many modes oscillates simultaneously in the cavity, giving rise to a speckle-like behavior of the field. In this situation, we observe extreme events appearing as anomalous waves of large amplitude over the lower amplitude and fluctuating background. These optical rogue waves occur erratically in space and time, and live for a typical time of the same order of the response time of the LCLV.

### 2.1. *The stretched exponential statistics*

The optical rogue wave phenomenon is characterized by measuring the PDFs of the light intensity in the high $F$ regime. The PDFs are determined by acquiring a large set of images (about 1000) and then performing the histograms of the intensity values on the whole image stack. At this purpose, a small fraction (4 %) of the field circulating in the cavity is extracted by a beam sampler and sent to a CCD camera ($1024 \times 768$ pixels and 12 bits depth).

While for relatively low pump intensity, $I_p = 2\,\mathrm{mW/cm^2}$, the cavity field shows a complex spatiotemporal dynamics, with the formation of many uncorrelated domains, at larger pump extreme events, rogue waves, appear as large amplitude peaks. An instantaneous experimental profile showing the transverse intensity distribution recorded for $I_p = 4.2\,\mathrm{mW/cm^2}$ is shown in Fig. 3(a). A large event in the form of a high amplitude peak can be clearly distinguished. The optical rogue waves evolve spontaneously in space and time, appearing randomly and with a small probability. The typical time scale of the dynamical evolution is 100 ms, which is ruled by the response time of the liquid crystals [31].

In Fig. 3(b) the PDF of the cavity field intensity, $I = |E_c|^2$, are displayed for different values of the pump, $I_p/I_{\mathrm{th}} = 1.8, 4.0, 4.2$, and 6.4, where $I_{\mathrm{th}} = 1.2\,\mathrm{mW/cm^2}$

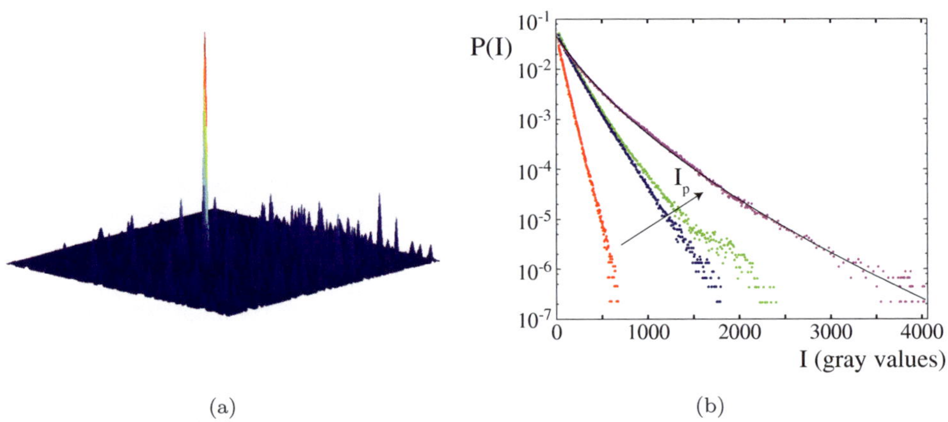


Fig. 3. (a) Instantaneous experimental profile of the transverse intensity distribution; an extreme event clearly appears over a speckle-like background; (b) Experimental PDF of the cavity field intensity; the pump is varied from $I_p/I_{\mathrm{th}} = 1.8, 4.0, 4.2$, and 6.4 from the steepest to the shallowest distribution; $I_{\mathrm{th}} = 1.2$ mW/cm$^2$ is the threshold intensity for the oscillation to start in the cavity. The steepest distribution is exponential. The solid black line is the stretched exponential function that fits the shallowest distribution, with $c_1 = 29.5$ (from [15]).

is the threshold for the optical oscillations to start in the cavity. The resulting PDF are plotted in log-linear scale, as shown in Fig. 3(b). Increasingly large deviations from the exponential behavior are observed as we increase the pump intensity. We have found that all the distributions are well fitted by the stretched exponential function $N\mathrm{e}^{-\sqrt{c_1+c_2 I}}$, where $1/c_1$ is a parameter that measures the deviation from the exponential behavior. The black line is the fitting function of the PDF with $I_p/I_{\mathrm{th}} = 6.4$.

Note that an exponential statistics for the intensity corresponds to a Gaussian statistics for the field amplitude. Therefore, an exponential PDF of the intensity is characteristic of a speckle pattern, where each point receives the uncorrelated contributions of many uncoupled modes. At low pump, indeed, this is the behavior displayed by the cavity field. However, when the pump increases, the increasing nonlinear coupling leads to a complex space–time dynamics and extreme events populate the tails of the PDF, providing large deviations from Gaussianity.

Another important point to note is that because the cavity is formed by three mirrors, there is an inversion symmetry around the transverse $y$ axis. This symmetry breaking is responsible for the cavity field $E_c$ to become inhomogeneous, hence, introducing a nonlocal coupling mechanism that appears as essential for inducing the extreme events.

Numerical simulations have been performed by using the full model equations of the cavity, as described in [32]

$$\tau \frac{\partial n_0}{\partial t} = -n_0 + \alpha |E_c|^2,$$
$$\tau \frac{\partial n_1}{\partial t} = -n_1 + \alpha E_c E_p^*, \tag{1}$$

where $\tau$ is the response time of the LCLV, $n_0$ and $n_1$ are, respectively, the amplitude of the homogeneous refractive index and the amplitude of the refractive index grating at the spatial frequency $k_p - k_c$, $k_p$ and $k_c$ being the optical wave numbers of the pump and cavity field, $\alpha$ is the nonlinear coefficient of the LCLV, and the diffusion length due to elastic coupling in the liquid crystal has been neglected. The dynamics of the liquid crystal is much slower than the settling of the cavity field; thus, $E_c$ adiabatically follows the evolution of $n_0$ and $n_1$. By taking the wave propagation equation with the cavity boundary conditions, we obtain

$$E_c = i \sum_{k=0}^{\infty} [\hat{C}\mathrm{e}^{i n_0} J_0(2|n_1|)]^k \hat{C}\mathrm{e}^{i n_0} \frac{n_1}{|n_1|} J_1(2|n_1|) E_p, \tag{2}$$

where $J_m$ is the Bessel function of the first kind and of order $m$, and $\hat{C}$ is an operator accounting for the geometry of the cavity and losses, $\hat{C} = \Gamma^{1/2}\mathrm{e}^{i\delta}\hat{S}_x \mathrm{e}^{i(L_0\nabla_\perp^2/2k_p)}\mathrm{e}^{-i(k_p \mathbf{r}_\perp^2/2f)}\mathrm{e}^{i(L_1\nabla_\perp^2/2k_p)}$, with $1-\Gamma$ the photon losses and $\delta$ the phase retardation in a round-trip. $\hat{S}_x$ is a symmetry operator that inverts the $x$ axis, thus accounting for the odd number of mirrors, $\nabla_\perp^2$ is the transverse Laplacian and $\mathbf{r}_\perp$ the position in the transverse $(x, y)$ plane.

In order to account for the finite size of the LCLV and the diaphragm inserted near the lens, we have used in the simulations a diaphragm both in the space and in the Fourier plane. In particular, we have used a spatial filter with radius equal to 0.4 cm and a Fourier filter with radius $k = 6.4 \times 10^{-3}$ rad. As for the other parameters, we have set the liquid crystal response time $\tau = 0.1$ s, the photon loss fraction $1 - \Gamma$ between 0.7 and 0.8, and $\alpha$ is chosen in such a way that the unit is the pump threshold for the activation of the cavity field.

Starting with a random initial condition, we observe a transient speckle-like behavior, then a symmetry breaking occurs, with an inversion symmetry around the $y$ axis. The symmetry breaking becomes evident by plotting the numerical intensity $|E_c|^2$ in the $z - y$ plane, as shown in Fig. 4(a). The yellow line in $z = 0$ represents the position of the LCLV inside the cavity. The spatial symmetry breaking seems to play an essential role in inducing large deviations from Gaussianity. The numerical PDFs of the cavity field intensity are displayed in Fig. 4(b) for different values of the pump. The intensities are rescaled in such a way that the PDFs have the same slope at the origin. The tails are increasingly populated as the pump increases, in agreement with the experimental observations.

### 2.1.1. *The hyper-cycle model*

The symmetry breaking introduces a nonlocal coupling between the cavity modes. This nonlocality couples local domains of the cavity field with other and distant domains, thus introducing large loops of amplification, a mechanism similar to the hyper-cycle chain of reactions in the catalytic processes [17]. Because of this mechanism, there is a sort of focusing, for which at some space locations the cavity field grows much more with respect to the surrounding places, giving rise to large amplitude peaks, hence, to large deviations from the Gaussian statistics.

To elucidate this mechanism, we consider a simple two-mode model, where only the evolution of the average refractive index $n_1$ is kept, the average being performed

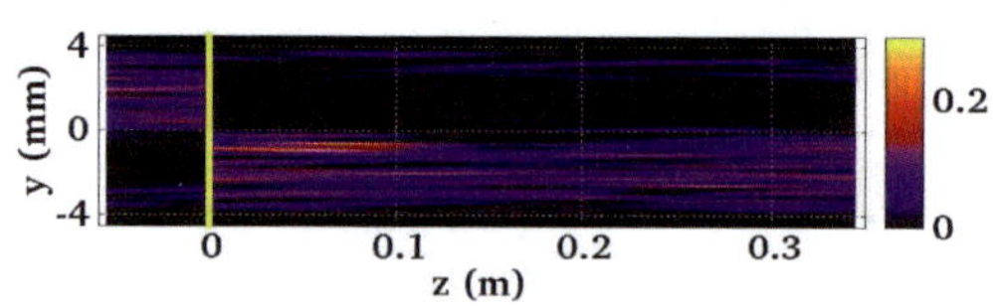


(a)

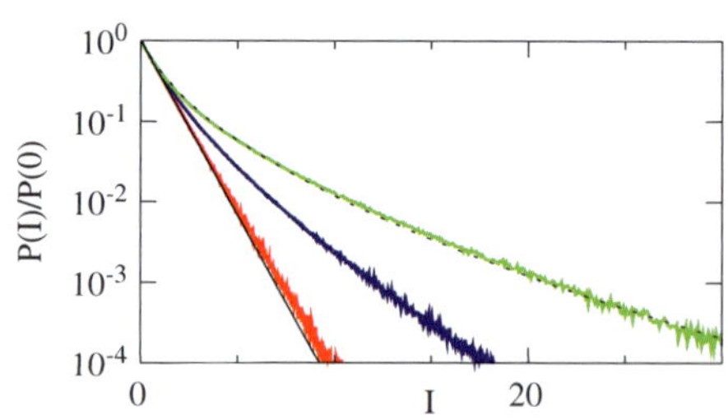


(b)

Fig. 4. (Color online.) Numerical simulations. (a) Cavity field intensity in the $y$–$z$ plane; the yellow vertical line indicates the position of the LCLV; $I_p = 10$, $1-\Gamma = 0.7$. (b) PDF of the cavity field intensity for different values of the pump $I_p/I_{\text{th}} = 6$ (black), 8 (red), 10 (green); $1 - \Gamma = 0.7$ (from [15]).

over the transverse plane. For a nearly plane cavity, $\bar{n}_1$ satisfies an equation of the form

$$\tau\frac{\partial \bar{n}_1}{\partial t} = -\bar{n}_1 + I_p F(\bar{n}_1), \tag{3}$$

where the first and second terms account for the liquid crystal relaxation and, respectively, the grating feeding provided by the pump and cavity fields. As an ansatz for the function $F$, we take a cubic function, which describes a linear growth followed by saturation, this last one being mainly due to multiple scattering processes [32].

When the nearly plane cavity is replaced by a cavity nearly spherical, the above mean-field picture is modified by the symmetry breaking and, thus, we have to consider two mean fields $\bar{n}_+$ and $\bar{n}_-$, where the averages are performed over the upper and lower half-planes, respectively. Accounting for the inversion $y \to -y$ of the cavity field, we have that, after a round trip, the grating $\bar{n}_\pm$ is fed by the grating $\bar{n}_\mp$ at the opposite side, i.e., the previous equation is replaced by the two following ones,

$$\begin{aligned}\tau\frac{\partial \bar{n}_+}{\partial t} &= -\bar{n}_+ + I_p F(\bar{n}_-),\\ \tau\frac{\partial \bar{n}_-}{\partial t} &= -\bar{n}_- + I_p F(\bar{n}_+),\end{aligned} \tag{4}$$

where we have assumed implicitly that the cavity losses are very high, i.e., the cavity field is negligible after more than one round trip.

The main features of $F$ can be captured by the cubic function $F(\bar{n}_\pm) = \bar{n}_\pm - \bar{n}_\pm^3$. It is easy to show that for $1 < I_p < 2$, Eq. (5) has only one stable solution, $\bar{n}_- = \bar{n}_+ = 0$, whereas for $I_p \geq 2$ a bifurcation occurs with the birth of the two stable antisymmetric states

$$\begin{aligned}n_+ &= \frac{1}{\sqrt{2}}\sqrt{1 \pm \sqrt{\frac{I_p^2-4}{I_p^2}}},\\ n_- &= \frac{1}{\sqrt{2}}\sqrt{1 \mp \sqrt{\frac{I_p^2-4}{I_p^2}}},\end{aligned} \tag{5}$$

which breaks the $y \to -y$ symmetry, thus qualitatively accounting for the experimental and numerical observations.

In order to stress the role of inhomogeneity, we define as an indicator of the deviations from the exponential statistics the nondimensional parameter [18]

$$D(\mathbf{r}) = \frac{\langle I^2(\mathbf{r})\rangle}{\langle I(\mathbf{r})\rangle^2} - 2, \tag{6}$$

which is equal to zero for an exponential statistics and, numerically, we evaluate it for different space regions of the cavity field. In Fig. 5 we plot the numerically

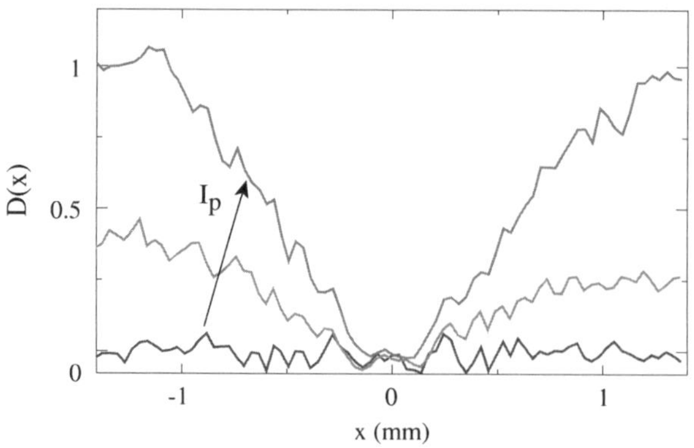


Fig. 5. (Color online.) Numerically calculated nondimensional indicator of rogue wave statistics for a nonlinear optical cavity versus the transverse coordinate $x$ and for increasing pump intensity; $I_p/I_{\text{th}} = 4, 6, 8$ for the blue (lower), green (middle) and red (higher) curve (from [18]).

calculated $D(x)$, which is $D(\mathbf{r})$ integrated along the vertical axis, as a function of the horizontal transverse coordinate $x$ and for $I_p/I_{\text{th}} = 4, 6, 8$. We see that $D(x)$ is an increasing function of both $I_p$ and $|x|$. While for increasing $I_p$ the nonlinearity increases, moving away from $x = 0$ (that is the inversion symmetry axis of the system), the inhomogeneity also increases because of the symmetry breaking introduced by the cavity configuration.

## 2.2. *The role of the symmetry breaking*

The role of symmetry breaking in inducing a stretched exponential statistics of the intensity has been experimentally investigated by building a cavity with four instead of three mirrors. In this configuration there is no inversion symmetry. A direct comparison is made for the resulting PDF of the intensity of the four-mirror cavity with the case of the three-mirror cavity. In both cases the Fresnel number is fixed to $F = 125$ and the pump intensity to $I_p = 5\,\text{mW/cm}^2$. The voltage applied to the LCLV is fixed to $V_0 = 40\,\text{V}$ *r.m.s.*, $f = 30\,\text{Hz}$, for the three-mirror cavity and to $V_0 = 32\,\text{V}$ *r.m.s.*, $f = 30\,\text{Hz}$, for the four-mirror cavity. These values have been chosen in order to obtain that, for low pump intensity, the PDFs of the three- and four-mirror cavity show the most similar exponential behavior.

For a qualitative comparison, Figs. 6(a) and 6(b) show two intensity profiles obtained from two instantaneous images recorded for the three- and four-mirror cavity, respectively. Figures 6(c) and 6(d) display two corresponding one-dimensional profiles. Rogue waves can be easily distinguished on the profiles recorded for the three-mirror cavity while they are not observed in the typical profiles recorded for the four-mirror cavity, showing that rogue waves become much less probable in the absence of symmetry breaking.

The PDF of the intensity are recorded in a similar way as described above, both for the three- and the four-mirror cavity. The results are plotted in Fig. 7. As it

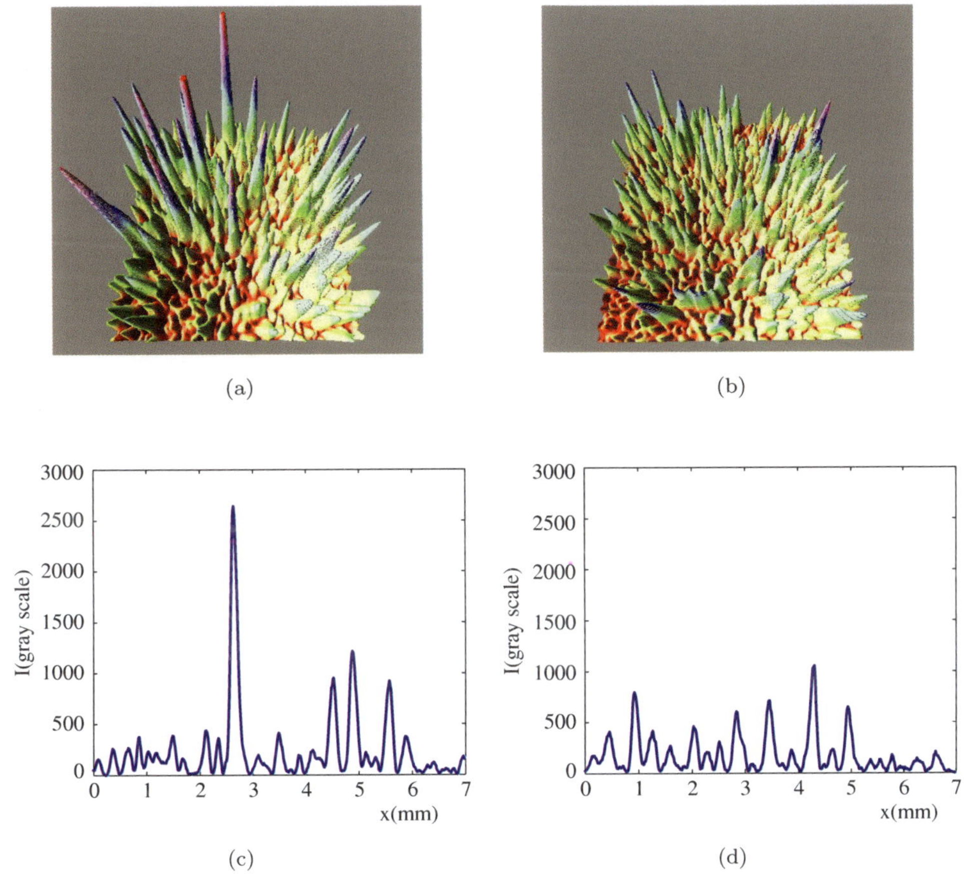


Fig. 6. Experimental intensity profiles for (a) the three-mirror cavity and (b) the four-mirror cavity; (c) and (d) are two corresponding one-dimensional profiles.

can be appreciated from the fitting curves, while for the three-mirror cavity the behavior is well described by a stretched exponential, for the four-mirror cavity the data are quite well fitted by an exponential function. We, therefore, recover a Gaussian field statistics when removing the symmetry breaking brought forth by the cavity configuration.

## 3. Optical Rogue Waves in a Linear Experiment

In order to outline the two key ingredients for the appearance of optical rogue waves, namely, granularity and inhomogeneity, we have set up a linear experiment. Here, granularity, that is, fragmentation of the wave field into a large number of elementary objects of still finite size, let us call them speckles, and inhomogeneity, that is, clustering of speckles in spatial domains with different average intensities, can be separately controlled.

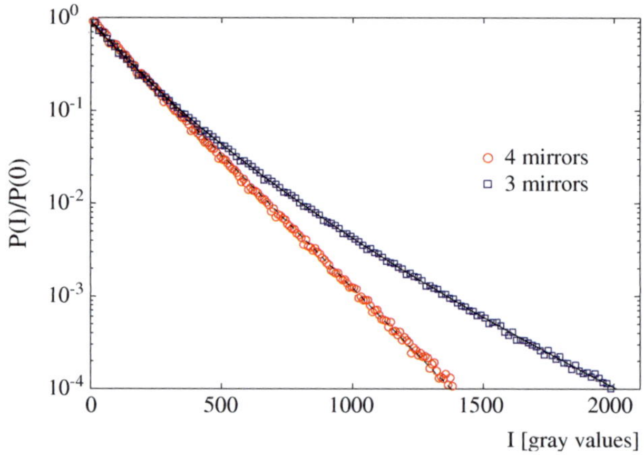


Fig. 7. (Color online.) Experimental PDF of the intensity for the four-mirror cavity (red circles) and the three-mirror cavity (blue squares). The four-mirror cavity distribution is practically exponential (dashed black line) while the three-mirror cavity is fitted by a stretched exponential (solid black line).

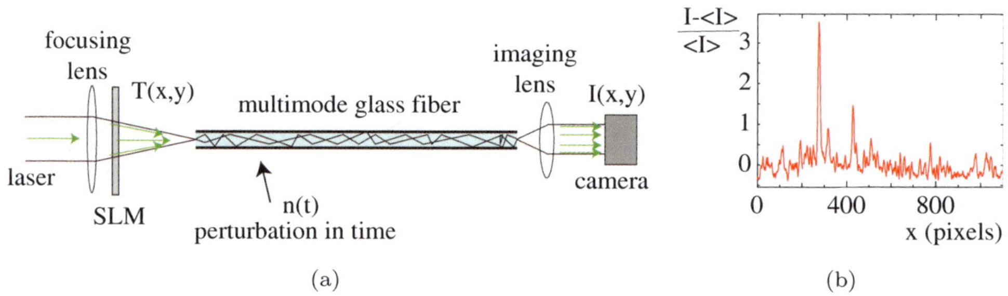


Fig. 8. (a) Experimental setup for the linear excitation of optical rogue waves: a laser input beam is focused into a multimode glass fiber; the output field is imaged onto a CCD camera; a SLM control the input beam profile through variable transmittance masks $T(x, y)$; a lateral perturbation $n(t)$ rules the fiber bending; (b) 1D intensity profile $I(x)$ taken from a corresponding image; $\langle I \rangle$ is the intensity averaged over the whole image (from [18]).

A schematic representation of the experimental setup is shown in Fig. 8(a). It consists of a multi-mode glass optical fiber (0.4 mm diameter and 2 m length) in which a laser beam is let to propagate. The input beam comes from a frequency doubled solid-state laser, wavelength 532 nm. Less than 1 mW/cm$^2$ is coupled inside the fiber and the input profile is controlled by a spatial-light-modulator, SLM. This is a computer driven liquid crystal display, 768 × 1024 pixels, one inch diagonal size. By setting onto the SLM appropriate transmittance masks $T(x, y)$, arbitrary profiles of the input intensity distribution can be introduced.

While an uniform mask $T(x, y)$ allows the whole cone of input wave vectors to be coupled into the fiber, an inhomogeneous mask with a black hole prevents the wave vectors passing in that direction to be coupled into the fiber, thus inducing at the output domains of different average intensity. This symmetry breaking mechanism introduces high amplitude peaks on the output intensity distributions $I(x, y)$. In Fig. 8(b) a one-dimensional intensity profile taken along an $x$-line of a corresponding image is displayed for the inhomogeneous case. A large amplitude peak, or optical rogue wave, can be clearly distinguished.

The average size of the grains (single speckle) is given by the effective aperture of the fiber, the total number of modes supported by the fiber being proportional to $(a/\lambda)2$, with $a$ the radius of the fiber core and $\lambda$ the wavelength of light in air [33]. For the typical experimental conditions, the total number of modes can be estimated as $\sim 10^6$. On the other hand, the inhomogeneous mask induces at the exit of the fiber an inhomogeneous distribution of the average intensity when this is evaluated over domains of adjacent grains. We determine the PDF of the intensity by acquiring with a CCD camera ($768 \times 1024$ pixels and 16 bits depth) a large set of images (about one thousand) and then performing the histograms of the intensity values on the whole image stack.

The results are shown in Fig. 9(a), where the PDF recorded for a uniform $T(x, y)$ is compared with the PDF recorded in the presence of an inhomogeneous mask. The first one (red solid line) is well fitted by an exponential function (black dashed line), as expected for speckles [33]. The second one (empty black dots) shows appreciable, though not too large, deviations from the exponential. In the same figure we plot also a PDF with strong deviations from the exponential; it is obtained by applying to the fiber a lateral perturbation $n(t)$ with a Gaussian distribution of the amplitude, as displayed in Fig. 9(b). The perturbation is realized by placing a piezoelectric emitter in contact with a side of the fiber (see Fig. 8(a)).

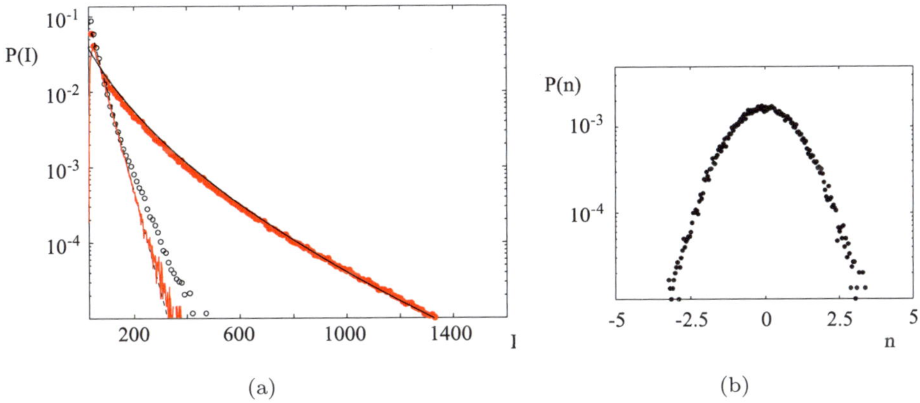


Fig. 9. (Color online.) (a) PDF of the intensity at the output of the fiber; solid (red) line: homogeneous mask; (black) empty dots: inhomogeneous mask; (red) filled dots: inhomogeneous mask and temporal perturbations; the black dashed line is a fit with an exponential function; the black solid line is a fit with a stretched exponential fit ($c_1 = 10$, $c_2 = 0.092$). (b) Gaussian PDF of the perturbation $n(t)$ used to drive the fiber and producing the large-tailed PDF (from [18]).

Thanks to the elasto-optic effect, when the piezo emits a low frequency acoustic wave, it locally modifies the optical paths inside the fiber, giving rise to a different spatial distribution of the output intensity, therefore, the detector collects events over different speckle configurations. Note that in the nonlinear experiment the multimode dynamics spontaneously introduces a continuous change of the cavity field, hence, allowing to explore a large number of configurations.

The effect of the temporal modulation greatly enhances the stretched exponential character of the intensity PDF. Note, however, that in the inhomogeneous case the PDF is a stretched exponential even without temporal perturbations (even though small-tailed when compared to the perturbed case), while in the homogeneous case the PDF remains exponential even in the presence of temporal modulations.

When inhomogeneity and temporal perturbations are simultaneously present, the resulting large-tailed PDF of the intensity is well described by a stretched exponential distribution, that is, $P(I) = \mathrm{e}^{-\sqrt{c_1+c_2 I}}$, where $c_2$ is a scale factor and $c_1$ a form parameter. In the limit $c_1 \to \infty$, the stretched distribution becomes an exponential. To account for the stretched exponential character of the PDF let us consider the role of inhomogeneity. At the fiber exit the intensity is distributed as a collection of speckle patterns made of various domains with different average intensity. Within a single domain the statistics is exponential but the variance changes from domain to domain. As a consequence, the overall PDF becomes a stretched exponential as the events are counted over the whole field. More precisely, the stretched distribution can be obtained as a statistical mixture of many exponential distributions with different variance,

$$P(I) = \int_0^\infty d\sigma \rho(\sigma) \frac{\mathrm{e}^{-I/\sigma}}{\sigma}, \tag{7}$$

where $\sigma$ is the variance of the PDF in a single domain. The stretched distribution is exactly obtained when the distribution of $\sigma$ is

$$\rho(\sigma) = \sqrt{\sigma} \mathrm{e}^{-\frac{c_2 \sigma}{4} - \frac{c_1}{c_2 \sigma}}. \tag{8}$$

Qualitatively, this prediction is well confirmed by the experimental results, where $\rho(\sigma)$ has been evaluated by performing the PDF analysis over small spatial domains and determining the local variance $\sigma$ of the associated fit [18].

### 3.1. *Statistics of the return times*

We consider, now, the statistics of the waiting times between successive rogue waves occurring at a given space position. As we limit our observations to a single homogeneous domain, we classify a Poisson distribution of event separation. However, as $n(t)$ jumbles the different domains within the detector aperture, we expect different rates associated with different domains. Interpolation of the different Poissonian should yield a log-Poisson statistics for the waiting times between two successive events [34, 35].

In the experiment, the fiber is driven by the piezoelectric perturbation $n(t)$, by sending an impulse and, then, letting the dynamics to relax. A typical spatio-temporal plot is shown in Fig. 10(a), where, after each impulse, a new configuration, followed by a slowing-down readjustment, is observed. In Fig. 10(b) the spatially averaged intensity is plotted as a function of time, with arrows indicating the instant times when the $n(t)$ impulse is released.

The waiting time between successive events is defined as $w = \ln(t_k) - \ln(t_{k-1}) = \ln(t_k/t_{k-1})$, where $t_k$ is the occurrence time of the $k$-th event. To obtain the PDF of the waiting times, the events are recorded by placing a photodiode in a fixed point and then counting a rogue wave each time the intensity is above a given threshold,

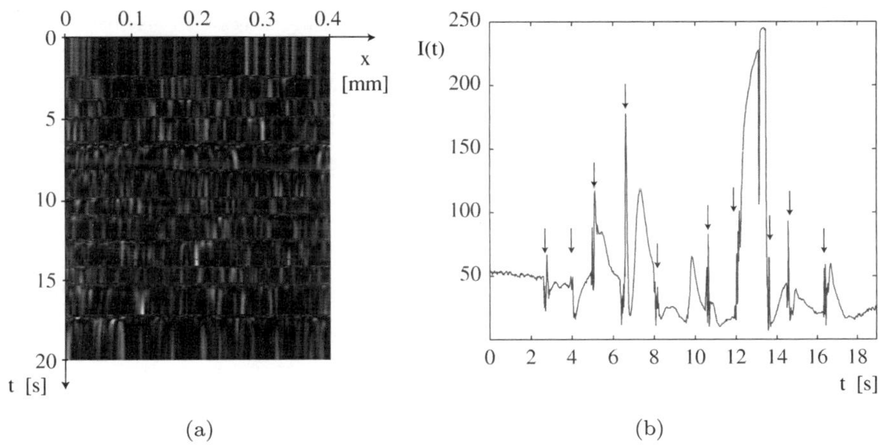


Fig. 10. (a) Spatiotemporal plots showing the evolution of the intensity when pulse-like perturbations $n(t)$ are applied to the fiber. (b) Corresponding spatially averaged intensity as a function of time; arrows indicate the instant times when the $n(t)$ impulses are released.

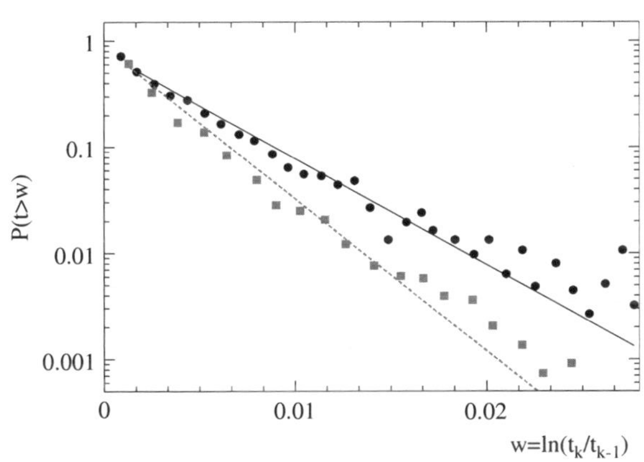


Fig. 11. Log-Poisson distribution of the waiting times between successive rogue waves; (red) squares, (black) dots, correspond to low, respectively, high amplitude of $n(t)$ (from [18]).

here taken as four times the average intensity. By analyzing the recorded data over a large number of configurations, each one triggered by the application of a $n(t)$ impulse, we obtain a log-Poisson distribution of the waiting time between successive rogue waves, as depicted in Fig. 11.

## 4. Conclusion

From the above results and considerations, we see that granularity and spatial inhomogeneity are joint generators that induce rogue waves even without nonlinearity. Nevertheless, the temporal scrambling of the spatially uncorrelated domains largely amplifies the deviations from the exponential statistics. We can, therefore, expect that in a nonlinear system where a spatiotemporal dynamics takes place spontaneously, the rogue wave phenomenon is very likely to be activated whenever clustering of grains occurs in inhomogeneous domains.

In this perspective, it is useful to draw a comparison between the linear and nonlinear case. In the nonlinear optical cavity, granularity is induced by the nonlinearity, which yields transverse spatial grains whose average size is $(\lambda L)/a$, where $L$ is the free propagation length in the cavity and $a$ is the size of the most limiting aperture. The same scaling of the grain size was previously obtained for a photorefractive cavity characterized by a space–time chaotic behavior [36, 37]. Inhomogeneity is then introduced by the chosen geometrical configuration of the resonator: if the coordinate system is taken such that $z$ is along the cavity axis and $x, y$ are on the transverse plane, there is an inversion symmetry around the $y$ axis, which provides a nonlocal coupling between distant domains.

In the linear experiment, granularity comes from the superposition (interference) of many spatial modes, giving rise to a speckle pattern. Then, inhomogeneity is introduced by tilting the angle of incidence of the input beam. However, in both systems large deviations from Gaussianity are obtained when a temporal scrambling of spatial domains occurs, either because of an externally imposed perturbation, as in the linear experiment, or because of an intrinsic spatiotemporal dynamics, as in the nonlinear experiment.

Finally, the log-Poisson distribution of the waiting time between successive events, as it has been measured in the linear experiment, suggests analogies between optical rogue waves and other types of rare events occurring in complex and cooperative systems, as earthquakes or social phenomena [34, 35].

## Acknowledgments

This work has been partially supported by the ANR international program, project ANR-2010-INTB-402-02, "COLORS."

## References

[1] P. Müller, Ch. Garrett and A. Osborne, *Oceanography* **18**(3) (2005) 66–75.
[2] B. S. White, *J. Fluid Mech.* **355** (1998) 113–138.

[3] N. Akhmediev and E. Pelinovsky, Discussion & Debate: Rogue Waves — Towards a Unifying Concept? *Eur. Phys. J. Special Topics* **185** (2010) 1, and articles therein.
[4] M. Onorato, A. R. Osborne and M. Serio, *Phys. Rev. Lett.* **96** (2006) 014503–4.
[5] N. Akhmediev, J. M. Soto-Crespo and A. Ankiewicz, *Phys. Lett. A* **373** (2009) 2137–2145.
[6] A. I. Dyachenko and V. E. Zakharov, *JETP Lett.* **88** (2008) 307–311.
[7] B. Eliasson and P. K. Shukla, *Phys. Rev. Lett.* **105** (2010) 014501–4.
[8] K. Hammani, B. Kibler, C. Finot and A. Picozzi, *Phys. Lett. A* **374** (2010) 3585–3589.
[9] V. Zakharov, V. Lvov and G. Falkovich, *Kolmogorov Spectra of Turbulence I: Wave Turbulence* (Springer, 1992).
[10] U. Bortolozzo, J. Laurie, S. Nazarenko and S. Residori, *J. Opt. Soc. B* **26** (2009) 2280–2284.
[11] A. I. Dyachenko, V. E. Zakharov, A. N. Pushkarev, V. F. Shvets and V. V. Yan'kov, *Sov. Phys. JETP* **69** (1989) 1144–1147.
[12] M. Onorato, T. Waseda, A. Toffoli, L. Cavaleri, O. Gramstad, P. A. E. M. Janssen, T. Kinoshita, J. Monbaliu, N. Mori, A. R. Osborne, M. Serio, C. T. Stansberg, H. Tamura and K. Trulse, *Phys. Rev. Lett.* **102** (2009) 114502–4.
[13] D. R. Solli, C. Ropers, P. Koonath and B. Jalali, *Nature* **450** (2007) 1054–1057.
[14] A. N. Ganshin, V. B. Efimov, G. V. Kolmakov, L. P. Mezhov-Deglin and P. V. E. McClintock, *Phys. Rev. Lett.* **101** (2008) 065303–4.
[15] A. Montina, U. Bortolozzo, S. Residori and F. T. Arecchi, *Phys. Rev. Lett.* **103** (2009) 173901–4.
[16] R. Höhmann, U. Kuhl, H. J. Stoc̆kmann, L. Kaplan and E. J. Heller, *Phys. Rev. Lett.* **104** (2010) 093901–4.
[17] M. Eigen and P. Schuster, *The Hypercycle: A Principle of Natural Self-Organization* (Springer, Berlin, 1979).
[18] F. T. Arecchi, U. Bortolozzo, A. Montina and S. Residori, *Phys. Rev. Lett.* **106** (2011) 153901–4.
[19] J. M. Dudley, G. Genty and B. J. Eggleton, *Opt. Exp.* **16** (2008) 3644–3651.
[20] K. Hammani, A. Picozzi and C. Finot, *Opt. Commun.* **284** (2011) 2594–2603.
[21] B. Kibler, J. Fatome, C. Finot, G. Millot, F. Dias, G. Genty, N. Akhmediev and J. M. Dudley, *Nat. Phys.* **6** (2010) 790–795.
[22] A. Chabchoub, N. P. Hoffmann and N. Akhmediev, *Phys. Rev. Lett.* **106** (2011) 204502–4.
[23] H. Bailung, S. K. Sharma and Y. Nakamura, *Phys. Rev. Lett.* **107** (2011) 255005–4.
[24] C. Bonatto, M. Feyereisen, S. Barland, M. Giudici, C. Masoller, J. R. Rios Leite and J. R. Tredicce, *Phys. Rev. Lett.* **107** (2011) 053901–4.
[25] D. Boggavarapu, R. Jin, J. Grantham, Y. Z. Hu, F. Brown de Colstoun, C. W. Lowry, G. Khitrova, S. W. Koch, M. Sargent Im and H. M. Gibbs, *Opt. Lett.* **18** (1993) 1846–1848.
[26] J. M. Soto-Crespo, P. Grelu and N. Akhmediev, *Phys. Rev. A* **84** (2011) 016604–7.
[27] P. Aubourg, J. P. Huignard, M. Hareng and R. A. Mullen, *Appl. Opt.* **21** (1982) 3706–3712.
[28] U. Bortolozzo, S. Residori, A. Petrosyan and J. P. Huignard, *Opt. Commun.* **263** (2006) 317–321.
[29] U. Bortolozzo, A. Montina, F. T. Arecchi, J. P. Huignard and S. Residori, *Phys. Rev. Lett.* **99** (2007) 023901–4.
[30] A. E. Siegman, *Lasers* (Oxford University Press, Oxford, 1986).
[31] P. G. De Gennes and J. Prost, *The Physics of Liquid Crystals*, 2nd edn. (Oxford Science Publications, Clarendon Press, 1993).

[32] A. Montina, U. Bortolozzo, S. Residori, J. P. Huignard and F. T. Arecchi, *Phys. Rev. A* **76** (2007) 033826–14.
[33] J. W. Goodman, *Speckle Phenomena in Optics: Theory and Applications* (Roberts and Company Publishers, Greenwood Village, CO 80111, USA, 2007).
[34] P. Sibani and P. B. Littlewood, *Phys. Rev. Lett.* **71** (1993) 1482–1485.
[35] T. O. Richardson, E. J. H. Robinson, K. Christensen, H. J. Jensen, N. R. Franks and A. B. Sendova-Franks, *PLoS one* **5** (2010) e9621–7.
[36] F. T. Arecchi, G. Giacomelli, P. L. Ramazza and S. Residori, *Phys. Rev. Lett.* **65** (1990) 2531–2534.
[37] F. T. Arecchi, G. Giacomelli, P. L. Ramazza and S. Residori, *Phys. Rev. Lett.* **67** (1991) 3749–3752.

# 4

# Noise in Biological Systems

# NOISE ANALYSIS IN STUDIES OF PROTEIN DYNAMICS AND MOLECULAR TRANSPORT

SERGEY M. BEZRUKOV
*Laboratory of Physical and Structural Biology, NICHD, National Institutes of Health, Bethesda, MD 20892-0924, USA*
*bezrukov@helix.nih.gov*

Received 26 January 2004
Revised 17 February 2004
Accepted 17 February 2004

Understanding the role of noise at cellular and higher hierarchical levels depends on our knowledge of the physical mechanisms of its generation. Conversely, noise is a rich source of information about these mechanisms. Using channel-forming protein molecules reconstituted into artificial 5-nm-thick insulating lipid films, it is possible to investigate noise in single-molecule experiments and to relate its origins to protein function. Recent progress in this field is reviewed with an emphasis on how this experimental technique can be used to study low-frequency protein dynamics, including not only reversible ionization of sites on the channel-forming protein molecule, but also molecular mechanisms of 1/f-noise generation. Several new applications of the single-molecule noise analysis to membrane transport problems are also addressed. Among those is a study on antibiotic translocation across bacterial walls. High-resolution recording of ionic current through the single channel, formed by the general bacterial porin, OmpF, enables us to resolve single-molecule events of antibiotic translocation.

*Keywords*: Ion channels; single molecules; fluctuations; bacterial porins.

## 1. Introduction

The last decade saw an explosion in biologically motivated noise research. The 2002 conference 'Unsolved Problems of Noise in Physics, Biology, and High Technology' held at the National Institutes of Health campus in Bethesda demonstrated significant progress in our understanding of noise at the many levels of biological organization, from single protein molecules, to neurons, to the human brain [1]. One of the leading themes of this research is the constructive role of noise in biology, the theme whose recent successful elaboration would have been impossible without landmark contributions of its ardent proponent and pioneer, Professor Frank Moss, and his colleagues [2–6].

The comprehensive picture of the constructive role of noise necessarily includes quantitative understanding of the physical mechanisms of noise generation. It is widely accepted that for neuronal systems one of the most important noise sources is related to the probabilistic gating of membrane ion channels [7, 8]. In retrospect, noise analysis of membrane ion currents played a crucial role in verifying the existence of protein ion channels — key membrane-bound structures that allow cells to sense changes in the chemical composition of the media, communicate with each other, and maintain homeostasis. The original measurements revealed channel conformational kinetics and conductance and, therefore, the channel density in the cell membrane [9–12]. Today, with the development of the single-channel reconstitution technique, noise and fluctuation analysis is opening new possibilities in studies of molecular dynamics.

Advances in studies of the dynamics of complex biological molecules [13, 14] rely on the development of new experimental techniques and the continuous refinement of existing methods. Optical studies of single protein molecules [15] are among most powerful approaches. Observations of enzymatic activity at the single-molecule level reveal rich behavior and conformational memory in enzyme operation. This is seen as a non-Markovian fluorescence signal with non-exponential state transition probabilities [16]. Noise analysis of the ionic current flowing through a single ion channel [17] — a nano-contact formed by a single protein molecule — allows one to address similar questions using resolution that is often higher than in single-molecule fluorescence experiments.

Protein channels are 'soft' structures in the sense that their biological function depends on protein ability to respond to moderate changes in its immediate environment. Channels respond to changes in applied voltage, proton concentration, temperature, applied mechanical tension, etc. by changing their conformation. To be able to respond, the difference in free energies of functionally important conformational states of the protein should be comparable to $kT$, where $k$ is the Boltzman constant and $T$ is the absolute temperature. The activation barrier for such a conformational transition should also be small to provide for a reasonable response rate.

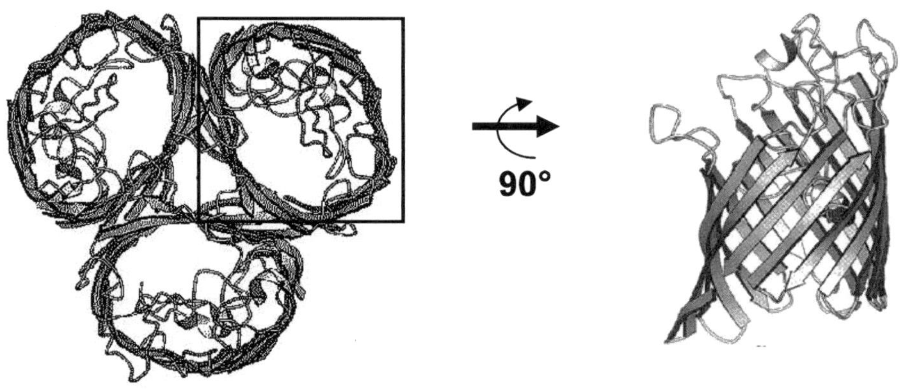


Fig. 1. High resolution X-ray structure of the channel formed by general bacterial porin, OmpF, viewed from out of the plane (left) and in the plane of the lipid film (right, only one monomer shown). The protein trimer has an overall diameter of about 10 nm. The main elements of the structure are three rigid β-barrels with several flexible loops folding into them (after [18]).

Even a brief glance at the channel architecture suggests a multitude of different physical processes that may contribute to the overall noise in the ionic current flowing through the channel. Figure 1 shows the structure of the main bacterial protein channel, OmpF, that makes the outer membrane of bacteria permeable for metabolites and other small solutes. OmpF is a trimer with three water-filled pores. In each monomer, sixteen β-strands span the outer membrane and form a barrel with short turns at one side of the membrane and large loops at the other. Several of the loops are packed inside the channel. One of the loops folds into the barrel to form a constriction zone at half the height of the channel.

Table 1 lists noise sources in their approximate hierarchical order. It combines the well-known results from different laboratories [9–12] with those obtained in our studies. The spectral power ranges in the table give typical orders of magnitude only; they are not intended to set any limitations or serve as definitions. Below we discuss several noise generation mechanisms in more detail.

Table 1. Hierarchy of noise in mesoscopic ion channels.

| Noise Source | Spectral Type | Amplitude Range*, $A^2$/Hz |
|---|---|---|
| Conformational Dynamics | Lorentzian, 1/f | $10^{-21} - 10^{-29}$ |
| Blocker Binding | Lorentzian | $10^{-22} - 10^{-27}$ |
| Residue Ionization | Lorentzian | $10^{-23} - 10^{-29}$ |
| Neutral Solute Transport | Diffusional, Lorentzian | $10^{-24} - 10^{-29}$ |
| Lipid Ionization | Diffusional, Lorentzian | $10^{-29} - 10^{-30}$ |
| Ion Shot Noise | White | $10^{-29} - 10^{-30}$ |
| Johnson Noise | White | $10^{-29} - 10^{-30}$ |

*Given for a single 'typical' mesoscopic channel of 0.1 to 1.0 nS conductance at 100 mV transmembrane voltage in the low-frequency decade of the sonic range.

## 2. Reversible Residue Ionization

Even for large protein channels, the water-filled pores are only about one nanometer in radius. This is the scale of the Debye screening length in water solutions with salt concentrations in the physiological range. Therefore, the charge state of the protein side chains that line the surface of the water-filled pore is crucial for channel ionic conductance. Strong electric fields produced by the side chain charges perturb the local distribution of ions and their mobility in the vicinity of the charge.

The charge state of the protein side chains depends on the proton concentration in the bulk solution, $[H^+]$. Bulk protons are in a dynamical equilibrium with side chain protons. For example, the average charge of the water-exposed carboxyl groups of glutamic and aspartic acids can be close to one half of the electron charge for a bulk proton concentration in the vicinity of 10 μM. Bulk protons hop on and off these side chains, thus modulating channel conductance and producing measurable fluctuations in the ionic currents through open channels [19–21].

Each monomer of the OmpF channel (Fig. 1) contains several dozen ionizable side chains, so it is natural to anticipate that their reversible ionization would be seen in its current noise. Indeed, this is the case. The results of noise measurements [22] are presented in Fig. 2. As expected, the open channel noise varies with the bulk proton concentration (pH = -$log$[H$^+$]).

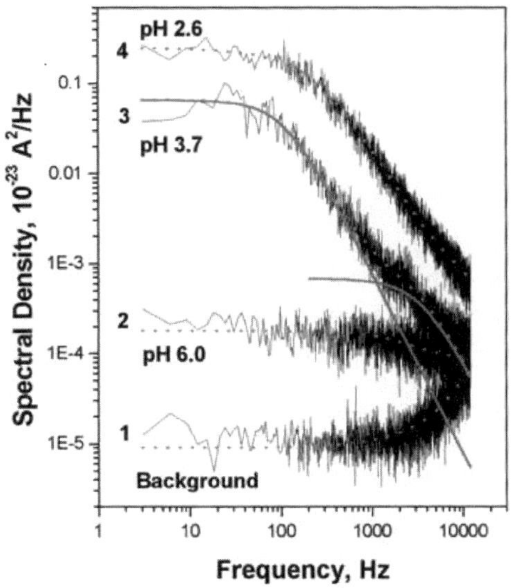


Fig. 2. Power spectral density of noise in ionic current through a single open channel of OmpF (Fig. 1) at different proton concentrations reveals dynamics of reversible ionization of the sites on the channel-forming molecule. The applied voltage was 100 mV; the OmpF was reconstituted into lipid films formed by opposition of two diphytanoyl phosphatidylcholine monolayers in aqueous solutions of 1 M potassium chloride. The background spectrum is measured at zero applied voltage [22]. The dotted lines are drawn to guide the eye; the solid lines represent an attempt of decomposition of the pH 3.7 spectrum into slow and fast Lorentzians.

The dominant low-frequency components are Lorentzians that allow us to determine the main characteristics of the residue ionization processes, such as the 'on' and 'off' rate constants and the influence of a single ionization event on channel conductance [19, 22]. This knowledge, taken together with the high-resolution structural information (Fig. 1), is very helpful in understanding electrostatics of the protein's nanoscale world.

### 3. 1/f Noise at the Single-Molecule Level

Our approach allows us to study 1/f noise at the level of a single molecule. Recently we have shown [23] that 1/f noise measured in the currents through multi-channel membranes [24] can be traced down to the step-wise conductance changes in a single 'fluctuator'. These step-wise changes have broad distributions of amplitudes and durations and reflect the complexity of protein structural organization and dynamics [13–16].

Figure 3 surveys the results of this study for the Maltoporin channel, whose structure is similar to the OmpF channel structure shown in Fig. 1, although its conductance is much smaller. The Maltoporin channel also exhibits 1/f noise that is absent in the case of OmpF (Fig. 2). Using high-resolution ionic current recording we have shown that the nanocontact formed by Maltoporin stochastically switches between sub-states of slightly differing conductance. The dynamics of these transitions are only weakly voltage-dependent, which suggests their equilibrium nature. We have found that the current through the contact is virtually 1/f-noise-free in any of the sub-states and concluded that the 1/f noise is generated by random transitions between them.

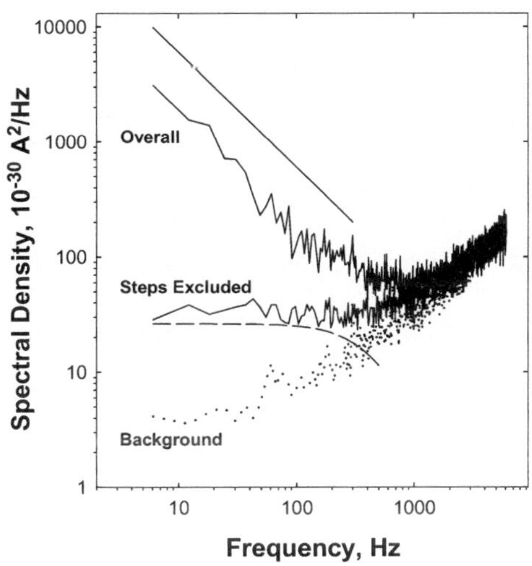


Fig. 3. Noise of a single open Maltoporin channel changes from 1/f-like behavior (upper curve) to 'white' one (middle curve) when step-wise transitions in the channel current are excluded [23]. The straight solid line shows the 1/f slope. The dashed line is a Lorentzian fit to the spectrum represented by the middle curve with the background subtracted. Lipid, salt solution, and applied voltage are as in Fig. 2; solution pH is 7.4.

Excision of the step-wise changes in the channel current transformed its low-frequency 1/f-like noise to noise with a flat spectrum. Importantly, very similar results could also be obtained using a selection of transition-free fragments of 1 s and longer by visual examination of current recordings. After subtracting the background, this noise component could be roughly approximated by a Lorentzian with a cut-off frequency of about 300 Hz. The intensity of its frequency-independent part is about $2.8 \cdot 10^{-29}$ A$^2$/Hz. This is still several times higher than the corresponding values for shot-noise, $2e\langle i\rangle = 5.8 \cdot 10^{-30}$ A$^2$/Hz (where $\langle i\rangle = 1.8 \cdot 10^{-11}$ A is the current through the single channel and $e$ is elementary charge), or Johnson noise, $4kTg = 2.9 \cdot 10^{-30}$ A$^2$/Hz (where $g = 1.8 \cdot 10^{-10}$ S is the channel conductance) expected in this case. The origin of this excess noise is not understood presently. It may be related to the still unresolved open channel conformational fluctuations or to the reversible recharging of ionizable sites in the channel-forming molecule that was discussed above. In any case, the 1/f component observed in the analysis of overall recordings is absent here to a high accuracy. The upper estimate for the Hooge parameter obtained from the data in Fig. 3 is $\alpha \leq 3 \cdot 10^{-7}$.

Qualitatively similar conclusions were recently reached in a study of ion current noise through nanofabricated synthetic pores [25]. Noise with the spectral distribution of $1/f^{\alpha}$ was attributed to the motions of channel wall constituents.

## 4. Channel-Facilitated Molecular Transport

Fluctuations in the channel's ionic current may be induced by the particles passing through the channel. The idea is that nanometer-scale particles going through a structure of the type shown in Fig. 1 create transient interruptions in the ionic conductance [17, 26-28]. These transients generate measurable excess noise and, sometimes, can be resolved as single-molecule events [17, 29–31]. Recently we have applied this idea to probe the mechanisms of antibiotic transport across bacterial walls [32].

Figure 4 shows that addition of a β-lactam antibiotic, ampicillin, introduces excess noise whose power spectra can be described by single Lorentzians. These spectra can be used to quantify the thermodynamics and kinetics of drug translocation [26–28, 32–34].

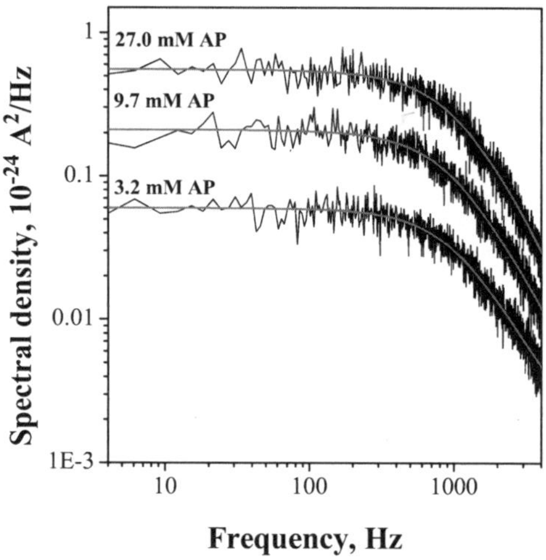


Fig. 4. Addition of the antibiotic ampicillin generates an excess conductance noise that can be used to characterize drug translocation through OmpF channels [32]. Power spectral densities of current fluctuations at three different ampicillin concentrations demonstrate that the process is well described by a simple two-state Markovian process. Solid lines through the spectra are drawn as Lorentzians with the relaxation time of $1.41 \cdot 10^{-4}$ s. The applied voltage was –100 mV, pH 6.0; other parameters as in Fig. 2.

Moreover, the characteristic time-scale of the process inferred from noise analysis suggests that the underlying single-molecule events can be time resolved. Figure 5 illustrates these time-resolved events.

Using this approach, we find that ampicillin and several other efficient penicillins and cephalosporins strongly interact with the residues of the constriction zone of OmpF channel. We conclude that, in analogy to the substrate-specific channels that evolved to bind certain metabolite molecules, antibiotics have 'evolved' to be channel-specific.

Molecular modeling [32] suggests that the charge distribution on the ampicillin molecule complements the charge distribution at the narrowest part of the bacterial porin. The interaction of these charges creates a region of attraction inside the channel that compensates for the entropy loss and facilitates drug translocation through the constriction zone. Therefore, the molecular mechanism that accounts for higher permeability rates of certain drugs can be traced down to their charge patterns as well as their 3D structure.

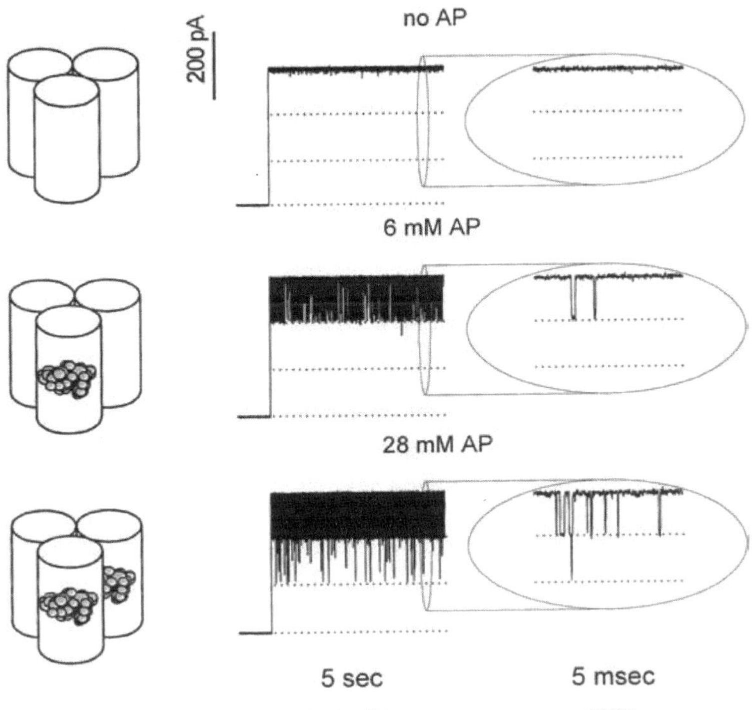


Fig. 5. Penetrating ampicillin (AP) molecules modulate ion current through a single *OmpF* channel [32]. Membrane bathing solution was 1M KCl, pH 6.0, the applied voltage was -100 mV. Top: In the absence of antibiotic the ion movement is mainly determined by the geometry and the surface properties of the channel pore. The ion current is stable, no interruptions are seen even at the high resolution recording shown on the right. Middle: In the presence of ampicillin in small concentrations one of the three *OmpF* pores gets spontaneously blocked by a translocating drug molecule. At the time resolution of 0.15 ms (left) blockage events look like downward spikes. At the higher resolution of 0.015 ms (right) they are seen as well-defined steps to 2/3 of the open channel current and back. Time between blockage events and their width provide both thermodynamic and kinetic parameters of antibiotic-pore interactions. Bottom: At higher ampicillin concentrations channel blockades are more frequent. Sometimes they overlap in time, leading to a transient reduction of ion current to 1/3 of its initial value.

## 5. Conclusions

Noise and fluctuation analysis of single protein nanocontacts provides important information about their structure and function. It advances our understanding of protein enzymatic activity by allowing us to study protein dynamics and interactions with different solutes at the single molecule level. Essential practical consequences include deeper understanding of molecular mechanisms of human illnesses and disorders, and further insights into the efficacy of newly designed drugs.

## References


[1] *Third International Conference on Unsolved Problems of Noise.* S. M. Bezrukov, editor, AIP Conference Proceedings, v.665, American Institute of Physics (2003).

[2] J. K. Douglass, L. Wilkens, E. Pantazelou and F. Moss, *Noise enhancement of information transfer in crayfish mechanoreceptors by stochastic resonance. Nature* **365** (1993) 337–340.
[3] Z. Gingl, L. B. Kiss and F. Moss, *Non-dynamical stochastic resonance: theory and experiment with white and arbitrarily coloured noise. Europhys. Lett.* **29** (1995) 191–196.
[4] E. Simonotto, M. Riani, C. Seife, M. Roberts, J. Twitty and F. Moss, *Visual perception of stochastic resonance. Phys. Rev. Lett.* **78** (1997) 1186–1189.
[5] D. Russell, L. A. Wilkens and F. Moss, *Use of behavioral stochastic resonance by paddlefish for feeding. Nature* **402** (1999) 219–223.
[6] L. W. Ward, A. Neiman and F. Moss, *Stochastic resonance in psychophysics and animal behavior. Biol. Cybern.* **87** (2002) 91–101.
[7] H. Lecar and R. Nossal, *Theory of threshold fluctuations in nerves. II. Analysis of various sources of membrane noise. Biophys. J.* **11** (1971) 1068–1084.
[8] J. A. White, J. T. Rubinstein and A. R. Kay, *Channel noise in neurons. Trends Neurosci.* **23** (2000) 131–137.
[9] F. Conti and E. Wanke, *Channel noise in nerve membranes and lipid bilayers. Quart. Rev. Biophys.* **8** (1975) 451–506.
[10] E. Neher and C.F. Stevens, *Conductance fluctuations and ionic pores in membranes. Ann. Rev. Biophys. Bioeng.* **6** (1977) 345–381.
[11] L. J. DeFelice, *Introduction to Membrane Noise*, Plenum Press, New York (1981).
[12] H. M. Fishman and H. R. Leuchtag, *Electrical noise in physics and biology. Curr. Topics Membr. Transp.* **37** (1990) 3–35.
[13] H. Frauenfelder, S. G. Sligar and P. G. Wolynes, *The energy landscapes and motions of proteins. Science*, **254** (1991) 1598–1603.
[14] W. A. Eaton, E. R. Henry, J. Hofrichter and A. Mozzarelli, *Is cooperative oxygen binding by hemoglobin really understood? Nature Struct. Biol.* **6** (1999) 351–358.
[15] X. S. Xie, *Single-molecule approach to dispersed kinetics and dynamic disorder: Probing conformational fluctuation and enzymatic dynamics. J. Chem. Phys.* **117** (2002) 11024–11032.
[16] S. Weiss, *Fluorescence spectroscopy of single biomolecules. Science* **283** (1999) 1676–1683.
[17] S. M. Bezrukov, *Ion channels as molecular Coulter counters to probe metabolite transport. J. Membrane Biol.* **174** (2000) 1–13.
[18] S. W. Cowan, T. Schirmer, G. Rummel, M. Steiert, R. Ghosh, R. A. Pauptit, J. N. Jansonius and J. P. Rosenbusch, *Crystal structures explain functional properties of two E.coli porins. Nature* **358** (1992) 727–733.
[19] S. M. Bezrukov and J. J. Kasianowicz, *Fluctuations in current through a single open ion channel reveal titration kinetics of ionizable residues. Phys. Rev. Lett.* **70** (1993) 2352–2355.
[20] J. J. Kasianowicz and S. M. Bezrukov, *Protonation dynamics of the alpha-toxin channel from spectral analysis of pH dependent current fluctuations. Biophys. J.* **69** (1995) 94–105.
[21] T. K. Rostovtseva, T.-T. Liu, M. Colombini, V. A. Parsegian and S. M. Bezrukov, *Positive cooperativity without domains or subunits in a monomeric membrane channel. Proc. Natl. Acad. Sci. USA* **97** (2000) 7819–7822.
[22] E. M. Nestorovich, T. K. Rostovtseva and S. M. Bezrukov, *Residue ionization and ion transport through OmpF channels. Biophys. J.* **85** (2003) 3718–3729.
[23] S. M. Bezrukov and M. Winterhalter, *Examining noise sources at the single-molecule level: 1/f noise of an open maltoporin channel. Phys. Rev. Lett.* **85** (2000) 202—205.
[24] F. Wohnsland and R. Benz, *1/f noise of open bacterial porin channels. J. Membr. Biol.* **158** (1997) 77–85.
[25] Z. Siwy and A. Fulinsky, *Origin of $1/f^{\alpha}$ noise in membrane channel currents. Phys. Rev. Lett.* **89** (2002) 158101.
[26] S. M. Bezrukov and I. Vodyanoy, *Probing alamethicin channels with water-soluble polymers: Effect on conductance of channel states. Biophys. J.* **64** (1993) 16–25.
[27] S. Nekolla, C. Andersen and R. Benz, *Noise analysis of ion current through the open and the sugar-induced closed state of the LamB channel of Escherichia coli outer membrane:*

*Evaluation of the sugar binding kinetics to the channel interior. Biophys. J.* **66** (1994) 1388–1397.

[28] S. M. Bezrukov, I. Vodyanoy and V. A. Parsegian, *Counting polymers moving through a single ion channel. Nature* **370** (1994) 279–281.

[29] J. J. Kasianowicz, E. Brandin, D. Branton and D. W. Deamer, *Characterization of individual polynucleotide molecules using a membrane channel. Proc. Natl. Acad. Sci. USA* **93** (1996) 13770–13773.

[30] S. M. Bezrukov, L. Kullman and M. Winterhalter, *Probing sugar translocation through maltoporin at the single channel level. FEBS Lett.* **476** (2000) 224–228.

[31] A. Meller, *Dynamics of polynucleotide transport through nanometer-scale pore. J. Phys. Condens. Matter.* **15** (2003) R581–R607.

[32] E. M. Nestorovich, C. Danelon, M. Winterhalter and S. M. Bezrukov, *Designed to penetrate: Time-resolved interaction of single antibiotic molecules with bacterial pores. Proc. Natl. Acad. Sci. USA* **99** (2002) 9789–9794.

[33] S. M. Bezrukov, A. M. Berezhkovskii, M. A. Pustovoit and A. Szabo, *Particle number fluctuations in a membrane channel. J. Chem. Phys.*, **113** (2000) 8206–8211.

[34] A. M. Berezhkovskii, M. A. Pustovoit and S. M. Bezrukov, *Effect of binding on particle number fluctuations in a membrane channel. J. Chem. Phys.* **116** (2002) 6216–6220.

# NONEQUILIBRIUM RATE THEORY FOR CONDUCTION IN OPEN ION CHANNELS

R. TINDJONG*, I. KAUFMAN† and P. V. E. McCLINTOCK‡

*Department of Physics*
*Lancaster University, Lancaster, LA1 4YB, UK*
**r.tindjong@lancaster.ac.uk*
*†i.kaufman@lancaster.ac.uk*
*‡p.v.e.mcclintock@lancaster.ac.uk*

D. G. LUCHINSKY

*Mission Critical Technologies Inc., 2041 Rosecrans Ave. Suite 225 El Segundo, CA 90245, USA*

*Department of Physics*
*Lancaster University, Lancaster, LA1 4YB, UK*
*dmitry.g.luchinsky@nasa.gov*

R. S. EISENBERG

*Department of Molecular Biophysics and Physiology*
*Rush Medical College*
*1750 West Harrison, Chicago, IL 60612, USA*
*beisenbe@rush.edu*

Received 23 September 2011
Revised 29 December 2011
Published 31 March 2012

Communicated by Richard Haley

We present a nonequilibrium reaction rate model of the ionic transition through an open ion channel, taking account of the interaction between an ion at the entrance of the channel and an ion at the binding site in a self-consistent way. The electrostatic potential is calculated by solution of the Poisson equation for a channel modeled as a cylindrical tube. The transition rate, and the binding site occupancy as a function of the left bulk concentration are compared to 1D Brownian dynamics simulations. The analysis is performed for a single binding site of high-affinity, with the exit rate influenced by barrier fluctuations at the channel exit. The results are compared with experimental data for the permeation of the $Na^+$ ion through the Gramicidin A channel, with which they are shown to be in good agreement.

*Keywords*: Ion channels; permeation; nonequilibrium rate; stochastic dynamics; fluctuating barrier.

## 1. Introduction

Ion channels are pathways down the centres of proteins embedded into the membranes of biological cells [1, 2], facilitating the diffusion of ions across the membranes. They are complicated devices [3], built of thousands of atoms, flexible, and filled with ions and water dipoles that coordinate their movements to allow ions to pass one by one across the membrane. To support life, channels have to be precise often selecting between similar ions with an accuracy of 1:1000 [4], yet able to deliver millions of ions per second, i.e., almost at the rate of free diffusion. In recent years impressive progress has been achieved both in understanding the structure of ion channels [3] and in modeling their properties [5]. Yet the mechanism that enables channels to be highly selective while still passing millions of ions per second remains a tantalizing paradox.

Channels displaying counter-intuitive behavior of this kind often have one singly occupied selectivity site [6] with a high affinity for a specific type of ion. The fact that a satisfactory explanation of the paradox has yet to be found, even in the most clearly characterized cases [7] of channel conduction, suggests that some basic physical phenomenon has been overlooked. The difficulties in modeling the system stem from the presence of the long-range Coulomb interaction and regions with low- and high-dielectric constants. The stochastic dynamics of ionic motion needs to be solved simultaneously with the Poisson equation for all the charges, which often results in strongly nonequilibrium stochastic dynamics. Therefore, only an approximate semi-analytic treatment of the problem is possible.

The simplifying assumptions that are frequently made to overcome these difficulties are sometimes not entirely self-consistent. First note that, in traditional [7–9] reaction rate theory (RRT), transitions of an ion from state $i$ to state $j$ associated with a potential barrier $E_{ji}$ are assumed to proceed at thermal equilibrium with constant rate $k_{ji}$

$$k_{ji} = \kappa_0 \exp\left(-\frac{E_{ji}}{k_B T}\right), \tag{1}$$

equal to the product of the frequency factor $\kappa_0$ and the Arrhenius factor $\exp(-\frac{E_{ji}}{k_B T})$, where $k_B$ is the Boltzman constant and $T$ is the temperature. At the same time to account for the Born self-energy it is often postulated [10] that an ion can only enter an empty channel: the effect of inter-ion interactions on the escape probabilities is ignored. Secondly, we note that the description of the current of ions through an open ion channel is often described by equilibrium Poisson–Nernst–Planck (PNP) theory, which also neglects the ion–ion interaction. The discrepancy between the PNP results and the results of Brownian dynamics (BD) simulations has been the subject of vigorous discussion [11, 12]. Yet to explain large ion fluxes through the channels with single high-affinity binding sites, the assumption of multiple-occupancy of the channel is sometimes made [13], ignoring the Born dielectric self-energy altogether. Although the assumptions mentioned may be valid in specific limits of physiological parameters, and helpful in explaining some of the observed experimental data, they are not mutually self-consistent. Being just postulated, rather than related to the channel parameters, they may even be misleading where a different range of physiological parameters is to be considered.

In standard RRT, ion channels are modeled as a sequence of energy barriers over which the ion must jump. The physical basis of the standard three-well model of an ion channel is based on Eyring's transition state theory. The binding sites model postulates that ions pass through channels by binding with one or more particular sites in the channel pore [10]. Läuger went a step further than the Eyring analysis by allowing for the phenomenon of saturation to be observed in his rate theory analysis of the transport [14]. In the binding site model, the net unidirectional flux can be calculated, taking into account the probability that the binding site well is occupied. An ion can enter the well only if it is not already occupied by another ion, and an ion can only leave a well if it is there already, i.e., if the well is occupied. Knowing the probability for the occupied and unoccupied wells is therefore enough to characterize the current.

In this paper we introduce a modified RRT that takes into account the dependence of the escape probabilities on the ion–ion interaction. The theory also illuminates the relation between the PNP and BD results. The electrostatic potential is calculated using the solution of the Poisson equation for a channel modeled as a cylindrical tube. The transition rate and the binding site occupancy as a function of the left bulk concentration are compared to 1D BD simulations, taking into account particle injection. The analysis is performed for a single binding site of high-affinity, with the exit rate governed by barrier fluctuations. We will compare the model results with experimental measurements of the permeation of $Na^+$ ions through the Gramicidin A channel, with which we will show that they are in good agreement.

## 2. Rate Model Description

We consider an archetypal model [1, 10] of an ion channel with a single high-affinity binding site, as shown in Fig. 1. It is considered as a 30 Å long cylinder of radius 2 Å with a negatively charged ring at its center. Assuming cylindrical symmetry of the channel, the potential energies for a $Na^+$ ion moving along the channel axis for the two state configurations are calculated by solution of Poisson's equation. The transition of an ion e.g., from the left to the right can be described within this model via the following steps (cf. [10]). To be able to enter the channel, an ion must arrive at the channel mouth. It does so with an arrival rate $j^l_{\mathrm{ar}}$ that is a function of the bulk concentration on the left $C_l$, and can be expressed as the average number of arrivals per unit time.

The probability of the ion being in the potential well must be connected to the probability of gaining entrance to the channel when it is empty. This is found by connecting the ion's arrival at the channel mouth to the probability of penetrating the empty channel. An estimate of the arrival time at the channel mouth, assuming

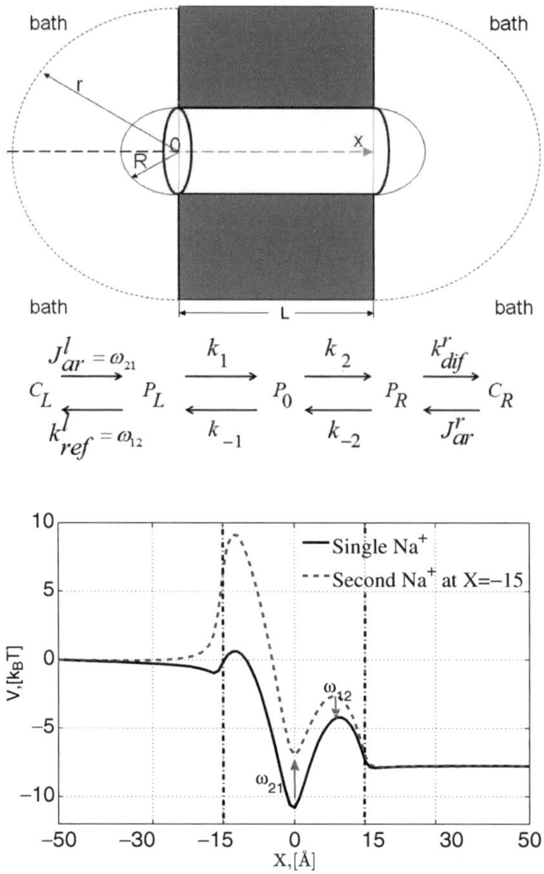


Fig. 1. Top panel: Geometry of the model system under consideration. It is comprised of a cylindrical channel with hemispheres at its left- and right-hand mouths. There is a fixed ring of negative charged (not shown) at its middle. This was the geometry used to solve Poisson's equation. Bottom: Potential energy along the channel axis, for 200 mV applied voltage. The full black line is the potential of a single $Na^+$ ion moving along the channel axis. This corresponds to State-1. The dashed line is the potential acting on a $Na^+$ ion moving along the channel axis when there is one $Na^+$ ion at the mouth of the channel, corresponding to State-2. The rate of jumping from state-1 to state-2 is $\omega_{21} = J^l_{\rm ar}$, and the rate of jumping from state-2 to state-1 is $\omega_{12} = k^r_{\rm dif}$.

charge neutrality and no applied field, can be obtained by considering pure diffusion of ions through an imaginary hemisphere at the mouth of the channel of radius equal to that of the channel. This problem is thus reduced to the determination of the ionic flux through a hemispherical sink at the channel entrance where all ions arriving at the boundary immediately vanish such that the ion concentration there goes to zero. A first approximation of the rate of arrival at the channel mouth is given by the Smoluchowski rate $1/\tau_{\rm arrival} = 2\pi CDR$, where $C$ is the bulk concentration far away from the channel mouth, $D$ is the ion diffusion coefficient and $R$ is the radius of the channel.

The bulk concentration is introduced into the model using the injection rate. The fluctuations in the potential energy barrier are the result of ion–ion interactions. We are specifically interested in mechanisms of strong modulation of the potential energy barrier that do not require conformational changes. The rate of barrier modulation is directly related to the rate of ion arrival at the channel mouth. From the left mouth the ion can either diffuse away at diffusion rate $k^l_{\text{dif}}$, or bind to the binding site with the rate $k_1$. Finally, the ion at the binding site can either return to the left compartment with the rate $k_{-1}$, or exit to the right compartment with the rate $k_2$. Similarly, the ion transition from the right to the left is characterized by the set of rates $\{j^r_{\text{ar}}, k^r_{\text{dif}}, k_2, k_{-2}, k_{-1}\}$ and concentration $C_r$. The formulation is completed by providing a model for the transition rates and writing a set of kinetic equations, which can readily be solved under conditions of steady current. Therefore, the kinetic equations for the current are given by:

$$\frac{dP_l}{dt} = J^l_{\text{ar}} - k^l_{\text{ref}} P_l - k_1 P_l (1 - P_0) + k_{-1} P_0, \tag{2}$$

$$\frac{dP_0}{dt} = k_1 P_l (1 - P_0) - (k_{-1} + k_2) P_0 + k_{-2} P_r (1 - P_0), \tag{3}$$

$$\frac{dP_r}{dt} = J^r_{\text{ar}} - k^r_{\text{dif}} P_r + k_2 P_0 - k_{-2} P_r (1 - P_0), \tag{4}$$

where $J^{l,r}_{\text{ar}}$ are the constant arrival rates at which ions are injected respectively into the left and right volumes located at the mouths of the channel. $P_{l,r}$ are respectively the occupation probabilities of these volumes. $P_0$ is the probability of occupation of the charged binding site in the middle of the channel. $k^l_{\text{ref}}$ and $k^r_{\text{dif}}$ are respectively the diffusive reflection rates on the left and right of the channel. From Eqs. (2) and (4), the currents to the right and to the left of the channel are given respectively by $J = k_1 P_l (1 - P_0) - k_{-1} P_0$ and $J = k_2 P_0 - k_{-2} P_r (1 - P_0)$. We are interested in the steady-state current.

The ion injection rate on the left is given by: $J^l_{\text{ar}} = 2\pi R C_l D$, where $C_l$ is the concentration in the left bulk. The diffusive and reflection rates at the left and right mouths, $k^r_{\text{dif}}$ and $k^l_{\text{ref}}$, are calculated from the stationary solution when the channel is closed ($J = 0$). Therefore, the occupation probability of the left hemisphere at the channel mouth is given by $P_l = 2\pi R^3 C_l / 3$; corresponding to the number of ions in the left hemisphere. Similar expressions can be written for the ion injection rate and occupation probability of the right mouth, leading to $k^l_{\text{ref}} = k^r_{\text{dif}} = 3D/R^2$. $P_0$ is calculated by combining elements of the steady-state equations. The corresponding equation for $P_0$ is quadratic. We are interested in the case of small $k_2$ parameters or, in other words in the case when $J \ll J^l_{\text{ar}}$. The current can then be calculated from:

$$J = \frac{k_1 (1 - P_0) J^l_{\text{ar}} - k^l_{\text{ref}} k_{-1} P_0}{k^l_{\text{ref}} + k_1 (1 - P_0)}. \tag{5}$$

Considering the case of a unidirectional current, the concentration on the right-hand side of the channel is set to $C_r = 0$. Since there is no backflow from the right bulk to the left bulk, we set $k_{-2} = 0$. The occupation probability at the channel binding site is given by:

$$P_0 = \frac{k_1 J^l_{\mathrm{ar}}}{k_1 J^l_{\mathrm{ar}} + (k_2 + k_{-1})k^l_{\mathrm{ref}}}. \tag{6}$$

The corresponding current is therefore reduced to $J = k_2 P_0$. For a permeating ion of charge $ze$, the current is given by $I = zeJ$. The occupation probabilities of the left and right mouth of the channel are given respectively by $P_l = (J^l_{\mathrm{ar}} - J)/k^l_{\mathrm{ref}}$ and $P_r = J/k^r_{\mathrm{dif}}$. Next, we calculate the rate $k_2$ of escape from the single binding site within the framework of fluctuating barrier theory.

### 2.1. *The ion–ion interaction*

The PNP approach [15, 16] has shed light on how an ion permeates the channel. Nevertheless, being an equilibrium approach, it ignores the ion-ion interaction. To take this into account we consider the changes in potential profile for the ion at the binding site induced by the presence of a second ion at the channel mouth. In this approximation the ion–ion interaction enters the dynamics of the ion inside the channel in the form of a Markovian dichotomous noise. The proposed reaction rate model introduces the rate $k_2$ calculated for the case when the barrier fluctuates between two states due to ion–ion interactions. Once the first ion is trapped at the binding site, conduction may occur when a second positive ion arrives at the channel mouth and the system switches to State-2. The latter corresponds to the reduced barrier needing to be overcome by the trapped ion in order to exit to the right bath.

Thus, as the second ion arrives at the channel mouth, the first ion initially at the binding site will have a higher chance of crossing the small barrier and exiting to the right-hand side of the channel, and a much smaller chance of diffusing back to the left bath. Hence the net transport to the right is caused directly by the random (fluctuational) arrival of ions at the left mouth of the channel. To a good approximation, no backflow of ions need be considered and we therefore calculate the unidirectional current.

To describe the two-state potential model, we consider the equation of a pulsating ratchet

$$m\gamma \frac{dx}{dt} = -\frac{dV(x, \zeta(t))}{dx} + \sqrt{2m\gamma k_B T}\xi(t). \tag{7}$$

We thus consider the overdamped unidirectional motion of a $Na^+$ ion of mass $m$, with a friction coefficient $\gamma$ which is related to its diffusion coefficient through the fluctuation dissipation relation. The thermal energy of the particle is $k_B T$. The potential $V(x, \zeta(t))$ in this approximation is obtained by solving Poisson's equation. It takes account of the following three contributions: (i) the potential of the Coulomb interaction with ions in the bulk; (ii) the electrostatic potential induced by interaction with the protein; (iii) the potential of the Coulomb interaction with the wall charge at the selectivity site. This mixed (hybrid) process is governed by

$\zeta(t)$; a dichotomous, stationary, Markovian noise and $\xi(t)$; a normalized white noise. For each state of the noise $\zeta(t)$ there corresponds a potential $V_i$. The system can be described by the probability distribution $P_i(x,t)$, which is the joint probability density for the position $x$ of the particle and the state of the potential $i$. The corresponding Chapman–Kolmogorov equation for this distribution is:

$$\frac{\partial P_i(x,t)}{\partial t} = \frac{\partial}{\partial x}\left(\left[\frac{1}{m\gamma}\frac{dV_i}{dx} + \frac{k_BT}{m\gamma}\frac{\partial}{\partial x}\right]P_i(x,t)\right) + \sum_j(\omega_{ij}P_j(x,t) - \omega_{ji}P_i(x,t)), \tag{8}$$

where $\omega_{ij}$ are the transition rates between states $j$ and $i$, given respectively by $\omega_{21} = J^l_{\text{ar}}$ and $\omega_{12} = k^l_{\text{ref}}$. For the two states 1 and 2, Eq. (8) is reduced to:

$$\frac{\partial}{\partial t}\begin{pmatrix} P_1(x,t) \\ P_2(x,t)\end{pmatrix} = \begin{pmatrix} \widehat{L}_1 - \omega_{21} & \omega_{12} \\ \omega_{21} & \widehat{L}_2 - \omega_{12}\end{pmatrix}\begin{pmatrix} P_1(x,t) \\ P_2(x,t)\end{pmatrix}, \tag{9}$$

where the operators $\widehat{L}_i$ are given by $\widehat{L}_i = \frac{1}{m\gamma}\partial_x\frac{dV_i}{dx} + \frac{k_BT}{m\gamma}\partial^2_x$. The particle starts at the bottom of the potential (corresponding to the channel binding site $x = x_0$), with probability $P_i(x,0) = \delta(x - x_0)$. Reflecting boundary conditions are chosen at the bottom of the potential ($x = x_0$) and absorbing boundary conditions at its top ($x = x_1$): $\frac{\partial P_i}{\partial x}(x = x_0, t) = 0$; $P_i(x = x_1, t) = 0$. Following the standard method described in Gardiner [17–19], the system of first-order ordinary differential equations giving the mean first passage time (MFPT) can be written:

$$\begin{cases} \dfrac{dT_1}{dx} = S_1, \\ \dfrac{k_BT}{m\gamma}\dfrac{dS_1}{dx} = \dfrac{1}{m\gamma}\dfrac{dV_1}{dx}S_1 + \omega_{21}(T_1 - T_2) - 1, \\ \dfrac{dT_2}{dx} = S_2, \\ \dfrac{k_BT}{m\gamma}\dfrac{dS_2}{dx} = \dfrac{1}{m\gamma}\dfrac{dV_2}{dx}S_2 + \omega_{12}(T_2 - T_1) - 1, \end{cases} \tag{10}$$

with the boundary conditions $\frac{dT_i}{dx}(x = x_0, t) = 0$; $T_i(x = x_1, t) = 0$.

### 2.2. *Solution of the boundary value problem*

The system Eq. (10) obtained from the fluctuating barrier model is solved numerically as a boundary value problem. The advantages of this method are that it does not require linear approximation of the potential energy and it can easily be extended to a larger number of states; specifically, one needs five states for two types of ions at each side of the channel. Each of the states corresponds to a different potential energy conformation. It is an approach that was not discussed earlier, and one that is easily accessible to a broad audience of potential users.

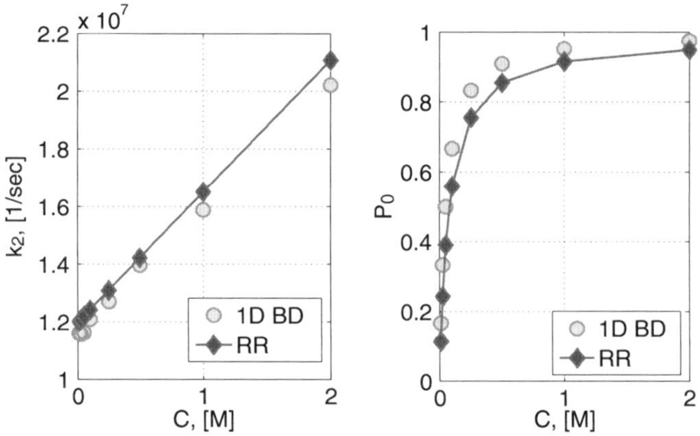


(a) (b)

Fig. 2. (a) Escape rate as a function of bulk concentration. The circles are obtained using 1D BD simulation for a $Na^+$ inside the channel with particle injection at the channel's left mouth. The lozenges are calculated using fluctuating barrier theory. (b) Channel binding site occupancy as a function of bulk concentration. The lozenges are calculated using the modified reaction rate (RR) theory and the circles are obtained from a 1D BD simulation with ion injection at the channel's left mouth.

The MFPT for a particle that starts from the bottom at $x = x_0$ is $T = \sum_{i=1}^{2} p^{(i)} T_i(x = x_0)$ [20], where $p^{(i)}$ is the occupation probability of the state $(i)$, such that $\sum_{i=1}^{2} p^{(i)} = 1$. The rate of escape from the channel binding site to the right bath is given by $k_2 = 1/T$.

## 3. Current Concentration Characteristics

We have calculated the modified rate $k_2$ as a function of the left bulk concentration for comparison with the rate obtained from direct BD simulations, taking into account particle injection at the mouth of the channel with the same Smoluchowski rate. As shown in Fig. 2(a), the two rates follow a similar linear dependence as a function of the concentration and are in reasonable agreement (cf. the Kramers rate over a nonfluctuating energy barrier which is independent of the concentration). The channel occupancy is also calculated and compared to that obtained from BD simulations: see Fig. 2(b). We compare the results with Andersen's experimental data [21] for the current of $Na^+$ ions in a gA channel in Fig. 3; we used the rates $k_1 = 3.3 \times 10^{10} \mathrm{s}^{-1}$, $k_{-1} = 1.2 \times 10^{7} \mathrm{s}^{-1}$ as fitting parameters and a diffusion coefficient of $D = 1.17 \times 10^{-9} \mathrm{m}^2\mathrm{s}^{-1}$ for the $Na^+$ ion.

## 4. Summary and Perspectives

We have introduced a nonequilibrium rate model of the ionic transition through an open ion channel within a self-consistent framework. The method takes explicit account of the interaction between an ion at the left mouth and an ion at the channel

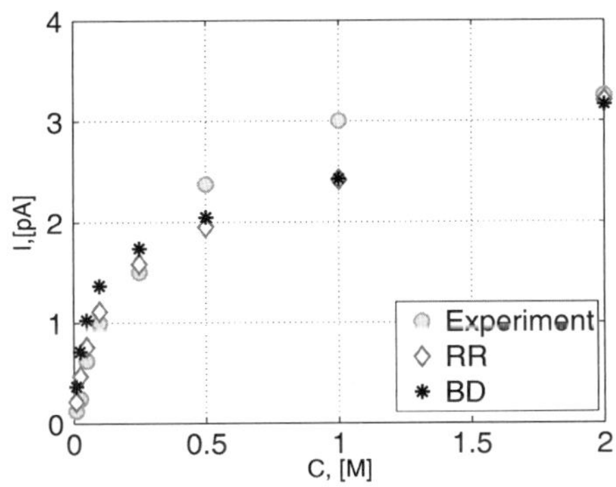


Fig. 3. Current as a function of concentration for a $Na^+$ ion in the gA channel. The circles are experimental data from Andersen *et al.* [21]. The lozenges represent the results of calculations based on the modified rate model. The stars are obtained from a 1D BD simulation for a $Na^+$ inside the channel with particle injection at the channel left mouth.

binding site. These ion–ion interactions result in fluctuations of the potential energy barrier of an ion at the binding site calculated from the channel structure by solving Poisson's equation. We calculated the transition rate and binding site occupancy of an $Na^+$ ion as a function of the left bulk concentrations, and compared the result with 1D BD simulations using ions injection at the channel mouth. The model was applied to experimental permeation rate data for $Na^+$ ions in the Gramicidin A channel, yielding satisfactory agreement. Note that, in the case of Gramicidin, the ion–ion interaction is not particularly important, so that equilibrium PNP theory may be sufficient. We expect, however, that the effect will be of crucial importance in channels that display both high conductivity and high selectivity, e.g., the $K^+$ and $Ca^{2+}$ channels. These will be analyzed in our future work. Note also that the boundary value technique used for solution of the MFPT system of equations can readily be extended to encompass a larger number of states. This feature will be needed, e.g. for the case when there are ions at both channel mouths, which will require five states.

## Acknowledgment

The research was supported by the EPSRC, UK.

## References


[1] B. Hille, *Ionic Channel of Excitable Membranes*, 3rd edn. (Sinauer Associates, Sunderland, MA, 2001).

[2] R. S. Eisenbeg, Ions in fluctuating channels: Transistors alive, *Fluct. Noise Lett.* **11** (2012) 1240001-1–1240001-20.

[3] J. D. Stockand and M. S. Shapiro, *Ion Channels: Methods and Protocols* (Humana Press, Totowa, NJ, 2006).

[4] D. A. Doyle, J. M. Cabral, R. A. Pfuetzner, A. Kuo, J. M. Gulbis, S. L. Cohen, B. T. Chait and R. MacKinnon, The structure of the potassium channel: Molecular basis of $K^+$ conduction and selectivity, *Science* **280** (1998) 69–77.

[5] S.-H. Chung, V. Krishnamurthy and O. S. Andersen, *Biological Membrane Ion Channels Dynamics, Structure, and Applications* (Springer, 2007).

[6] P. Hess and R. W. Tsien, Mechanism of ion permeation through calcium channels, *Nature* **309** (1984) 453–456.

[7] E. W. McCleskey, Calcium channel permeation: A field in flux, *J. Gen. Physiol.* **113** (1999) 765–772.

[8] P. C. Jordan, Ion permeation and chemical kinetics, *J. Gen. Physiol.* **114** (1999) 601–604.

[9] O. S. Andersen, Perspectives on ion permeation, *J. Gen. Physiol.* **113** (1999) 763–764.

[10] D. J. Aidley and P. R. Stanfield, *Ion Channels: Molecules in Action* (Cambridge University press, Cambridge New York Melbourne, 1996).

[11] G. Moy, B. Corry, S. Kuyucak and S.-H. Chung, Tests of continuum theories as models of ion channels. I. Poisson-Boltzmann theory versus Brownian dynamics, *Biophys. J.* **78** (2000) 2349–2363.

[12] B. Corry, S. Kuyucak and S.-H. Chung, Tests of continuum theories as models of ion channels. II. Poisson-Nernst-Planck theory versus Brownian dynamics, *Biophys. J.* **78** (2000) 2364–2381.

[13] C. M. Armstrong and J. Neytonc, Ion permeation through calcium channels, *Ann. New York Acad. Sci.* **635** (1991) 18–25.

[14] P. Läuger, Ion transport through pores: A rate-theory analysis, *Biochim. Biophys. Acta, Biomembr.* **311** (1973) 423–441.

[15] M. G. Kurnikova, R. D. Coalson, P. Graf and A. Nitzan, A lattice relaxation algorithm for 3d poisson-nernst-planck theory with application to ion transport through the gramicidin a channel, *Biophys. J.* **76** (1999) A211–A211.

[16] B. Corry, S. Kuyucak and S.-H. Chung, Tests of Poisson-Nernst-Planck theory in ion channels, *J. Gen. Physiol.* **114** (1999) 597–599.

[17] C. Gardiner, *Handbook of Stochastic Methods, for Physics, Chemistry and the Natural Sciences*, 2nd Edn. (Springer, Berlin Heidelberg New York Barcelona Hong Kong London Milan Paris Tokyo, 1983).

[18] M. Bier and R. D. Astumian, Matching a diffusive and a kinetic approach for escape over a fluctuating barrier, *Phys. Rev. Lett.* **71** (1993) 1649–1652.

[19] C. R. Doering and J. C. Gadoua, Resonant activation over a fluctuating barrier, *Phys. Rev. Lett.* **69** (1992) 2318–2321.

[20] J. H. Li, D. Y. Xing, J. M. Dong and B. Hu, Resonant activations for a fluctuating barrier system driven by dichotomous noise and Gaussian white noise. *Phys. Rev. E.* **60** (1999) 1324–1328.

[21] O. S. Andersen, R. E. Koeppe and B. Roux, Gramicidin channel, *IEEE Trans. Nanosci.* **4** (2005) 10–20.

# ESTIMATION OF DETECTION LIMITS OF THE PHAGE-INVASION BASED IDENTIFICATION OF BACTERIA

L. B. KISH, M. CHENG, J. U. KIM and S. SEO
*Department of Electrical Engineering, Texas A&M University, College Station, TX 77843-3128, USA*

M. D. KING and R. YOUNG
*Department of Biochemistry and Biophysics, Texas A&M University*
*2128 TAMU, College Station TX 77843-2128*

A. DER
*Institute of Biophysics, Biology Research Center Hungarian Academy of Science*
*P.O.B. 521, Szeged, H-6701 Hungary*

G. SCHMERA
*Space and Naval Warfare System Center, Signal Exploitation & Information Management*
*San Diego, CA 92152-5001, USA*

Received 26 January 2005
Revised 21 February 2005
Accepted 21 February 2005
Communicated by Robert Vajtai

Recently a new method, SEnsing of Phage-Triggered Ion Cascade (SEPTIC) was proposed for the rapid detection and identification of bacteria via the electrical field caused by the stochastic emission of ions during phage infection. In this Letter, we present linear network theoretical considerations about the detection limits of the method. The considerations are based on our published data of the *E. coli* detection experiments and on the assumption of a linear response between the number of bacteria and the measured power density spectrum of the fluctuation-signal. Some practical limits of the detectability of the present agents with possible noise measurement arrangements are discussed in this paper. The calculations indicate that the detection and identification of a *single* bacterium can be achieved with natural (wild) phages with reasonable efforts within a time window of 10 minutes.

*Keywords*: Prompt identification of bacteria; detection limits; nano-bio chip; phages; noise.

## 1. Introduction

Very recently, a method of prompt detection and identification of bacteria was proposed [1]. The new method, SEnsing of Phage-Triggered Ion Cascade (SEPTIC), is detecting and analyzing the electrical field caused by the stochastic emission of ions during phage infection. The detection and identification is done by measuring the microscopic voltage fluctuations in a nano-well device [1,2], and it takes less that 10 minutes. Because the phage infection is a very selective process, where only bacteria of a specific strain are infected, a SEPTIC-biochip [2] containing an array of sensors where each sensor is sensitized with a different phage can detect and identify all relevant bacteria with the extraordinary speed and selectivity. In this paper, we present some considerations about the sensitivity of the method.

## 2. Noise Circuitry of the Measurement Setup

The noise circuit diagram of the SEPTIC measurement setup is shown in Fig. 1. This model circuitry is simplified in accordance with the low-frequency range of the investigations (<10 Hz). The resistance $R_{nw}$ of the nano-well is about 1 MOhm. $S_{nw}$ and $S_{u,n}$ are the power density spectra of the nano-well signal voltage and the input noise voltage generator of the preamplifier, respectively. $S_{i,n}$ is the power density spectrum of the input current noise of the preamplifier. During the experiments and at any foreseeable/reasonable low-frequency arrangements in the future, the input impedance of the amplifier can be neglected. Thus, the equivalent input voltage noise spectrum [2] can be given as:

$$S_{u,eq}(f) = S_{nw}(f) + 4kTR_{nw} + S_{u,n}(f) + R_{nw}^2 S_{i,n}(f), \tag{1}$$

where the $S_{nw}$ has a $1/f^2$ shape. In the relevant frequency range (<10 Hz), $S_{u,n}$ and $S_{i,n}$ expected to have $1/f$ shape and $R_{nw}$ is independent of the frequency.

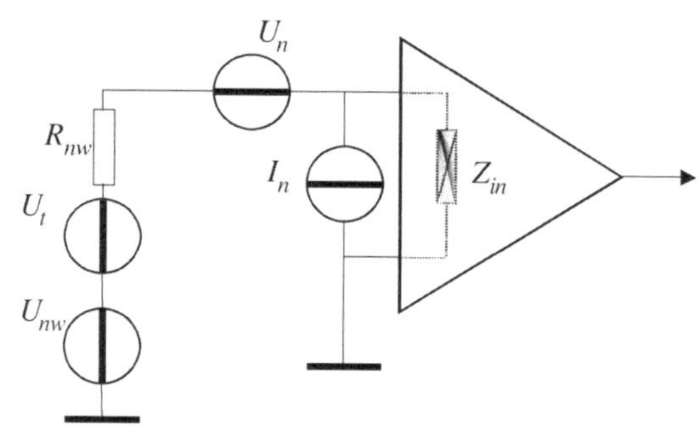


Fig. 1. Simplified noise circuitry of the nano-well arrangement. $U_{nw}$, $U_t$ and $R_{nw}$ are the signal voltage generator, the thermal noise voltage generator and the internal resistance (1 MOhm in the experiments) of the nano-well, respectively. $U_n$ and $I_n$ are the input noise voltage and input noise current generators of the preamplifier, respectively. $Z_{in}$ is the input impedance of the preamplifier (100 MOhm in the experiments, thus it can be disregarded).

## 3. Estimation of Detection Limits

The results are presented in Fig. 2. During the experiments [1], $10^7$ bacteria were present in the total sample volume of $10\text{mm}^3$. Based on Eq. 1 and the experimental data [1], the detection limits of the number of bacteria are estimated at frequency 1 Hz. The data published in [1] are obtained at the total bacterium count of $10^7$ and at 5 minutes incubation time.

The rest of the curves are calculated by assuming a linear response in the power density spectrum of the signal against the number of bacteria. This assumption is justified by the facts that the signal is stochastic and the bacteria act as independent sources of the total fluctuations. It is however an important precondition that, while the bacterium concentration may be low, the probability of infection of a given bacterium should be the same as at high bacterium concentrations. This important condition is naturally satisfied by using a fixed number $N_p$ of phages, where $N_p$ is much greater than the highest possible number $N_{b,max}$ of bacteria. At the experiments, the $N_p$ was $10*N_{b,max}$. The numbers of bacteria are valid for 10 mm$^3$ fluid, which is the practical experimental condition set by the smallest drop volume (5 mm$^3$) and the need of mixing two fluids (bacterium and phage containing solutions).

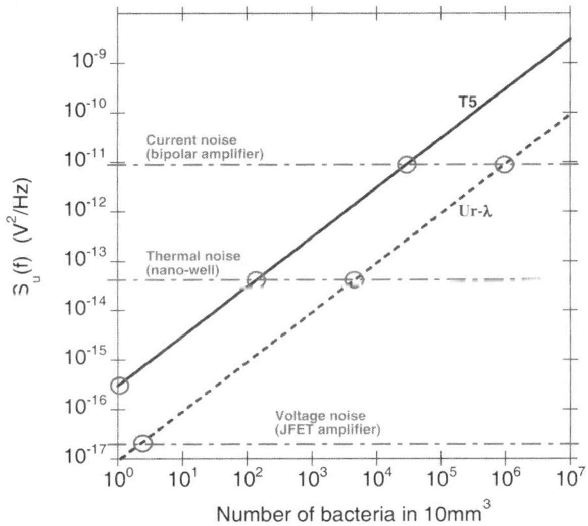


Fig. 2. Estimation of the ultimate limits of detectability of *E. coli* with the tested phages and various measurement arrangements. A linear response is assumed. The various detection limits are indicated by circles where the X coordinate of the center of the circle is the limit.

The highest limit is set by the actual conditions of the published experiments. This limit is provided by the input current noise of the bipolar preamplifier which makes the last term of Eq. 1 dominant among terms 2-4. We used the highest 1/f noise level measured at the negative experiments to determine this limit. The corresponding input current noise is about $10^{-23}$ $A^2/Hz$ which is consistent with the amplifier data at 1 Hz. This mechanism sets the detection limits to about 30,000 bacteria with the T5 and 1 million bacteria with the $Ur-\lambda$ phages.

A significantly lower limit is set by the thermal noise. This limit can easily be reached by a JFET preamplifier which has negligible current noise (about $10^{-30}$ $A^2/Hz$). That makes the last term of Eq. 1 the smallest for JFET low-noise preamplifiers. In this case, the dominant term among terms 2-4 in Eq. 1 is the 3rd one. The limit is obtained supposing 1 MOhm for $R_{nw}$. This mechanism sets the detection limits to about 200 bacteria with the T5 and 5000 bacteria with the $UR-\lambda$ phages.

The most enhanced sensitivity could be achieved using thermal noise reduction techniques, such as cross-correlation type nano-well and preamplifier arrangements and reducing the conductivity of the fluids thus reducing $R_{nw}$. The last action may decrease the sensor response too; however, the probability of this is vague because the Debye screening length of deionized water is only 1 micron and the mean distance between bacteria in the experiments is 10 micron, which indicates that the observed effect is not a long-range one. Thus decreasing the Debye length by increasing salt concentrations will most probably have a negligible effect on the induced electrical field fluctuations. In the most optimal case, if the thermal noise can be eliminated by conductance increase so much that the second term of Eq. 1 becomes the dominant, we can reach the highest sensitivity shown in Fig. 2. Typical voltage noise data of low-noise JFET amplifiers are used. This mechanism decrease the detection limits to 1 bacterium with the T5 and 2 bacteria with the $\mathrm{Ur}-\lambda$ phages. By using cross-correlation technique this limit can further decreased.

Note, if the conductivity of the fluid cannot be increased, it is still reasonable to suppose that we can reduce the thermal noise by cross-correlation technique and optimized nano-well geometry by a factor of 100. That achievement would provide detection limits of 1 bacterium with the T5 and about 30 bacteria with the $\mathrm{Ur}-\lambda$ phages.

## 4. Conclusions

It was shown by linear circuit theory and supposing linear response that, by using low-noise JFET preamplifiers and 10 $\mathrm{mm}^3$ sample droplet, the detection limits of *E. coli* are about 200 bacteria with the T5 and 5,000 bacteria with the $\mathrm{Ur}-\lambda$ phages, respectively. It was also shown that we could reasonably suppose that the application of increased fluid conductivity and/or more sophisticated techniques, such as cross-correlation arrangements, can decrease the detection limit to 1 bacterium.

The significant difference between the detection limits of E. coli using T5 versus Ur-lambda phages, respectively, may be due to the fact that the adsorption of Ur-lambda is temperature sensitive. However, all the measurements, except for the preincubation time, were carried out at room temperature, below the optimal 37°C. Thus, in a practical sensing arrangement, keeping the chip temperature at 37°C may imply a significant enhancement of sensitivity.

Note, there are also other tools, which may enhance sensitivity, such as methods of driving the bacteria to the nano-well surface, or nano/micro-fluidics to accelerate the dynamics of the system. These techniques may be needed if, at low concentrations, the dynamics of the response slows down and too long measuring times are necessary to get stationary data at low concentrations. This possibility is an open question at the moment and further experimental and theoretical efforts are necessary to clarify the situation.

## Acknowledgments

Discussions with Sergey Bezrukov and Bob Biard are appreciated.

## References

[1] M. D. King, S. Seo, J. U. Kim, R. F. Young, M. Cheng and L. B. Kish, Rapid Detection and Identification of Bacteria: SEnsing of Phage-Triggered Ion Cascade (SEPTIC) *J. Biological Physics and Chemistry* **5** (1), in press.

[2] C. D. Motchenbacher & J. A. Connelly, Low-noise electronic system design, *John Wiley & Sons*, New York (1993).

# FLUCTUATIONS, NOISE AND SCALING IN THE CARDIO-PULMONARY SYSTEM

BÉLA SUKI*, ADRIANO M. ALENCAR
*Department of Biomedical Engineering, Boston University, Boston, MA 02215 USA*
**bsuki@bu.edu*

URS FREY
*Department of Pediatrics, University Hospital of Berne, CH-3010 Switzerland*

PLAMEN CH. IVANOV, SERGEY V. BULDYREV, ARNAB MAJUMDAR, H. EUGENE STANLEY
*Center for Polymer Studies and Department of Physics, Boston University, Boston, MA 02215 USA*

CHRISTOPHER A. DAWSON
*Department of Physiology, Medical College of Wisconsin, 53226, Department of Biomedical Engineering, Marquette University, 53201, Research Service, Zablocki V A Medical Center, 53295*

GARY S. KRENZ
*Department of Mathematics, Statistics, and Computer Science, Marquette University 53201*

MICHIAKI MISHIMA
*Department of Respiratory Medicine, Graduate School of Medicine, Kyoto University, Japan*

Received 29 July 2002
Revised 3 February 2003
Accepted 4 February 2003

The structure and the functioning of cardio-pulmonary system is complex and statistical physics appear to be suitable for their characterization. In this review, we examine scaling in cardio-pulmonary physiology. The focus will be on the interpretation of scaling behaviors and their relation to structure-function in the normal and diseased cardio-pulmonary system. First, we overview fluctuations and scaling in respiratory rate variability in terms of a neural network model. Next, we analyze fluctuations in human heartbeat dynamics under healthy and pathologic conditions using wavelets and multifractal approaches. We then discuss avalanche behavior of airway openings as well as scaling behavior of crackling sound generated during the process of airway openings. We also examine the relationship between the observed scaling properties and the design features of the pulmonary vascular tree. Finally, we show how the network failure of lung tissue structure leads to emphysema, a leading cause of respiratory disability and death worldwide.

*Keywords:* Fractal; respiratory rate; heart rate; blood flow; airway opening; crackle sound; lung tissue.

## 1. Introduction

The central task of statistical physics is to study macroscopic phenomena that result from microscopic interactions among many individual components. This problem is akin to many investigations undertaken in physiology. In particular, physiological systems under neuroautonomic regulation, such as regulation of heart or breathing, are good candidates for such an approach since: 1) the systems often include multiple components, thus leading to very large numbers of degrees of freedom, and 2) the systems are usually driven by competing forces. Therefore, it seems reasonable to consider the possibility that physiological systems may exhibit temporal dynamics that are similar to those found in physical systems. Indeed, concepts and techniques originating in statistical physics are promising for quantitative analysis of complex physiological systems. In this review, we apply various tools from statistical physics to problems in cardio-vascular physiology.

The essential physiological functions of the cardio-pulmonary system are to maintain gas exchange and supply the organs and cells of the body with oxygenated blood. Exchange of oxygen and carbon dioxide occurs in the lungs at the level of the air sacs called the alveoli. Fresh air flows into the lung through the airways which constitute a 3-dimensional branching structure. The air diffuses through the thin walls of the capillaries into the blood. The oxygenated blood then enters the heart through the pulmonary circulation. The coordinated electrical activity of the heart provides a rhythmic contraction of the heart muscles and the oxygenated blood is pumped through the arteries into the body including the brain and the extremities [1].

The structure as well as the functioning of the cardio-pulmonary system are complex and include subsystems that are themselves inhomogeneous and irregular. The three-dimensional structure of the airways is a well-known example of a complex fractal structure where the branching pattern repeats itself over multiple length scales [2]. In disease, segments of the airway tree can become blocked by fluid plugs during expiration. The reopening of the closed airway segments during inspiration occurs in avalanches and the size distribution of the avalanches follows a power law [3]. The pulmonary vascular tree running parallel to the airway tree is also a fractal structure [4] and the resistance to blood flow of the tree shows scaling behavior [5]. Another example of tree-like structures is the electrical fiber network of the heart tissue, the His-Purkinje conduction system, through which propagating voltage pulses generate complex patterns with fractal properties [6].

Physical measurements on various parts of the cardio-pulmonary system display fluctuations at scales spanning several orders of magnitude. However, not only the structure, but various physiological time series obtained in the cardiopulmonary system are extremely inhomogeneous and non-stationary, fluctuating in an irregular and complex manner. In the medical literature, these fluctuations are often regarded as noise and neglected. However, these fluctuations often carry important information about the structure and function of the heart and lungs. Examples include fluctuations in heart rate [7], respiratory rate [8], lung volume [9], ventilation and perfusion [10], tidal volume, end-expiratory oxygen and carbon dioxide levels [11], and blood flow [12, 13].

In the following sections, we examine fluctuations and scaling relationships in cardio-pulmonary physiology in terms of distributions and correlations. Specifically, we explore the degree to which concepts developed in statistical physics can be usefully applied to physiological structures and signals. The focus will be on the interpretation of the various scaling behaviors in the normal and diseased states of the cardio-pulmonary system. First, we overview fluctuations and scaling respiratory rate variability in terms of a neural network model. Next, we illustrate the problems related to physiologic signal analysis with representative examples of human heartbeat dynamics under healthy and pathologic conditions. We then examine the avalanche behavior of airway openings as well as the scaling behavior of crackling sound generated during the process of airway openings. We also examine the relationship between the scaling properties and the design features of the pulmonary vascular tree. Finally, we show how the network failure of lung tissue structure leads to emphysema, a leading cause of respiratory disability and death worldwide.

## 2. Breathing Irregularities in Infants and the Singularity of the Respiratory Oscillator

Newborns and premature infants often develop irregular breathing patterns suggesting that there are important developmental differences in respiratory regulation during the postnatal period compared with later life [14]. Such irregularities may potentially lead to life-threatening periods of insufficient breathing (hypopneas) that occur in early life in preterm infants and which may also occur in infants at risk for sudden infant death syndrome (SIDS). The major underlying factors responsible for these irregularities may include immaturity of the brain stem rhythm generators [15] and immature central and peripheral chemoreceptors [16]. Scaling laws and modeling can have a significant impact on the characterization and understanding of the fluctuations in breathing irregularities.

### 2.1 *Irregularities and scaling in inter-breath intervals*

Recently, Frey *et al.* [8] reported long-term measurements of breathing pattern in term and preterm infants. Figure. 1 shows a typical example of inter-breath intervals of a preterm baby at post-conceptional ages of 39 (Fig. 1(A)) and 61 (Fig. 1(B)) weeks. The inter-breath intervals were calculated from abdominal movements using a threshold algorithm. Each inter-breath interval includes an apneic period (time period between two full breaths) and potentially several small insufficient tidal excursions (hypopneas) which do not reach a certain threshold level and hence do not provide adequate gas exchange. From Fig. 1, it is apparent that this particular baby's breathing pattern is highly irregular at 39 weeks, but the fluctuations are significantly reduced by 61 weeks

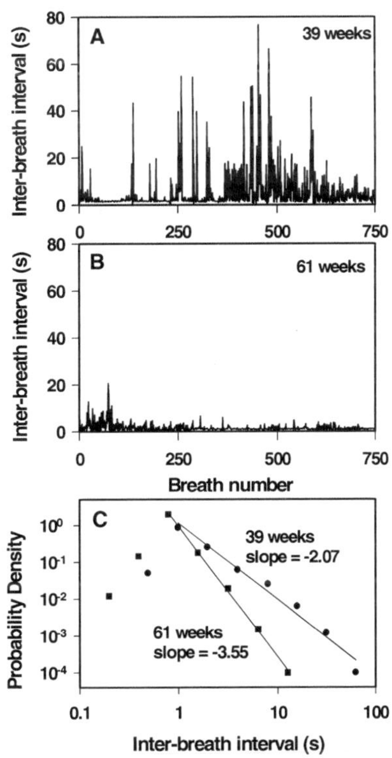


Fig. 1. Time series of inter-breath intervals in a baby at post-conceptional ages of 38 (A) and 61 (B) weeks. (C) Distribution of inter-breath intervals and regression lines. From Ref. 8 with permission.

when the baby's breathing pattern is more mature. The mean and standard deviation (SD) of inter-breath intervals are 5.2 ± 8.2 s and 1.4 ± 0.9 s at 39 and 61 weeks, respectively. There is a four fold decrease in the mean, but the decrease in SD is nearly 10 fold suggesting that there is critical information in the fluctuations. Indeed, displayed on a double logarithmic graph, both distributions have a peak around 1 s followed by a linear decrease (Fig. 1(C)) which implies that the distributions have a power law tail. Interestingly, while maturation preserved the power law form, the negative slope of a straight line fit to the data, or exponent $\alpha$, increased from 2.07 to 3.55, and hence is sensitive to maturational changes of respiratory control in babies. In a group of 15 normal term babies at postnatal age of 8.8 ± 4.1 weeks, $\alpha$ was 2.9 ± 0.4 whereas in a group of 10 preterm infants at postnatal age of 6.6 ± 4.3 weeks, $\alpha$ was 2.5 ± 0.2. The postnatal age was not statistically different in the two groups, however, $\alpha$ was statistically significantly smaller ($p < 0.02$) in the preterm infants. In 10 preterm infants, longitudinal recordings were taken at postnatal age of 9.3 ± 5.1 and 12.6 ± 5.6 weeks. The corresponding $\alpha$ increased statistically significantly ($p < 0.002$) from 2.56 ± 0.4 to 3.2 ± 0.4. These data imply that 1) preterm babies have larger fluctuations with smaller exponents than term babies at the same postnatal age and 2) the exponent is well suited to characterize the fluctuations and maturations with advancing age.

## 2.2 *Neural network model of breathing*

To interpret these findings, Frey *et al.* [8] used a neural network model originally proposed by Botros and Bruce [17] which offers a possible explanation of the statistical characteristics of the breathing pattern shown in Fig. 1. The model is based on the assumptions that there is noise in the neural network regulating the control of breathing and that this noise level is changing with maturation. Possible biological evidence for this assumption is given below.

To reproduce the observed irregularities, noise was added to the neural oscillator model of Botros and Bruce [17], which transforms tonic neural inputs (TNI) into a regular rhythm and hence breathing. The model consists of five coupled nonlinear differential equations corresponding to the activities of five neuron groups in the respiratory center. The ramp-inspiratory neuron group provides periodic outputs to the phrenic nerve similar to measured data. Therefore, the time interval between the peaks of the output of the ramp-inspiratory neurons can be taken as proportional to inter-breath interval. As a first step, uniformly distributed noise with small SD was added to the TNI of the first or ramp-inspiratory group (TNI1). With suitable choice of parameters, the obtained time sequence reproduces the occurrences of high spikes similar to those in Figs. 1(A) and (B). The time series of the model has a power law distribution with $\alpha = 2.28$. The power law distributed inter-breath interval sequence is generated within the respiratory oscillator. First, fluctuations in TNI1 lead to irregularities in inter-breath intervals. Second, as TNI1 decreases, the inter-breath-intervals diverge (Fig. 2) with a singularity. If a uniformly

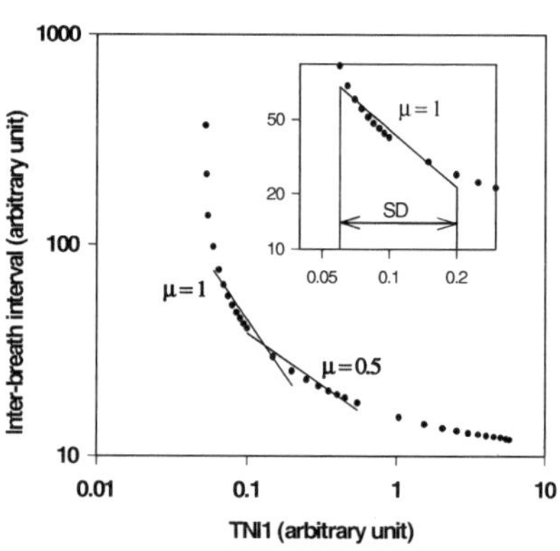


Fig. 2. Singularity of the respiratory neural network. Inter-breath interval diverges as TNI1 decreases. From [8] with permission.

distributed noise explores such a singularity, i.e., it is subject to a power law transformation with exponent $\mu$, then the resulting noise will have a power law distribution with an exponent $\alpha = 1 + 1/\mu$. While the singularity curve in Fig. 2 is not exactly a power law, if the noise SD is small, then, in the vicinity of the mean of TNI1, a power law fit of the form log(IBI) ~ $\mu$ log(TNI1) is a reasonable approximation (Fig. 2 inset). Thus, for small SD the uniform noise in TNI1 is transformed into a power law distributed noise due to the singularity. However, the exponent obtained from the simulations may be slightly different from the theoretical $\alpha = 1 + 1/\mu$ relationship, being determined by the average of the local slopes on the singularity curve sampled by TNI1. This model is also capable of accounting for maturation. Increasing the mean of TNI1 results in a decreasing $\mu$ (Fig. 2), which in turn increases $\alpha$. Accordingly, we conclude that maturation corresponds to a shift of the mean of TNI1 away from the singularity. Indeed, if we increase the mean TNI1 in the model, we find again a power law distributed IBI time series with $\alpha = 3.57$ which is in accord with the distribution in Fig. 1C at PCA = 61 weeks. Thus, by examining the effects of only TNI1, we can demonstrate a mechanism that generates a power law distribution of fluctuations in breathing pattern which also explains maturation. However, in reality, all TNIs can be noisy and the measured distribution of the inter-breath intervals will be determined by their combined effects. Indeed, the singular behavior in Fig. 2 is robust since the inter-breath intervals also diverge with respect to several other TNIs.

The correspondence between the various neural functions and the parameters of the model are relatively well understood [17, 18] and the original model accounts for much of the neurophysiology of respiratory control in newborns [19]. This allows us to establish three key ingredients necessary to qualitatively and quantitatively reproduce the observed irregularities. First, the existence of inherent noise is responsible for the irregularities. Besides the noisy operation of neurons, the heterogeneity of maturation of the vagal nerve (myelination) may play a role in infants [20]. Since this myelination process mainly determines the speed of propagation of action potentials, noise appears due to the heterogeneity of transmission times in a nerve consisting of a bundle of parallel neurons. Second, the existence of a singularity that produces large variations in inter-breath intervals for small variations in a TNI is needed to reproduce the statistical properties of irregularities. If the average value of one or more TNIs is sufficiently close to a singularity of the oscillator, then the irregularities will be distributed according to a power law. The existence of a certain type of singularity in the respiratory oscillator has indeed been experimentally observed [21]. Nevertheless, the precise nature of the singularity in the infant respiratory oscillator is largely unknown. Third, with maturation, a shift in TNI away from this singularity significantly reduces the irregularities that appear in the breathing pattern of infants and accounts for the changes in the exponent with increasing age. Finally, since the exponent is sensitive to maturation, it has the potential to be used as a simple index for evaluating the likelihood of very long insufficient breathing periods in various groups of infants such as preterms or those who are at risk for SIDS.

## 3. Scaling Behavior in Heart Rate Variability

In recent years the study of the statistical properties of heartbeat interval sequences has attracted the attention of researchers from different fields [22–25]. Analysis has focused extensively on interbeat interval variability as an important quantity to help elucidate possibly nonhomeostatic physiologic variability because: (i) the heart rate is under direct neuroautonomic control, (ii) interbeat interval variability is readily measured by noninvasive means, and (iii) analysis of these heart rate dynamics may provide important practical diagnostic and prognostic information. Figure 3 top shows a cardiac interbeat time series — the output of a spatially and temporally integrated neuroautonomic control system. The time series shows "erratic" fluctuations and "patchiness". These fluctuations are usually ignored in conventional studies which focus on averaged quantities. In fact, these fluctuations are often labeled as "noise" to distinguish them from the true "signal" of interest. Generally, in the conventional approach it is assumed that there is no meaningful structure in apparent noise and, therefore, one does not expect to gain any understanding about the underlying system through the study of these fluctuations. However, by adapting and extending methods developed in modern statistical physics and nonlinear dynamics [26], we find that the physiologic fluctuations shown in Fig. 3 top exhibit an unexpected hidden scaling structure [7, 27, 28, 29, 30, 31]. Furthermore, the dynamical patterns of these fluctuations and the associated scaling features change with pathological perturbations. These findings raise the possibility that understanding the origin of such temporal structures and their alterations with disease may elucidate certain basic aspects of heart rate control mechanisms, and may have potential for clinical monitoring.

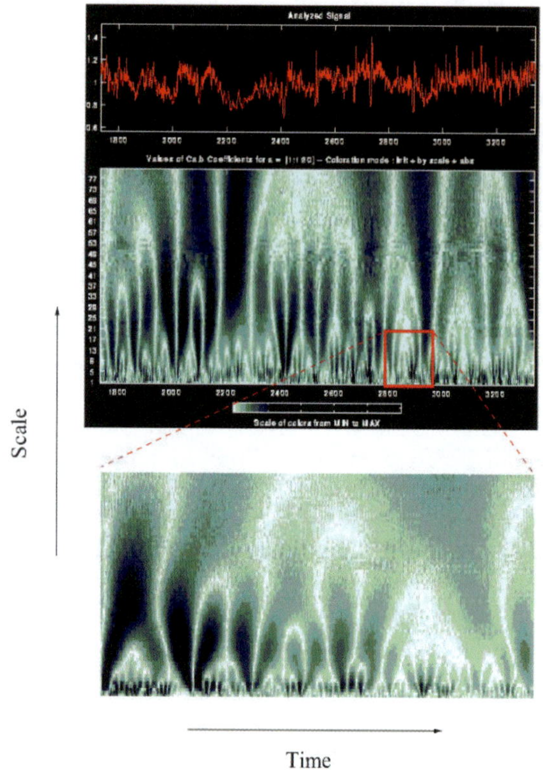


Fig. 3. (Top) Consecutive heartbeat intervals measured in seconds are plotted vs beat number from a record of a healthy subject. The time series exhibits very irregular and nonstationary behavior. (Middle) Color coded wavelet analysis of a heartbeat interval signal. The $x$-axis represents time (~1700 beats) and the $y$-axis indicates the scale of the wavelet used ($a$=1,2,…,80; i.e. from ~5 seconds to 5 minutes) with large scales at the top. This wavelet decomposition reveals a self-similar fractal structure in the healthy cardiac dynamics - a magnification of the central portion of the top panel with 200 beats on the $x$-axis and wavelet scale $a$=1,2,…,20 on the $y$-axis shows similar branching patterns (bottom). From Ref. 30 with permission.

### 3.1 *Generalized distributions of interbeat intervals*

Quantifying the probability distribution of heartbeat fluctuations can provide insights into the underlying dynamical processes because the distribution of interbeat intervals is directly related to the mechanisms which control heart rate variability. Therefore, by finding consistent features of the distribution which are robust with respect to different healthy subjects, we can quantify physiologic dynamics. However there are important technical difficulties which must be overcome first. Among the possible reasons why an interbeat interval histogram can differ from case to case are: i) histograms can differ because they have different means and standard deviations but follow the same functional form ii) histograms are described by different functional forms i.e., they belong to different classes of processes. The first type of difference is commonly observed in physiological data and should be taken care of by properly "renormalizing" (with respect to the mean and standard deviation) the histogram. If we assume that heart rate control mechanisms in healthy subjects follow the same general set of dynamical rules, then we expect that some variables of the system's output will be described by a single, well-defined distribution function. Functional differences between distributions, on the other hand, can be a result of altered mechanisms, and could be indicative of pathological behavior.

Direct analysis of heartbeat interval histograms does not lead to separation between healthy and abnormal cardiac dynamics leading to the incorrect conclusions that no pertinent information related to the underlying dynamics can be extracted [32]. Such histograms measured directly for each subject do not converge to a single representative curve describing healthy dynamics [27]. Even rescaling the time series to give all histograms identical means and variances does not lead to a common curve and does not distinguish clearly between signals from healthy and sick subjects. The problem is that: i) masking effects of non-stationarities in the signal have to be first properly reduced; and ii) different time scales of analysis have to be used to observe dynamical patterns corresponding to the actual time scales of the underlying physiological processes.

To address these problems, the cumulative variation amplitude analysis (CVAA) method was developed [7]. This method comprises sequential application of a set of algorithms based on wavelet and Hilbert transform analysis. The wavelet transform allows one to "extract" from the data particular features — it probes the fluctuations in the heart rate signal at different time scales. In addition to extracting the variations over given time-scales in the heart rate signal, the wavelet transform reduces masking effects of the non-stationarities, since the analyzing wavelet can be chosen orthogonal to local polynomial trends [33]. The wavelet transform is thus a cumulative measure of the variations in the heart rate signal over a region proportional to the wavelet scale $a$, so the study of the behavior of

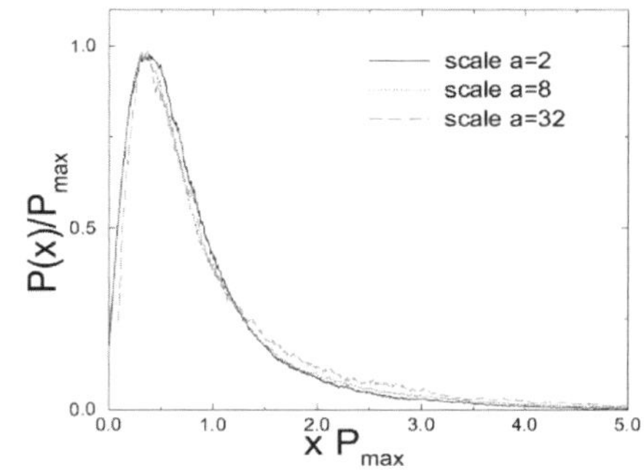


Fig. 4. Group average of the rescaled distributions of the cumulative variation amplitudes for the healthy individuals during nocturnal hours. The observed Gamma scaling remains stable for a wide range of the wavelet transform scales $a$. From Ref. 7 with permission.

the wavelet values can reveal intrinsic properties of the dynamics masked by non-stationarity. The next step of the CVAA is to extract the amplitudes of the variations in the beat-to-beat signal by means of an analytic signal approach [34] which also does not require stationarity.

Inspection of the distribution functions of the amplitudes of the cumulative variations for the healthy subjects reveals that these distributions are well fit by the generalized homogeneous form [36] - the Gamma distribution. Our results show that a common scaling function defines the probability density of the magnitudes of the variations in the beat-to-beat intervals for each healthy subject [7, 27]. To test the hypothesis that there is a hidden, possibly universal, structure to these heterogeneous time series, we rescale the distributions and find for all healthy subjects that the data conform to a single scaled plot ("data collapse") [7, 37]. Such behavior is reminiscent of a wide class of well-studied physical systems with universal scaling properties [37]. In contrast, subjects with sleep apnea, a common cardio-pulmonary disorder, show individual probability distributions that fail to collapse [7, 36]. The collapse of the individual distributions for all healthy subjects after rescaling their individual parameter is indicative of a "universal" structure. The term "universal" is used in the sense that a closed mathematical scaling form is established describing in a unified quantitative way the cardiac dynamics of all healthy subjects.

An analysis of the heart rate dynamics for healthy subjects during the daytime (noon – 6 p.m.) and nocturnal (12 a.m. – 6 a.m.) hours indicates that the observed universal behavior holds not only for the night phase but for the day phase as well. Thus the observed feature is independent of apparent external activity-related influences but rather relates to the intrinsic mechanisms or cardiac regulation. Semilog plots of the averaged distributions show a systematic deviation from the exponential form (slower decay) in the tails of the night-phase distributions, whereas the day-phase distributions follow the exponential form over practically the entire range [27]. Note that the tail of the distribution for the night phase surprisingly indicates higher probability of larger variations in the healthy heart dynamics during sleep hours in comparison with the daytime dynamics. Moreover, we observe for the healthy group good data collapse with a stable scaling form for a broad range of wavelet (time) scales $a = 2$ up to $a = 64$ corresponding to time scales from few seconds to 5–6 minutes (Fig. 4). The stability of this scaling form indicates that the underlying dynamical mechanisms regulating the healthy heart beat have similar statistical properties on different time scales.

### 3.2 *Correlation properties of interbeat intervals*

A quantity widely used to measure correlations in a time series is the power spectrum, which measures the relative frequency content of a signal. The analysis of heart beat fluctuations focused initially on short time oscillations associated with breathing and blood pressure as well as other control [22]. Studies of longer heartbeat records revealed *1/f*-like scale-free behavior [38]. A power spectrum calculation assumes that the signal studied is stationary [34], and when applied to non-stationary time series can lead to misleading results. However, time series of beat-to-beat (RR) heart rate intervals obtained from digitized electrocardiograms are typically non-stationary (Fig. 3 top). Recently, the *detrended fluctuation analysis* (DFA) method [39] was introduced to accurately quantify long-range correlations in physiological fluctuations when these are embedded in a seemingly non-stationary time series. The advantage of the DFA method over conventional methods is that it avoids the spurious detection of apparent long-range correlations that are an artifact of non-stationarity related to linear and higher order polynomial trends in the data.

We analyze the nocturnal and diurnal fractions of the dataset of each subject, which correspond to the 6h ($n \approx 22{,}000$ beats) from midnight to 6am and noon to 6pm. These periods include segments with the lowest and highest heart rates in the time series, which we and others found to be the best indirect marker of sleep. We find that the data during wake hours display long-range power-law correlations over several decades with average exponent $\alpha_W \approx 1.05$ for the healthy group [28], suggesting fractal organization in heartbeat fluctuations. These correlations change for the heart failure patients which are characterized by $\alpha_W \approx 1.2$ during wake hours. For the sleep data, we find a systematic crossover at scale $n \approx 60$ beats followed by a scaling regime extending over two decades characterized by a smaller exponent: $\alpha_S \approx 0.85$ for the healthy and $\alpha_S \approx 0.95$ for the heart failure group (Fig. 5) [40]. We also find that for all individuals studied, the heartbeat dynamics during sleep are characterized by a smaller exponent [40]. To test the robustness of our results, we analyze datasets from cosmonauts during long-term orbital flight on the Mir space station under the extreme conditions of zero gravity and high stress activity. Each dataset contains continuous periods of 6h data under both sleep and wake conditions. We find that for all cosmonauts the heartbeat interval series exhibit long-range correlations with scaling exponents consistent with those found for the healthy terrestrial group: $\alpha_W \approx 1.04$ for the wake phase and $\alpha_S \approx 0.82$ for the sleep phase [40]. The values of these exponents indicate that the fluctuations in the interbeat intervals are anticorrelated for the wake phases and even stronger anticorrelated for the sleep phase. This sleep-wake scaling difference is observed not only for the group averaged exponents but for each individual cosmonaut dataset (Fig. 5(b)). Moreover, the scaling differences are persistent in time, since records of the same cosmonaut taken on different days (ranging from the 3rd to the 158th day in orbit), exhibit a higher degree of anticorrelation in sleep. Thus, the larger values for the wake phase scaling exponents observed for healthy subjects cannot be a trivial artifact of activity. Furthermore, the larger value of the average wake exponent for the heart failure group compared to the other two groups cannot be attributed to external stimuli either, since patients with severe cardiac disease are strongly restricted in their physical activity. We note, however, that the average sleep-wake scaling difference remains the same ($\sim 0.2$) for all three groups. Such sleep-wake changes in the scaling characteristics may indicate different regimes of intrinsic neuroautonomic regulation of the cardiac dynamics, which may "switch" on and off in accordance with circadian rhythms. A recent study

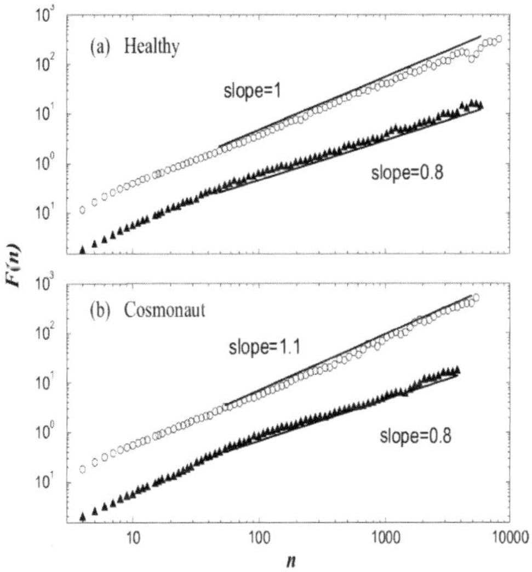


Fig. 5. Log-log plots of the average root-mean-square fluctuation function $F(n)$ vs. number of beats $n$ for 6h wake (open circles) and sleep records (filled triangles) of (a) one typical healthy subject; and (b) one cosmonaut (during orbital flight). Note the systematic lower exponent for the sleep phase (filled triangles), indicating stronger anticorrelations in the heartbeat fluctuations. From Ref. 40 with permission.

confirms our finding of lower value for the scaling exponent during sleep and shows that different stages of sleep (e.g. light sleep, deep sleep, rapid eye movement stages) could be associated with different correlations in the heartbeat fluctuations [41]. The findings of *stronger* anticorrelations [40], as well as higher probability for larger heartbeat fluctuations during sleep [7, 27, 36] are of interest from a physiological viewpoint, since they suggest that the above dynamical characteristics are related to intrinsic mechanisms of neuroautonomic control, and support a reassessment of the sleep as a surprisingly *active dynamical* state.

The finding of scaling features in the human heartbeat and their change with disease or sleep-wake transition have motivated new modeling approaches to better understand heart rate regulation [42]. The statistical self-similarity, as observed in the stable form of the probability distributions (Fig. 4) as well as in the power-law long-range correlations of the heart beat fluctuations (Fig. 5), is an important characteristic of fractal objects [43]. The wavelet decomposition of beat-to-beat heart rate signals can be used to provide a visual representation of this fractal structure (Fig. 3 middle and bottom). The brighter colors indicate larger values of the wavelet amplitudes (corresponding to large heartbeat fluctuations) and white tracks represent the wavelet transform maxima lines. The structure of these maxima lines shows the evolution of the heartbeat fluctuations with scale and time revealing a self-similar cascade — a magnification of the central portion of the top panel shows similar branching patterns (Fig. 3 bottom). Such fractal cascade results from the interaction of many nonlinearly coupled physiological components, operating on different scales (polynomial trends due to daily activity are filtered out).

### 3.3 *Multifractal behavior in heart rate control*

Monofractal signals are homogeneous in the sense that they have the same scaling properties, characterized locally by a single singularity exponent $h_0$, throughout the entire signal. Therefore, monofractal signals can be indexed by a single global exponent — the Hurst exponent $H \equiv h_0$ [44] — which suggests that they are stationary from the viewpoint of their local scaling properties. However, there is growing evidence that heartbeat dynamics exhibits nonlinear properties [7,23,25] which are often associated with multifractal behavior. Multifractal signals can be decomposed into many subsets — possibly infinitely many — characterized by different local Hurst exponents $h$, which quantify the local singular behavior and hence relate to the local scaling of the time series. Thus, multifractal signals require many exponents to fully characterize their scaling properties [43]

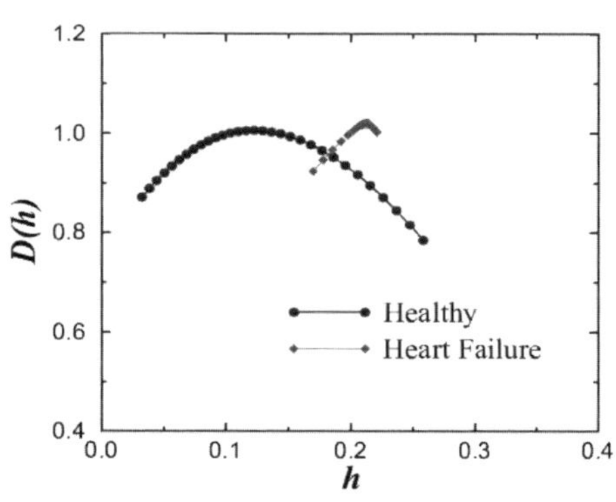


Fig. 6. Multifractal spectrum $D(h)$. The shape of $D(h)$ for the individual records and for the group average is broad ($\Delta h \approx 0.25$), indicating multifractal behavior for healthy heart beat dynamics. On the other hand, $D(h)$ for the heart failure group is very narrow ($\Delta h \approx 0.05$), indicating loss of multifractality. The different form of $D(h)$ for the heart failure group may reflect perturbation of the cardiac neuroautonomic control mechanisms associated with this pathology. Note, that for $q$=2 the heartbeat fluctuations of healthy subjects are characterized by $h \approx 0.1$, which corresponds to $\alpha \approx 1.1$ for the interbeat interval series obtained from DFA analysis. From Ref. 29 with permission.

and are intrinsically more complex, and inhomogeneous, than monofractals. The statistical properties of the different subsets characterized by these different exponents $h$ can be quantified by the function $D(h)$, where $D(h_o)$ is the fractal dimension of the subset of the time series characterized by the local Hurst exponent $h_o$. Thus, the multifractal approach has the potential to describe a wide class of signals that are more complex than those characterized by a single fractal dimension.

In a recent study, by means of a wavelet-based multifractal formalism [33], we establish the relevance of the multifractal formalism for the description of a physiological signal - the human heartbeat. We have shown that healthy human heartbeat dynamics exhibit higher complexity (than previously expected from the finding of fractal $1/f$ scaling) which is characterized by a broad multifractal spectrum (Fig. 6) [29]. In particular, we find that that for healthy subjects, $D(h)$ has nonzero values for a broad range of local Hurst exponents $h$ (Fig. 6). The multifractality of healthy heartbeat dynamics cannot be explained by activity, as we analyze data from subjects during nocturnal hours. Furthermore, this multifractal behavior cannot be attributed to sleep-stage transitions, as we find multifractal features during daytime hours as well [45]. The range of scaling exponents — $0 < h < 0.3$ — with nonzero fractal dimension $D(h)$, suggests that the fluctuations in the healthy heartbeat dynamics exhibit anti-correlated behavior ($h = 1/2$ corresponds to uncorrelated behavior while $h > 1/2$ corresponds to correlated behavior).

In contrast, we find that heart rate data from subjects with a pathological condition — congestive heart failure — show a clear loss of multifractality. For the heart failure subjects $D(h)$ is non-zero only over a very narrow range of exponents $h$ indicating monofractal behaviour (Fig. 6). Furthermore, our tests indicate that the observed multifractality is related to nonlinear features of the healthy heartbeat dynamics — we find an explicit relation between the nonlinear features (represented by the Fourier phase interactions) and the multifractality of healthy cardiac dynamics [29, 31]. The origin and nature of these Fourier phase interactions remain an open question. Nevertheless, the detection of robust fractal and multifractal scaling in the heart rate dynamics is of interest because it raises the intriguing possibility that the control mechanisms regulating the heartbeat interact as part of a coupled cascade of feedback loops in a system operating far from equilibrium — an extraordinarily complex behavior which in physical systems has been connected with turbulence and related multiscale phenomena [46].

## 4. Dynamics of Airway Opening and Crackles

The ventilation of the lung occurs via the airway tree which is a 3-dimensional binary tree forming a fractal structure [2]. In the normal lung and under natural breathing conditions, all airways are open providing only airflow resistance to breathing. The airways and alveoli are coated with a thin liquid layer containing surfactant. When the outward elastic tethering forces become smaller than the inward surface tension generated forces, the segment closes either via developing a liquid bridge [47, 48]. Thus, airway segments can develop closure when lung volume is lowered below the so called closing volume. Additionally, in diseases such as asthma, airway closure can also develop during normal breathing [49] due to the inward force generated by the contractile apparatus of airway smooth muscle cells. If the closed segment do not reopen during inspiration, then ventilation and gas exchange will be seriously impaired leading to potentially lethal situations such as an asthma attack. However, if a closed segment reopens, a short "explosive" transient waves, called crackle, is generated. Forgacs was the first to propose that crackles are associated with sudden opening of closed airways [50]. The discrete emission of crackles from the lung tissue has also been modeled as a stress-relaxation quadrupole [51], and this work has received experimental support [52]. Despite the fact that characteristics of individual crackles have long been used as diagnostic tools [53, 54], the statistical properties of crackles have only been recently analyzed [55, 56]. In the next sections, we examine the avalanche-like re-opening phenomenon and the statistical properties associated crackle sound.

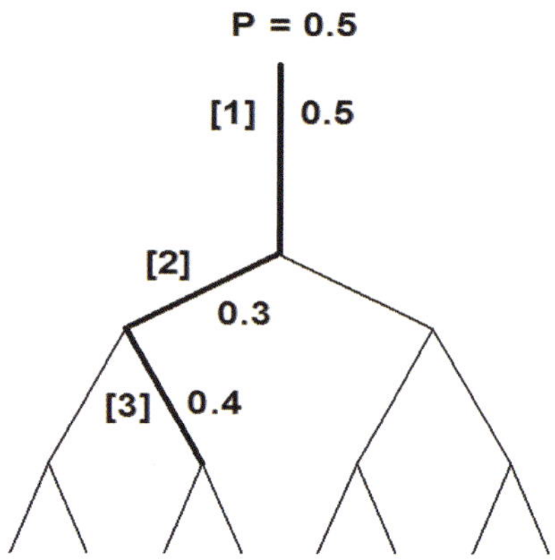


Fig. 7. A 4-generation tree model of the airways. The numbers in brackets are segments opening in an avalanche (thick lines) when pressure P=0.5. Thin lines are segments that have not opened. The numbers to the right of the segments are the normalized opening pressures.

### 4.1 *Avalanches in airway opening*

We examine the phenomenon of airway closure and reopening using a simple model of the periphery of the airway tree (Fig. 7). To mimic closure, we assume that every segment of the tree is collapsed. An airway reopens when the pressure across it reaches a critical threshold, the opening pressure Pth [48]. It is unlikely that two airways would have the same opening pressure and we assume that Pth is uniformly distributed between 0 and 1 [3, 57]. Here 0 corresponds to the pressure at the lowest lung volume (residual volume) whereas 1 corresponds the pressure at total lung capacity. During inflation, the pressure P is increased at the root of the tree (segment #1) at a constant rate. If the pressure behind the closures is 0 then the pressure across a collapsed segment is also P. When P reaches the opening pressure Pth of an airway, the airway suddenly opens. The local process of opening is an sudden rupture of the liquid film between the walls of the airway and consequently a short sound burst, called crackle, is generated locally [50, 55, 56]. As inflation proceeds, segment #2 would open at P = 0.3. However, segment #2 will not experience this pressure until the root opens. When the root opens, segment #2 suddenly becomes exposed to a higher pressure than its own opening pressure and opens in a short time. Once segment #2 opened, segment #3 also becomes exposed to P = 0.5 and it too opens with segment #2. This process leads to an avalanche of openings [3]. The avalanche-like opening is a

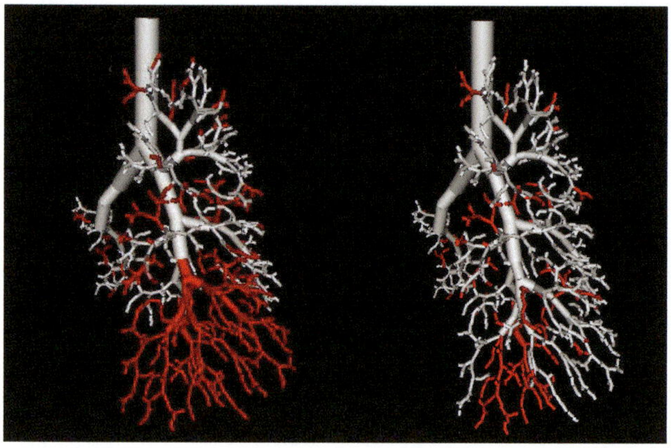

Fig. 8. 3-dimensional model of the airway tree before (left) and after (right) an avalanche suddenly opens many segments. White and red are open and closed segments, respectively.

consequence of the hierarchical structure of the tree because the connectivity of the tree imposes constraints on the possible sequence of openings [3,57]. An avalanche in a 3-dimensional tree structure [2] is visualized in Fig. 8.

## 4.2 *Scaling in crackling sound*

To characterize the avalanche process, one can analyze the time series of crackles emitted during opening. A crackle consists of an initial acoustic spike corresponding to the rupture of the liquid film followed by a damped acoustic ringing . The presence of the initial spike makes it easy detect a crackle. The time series of inter-crackle intervals measured in a real lung at the root of the tree (trachea) during inflation shows a complex behavior. Indeed, recent experiments have provided evidence of two separate scaling regions in the distribution of inter-crackle intervals (Fig. 9). The first regime, between time intervals of 10-4 and 10-3 s, is due to the fact that the avalanche propagation speed is finite. Numerical simulations suggest that this first power law region is related to the distribution of opening time delays, which in turn is related to the length distribution of the airway segments [56]. Since the airway tree is self-similar [2, 58] the length distribution is a power law. Thus, the exponent is likely to be related to the exponent of the airway length distribution. The second regime, between time intervals of 0.1 and 10 s, is due to inter-avalanche timing. This process is related to the propagation of an "active surface", defined as the closed segments connected to the root via open segments, on a Cayley tree. It has been shown using mean field calculations that the waiting times between avalanches are distributed according to a power law with an exponent that is close to 2 with a correction factor inversely proportional to the size of the tree [56]. Figure 9 shows that the numerical model developed by Alencar *et al.* [56] is able to fit the data over a range of 5 orders of magnitude in inter-crackle intervals. Additionally, as crackles propagate up the tree, the sound amplitude is attenuated at successive bifurcations. Interestingly, the distribution of crackle amplitudes measured in the bronchi is also a power law with an exponent that is related to airway geometry at bifurcations [55]. An analytical scaling relation has also been developed for the amplitude distribution [55]. We conclude that by measuring crackles in the lung and analyzing the distributions of the amplitudes and time intervals,

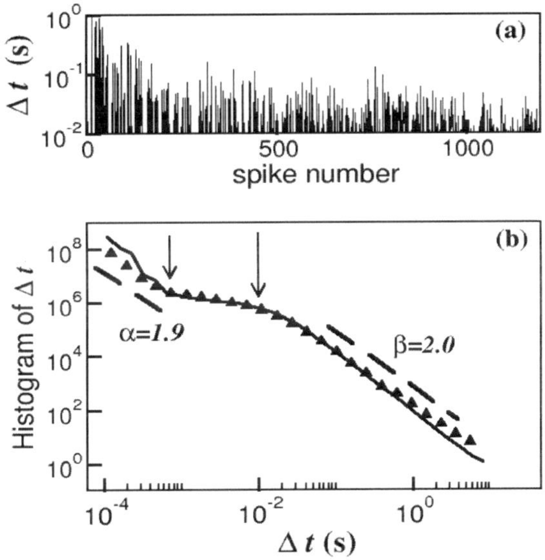


Fig. 9. Data analysis. (a) Linear-log plot of one example of inter-crackle intervals, in seconds, against consecutive spike numbers. (b) Histogram of inter-crackle intervals measured during the inflation of dog lungs (symbols). Dashed lines show linear regression with exponents. The arrows indicate the beginning and ending of the plateau region. Solid line is model simulation. From Ref. 56 with permission.

significant structural information can be obtained about the airway tree structure and the nature of airway closure [59]. Further studies are needed to assess whether these exponents are sensitive for the presence of disease in the lung.

## 5. Pulmonary Arterial Tree Design

The pulmonary arterial tree, like many other biological distribution networks, is a complex structure (Fig. 10(A)). Its primary functions are to deliver the blood to the pulmonary capillaries in proportion to local ventilation while providing the appropriate impedance for the output of the right ventricle. Its complex appearance tends to obscure a direct intuitive assessment of its key design features with respect to these functions. The following represents one approach to this question.

### 5.1 *Morphometry*

The initial step is a quantitative summary of the tree structure referred to as morphometry. Morphometric data have generally been obtained from plastic corrosion casts [60,63,66,67,68,76,79,82], and data obtained using imaging methods are becoming available [62,70,75]. The data consist of measurements of diameters lengths and numbers of vessels, and are commonly summarized by binning according to some ordering scheme [4,66,68].

Figure 10(B) is one graphic representation of the resulting binned or ordered data, wherein the number, N, of vessels in order j is plotted against mean diameter, $\overline{D}$, in order j. Despite differences in branching patterns observable at a macroscopic level, these graphs are remarkably similar for all the species studied, with the most notable difference being in the intercept values reflecting the differences in lung size. The value of β (defined as minus the slope of the log Nj vs. log $\overline{D}$) is a measure of the taper between bifurcations along a pathway through the tree. The range in the values of β obtainable from morphometric analysis of pulmonary arterial trees provided in Table 1 is rather small, suggesting that deviations beyond this range incur a significant penalty in the evolutionary sense.

The most common ordering system used to provide summary data such as in Fig. 10(B) and Table 1 is the Strahler system. The Strahler system (Fig. 10(C)) starts with the terminal arterioles designated as order 1 [4,66,68]. When two vessel segments of the same order meet at a bifurcation, they converge into a parent vessel segment of the next higher order number. When two vessel segments of different orders meet, the parent order number is the higher of the two daughter orders. Once the vessel segment orders have been assigned, contiguous segments of a common order are combined into vessels having the mean diameter and sum of the lengths of the combined segments. Then the number, and the averages of the diameters and lengths of the vessels comprising each order make up the morphometric summary of the tree, which can be represented in graphical form as in Fig. 10(B). Strahler ordering also captures the asymmetry of the tree as shown in Fig. 10(C). That is, the average number ratio, B, between orders (reflected in the vertical differences between the points on a given plot in Fig. 10(B) is a measure of asymmetry defined as inequality in the numbers of bifurcations from inlet artery to terminal arterioles among the various pathways through the tree.

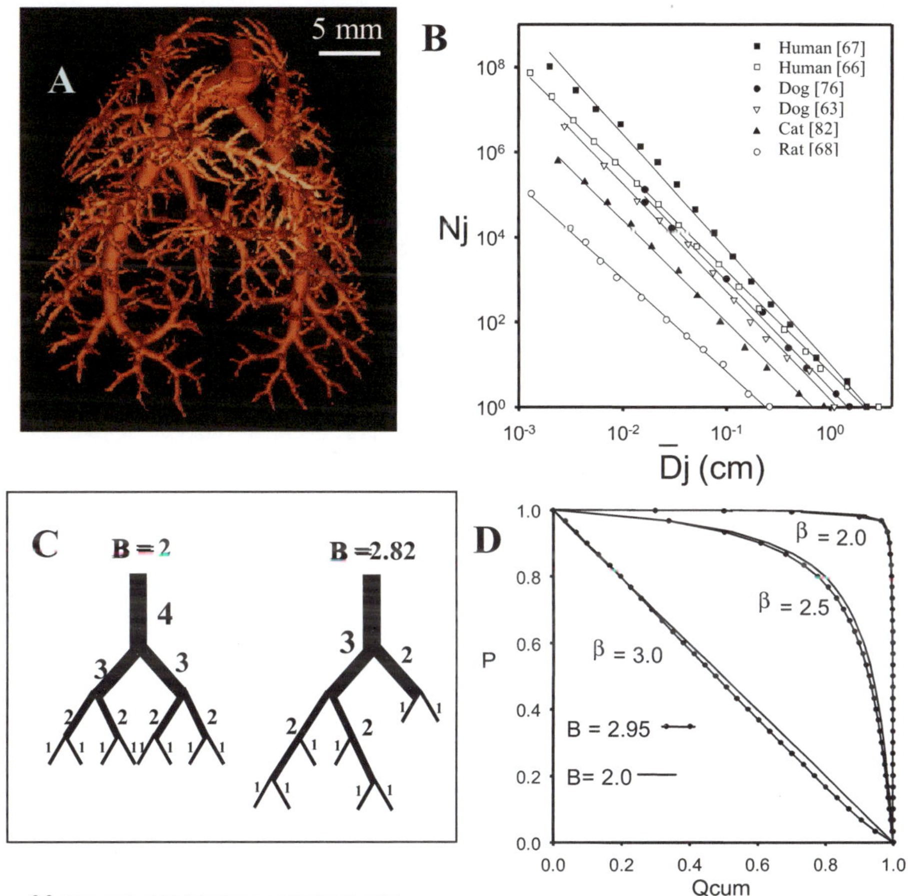


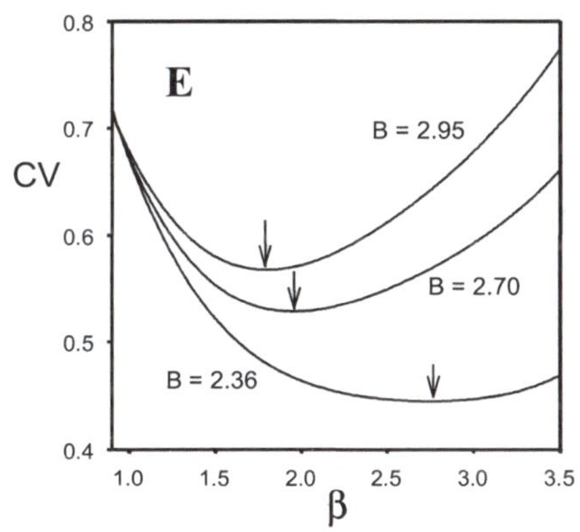


Fig. 10. (A) A surface shaded rendering of a CT scan of a rat pulmonary arterial tree. (B) Graph of the number of vessels, N, in order j vs. the average diameter, $\overline{D}$, of the vessels in j for the pulmonary arterial trees from several species. (C) A diagrammatic representation of Strahler ordering of a symmetrical tree on the left, for which B = 2, and an asymmetrical tree on the right, for which B = 2.85. The numbers beside the branches are the Strahler order numbers. (D) Model simulation intravascular pressure, P (normalized to the difference between the arterial inlet pressure and the pressure at the outlet of the terminal arteries) vs. the volume upstream from all of the locations within the tree at which pressure = P, Qcum (normalized to the total arterial volume), for three values of β (2, 2.5, and 3) and two values of B (2.0 and 2.95). (E) Model simulation coefficient of variation in terminal flows, CV, vs. β for three values of B (2.36, 2.7, and 2.95). The arrows identify the minimum CV for each value of B. Panels B and C reprinted from [4, pp. 407 and 410], by courtesy of Marcel Dekker. Inc.

For a symmetrical bifurcating tree, B = 2; B increases above 2 as the tree becomes more asymmetrical [4]. Values of B obtainable from the pulmonary arterial morphmetric literature are also presented in Table 1.

Table 1.

| *Species* [ref] | $\beta$ | *B* |
|---|---|---|
| Dog [76] | *2.52* | |
| Dog [63] | *2.55* | *3.67* |
| Cat [82] | *2.35* | *3.58* |
| Rat [78] | *2.15* | *2.76* |
| Rabbit [61] | *2.76* | |
| Human [66] | *2.40* | 3.10 |
| Human [67] | *2.71* | 3.36 |

### 5.2 *Model*

An approach for examining the functional implications of the values of these morphometric parameters is to construct a model that is consistent with obvious features of the real structure and in which the morphometric parameters obtainable from the data can be systematically varied to examine their effects [4,62,72]. In other words, with such a model, one can examine what deviations from the observed parameters values might produce a functional error signal upon which signal transduction mechanisms or evolutionary forces might act to constrain the network structure.

### 5.3 *Assignment of model vessel dimensions*

To construct a model tree for this purpose, we begin with the assumption that no two vessel segments have exactly the same diameter. Each vessel segment that will comprise the tree is assigned a number (or rank) Ncum, that is 1 plus the number of vessels having diameter larger than D(Ncum). Then, the diameter D of vessel Ncum is assigned by Eq. (1) [69].

$$D(\mathrm{Ncum}) = D(1)\left(\frac{\mathrm{Ncum}+0.5}{1.5}\right)^{-1/\beta} . \qquad (1)$$

Another characteristic of the pulmonary arterial network established by the morphometric data is that the average lengths, L, of the vessels are in nearly constant proportion to their diameters [62, 78]. Thus, in the model, vessel lengths are also determined by equation 1 as a fixed multiple of diameter, D.

### 5.4 *Connecting vessels into a model tree structure*

The vessel segments are then connected together into a tree structure. Asymmetry is controlled during the process of connecting the vessel segments by a parameter $\phi$ as follows. The algorithm begins with vessel segment Ncum = 1 and proceeds through Ncum = (Ntot - 1)/2 (where Ntot is the total number of segments comprising the tree). Each vessel segment is randomly assigned one daughter (which will ultimately be the larger of the two daughters) from the unattached segments left in the sequence Ncum —

1 to 2Ncum. After each Ncum through (Ntot - 1)/2 vessel has had one daughter attached, the remaining vessel segments (each of which will be the smaller of the two daughters at a bifurcation) are assigned to parents in the sequence of largest remaining daughter attached to largest remaining parent. To vary the asymmetry of the tree, limits are placed on the initial daughter assignment so that the choice of the largest daughter is among vessel segments having Ncum less than or equal to, 2 times the parent Ncum but larger than the closest integer equal to or smaller than $\phi$ times the parent Ncum, where $1 < \phi < 2$. Thus, as $\phi$ approaches 1 or 2 the tree will be, respectively, more or less asymmetrical. Regardless of the degree of asymmetry the terminal vessels are from the Ncum sequence from (Ntot + 1)/2 to Ntot. In other words, the terminal arterioles, i.e., those that would connect to the capillaries, are the smallest vessel segments in the tree. This is a general characteristic of arterial trees, which ultimately have to connect to capillaries that have virtually a common diameter.

When such a tree is Strahler ordered, the log Nj vs. $\log \overline{D}$ is approximately linear, with the model input value of $-\beta$ as its slope [69]. Thus, an asymmetrical tree is constructed with this morphometric relationship predetermined. In addition, as $\phi$ approaches 1, the average number of vessels comprising an order, $B^{n-j}$, approaches $3^{n-j}$ (where n is the total number of orders) rather than the $2^{n-j}$ in a symmetrical tree, as in the Strahler ordered morphometric data from the human lung reported by Horsfield [66] and Table 1.

### 5.5 *Pressure and flow calculations*

Having constructed the simulated tree as indicated above, the individual vessel segment flow rate is determined by first calculating the Poiseuille resistance of each segment. The total downstream resistance at each branch point is then calculated successively beginning with the subtended terminal vessel segments. Then for a given total flow and a common terminal outlet pressure, the pressure and the flow division at each bifurcation can be recursively calculated for each bifurcation from the two parallel total downstream resistances. Recognizing that in a real arterial tree the vessel diameters are pressure dependent, an implicit assumption is that the model vessel diameters are those that would exist at the vascular pressures output by the simulations.

### 5.6 *Simulations*

For the simulations presented herein the model trees are comprised of approximately one million vessel segments. The local pressure, flow, volume, and diameter outputs are normalized to the total arterial pressure drop, inlet flow, and total arterial volume, and inlet diameter, respectively, because particular values can be achieved by adjusting scaling factors: total inlet flow, viscosity, D(1), and L/D. In addition, the longitudinal (inlet artery to terminal arteriole) distribution of intravascular pressure is expressed as a function of cumulative intravascular volume, Qcum (defined as the total volume upstream from all those sites within the arterial tree where the pressures are equal). The intravascular pressure, P(Qcum), is plotted as a function of Qcum rather than diameter or distance from the arterial inlet, because, in the asymmetrical heterogeneous tree, there is no common isobaric diameter or distance. Thus, Qcum is a more concisely stated independent variable in this context. The parallel distribution of terminal flows is represented by its coefficient of variation.

Model simulations presented in Figs. 10(D) and 10(E) show how the longitudinal distribution of intravascular pressure (the longitudinal distribution of resistance being one determinant of the arterial impedance) and the distribution of terminal flows (the capillary perfusion distribution being one determinant of efficency of transport between blood and tissue or between blood and alveolar gas), respectively, are affected by variations in β and B. It can be seen that β is a key determinant of the longitudinal pressure profile, P vs Qcum, and that P vs Qcum is robust with respect to the asymmetry of the tree reflected in B. On the other hand, the parallel distribution of terminal flows is affected by both β and B (Fig. 10(E)). It is not surprising that the coefficient of variation, CV, in terminal flows would depend on the asymmetry of the tree, but the effect of β on the flow distribution may be less easily anticipated. For a given B, there is a β that minimizes the variation in flows.

The model simulations suggest that in a vascular tree that is consistent with the model assumptions, there is an optimal value of the taper reflected by β resulting in a minimum variation in terminal flows, the greater the asymmetry (larger B) the greater the taper (smaller β) will need to be. The reason for the impact of the value of β on the distribution of terminal flows is apparently as follows. The smaller the value b, the larger is the fraction of the total vascular the resistance concentrated in the small vessels and the more the larger arteries act as a relatively constant pressure manifold through which flow can be distributed among parallel pathways with relatively small resistance over a relatively large fraction of the distance along each pathway. This would tend to reduce the impact of the variation in distances traveled on terminal flows. However, as β gets smaller the variation in diameters at a given level in the tree increases, tending to increase the heterogeneity in flows among asymmetrically arranged parallel vessels. Thus, if β is too small, the variation in diameters among parallel vessels begins to have a more dominant effect on the distribution of terminal flows.

Of course the model is a simplified version of any real vascular tree. For example, values of B obtainable from available morphometric studies fall between about 2.8 and 3.7 (Table 1), whereas the model maximum is 3. However, the trends suggest that the results with more asymmetrical trees are predicable. While it is clear that the local flows can be quite heterogeneous in normal lungs [64], there are limits on the heterogeneity in local flows consistent with optimal gas exchange efficiency, and, in general, the efficiency of solute transport between blood and tissue is inversely proportional to the variance in the distribution of flows in the vessels involved in the transport [65]. Thus, it is conceivable that, given the asymmetry needed to distribute flow throughout an odd shaped structure like the lungs, efficiency with regard to the transport within the microvasculature may be part of the evolutionary pressure determining the taper expressed by β. Other optimality criteria have been evaluated as they might apply to the vascular structure-function relationship. Probably the most extensively studied is that of Murray's Law, which links structure, energy cost, and shear stress distribution [4,71,73,74,76,78,83]. Deviations from Murray's law in real vascular trees [4,60,80,81] suggests that other aspects of organ function may need to be considered as well. These may include those affected by the heterogeneity of the flow distribution.

## 6. Alveolar Tissue Structure in Emphysema

Emphysema, one type of chronic obstructive pulmonary disease (COPD) mostly induced by smoking, is a disease of the elastic fiber network of the lung tissue which is slowly destructed over a time period of 5-20 years [84]. High resolution computed tomography

(CT) is a sensitive method to examine lung structure and its alterations induced by COPD or other diseases such as fibrosis [85, 86].

### 6.1 *Distribution of low attenuation area clusters in CT images*

Characterization of CT images can be achieved as follows. First, we generate low attenuation area (LAA) clusters. Pixels in the image with density less than a selected threshold (-960 Hounsfield units) are designated as low density tissue (containing mostly air) and assigned a value of 1, whereas pixels with density larger than the threshold are designated as tissue assigned a value of 0. A LAA cluster is then defined as a contiguous region with values of 1. The total number of pixels of such a LAA cluster provides the cluster size. These binary maps exhibit spatial heterogeneity with many irregular clusters of widely varying sizes (Fig. 11). In clinical applications, the sum of the clusters, the total LAA, is used to characterize the progression of the disease. The total LAA is, however, the same as the average cluster size which does not reflect the irregularities in cluster size and shape. Thus, it is important to examine the probability distribution of LAA cluster sizes. Indeed, it has been reported that the size distribution of LAA clusters follows a power law for both normal subjects and patients with chronic obstructive pulmonary disease (COPD) [87]. The exponents of the LAA cluster distribution are sensitive to the progression of emphysema because as Fig. 12 shows, the exponent decreases with disease severity [87]. Thus, the likelihood of finding a large LAA cluster, where elastic recoil of the tissue and gas exchange are compromised, is much higher in COPD than in normal subjects. Interestingly, normal subjects and patients with early emphysema have similar total LAA. However, the exponent is smaller in early emphysema (Fig. 13). This suggests that the spatial organization of LAA clusters is different in COPD than in normal subjects [87]. Thus, the exponent is sensitive to early microstructural reorganization of the tissue due to the progression of emphysema.

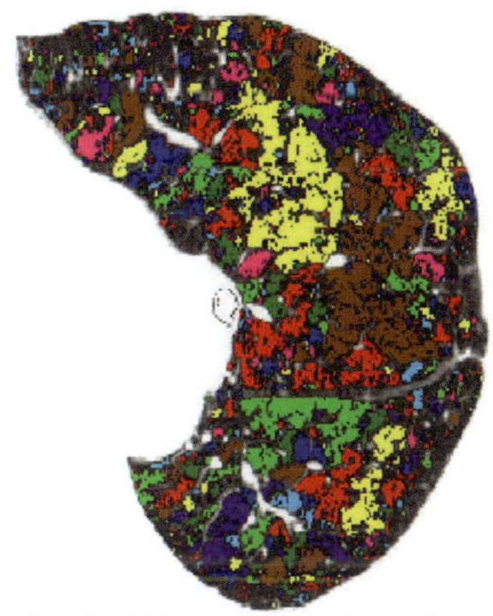

Fig. 11. CT image of the lung of an emphysematous patient. Color clusters represent contiguous Low Attenuation Areas. From Ref. 87 with permission.

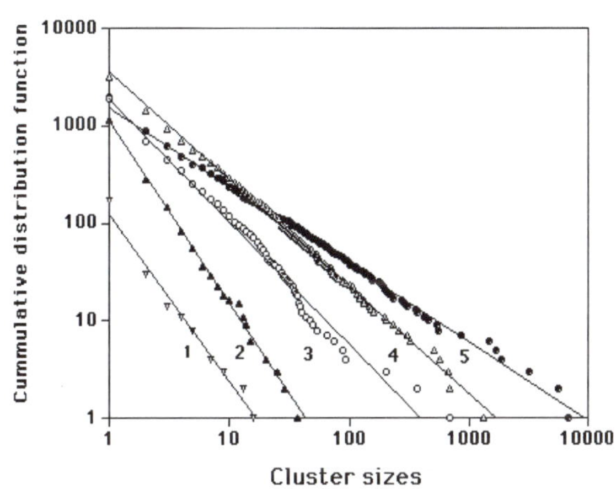


Fig. 12. Cumulative distribution of cluster sizes in 5 emphysematous patients. Severity of emphysema increase from 1 to 5. The exponent of the distributions decreases with increasing severity. From Ref. 87 with permission.

### 6.2 *An elastic network model of CT images*

The relative total LAA (total LAA normalized by the number of pixels in the lung filed) is similar to the probability of occupying a site in percolation. One might be attempted to use percolation models [88] to describe the process of tissue breakdown. However, since the exponent of the distribution varies from subject to subject even at the same relative total LAA, general percolation models are not appropriate. In order to better understand the breakdown of lung tissue structure, a large 2-dimensional elastic network model representing a slice of lung tissue has been developed [87]. The network consists of a 500 x 500 square lattice of Hookean springs having identical spring constants. The border nodes of the lattice are fixed, but the internal nodes are free to move. To account for an LAA cluster in a CT image, we first randomly remove a node from the lattice. Next, we perform a random walk starting from the removed node. Along this random walk, we remove S adjoining nodes, each time cutting all springs that are connected to the nodes. This produces a cluster of S+1 points removed from the original lattice. To create multiple LAA clusters, we repeat this process NP times with a different randomly selected seed point and vary S uniformly between 0 and Smax. The equilibrium configuration of the remaining nodes can be found by simulated annealing [87, 89]. The coordinates of each node are changed in small steps along the direction of the resultant force acting on it by its connecting springs, until the elastic potential energy of the entire network is minimized. The area enclosed by the LAA clusters are measured numerically and the cluster size distribution and the total LAA are calculated.

### 6.3 *Network breakdown*

The elastic network model has only two parameters: Smax and NP. The total LAA can be increased by increasing Smax and/or increasing NP. We choose Smax to be either 10 or 20 and varied NP between 1% and 16% of the total number of nodes. The cluster distributions from the model simulations also follow power laws (not shown) similar to those in Fig. 12. As NP is increased the distribution becomes wider, covering almost 4 orders of magnitude when NP=16%. The solid line in Fig. 13 shows that the variation of the exponents with the normalized total LAA from the model is in quantitative agreement with those obtained from the CT images.

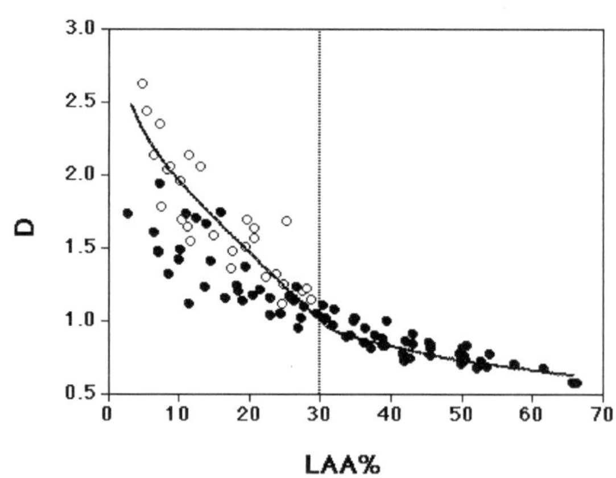


Fig. 13. Exponent D of the cumulative distributions of cluster sizes from Ref. [89] as a function of the percent Low Attenuation Area (LAA%) in normal (open symbols) and emphysematous (filled symbols) subjects. Solid line is model simulation. Based on Ref. 87 with permission.

The process of tissue breakdown is based on the assumption that mechanical forces can rupture the tissue [87]. We set up a threshold force and eliminate all the springs from the network which carries a force higher than the threshold. Next, the threshold is lowered and more springs are eliminated from the network. The underlying assumption is that with the progression of the disease, enzymes chemically weaken the tissue and the tissue breaks at points of high force concentration. This model predicts that initially an LAA cluster grows slowly around its perimeters, but soon significant stress concentration develops on alveolar walls separating two neighboring LAA clusters. When the wall separating two clusters breaks, the clusters coalesce which results in a sudden rearrangement of the structure. Next, the forces are redistributed along the perimeter of the new cluster which starts growing slowly until it coalesces with another cluster. This process is similar to crack propagation in plastic materials which occurs in avalanches following a power law distribution [90]. Thus, it is not surprising that the shape of the LAA distribution remains a power law following the propagation of cracks. However, the coalescence of clusters decreases the exponent because essentially we take away two small clusters (decreasing their probability) and add one larger cluster (increasing their probability) so the distribution flattens. Additionally, following a coalescence of two neighboring clusters, only a narrow region of non-LAA pixels separating the clusters change to LAA pixels. Therefore, the total LAA increases minimally. This explains that the total LAA in early emphysema is nearly normal, but the exponent is sensitive to small changes in cluster reorganization [87]. Finally, we also note that experiments have recently been reported supporting the notion that mechanical forces during breathing are capable of rupturing lung tissue structure [91].

## 7. Conclusion

Fluctuations in the cardiopulmonary system carry important information about its structure and/or functioning. Quantifying the fluctuations using statistical mechanical tools can reveal hidden dynamics and may help early detection of diseases or improve prognosis of exacerbations. Analytical and numerical modeling can help better understand these scaling laws which may shed light on the origins of the fluctuations and perhaps the mechanism of how a particular disease develops or propagates.

## Acknowledgement

This study was supported by grants NSF BES-0114538, and HL59215-04 and HL 19298 from the National Heart, Lung and Blood Institute, and the Department of Veterans' Affairs. P.Ch.I. and H.E.S. thank NIH/National Center for Research Resources for support.

## References

[1] E. R. Weibel, *The pathway for oxygen: structure and function in the mammalian respiratory system,* Harvard University Press, Cambridge (1984).

[2] H. Kitaoka, R. Takaki and B. Suki, *A three-dimensional model of the human airway tree, J. Appl. Physiol.* **87** (1999) 2207–2217.

[3] B. Suki B, A. L. Barabási, Z. Hantos, F. Peták and H. E. Stanley, *Avalanches and power law behaviour in lung inflation, Nature* **368** (1994) 615–618.

[4] C. D. Dawson, G. S. Krenz and J. H. Linehan, *Complexity and structure-function relationships in the pulmonary arterial tree*, Chapter 13, In: *Lung biology in health and disease, complexity in structure and function of the lung,* eds. M. P. Hlastala and H. T. Robertson, New York: Marcel Dekker, inc. (1998) 401–427.

[5] R. W. Glenny, S. L. Bernard and H. T. Robertson, *Pulmonary blood flow remains fractal down to the level of gas exchange, J. Appl. Physiol.* **89** (2000) 742–748.

[6] B. J. West, *Physiology in fractal dimensions - error tolerance, Ann. Biomed. Eng.* **18** (1990) 135–149.

[7] P. Ch. Ivanov, M. G. Rosenblum, C. K. Peng, J. Mietus, S. Havlin, H. E. Stanley and A. L. Goldberger, *Scaling behaviour of heartbeat intervals obtained by wavelet-based time-series analysis, Nature* **383** (1996) 323–327.

[8] U. Frey, M. Silverman, A. L. Barabási and B. Suki, *Irregularities and power law distributions in the breathing pattern in preterm and term infants, J. Appl. Physiol.* **86** (1998) 789–797.

[9] X. B. Zhang and E. N. Bruce, *Fractal characteristics of end-expiratory lung volume in anesthetized rats, Ann. Biomed. Engin.* **28** (2000) 94–101.

[10] H. T. Robertson, W. A. Altemeier and R. W. Glenny, *Physiological implications of the fractal distribution of ventilation and perfusion in the lung. Ann. Biomed. Engin.* **28** (2000) 1028–1031.

[11] M. Cernelc, B. Suki, B. Reinmann, G. L. Hall and U. Frey, *Correlation properties of tidal volume, end-tidal 02 and C02 concentrations in healthy infants. J. Appl. Physiol.* **92** (2002) 1817–1827.

[12] R. W. Glenny, N. L. Polissar, S. Mckinney and H. T. Robertson, *Temporal heterogeneity of regional pulmonary perfusion is spatially clustered, J. Appl. Physiol.* **79** (1995) 986–1001.

[13] J. G. Venegas and G. G. Galletti, *Low-pass filtering, a new method of fractal analysis: application to PET images of pulmonary blood flow, J. Appl. Physiol.* **88** (2000) 1365–1373.

[14] J. E. Hodgman, F. Gonzales, T. Hoppenbroewers and L. A. Cabal, *Apnea, transient episodes of bradycardia, and periodic breathing in preterm infants, Am. J. Dis. Child.* **144** (1990) 54–57.

[15] D. J. Henderson-Smart, A. G. Pettygrew and D. J. Campell, *Clinical apnea and brainstem neural function in pretem infants, N. Engl. J. Med.* **308** (1983) 353–357.

[16] T. Gerhardt and E. Bancalari, *Apnea of prematurity: lung function and regulation of breathing, Pediatrics* **74** (1984) 58–62.

[17] S. M. Botros and E. N. Bruce, *Neural network implementation of a three phase model of respiratory rythm generation, Biol. Cybern.* **63** (1990) 143–153.

[18] D. W. Richter and D. Ballantyne. *A three phase theory about the basic respiratory pattern generator,* In: *Central Neurone Environmen*t, eds. M. E. Schlafke, H. P. Koepchen and W. R. See, Berlin: Springer-Verlag, (1983) 165–174.

[19] J. P. Farber, *Medullary inspiratory activity during opossum development, Am. J. Physiol.,* **23** (1988) R578–R584.

[20] P. N. Sachis, D. L. Armstrong, L. E. Becker and A. C. Bryan, *The vagus nerve and sudden infant death syndrome: a morphometric study, J. Neuropathol. Exp. Neurol.* **41** (1982) 466–472.

[21] D. Paydarfar and D. M. Buerkel, *Dysarythmias of the respiratory oscillator, Chaos* **5** (1995) 18–29.

[22] S. Akselrod, D. Gordon, F. A. Ubel, D. C. Shannon, A. C. Barger and R. J. Cohen, *Power spectrum analysis of heart rate fluctuation: a quantitative probe of beat-to-beat cardiovascular control, Science* **213** (1981) 220.

[23] J. Lefebvre, D. A. Goodings, M. V. Kamath and E. L. Fallen, *Predictability of normal heart rhythms and deterministic chaos, Chaos* **3** (1993) 267.

[24] L. Glass, P. Hunter and A. McCulloch, eds., *Theory of heart,* Springer Verlag, New York (1991).

[25] C.-S. Poon and C. K. Merrill, *Decrease of cardiac chaos in congestive heart failure, Nature* **389** (1997) 492–495.

[26] M. F. Shlesinger and B. J. West, *Random fluctuations and pattern growth: Experiments and models*, Kluwer Academic Publishers, Boston (1988).

[27] P. Ch. Ivanov, M. G. Rosenblum, C.-K. Peng, J. Mietus, S. Havlin, H.E. Stanley and A. L. Goldberger, *scaling and universality in heart rate variability distributions, Physica A* **249** (1998) 587–593.

[28] C.-K. Peng, S. Havlin, H. E. Stanley and A. L. Goldberger, *Quantification of scaling exponents and crossover phenomena in nonstationary heartbeat time series,* in *Proc. NATO dynamical disease conference*, eds. L. Glass, *Chaos* **5** (1995) 82–87.

[29] P. Ch. Ivanov, L. A. N. Amaral, A. L. Goldberger, S. Havlin, M. G. Rosenblum, Z. Struzik and H. E. Stanley, *Multifractality in human heartbeat dynamics, Nature* **399** (1999) 461–465.

[30] P. Ch. Ivanov, L. A. N. Amaral, A. L. Goldberger, S. Havlin, M. G. Rosenblum, H. E. Stanley and Z. Struzik, *From 1/f noise to multifractal cascades in heartbeat dynamics, Chaos* **11** (2001) 641–652.

[31] Y. Ashkenazy, P. Ch. Ivanov, S. Havlin, C.-K. Peng, A. L. Goldberger and H. E. Stanley, *Magnitude and sign correlations in heartbeat fluctuations, Phys. Rev. Lett.* **86** (2001) 1900–1903.

[32] A. A. Aghili, X. X. Rizwan-uddin, M. P. Griggin and J. R. Moorman, *Scaling and ordering of neonatal rate variability, Phys. Rev. Lett.* **74** (1995) 1254–1257.

[33] J. F. Muzy, E. Bacry and A. Arneodo, *The multifractal formalism revisited with wavelets, Int. J. Bifurc. Chaos* **4** (1994) 245–302.

[34] D. Panter, *Modulation, noise and spectral analysis,* McGraw-Hill, New York (1965).

[35] D. Stauffer and H. E. Stanley, *From Newton to Mandelbrot: a primer in theoretical physics,* 2nd Edition, Springer-Verlag, Heidelberg & New York (1996).

[36] P. Ch. Ivanov M. G. Rosenblum, C.-K. Peng, S. Havlin, H. E. Stanley and A. L. Goldberger, *Wavelets in medicine and physiology* in *Wavelets in physics*, ed. H. van der Berg, Cambridge University Press, Cambridge (1998).

[37] H. E. Stanley, *Introduction to phase transitions and critical phenomena,* Oxford University Press, London (1971).

[38] M. Kobayashi and T. Musha, *1/f fluctuation of heartbeat period, IEEE Trans. Biomed. Eng.* **29** (1982) 456.

[39] C.-K. Peng, S. V. Buldyrev, S. Havlin, M. Simons, H. E. Stanley and A. L. Goldberger, *Mosaic organization of DNA nucleotides, Phys. Rev. E* **49** (1994) 1685–1689.

[40] P. Ch. Ivanov, A. Bunde, L. A. N. Amaral, S. Havlin, J. Fritsch-Yelle, R. M. Baevsky, H. E. Stanley and A. L. Goldberger, *Sleep-wake differences in scaling behavior of the human heartbeat: analysis of terrestrial and long-term space flight data, Europhysics Letters* **48** (1999) 594–600.

[41] A. Bunde, S. Havlin, J. W. Kantelhardt, T. Penzel, J. H. Peter and K. Voigt, *Correlated and uncorrelated regions in heart-rate fluctuations during sleep, Phys. Rev. Lett.* **85** (2000) 3736.

[42] P. Ch. Ivanov, L. A. N. Amaral, A. L. Goldberger and H. E. Stanley, *Stochastic feedback and the regulation of biological rhythms, Europhys. Lett.* **43** (1998) 363.

[43] A. Bunde and S. Havlin, *Fractals in science,* Springer-Verlag, Berlin (1994).

[44] H. E. Hurst, *Long-term storage capacity of reservoirs, Trans. Am. Soc. Civ. Eng.* **116** (1951) 770.

[45] L. A. N. Amaral, P. Ch. Ivanov, N. Aoyagi, I. Hidaka, S. Tomono, A. L. Goldberger, H. E. Stanley and Y. Yamamoto, *Behavioral-independent features of complex heartbeat dynamics, Phys. Rev. Lett.* **86** (2001) 6026.

[46] U. Frisch, *Turbulence*, Cambridge University Press, Cambridge UK (1995).

[47] R. D. Kamm and R. C. Schroter, *Is airway closure caused by a liquid film instability?, Respir. Physiol.* **75** (1989) 141–156.

[48] D. P. Gaver III, R. W. Samsel and J. Solway, *Effects of surface tension and viscosity on airway reopening, J. Appl. Physiol.* **69** (1990) 74–85.

[49] K. R. Lutchen, D. W. Kaczka, E. Israel, B. Suki and E. P. Ingenito, *Airway constriction pattern is a central component of asthma severity: The role of deep inspirations, Am. J. Respir. Crit. Care Med.* **164** (2001) 207–215.

[50] P. Forgacs, *Crackles and wheezes, Lancet* **2** (1967) 203–205.

[51] J. J. Fredberg and S. K. Holford, *Discrete lung sounds: crackles (rales) as stress relaxation quadrupole, J. Acoust. Soc. Am.* **73** (1983) 1036–1046.

[52] M. Munakata, Y. Homma, M. Matsuzaki, H. Ogasawara, K. Tanimura, H. Kusaka and Y. Kawakami, *Production mechanism of crackles in excised normal canine lungs, J. Appl. Physiol.* **61** (1986) 1120–1125.

[53] R. L. H. Murphy, S. K. Holford and W. C. Knowler, *Visual lung-sound characterization by time expanded wave-form analysis, N. Eng. J. Med.* **296** (1977) 968–971.

[54] H. Pasterkamp, S. S. Kraman and G. R. Wodicka, *Respiratory sounds – Advances beyond the stethoscope, Am. J. Respir. Crit. Care Med.* **156** (1997) 974–987.

[55] A. M. Alencar, Z. Hantos, F. Petak, J. Tolnai, T. Asztalos, S. Zapperi, J. S. Andrade Jr, S. V. Buldyrev, H. E. Stanley and B. Suki, *Scaling behavior in crackle sound during lung inflation, Phys. Rev. E* **60** (1999) 4659–4663.

[56] A. M. Alencar, S. V. Buldyrev, A. Majumdar, H. E. Stanley and B. Suki, *Avalanche dynamics of crackle sound in the lung, Phys. Rev. Lett.* **8708** (2001) 8101–8104.

[57] A. L. Barabási, S. V. Buldyrev, H. E. Stanley and B. Suki, *Avalanches in the lung: A statistical mechanical model, Phys. Rev. Lett.* **76** (1996) 2192–2195.

[58] M. F. Shlesinger and B. J. West, *Complex fractal dimension of the bronchial tree, Phys. Rev. Lett.* **67** (1991) 2106–2109.

[59] B. Suki, *Fluctuations and power laws in pulmonary physiology, Am. J. Resp. Crit. Care. Med.* (in press).

[60] S. H. Bennett, B. W. Goetzman, J. M. Milstein and J. S. Pannu, *Role of arterial design on pulse wave reflection in a fractal pulmonary network, J. Appl. Physiol.* **80** (1996) 1033–1056.

[61] C. G. Caro and P. G. Saffman, *Extensibility of blood vessels in isolated rabbit lungs, J. Physiol. (London)* **178** (1965) 193–210.

[62] C. A. Dawson, G. S. Krenz, K. L. Karau, S. T. Haworth, C. C. Hanger and J. H. Linehan, *Structure-function relationships in the pulmonary arterial tree, J. Appl. Physiol.* **86** (1999) 569–583.

[63] R. Z. Gan and R. T. Yen, *Vascular impedance analysis in dog lung with detailed morophometric and elasticity data, J. Appl. Physiol.* **77** (1994) 706–717.

[64] R. W. Glenny, S. L. Bernard and H. T. Robertson, *Pulmonary blood flow remains fractal down to the level of gas exchange, J. Appl. Physiol.* **89** (2000) 742–748.

[65] J. M. Gonzalez-Fernandez and S. E. Atta, *Maximal substrate transport in capillary networks, Microvascular Res.* **5** (1973) 180–198.

[66] K. Horsfield, *Morphometry of small pulmonary arteries in man, Circ. Res.* **42** (1978) 593–597.

[67] W. Huang, R. T. Yen, M. Mclaurine and G. Bledsoe, *Morphometry of the human pulmonary vasculature, J. Appl. Physiol.* **81** (1996) 2123–2133.

[68] Z. Jiang, G. S. Kassab and Y. C. Fung, *Diameter-defined Strahler system and connectivity matrix of the pulmonary arterial tree, J. Appl. Physiol.* **76** (1994) 882–892.

[69] K. L. Karau, G. S. Krenz and C. A. Dawson, *Branching exponent heterogeneity and wall shear stress distribution in vascular trees, Am. J. Physiol. Heart Circ. Physiol.* **280** (2001) H1256–H1263.

[70] K. L. Karau, R. C. Molthen, A. H. Dhyani, S. T. Haworth, C. C. Hanger, D. L. Roerig, R. H. Johnson and C. A. Dawson, *Pulmonary arterial morphometry from microfocal x-ray computed tomography, Am. J. Physiol. Heart Circ. Physiol.* **281** (2001) H2747–H2756.

[71] G. S. Kassab and Y. B. Fung, *The pattern of coronary arteriolar bifurcations and the uniform shear hypothesis, Ann. Biomed. Engr.* **23** (1995) 13–20.

[72] G. S. Krenz, J. Lin, C. A. Dawson and J. H. Linehan, *Impact of parallel heterogeneity on a continuum model of the pulmonary arterial tree, J. Appl. Physiol.* **77** (1994) 660–670.

[73] M. Labarbera, *Principles of Design of Fluid Transport Systems in Zoology, Science* **249** (1990) 992–1000.

[74] H. H. Lipowsky, *Shear stress in the circulation. In: Flow-Dependent Regulation of Vascular Function*, edited by J. A. Bevan, Kaley, G., and G. M. Rubanyi, New York: Oxford University Press, (1995) 28–45.

[75] Y.-H. Liu and E. L. Ritman, *Branching pattern of pulmonary arterial tree in anesthetized dogs, J. Biomech. Eng.* **108** (1986) 289–293.

[76] W. S. Miller, *The structure of the lung, J. Morphol.* **8** (1983) 165–188.

[77] D. C. Murray, *The physiological principle of minimum work. I. The vascular system and the cost of blood volume, Proc. Natl. Acad. Sci. USA* **12** (1926) 207–214.

[78] T. F. Sherman, A. Koller, P. C. Johnson and A. S. Popel, *The cost of departure from optimal radii in microvascular network, J. Theor. Biol.* **136** (1989) 245–265.

[79] N. Suwa, T. Niwa, H. Fukasawa and Y. Sasaki, *Estimation of intravascular blood pressure gradient by mathematical analysis of arterial casts, Tohoku J. Exper. Med.* **79** (1963) 168–198.

[80] H. B. M. Uylings, *Optimization of diameters and bifurcation angles in lung and vascular tree structures, Bull. Math. Biol.* **39** (1977) 509–520.

[81] G. B. West, J. H. Brown and B. J. Enquist, *The fourth dimension of life: Fractal geometry and allometric scaling of organisms, Science* **284** (1999) 1677–1679.

[82] R. T. Yen, F. Y. Zhuang, Y. C. Fung, H. H. Ho, H. Tremer and S. S. Sobin, *Morphometry of cat's pulmonary arterial tree, J. Biomech. Eng.* **106** (1984) 131–136.

[83] M. Zamir, *Shear forces and blood vessel radii in the cardiovascular system, J. Gen. Physiol.* **69** (1977) 449–461.

[84] P. J. Barnes, *Chronic obstructive pulmonary disease, N. Engl. J. Med.* **343** (2000) 269–80.

[85] R. Uppaluri, T. Mitsa, M. Sonka, E. A. Hoffman and G. McLennan, *Quantification of pulmonary emphysema from lung computed tomography images, Am. J. Respir. Crit. Care Med.* **156** (1997) 248–254.

[86] K. Shimizu, T. Johkoh, J. P. Ikezoe, K. Ichikado, J. Arisawa, H. Nakamura, S. Tamura, T. Nagareda, *Fractal analysis for classification of ground-glass opacity on high-resolution CT: An in vitro study, J. Comp. Assist. Tomog.* **21** (1997) 955–961.

[87] M. Mishima, T. Hirai, H. Itoh, Y. Nakano, H. Sakai, S. Muro, K. Nishimura, Y. Oku, K. Chin, M. Ohi, T. Nakamura, J. H. T. Bates, A. M. Alencar and B. Suki, *Complexity of terminal airspace geometry assessed by lung CT in normal subjects and patients with chronic obstructive pulmonary disease, Proc. Natl. Acad. Sci.* **96** (1999) 8829–8834.

[88] D. Stauffer and A. Aharony, *Introduction to percolation theory,* Taylor and Francis Inc., London (1994).

[89] S. Kirkpatrick, C. D. Gelatt and M. P. Vecchi Jr, *Optimization by simulated annealing, Science* **220** (1983) 671–680.

[90] S. Zapperi S, A. Vespignani and H. E. Stanley, *Plasticity and avalanche behavior in microfracturing phenomena, Nature* **388** (1997) 658–660.

[91] S. Kononov, K. Brewer, H. Sakai, F. S. A. Cavalcante, C. R. Sabayanagam, E. P. Ingenito and B. Suki, *Roles of mechanical forces and collagen failure in the development of elastase-induced emphysema, Am. J. Resp. Crit. Care. Med.* **164** (2001) 1920–1926.

# THE APPLICATION OF SUPRATHRESHOLD STOCHASTIC RESONANCE TO COCHLEAR IMPLANT CODING

N. G. STOCKS* and D. ALLINGHAM
*School of Engineering, University of Warwick, Coventry, CV4 7AL, UK*
**ngs@eng.warwick.ac.uk*

R. P. MORSE
*MacKay Institute of Communication and Neuroscience, University of Keele*
*Keele, ST5 5BG, UK*

Received 21 June 2002
Accepted 20 September 2002

In this paper we explore the possibility of using a recently discovered form of stochastic resonance - termed suprathreshold stochastic resonance - to improve speech comprehension in patients fitted with cochlear implants. A leaky-integrate-and-fire (LIF) neurone is used to model cochlear nerve activity when subject to electrical stimulation. This model, in principle, captures key aspects of temporal coding in analogue cochlear implants. Estimates for the information transmitted by a population of nerve fibres is obtained as a function of internal (neuronal) noise level. We conclude that SSR does indeed provide a possible mechanism by which information transmission along the cochlear nerve can be improved - and thus may well lead to improved speech comprehension.

*Keywords*: Suprathreshold stochastic resonance; cochlear nerve; neural coding; cochlear implants; information transmission; noise.

## 1. Introduction

In a recent series of papers [1–6] a novel form of stochastic resonance (SR) – termed suprathreshold stochastic resonance (SSR) – was introduced and discussed. Similar to 'conventional' SR [7,8], the SSR effect gives rise to a nonmonotonic dependence of the transinformation (*trans*mitted *information*) on noise intensity - thus, maximum transinformation occurs at a non-zero level of noise intensity. However, SSR has an important difference; unlike conventional SR it can occur for signal strengths that are above threshold. Indeed, the SSR effect is optimised precisely when the signals are suprathreshold [1,2]. It was for this reason the effect was termed *suprathreshold* SR [1]. The fact that SSR occurs for suprathreshold signal strengths gives it potential advantages over conventional SR; SSR gives rise to much larger noise-induced information gains than can be obtained with conventional SR, and it does not rely on the signal being weak.

SSR occurs in parallel arrays of threshold elements like that shown in Fig. 1; it can not occur in a single element. Each unit in the array is subject to the common input signal $s(t)$ but evolves under the influence of its own internal (independent) noise source. The SSR effect was first studied in an array comprising of simple threshold devices (comparators) [1,2,6] but has more recently been applied to neural populations [3–5] where each unit in the array was modelled as a FitzHugh-Nagumo neurone [9]. It has been demonstrated that global noise-induced information gains can be obtained regardless of the input signal magnitude or the value of the common threshold(all units have the same threshold). In short, in a population of similar neurones, the information can always be maximized by an optimal level of internal noise [3]. It is the possible exploitation of this idea that we explore in the context of cochlear implants.

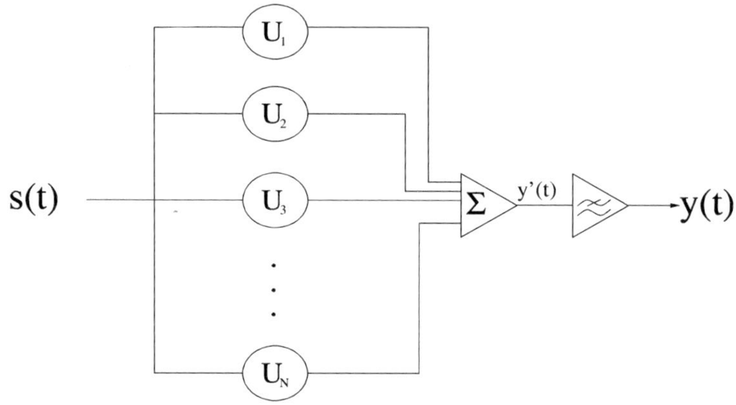


Fig. 1. Parallel array of N identical 'units'denoted $U_1$ to $U_N$. Each unit has a common input signal $s(t)$ but evolves under the influence its own internal noise source. The internal noise intensity of each unit is assumed to be the same. The response $y(t)$ is obtained by summing the units' responses and low-pass filtering.

In many profoundly deaf people the hair cells of the inner ear that transduce mechanical sound vibration into electrical signals are damaged or missing [10–12]. A cochlear implant can often be surgically implanted into the cochlea to replace the function of the hair cells [13–15] and partially restore hearing. Cochlear implants work by direct electrical stimulation of the cochlear nerve fibres. The purpose of the implant is to evoke, with electrical stimulation, a similar pattern of nerve activity to that expected in a healthy ear when stimulated acoustically. A schematic diagram of a cochlear implant is shown in Fig. 2. Cochlear implants consist of a number of electrodes (typically about twenty) that each stimulate a large group (population) of nerve fibres. In principle, it is possible to excite the whole population with a common signal just as in Fig. 1. However, in practice, this is not done. It is usual to first pass the signal through a bank of passband filters (channels) - each channel having a different centre frequency. The output from each channel is then connected to just one electrode. This process is undertaken to reflect the different tuning characteristics of the hair cells and hence mimic the effect of 'place coding' in the ear. The array in Fig. 1 can be used to model the cochlear implant in

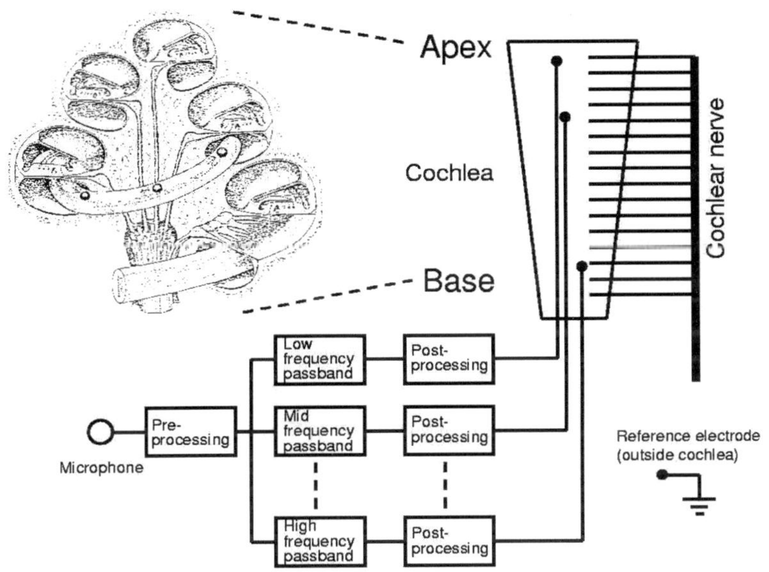


Fig. 2. Diagram of a cochlear implant. The sketch of the the cochlear has been taken from [16]

two possible ways. First, each electrode stimulates a (sub)population of nerve fibres with the same signal. In this case the parallel array models information propagation in this subpopulation. Second, in principle, it is possible to connect more than one electrode to each filter channel - thus again stimulating a population with the same signal. Thus, in effect, we are considering information transmission in only a subpopulation of the total number of nerve fibres that the cochlear implant stimulates. We make the initial simplifying assumption that all neurones are identical.

To understand how noise may be used to improve speech comprehension in cochlear implants it is important to first understand what gives rise to SSR. The SSR effect, over the range of noise intensities that increase transinformation, can be understood as follows; if there is no internal noise, and each unit is identical, then the response of all the units will also be identical (because they each receive the same input signal). The implication of this in information-theoretic terms is (rather obviously) that each unit will carry *identical* information about the signal and, hence, the total information transmitted by the array will be simply equal to the information transmitted by any one unit. However, this is not the case if internal noise is present. The presence of the noise results in a desynchronization [2] of the individual responses that can lead to an increase in the output entropy[a] of the array. The increase in output entropy occurs because the noise introduces a degree of independence in the information transmitted by individual units by slightly randomising the firing times of the units. This increase in output entropy gives rise to a net increase in the transinformation (because the output entropy rises faster than the information lost due to noise [2]).

[a]The output entropy ($H(y)$ in Eq. (1)) is the total information (signal and noise), it should not be confused with transinformation which is the information at the output about the signal.

These information gains are only observed over a limited range of noise intensities. At sufficiently larger noise the situation reverses and the information loss due to noise increases faster than the output entropy - this leads to a reduction in the transinformation.

It is important to note that the net global gain in information [3] is obtained by summing information over all units - noise always lowers the amount of information carried by each individual unit (assuming that the signal is suprathreshold). Indeed, the SSR effect has other non-intuitive features. For example, although the transinformation goes through a maximum with increasing noise, no such maximum is observed in the output signal-to-noise ratio(SNR). The SNR decreases monotonically with noise (see Ref. [2] for the definition of SNR). This is another major difference in the phenomenology of SSR compared to conventional SR. This difference has implications for signal detection - it means that SSR cannot be used to improve detection in a hypothesis testing sense (i.e. determining whether a signal is present) but can be used to improve the fidelity of the transmitted signal. It is the latter of these tasks that is important for cochlear implants.

The idea that noise may be useful as a means of desynchronizing neural responses across a population has already begun to receive attention in the context of cochlear implant coding. This is because it is known that conventional cochlear implants produce temporal cochlear nerve activity that differs significantly from that observed in a healthy ear. The main difference is that the cochlear implant produces abnormally high levels of across-fibre synchrony. Work by Morse and Evans [17] demonstrated that, because of high synchronization, the temporal pattern of nerve discharges evoked by conventional analogue cochlear implants would not be expected to convey an essential cue for both vowel identification and the identification of some consonants. This abnormally high level of across-fibre synchronization has also been noted in other studies using periodic stimuli [18–22].

It has been postulated that the abnormally high level of synchronization is due to lower than normal neuronal noise levels. In a healthy ear the hair cells appear to be a major generator of noise - the noise originates from Brownian motion of the hair cells and from synaptic connections to the cochlear nerve fibres. Therefore, in cochlear implant patients, the absence of hair cells is likely to lead to lower than normal stochastic nerve activity. This postulate is backed up by studies that have demonstrated that the healthy mammalian cochlea has a significant amount of spontaneous activity (neural firing) 'in quiet' [23] that is largely not observed if the hair cells and presynaptic connections are damaged or absent [10,24]. This has led to the suggestion that noise should be re-injected back into the deafened ear to reduce the across-fibre synchrony [17,22,25]. Indeed, the addition of noise has been shown to enhance the coding of vowel formant cues [17], reduce across-fibre synchrony [22], and lead to improved sensitivity to envelope modulation [25]. However, it has not been established that internal noise can actually lead to enhanced information transmission. It would appear that SSR offers a mechanism by which this can be achieved. Indeed, cochlear implants are normally operated at suprathreshold stimulus intensities – a parameter range, as discussed above, that precludes the occurrence of conventional SR [1,2], but not SSR. The key finding in this paper is that SSR does appear to offer a means of improving information transmission in cochlear implants.

## 2. The Transinformation

The global information transmission through the array is characterised by the transinformation $I$, which, for a continuous channel, is defined as [26]

$$\mathrm{I} = H(y) - H(y|s) = -\int_{-\infty}^{\infty} P_y(y)\log_2 P_y(y)dy$$
$$-\left(-\int_{-\infty}^{\infty} P_s(s)ds\int_{-\infty}^{\infty} P(y|s)\log_2 P(y|s)dy\right) \tag{1}$$

where $H(y)$ is the information content (or entropy) of $y(t)$ and $H(y|s)$ can be interpreted as the amount of encoded information lost in the transmission of the signal. $P_y(y)$ and $P_s(s)$ are the probability density functions (pdfs) of $y(t)$ and $s(t)$ respectively and $P(y|s)$ is the conditional pdf. All the relevant distributions can easily be obtained from simulation of the model of the array.

In the absence of noise $H(y|s) = 0$ and hence $I = H(y)$, and for sufficiently large noise $H(y|s) = H(y)$ and hence $I = 0$. Therefore, $I$ can only increase in the presence of noise if $H(y)$ increases faster (with increasing noise) than $H(y|s)$. This is precisely what happens in SSR [2].

## 3. Model of Electrically Stimulated Auditory Neurones

We use a leaky-integrate-and-fire (LIF) [9] model to capture the behaviour of cochlear nerve fibres subject to electrical stimulation. The LIF model has been developed directly from physiological studies of the electrically stimulated sciatic nerve of the toad *Xenopus laevis*. The use of the toad sciatic nerve as a physiological model for the mammalian cochlear nerve will be addressed in detail elsewhere [27], as will the detailed comparison between the LIF model and the sciatic nerve data [28]. Here we simply present the model and show some representative comparisons between the LIF model and physiological data.

Each unit in the array was modelled using the LIF neurone

$$\tau_m \dot{x}_i = -x_i + s(t) + \sqrt{2D}\xi_i(t)\,, \tag{2}$$
$$\langle \xi_i(t)\xi_j(t')\rangle = \delta_{ij}\delta(t-t');\ \langle \xi_i\rangle = 0$$

where $x_i$ represents the membrane voltage of the $i$th neurone, $\tau_m$ is the membrane time constant, $D$ is the common noise intensity, $s(t)$ is the input signal, $\delta_{ij}$ is the Kronecker delta, $\delta(..)$ is the Dirac delta function, and $i = 1...N$, where $N$ is the total number of neurones in the array. Each neurone evolves under the influence of its own independent noise source $\xi_i(t)$. In the physiological experiments this noise term originates from the fluctuations in the opening and closing of the ion channels (membrane noise) but in this study we treat its intensity as an adjustable parameter.

The membrane voltage is allowed to evolve according to (2) until it reaches the common threshold value $\theta$, at which point a spike (represented by a delta function) is deemed to have occurred and the membrane voltage is reset to zero. An absolute refractory period, $\tau_{abs}$, is then imposed during which time the membrane voltage is held fixed at zero. After a duration $\tau_{abs}$ the neurone then enters a relative refractory

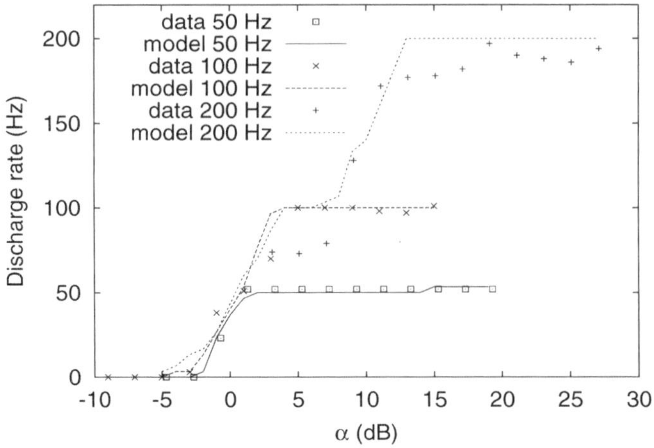


Fig. 3. Discharge rates vs signal-to-threshold ratio in decibels. Three sets of curves are shown, representing the response to a sinusoidal signal with a frequency of 50 Hz, 100 Hz or 200 Hz. The results from the LIF model were obtained using the parameter values $\tau_m = 4.8$ms, $\tau_{abs} = 3.3$ms, $\tau_{rel} = 5.5$ms and $D = 8 \times 10^{-5}$. All these values are consistent with the known physiological values [27]. All the experimental data was obtained from the same neurone.

where the threshold is increased. This behaviour is modelled by multiplying the signal and noise in (2) by the factor $(1 - \exp(-t/\tau_{rel}))$, where $\tau_{rel}$ is the relative refractory time constant. The spike trains from all units are then summed to give $y'(t)$, and the final response $y(t)$ is obtained from the convolution

$$y(t) = \int_{-\infty}^{\infty} y'(s) f(t-s) ds \tag{3}$$

where,

$$f(t) = \begin{cases} A(1 - \exp(-t/\tau_s)) & \text{if } t < T_p, \\ A(1 - \exp(-T_p/\tau_s)) \exp(-(t - T_p)/\tau_s) & \text{if } t \geq T_p. \end{cases} \tag{4}$$

Convolving with the function $f(t)$ is equivalent to assuming the original action potential has a rectangular pulse of width $T_p$ and amplitude $A$ that is then passed through a single-pole low-pass filter with a cut-off frequency of $\omega = 1/\tau_s$. In physical terms, the total response $y(t)$ can be viewed as the membrane voltage of a summing neurone (with membrane time constant $\tau_s$) to which all the other neurones are connected (although in this picture synaptic filtering effects have been neglected). The filter output provides an estimate of the input signal - it removes the high frequencies that are largely dominated by noise. The value of $T_p = 2.5$ms and $\tau_s = 1$ms was chosen to maximise the transinformation.

To demonstrate that (2) captures the important dynamics of electrically stimulated nerve fibres, we show a comparison between the results obtained from the LIF model and results from physiological experiment. Fig. 3 shows results for the discharge rates (spikes per second) of a single fibre (from the sciatic nerve of the toad)

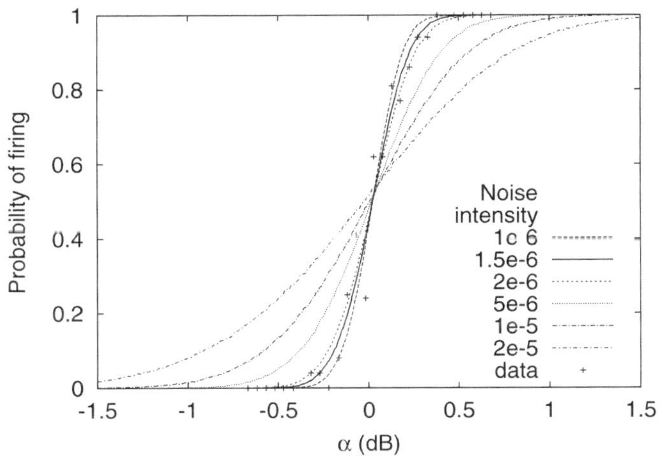


Fig. 4. Probability of firing vs the signal-to-threshold ratio in decibels for various noise intensities. Smooth curves are from simulation of the model and data points are from the experiments. The value of $1.5 \times 10^{-6}$ matches the relative spread as calculated from experimental data. Parameters for fitting an error function to each of these curves are shown in Table 1. A pulse width of $500\mu s$ was used.

when subject to a sinusoidal current for three different frequencies of the driving current. The results are plotted against $\alpha$ - the stimulus-amplitude-relative-to-threshold ratio expressed in decibels; 0dB corresponds to the case when the signal is just sufficient (in the absence of noise) to cause the neurone to spike. In effect, $\alpha$ quantifies the amplitude of the sinusoidal forcing (the mean of signal plus noise is zero). The parameters $\tau_m$, $\tau_{abs}$, $\tau_{rel}$ and $D$ were adjusted to give the best fit to the data by eye. Excellent agreement between the model and experiments is observed for stimulus frequencies of 50 Hz and 100 Hz, although some discrepancy is seen at 200 Hz. However, even at 200 Hz the model captures the important qualitative features such as the small step at a discharge rate of 100 Hz (approximately 75 Hz in the data).

The membrane noise level is estimated in the experiment by measuring the RS. This is measured by replacing the sinusoidal forcing with a pulse of fixed width and repeatedly applying the pulse to obtain the probability of firing. This process is repeated for various pulse heights and the results plotted as in Fig. 4. In this case $\alpha$ is defined to be the pulse height-to-threshold distance. An integrated Gaussian (error function) with mean $\mu_{fit}$ and standard deviation $\sigma_{fit}$ is then fitted to the data. The relative spread is then defined to be [29, 30],

$$RS = \frac{\sigma_{fit}}{\mu_{fit}} . \tag{5}$$

The noise intensity in the model (2) can then be adjusted to give the same RS value. A family of curves is produced by varying the noise intensity $D$; this enables a table (Table 1) to be constructed that can be used to map the experimentally measured values of RS onto an equivalent noise intensity.

Table 1. Relative spread and parameters of fit to the error function for various noise intensities. All other parameters were the same as those used for the discharge rate curves in Fig. 3.

| Noise intensity | $\mu_{fit}$ | $\sigma_{fit}$ | Relative spread |
|---|---|---|---|
| 5e-4 | 0.875 | 1.6387 | 1.873 |
| 2e-4 | 0.963 | 0.6045 | 0.6277 |
| 1e-4 | 0.976 | 0.3189 | 0.3267 |
| 5e-5 | 0.988 | 0.1930 | 0.1953 |
| 2e-5 | 0.997 | 0.1074 | 0.1077 |
| 1e-5 | 1.001 | 0.0721 | 0.0720 |
| 5e-6 | 1.002 | 0.0498 | 0.0497 |
| 2e-6 | 1.003 | 0.0310 | 0.0309 |
| **1.5e-6** | **1.003** | **0.0268** | **0.0267** |
| 1e-6 | 1.003 | 0.0218 | 0.0217 |

## 4. Results

All results presented in the section were obtained using an input signal that was Gaussianly distributed and exponentially correlated with a correlation time of 0.2s. The 0dB signal strength relative to threshold was defined as $\theta = 2\sigma_s$ where $\sigma_s$ is the signal strength measured at the *output* of one of the neurones i.e. it is the standard deviation of the membrane voltage $x_i$ (before the threshold is taken into account) in the absence of noise. With this definition, at 0dB the system (in the absence of noise) spends 99% of its time below threshold and is thus predominantly subthreshold. In general the signal intensity in decibels, $S_{dB}$, is found from the formula,

$$S_{dB} = 20 \log \frac{2\sigma_s}{\theta}\,.$$

In Fig. 5 the transinformation against scaled noise intensity, $\sigma$, is plotted for two different signal strengths and various $N$. The scaled noise intensity is defined as $\sigma = \sigma_n^2/\sigma_s^2$, where $\sigma_n^2 = D/\tau_m$ is the variance of the membrane voltage $x_i$ measured in the absence of the signal (i.e. variance due to noise). Consequently, $\sigma$ is equal to the inverse of the SNR measured relative to the output. The results presented were obtained with $\theta = 1$.

Clearly, all curves in Fig. 5 (except $N = 1$) show a non-monotonic dependence on noise intensity and thus display an SR type effect. The height of the maximum is seen to increase as the number of neurones in the array is increased. Note that the optimal noise intensity (position of maximum) is located in the range $-10\text{dB} < \sigma < 0\text{dB}$ - this implies that the information is maximised when the variance of the noise is of the same order as the variance of the signal. All these observations are consistent with the SSR effect.

The arrow on the abscissa (Fig. 5) indicates the estimated value of the noise intensity based on the physiological experiments. Clearly this lies well below (by more than two orders of magnitude) the optimal noise intensity, and therefore suggests that improved information transmission in electrically stimulated neural populations could be achieved by increasing the level of independent noise at each neurone.

In Fig. 6 the transinformation is plotted for a number of different signal strengths. The signal strengths are representative of the values used in a cochlear

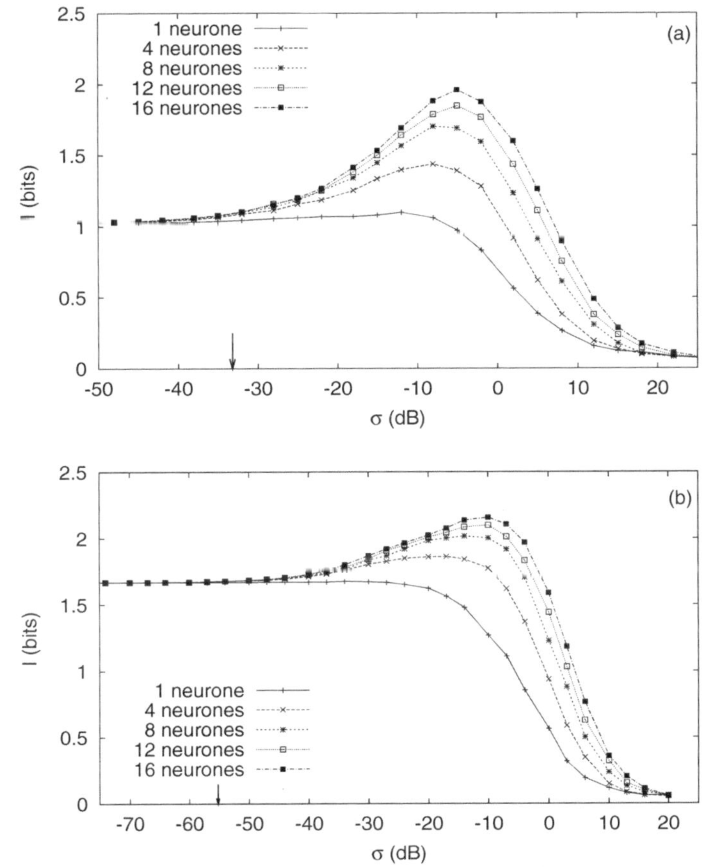


Fig. 5. Average mutual information vs scaled noise intensity (see text for definition of $\sigma$) for arrays of neurones of various sizes, for a signal strength (a) 8 and (b) 30 decibels above threshold. The signal strengths were chosen to coincide with those typically employed in cochlear implants. The arrow indicates the approximate average value of internal noise in the experimental data.

implant [15]. Considering Fig. 6(a) first, it is observed that the height of the maximum increases with signal strength, i.e. the more suprathreshold the signal, the greater the information transfer. It is also evident that the optimal noise intensity increases with increasing signal strength. Again these observations are consistent with the SSR effect [1–4]. The data in Fig. 6(b) is identical to that shown in Fig. 6(a) except it has been plotted against the scaled noise intensity $\sigma$. This rescaling again demonstrates that the optimal noise intensity occurs when the signal and noise intensities are of similar size – i.e. when the output SNR is in the range $0 - 10$dB (recall $\sigma$=1/SNR.)

## 5. Discussion and Conclusions

The results shown in the previous section clearly indicate that SSR can be observed in a model population of electrically stimulated cochlear nerves. This gives hope

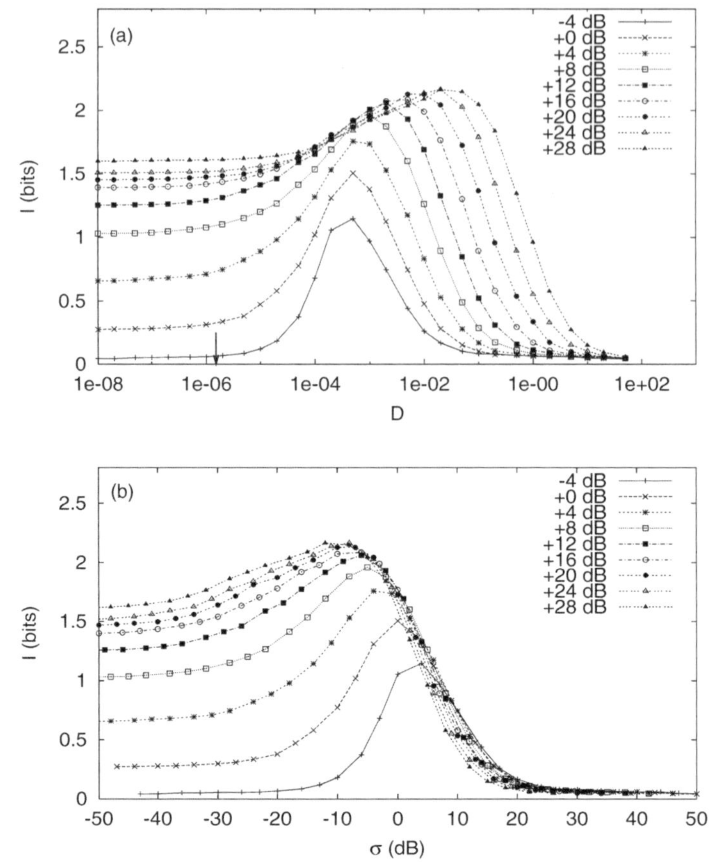


Fig. 6. Average mutual information vs scaled noise intensity for an array of 16 neurones for a variety of signal strengths (shown in the legend). The arrow indicates the approximate average value of internal noise in the experimental data. In (b) the results have been replotted against scaled noise intensity $\sigma$.

that SSR could be exploited in cochlear implants to improve speech comprehension. It was also observed that the noise intensity estimated from the physiological studies [27] fell several orders of magnitude below that required to optimise information transmission via SSR. Although the level of noise observed in the cochlear nerve is twice that observed in the sciatic nerve [27], this level of noise is still well below the optimal noise intensity. This suggests that deafened cochlear nerve fibres are sufficiently noise-free that the addition of independent noise to cochlear implant electrodes could lead to improved information transmission along the auditory nerve.

However, this statement needs additional qualification. First, our model assumes that all neurones are identical - clearly this is not the case in reality. Nevertheless, given that there are approximately 30,000 cochlear nerve fibres in the human ear [31], subpopulations of fibres with similar characteristics would be expected. Hence, our results may well be applicable to these subpopulations. A second, related, point is it that our model does not bandpass the signal before application to the neurones and therefore fails to capture the tuning characteristics (place-coding effects) of the auditory system. However, this fact opens up some interesting questions. Current cochlear implants filter the signal through a bank of bandpass filters before applying the filtered signal to the electrodes of the cochlear implant (see Fig. 2). Each electrode receives a filtered signal with a different centre frequency. Our results suggest that this may not be the most efficient design strategy; benefit could be obtained by connecting the same channel to multiple electrodes. Providing the electrodes stimulate neurones with similar properties, SSR could be employed to enhance information transmission. The question as to how channels should be split between different electrodes is something that needs further consideration. This strategy would also enable emphasis/de-emphasis of specific frequency ranges by varying the number of electrodes connected to each channel.

Finally, we note that the SSR effects we observe are not small. At a signal intensity of 8dB the transinformation improved by 1 bit when the noise was increased from the physiologically estimated value to the optimal value. An increase of 1 bit is equivalent to a doubling of the accuracy to which the transmitted signal is known. The results also suggest that the effect does not need large populations for it to be useful - even a population of four neurones can display a 40% gain in information. Therefore, the implementation of such a coding scheme is easily achievable using current cochlear implant technology.

## Acknowledgements

This work was funded in part by the MRC (grant G0001114) and the EPSRC (grant GR/35650/01).

## References

[1] N. G. Stocks, *Suprathreshold stochastic resonance in multilevel threshold systems*, *Phys. Rev. Lett.* **84** (2000) 2310–2314.

[2] N. G. Stocks, *Information transmission in parallel arrays of threshold elements: suprathreshold stochastic resonance*, *Phys. Rev. E* **63** art. num. 041114 (2001) 1–11.

[3] N. G. Stocks and R. Mannella, *Generic noise-enhanced coding in neuronal arrays*, *Phys. Rev. E* **64** art. num. 030902 (2001) 1–4.

[4] N. G. Stocks and R. Mannella, *Suprathreshold stochastic resonance in a neuronal network model: A possible strategy for sensory coding*, in *Future Directions for Intelligent systems and Information Sciences*, ed. N. Kosabov (Physica-Verlag, 2000) 236–246.

[5] N. G. Stocks, *Optimising information transmission in model neuronal ensembles*, in *Stochastic Processes in Physics, Chemistry and Biology*, ed. Jan A. Freund and Thorsten Poschel, Lecture Notes in Physics, LNP 557, (Springer-Verlag, Berlin Heidelberg, 2000) 150–159.

[6] N. G. Stocks, *Suprathreshold stochastic resonance: an exact result for uniformly distributed signal and noise*, *Phys. Lett. A* **279**(2001) 308–312.

[7] L. Gammaitoni, P. Hanggi, P. Jung and F. Marchesoni, *Stochastic resonance,Rev. of Mod. Phys.* **70**,(1998) 223–287.

[8] *Proceedings of the International Workshop on Fluctuations in Physics and Biology: Stochastic Resonance, Signal Processing and Related Phenomena*, ed. R. Mannella, et al, *Nuovo Cimento Soc. Ital. Fis.* **17D**, (1995).

[9] H. C. Tuckwell, *Introduction to Theoretical Neurobiology: Volume 2, Nonlinear and Stochastic Theories*, (CUP, Cambridge 1988).

[10] N. Y. S. Kiang, E. C. Moxon and R. A. Levine, *Auditory nerve activity in cats with normal and abnormal cochleae*, in *Sensorineural Hearing Loss*, eds. G. E W. Wolstenholme and J. Knight, (Churchill, London, 1970) 241–273.

[11] R. Hinjosa and J. R. Lindsay, *Profound deafness: associated sensory and neural degeneration*, *Archives of Otolaryngology* **106** (1980) 193–209.

[12] J. B. Nadol, Jr., *Histopathology of human aminoglycoside ototoxicity*, in *Aminoglycoside Ototoxicity* eds. S. A. Lerner, G. J. Matz and J. E. Hawkins (Little and Brown and Co., Boston, 1981) 409–434.

[13] B. S. Wilson, *The future of cochlear implants*, *Brit. J. of Audiology* **31** (1997) 205–225.

[14] P. C. Loizou, *Signal processing techniques for cochlear implants - A review of progress in deriving electrical stimuli for the speech signal*, *IEEE Eng. Med. Biol. Mag.* **18** (1999) 34–46.

[15] S. Rosen, *Cochlear implants*, in *Scott-Brown's Otolarangology( 6th edition) Volume 2*, ed. D. Stephens (Butterworth Heinemann, Oxford, 1996).

[16] G. E. Loeb, *The functional replacement of the ear*, *Scientific American* **252** (1985) 86–92.

[17] R. P. Morse and E. F. Evans, *Enhancement of vowel coding for cochlear implants by addition of noise*, *Nature Med.* **2**, No. 8 (1996) 928–932.

[18] N. Y. S Kiang and E. C. Moxon, *Physiological consideration in artificial stimulation of the inner ear*, *Ann. Otol. Rhinol. Laryngol.* **81** (1972) 714–730.

[19] R. Hartmann, G. Topp and R. Klinke, *Discharge patterns of cat primary auditory nerve fibres with electrical stimulation of the cochlea*, *Hear. Res.* **13** (1984) 46–62.

[20] C. W. Parkins, *Temporal response patters of auditory nerve fibres to electrical stimulation in deafened squirrel monkeys*, *Hear. Res.* **41** (1989) 137–168.

[21] A. J. Matsuoka, P. J. Abbas, J. T. Rubinstein and C. A. Miller, *The neural response to electrical constant-amplitude pulse train stimulation: evoked compound action potential recordings*, *Hear. Res.* **149** (2000) 115–128.

[22] A. J. Matsuoka, P. J. Abbas, J. T. Rubinstein and C. A. Miller, *The neural response to electrical constant-amplitude pulse train stimulation: additive Gaussian noise*, *Hear. Res.* **149** (2000) 129–137.

[23] N. Y. S. Kiang, T. Watanabe, E. C. Thomas and L. F. Clark, *Discharge Patterns of Single Fibres in the Cat's Auditory Nerve*, (MIT Press, Cambridge, MA, 1995).

[24] M. C. Liberman and L. W. Dodds, *Single-neuron labelling and chronic cochlear pathology II. Stereocilia damage and alterations of spontaneous discharge rates*, *Hear. Res.* **16** (1984) 43–53.

[25] M. Chatterjee and M. E. Robert, *Noise enhances modulation sensitivity in cochlear implant listeners: stochastic resonance in a prosthetic sensory system?* *JARO* **2** (2001) 159–171.

[26] F. M. Reza, *An Introduction to Information Theory*, (Dover Publications, New York, 1994).

[27] R P Morse and E F Evans, *The sciatic nerve of the toad* Xenopus laevis *as a physiological model of the human cochlear nerve*, submitted.

[28] D. Allingham, N. G. Stocks and R. P. Morse, *A model of the electrically stimulated cochlear nerve based on physiological studies of the toad* Xenopus laevis, to be submitted to *IEEE Tran. on Biomed. Eng.*

[29] A. A. Verveen, *Axon diameter and fluctuation in excitability*, *Acta Morphologica* **5**, (1962) 79–85.

[30] I. C. Bruce, M. W. White, L. S. Irlicht, S. J. O'Leary, S. Dynes and G. M. Clark, *A stochastic model of the electrically stimulated auditory nerve: A single-pulse response*, *IEEE Trans. Biomed. Eng.* **46** (1999) 617–629.

[31] H. Spoendlin and A. Schrott, *Analysis of the human auditory nerve*, *Hear. Res.* **43** (1989) 25–38.

# NOISE EFFECTS ON THE ELECTROSENSE-MEDIATED FEEDING BEHAVIOR OF SMALL PADDLEFISH

DAVID F. RUSSELL, ANNA TUCKER, BARBARA A. WETTRING, ALEXANDER B. NEIMAN, LON WILKENS, and FRANK MOSS

*Center for Neurodynamics, Univ. of Missouri at St. Louis, 63121-4499, USA*
*drussell@admiral.umsl.edu, neiman@neurodyn.umsl.edu, mossf@umsl.edu*

Received 14 February 2001
Revised 15 May 2001
Accepted 30 May 2001

Weak electrical noise applied in the water around small paddlefish, *Polyodon spathula*, increases the spatial range over which they can detect and capture planktonic prey (*Daphnia*), demonstrating stochastic resonance at the level of an animal's feeding behavior. Here we show that optimal-amplitude (~0.5 $\mu V \cdot cm^{-1}$) noise causes a fish to prefer more vertical angles of attack when striking at prey, as revealed in polar graphs. Increased spatial range is also seen in horizontal directions, as outlying shoulders in the probability distribution of horizontal strike distances. High levels of noise increased the distance that approaching prey travelled along the rostrum (an elongated appendage anterior to the head, functioning as an electrosensitive antenna), before the fish first showed a visible fin or body motion in response. There was no significant effect of optimal-amplitude noise on the rate of strikes, although high-amplitude noise reduced the strike rate. The behavioral data were confirmed in neurophysiological experiments demonstrating that stochastic resonance occurs in individual electroreceptors, and in fact occurs at a similar optimal noise level as in behavioral experiments. We conclude that stochastic resonance can be demonstrated in the behavior of animals, and that animals can make use of the increased sensory information available during near-threshold environmental noise.

*Keywords:* Stochastic resonance; External noise; Neuroscience; Foraging behavior; Electroreceptor.

## 1. Introduction

Stochastic resonance (SR) [1–5], as applied to (biological) sensory nervous systems, has mainly been demonstrated at the level of peripheral sensory receptors. At an individual sensory receptor, SR can affect the transduction of stimulus information into a time series of action potentials, travelling to the central nervous system. This has been shown for several types of peripheral mechanoreceptors, including mechanoreceptive hairs on the tailfan of crayfish for near-field hydrodynamic stimuli (weak water waves) [6], mechanoreceptors on the tail of crickets for sensing vibrations and air currents [7], touch receptors in the skin of rats [8, 9], and muscle spindle proprioceptors in humans [10]. SR has also been demonstrated in auditory hair cells of frogs [11]. In all of these cases, the firing activity of a single sensory receptor was recorded while applying a weak external signal along with external noise. The only variable was the r.m.s. amplitude of the external noise. Computer analysis of the resulting spike train or membrane currents then revealed stochastic resonance as a local maximum in a graph of signal-to-noise ratio (SNR), or other information measure, as a function of the r.m.s. noise amplitude. The noise level at which the local maximum occurs is termed "optimal noise". Higher levels of noise degrade the transduction and SNR. A longstanding criticism of such research has been the lack of evidence that living creatures actually utilize the increased sensory information available from stochastic resonance. This point has been addressed in psychophysical experiments on humans [12, 13].

Recent experiments on the feeding behavior of paddlefish have answered this criticism [14]. Small paddlefish (<20 cm) use their passive electrosense to locate and capture individual zooplankton prey such as *Daphnia*. The juveniles feed in a "particulate" manner, that is, by capturing zooplankton one-by-one. This presents an advantageous system for quantifying the role of sensory noise in an animal's behavior, measured from the ability to spatially locate prey.

Paddlefish have tens of thousands of electroreceptors along the "rostrum", a long appendage projecting in front the head, organized into four linear arrays. Additional arrays are found on the gill covers, the top of the head, and around the mouth [15]. The electroreceptors are of the ampullary type [16, 17]. They can indeed detect the weak electrical signals emitted by *Daphnia* prey [18], the preferred food for paddlefish. These crustacean zooplankton are only 1–3 mm long, but are consumed in large numbers. The signals from *Daphnia* include weak AC oscillatory signals of 5–50 μV peak-peak amplitude, arising from swimming motions of the antennae or rhythmical beating motions of the feeding legs [14, 18, 19]. *Daphnia* also produce large (50–200 μV) one-shot electrical transients in the nearby water, due to a continuous electrical current between the head and tail, forming an electric dipole, whose vector points in different directions as a *Daphnia* moves or reorients in the water [19]. Which of these signals from *Daphnia* a paddlefish pays attention to during feeding remains unclear. However it is clear that small paddlefish will turn acrobatically to easily capture single *Daphnia* even in the absence of visible light, e.g. when viewed by infrared video, or when other possible sensory cues such as hydrodynamic signals or chemicals from prey are masked, indicating that electrosense is the primary sensory modality employed to locate prey [18, 20].

As shown previously [14], presentation of an optimal amplitude of noise will increase the spatial range over which small paddlefish can detect and strike at *Daphnia* prey, a phenomenon we have termed "behavioral stochastic resonance". Increased sensory information, as measured by the Fisher information, is used by fish for feeding during presentation of an optimal amplitude of environmental noise [21]. Theoretical models of the noise produced by populations of *Daphnia* have been developed [22]. The present article presents additional experimental data on how noise affects the feeding behavior. Also, SR is demonstrated in single paddlefish electroreceptors, and then compared to the behavioral results.

## 2. Experimental Methods

Individual small (18–20 cm) paddlefish were videotaped while capturing *Daphnia* prey one-by-one, swimming in the viewing chamber of a recirculating stream of water ("swim mill", Fig. 1), located in a closed dark temperature-regulated environmental room [14, 18, 20, 23]. Near-infrared illumination, invisible to these fish [24], was from arrays of light-emitting diodes. Water passed through a flow collimator before entering the viewing channel, with nearly laminar flow. Feeding behavior was videotaped by combining the signals from two CCD cameras, aimed at the side and bottom (via a mirror) of the viewing channel in which the fish swam. In offline analysis, feeding events were located, and then the videotape was backed up until the captured plankton was in a vertical plane aligned with the tip of the fish's rostrum, chosen to be far enough forward that the fish had not yet reacted usually. The criterion for data acceptance of a feeding event was that a fish had to be swimming approximately straight ahead into the water flow, not turned in any direction by more than 15°, such that the fish and prey followed nearly parallel paths, as required for valid measurement of the distance between the *Daphnia* and the midline long axis of the rostrum. With the videotape stopped, the selected frame was digitized, and then measured using software cursors. The side-view camera image gave the vertical distance to the approaching *Daphnia*, while the bottom-view camera gave the horizontal (side-to-side) distance. Distances were corrected <2% for perspective ("parallax") error.

"Strikes" were defined as attempts by a fish to capture a nearby *Daphnia*, during which the fish would open its mouth wide, expand the head, and lunge at the approaching prey. Fish rarely performed such motions except during attempts to capture prey. Most strikes did culminate in prey capture. However, some unsuccessful attempts were also included in the analysis, because we were primarily interested in the detection of prey, which the fish had clearly achieved. The fish lunged in the correct direction, but would fail to engulf the prey due to vagaries of water flow around the mouth during the final approach. Hence we use the term "strike" rather than "capture".

Electrical noise was applied between two large Ag plate electrodes, chlorided by immersion in bleach, and coated with agarose to stabilize the surface, placed as far as possible in front of the fish and behind it, 95 cm apart (Fig. 1). Water conductivity was adjusted to be 760 ± 10 $\mu S \cdot cm^{-1}$ by adding NaCl or purified water, as measured using an Orion (USA) model 115 meter, calibrated with a commercial standard solution; this conductivity is in the normal range for paddlefish [24]. The resistance between the plate electrodes was measured as 1200–2000 Ω, arising partly from the water, and partly from the surface properties of the Ag/AgCl plates. A third Ag/AgCl plate electrode in the water led to ground. The noise was from a vacuum tube source (Fig. 1). Its output was

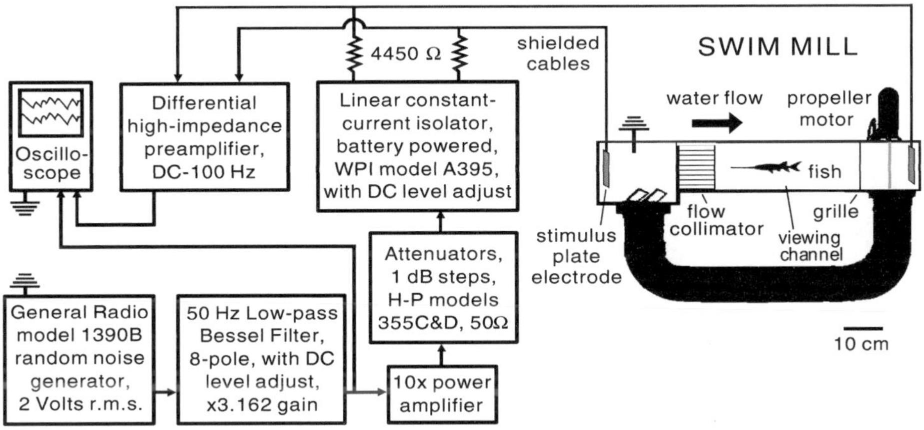


Fig. 1. Instrumentation for presenting electrical noise in the water around a small paddlefish, feeding on plankton in a recirculating stream (swim mill), drawn to scale, abstracted from a photo. The viewing channel in which the fish swam between the flow collimator and the grille near the propeller was photographed by two video cameras.

low-pass filtered at 50 Hz, amplified at low impedance, and attenuated to the selected amplitude. It then commanded a battery-powered linear constant-current isolator, having ±75 V compliance, which was factory-modified to generate a ±10 μA range of output current (to the plate electrodes) for a ±10 V range of command input. The internal noise of the isolator was low, measured as 0.42 nA (r.m.s.) for the 0.1–100 Hz frequency band. The 4.45 kΩ series resistors served to boost the ouput voltage excusion and SNR of the constant-current isolator. Long shielded cables went into the environmental room, to the plate electrodes in the swim mill. During experiments, we monitored on an oscilloscope the noise signal falling across the resistance of the water, i.e. the actual noise voltage gradient being presented to the fish, using a high impedance differential amplifier connected between the stimulus plate electrodes. This also revealed any 60 Hz interference from the power mains, which was minimized by unplugging other nearby equipment during experiments. Even the lowest amplitude of noise had an adequate SNR.

The r.m.s. amplitude of the applied noise, and the uniformity of the noise amplitude in different regions of the viewing channel in which the fish swam, were directly verified in a series of calibration tests using two focal pipet electrodes. Each consisted of a chlorided Ag wire immersed in water/agarose, sealed inside a horizontal glass pipet having a 2 mm tip opening. One electrode was positioned where a fish would swim, and was movable in a vertical plane. The other electrode was fixed in the center of the channel, 30 cm downstream. The amplified signal recorded differentially between the pipet electrodes, due to the external noise current falling across the resistance presented by a 30 cm path through the water channel, was digitized and its r.m.s. amplitude computed. The noise amplitude varied less than 3% over the cross-section of the viewing channel, except for falling 9% in one corner.

The juvenile paddlefish were obtained once a year, usually in June, from a hatchery operated by the Missouri Dept. of Conservation. Fish were habituated to the swim mill in several "adaptation" sessions before use. A key requirement was to fill the swim mill with "conditioned" water from the holding aquarium. Some fish were better than others at feeding within the confines of the viewing channel, and ignoring unusual stimuli such as vibrations from the propeller motor. Inevitably we selected for cooperative fish. Nevertheless their feeding behavior was "not learned". Juveniles typically grew to ~18 cm by the time they were "adapted" and had recovered good health in our aquarium system. The most important parameter in these experiments was the healthy condition of the fish. Data were from n=4 fish. Experiments were replicated with fish from two different years.

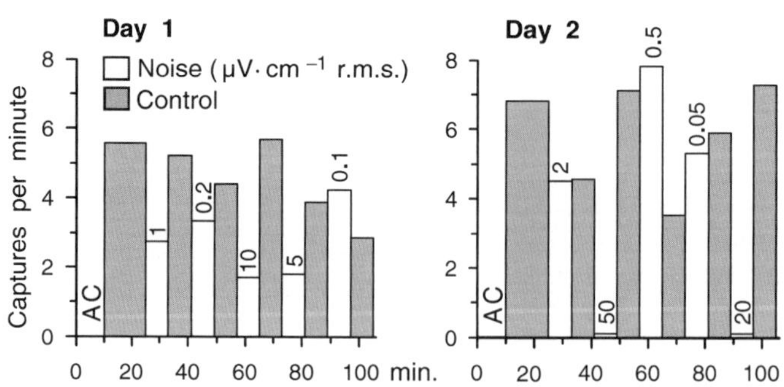


Fig. 2. Example of the experimental protocol. Time zero was when a fish was placed in the swim mill. AC: Acclimation time, when a fish was initially put in the swim mill, without *Daphnia* present.

The elapsed time on videotapes was used to keep track of the start and end of noise presentations, each 8 min in duration. The experimental protocol (Fig. 2) was to place one fish in the swim mill for 10 min, then add 50–60 *Daphnia* and allow the fish to feed for 15 min, adding additional *Daphnia* as needed to keep the density of plankton in the water approximately constant at ~2 $L^{-1}$. Hence the total habituation time on a given day was 25 min, allowing the fish to adjust to the swim mill before collecting data.

Then, noise of a selected amplitude was presented for 8 min, followed by control feeding (without noise) for 8 min. This alternation of noise and control epochs continued until the total feeding time (i.e., from time=10 min) exceeded 95 min, at which point the session was terminated to limit satiation. During data analysis, the number of *Daphnia* consumed during an 8 min noise presentation was always normalized to the number eaten during the subsequent 8 min control. This normalization procedure compensated for any nonstationarity over the course of the session due to changes in motivation, satiety, etc. The 8 min duration for noise or control epochs was chosen to be long enough that short-term variation in the feeding rate averaged out, yet short enough that up to five different noise amplitudes could be presented on a given day. To present an entire sequence of 10–12 noise amplitudes required 2–4 experimental sessions, each on a different day. Sometimes a session was terminated early if a fish was reluctant to feed.

The plankton density was monitored during feeding sessions using a specially constructed electronic particle counter, based on a linear array of photodetectors applied vertically to the screen of a TV monitor displaying an image of the swim mill from the side-view video camera. A *Daphnia* appeared as a bright dot against a black background, and as it flowed downstream, its image crossed the photodetector array and was counted. *Daphnia* were added remotely, in groups of 10 in 50 mL aliquots of water, via a small plastic tube going to the swim mill from an adjacent room. An overflow vent on the swim mill prevented its water level from rising as *Daphnia* were added repeatedly. *Daphnia* body lengths were controlled at 2.64 ± 0.18 mm by manually sorting them using mesh nets and pipets. Body lengths were measured from the head shield to the base of the caudal spine, from samples of *Daphnia* recovered from the swim mill after the feeding sessions, and preserved in formalin.

Between experiments, fish were returned to a large aquarium and fed daily, but then deprived of food for 24–48 h before the next experimental session. Experiments were run at night (20:00 to 01:00) since young paddlefish feed much more briskly at night than during the day [14]. For example, a fish that consumed 350–800 *Daphnia* in 95 min at night would eat only ~50 *Daphnia* in 95 min during the daytime.

As conventionally, the probability ($P$) values listed for statistical tests are $P(H_0)$, i.e. the probability of the null hypothesis $H_0$ that data from control and experimental conditions, or from two different experimental conditions, have the same distribution (are not different).

## 3. Experimental Results

The initial aim was to test the hypothesis that paddlefish use electrosense to locate prey, by showing that a fish's ability to locate prey is degraded during presentations of environmental electrical noise, applied between large plate electrodes, in front of the fish and behind it (Fig. 1). The amplitude spectrum of the low-frequency noise [14], low-pass filtered at 50 Hz, matched the 0.1–30 Hz bandwidth of paddlefish electroreceptors [18]. As expected, feeding did slow or stop during high-amplitude noise, 5–50 $\mu V \cdot cm^{-1}$ r.m.s. However, it was noticed that the feeding rate for some fish actually increased during presentations of weak-amplitude noise, ~1 μV·cm$^{-1}$, compared to controls without noise, suggesting stochastic resonance.

### 3.1 *Effects of external noise on strike distance*

The goal was to measure any effects of noise on the spatial range over which paddlefish could detect and strike at prey. Initially this was attempted by measuring the radial distance to prey from the rostrum. However, such analysis gave one-tailed probability distributions, since the origin (the rostrum's midline long axis) is at zero radial distance. Most statistical methods are for two-tailed distributions. A suggestion by G.V. Welland that we analyze separately the vertical and horizontal components of strike distance proved useful, since this approach gave two-tailed probability distributions centered around the location of the fish, specifically the midline long axis of the rostrum. An increase in spatial range could then be measured as an increase in the variance or "width" of the probability distribution of strike distances, for the vertical or horizontal component.

The noise amplitudes used, 0.05–50 μV·cm$^{-1}$ r.m.s., were determined empirically. Since feeding nearly stopped during 50 μV·cm$^{-1}$ noise, higher levels were not used, while levels below 0.2 μV·cm$^{-1}$ had little effect. The different amplitudes were spaced regularly on a logarithmic scale, since the responses of sensory nervous systems are typically scaled to the logarithm of the stimulus amplitude.

#### 3.1.1 *Vertical component of strike distance during optimal noise*

During low-amplitude noise, 0.2–1 μV·cm$^{-1}$, the probability distribution for the vertical component of spatial range changed shape: the peak shifted ~10 cm downward, and the shape became asymmetrical, and wider (Fig. 3A), even though the mean or median shifted downward only slightly [14]. That is, the fish detected prey farther away, along the vertical axis, and preferred the region below the rostrum. This effect was maximal for 0.5 μV·cm$^{-1}$ noise, which we therefore define as "optimal noise". Lower levels of noise, 0.05–0.1 μV·cm$^{-1}$, had no consistent effect (that is, gave noise/control ratios approximately equal to 1).

The increased vertical spatial variance during presentations of optimal (0.5 μV·cm$^{-1}$) noise, compared to controls (without noise) immediately afterwards, was confirmed using

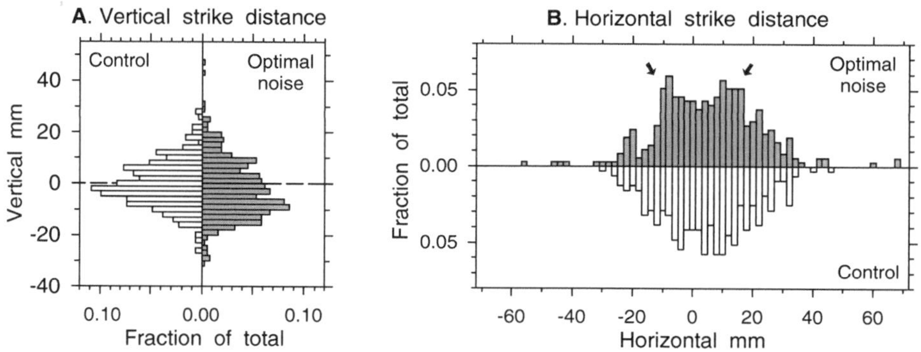


Fig. 3. Broadening of the probability distribution of fish-to-prey strike distances during optimal noise (0.5 μV·cm$^{-1}$ r.m.s.; solid bars), compared to controls without noise (clear bars). The vertical (**A**) and horizontal (**B**) components of fish-to-prey strike distances were analyzed separately. "Zero" distance represents the midline long axis of the paddlefish rostrum. The optimal noise data included 373 strikes from a total of 4 fish. Control data included 312 strikes from the same fish. Arrows, **B**: Outlying shoulders during optimal noise.

a nonparametric statistical test based on the summed variance of ranks, sometimes called the Mood or Mood-Hollander test for "scale", the nonparametric equivalent of "variance" [25]. $P < 0.0001$ by this test [14]. The noise and control data had similar median values, as required for this test [25], but to make them identical the values during noise were all translated by 0.48 mm.

### 3.1.2 *Horizontal component of strike distance during optimal noise*

In control data, the probability distribution of horizontal strike distances was always wider than the distribution of vertical strike distances. The effects of optimal noise on the horizontal components of fish-to-prey strike distances, compared to "subsequent control" data, were not significant by the Mood test for scale, nor by a Kolmogorov-Smirnov (K-S) two-sample test ($P = 0.94$). Nevertheless, there was a shape change in the probability distribution of horizontal strike distances during 0.5 $\mu V{\cdot}cm^{-1}$ noise, seen as splitting of the central peak of the distribution to have a distinct left and right shoulder (arrows, Fig. 3B).

### 3.1.3 *Strike angles during optimal noise*

Does optimal noise affect the preferred angle of attack for prey capture? This was examined by comparing polar plots of the angles to approaching prey, in a transverse plane relative to the rostrum, for strikes during optimal noise or controls (Fig. 4A). Data were grouped from the left and right sides, and plotted as 0° = horizontal, 90° = vertical above the rostrum, –90° = vertical below the rostrum. It can be seen that angles around 0° were preferred in the control data (Fig. 4A, right side), whereas more vertical angles were preferred during optimal noise (Fig. 4A, left side). That is, optimal noise caused the fish to make increased use of attack angles in the range from 60° below to 30° above the rostrum. This is seen as a region of steeper slope in the cumulative probability distribution of strike angles during optimal noise (arrow, Fig. 4B), which diverged significantly from the control curve ($P = 0.014$, K-S test). Both curves were significantly different from a uniform distribution ($P < 0.001$, K-S test).

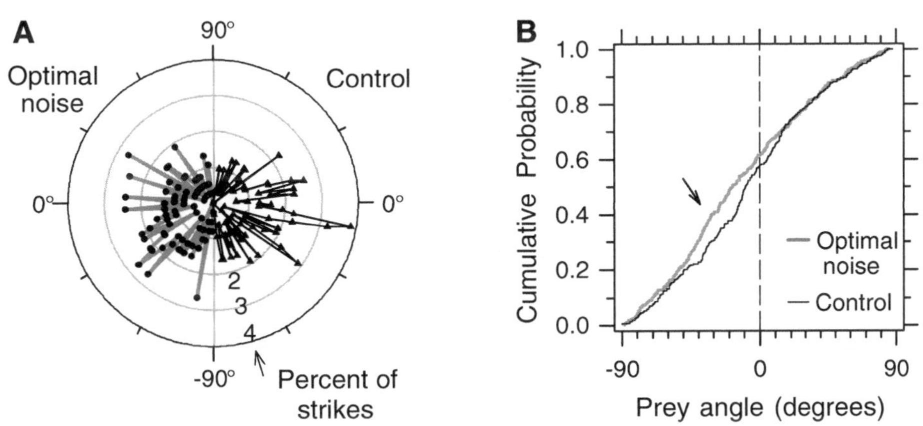


Fig. 4. Strike angles in a transverse plane, to approaching *Daphnia* which the fish eventually attacked. 0° = horizontal, 90° = vertical above the rostrum, –90° = vertical below the rostrum. Data were grouped from the left and right sides. **A**: Polar graph from the perspective of a fish "looking" forward; the center of the graph corresponds to the midline long axis of the rostrum. **Left half**: Strikes during presentations of optimal noise. **Right half**: Control. **B**: Cumulative probability as a function of strike angle, for optimal noise (373 strikes) and control data (312 strikes).

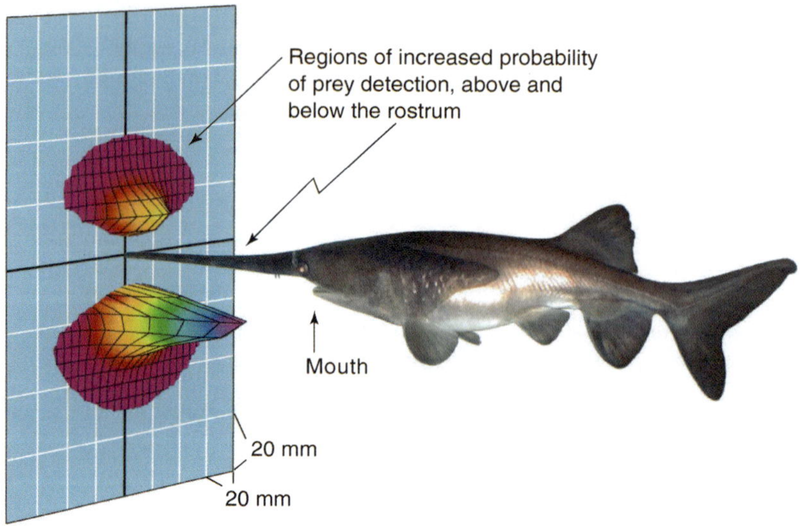


Fig. 5. Regions of increased strike probability during optimal noise, in relation to the rostrum of a paddlefish. Regions of increased strike probability were calculated by subtracting the spatial probability distribution for control data, in a transverse plane at the tip of the rostrum (where data measurements were made), from the distribution during presentations of optimal noise. Only positive differences are shown. P = +0.0043 at the maximum below the rostrum. An interpolated mesh surface was fitted to the data points.

### 3.1.4 *Regions of increased strike probability during optimal noise*

To visualize the regions in space around the rostrum where optimal noise enhances the probability of prey detection, a spatial distribution with three dimensions (probability, horizontal distance, and vertical distance) was constructed for strikes during optimal

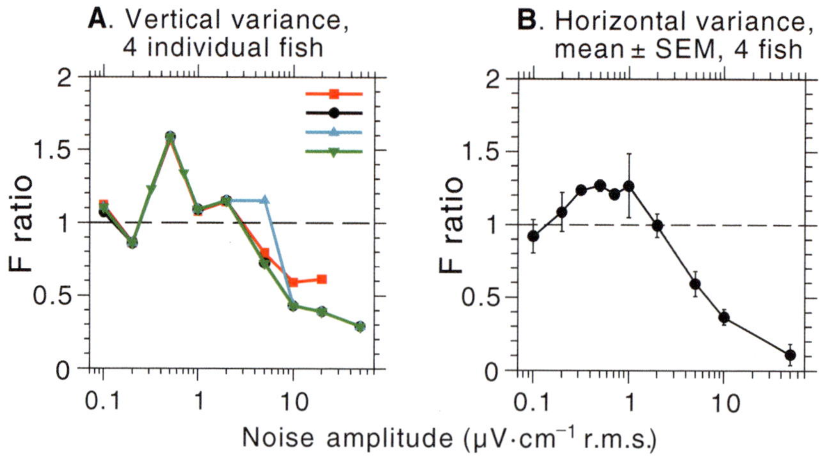


Fig. 6. Effects of different noise amplitudes on strike distances, quantified as noise/control variance ratios ($F$ ratios). For each fish, data were pooled from two runs (series of noise amplitudes). However data at 0.315 and 0.708 $\mu V \cdot cm^{-1}$ noise were from only one fish. **A**: Vertical component of strike distances, for individual fish (different colors). At each noise amplitude, the variance during noise was divided by the variance during the immediately subsequent control. **B**: Horizontal components of strike distances. SEM: standard error of the mean bars show variation between the four fish. The error bar at 0.5 $\mu V \cdot cm^{-1}$ noise is too small to be visible.

noise, and then the corresponding control distribution was subtracted from it (Fig. 5). The graph reveals two regions of increased probability of strikes during optimal noise, one above the rostrum, and a larger region below the rostrum. The main effect of optimal noise was to increase the vertical range of strike distances.

### 3.1.5 *Strike distances at different noise amplitudes*

As seen in Fig. 6, high amplitudes of noise, 10–50 μV·cm$^{-1}$ r.m.s., reduced both the vertical and the horizontal spatial range, such that only nearby plankton were captured [14]. Moderate noise amplitudes of ~2 μV·cm$^{-1}$ had little effect on the spatial range. As already described, 0.3–0.7 μV·cm$^{-1}$ noise increased the vertical component of strike distances (Fig. 6A), significantly so at 0.5 μV·cm$^{-1}$ noise ($P < 0.0001$, Mood test). The results from the four individual fish were remarkably consistent (Fig. 6A). For one of the fish, additional noise levels at 0.315 and 0.708 μV·cm$^{-1}$ were tested to explore the limits of the optimal range of noise amplitudes. The horizontal variance was also elevated at 0.3–0.7 μV·cm$^{-1}$ noise (Fig. 6B). The lowest noise levels, 0.2 μV·cm$^{-1}$ or less, had little effect. Hence the "optimal" range was quite narrow, 0.2–1 μV·cm$^{-1}$, with the largest effects at 0.3–0.7 μV·cm$^{-1}$.

## 3.2 ***Rate of strikes***

The effects of different noise amplitudes on the strike rate paralleled the effects on strike distance. That is, the rate declined at high levels of noise, to almost zero at 50 μV·cm$^{-1}$. In two of the fish, 0.2–1 μV·cm$^{-1}$ noise did cause the strike rate to increase (Fig. 7A), as much as 2.5-fold in one run. However in the other two fish, optimal noise had little effect on the normalized strike rate ratio, which was close to 1. Over the sample of four fish, there was no statistically significant effect of 0.5 μV·cm$^{-1}$ "optimal" noise on the mean strike rate (Fig. 7B), despite the small peak there.

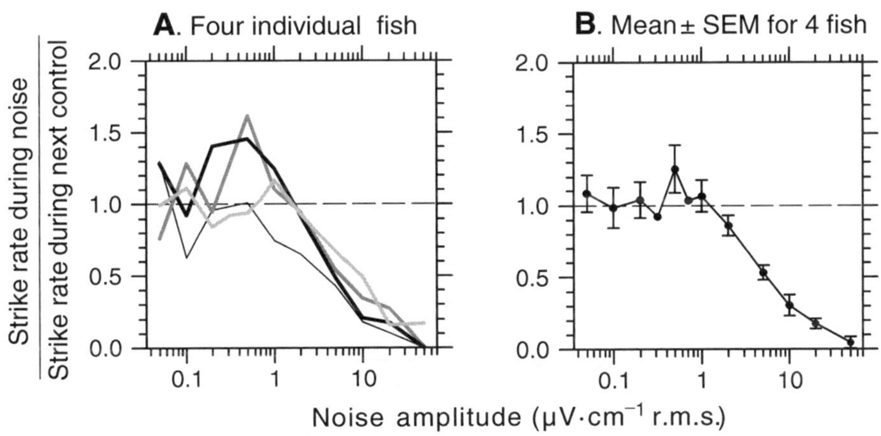


Fig. 7. The strike rate was not significantly affected by optimal noise. **A**: Normalized rates at different levels of noise, for 4 individual fish (separate curves). Each curve was an average of 2 runs (complete series of noise amplitudes) per fish. Two of the fish showed an elevated strike rate during "optimal noise", but the other two fish did not. **B**: When all the data were averaged (from 4 fish, 2 runs per fish), the average strike rate for the group was not significantly elevated during optimal noise (P [null hypothesis] = 0.182, analysis of variance with two replicates). Bars: standard error of the mean (SEM).

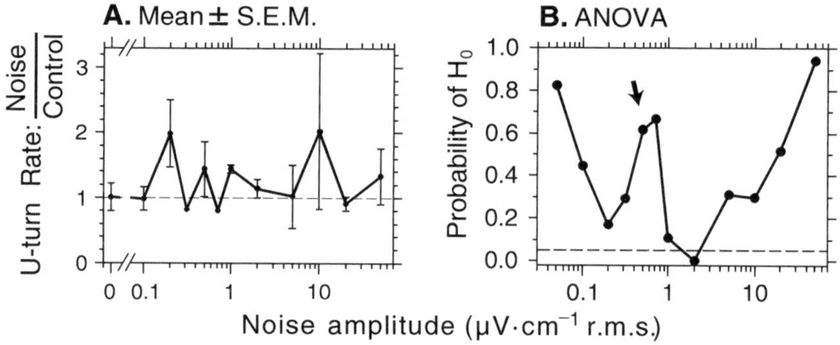


Fig. 8. U-turns were events when a fish turned to swim downstream. U-turns were used as an alternate type of behavior, for comparison to feeding behavior. **A**: The normalized average rate of U-turns for 4 fish appears to be independent of the noise amplitude. SEM: standard error of the mean bars. **B**: In the same data, an effect of optimal noise on U-turns was revealed by plotting the probability of the null hypothesis ($H_0$; see text) from a two-way analysis of variance (ANOVA) with two replicates (series of noise amplitudes) per fish, which revealed a local maximum (arrow).

### 3.3 *Did noise cause arousal?*

The increased feeding rates observed for certain fish during optimal noise could simply be due to excitation or arousal of the fish [26]. However the fish did not visibly react when any of the noise amplitudes were switched on: they did not show special fin motions or startle flexures. For example, even at the highest noise levels, the fish continued swimming normally when the noise was switched on, but simply stopped feeding. Of course the fish were probably already in a state of stress, due to netting them, disturbances during transport, and vibration from the swim mill, although we tried to minimize such arousing stimuli. As mentioned, we inevitably selected for fish which were cooperative and eager to feed under even novel circumstances.

As confirmation, the effects of noise on a different type of behavior, "U-turns", were also studied, to test whether noise might cause arousal. All fish occasionally performed U-turns, in which a fish would turn and swim downstream, eventually bump into the downstream grille, then return to its normal upstream position close behind the flow collimator (Fig. 1). U-turns appeared to signify a lapse in the motivation to feed. The frequency of such U-turns was scored during 8-min presentations of noise at different amplitudes, normalized to the rate during the immediately subsequent 8-min controls (Fig. 8A). In general, U-turns did not occur significantly more frequently at most levels of noise, including optimal noise or high levels, unlike what would be expected if noise were to cause arousal.

Nevertheless, an effect of optimal noise on the rate of U-turns could be discerned by plotting $P(H_0)$ (the probability of the null hypothesis that the U-turn rate during noise was the same as during the subsequent control), at different noise amplitudes (Fig. 8B). The curve crossed the $P = 0.05$ significance line only at 2 $\mu V \cdot cm^{-1}$, which the fish apparently finds significantly distracting. Interestingly, the graph shows a local maximum at "optimal" noise amplitudes (arrow). This presumably relates to questions of attention: the fish may mistakenly perceive the near-threshold fluctuations of external noise as signifying the presence of prey, and as a result become more motivated to feed, engaging less often in U-turns.

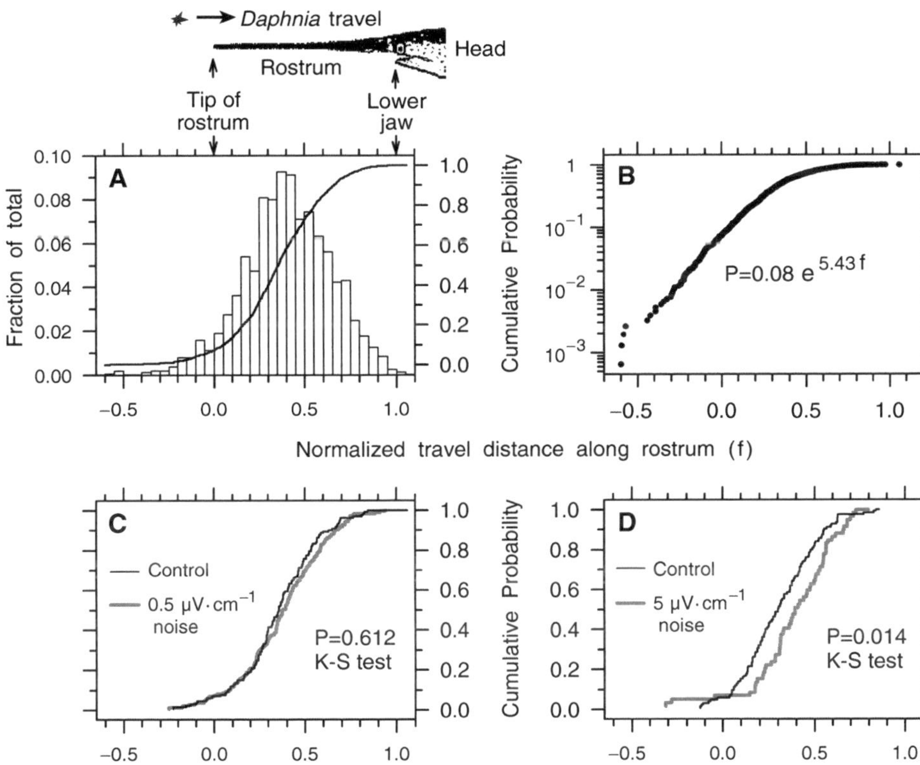


Fig. 9. Noise effects on the electrosensory integration time needed for detection of prey. The horizontal axis for each graph is the distance along the rostrum that approaching *Daphnia* had travelled when a fish first reacted (see text), normalized to the mean rostrum length (81 mm). **A**: Control data without noise, showing a histogram (left vertical axis) and cumulative probability distribution (right vertical axis) for 1536 *Daphnia* that the 4 fish eventually struck at. **B**: Semilog plot of the cumulative probability distribution from (A), revealing exponential growth of the response probability over the range of –0.44 to +0.3 rostrum length; the equation shown was fitted to this linear region. **C**: Optimal noise had little effect on the travel distance needed for prey detection: n=186 *Daphnia* during noise, n=125 for control. **D**: Higher-amplitude noise significantly delayed the detection of prey; n= 59 for noise, n=133 for control. A parabolic correction was applied to distances along the length of the swim mill's viewing channel, to compensate for optical distortion in which objects appeared longer at locations to the side (upstream or downstream) than in the center. K-S: Kolmogorov-Smirnov two-sample test.

### 3.4 *Effects of noise on reaction times*

Possible effects of external noise on the integration time (reaction time) for prey detection were characterized, to test the hypothesis that if optimal noise increases the available electrosensory information, by stochastic resonance, the fish might react sooner during optimal noise, compared to controls without noise. As an individual *Daphnia* approached a fish, the *Daphnia* traveled along a path approximately parallel to the long axis of the rostrum. As it traveled down the rostrum, at some point the fish responded with a fin or body motion. Most frequently the earliest visible motor response was to bring the pectoral fins forward, but in some cases the earliest response was arching of the spine.

In control data (without noise), 1536 such events were measured, in terms of the fraction (normalized) of the rostrum length that a *Daphnia* had travelled when a fish first showed an externally visible response movement (Fig. 9A). The *Daphnia* approached from upstream. The mean rostrum length of the four fish was 81 mm, defined here as the distance from the rostrum tip (normalized travel distance = 0.0) to the leading edge of the lower jaw (normalized travel distance = 1.0). Data were grouped for prey at all radial angles and distances. Half (50%) of the prey were detected by 0.35 rostrum lengths, the histogram peaked around 0.4 rostrum lengths, and most prey were detected by the time they reached the level of the mouth (Fig. 9A). Of course there may be even earlier responses internally, e.g of the heartbeat rate [27], but the present study was restricted to externally visible reactions of the fish. This control distribution was not significantly different from a Gaussian distribution with a mean of 0.35 rostrum lengths and a standard deviation of 0.24 rostrum lengths, fitted to the data ($P$ = 0.189, K-S test). In ~8% of strikes, the fish reacted before the prey reached the rostrum tip. Another view of the control data is that the probability of reacting increased exponentially as prey approached and passed the rostrum tip, i.e. over the range of –0.44 to 0.3 rostrum lengths (Fig. 9B). With a water velocity $v$ of 100 mm·s$^{-1}$, a mean rostrum length $m$ of 81 mm, and an exponential growth constant $b$ of 5.43 (from least-squares curve fitting; correlation coefficient $r$ = 0.999), an estimate of an "integration time" $t$ needed for detection of *Daphnia* can be calculated as $t = m / bv$ = 0.149 s. That is, the cumulative probability of detection increased $e$-fold every 149 ms or 14.9 mm as a *Daphnia* approached and travelled along the anterior third of the rostrum. This definition of "integration time" avoids having to decide at what point detection begins.

A few prey were not detected until after they had passed the mouth. Detection in this case is probably mediated by electroreceptors on the gill covers, which are quite long, serving as auxiliary antennas. In fact the length of the electroreceptor arrays on the head plus gill covers approximately equals the length of the arrays on the rostrum.

However, after extended analysis, we were not able to demonstrate a statistically significant effect of optimal noise on the integration time, since the "travel distance" curves for optimal noise presentations and the immediately subsequent controls were not distinguishable (Fig. 9C). This was also true when the data were restricted to strike locations directly above or below the rostrum, i.e. not to the sides. Higher-amplitude noise, 5 μV·cm$^{-1}$, did significantly retard the detection of *Daphnia* by 0.1 rostrum length, equivalent to 8.1 mm extra travel distance or 81 ms extra time required for prey detection (Fig. 9D). It was difficult to test statistically the effects of higher noise levels, 10–50 μV·cm$^{-1}$, because the number of strikes became small, but these higher noise levels also appeared to retard the detection of prey.

### 3.5 *Stochastic resonance in individual electroreceptors*

Do individual electroreceptors display stochastic resonance? To show this, neurophysiological experiments were conducted on immobilized fish, using a metal microelectrode to record the "single unit" spikes of an individual electrosensory neuron's axon (afferent) [18]. A weak 5 Hz sine wave stimulus of fixed amplitude was presented from a local dipole electrode, having a dipole axis length of 2 mm, placed near the receptive field of the afferent from the rostrum. The electroreceptors are most sensitive at ~5 Hz [18]. External noise of selected amplitude was presented simultaneously, between plate electrodes in front of the fish and behind it. The signal-to-noise ratio (SNR) of the response at 5 Hz was calculated from power spectra of the afferent spike trains [6].

Figs. 10A and B illustrate raw data for one electroreceptor (further methods are given in the legend). It showed an increased SNR for the 5 Hz weak sine wave stimulus when 0.33 μV·cm$^{-1}$ noise was presented simultaneously, as expected for stochastic resonance

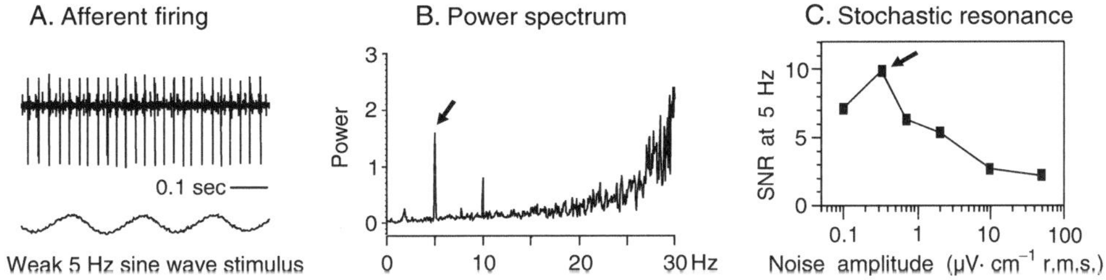


Fig. 10. Stochastic resonance in a single electroreceptor. **A**: The low-amplitude 5 Hz sine wave stimulus did not cause visible rate-modulation of the spontaneous firing of the afferent. **B**: The afferent response could be detected as a narrow peak at 5 Hz (arrow) in a power spectrum of the afferent spike time series. **C**: When external noise of selected r.m.s. amplitude was presented along with a 5 Hz sine wave, this receptor showed a local maximum (arrow) in the SNR at 5 Hz, and lower SNR at higher levels of noise.

(arrow, Fig. 10C). Higher levels of noise degraded the SNR. The "optimal noise" level for three electroreceptors, 0.33 $\mu V{\cdot}cm^{-1}$ (black line and left-side vertical axis, Fig. 11) was similar to the optimal noise level from behavioral experiments on feeding, 0.5 $\mu V{\cdot}cm^{-1}$ (gray line and right-side vertical axis, Fig. 11).

## 4. Discussion

The significance of this work is in showing that an animal can make use of the increased sensory information provided by stochastic resonance in sensory nervous systems, as demonstrated for a vital unlearned natural behavior in paddlefish, namely feeding.

The key hypothesis that weak electrical noise can increase a fish's electrosensitivity is formalized in Fig. 12, showing hypothetical distributions of fish-to-prey capture distances (abscissa). The fish would be at the origin; positive and negative signs would correspond to locations above and below the fish, respectively. The vertical axis is intended to be a measure of feeding behavior *H*, for example: $H$ = *(strike probability)* $*$ *(strike rate)*. This measure allows the area under the curve to change. Compared to the control distribution without noise (Fig. 12A), we expected a wider distribution, with capture of outlying prey at larger distances, if optimal-amplitude noise were to induce stochastic resonance, since SR would increase the sensitivity of electroreceptors, and allow a fish to detect the near-

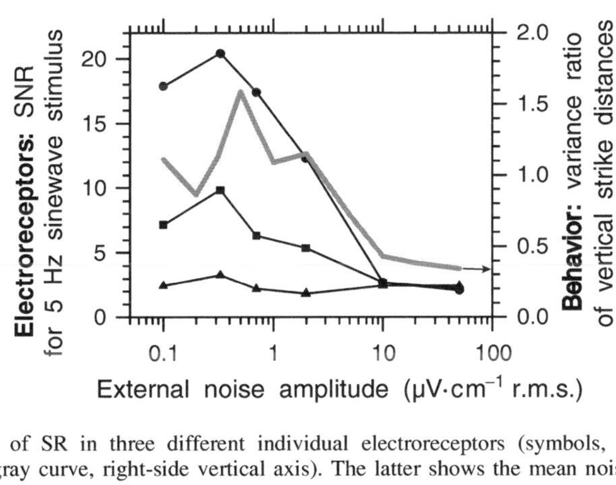


Fig. 11. Comparison of SR in three different individual electroreceptors (symbols, left vertical axis) to behavioral SR data (gray curve, right-side vertical axis). The latter shows the mean noise/control ratio of the vertical variance of strike distances (see Fig. 6A). SR occurs at similar amplitudes of noise in both cases.

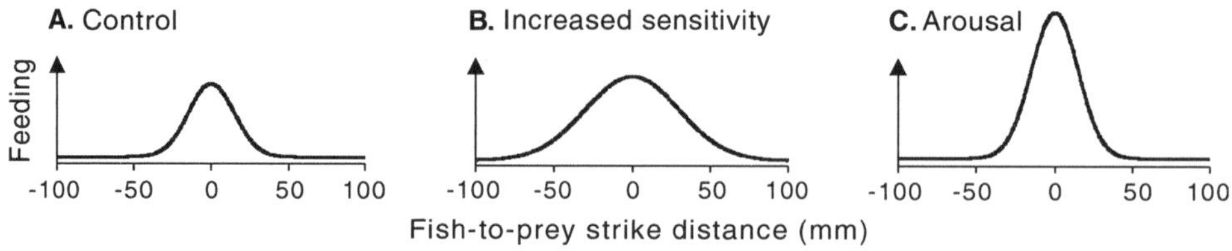


Fig. 12. Three hypothetical distributions of fish-to-prey strike distances.

threshold signals from more distant prey. That is, we expected increased variance of the spatial distribution (Fig. 12B). An alternate hypothesis, that noise might simply excite the fish ("arousal"; Fig. 12C), without altering its sensitivity, predicts that the spatial variance would remain the same as in Fig. 12A, because the fish would not be able to detect prey farther away. However the rate of capture (number of plankton captured per minute) might be higher, because the fish might be aroused, shown as increased area under the curve in Fig. 12C. Our experimental data conform to model B, with increased variance during optimal noise, indicating that the fish become more sensitive during presentations of external optimal-amplitude noise.

The inconsistent effect of optimal noise on the rate of strikes (Fig. 7) allows us to exclude arousal as a mechanism. However Fig. 7 raises the question of what "behavioral stochastic resonance" accomplishes, since on average the fish did not capture significantly more plankton per unit of time during optimal-amplitude noise. From viewing the videotapes of feeding behavior, it was apparent that fish would often ignore prey which were in easy strike range. That is, the strike rate appeared to be mainly affected by a fish's hunger level and the time of day, more than by prey availability. Under the experimental conditions used, the prey concentration was held approximately constant at a low value of ~2 *Daphnia* per liter [28]. The photocounter that we used showed that 60–100 *Daphnia* passed the fish per minute, given a water velocity of ~100 mm·s$^{-1}$. There were abundant opportunities to strike at prey, well beyond the ~6 captures per minute observed (Fig. 2). Hence the strike rate may have been set by an internal clock. However, under natural conditions in the wild, it may instead be the low availability of prey which limits the capture rate. Behavioral stochastic resonance may contribute significantly to feeding success and the capture rate under conditions of low prey availability.

Electroreceptor SR occurs over the same range of noise amplitudes as behavioral SR, 0.3–0.7 μV·cm$^{-1}$ (Fig. 11). This range of r.m.s. noise amplitude corresponds well to values reported for the limit of sensitivity of ampullary-type electroreceptors in freshwater fish, approximately 1 μV·cm$^{-1}$ [16, 29]. That is, our experiments demonstrate that SR in paddlefish electrosense is a near-threshold phenomenon, as expected theoretically [1–4]. This argument can be inverted to suggest that SR could be used as a general criterion for defining the "threshold" stimulus amplitude of sensory receptors [14]. The data in Fig. 11 identify the peripheral electroreceptors as one site where SR occurs. For the three receptors, the average increase in SNR during optimal noise was 33%, relative to responses at adjacent noise levels. However SR might also occur at higher levels of information processing in the electrosensory nervous system, in the brain.

The effects of synaptic noise on neuronal spike timing have been modelled extensively [30]. Harvey and Bruner modelled the spike generation zone of ampullary electroreceptors, with injections of "synaptic noise", and concluded that the minimum resolvable change of a stimulus increases in direct proportion to the r.m.s. amplitude of the noise [31].

Why do paddlefish electroreceptors, as well as the electrosense-mediated feeding behavior, still respond to external noise? It would instead seem advantageous to incorporate into electroreceptors an internal source of noise to induce SR. Internal noise has the advantage of making the receptor sensitivity less susceptible to the vagaries of external noise amplitude. For example, the auditory system of frogs has been shown to be already optimized for stochastic resonance using "internal noise", that is, arising from sources internal to the receptor system [11]. Fossil paddlefish date from the Cretaceous period [32]. With such a long evolutionary history, 65 x $10^6$ years, it can be presumed that present-day paddlefish have become optimized by natural selection, especially for vital functions like feeding, requiring prey detection. Paddlefish may have retained a dependence on external noise because swarms of planktonic prey produce electrical noise [14]. A fish may be able to adjust the amplitude of the ambient electrical noise it encounters by swimming around the edges of a plankton swarm [22]. Paddlefish readily swim in circles [24], and are known anecdotally to circle around patches of plankton.

We have shown recently that paddlefish electroreceptors have internal stochastic ~25 Hz oscillators [33], and have suggested that they may serve as an internal source of harmonic noise to induce SR, among other possible functions. Nevertheless, external noise of near-threshold amplitude can also induce SR in the electroreceptors (Figs. 10, 11). Further studies are needed to define the respective roles in electroreception of internal and external noise.

### Acknowledgements

Supported by the U.S. Office of Naval Research - Physics Division (F.M.), the Whitehall Foundation (D.R.), and the Fetzer Institute (A.N.). Eva Wagner participated in preliminary experiments on sensory modalities. We thank Wayne Garver for designing and constructing the photocounter for plankton.

### References

[1] K. Wiesenfeld and F. Moss, *Stochastic resonance and the benefits ofnoise: from ice ages to crayfish and SQUIDs, Nature* **373** (1995) 33–36.

[2] A. Bulsara and L. Gammaitoni, *Tuning in to noise, Physics Today* **49** (1996) 39–45.

[3] L. Gammaitoni, P. Hanggi, P. Jung, and F. Marchesoni, *Stochastic resonance, Rev. Mod. Phys.* **70** (1998) 223–288.

[4] V. S. Anishchenko, A. B. Neiman, F. Moss and L. Schimansky-Geier, *Stochastic resonance: noise-enhanced order, Physics - Uspekhi* **42** (1999) 7–36. (available from: www.ufn.ru)

[5] S.M. Bezrukov and I. Vodyanoy, *Noise-induced enhancement of signal transduction across voltage-dependent ion channels, Nature* **378** (1995) 362–365.

[6] J. K. Douglass, L. Wilkens, E. Pantazelou and F. Moss, *Noise enhancement of information transfer in crayfish mechanoreceptors by stochastic resonance, Nature* **365** (1993) 337–304.

[7] J. E. Levin and J. P. Miller, *Broadband neural encoding in the cricket cercal sensory system enhanced by stochastic resonance, Nature* **380** (1996) 165–168.

[8] J. J. Collins, T. T. Imhoff and P. Grigg, *Noise-enhanced information transmission in rat SA1 cutaneous mechanoreceptors via aperiodic stochastic resonance, J. Neurophysiol.* **76** (1996) 642–645.

[9] C. Ivey, A. V. Apkarian and D. Chialvo, *Noise-induced tuning curve changes in mechanoreceptors, J. Neurophysiol.* **79** (1998) 1879–1890.

[10] P. Cordo, J. T. Inglis, S. Verschueren, J. J. Collins, D. M. Merfeld, S. Rosenblum, S. Buckley and F. Moss, *Noise in human muscle spindles, Nature* **383** (1996) 769–770.
[11] F. Jaramillo and K. Wiesenfeld, *Mechanoelectrical transduction assisted by Brownian motion: a role for noise in the auditory system, Nature Neurosci.* **1** (1998) 384–388.
[12] J. J. Collins, T. T. Imhoff and P. Grigg, *Noise-enhanced tactile sensation, Nature* **383** (1996) 770.
[13] E. Simonotto, M. Riani, C. Seife, M. Roberts, J. Twitty and F. Moss, *Visual perception of stochastic resonance, Phys. Rev. Letts.* **78** (1997) 1186–1189.
[14] D. F. Russell, L. A. Wilkens, F. Moss, *Use of behavioral stochastic resonance by paddle fish for feeding, Nature* **402** (1999) 291–294.
[15] J. M. Jørgensen, Å. Flock and J. Wersäll, *The Lorenzinian ampullae of Polyodon spathula, Z. Zellforsch.* **130** (1972) 362–377.
[16] T. H. Bullock, *Electroreception, Ann. Rev. Neurosci.* **5** (1982) 121–170.
[17] H. H. Zakon, *The electroreceptive periphery*, in Electroreception, ed. T. H. Bullock and W. Heiligenberg. Wiley, New York (1986) 103–156.
[18] L. A. Wilkens, D. F. Russell, X. Pei and C. Gurgens, *The paddlefish rostrum functions as an electrosensory antenna in plankton feeding, Proc. Roy. Soc. Lond.* B **264** (1997) 1723–1729.
[19] A. B. Neiman, D. F. Russell, X. Pei, W. Wojtenek, J. Twitty, E. Simonotto, B. A. Wettring, E. Wagner, L. A. Wilkens and F. Moss, *Stochastic synchronization of electroreceptors in the paddlefish, Int. J. Bifur. Chaos* **10** (2000) 2499–2517.
[20] L. Wilkens, B. A. Wettring, E. Wagner, W. Wojtenek and D. F. Russell, *Prey detection in selective plankton feeding by the paddlefish: is the electric sense sufficient?, J. Exp. Biol.* **204** (2001) 1381–1389.
[21] P. E. Greenwood, L. M. Ward, D. F. Russell, A. Neiman and F. Moss, *Stochastic resonance enhances the electrosensory information available to paddlefish, Phys. Rev. Letts.* **84** (2000) 4773–4776.
[22] J. J. Freund, J. Kienert, L. Schimansky-Geier, B. Beisner, A. Neiman, D. F. Russell, T. Yakusheva and F. Moss, *Behavioral stochastic resonance: how a noisy army betrays its outpost, Phys. Rev. E* **63** (2001) 031910.
[23] W. W. Burggren and W. E. Bemis, *Metabolism and ram gill ventilation in juvenile paddlefish, Polyodon spathula (Chondrostei: Polyodontidae), Physiological Zool.* **65** (1992) 515–539.
[24] C. Gurgens, D. F. Russell and L. A. Wilkens, *Electrosensory avoidance of metal obstacles by the paddlefish, J. Fish Biol.* **57** (2000) 277–290.
[25] L. A. Marascuilo and M. McSweeney, *Nonparametric and Distribution-Free Methods for the Social Sciences*, Brooks/Cole, Monterey (1977).
[26] P. R. Laming, *The physiological basis of alert behaviour in fish*, in Brain Mechanisms of Behaviour in Lower Vertebrates, ed. P.R. Laming, Cambridge Univ. Press, Cambridge U.K. (1981) 203–222.
[27] A. J. Kalmijn, *Electro-perception in sharks and rays, Nature* **212** (1966) 1232–1233.
[28] R. J. H. Leong and C. P. O'Connell, *A laboratory study of particulate and filter feeding of the northern anchovy (Engraulis mordax), J. Fisheries Res. Bd. Canada* **26** (1969) 557–582.
[29] R. C. Peters and F. Wijland, *Electro-orientation in the passive electric catfish, Ictalurus nebulosus, J. Comp. Physiol.* **92** (1974) 273–280.
[30] R. B. Stein, *Some models of neuronal variability, Biophys. J.* **7** (1967) 37–68.
[31] J. R. Harvey and L. J. Bruner, *The spike generation zone of the ampullary electroreceptor. II. Oscillator period noise and the limits of sensitivity, Biol. Cybern.* **72** (1995) 379–387.
[32] L. Grande and W. E. Bemis, *Osteology and phylogenetic relationships of fossil and recent paddlefishes (Polyodontidae) with comments on the interrelationships of Acipenseriformes. J. Vert. Paleontol.* **11** (Memoir 1) (1991) i–121.
[33] A. Neiman and D. F. Russell, *Stochastic biperiodic oscillations in the electroreceptors of paddlefish, Phys. Rev. Letts.* **86** (2001) 3443–3446.

# STOCHASTIC RESONANCE AND OPTIMAL CLUSTERING FOR ASSEMBLIES OF ION CHANNELS

G. SCHMID, I. GOYCHUK and P. HÄNGGI
*Institut für Physik, Universität Augsburg*
*86135 Augsburg, Germany*

S. ZENG and P. JUNG
*Department of Physics and Astronomy and Quantitative Biology Institute, Ohio University*
*Athens, OH 45701*

Received 19 January 2004
Revised 23 February 2004
Accepted 23 February 2004

We consider the statistical properties of action potentials generated by clusters of sodium and potassium channels due to channel noise and/or external stimulation. Since the fluctuations are related to the cluster size, a size-resonance effect — analog to stochastic resonance — is observed that facilitates optimal decoding of small external stimuli at optimal cluster sizes. Furthermore, in analogy to the coherence resonance effect, the channel-noise induced firing patterns exhibit a resonant-like temporal coherence as a function of the cluster size even in the absence of a periodic stimulus. In the presence of additional synaptic noise, SR occurs only for large cluster sizes which possess suboptimal internal noise levels.

*Keywords*: Excitable membranes; ion channel clusters; channel noise; spiking rate; intrinsic coherence resonance; intrinsic stochastic resonance.

## 1. Introduction

Clustering of ion channels is a frequently observed phenomenon in nature. It occurs naturally in myelinated neurons, where the active sodium channels are concentrated at the *nodes of Ranvier* acting as signal boosters. But it also occurs in neurons that are *not myelinated*, e.g. in retinal neurons, where myelination would interfere with vision [1]. The mechanisms for clustering of ion channels and receptors are subject of current investigations in neurobiology and cell-biology. Some theories are based on attractive interactions between the channel proteins [2], others on the formation of micro-domains ("rafts") in the membrane with higher affinity for signaling proteins, and still others the involvement of the cytoskeleton anchoring the channels by a sub-membrane undercoat [1]. In this paper we review recent papers that explore the novel idea that ion channel clustering enhances neuronal information encoding [3, 4]. In these papers, it has been shown that small signals may be better detected by smaller clusters of ion channels. The essential idea is based on the theory of *stochastic resonance* [5, 6]. A subthreshold signal can be encoded by an excitable system when an optimal amount of noise is added to the signal. In the case of ion channel clusters, the fluctuations are generated by the thermal opening and closing of ion channels. The number of ion channels determines the amplitude of the fluctuations of the membrane potential. Thus instead of tuning the noise, one can tune the size, resulting in a system-size resonance first reported in [3,4,7] (see also [8,9]). While the principle idea is straightforward, there are some more subtle issues in transferring the concept of excitable systems and stochastic resonance to *small systems.* For large clusters, action potentials are generated when the membrane potential exceeds a threshold value. As the ion channel cluster becomes smaller, the threshold softens and becomes conceptually irrelevant. The threshold-crossing criterion is replaced by a criterion based on channel kinetics [10]. Second, the relation between size and noise is not simple and monotonous. The spontaneous spiking rate as a function of the cluster size exhibits a maximum. A sequence of detailed studies of this novel form of stochastic resonance, for the particular case of clustered sodium and potassium channels, is being reviewed in this paper. In Sec. 1 we will briefly introduce the stochastic Hodgkin–Huxley model. In Sec. 2 we will discuss and evaluate some of the methods of solution. In Sec. 3 we review previous results and present some novel results.

## 2. The Stochastic Hodgkin–Huxley Model

We adopt the classic model for the ion channels introduced by Hodgkin and Huxley that mimics the potassium channel by four identical gates that stochastically switch between an open state and a closed state. The probabilities $p_n$ for the four gates $n = 1, 2, 3, 4$ to be open are described by the rate equations

$$\dot{p}_n(t) = -\left(\alpha_{\mathrm{K}}(v) + \beta_{\mathrm{K}}(v)\right) p_n(t) + \alpha_{\mathrm{K}}(v)\,, \tag{1}$$

where $\alpha_{\mathrm{K}}(v)$ and $\beta_{\mathrm{K}}(v)$ are the membrane-voltage $v$ dependent opening and closing rates

$$\alpha_{\mathrm{K}}(v) = \frac{0.01(10 - v)}{\exp\left((10 - v)/10\right) - 1}\,, \quad \beta_{\mathrm{K}}(v) = 0.125 \exp\left(-\frac{v}{80}\right). \tag{2}$$

The trans-membrane voltage $v$ is measured here and in all equations below in mV with respect to the physiologic cellular resting potential of $-65$ mV. The potassium channel is open only when all four gates are open, i.e. with probability $p_1p_2p_3p_4$.

The sodium channel consists of four gates. Three identical fast gates increase their opening probability $q_n$, $n = 1, 2, 3$ when the voltage $v$ becomes larger than the resting potential. The slower fourth deactivation gate decreases its open probability $q_4$ when the membrane potential increases. The gate variables obey the following rate equations:

$$\begin{aligned} \dot{q}_n(t) &= -\left(\alpha^f_{\mathrm{Na}}(v) + \beta^f_{\mathrm{Na}}(v)\right) q_n(t) + \alpha^f_{\mathrm{Na}}(v)\,, \\ \dot{q}_4(t) &= -\left(\alpha^s_{\mathrm{Na}}(v) + \beta^s_{\mathrm{Na}}(v)\right) q_4(t) + \alpha^s_{\mathrm{Na}}(v)\,, \end{aligned} \tag{3}$$

with the opening and closing rates

$$\alpha^f_{\mathrm{Na}}(v) = \frac{0.1(25-v)}{\exp((25-v)/10)-1}, \quad \beta^f_{\mathrm{Na}}(v) = 4.0\exp\left(-\frac{v}{18}\right),$$

$$\alpha^s_{\mathrm{Na}}(v) = 0.07\exp\left(-\frac{v}{20}\right), \quad \beta^s_{\mathrm{Na}}(v) = \frac{1}{\exp((30-v)/10)+1}. \tag{4}$$

The membrane voltage is measured in mV with respect to the resting potential.

Although each individual ion channel opens and closes independently, the opening and closing rates (for the $n$ gate, they are $\alpha_n$ and $\beta_n$) are regulated by the same membrane potential. As a consequence, all ion channels are globally coupled through the membrane potential. For the density of the sodium and potassium channels (number of channels per area) we use $\rho_{\mathrm{Na}} = 60\,\mu\mathrm{m}^{-2}$ and $\rho_{\mathrm{K}} = 18\,\mu\mathrm{m}^{-2}$, respectively. The single-channel conductances of the sodium and potassium channels are given by $\gamma_{\mathrm{Na}} = \gamma_{\mathrm{K}} = 20\,\mathrm{pS}$. Using a membrane capacitance of $c = 1\,\mu\mathrm{F/cm}^2$ [11] we end up with the following equation for the membrane potential

$$\dot{v} = -\left(\frac{N^{\mathrm{open}}_{\mathrm{K}}}{\tau_{\mathrm{K}} N_{\mathrm{K}}}(v - v^{\mathrm{rev}}_{\mathrm{K}}) + \frac{N^{\mathrm{open}}_{\mathrm{Na}}}{\tau_{\mathrm{Na}} N_{\mathrm{Na}}}(v - v^{\mathrm{rev}}_{\mathrm{Na}}) + \frac{1}{\tau_L}(v - v_l)\right) + i_{\mathrm{ext}}(t), \tag{5}$$

where $I_{\mathrm{ext}} = c\, i_{\mathrm{ext}}(t)$ denotes an externally applied current per area of the membrane and $v^{\mathrm{rev}}_{\mathrm{K}} = -12\,\mathrm{mV}, v^{\mathrm{rev}}_{\mathrm{Na}} = 115\,\mathrm{mV}, v_l = 10.6\,\mathrm{mV}$ the reversal potentials of the potassium systems, sodium system and leakage system, respectively. The time constants are given by

$$\begin{aligned} \tau_{\mathrm{K}} &= \frac{1}{36}\,\mathrm{ms}, \\ \tau_{\mathrm{Na}} &= \frac{1}{120}\,\mathrm{ms}, \\ \tau_L &= 3.3\,\mathrm{ms}. \end{aligned} \tag{6}$$

The numbers of open $\mathrm{K}^+$ and $\mathrm{Na}^+$ channels, $N^{\mathrm{open}}_{\mathrm{K}}$ and $N^{\mathrm{open}}_{\mathrm{Na}}$, respectively, have to be determined as a function of time by stochastic simulations with methods described in the next section.

The time scale of synaptic noise is about one order of magnitude smaller than channel noise (see e.g. in [12]). Thus we can consider synaptic noise as Gaussian white noise $\xi_s(t)$ with

$$\begin{aligned} \langle \xi_s(t) \rangle &= 0, \\ \langle \xi_s(t)\xi_s(t') \rangle &= \sigma_s^2 \delta(t - t'), \end{aligned} \tag{7}$$

where $\sigma_s$ describes the strength of the synaptic noise. Since synaptic noise leads to elementary events in the synapse that are integrated, it has to be added to the right hand side of Eq. 5, i.e.

$$\dot{v} = -\left(\frac{N^{\mathrm{open}}_{\mathrm{K}}}{\tau_{\mathrm{K}} N_{\mathrm{K}}}(v - v^{\mathrm{rev}}_{\mathrm{K}}) + \frac{N^{\mathrm{open}}_{\mathrm{Na}}}{\tau_{\mathrm{Na}} N_{\mathrm{Na}}}(v - v^{\mathrm{rev}}_{\mathrm{Na}}) + \frac{1}{\tau_L}(v - v_l)\right) + i_{\mathrm{ext}}(t) + \xi_s(t). \tag{8}$$

### 2.1. *Tracking the number of open ion channels*

In order to integrate Eq. 8, the number of open potassium and sodium channels needs to be determined. The simplest and most accurate method is to simulate each gate of each channel under the assumption that the opening and closing of the gates is a Markov-process. This method is inefficient since many simulated transitions of gates do not change the state of the channel (open or closed) and thus have no impact on the membrane voltage. More efficient schemes are based on representing the cluster of ion channels by the occupation numbers of channel-states. The potassium channel can be in five different states, $n_{0,1,2,3,4}$, which are characterized by $0, 1, 2, 3, 4$ gates open. Given the opening and closing rates of each gate by $\alpha_{\mathrm{K}}$ and $\beta_{\mathrm{K}}$ (see Eqs. 2), transition rates between the four states can be easily obtained. The kinetic scheme of a potassium channel is shown in Fig. 1. Similarly, the sodium channel, is characterized by the kinetic scheme in Fig. 2. The difference to the potassium channel is the slow deactivation gate that leads to a more complicated kinetic scheme.

$$[n_0] \underset{\beta_n}{\overset{4\alpha_n}{\rightleftarrows}} [n_1] \underset{2\beta_n}{\overset{3\alpha_n}{\rightleftarrows}} [n_2] \underset{3\beta_n}{\overset{2\alpha_n}{\rightleftarrows}} [n_3] \underset{4\beta_n}{\overset{\alpha_n}{\rightleftarrows}} [n_4]$$

Fig. 1. Kinetic scheme for a stochastic potassium channel.

$$\begin{array}{ccccccc}
[m_0h_1] & \underset{\beta_m}{\overset{3\alpha_m}{\rightleftarrows}} & [m_1h_1] & \underset{2\beta_m}{\overset{2\alpha_m}{\rightleftarrows}} & [m_2h_1] & \underset{3\beta_m}{\overset{\alpha_m}{\rightleftarrows}} & [m_3h_1] \\
\alpha_h\uparrow\downarrow\beta_h & & \alpha_h\uparrow\downarrow\beta_h & & \alpha_h\uparrow\downarrow\beta_h & & \alpha_h\uparrow\downarrow\beta_h \\
[m_0h_0] & \underset{\beta_m}{\overset{3\alpha_m}{\rightleftarrows}} & [m_1h_0] & \underset{2\beta_m}{\overset{2\alpha_m}{\rightleftarrows}} & [m_2h_0] & \underset{3\beta_m}{\overset{\alpha_m}{\rightleftarrows}} & [m_3h_0]
\end{array}$$

Fig. 2. Kinetic scheme of a sodium channel.

The states of the sodium and potassium channels in the cluster are fully specified by the numbers of potassium channels $[n_k]$ in the states $n_k$ and the number of sodium channels $[m_ih_j]$ in the states $m_ih_j$. The number of activated potassium and sodium channels is given by $[n_4]$ and $[m_3h_1]$, respectively. Let, for example, the transition rate between the two states $S_1$ and $S_2$ be $\gamma_{S_1S_2}$ and the populations of these states be $[n_1]$ and $[n_2]$. Then, the probability $p$ that a channel switches within the time interval $(t, t+\delta t)$ from state $S_1$ to $S_2$ is given by $p = \gamma_{S_1S_2}\delta t$. The probability that, $[\delta n_{12}] = [n_1(t)] - [n_1(t+\delta t)] = [n_2(t+\delta t)] - [n_2(t)]$, channels switch from state $S_1$ to state $S_2$ in the same time interval satisfies the binomial distribution

$$P([\delta n_{12}]) = \binom{[n_1]}{[\delta n_{12}]} p^{[\delta n_{12}]}(1-p)^{([n_1]-[\delta n_{12}])} . \tag{9}$$

Thus the number of switching channels between the states is sequentially drawn from binomial distributions. If the cluster of channels is large i.e. $[n_1]$ is large, the number of switching channels is also large in the average. Thus, in the time interval $\delta t$ larger channel clusters experience more transitions.

In order to update the state of the population of ion channels with time, we have to create rules in what sequence the states are updated. The simple stochastic methods, described in the previous section, does not require such rules. In order to enforce positive occupation numbers we update the occupation numbers sequentially, starting with the process with the largest rate and so forth.

### 2.2. *Langevin approach*

Based on a the truncation of a Kramers–Moyal expansion of the master equations for single gates, Fox and Lu [13, 14] have derived the following set of Ito–Langevin equations for the gating variables $n, m$ and $h$ for large ion channel clusters (i.e. when the number of channels in the cluster is large)

$$\begin{aligned}\frac{d}{dt}n &= \alpha_n(1-n) - \beta_n n + \overline{g}_n(t)\,,\\ \frac{d}{dt}h &= \alpha_h(1-h) - \beta_h h + \overline{g}_h(t)\,,\\ \frac{d}{dt}m &= \alpha_m(1-m) - \beta_m m + \overline{g}_m(t)\,,\end{aligned} \tag{10}$$

where the variables $\overline{g}_n(t)$, $\overline{g}_h(t)$, $\overline{g}_m(t)$ denote Gaussian, zero-mean white noise with

$$\begin{aligned}\langle\overline{g}_n(t)\overline{g}_n(t')\rangle &= \frac{2}{N_{\mathrm{K}}}\frac{\alpha_n\,\beta_n}{\alpha_n+\beta_n}\,\delta(t-t')\,,\\ \langle\overline{g}_m(t)\overline{g}_m(t')\rangle &= \frac{2}{N_{\mathrm{Na}}}\frac{\alpha_m\,\beta_m}{\alpha_m+\beta_m}\,\delta(t-t')\,,\\ \langle\overline{g}_h(t)\overline{g}_h(t')\rangle &= \frac{2}{N_{\mathrm{Na}}}\frac{\alpha_h\,\beta_h}{\alpha_h+\beta_h}\,\delta(t-t')\,.\end{aligned} \tag{11}$$

Here $N_{\mathrm{K}}$ and $N_{\mathrm{Na}}$ denote the total number of potassium and sodium channels. It is necessary to include restriction to guarantee that $m$, $n$ and $h$ do not leave the unit interval [0, 1]. The differential equation for the membrane potential is the classic Hodgkin–Huxley equations where $m^3h$ determines the fraction of open sodium channels and $n^4$ the fraction of open potassium channels, i.e.

$$\dot{v} = -\left(\frac{1}{\tau_{\mathrm{K}}}n^4\,(v-v_{\mathrm{K}}^{\mathrm{rev}}) + \frac{1}{\tau_{\mathrm{Na}}}m^3h\,(v-v_{\mathrm{Na}}^{\mathrm{rev}}) + \frac{1}{\tau_L}\,(v-v_l)\right) + i_{\mathrm{ext}}(t) + \xi_s(t)\,. \tag{12}$$

Equations (10), (11), (12) have to be integrated numerically in order to predict a neuronal spike train with the spikes occurrences at $t_i$, $i = 0, ..., N$.

## 3. Results

### 3.1. *Spike rates*

While in the original, deterministic Hodgkin–Huxley model action potentials occur only for external current stimuli that exceed a threshold, the intrinsic channel noise can initiate spontaneous spikes even in the absence of external stimuli [15–19]. The

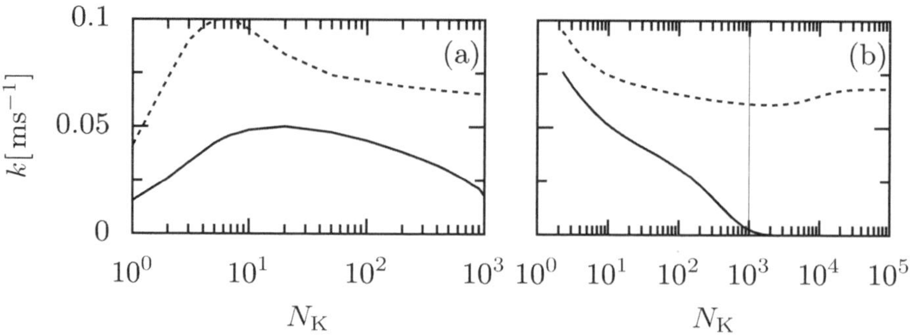


Fig. 3. The mean spiking rate $k$ which is given as the inverse of the mean interspike interval, cf. Eq. 13, is plotted against the number of potassium ion channels for $I_\mathrm{ext} = 0$ (solid lines) and $I_\mathrm{ext} = 10\,\mu\mathrm{A/cm}^2$ (dotted lines) using direct Markov simulations (a) and the Langevin approach (b). The vertical line in Fig. 3(b) corresponds to the maximal $N_\mathrm{K}$ in Fig. 3(a).

mean interspike interval, i.e.

$$\langle T \rangle = \lim_{N\to\infty} \frac{1}{N} \sum_{i=1}^{N} (t_i - t_{i-1}) \,, \tag{13}$$

becomes a function of the number of ion channels. The inverse mean interspike interval defines the spiking rate $k = 1/\langle T \rangle$. Fig. 3 shows the dependence of the spiking rate $k$ on the number of potassium ion channels $N_\mathrm{K}$ (the number of sodium ion channels $N_\mathrm{Na}$ is kept proportional, $N_\mathrm{Na} = \frac{60}{18} N_\mathrm{K}$, to the number of potassium ion channels). For $I_\mathrm{ext} = 10\,\mu\mathrm{A/cm}^2$ the kinetic scheme simulations and the Langevin approach shows pretty good agreement except for small clusters $N_\mathrm{K} \lesssim 10$. For vanishing external current ($I_\mathrm{ext} = 0$) the methods do not compare as well. Similar discrepancies between the master equation description and the corresponding Fokker–Planck approximation in a weak noise limit are known for a long time [20].

The spontaneous spiking rate $k$ becomes a non-monotonic function of the size of the cluster. It shows a distinct maximum near $N_\mathrm{K} = 7$, cf. Fig. 3(a), within the stochastic kinetic scheme description. For a small number of ion channels, the stochastic opening of a single sodium channel results with large likelihood in an action potential. Therefore, the rate of spontaneous action potentials initially increases with the number of channels. For a larger number of ion channels, however, the common membrane voltage couples the random opening and closing events more tightly. This results in the collective events stemming from a larger portion of ion channels available at the same time. In the corresponding regime, an exponential decrease of the spiking rate occurs with the increasing patch size for a deterministically subthreshold driving $I_\mathrm{ext} < 6.26\,\mu\mathrm{A/cm}^2$. Since the generation of spontaneous action potentials is caused merely by the channel noise stemming from the fluctuations of the number of open ion channels, a decreasing noise level or an increasing number of ion channels, respectively, leads to a less probable activation of the cell membrane. With an additional, externally applied positive and constant current, the barrier crossing to excitation becomes generally more probable. This results in increasing spiking rate. In the regime of super-threshold constant

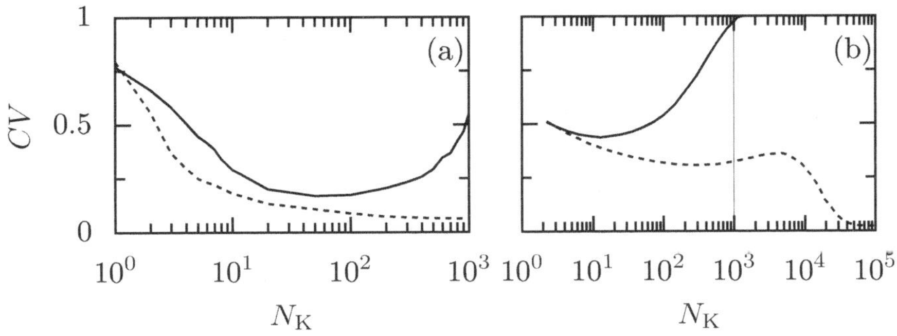


Fig. 4. Same figures as Fig. 3 but for the coefficient of variation $CV$, cf. Eq. 14.

current driving $I_{ext} \geq 9.762\,\mu\text{A/cm}^2$ a different effect takes place. Here, for large clusters, where the spiking events are determined by the excitable properties of the macroscopic Hodgkin–Huxley description of the neuron, the spiking rate decreases with increasing cluster sizes to approach an asymptotic value. For small cluster sizes it increases with increasing cluster size for firing events are determined by single-channel kinetics (see discussion above and Fig. 3(b)).

### 3.2. *Coherence resonance*

We next address the regularity of spontaneous action potentials. A proper measure is the coefficient of variation, $CV$, a measure of coherence, which is given as the ratio of the standard deviation of interspike time-intervals from the corresponding mean value to the mean value:

$$CV = \frac{\sqrt{\langle T^2 \rangle - \langle T \rangle^2}}{\langle T \rangle}. \tag{14}$$

In (14), $\langle T^2 \rangle := \frac{1}{N} \sum (t_i - t_{i-1})^2$ is the mean-squared interspike interval. For a fully disordered point process (the case of Poisson process) the coefficient of variation $CV$ assumes the value $CV = 1$, while for more ordered processes it assumes smaller values and for a deterministic signal $CV$ vanishes.

Figure 4 depicts the coefficient of variation for different constant driving currents. The $CV$ reveals the phenomenon of intrinsic coherence resonance [3, 4]. At an optimal dose of internal noise, i.e. an optimal number of ion channels, the $CV$ for $I_{ext} = 0$ exhibits a minimum, where the spike train becomes distinctly more ordered. This maximum in temporal coherence occurs at a cluster size at which the spontaneous firing rate exhibits the maximum. For increasing applied external constant current the spike generation becomes more rhythmic. For a strong driving $I_{ext} > 9.763\,\mu\text{A/cm}^2$, $CV$ tends to zero in the limit of infinite numbers of ion channels, cf. Fig. 4(b).

### 3.3. *Stochastic resonance*

After studying the impact of constant current driving we focus next our attention on the effect of stochastic resonance (SR). We address the effect of genuine *intrinsic*

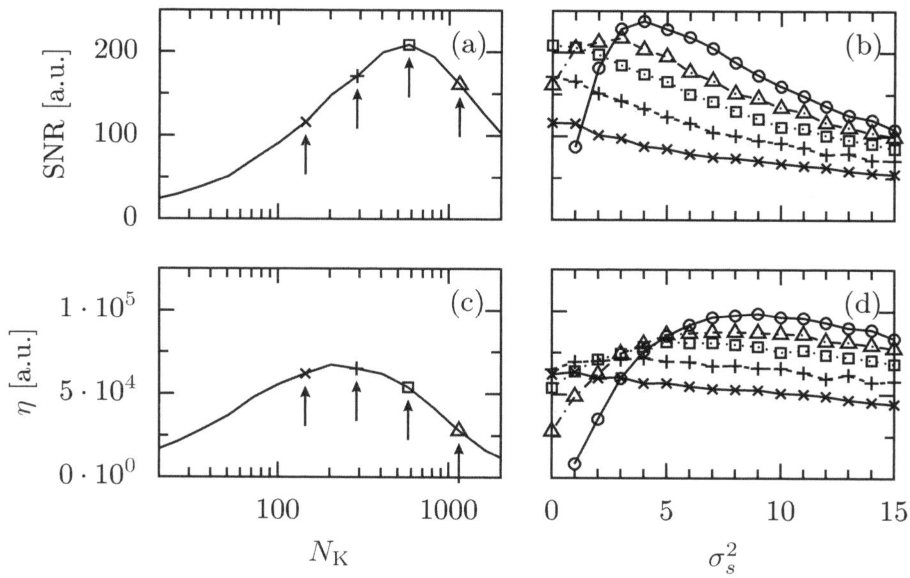


Fig. 5. The signal-to-noise ratio $SNR$ and the spectral amplification $\eta$ of an external subthreshold sinusoidal stimulus with amplitude $A = 1.0\,\mu\text{A/cm}^2$ and angular driving frequency $\Omega = 0.3\,\text{ms}^{-1}$ for different observation areas: (a) and (c) no synaptic noise; (b) and (d) the $SNR$ and $\eta$ versus the synaptic noise for the system sizes indicated by the arrows in (a) and (c): $N_K = 144$, solid lines through the crosses; $N_K = 288$, long dashed line connecting the pluses; $N_K = 576$, short dashed line through squares; $N_K = 1152$, dotted line connecting the triangles. The situation with no internal noise (i.e. formally $N_K \to \infty$) is depicted by the dotted line connecting the circles.

*stochastic resonance*, where the response of the system to a subthreshold external stimulus is optimized solely due to internal, ubiquitous channel noise. Therefore, we apply a sinusoidal driving current to the cell membrane with the ion channel cluster:

$$I_{\text{ext}}(t) = A \sin(\Omega t)\,, \tag{15}$$

with the driving amplitude $A$ and the angular driving frequency $\Omega$. Such a driving influences the spiking rate and the relative fluctuations of the spikes. A quantitative measure of the encoding of the sinusoidal signal in the spike train can be derived from the power spectrum $S(\omega)$ given as:

$$S(\omega) = \lim_{T\to\infty} \frac{1}{T} \left| \sum_n \mathrm{e}^{-i\omega t_n} \right|^2 , \tag{16}$$

with $T$ is the total integration time and the $t_n$ denote the times of spike occurrences. Most notably, we found a narrow peak at the driving frequency $\Omega$. Due to the finite length of the spike trains, this peak is not $\delta$-shaped, but is slightly broadened. From $S(\omega)$ we compute the signal-to-noise ratio $SNR$ and the spectral amplification $\eta$ (ratio of the powers at the frequency $\Omega$ of the spike train and the ac-signal).

In the absence of synaptic noise, cf. Fig. 5(a) and (c), we discover the effect of genuine intrinsic stochastic resonance, where the response of the system to the subthreshold external stimulus is optimized solely due to internal, ubiquitous noise.

For the given parameters, SR in the spectral amplification of signal occurs for $N_K \approx 180$ and in the signal-to-noise ratio for a different value $N_K \approx 580$. Starting from $N_K \approx 180$ the growing internal noise monotonically deteriorates the amplitude of system response at the driving frequency. Moreover, upon reaching $N_K \approx 580$ it deteriorates the quality of signal transduction which is measured by $SNR$.

Under such circumstances, one would predict that the addition of an external noise (which corresponds to the conventional situation in biological SR studies) cannot improve $\eta$ and $SNR$ further, i.e. the conventional SR will not be exhibited. In order to verify this prediction, we add synaptic noise. The corresponding results, depicted in Fig. 5(b) and (d) fully confirm the above prediction. Conventional SR therefore occurs only for large ion channel clusters beyond an optimal size and it reaches a saturation in the limit $N_K \to \infty$ (limit of the deterministic Hodgkin–Huxley model).

## 4. Conclusions

In conclusion, we have investigated stochastic resonance in a stochastic generalization of the Hodgkin–Huxley model. The spontaneous fluctuations of the ion channel conductance due to individual stochastic ion channel dynamics has been systematically taken into account by two methods: firstly, we have simulated the evolution of the membrane potential based upon a Markov-simulation of the kinetic of the potassium and sodium ion channels, secondly, we have followed a Langevin approach. Within the both methods we have shown that the excitable membrane patches exhibit a spontaneous spiking activity due to the omnipresent internal noise. The main results of this study refers to the phenomenon of intrinsic stochastic resonance. Here, the intrinsic channel noise alone give rise to SR (see Fig. 5). Moreover, SR induced by synaptic noise only occurs for sufficiently large clusters of ion channels, where the internal noise strength is not at its optimal value.

## Acknowledgement

It is a pleasure to dedicate this work to Frank Moss on the occasion of his 70th birthday together with our best wishes for many more fruitful collaborations and scientific discussions. As a matter of fact, Frank is still young enough to appreciate and to contribute great science. G. Schmid, I. Goychuk and P. Hänggi gratefully acknowledge the support of this work by the Deutsche Forschungsgemeinschaft, SFB 486 *Manipulation of matter on the nanoscale*, Project A10. This research (P. Jung and S. Zeng) was also supported by the National Science Foundation under Grant No. IBN-0078055.

## References

[1] C. Hildebrand and S. Waxman, *Regional node-like membrane specializations in non-myelinated axons of rat retinal nerve fiber layer*, *Brain Research* **258** (1983) 23–32.

[2] C. Guo and H. Levine, *A Thermodynamic Model for Receptor Clustering*, *Biophys. J.* **77** (1999) 2358–2365.

[3] P. Jung and J. W. Shuai, *Optimal size of ion channel clusters*, *Europhys. Lett.* **56** (2001) 29–35.

[4] G. Schmid, I. Goychuk and P. Hänggi, *Stochastic resonance as a collective property of ion channel assemblies*, *Europhys. Lett.* **56** (2001) 22–28.

[5] L. Gammaitoni, P. Hänggi, P. Jung and F. Marchesoni, *Stochastic resonance*, *Rev. Mod. Phys.* **70** (1998) 223–288; P. Hänggi, P. Talkner and M. Borkovec, *Reaction-rate theory: fifty years after Kramers*, *Rev. Mod. Phys.* **62** (1990) 251–342.

[6] K. Wiesenfeld and F. Moss, *Stochastic resonance and the benefits of noise — from ice ages to crayfish and squids*, *Nature (London)* **373** (1995) 33–36.

[7] G. Schmid, I. Goychuk and P. Hänggi, *Channel noise and synchronization in excitable membranes*, *Physica A* **325** (2003) 165–175.

[8] A. Pikovsky, A. Zaikin and M. A. Casa, *System size resonance in coupled noisy systems and in the Ising model*, *Phys. Rev. Lett.* **88** (2002), art. no. 050601.

[9] R. Toral, C. R. Mirasso and J. D. Gunton, *System size coherence resonance in coupled FitzHugh–Nagumo models*, *Europhys. Lett.* **61** (2003) 162–167.

[10] S. Zeng and P. Jung, *Mechanism for neuronal spike-generation by small and large ion channel clusters*, subm. to *Phys. Rev. E.*

[11] A. L. Hodgkin and A. F. Huxley, *A quantitative description of the membrane current and its application to conductance and excitation*, *J. Physiol. (London)* **117** (1952) 500–544.

[12] A. Manwani and C. Koch, *Detecting and estimating signals in noisy cable structures*, *Neural Computation* **11** (1999) 1797–1873.

[13] R. F. Fox and Y. N. Lu, *Emergent collective behavior in large number of globally coupled independently stochastic ion channels*, *Phys. Rev. E* **49** (1994) 3421–3431.

[14] R. F. Fox, *Stochastic versions of the Hodgkin–Huxley equations*, *Biophys. J.* **72** (1997) 2068–2074.

[15] J. A. White, J. T. Rubinstein and A. R. Kay, *Channel noise in neurons*, *Trends Neurosci.* **23** (2000) 131–137.

[16] C. C. Chow and J. A. White, *Spontaneous action potentials due to channel noise*, *Biophys. J.* **71** (1996) 3013–3021.

[17] J. R. Clay and L. J. DeFelice, *Relationship between membrane excitability and single channel open-closed kinetics*, *Biophys. J.* **42** (1983) 151–157.

[18] A. F. Strassberg and L. J. DeFelice, *Limitations of the Hodgkin–Huxley formalism: effects of single channel kinetics on transmembrane voltage dynamics*, *Neural Comput.* **5** (1993) 843–855.

[19] L. J. DeFelice and A. Isaac, *Chaotic states in a random world: relationship between the nonlinear differential equations of excitability and the stochastic properties of ion channels*, *J. Stat. Phys.* **70** (1993) 339–354.

[20] P. Hänggi, H. Grabert, P. Talkner and H. Thomas, *Bistable Systems: Master Equation Versus Fokker–Planck Modeling*, *Phys. Rev. A* **29** (1984) 371–378; P. Hänggi and P. Jung, *Bistability in Active Circuits: Application of a Novel Fokker-Planck Approach*, *IBM Journal of Research and Development* **32** (1988) 119–125.

# SYNCHRONIZATION APPROACH TO ANALYSIS OF BIOLOGICAL SYSTEMS

MICHAEL G. ROSENBLUM, ARKADY S. PIKOVSKY and JÜRGEN KURTHS
*Department of Physics, Potsdam University*
*Am Neuen Palais, 10, D-14415, Potsdam, Germany*

Received 3 December 2003
Revised 2 February 2004
Accepted 4 March 2004

In this article we review the application of the synchronization theory to the analysis of multivariate biological signals. We address the problem of phase estimation from data and detection and quantification of weak interaction, as well as quantification of the direction of coupling. We discuss the potentials as well as limitations and misinterpretations of the approach.

*Keywords*: Synchronization; multivariate data; interaction strength; directionality.

## 1. Introduction

Synchronization is a fundamental nonlinear phenomenon, and it plays an important role in various fields of science and engineering [1–5]. Very often it is found in live systems, being observed on a level of single cells, physiological subsystems, organisms and even on the level of populations [3, 6, 7]. Sometimes, this phenomenon is essential for a normal functioning of a system, e.g. for a coordinated motion of several limbs or for the performance of a pacemaker, where many cells fire synchronously, and in this way produce a macroscopic rhythm that governs respiration, heart contraction, etc. Sometimes, the onset of synchrony leads to a severe pathology, e.g. in case of the Parkinson's disease, when locking of many neurons results in the tremor activity. Quite often, the functional role of synchrony is yet unknown, e.g. in case of cardiorespiratory coordination [8–12] or in case of mutual entrainment of respiration and locomotion; possibly its appearance is just a manifestation of a general property of self-sustained oscillators — to adjust their rhythms due to a weak interaction. On the other hand, onset or cessation of synchrony reflects variation in the state of the complex system, and therefore may provide important physiological information.

In this paper we review the application of synchronization ideas to analysis of multivariate data measured from biological systems. We discuss and compare traditional and recently introduced techniques aimed to reveal and quantify the interaction between physiological subsystems from measurements. The main idea is to consider the inverse problem, i.e. to analyze the observed phase relations in order to obtain some information on the interaction of the systems generating the signals. In particular, we are interested in quantification of the strength and of the interaction and estimation of the directionality of coupling. We pay especial attention to the discussion of possible pitfalls and limitations of the method.

We note that such well-known effects as classical and stochastic resonance [13,14] in some respects are close to synchronization and cannot be distinguished from it within the framework of a passive experiment, i.e. when the system is simply observed, without any controlled variation of its parameters (for a discussion of active and passive experiments see [3]). A common feature of these effects is an appearance of a relation between phases and frequencies of two or many signals, and therefore the analysis of these relations can be used in case of a resonance as well. In particular, this approach has been successfully exploited for the investigation of stochastic systems [5,15–17].

## 2. Estimation of Instantaneous Phases

Prior to the analysis of phase relations we have to estimate phases from data. There exist three main approaches to the problem. One is based on the construction of the complex *analytic signal* $\zeta(t)$ [18] from a scalar experimental time series $s(t)$ via the Hilbert transform (HT)

$$\zeta(t) = s(t) + is_H(t) = A(t)e^{i\phi(t)}\,, \qquad s_H(t) = \pi^{-1}\text{P.V.}\int_{-\infty}^{\infty} \frac{s(\tau)}{t-\tau}d\tau\,, \tag{1}$$

where $s_H(t)$ is HT of $s(t)$. Equation (1) unambiguously provides an instantaneous phase $\phi(t)$ and an amplitude $A(t)$. Note that HT is parameter free. Practical hints for computation and usage of the HT, as well as further citations can be found in [3,19]. Here we briefly mention the crucial points:

- Mathematically, HT is defined for an arbitrary signal. However, $\phi(t)$ and $A(t)$ admit a clear physical interpretation only for narrow band signals. If the signal has no well-expressed peak in its power spectrum, then the computation of the phase and application of the synchronization approach is highly doubtful.
- A complex signal that can be considered as a mixture of several narrow band processes should be first decomposed into oscillatory components that can be considered as signals with slowly varying amplitude and frequency; as the next step, the phases of these components can be obtained via HT. (Note that sometimes it is difficult to decide whether a peak in the spectrum represents another process or a harmonic.) Decomposition can be done by means of a band-pass filter or by more sophisticated techniques like independent component analysis.
- Determination of $\phi(t)$ is very sensitive to low-frequency trends, what makes the preprocessing of the data a crucial step in the analysis.

The second approach exploits the wavelet analysis with a complex wavelet function and provides a phase (and an amplitude) as functions of time for a certain spectral frequency band [20, 21]:

$$A(t; f)e^{\imath\phi(t;f)} = \int_{-\infty}^{\infty} s(\tau)\Psi^*(t,\tau;f)d\tau\,, \tag{2}$$

where $\Psi(t,\tau;f)$ is the Morlet, or Gabor, wavelet:

$$\Psi(t,\tau;f) = \sqrt{f}\exp\left(\imath \cdot 2\pi f(\tau - t)\right)\exp\left(-\frac{(\tau-t)^2}{2\sigma^2}\right).$$

This procedure is equivalent to a band-pass filtration and subsequent HT of the signal $s(t)$ [22]. The central frequency of the filter is $f$, and its width is determined by the parameter $\sigma$.

Third, the phase can be very easily introduced for processes that can be treated as a series of well-defined events taking place at times $t_k$ (point processes). Examples include signals characterizing heart beat or neuron firing [23]. If the interval between two events can be considered as a cycle, then it is natural to say that the phase increment between the events is exactly $2\pi$. Hence, we can assign to the times $t_k$ the values of phase $\phi(t_k) = 2\pi k$, and for arbitrary instant of time $t_k < t < t_{k+1}$ take

$$\phi(t) = 2\pi k + 2\pi\frac{t - t_k}{t_{k+1} - t_k}\,. \tag{3}$$

Finally, we mention that definition and practical determination of a phase of a complex signal in the context of the synchronization analysis remains an open problem. One approach, called locking-based frequency measurement was suggested in [24]. The idea of this approach is to use the signal under study to drive a set of uncoupled limit cycle oscillators with different natural frequencies. A subset of these probe oscillators can be entrained by the common forcing, and therefore synchronize in between; the frequency and the phase of these locked oscillators can be taken as an estimate of the frequency and phase of the original signal.

## 3. Quantification and Interpretation of Phase Relations

As soon as the phases of two signals are estimated by appropriate methods, the relations between phases can be quantified. For this purpose we recall that the interaction of noisy oscillators is reflected in the distribution of the cyclic relative phase

$$\psi_{n,m}(t) = (n\phi_1(t) - m\phi_2(t)) \bmod 2\pi\,,$$

where $n, m$ are integers characterizing the order of locking (see, e.g. [3]). For uncoupled oscillators the distribution of $\psi_{n,m}$ is uniform, whereas the interaction makes this distribution unimodal. Hence, the strength of interaction between the systems can be estimated via quantification of the form of the distribution of $\psi_{n,m}$. The correspondent measures are called *synchronization indices*; usually they are normalized in such a way that zero index corresponds to complete independence, whereas unit index corresponds to perfect phase locking. An index can be based on the Shannon entropy of the distribution of $\psi_{n,m}(t)$ [25]. Alternatively, it can be taken equal to the intensity of the first Fourier mode of the distribution [19, 20]:

$$\rho_{n,m}^2 = \langle \cos \psi_{n,m} \rangle^2 + \langle \sin \psi_{n,m} \rangle^2 .$$

The advantage of this measure is the absence of parameters. Note that the integers $n, m$ can be evaluated from the estimation of the average frequencies of two signals (e.g. from their power spectra), or found by trial and error by looking for the maximum of $\rho_{n,m}$.

If one of the signals can be considered as a point (event) process, then the phase interrelation can be conveniently quantified with the help of the *phase stroboscope.* For example, this technique turned out to be useful in the analysis of cardiorespiratory interaction [3, 10, 19]. Phase stroboscope implies, that the second signal is observed at the moments when the events in the first signal occur, i.e. when $\phi_1(t) = 2\pi n$. So, in case of the cardiorespiratory data, the phase of the respiratory signal is determined at the instants of the appearance of the R-peak in the cardiogram. Interaction between the systems can be revealed by means of a graphical tool, called *synchrogram,* and quantified by means of characterization of the stroboscopically observed phase [11].

Concerning the interpretation of the phase relations we want to emphasize the difference between the approach developed in our previous papers (see [19] for a review) and the approach, suggested in [20, 21]. Our approach, based on the theory of coupled oscillators assumes that there exists a unique phase for each of the interacting oscillators. Therefore the preprocessing, e.g. by means of a band-pass filter, is aimed to extraction of a certain signal from its mixture with other signals and broad band noise. This extracted signal is assumed to be generated by one of the subsystems under study, and the phase of this subsystem is estimated and $n : m$ relation between phases is quantified. On the contrary, in the approach of [20, 21] the phase is computed subsequently for all spectral frequency bands by means of the continuous variation of the wavelet parameter corresponding to the central frequency of the band-pass filter. Then, a $1:1$ phase relation of two signals is quantified for each frequency band, providing the index $\rho_{1,1}$ as a function of the spectral frequency $f$. $\rho_{1,1}(f)$ can be considered in analogy to the classical tool of the cross-spectrum analysis, namely to the coherence function $\gamma(f)$. Similarly to $\gamma(f)$, $\rho_{1,1}(f)$ characterizes the interdependence between certain frequency components of two signals; however, in contrast to $\gamma(f)$, $\rho_{1,1}(f)$ takes into account only the phase information. Thus, the technique [20, 21] should be considered as an extension of the cross-spectrum analysis rather than in the framework of coupled oscillator approach.

Concluding this section, we comment on the relevance of possible relations between the amplitudes of the signals. Our approach is motivated by the synchronization phenomenon, where the phase (and frequency) relations are of primary importance, whereas the amplitudes are generally not much influenced by a weak interaction, and, e.g. in case of chaotic systems can be weakly correlated. However, the experimental data do not always comply with this model, and then the amplitude relations can be also important. An example is given in [26], where interhemispheric synchrony of spontaneous beta oscillations in humans was analyzed.

It was shown that the phase relations are amplitude-dependent, namely, the synchronization index increases with the amplitude of signals; it was hypothesized that this is due to the increase of the signal-to-noise ratio. However, a visual inspection of the signals shows a strong amplitude modulation of signals (that is why these oscillations are called spontaneous) which is typical for narrow band noise. So, we may assume that the systems under study cannot be considered as self-sustained oscillators, and, therefore, the notion of synchronization may not be applicable in this case. The observed phase relation should be rather interpreted in terms of a resonance, where the amplitude relations, certainly, are important.

## 4. Asymmetric Phase Relations

If an interaction between two oscillators is established, the next step in the analysis can be the quantification of the direction of coupling. This question is nontrivial because synchronization may be caused either by biderectional or by unidirectional coupling. We outline here two approaches. One, based on information theory, quantifies the direction of the information flow [27]; it operates with raw signals. This approach goes back to the mutual predictability concept of Granger [28] that we can briefly formulate as follows: if system 2 acts on system 1, then we can better predict the future of the system 1 if we take into account the information on the system 2; this can be also formulated in terms of entropy. The other approach [29] works with the phase data and is based on the coupled oscillator theory.

The main idea is that in case of weak coupling the description of two interacting oscillators can be reduced to the phase dynamics

$$\dot{\phi}_{1,2} = \omega_{1,2} + f_{1,2}(\phi_{2,1}, \phi_{1,2}) + \xi_{1,2}\,, \tag{4}$$

where $\phi_{1,2}$ are phases, $\omega_{1,2}$ are natural frequencies, and the terms $f_{1,2}$ describe the coupling between the systems. The random terms $\xi_{1,2}$ describe noisy perturbation and/or chaotic amplitude dynamics. Given the time series of phases, $\phi_{1,2}(t_k)$, we consider the phase increments during some fixed time $\tau$, $\Delta_{1,2}(k) = \phi_{2,1}(t_k + \tau) - \phi(t_k)$. These easily obtained new time series can be considered as generated by some unknown two-dimensional noisy map

$$\Delta_{1,2} = \omega_{1,2}\tau + \mathcal{F}_{1,2}(\phi_{2,1}, \phi_{1,2}) + \eta_{1,2}\,. \tag{5}$$

Deterministic parts of this map can be estimated by fitting the dependencies of $\Delta_1$ and $\Delta_2$ on $\phi_1$, $\phi_2$. As the seeked functions $\mathcal{F}$ are $2\pi$-periodic with respect to both arguments, the natural choice of the probe function is the Fourier-series.

As soon as the map (5) is identified, the information on the directionality is provided by the coefficients of two Fourier series approximating the unknown functions $\mathcal{F}_{1,2}$. However, this information is not easy to interpret. Here we face the problem how to quantify the direction of coupling by one number, what is definitely ambiguous. One way to introduce such a number, called *directionality index*, was suggested in [29]. Let $F_{1,2}$ be the fits to $\mathcal{F}_{1,2}$. The sensitivity of the phase dynamics of one system with respect to the variation of the phase of the second system can be quantified by the average of the derivative $\dfrac{\partial F_{1,2}}{\partial \phi_{2,1}}$. Namely, we characterize this by computing the coefficients

$$c_{1,2}^2 = \langle \frac{\partial F_{1,2}}{\partial \phi_{2,1}} \rangle ,$$

where $\langle\rangle$ stays for averaging over $\phi_{1,2}$. Finally, the directionality index is computed as

$$d^{(1,2)} = \frac{c_2 - c_1}{c_1 + c_2} .$$

This index varies from 1 in the case of unidirectional coupling $1 \to 2$ to $-1$ in the opposite case, $2 \to 1$. Intermediate values correspond to bidirectional coupling, and the case $d^{(1,2)} = 0$ can be denoted as symmetric coupling. Note, however, that except for unidirectional driving, the definition of directionality is ambiguous.

We note also that the predictability (information-theoretical) approach can be also applied to the phase data [30, 31]. For examples of application to the cardiorespiratory and brain data see [30,32]. Statistical properties of the directionality estimation are considered in [33].

## 5. Discussion: Potentials, Limitations and Pitfalls

Synchronization analysis proved to be a useful tool in experimental investigation of different biological systems. So, tendency to phase locking of different order $n : m$ was revealed in a group of young athletes [10, 34] and in normal adults [12]. Strength and directionality of the cardiorespiratory interaction in healthy newborns was studied in [11,30,35] in dependence on the age and sleep stages, what revealed the dependence of the interaction parameters on the respiratory frequency and provided the information on the role of the kinetics involved in the vagal-atrial transmission. As further examples we mention the prediction and localization of epileptic activity [36, 37], analysis of the role of interbrain synchronization and localization of pathological brain activity in Parkinson's disease [25, 38], or nephron-nephron interaction [39]. Stochastic phase synchronization in sensory systems and its role in mediating sensory responses was studied in a series of papers of the St.Louis group, in experiments with electrosensitive afferent neurons of the paddlefish and with light-sensitive and mechanosensitive neurons of the crayfish, as well as theoretically [15–17, 23, 40]. The algorithm for estimation of the directionality in coupling was exploited for understanding of the functional connectivity in the brain during a prescribed motor task [32, 41].

In what follows we discuss the limitations of this synchronization approach and problems of an interpretation of the results. First of all we emphasize again that blind application of the Hilbert transform to a broad band signal can provide physically meaningless results. The first step of the synchronization analysis — estimation of instantaneous phases requires a careful choice of the preprocessing and is crucial.

Next, we stress that the detection of phase relations between two signals does not always mean that we deal with a synchronization process. These relations can be, e.g. a manifestation of classical or stochastic resonance. Quantification and analysis of time dependencies of such relations can nevertheless provide useful information on the biological system, though one should be careful in the conclusions regarding underlying physical mechanisms.

The synchronization approach is based on the assumptions that we deal with interacting self-sustained oscillators, and that the signals we measure represent the dynamics of different systems. Violation of these assumptions, e.g. when two signals reflect different variables of the same system or are input and output of some passive transformer, makes the analysis hardly useful.

Finally, we mention that in the physiological literature, in particular in neuroscience, the term "synchrony (synchronization)" is very often used as a synonym to "correlation". Hence, the cross-spectral (cross-correlation) technique is used as a standard tool to quantify synchrony. If we understand synchronization in a physical sense, i.e. as an adjustment of phases and frequencies of interacting oscillating objects, then it is clear that this tool is not an appropriate one. Indeed, high coherence in the spectral domain can be a result of other types of interaction than weak synchronizing coupling, e.g. it can be observed in case of modulation. Moreover, the cross-spectral (cross-correlation) technique accounts for both amplitude and phase interrelations. Generally, correlation (coherence) and synchronization analysis address different aspects of interaction and can, e.g. lead to different results in localization of pathological brain activity [19, 25]. Next important point is that sometimes the conclusion on the driver-response relationship (i.e. on the directionality) is made on the basis of the phase of the cross-spectrum at a certain dominating frequency. It is concluded that the system that lead in phase is the driver. This concept, inherited from the input-output approach to passive transformers, definitely fails for coupled oscillators, where the phase shift in the synchronous regime depends on coupling functions and natural frequencies of both interacting systems, and these factors cannot be separated. [Indeed, considering the simplest model of two interacting oscillators, $\phi_{1,2} = \omega_{1,2} + \varepsilon_{1,2}\sin(\phi_{2,1} - \phi_{1,2})$ (Adler equations, cf. with Eqs. (4)), one easily obtains for the phase difference $\phi_1 - \phi_2 = \arcsin\left((\omega_1 - \omega_2)/(\varepsilon_1 + \varepsilon_2)\right)$.] The same is valid for the estimation of the delays by re-computation of the phase shift (obtained from the cross-spectrum) into conduction time (delay): justified in the treatment of passive systems, this approach leads to spurious results if two signals represent coupled oscillators.

As the last remark, we mention that different notions of synchronization, such as generalized and event synchronization are also used in the analysis of interdependencies from multivariate data [42–46].

## Acknowledgements

In our research on synchronization analysis of biological systems we have been highly inspired by fruitful discussions with Professor Frank Moss. We gratefully dedicate this paper to his 70th birthday.

## References

[1] I. I. Blekhman, *Synchronization in Science and Technology*, Nauka, Moscow (1981), (In Russian); English translation: ASME Press, New York (1988).

[2] P. S. Landa, *Nonlinear Oscillations and Waves in Dynamical Systems*, Kluwer Academic Publishers, Dordrecht–Boston–London (1996).

[3] A. Pikovsky, M. Rosenblum and J. Kurths, *Synchronization. A Universal Concept in Nonlinear Sciences*, Cambridge University Press, Cambridge (2001).

[4] E. Mosekilde, Yu. Maistrenko and D. Postnov, *Chaotic Synchronization. Applications To Living Systems*, World Scientific, Singapore (2002).

[5] V. Anishchenko, A. Neiman, V. Astakhov, T. Vadiavasova and L. Schimansky-Geier, *Chaotic and Stochastic Processes in Dynamic Systems*, Springer Verlag, Berlin (2002).

[6] L. Glass and M. C. Mackey, *From Clocks to Chaos: The Rhythms of Life*, Princeton Univ. Press, Princeton, NJ (1988).

[7] L. Glass, *Synchronization and rhythmic processes in physiology*, *Nature* **410** (2001) 277–284.

[8] H. Pessenhofer and T. Kenner, *Zur Methodik der kontinuierlichen Bestimmung der Phasenbeziehung zwischen Herzschlag und Atmung*, *Pflügers Archiv* **355** (1975) 77–83.

[9] T. Kenner, H. Pessenhofer and G. Schwaberger, *Method for the analysis of the entrainment between heart rate and ventilation rate*, *Pflügers Arch.* **363** (1976) 263–265.

[10] C. Schäfer, M. G. Rosenblum, J. Kurths and H.-H. Abel, *Heartbeat synchronized with ventilation*, *Nature* **392** (1998) 239–240.

[11] R. Mrowka, A. Patzak and M. G. Rosenblum, *Qantitative analysis of cardiorespiratory synchronization in infants*, *Int. J. of Bifurcation and Chaos* **10** (2000) 2479–2488.

[12] M. Bračič and A. Stefanovska, *Synchronization and modulation in the human cardiorespiratory system*, *Physica A* **283** (2000) 451–461.

[13] K. Wiesenfeld and F. Moss, *Stochastic resonance: From ice ages to crayfish and SQUIDs*, *Nature* **373** (1995) 33–36.

[14] F. Moss, D. Pierson and D. O'Gorman, *Stochastic resonance: Tutorial and update*, *Int. J. of Bifurcation and Chaos* **4** (1994) 1383–1397.

[15] A. Neiman, X. Pei, D. F. Russell, W. Wojtenek, L. Wilkens, F. Moss, H. A. Braun, M. T. Huber and K. Voigt, *Synchronization of the noisy electrosensitive cells in the paddlefish*, *Phys. Rev. Lett.* **82** (1999) 660–663.

[16] A. Neiman, L. Schimansky-Geier, A. Cornell-Bell and F. Moss, *Noise-enhanced phase synchronization in excitable media*, *Phys. Rev. Lett.* **83** (1999) 4896–4899.

[17] A. B. Neiman and D. F. Russell, *Synchronization of noise-induced bursts in noncoupled sensory neurons*, *Phys. Rev. Lett.* **88** (2002) 138103.

[18] D. Gabor, *Theory of communication*, *J. IEE London* **93** (1946) 429–457.

[19] M. G. Rosenblum, A. S. Pikovsky, J. Kurths, C. Schäfer and P. A. Tass, *Phase synchronization: From theory to data analysis*, in *Neuro-informatics and Neural Modeling*, volume 4 of *Handbook of Biological Physics*, ed. F. Moss and S. Gielen, Elsevier, Amsterdam (2001) 279–321.

[20] E. Rodriguez, N. George, J.-P. Lachaux, J. Martinerie, B. Renault and F. J. Varela, *Perception's shadow: Long distance synchronization of human brain activity*, *Nature* **397** (1999) 430–433.

[21] J.-P. Lachaux, E. Rodriguez, M. Le van Quyen, A. Lutz, J. Martinerie and F. J. Varela, *Studying single-trials of phase synchronous activity in the brain*, *Int. J. Bifurc. Chaos* **10** (2000) 2429–2440.

[22] R. Quian Quiroga, A. Kraskov, T. Kreuz and P. Grassberger, *Performance of different synchronization measures in real data: A case study on electroencephalographic signals*, *Phys. Rev. E* **65** (2002) 041903.

[23] S. Bahar, A. Neiman, L. A. Wilkens and Frank Moss, *Phase synchronization and stochastic resonance effects in the crayfish caudal photoreceptor*, *Phys. Rev. E* **65** (2002) 050901.

[24] M. Rosenblum, A. Pikovsky, J. Kurths, G. Osipov, I. Kiss and J. Hudson, *Locking-based frequency measurement and synchronization of chaotic oscillators with complex dynamics*, *Phys. Rev. Lett.* **89** (2002) 264102.

[25] P. Tass, M. G. Rosenblum, J. Weule, J. Kurths, A. S. Pikovsky, J. Volkmann, A. Schnitzler and H.-J. Freund, *Detection of $n : m$ phase locking from noisy data: Application to magnetoencephalography*, *Phys. Rev. Lett.* **81** (1998) 3291–3294.

[26] V. V. Nikouline, K. Linkenkaer-Hansen, J. Huttunen and R. J. Ilmoniemi, *Interhemispheric phase synchrony and amplitude correlation of spontaneous beta oscillations in human subjects: A magnetoencephalographic study*, *NeuroReport* **12** (2001) 2487–2491.

[27] T. Schreiber, *Measuring information transfer*, *Phys. Rev. Lett.* **85** (2000) 461–464.

[28] C. W. J. Granger, *Investigating causal relations by econometric models and cross spectral models*, *Econometrica* **37** (1969) 424–438.

[29] M. Rosenblum and A. Pikovsky, *Detecting direction of coupling in interacting oscillators*, *Phys. Rev. E* **64** (2001) 045202(R).

[30] M. G. Rosenblum, L. Cimponeriu, A. Bezerianos, A. Patzak and R. Mrowka, *Identification of coupling direction: Application to cardiorespiratory interaction*, *Phys. Rev. E* **65** (2002) 041909.

[31] M. Paluŝ and A. Stefanovska, *Direction of coupling from phases of interacting oscillators: An information-theoretic approach*, *Phys. Rev. E* **67** (2003) 055201.

[32] L. Cimponeriu, M. G. Rosenblum, T. Fieseler, J. Dammers, M. Schiek, M. Majtanik, P. Morosan, A. Bezerianos and P. A. Tass, *Asymmetric relations between interacting neuronal oscillators*, *Progress of Theoretical Physics Suppl.* **150** (2003) 22–36.

[33] D. A. Smirnov and B. P. Bezruchko, *Estimation of interaction strength and direction from short and noisy time series*, *Phys. Rev. E.* **68** (2003) 046209.

[34] C. Schäfer, M. G. Rosenblum, H.-H. Abel and J. Kurths, *Synchronization in the human cardiorespiratory system*, *Physical Review E* **60** (1999) 857–870.

[35] R. Mrowka, L. Cimponeriu, A. Patzak and M. G. Rosenblum, *Directionality of coupling of physiological subsystems - age related changes of cardiorespiratory interaction during different sleep stages in babies*, *American J. of Physiology (Regul Comp Integr Physiol)* **285** (2003) R1395–R1401.

[36] F. Mormann, K. Lehnertz, P. David and C. E. Elger, *Mean phase coherence as a measure for phase synchronization and its application to the EEG of epilepsy patients*, *Physica D* **144** (2000) 358–369.

[37] F. Mormann, R. G. Andrzejak, T. Kreuz, Ch. Rieke, P. David, Ch. E. Elger and K. Lehnertz, *Automated detection of a preseizure state based on a decrease in synchronization in intracranial electroencephalogram recordings from epilepsy patients*, *Phys. Rev. E* **67** (2003) 021912.

[38] P. A. Tass, T. Fieseler, J. Dammers, K. Dolan, P. Morosan, M. Majtanik, F. Boers, A. Muren, K. Zilles and G. R. Fink, *Synchronization tomography: A method for three-dimensional localization of phase synchronized neuronal populations in the human brain using magnetoencephalography*, *Phys. Rev. Lett.* **90** (2003).

[39] N.-H. Holstein-Rathlou, K.-P. Yip, O. V. Sosnovtseva and E. Mosekilde. Synchronization phenomena in nephron-nephron interaction, *Chaos* **11** (2001) 417–426.

[40] S. Bahar and F. Moss, *Stochastic phase synchronization in the crayfish mechanoreceptor/photoreceptor system*, *Chaos* **13** (2003) 138–144.

[41] J. Gross, L. Timmermann, J. Kujala, M. Dirks, F. Schmitz, R. Salmelin and A. Schnitzler, *The basis of intermittent motor control in humans*, *Proc. Nat. Acad. Sci.* **99** (2002) 2299–2302.

[42] S. J. Schiff, P. So, T. Chang, R. E. Burke and T. Sauer, *Detecting dynamical interdependence and generalized synchrony through mutual prediction in a neural ensemble*, *Phys. Rev. E* **54** (1996) 6708–6724.

[43] M. Le van Quyen, C. Adam, M. Baulac, J. Martinerie and F. J. Varela, *Nonlinear interdependencies of EEG signals in human intracranially recorded temporal lobe seizures*, *Brain Research* **792** (1998) 24–40.

[44] J. Arnhold, P. Grassberger, K. Lehnertz and C. E. Elger, *A robust method for detecting interdependences: Application to intracranially recorded EEG*, *Physica D* **134** (1999) 419–430.

[45] R. Quian Quiroga, J. Arnhold and P. Grassberger, *Learning driver-response relationships from synchronization patterns*, *Phys. Rev. E* **61** (2000) 5142–5148.

[46] R. Quian Quiroga, T. Kreuz and P. Grassberger, *Event synchronization: A simple and fast method to measure synchronicity and time delay patterns*, *Phys. Rev. E* **66** (2002) 041904.

# COHERENCE BETWEEN FLUCTUATIONS IN BLOOD FLOW AND OXYGEN SATURATION

ALAN BERNJAK*, ANETA STEFANOVSKA† and PETER V. E. McCLINTOCK‡

*Department of Physics*
*Lancaster University, Lancaster, LA1 4YB, UK*
**a.bernjak@lancaster.ac.uk*
*†aneta@lancaster.ac.uk*
*‡p.v.e.mcclintock@lancaster.ac.uk*

P. JANE OWEN-LYNCH

*Biomedical and Life Sciences Division*
*School of Health and Medicine*
*Lancaster University, Lancaster, LA1 4YQ, UK*
*j.owen-lynch@lancaster.ac.uk*

PETER B. M. CLARKSON

*Cardiology Department*
*Raigmore Hospital, Old Perth Road*
*Inverness, IV2 3UJ, UK*
*peter.clarkson1@nhs.net*

Received 17 September 2011
Revised 13 December 2011
Published 31 March 2012

Communicated by Igor Goychuk

We have determined the wavelet phase coherence between simultaneously recorded microvascular blood flow and oxygen saturation signals from 88 healthy subjects, thus enabling us to study their common fluctuations. Measurements were taken for 30 min from the arm and leg, at two depths. In the skin, blood flow and oxygen saturation were found to be coherent both at the cardiac frequency and below 0.1 Hz down to about 0.01 Hz. Coherence in the arm extends to lower frequencies than that in the leg. From the deeper recordings, no coherence was found on either limb. The existence of coherence between skin blood flow and oxygen saturation demonstrates causal connections between them within certain frequency ranges. The method has yielded the first detailed insight into the dynamics of blood oxygenation.

*Keywords*: Wavelet phase coherence; blood flow; oxygen saturation; microcirculation.

## 1. Introduction

Normal cellular function requires a continuous supply of oxygen. In higher animals a cardiovascular system is used to provide for its convective delivery to the smallest branches of the network of blood vessels — the microcirculation — from which oxygen then passes to the cells through passive diffusion. The oxygen is transported through the cardiovascular system by hemoglobin (Hb), an iron-containing protein in the red blood cells. The Hb oxygen saturation in the blood consists of a steady level that has been extensively studied in terms of time-averages, in both the resting state and during exercise. We discuss here oscillatory fluctuations around this level, similar to those that occur in blood flow. In the latter case, studies of the oscillations measured by noninvasive laser-Doppler flowmetry (LDF) [1–4], have brought much insight into the underlying physiological processes. Oscillations have also been observed in recordings of tissue oxygenation [5, 6]. Vasomotion might have a beneficial effect on tissue oxygenation, but their relationship is unclear.

In this paper we investigate the interactions between local microvascular blood flow and oxygen saturation by applying the wavelet phase coherence method to simultaneously recorded signals. Techniques for data acquisition are described in Sec. 2. The wavelet transform and wavelet phase coherence methods are summarised in Sec. 3, followed in Sec. 4 by their application to data recorded from 88 healthy subjects. We discuss the results and draw conclusions in Secs. 5 and 6, respectively.

## 2. Methods for Measurement of Skin Blood Perfusion and Oxygenation

### 2.1. *Laser-Doppler flowmetry*

Blood flow in the skin was measured by LDF, based on the technique introduced by Stern [7] and developed by Nilsson *et al.* [8]. Infrared light is delivered to the skin via a flexible optical fiber. The light backscattered from red blood corpuscles is collected, and the distribution of the Doppler shift indicates their average speed and concentration. Strictly, the output value from the LDF device represents flux. However, in the literature flux is often used interchangeably with perfusion or blood flow, assuming a constant concentration of blood corpuscles. The LDF measurements in the present work were made with the O2C ("oxygen to see") device manufactured by LEA Medizintechik, Germany. In addition to blood flow, the device (used with the LF-2 flat probe) enables simultaneous acquisition of Hb oxygen saturation (see below) in the same location. A near-infrared laser diode (830 nm, 30 mW) is the source of light used to measure blood flow. The light is brought to the skin through optical fibers and the collector fibers are set at two different separations to allow for recordings from two depths. The exact measurement depth depends on tissue properties but the shallow recording (fiber separation = 2 mm) reaches to around 1 mm in the skin (typical for LDF devices) while the deeper recording (fiber separation = 8 mm) can reach several mm deep. In the text we refer to the two depths as "Skin" and "Deep". LDF recordings provide flux expressed in arbitrary perfusion units (AU or PU) on account of the so-called "biological zero": even during occlusion, when the blood flow is completely blocked, the instrument yields finite readings due to Brownian motion of the red blood cells and vascular wall movement, as well as movement due to mechanical contraction of the surrounding tissue.

### 2.2. *Oxygenation measurements*

Tissue oxygenation depends on the balance between oxygen delivery and utilization [9], but oxygen saturation does not reflect tissue oxygenation directly because the affinity of oxygen to hemoglobin is affected by physiological factors such as e.g., pH or body temperature. In the current study Hb oxygen saturation was measured by optical reflectance spectroscopy, a form of near-infrared spectroscopy (NIRS). In general, NIRS enables noninvasive, *in vivo*, continuous detection of changes in oxygenated hemoglobin, deoxygenated hemoglobin or total hemoglobin (blood volume). It is based on the Beer–Lambert law [10] which relates the absorption of light to the properties of the tissue through which it passes. Oxygenated and deoxygenated hemoglobin have different absorption characteristics in the near-infrared range (700–1000 nm) and the measured absorption can be used to determine the local average oxygen saturation of the hemoglobin. Absolute concentrations are difficult to calculate accurately because the exact geometrical parameters (path of the scattered light) in the tissue are not known. Mean baseline values of oxygenation saturation and responses to provocation have been intensively investigated in the muscles [10] and brain [11]. Recent studies report, analyze and discuss spontaneous fluctuations of the Hb oxygen saturation in time [5, 6, 12, 13].

In the experiments to be reported below, we measured Hb oxygen saturation with the O2C device, using the LP-2 probe. A difference from standard NIRS is that the O2C is not restricted to specific frequencies. White light (halogen lamp) is shone onto/into the skin and a wide spectrum of back-reflected light (650–795 nm from the skin and 500–630 nm from the deeper tissue) is analyzed to estimate oxygen saturation. By using the same probe separations as for the LDF, the sampled areas of tissue for the oxygen saturation and LDF recordings overlap each other. All signals are sampled at 40 Hz, but oxygen saturation data are refreshed once per second. The output values of all data are discretized to integers.

Using the O2C, the signals are recorded from venules — at the venous end of the capillaries. Oxygen saturation in these vessels can be much lower than in the large veins, because the venous blood includes "shunt-blood" which does not pass oxygen consuming cells. To determine local oxygen consumption accurately, oxygen saturation should be known on both the arterial and venous sides of the capillary bed, together with the hemoglobin level. The fact that oxygen saturation and blood flow are measured simultaneously in the same volume enables us to investigate the relationship between the two processes and promises to yield improved understanding of the mechanisms of oxygen delivery within the microcirculation.

## 3. Wavelet Transform and Wavelet Phase Coherence

The cardiovascular system can be described as a set of at least six coupled oscillators whose characteristic frequencies vary in time [1, 3]. The associated physiological processes generate complex oscillatory patterns in the blood flow, especially at the level of capillaries. Wavelet analysis with the Morlet mother wavelet can detect these oscillations with logarithmic frequency resolution, and follow the variations of their frequencies and amplitudes in time [1, 2]. By this means, six frequency intervals were defined: I: 0.6–2 Hz (cardiac activity), II: 0.145–0.6 Hz (breathing), III: 0.052–0.145 Hz (myogenic activity), IV: 0.021–0.052 Hz (neurogenic activity), V: 0.0095–0.021 Hz (NO-related endothelial) and VI: 0.005–0.0095 Hz (non-NO endothelial activity) [3, 4]. The blood flow signals thus reflect the nonautonomous character of the underlying oscillatory processes and we expect similar modulations to be present in the oxygen saturation recordings.

The wavelet transform is a method that provides for the complex transformation of a time series from the time to the time-frequency domain. It involves convolving the time series $g(u)$ with a family of generally nonorthogonal basis functions, generated from the mother wavelet:

$$g(s,t) = \frac{1}{\sqrt{s}} \int_{-\infty}^{\infty} \psi\left(\frac{u-t}{s}\right) g(u)du. \tag{1}$$

In the above expression $\psi$ is the Morlet mother wavelet, scaled by the factor $s$ and translated in time by $t$. The Morlet wavelet is a complex sinusoid modulated by the Gaussian function:

$$\psi(u) = \frac{1}{\sqrt[4]{\pi}} \cdot \mathrm{e}^{-i2\pi u} \cdot \mathrm{e}^{-u^2/2}, \tag{2}$$

where $i = \sqrt{-1}$. Scaling of the wavelet enables the detection of oscillations with different frequencies, while its translation in time allows one to follow how the spectra evolve in time. The rationale for using the Morlet wavelet is that its Gaussian envelope provides good localization of events in both time and frequency; in time, approximately 6–7 periods are used to detect the power of each oscillatory component. Moreover, there is a straightforward inverse relationship between the scaling factor $s$ and its corresponding frequency: $f = 1/s$.

As noted above, oscillations in the cardiovascular system manifest over the wide frequency range from 0.005 Hz to 2 Hz in recordings of around 30 min duration. For longer recordings, even lower frequency oscillations would be expected to appear if problems of movement artifacts and nonstationarity could be overcome. To accommodate both high and low frequencies, logarithmic frequency resolution is used. Starting with scale $s = 0.5$ (corresponding to the highest frequency 2 Hz), the wavelets are stretched by factor 1.05, until $s = 200$ is reached (corresponding to the lowest frequency 0.005 Hz). The resolution in time is set to 1 s. The two most important aspects of analyzing cardiovascular signals with wavelets are the adjustable window length, so that each oscillatory component is detected with its corresponding window length, and the logarithmic frequency resolution, which enables accurate characterization of the low frequency oscillations.

### 3.1. *Wavelet phase coherence*

With the complex Morlet wavelet, the wavelet coefficients are complex numbers in the time-frequency plane. They define the absolute amplitude and instantaneous relative phase for each frequency and time, on a grid. The phase information can be used to investigate relationships between the oscillations from different signals. For example, two noisy signals can contain oscillations at the same frequency but it may not be known whether they are causally related. Wavelet phase coherence can uncover possible relationships by evaluating how well the instantaneous phases from the two signals are matched [14, 15].

First, instantaneous phases are calculated at each time $t_n$ and frequency $f_k$ for both signals: $\phi_{1k,n}$ and $\phi_{2k,n}$. Their relative phase difference is then obtained as $\Delta\phi_{k,n} = \phi_{2k,n} - \phi_{1k,n}$. The sine and cosine components of the phase differences are calculated and averaged in time for the whole length of the signal. Finally, the phase coherence function is defined as:

$$C_\phi(f_k) = \sqrt{\langle \cos \Delta\phi_{k,n} \rangle^2 + \langle \sin \Delta\phi_{k,n} \rangle^2}. \tag{3}$$

The phase coherence function $C_\phi(f_k)$ has a value between 0 and 1. It quantifies the tendency for the phase difference between the two signals at a particular frequency to remain constant (where the phase coherence value would be close to 1). When two oscillations are unrelated, their phase difference changes continuously in time and so their phase coherence is close to zero.

Phase coherence values must, however, be interpreted with care. In finite-length signals, low-frequency components are represented by fewer periods than high-frequency components. Consequently, less variation of phase difference occurs at low-frequencies, resulting in an artificially increased phase coherence. In fact, the phase coherence between two completely unrelated signals increases monotonically towards lower frequencies. To identify a frequency-dependent floor representing truly significant coherence, we employed surrogate analysis [16]. Amplitude-adjusted Fourier transform (AAFT) surrogate signals were generated by shuffling the phases of the original time series to create new time series with the same means, variances, autocorrelation functions (and therefore, the same power spectra) as the original sequences, but without any phase relations. We averaged 100 wavelet phase coherences calculated from surrogate signals, and considered that wavelet coherence from the original recording was statistically significant when it was two standard deviations above the mean surrogate coherence.

## 4. Experiments on Coherence Between Blood Flow and Oxygen Saturation

Using the O2C with two probes (LF-2), simultaneous measurements of blood flow and oxygen saturation were recorded for 30 min from 88 healthy normotensive subjects (50 females, 38 males, aged $48 \pm 17$ years, age range 18–90 years, systolic blood pressure $< 150$ mmHg, diastolic blood pressure $< 90$ mmHg), lying relaxed in a supine position. All the signals were sampled at 40 Hz. The probes were positioned on the right arm (forearm, inner side and center) and right leg (middle of shinbone). Skin temperature was measured using a temperature sensor (Thermilinear, YSI Inc., USA) placed on the right forearm and the group skin temperature was $31.8 \pm 1.3$°C. Simultaneous recordings of blood flow and oxygen saturation in the skin are shown in Fig. 1.

Group median values and ranges of blood flow and oxygen saturation are presented in Table 1. Data are given for both depths and locations. For each subject, the average across the whole recording was calculated.

Median wavelet power spectra of blood flow and oxygen saturation signals from the arm are presented for the group in Fig. 2. Spectra of signals recorded at two

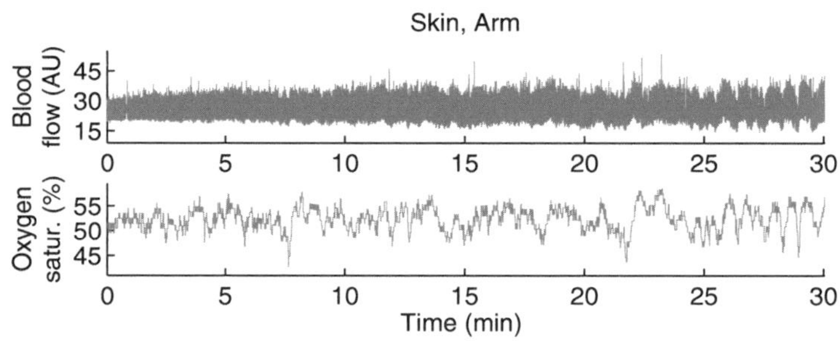


Fig. 1. Simultaneous recordings of blood flow using LDF (top panel), and oxygen saturation using optical reflectance spectroscopy (bottom). The signals were simultaneously recorded with the O2C instrument on the same location from the skin of the right arm.

Table 1. Median values and ranges [25th percentile and 75th percentile] of blood flow and oxygen saturation ($N = 88$).

| Signal, probe position | Skin | Deep |
|---|---|---|
| Blood flow (AU), Arm | 18.85 [11.70 34.81] | 77.31 [54.50 128.06] |
| Blood flow (AU), Leg | 21.80 [16.62 28.73] | 73.44 [49.00 112.36] |
| Oxygen saturation (%), Arm | 46.68 [41.20 55.74] | 87.54 [84.82 89.34] |
| Oxygen saturation (%), Leg | 43.35 [36.35 53.70] | 75.54 [65.58 80.65] |

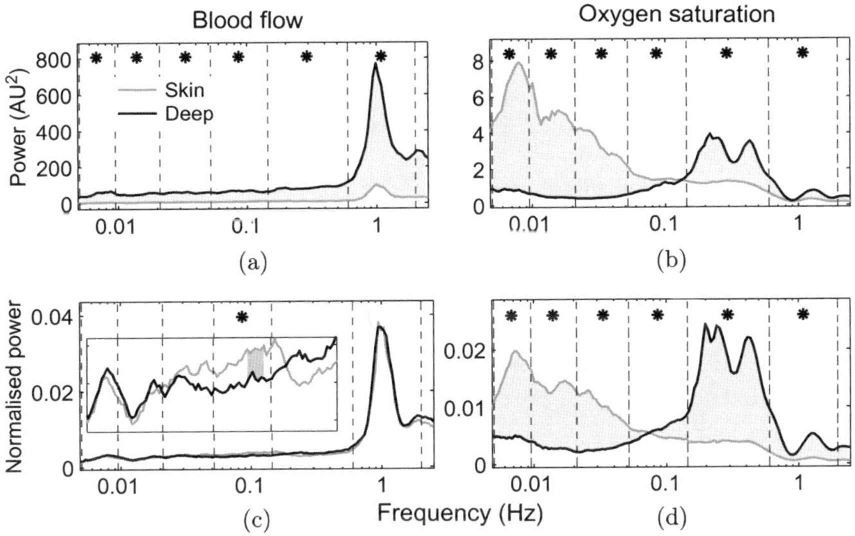


Fig. 2. Wavelet power spectra of signals recorded from the skin and deeper tissue of the arm are compared within the 0.0095–2 Hz frequency range. Group median spectra of blood flow ((a) and (c)) and oxygen saturation ((b) and (d)) are presented in terms of absolute ((a) and (b)) and normalised ((c) and (d)) power spectra. A paired test was used to calculate significant differences ($P < 0.05$) at specific frequencies (light grey shading) or within the defined frequency intervals (indicated by asterisks). Blood flow spectra are similar for recordings from the skin and deeper tissue ((a) and (c)), whereas the oxygen saturation spectra are clearly different ((b) and (d)). The inset in (c) is to illustrate the low frequency contribution in greater detail.

depths (skin and deeper) are compared in terms of (a, b) absolute and (c, d) normalized power. Normalized spectra were calculated by dividing the power spectra by the total power within the whole frequency range from 0.005 Hz to 2 Hz for each subject. A paired signed rank test was used to determine statistically significant differences ($P < 0.05$) between the two depths for each specific frequency (indicated by light grey shading). Average power and normalized power were also calculated within each of the six frequency intervals. Significant differences within the intervals are indicated by asterisks.

Blood flow recorded from the deeper tissue has significantly more power across the whole frequency range compared to blood flow from the skin (Fig. 2(a)). The comparison of normalized power spectra indicates that the distribution of power is very similar (Fig. 2(c)) in both cases. Note that the significant difference around 0.1 Hz (asterisk) is obtained by a paired test. In contrast to blood flow, spectra of oxygen saturation from the skin and from the deeper tissue are clearly different. In the deeper tissue, most of the power is concentrated around the respiratory frequency while oxygenation spectra from the skin contain most of the power at frequencies below 0.1 Hz (Figs. 2(b) and 2(d)). The findings are similar for both arm and leg.

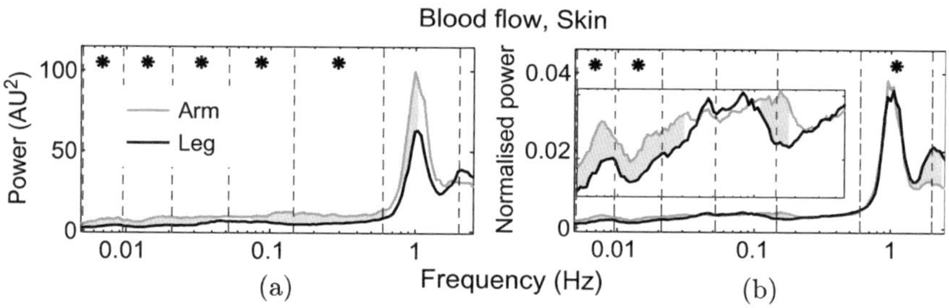


Fig. 3. The wavelet power spectra of skin blood flow signals from the arm and leg are compared within the 0.0095–2 Hz frequency range. A paired test was used to reveal significant differences ($P < 0.05$) at specific frequencies (grey shading) or within the defined frequency intervals (indicated by asterisks). Median group power is significantly higher in the arm compared to the leg (a), while the normalized spectra show that in terms of proportion of power the differences are significant within the cardiac and endothelial-related intervals (b). The inset provides greater detail of the low frequency components.

A comparison of skin blood flow power spectra from the arm and leg reveals significantly higher power at frequencies below 1 Hz for the arm compared to the leg (Fig. 3). In terms of distribution of power, blood flow in the leg contains less power in the cardiac and the endothelial-related frequency intervals compared to the arm (Fig. 3(b)). Note that in the current study, LDF recordings were taken from the forearm and on the shin bone in contrast to previous work [1–4], where the probes were positioned on the bony prominences on the wrist and the ankle.

Wavelet phase coherence between blood flow and oxygen saturation was calculated for both arm and leg, at both depths. The median coherence for 88 subjects is presented in Fig. 4. Results from skin on the arm and leg are shown in panels (a) and (b), respectively; results from the deeper measurements on the arm and leg are shown in (c) and (d). For each pair of recordings, 100 AAFT surrogate signals and their phase coherence were calculated. The mean coherence is indicated by the lower, dash-dotted line. The higher dashed line indicates a level two standard deviations above the mean for each frequency. On the arm, there is a statistically significant phase coherence between blood flow and oxygen saturation across a wide frequency range from interval III down to interval VI, and also at the cardiac frequency in interval I (Fig. 4(a)). On the leg, the coherence is higher at the cardiac frequency compared to the arm, but it is lower or nonsignificant within intervals V and VI (Fig. 4(b)). There is no coherence between blood flow and oxygen saturation recorded from deeper tissue, neither on the arm nor on the leg (Figs. 4(c) and 4(d)).

## 5. Discussion

Coherence between simultaneously measured blood flow and oxygen saturation in the same catchment volume in the microcirculation has not previously been

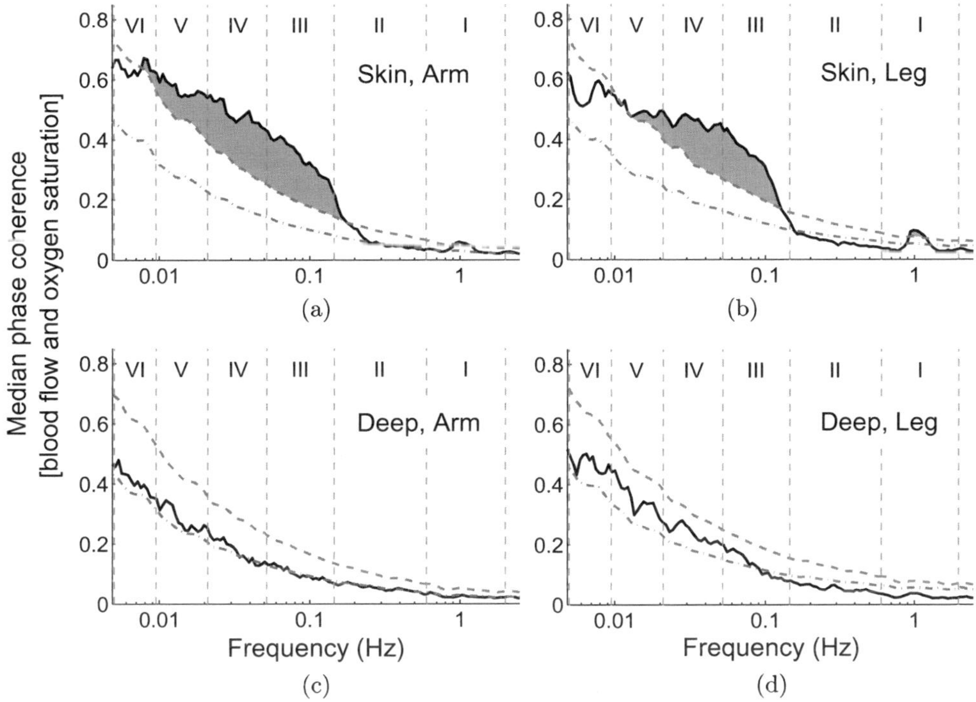


Fig. 4. Wavelet phase coherence between blood flow and oxygen saturation recorded from the arm ((a) and (c)) and the leg ((b) and (c)) and from two depths, skin ((a) and (b)) and deeper tissue ((c) and (d)). In each case, the median coherence for 88 subjects is indicated by the full line and the dashed lines show the mean, and two standard deviations above the mean, respectively, for the coherence calculated from 100 surrogate signals per subject. The coherence is significant across a wide frequency intervals for signals recorded from the skin ((a) and (b)), but not so from the deeper tissue ((c) and (d)).

reported. Following a theoretical work [17] suggesting that oscillations in microcirculatory flow can have a significant effect on tissue oxygenation, Goldman and Popel [18] used their capillary network model to identify the largest effect in the frequency range between 0.025 Hz and 0.05 Hz. This interval coincides with our interval IV, where we observe significant coherence for skin on both arm and leg. Studies of interactions between simultaneously measured signals are few and mostly involve cerebral circulation. Apart from this, the approaches used to measure flow or oxygenation are often invasive and not focused on the same tissue. To our knowledge, the O2C is unique in its ability to measure both signals from the same location, noninvasively, and with a high temporal resolution. Using an O2C, Thorn *et al.* [12] performed measurements similar to ours and identified oscillations in skin oxygen saturation below 0.02 Hz. Based on parameters, defined in the time domain, they associated them with oscillations in skin blood flow on the arm. We show significant coherence at this frequency (interval V) for the arm, but only partly so for the leg.

The difference between the limbs in terms of coherence in the skin might be related to the relatively higher contribution of blood flow oscillations at low frequencies for the arm compared to the leg (Fig. 3). The specific effect of ageing on the parameters discussed in this paper will be presented elsewhere.

There is no coherence between blood flow and Hb oxygen saturation at the deeper recording, neither for the arm nor for the leg. We assume that this is due to the very different power spectra of oxygen saturation in the deeper tissue compared to the skin, together with the observation that the corresponding blood flow spectra are relatively similar. Whether this has a physiological explanation, or is related to the measurement technique, is beyond the scope of this study and will have to be further investigated.

Cross-spectra are commonly used to investigate relationships between signals by identifying frequency regions with high common power. Using wavelets, this can be further extended to determine these regions in the time-frequency plane [19]. However, common power at a particular frequency and time does not necessarily imply common cause. On the other hand, a consistent phase relationship is much more suggestive of causality between signals [20]. In the current study, the wavelet transform with logarithmic frequency resolution was used to extract phase information from the signals, and wavelet phase coherence was applied to identify oscillations with consistent phase differences. This approach allowed us to find significant coherence even when the common power was low, which is important in the case of cardiovascular signals where low-frequency components make important, but not necessarily large, contributions. To determine instantaneous phases of oscillations, the wavelet transform applies the same approach as the Hilbert transform. The difference is that, with wavelets, the window is localized and slides within the signal, thus enabling one to trace the variations of phase in time.

Spectroscopic methods, such as used in NIRS or the O2C device, determine oxygen saturation in the capillary-venous part of the microcirculation. Thus the amount of oxygen bound to hemoglobin is assessed after distribution of oxygen into the tissue and the oxygen saturation signal therefore illuminates local oxygen utilization. Oscillations can be identified in the time series but, due to the complex nature of the blood flow in the capillaries, oxygen saturation on its own is insufficient to explain their origin. It is the O2C facility for simultaneous measurements of flow and saturation in the same location that allows an analysis of their interactions to reveal further information about the mechanisms of oxygen consumption within the microcirculation. Wavelet phase coherence has identified a matching of phase differences over a wide frequency interval, clearly implying causal relationships between oscillations in blood flow and in oxygen saturation.

## 6. Conclusion

In conclusion, we have demonstrated for the first time the existence of coherence between simultaneously measured blood flow and oxygen saturation in the skin microcirculation of healthy subjects; no such coherence was observed in deeper layers containing skeletal muscle and larger vessels. The dynamical pattern of blood flow differs between these two layers but is scaled so that the normalized power of each oscillatory component is comparable. The dynamics of blood oxygenation is completely different between the two layers. While the power of the skin oxygenation signal comes mainly from low frequency components, that in deeper tissue comes mainly from oscillations at around the respiratory frequency. This is a novel and interesting result, and its physiological meaning has yet to be determined. It may be useful to establish the pattern of coherence in cardiovascular diseases and skin cancer.

## Acknowledgments

We thank the subjects who volunteered for and participated in this research. The research was supported by the ESRC, UK "New Dynamics of Ageing" programme.

## References

[1] A. Stefanovska, M. Bračič and H. D. Kvernmo, Wavelet analysis of oscillations in the peripheral blood circulation measured by laser Doppler technique, *IEEE Trans. Bio. Med. Eng.* **46** (1999) 1230–1239.

[2] A. Stefanovska and M. Bračič, Physics of the human cardiovascular system, *Contemp. Phys.* **40** (1999) 31–55.

[3] P. Kvandal, S. A. Landsverk, A. Bernjak, A. Stefanovska, H. D. Kvernmo and K. A. Kirkebøen, Low frequency oscillations of the laser Doppler perfusion signal in human skin, *Microvasc. Res.* **72** (2006) 120–127.

[4] Y. Shiogai, A. Stefanovska and P. V. E. McClintock, Nonlinear dynamics of cardiovascular ageing, *Phys. Rep.* **488** (2010) 51–110.

[5] Z. Y. Li, Y. H. Wang, Y. Li, Y. Wang, J. P. Li and L. L. Zhang, Wavelet analysis of cerebral oxygenation signal measured by near infrared spectroscopy in subjects with cerebral infarction, *Microvasc. Res.* **80** (2010) 142–147.

[6] H. Obrig, M. Neufang, R. Wenzel, M. Kohl, J. Steinbrink, K. Einhaupl and A. Villringer, Spontaneous low frequency oscillations of cerebral hemodynamics and metabolism in human adults, *Neuroimage* **12** (2000) 623–639.

[7] M. D. Stern, In vivo observation of microcirculation by coherent light scattering, *Nature* **254** (1975) 56–58.

[8] G. E. Nilsson, T. Tenland and P. L. Öberg, Evaluation of a laser Doppler flowmeter for measurement of tissue blood flow, *IEEE Trans. Biomed. Eng.* **27** (1980) 597–604.

[9] D. M. Mancini, L. Bolinger, H. Li, K. Kendrick, B. Chance and J. R. Wilson, Validation of near-infrared spectroscopy in humans, *J. Appl. Physiol.* **77** (1994) 2740–2747.

[10] R. Boushel and C. A. Piantadosi, Near-infrared spectroscopy for monitoring muscle oxygenation, *Acta Physiol. Scandinav.* **168** (2000) 615–622.

[11] A. Villringer and B. Chance, Non-invasive optical spectroscopy and imaging of human brain function, *Trends. Neurosci.* **20** (1997) 435–442.

[12] C. E. Thorn, H. Kyte, D. W. Slaff and A. C. Shore, An association between vasomotion and oxygen extraction, *Am. J. Physiol.: Heart Circ. Physiol.* **301** (2011) H442–H449.

[13] A. Stefanovska, Dynamics of blood oxygenation gives better insight into tissue hypoxia than averaged values, *Am. J. Physiol.: Heart Circ. Physiol.* **296** (2009) H1224–H1226.
[14] A. Bandrivskyy, A. Bernjak, P. V. E. McClintock and A. Stefanovska, Wavelet phase coherence analysis: Application to skin temperature and blood flow, *Cardiovasc. Eng.* **4** (2004) 89–93.
[15] L. W. Sheppard, V. Vuksanović, P. V. E. McClintock and A. Stefanovska, Oscillatory dynamics of vasoconstriction and vasodilation identified by time-localized phase coherence, *Phys. Med. Biol.* **56** (2011) 3583–3601.
[16] T. Schreiber and A. Schmitz, Improved surrogate data for nonlinearity tests, *Phys. Rev. Lett.* **77** (1996) 635–638.
[17] A. G. Tsai and M. Intaglietta, Evidence of flowmotion induced changes in local tissue oxygenation, *Int. J. Microcirc. Clin. Exp.* **12** (1993) 75–88.
[18] D. Goldman and A. S. Popel, A computational study of the effect of vasomotion on oxygen transport from capillary networks, *J. Theor. Biol.* **209** (2001) 189–199.
[19] C. Torrence and G. P. Compo, A practical guide to wavelet analysis, *Bull. Am. Meteorol. Soc.* **79** (1998) 61–78.
[20] A. Grinsted, J. C. Moore and S. Jevrejeva, Application of the cross wavelet transform and wavelet coherence to geophysical time series, *Nonlin. Process Geophys.* **11** (2004) 561–566.

# 5

# Noise in Materials, Circuitry, Devices and Sensing

# Memristor Equations: Incomplete Physics and Undefined Passivity/Activity

Kyle M. Sundqvist*, David K. Ferry† and Laszlo B. Kish‡

*Department of Physics, San Diego State University
5500 Campanile Drive, San Diego, CA 92182-1233, USA

†School of Electrical, Computer, and Energy Engineering and
Center for Solid State Electronic Research
Arizona State University, Tempe, AZ 85287-5706, USA

‡Department of Electrical and Computer Engineering
Texas A&M University, TAMUS 3128
College Station, TX 77843 3128, USA

Received 12 July 2017
Accepted 5 August 2017
Published 20 September 2017

Communicated by Igor Goychuk

In his seminal paper, Chua presented a fundamental physical claim by introducing the memristor, "The missing circuit element". The memristor equations were originally supposed to represent a passive circuit element because, with active circuitry, arbitrary elements can be realized without limitations. Therefore, if the memristor equations do not guarantee that the circuit element can be realized by a passive system, the fundamental physics claims about the memristor as "missing circuit element" loses all its weight. The question of passivity/activity belongs to physics thus we incorporate thermodynamics into the study of this problem. We show that the memristor equations are physically incomplete regarding the problem of passivity/activity. As a consequence, the claim that the present memristor functions describe a passive device lead to unphysical results, such as violating the Second Law of thermodynamics, in infinitely large number of cases. The seminal memristor equations cannot introduce a new physical circuit element without making the model more physical such as providing the Fluctuation–Dissipation Theory of memristors.

*Keywords*: Memristors; incomplete models; activity; passivity.

## 1. Introduction

### 1.1. *On Chua's memristor model*

In 1971, Chua introduced the memristor, a new mathematical circuit theory element [1] that, to be fundamentally interesting, was supposed to be a passive device.

‡Corresponding author.

The mathematical model interrelates the time-integral of the voltage (voltage flux) on the memristor

$$\Phi = \int_0^t U(\tau)d\tau \tag{1}$$

with the time-integral of the current (charge) through the device

$$q = \int_0^t I(\tau)d\tau \tag{2}$$

as follows:

$$M(q) = \frac{d\Phi}{dq}, \tag{3}$$

$$U(t) = M[q(t)]I(t), \tag{4}$$

where $M(q)$ is the memristor function see Fig. 1. It is claimed that for the memristor to be a *passive* circuit element the following condition must be satisfied:

$$M(q) \geq 0. \tag{5}$$

(Note, the lower boundary of the integral is often taken from minus infinity, which is unphysical. The more physical time coordinate "zero" can be the moment of time when the memristor comes into existence, and an initial value of the integral can be given, if needed. However, this issue has no importance in our present paper).

If the memristor function is constant, $M_0$, then the memristor and Eq. (4) represents a resistor of resistance $R_0$, that is, $M(q) = M_0$, where $R_0 = M_0$. Then, Eq. (4) becomes

$$U(t) = R_0 I(t). \tag{6}$$

One of the most fundamental questions about Chua's memristor model is:

*Do Eqs.* (1)–(5) *indeed always represent a passive device, or there are situations where the realization of a memristor function requires an active device?*

With active circuitry, arbitrary elements can be realized without limitations. Therefore, *if the memristor equations do not guarantee that the circuit element can be realized by a passive system, the fundamental physics claim about the memristor as "the missing circuit element" loses all its weight.*

It is important to note that many nonlinear impedances with memory effects exist since the existence of electronics that are passive devices and can be called memristors (for recent ones, see e.g., [2, 3]). These devices are not of our interest because they are examples for practical passive devices and they do not answer our fundamental physical question about the generality of the memristor claim.

The question of passivity/activity is a physical problem and former engineering definitions of this matter were shown self-contradictory and unphysical at certain practical conditions where they lead to perpetual motion machines [4]. Therefore, we will use the most advanced, statistical-thermodynamic definition of passivity/activity based on statistical physics that works correctly even with thermal noise (the quoted text below is from [4]):

(a) The device-in-question is active if the following condition holds. Suppose, we have a hypothetical-device, which does not require an external energy source to function and has the same signal-response characteristics as the device-in-question. In an isolated system with thermal equilibrium, such a hypothetical-device in a proper circuit would be able to produce steady-state entropy reduction in the system that is originally in thermal equilibrium, where the other elements are all passive. In other words, such hypothetical device would violate the Second Law of Thermodynamics.
Note for example, such an entropy reduction could be a steady-state positive or negative temperature gradient in the device's environment; etc.; anything that can break thermal equilibrium conditions and persists over the duration of operation of the device.
(b) Physical implication: Such a hypothetical-device cannot exist in practice, thus an active physical device always requires an external energy source to execute its response characteristics of activity.
(c) The device-in-question is passive if it is not active.

The structure of the paper is as follows:

In Sec. 2, we briefly outline the Fluctuation–Dissipation Theorem (FDT) of classical statistical physics.

In Sec. 3, we show why the above memristor equations are only abstract circuit theoretical models that are unphysical and/or incomplete in their present form and that the present model requires an active device in an infinite number of situations including the case when the memristor emulates a resistor.

## 2. FDT and Thermal Noise of Impedances and Resistors

Contrary to Chua's memristor model (Eqs. (1)–(6)), the statistical thermodynamics of the three fundamental circuit elements, that is, the resistor, capacitor and inductor, is sufficiently defined via the FDT [5–8]. It states that, in the classical

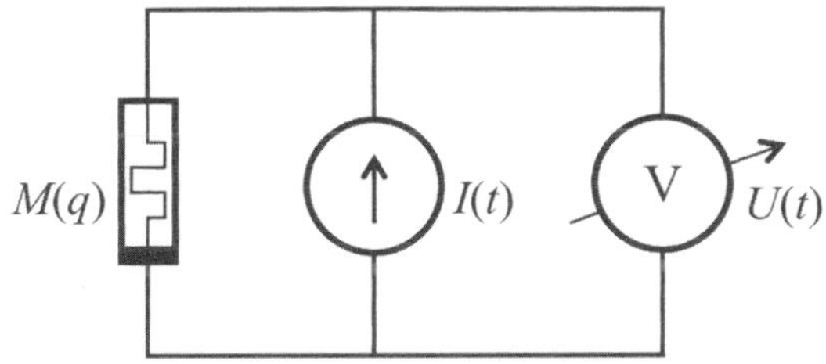


Fig. 1. Memristor driven by a current generator while the voltage on it is monitored. For definitions, see Eqs. (1)–(5) in the text.

physical limit, any impedance combined from these elements will act as a thermal noise generator with an internal impedance value identical to this particular impedance.

The Second Law of Thermodynamics requires [5–8] that, in thermal equilibrium, the time-average of the instantaneous power flow between two parallel impedances is zero, i.e.,

$$\langle P_{a\rightarrow b}(t,T)\rangle_t = 0, \tag{7}$$

where $t$ is time and $P_{a \Longleftrightarrow b}(t,T)$ denotes instantaneous power flow between resistors $Z_a$ and $Z_b$, as illustrated in Fig. 2. This expression holds for any frequency range.

For a *passive* impedance $Z(f)$ in thermal equilibrium, Eq. (7) demands that the functional form of the power density spectral function of noise voltage $U(t)$ is [8]:

$$S_u(f,T) = \mathrm{Re}[Z(f)]Q(f,T) = R(f)Q(f,T), \tag{8}$$

where $R(f)$ is the real part of the impedance and $S_u(f)$ represents the power density spectrum of thermal (Johnson) noise voltage. $Q(f,T)$ is a universal function of frequency and temperature and it is independent of material properties, geometry, and mechanism for electrical conduction [5–8].

In the low-frequency (classical physical limit):

$$Q(f,T) = 4kT. \tag{9}$$

For active impedances, there is no related fundamental law and the noise is determined by also other factors not visible in Eq. (8).

For inductors and capacitors, Eq. (8) implies zero thermal noise. For a resistor, Eqs. (8) and (9) imply that the voltage noise spectrum is $S_U(f,T) = 4kTR$ and its Norton-equivalent, the thermal noise spectrum is $S_I(f,T) = 4kT/R$.

Let us assume that the impedance is a resistor $R_0$. Then, the current driven resistor will produce the following voltage:

$$U(t) = R_0 I(t) + U_n(t), \tag{10}$$

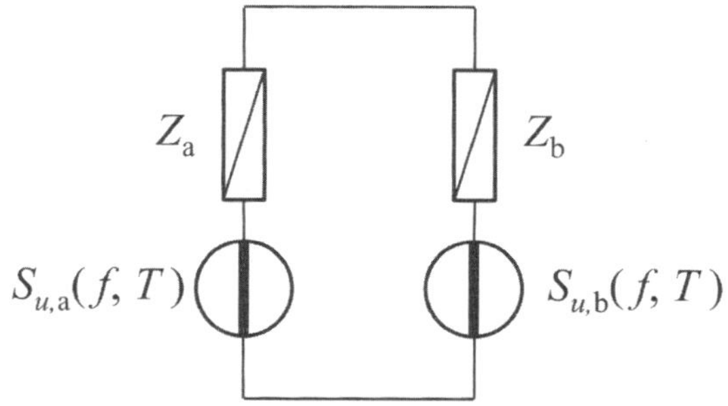


Fig. 2. Circuit for discussing Johnson noise and representing net power flow between resistors $Z_a$ and $Z_b$. The voltage generators produce $S_u(f,T)$ power density spectra at frequency $f$ and temperature $T$.

where $U_n(t)$ is a Gaussian thermal noise with mean-square value $\int_{f_L}^{f_H} 4kTR_0 df$, where $f_L$ and $f_H$ are the low and high frequency cutoffs of the frequency band of observation, respectively.

## 3. The Memristor Equations are Incomplete and Unable to Define a Physical Device

### 3.1. *The case of linear memristors*

There is a striking difference between Eqs. (6) and (10). Equation (6) with zero thermal noise for the memristor emulating a resistor is unphysical unless the memristor is an active device or, if it is passive, it must be at zero Kelvin absolute temperature to have zero noise. (Note, even the assumptions that an active impedance can have zero noise or that a physical body has zero absolute temperature are unphysical but let us disregard this problem).

To prove the above statement, let us suppose that the memristor is passive. Then, in the arrangement shown in Fig. 2, suppose that $Z_a$ and $Z_b$ are a memristor and a resistor, defined by Eqs. (6) and (10), respectively and both are at the same temperature $T$. As it is well-known, it follows from Eqs. (6) and (8)–(10), that in the frequency band $[f_L, f_H]$, the mean power flow between two identical resistors of $R_0$ at temperatures $T_1$ and $T_2$ is given as (e.g., [5–7]):

$$\langle P_{1\to 2}\rangle_t = 4k(T_1 - T_2)(f_H - f_L)\frac{R_0^2}{4R_0^2} = k(T_1 - T_2)(f_H - f_L). \tag{11}$$

Then, if the memristor model provides a full description, there would be a nonzero mean power flow $\langle P_{R\to M}(T)\rangle_t$ from a passive resistor at temperature $T > 0$ to a parallel memristor with zero thermal noise at the same temperature:

$$\langle P_{R\to M}(T)\rangle_t = kT(f_H - f_L) > 0. \tag{12}$$

However, both the memristor and the resistor are at the same temperature $T$ therefore a nonzero power flow violate the Second Law of Thermodynamics unless there is an external energy source that supplies this power. *The physical conclusion is* [4], *that a memristor described by Eq.* (6) *must be an active device.*

Alternatively, if the given memristor is actually a passive device then we must realize that the memristor equations are incomplete to address the statistical thermodynamics of the system. The statistical physics (FDT) of the memristor is undefined and, without that, Eqs. (1)–(6) are not only incomplete but they are also incorrect if they are used to decide about the passivity/activity of the memristor.

### 3.2. *The case of nonlinear memristors*

The violation of the Second Law due to assuming passivity of the memristors and the validity of Eqs. (1)–(5) goes much beyond the case of linear memristors. Both from the above considerations and from [4], it is obvious, that assuming a thermal

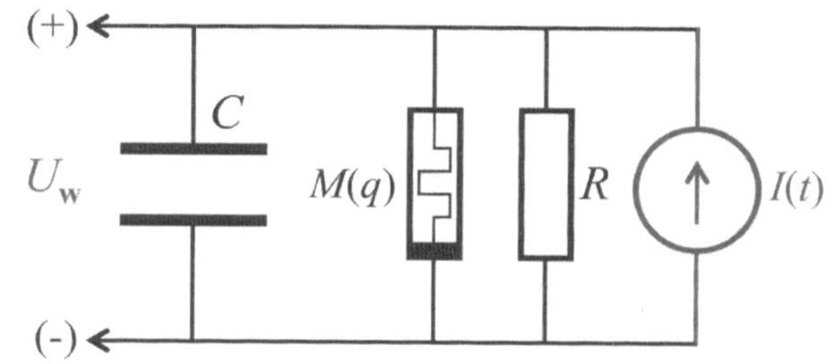


Fig. 3. A DC voltage source by a rectifying-type memristor with the circuit model of the thermal noise current $I(t)$ of a shunt resistor.

noise-free memristor, as Eqs. (1)–(6) do, will break the Second Law whenever the memristor function describes a dissipative device. Then, similarly to the linear case above, connecting a resistor parallel to the memristor will transfer non-zero net thermal noise power to the memristor from the resistor. As a consequence, the resistor is cooled and the memristor is heated.

Here, as the demonstration of infinite number of cases of activity, instead of investigating the general question of dissipativity of memristor functions, we point out that a smaller yet still infinitely large number of cases exists when the nonlinear memristor described by Eqs. (1)–(5) must be active: the case when an asymmetric memristor function shows rectifying characteristics, even at the slightest level, see Fig. 3.

If the thermal noise current flowing through the memristor and due to the proper asymmetry of the $\Phi$ and $M$ functions, the $U_w$ voltage on the capacitor has non-zero

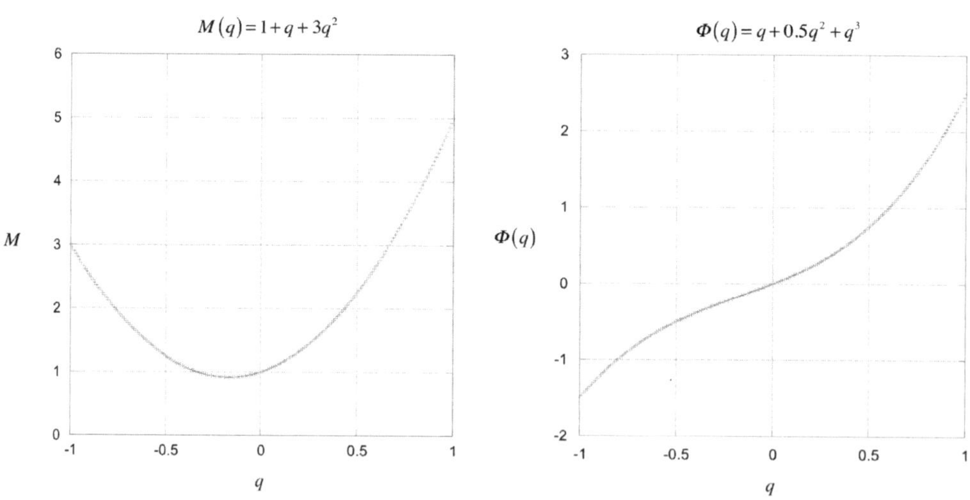


Fig. 4. Memristor functions with rectifying characteristics. The asymmetry of $M$ provides an asymmetry in $\Phi$ and a nonzero DC component in $U_w$ when driven by an AC current. In physical cases, such $M$ and $\Phi$ functions will be bounded at higher values however it is unnecessary to deal with this issue for thermal noise, which is the weakest signal in the system.

mean (DC) value, even if it has only an infinitesimally small DC component, a perpetual motion machine can be built. The same situation for an unphysical (that is, noise-free) diode was already pointed out by Brillouin [9].

We gave a proof of Brilluoin's paradox in [4] by showing that putting a large number $N$ of circuits shown in Fig. 3 into a series circuitry results in an unbounded available power proportional to $N$ due to the DC components that are always positively correlated. In conclusion, any passive circuitry that can rectify thermal noise, even in the slightest way, in unphysical because it breaks the Second Law. Such circuit must contain an active element.

As an illustrative example, we created simple memristor functions that are able to rectify, see Fig. 4. It is obvious that these simple examples represent an infinite set of rectifying memristor functions

$$\Phi(q) = aq + bq^2 + cq^3, \tag{13}$$

if we choose the coefficients of $q$ so that Eq. (5) is satisfied and each of these coefficients are greater than zero.

## 4. Conclusion

In their present noise-free form, the memristor Eqs. (1)–(5) require the presence of an active device in infinitely many cases of memristor functions.

Chua's general proof that the memristor is passive based on Eqs. (1)–(5) is invalid because the memristor equations do not provide a complete description of the physics of memristors. A relevant statistical thermodynamics is needed for the memristors however the memristor equations do not contain sufficient physics to deduce that.

While the response functions of passive resistors, capacitors and inductors determine their FDT, the memristor model lacks sufficient physics for a FDT. To create the noise theory (FDT) of a memristor, it is essential to know the internal material structure of the particular memristor thus its general formulation, like it is done with the three basic circuit elements, is impossible.

Finally, we note that, in a similar, nonlinear situation Nico van Kampen had already showed [10] that the vacuum diode equations lead to unphysical results if the thermal noise of the diode is not included. He deduced the correct theory of noise in freestanding diodes, which is a relevant illustration of the essentially missing component of memristor theories.

## References

[1] L. Chua, Memristor — The missing circuit element, *IEEE Trans. Circuit Theory* **18** (1971) 507–519.

[2] D. B. Strukov, G. S. Snider, D. R. Stewart and R. S. Williams, The missing memristor found, *Nature* **453** (2008) 80–83.

[3] L. Chua, Resistance switching memories are memristors, *Appl. Phys. A* **102** (2011) 765–783.

[4] K. Sundqvist, D. K. Ferry and L. B. Kish, Second Law based definition of passivity/activity of devices, accepted for publication in *Phys. Lett. A*, doi:10.1016/j.physleta.2017.08.039; arXiv:1705.08750.
[5] L. B. Kish, G. A. Niklasson and C. G. Granqvist, Zero-point term and quantum effects in the Johnson noise of resistors: A critical appraisal, *J. Stat. Mech.* **2016** (2016) 054006.
[6] L. B. Kish and T. Horvath, Notes on recent approaches concerning the Kirchhoff-Law-Johnson-Noise-based secure key exchange, *Phys. Lett. A* **373** (2009) 2858–2868.
[7] L. B. Kish, *The Kish Cypher* (World Scientific, Singapore, 2017).
[8] L. B. Kish, G. A. Niklasson and C. G. Granqvist, Zero thermal noise in resistors at zero temperature, *Fluct. Noise Lett.* **15** (2016) 1640001.
[9] L. Brillouin, Can the rectifier become a thermodynamical demon?, *Phys. Rev.* **78** (1950) 627–628.
[10] N. G. van Kampen, Non-linear thermal fluctuations in a diode, *Physica* **26** (1960) 585–604.

# USE OF THE ONE-AND-A-HALF-SPECTRUM TO DETERMINE BARKHAUSEN NOISE ASYMMETRY

A. C. MILLS* and M. B. WEISSMAN†
*Physics Department, University of Illinois at Urbana-Champaign*
*1110 West Green St., Urbana, IL, 61801-3080*
†*mbw@uiuc.edu*

G. DURIN
*Istituto Nazionale di Ricerca Metrologica*
*str. delle Cacce 91, 10135 Torino, Italy*

Received 9 October 2006
Revised 26 December 2006
Accepted 4 January 2007
Communicated by Edmunk Nowak

We demonstrate the use of the one-and-a-half spectrum as a way of looking at the time-asymmetry of Barkhausen noise. This method is of particular value because it allows us to look in the regime where individual Barkhausen pulses are no longer distinguishable. We use the one-and-a-half spectrum on Barkhausen data from two different ferromagnetic samples and find asymmetries in the noise of both. The magnitude of the asymmetry is different in the two samples and we find no clear change in the asymmetry when we move from the regime of well-separated jumps at low field rate to the continuous fluctuating signal at higher rates.

*Keywords*: Barkhausen noise; domain wall motion; higher-order spectra; time-asymmetry.

## 1. Introduction

Testing models of Barkhausen noise and other driven 'crackling' noises generally requires use of statistical properties beyond the ordinary power spectrum, since such noises typically have rather non-informative spectra with few distinct features. Most theoretical discussion of higher-order statistics have been framed in terms of properties of individual events- 'pulses' or 'avalanches'. Such characterizations are usually adequate so long as such discrete events actually occur and so long as it is practical to isolate them experimentally. Neither condition holds in general, hence it is valuable to have other statistical techniques which get at the same informa-

*Current address: 118 Chestnut St., Kitchener, Ontario N2H 1V1 Canada tion found in pulse-parameter distributions but without requiring the existence and detectability of discrete pulses.

In this paper we pick one well-established and now understood statistical feature, the time asymmetry of the Barkhausen pulses in metallic magnets, and show that it is measurable by higher-moment analysis of noise, in this case a third-moment analysis using modified Haar transform techniques. We show experimentally that this time asymmetry is little changed when the field sweep rate is too high to sustain separable pulses, in agreement with the current understanding that the asymmetry arises from inductive delay in the onset of eddy currents [1], rather than from some more subtle property of avalanche dynamics.

## 2. Experiment

We measure the Barkhausen noise in the usual way, by finding the voltage across a pick-up coil surrounding a ferromagnetic sample undergoing slow changes in external magnetic field. The sample is placed in a solenoid which drives the sample to its saturation magnetization by a triangle wave along its long axis. The ends of the sample are connected by a high-permeability yoke to reduce the demagnetising field. The Barkhausen noise is measured by a copper pick-up coil wound around the middle of the sample. The voltage from the pick-up coil is amplified, then anti-alias filtered and digitised. To prevent pick-up from ambient fields, the set-up is shielded by a high-permeability canister. Since domain wall motion dominates over other means of magnetisation in the steepest part of the hysteresis loop, data are generally collected in this range, near the centre of the hysteresis loop.

The $Fe_{21}Co_{64}B_{15}$ sample is an amorphous material under no tensile stress measuring 21 cm x 1 cm x 30$\mu$m. We use the dimensionless parameter $c$ introduced by Alessandro and co-workers [2] to describe the field rate. This is convenient because it divides the data into two regimes: for $c << 1$ pulses are well-separated while for $c > 1$ individual pulses are not distinguishable. The $Fe_{21}Co_{64}B_{15}$ sample was driven at frequencies of 0.005, 0.01, 0.1 and 0.173 Hz, corresponding to $c = 0.045$, $c = 0.09$, $c = 0.9$ and $c = 1.5$, respectively. The FeSi 7.8% wt sample is polycrystalline and has been shown to have strong long-range interactions [3]; it measures 30 cm x 0.5 cm x 60 $\mu$m. It was driven at frequencies of of 0.005, 0.01, 0.02 and 0.05 Hz, corresponding approximately to $c = 0.11$, $c = 0.22$, $c = 0.65$ and $c = 1$, respectively.

## 3. $S_{1.5}(f_1, f_2)$: The One-and-A-Half Spectrum

We use a variation on the 1.5-spectrum developed in references [4] and [5] to find the frequency-dependent time-asymmetry. The idea is to find the cross-correlation function of the signal itself with its noise power (at a higher frequency). Following Ref. [4], we consider a time series

$$\{v[t_i] : 0 \le i \le 2^m - 1\}$$

where $t_i - t_{i-1} = 1/f_s$ and $f_s$ is the sampling frequency.

We can coarsen the voltage signal by making running time averages of the signal over different time scales:

$$V_n(j) \equiv \sum_{i=2^n j}^{2^n(j+1)-1} v(t_i)/2^n \tag{1}$$

for $0 \leq n < m$ and $0 \leq j \leq 2^{m-n} - 1$. In this way, we average non-overlapping collections of $2^n$ points to get a sequence of voltages with an effective sampling rate that we define as $f_1^{(n)} = f_s/2^n$. To capture the noise power, we create a new sequence for each $n$ called the Haar power:

$$P_{n+1}(j) \equiv [V_n(2j+1) - V_n(2j)]^2 \tag{2}$$

for $0 \leq j \leq 2^{m-n-1} - 1$.

We now need to find the cross-spectrum between $V_n(j)$ and $P_n(j)$. In reference [4], this was done by finding the Fourier transforms of both and then multiplying one by the complex conjugate of the other. However, Haar transforms offer the advantage of computational simplicity, and we will use them below. The Haar function is defined as [6]:

$$\psi(x) = \begin{cases} 1\, 0 \leq x < 1/2 \\ -1\; 1/2 \leq x < 1 \\ 0 \text{ otherwise.} \end{cases} \tag{3}$$

The basis functions of the Haar transform are the collection of functions $\psi_{g,h}(x)$ where $\psi$ is coarsened by $g$ and shifted by $h$ in the following way [6]:

$$\psi_{g,h}(x) = 2^{-g/2}\psi(2^{-g}x - h). \tag{4}$$

We use an unconventional variation on the Haar transform in order to look for time asymmetry. We first define $\overrightarrow{H_k}$ of a sequence $f$ (of length $2^N$) by

$$\overrightarrow{H_k(f)} = (H_k(f)_0, H_k(f)_1, \ldots, H_k(f)_{2^{N-k}}) \tag{5}$$

for $0 \leq j < 2^{N-k}$ where $k$ represents the frequency octave. What we call the real part of the complex vector $\overrightarrow{H_k}$ is the usual Haar transform. The imaginary part of $\overrightarrow{H_k}$ comes from using a shifted Haar tranform, as described below. Essentially, we divide $f(i)$ up into disjoint "packages" of length $2^k$. Within each package, we calculate the real part of the transform by adding up the values of $f$ in the first half of the package and subtracting the values of $f$ in the second half of the package. In this way, the $j$-th element of the real part of the vector is given by:

$$Re(H_k(f))_j = \sum_{i=2^k j}^{2^k j+2^{k-1}-1} f(i) \quad - \sum_{i=2^k j+2^{k-1}}^{2^k(j+1)-1} f(i) \tag{6}$$

for $1 \leq k \leq n-1$. This is the usual Haar transform.

Analogous to the relationship between cosine and sine, the imaginary part of $\overrightarrow{H_k}$ is generated by shifting the regular Haar function by one quarter of the "package length" (by $2^{k-2}$). That is, this time we will add up the values of the first and last quarters of the package and subtract the values of the second and third quarters of the package [7]. For $k = 1$, the imaginary part is zero, analogous to the Nyquist frequency.

To calculate the cross-spectrum, we find

$$S_{1.5}(n,k) = \overrightarrow{H_k(V_n(j))} \cdot \overrightarrow{(H_k(P_n(j))}^{\star} \tag{7}$$

To return to more familiar notation, we need only note that $f_1^{(n)} = f_s/2^n$ and that $f_2^{(k)} = f_s/2^{k+1}$. Then,

$$S_{1.5}(n,k) = S_{1.5}(f_1^{(n)}, f_2^{(k)}) \tag{8}$$

where we usually drop the superscripts on the frequencies.

Since the 1.5-spectrum measures the correlation between the signal and the noise power (at a higher frequency), we can look at

$$Im(S_{1.5}(f_1, f_2)/Re(S_{1.5}(f_1, f_2)) \tag{9}$$

to learn about the phase difference. For time-symmetrical signals, this quotient is zero. For positive pulses, negative results correspond to high-frequency pulses preceding low-frequency pulses, while positive results indicate that low-frequency pulses precede high-frequency pulses. For negative pulses, the signs are reversed.

## 4. Results

Figure 1 gives examples of the raw data on the $Fe_{21}Co_{64}B_{15}$ sample. Figure 1a shows $c = 0.045$ data, in which the pulses are well-separated, as expected for $c << 1$. Figure 1b shows $c = 1.5$ data; here we cannot distinguish between individual pulses.

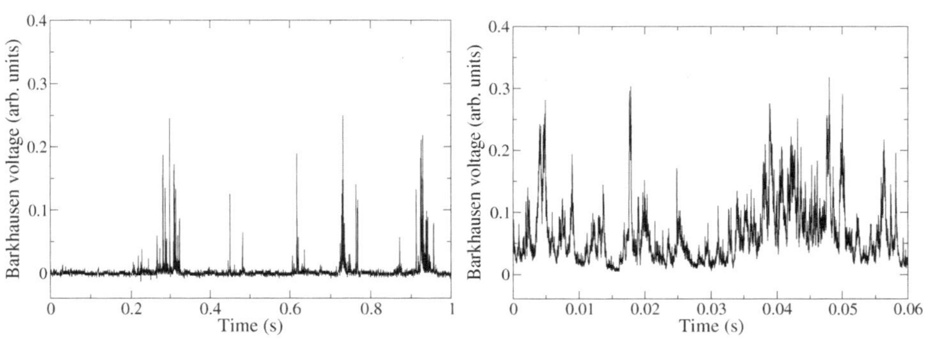


(a) $c = 0.045$ (b) $c = 1.5$.

Fig. 1. Examples of the measured Barkhausen voltage on the $Fe_{21}Co_{64}B_{15}$ sample. On the left, the sweep rate was slow ($c = 0.045$), while the right plot shows higher sweep rate data ($c = 1.5$)

Figure 2 shows plots of $Im(S_{1.5})/Re(S_{1.5})$ versus frequency $f_2$ for the two samples. We see a similar effect in each sample. $Im(S_{1.5})/Re(S_{1.5})$ is near zero at low frequencies ($f_2$); that is, the Barkhausen noise is symmetric in time in this frequency range. At higher frequencies, $Im(S_{1.5})/Re(S_{1.5})$ becomes more and more negative: here the Barkhausen noise is increasingly time-asymmetric.

The negative sign of $Im(S_{1.5})/Re(S_{1.5})$ indicates that the leading edge of the pulses have more high-frequency structure than the trailing edge. This is consistent with observations made by Petta and co-workers [4] and with the asymmetries seen in experimental pulse shape collapses [8–11]. The fact that the asymmetry shows up at high frequencies tells us that it comes from the asymmetry of the pulses themselves, and is not related to the spacing between pulses (also observed by Petta *et al.* [4]).

One can see that there is a significant difference in the asymmetry values. Here we show only $f_1 = f_s/2$, the highest-frequency (lowest $n$) spectrum. The plots show all the different driving rates, or $c$, used for each sample. Since there were many runs for each sample, we plot the averages. In some cases, we have also included error bars, computed from the standard deviation of the mean.

By comparing the asymmetry at the highest $f_2$ common to both samples, $f_2 =$ 6250 Hz, we can see that the $Fe_{21}Co_{64}B_{15}$ sample has a larger asymmetry (-0.30) than the iron-silicon sample (-0.15). In these samples, we see little evidence for a dependence of asymmetry on sweep rate, $c$. We have however, seen some evidence for growing asymmetry with reduced $c$ in other samples [7].

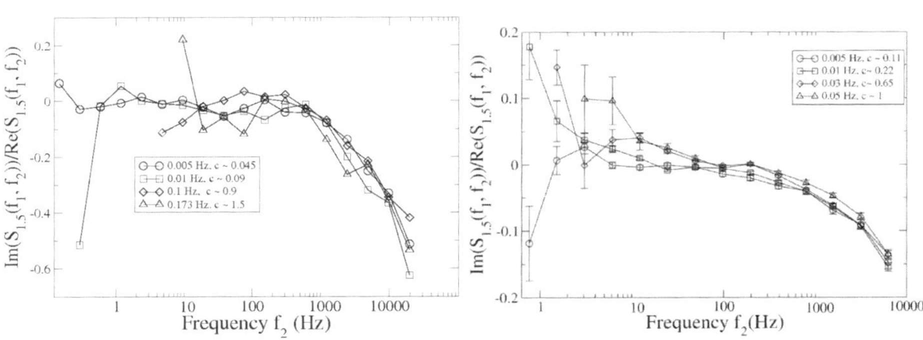


(a) $Fe_{21}Co_{64}B_{15}$ (b) FeSi 7.8 % wt.

Fig. 2. The imaginary part over the real part of the 1.5 spectrum for $Fe_{21}Co_{64}B_{15}$ ($f_1$ = 78 125 Hz) and FeSi 7.8 % wt. ($f_1$ = 25 000 Hz). Each data set represents an average over several files.

## 5. Conclusion

The time asymmetry found by the Haar $S_{1.5}$ method closely resembles that found in collections of pulse shapes. In particular, it does not show scaling over a range of frequencies but rather shows a characteristic frequency, shown elsewhere to correspond to the longest inductive delay, set by sample geometry and conductivity [1]. The $S_{1.5}$ method also shows that the existence of discrete avalanches is not necessary for this time asymmetry to exist nor for it to be measurable. A similar conclusion should apply to various other statistical properties of crackling noise, for which both $S_{1.5}$ and second spectra (fourth moments) can be useful.

## Acknowledgements

We acknowledge support from NSF DMR 02-40644.

## References

[1] S. Zapperi, C. Castellano, F. Colaiori and G. Durin, *Signature of effective mass in crackling-noise asymmetry*, *Nature Phys.* **1** (2005) 46–49.

[2] B. Alessandro, C. Beatrice, G. Bertotti and A. Montorsi, *Domain-wall dynamics and Barkhausen effect in metallic ferromagnetic materials. I. Theory*, *J. Appl. Phys.* **68** (1990) 2901–2908.

[3] G. Durin and S. Zapperi, *Scaling exponents for Barkhausen avalanches in polycrystalline and amorphous ferromagnets*, *Phys. Rev. Lett.* **84** (2000) 4705–4708.

[4] J. Petta, M. Weissman and G. Durin, *Barkhausen pulse structure in an amorphous ferromagnet: Characterization by high-order spectra*, *Phys. Rev. E* **57** (1998) 6363–6369.

[5] K. O'Brien and M. Weissman, *Statistical signatures of self-organization*, *Phys. Rev. A* **46** (1992) R4475–R4478.

[6] I. Daubechies, *Ten Lectures on Wavelets*, Society for Industrial and Applied Mathematics (1992).

[7] A. Mills, *Barkhausen noise studies: Pinning fields and time-asymmetries*, Ph. D. Thesis, University of Illinois at Urbana-Champaign (2004).

[8] D. Spasojevic, S. Bukvic, S. Milosevic and H. Stanley, *Barkhausen noise: Elementary signals, power laws, and scaling relations*, *Phys. Rev. E* **54** (1996) 2531–2546.

[9] G. Durin and S. Zapperi, *On the power spectrum of magnetization noise*, *J. Magn. Magn. Mater.* **242** (2002) 1085–1088.

[10] F. Colaiori, S. Zapperi and G. Durin, *Shape of a Barkhausen pulse*, *J. Magn. Magn. Mat.* **272-276** (2004) E533–E534.

[11] A. Mehta, A. Mills, K. Dahmen and J. Sethna, *Universal pulse shape scaling function and exponents: Critical test for avalanche models applied to Barkhausen noise*, *Phys. Rev. E* **65** (2002) 046139.

# IS THE REGIME WITH SHOT NOISE SUPPRESSION BY A FACTOR 1/3 ACHIEVABLE IN SEMICONDUCTOR DEVICES WITH MESOSCOPIC DIMENSIONS?

P. MARCONCINI*, M. MACUCCI,
D. LOGOTETA and M. TOTARO

*Dipartimento di Ingegneria dell'Informazione*
*Università di Pisa, Via Caruso 16*
*Pisa, I-56122, Italy*
**p.marconcini@iet.unipi.it*

Received 15 September 2011
Revised 29 December 2011
Published 31 March 2012

Communicated by Richard Haley

We discuss the possibility of diffusive conduction and thus of suppression of shot noise by a factor 1/3 in mesoscopic semiconductor devices with two-dimensional and one-dimensional potential disorder, for which existing experimental data do not provide a conclusive result. On the basis of our numerical analysis, we conclude that it is quite difficult to achieve diffusive transport over a reasonably wide parameter range, unless the device dimensions are increased up to the macroscopic scale where, however, shot noise disappears because the device length exceeds the Debye length. In addition, in the case of one-dimensional disorder, some mechanism capable of mode-mixing has to be present in order to reach or even approach the diffusive regime.

*Keywords*: Diffusive regime; shot noise suppression; Fano factor; disordered conductor; tunnel barriers; strong localization; magnetic field.

## 1. Introduction

In nanoscale devices shot noise (i.e., the noise due to the discreteness of the electron charge) is often suppressed [1, 2] with respect to the value of power spectral density given by the Schottky formula $S_I = 2e|I|$ (where $e$ is the elementary charge and $I$ is the average current), which would be expected in the absence of correlations between the carriers [3]. The ratio that quantifies this suppression, due to the correlations introduced by Coulomb interactions or (as in the cases discussed in the following) by Fermi exclusion, is called Fano factor.

Particularly interesting is the case of diffusive conductors. It is well known [4] that, if the length $L$ of the device is much less than the elastic mean free path $l$, the transport regime is ballistic and shot noise is strongly suppressed, while, if $L$ is greater than the localization length $L_l$ and the conductor is assumed to be phase coherent (as will be the case here), transport is characterized by strong localization, with the resistance increasing exponentially with the length of the device and the Fano factor approaching unity. The localization length is equal to $L_l = (\beta N + 2 - \beta)l$ [5–7], where $N$ is the number of propagating modes and the symmetry index $\beta$ is equal to 1 with no magnetic field, 2 with a finite magnetic field and 4 with a strong spin–orbit scattering or magnetic impurities; if $N \gg 1$, $L_l \approx \beta N l$. In the intermediate regime in which the values of $l$ (determined by the disorder inside the device), $L$ and $L_l$ satisfy the inequalities $l \ll L \ll L_l$, transport is diffusive. A study of the crossover between these different transport regimes can be found in [8]. In the diffusive regime the resistance increases linearly with length and, being the distribution of the transmission eigenvalues $T$ bimodal and proportional to $1/(T\sqrt{1-T})$ [9, 10], shot noise is suppressed by a factor 1/3, due to the presence of a few open quantum channels with nearly unit transmission probability [11, 12]. In the case of devices with disorder in more than one dimension, this result has been theoretically obtained using quantum-mechanical [13–15] or semiclassical [16] approaches, both of which include the effect of Fermi exclusion. The same result has been obtained in the case of one-dimensional disorder (i.e., a series of unevenly spaced tunnel barriers) using a semiclassical model [17].

In this paper we will try to answer the question whether in semiconductor devices with mesoscopic dimensions it is possible to reach the diffusive transport regime and thus the corresponding 1/3 suppression of shot noise over a significant parameter range.

From the experimental point of view, Henny *et al.* [18] measured the shot noise suppression factor in a thin metallic wire, finding the theoretically predicted 1/3 reduction by asymptotically widening the contact reservoirs and thus reducing the reservoir heating which otherwise would increase the measured noise.

Instead, in the case of semiconductor devices, the existing experimental data are not conclusive. Liefrink *et al.* [19] reported, for a wire obtained confining the two-dimensional electron gas of a GaAs/AlGaAs heterostructure, shot noise reduction factors varying between 0.2 and 0.45, depending on the width of the wire.

Song *et al.* [20] have measured the Fano factor for a semiconductor structure with one-dimensional (1D) disorder, in particular for a superlattice in which carriers have been injected into the conduction band through optical excitations from the valence band. Also in this case, for low applied fields, the measured Fano factor has values that strongly differ from 1/3, varying from one superlattice to another and in particular typically increasing with barrier width (i.e., decreasing barrier transparency).

Prompted by these results, we have performed a numerical investigation of the noise behavior of semiconductor structures characterized by 2-dimensional (2D) and 1D disorder, using different models, with varying degree of approximation. In the case of 2D disorder, our conclusion is that in a semiconductor device with mesoscopic dimensions it is very difficult to obtain a diffusive behavior over a significant range of the scatterer concentration or of the disorder strength: in order to reach such a result, we should increase the number of propagating modes up to several thousands, thereby reaching a macroscopic size. However, in general shot noise is not observable in macroscopic conductors, since their longitudinal extension is much larger than the Debye length (thereby preventing the observation of shot noise [21]). On the other hand, in the case of 1D disorder we have found that the absence of coupling among the modes propagating in the structure makes it inherently impossible to obtain a diffusive regime, and thus a 1/3 Fano factor, unless some additional contribution (such as a magnetic field) is introduced to create mode mixing.

## 2. Two-Dimensional Disorder

The structure we consider is a quantum wire obtained laterally confining, by means of negatively biased gates located on the surface, the two-dimensional electron gas (2DEG) of a GaAs/AlGaAs heterostructure. The ionized donors located inside the n-doped AlGaAs layer (together with other charged impurities present inside the heterostructure) determine potential fluctuations at the 2DEG level.

In a previous paper by our group [22] a self-consistent calculation of the average potential, combined with a semi-analytical formula for the effect of ionized donors, was used to fit the conductance measurement on a fabricated quantum wire and to numerically predict its noise behavior. In that case it was found that the Fano factor did not stabilize at 1/3, but it rather crossed it for a single value of the gate bias voltage.

In order to gain a better understanding of the problem, we have now studied the noise behavior of the structure for a larger range of parameters, for some choices of a model potential. In detail, we have considered a $4.9\,\mu$m long and $8.4\,\mu$m wide conductor with a hard-wall lateral confinement (since in our previous investigations the detailed shape of the confinement potential did not appear to play a significant role on the noise behavior). We have considered that all of the dopants are located 40 nm above the 2DEG and, for each considered dopant concentration (with a uniform random distribution), we have initially evaluated the effect, at the level of the 2DEG, by summing the contribution of each single dopant, evaluated with the semi-analytical expression given by Stern and Howard [23]. In order to have a neutral system with a potential landscape symmetric around zero, which simplifies the investigation and comparison of a large number of different cases, we have considered an artificial situation with an equal number of positive and negative charges. Then, different disorder strengths for each dopant concentration have been obtained simply by multiplying the thus obtained potential profile by a scale factor $K$. As an example, in Fig. 1 we show a map of the potential obtained at the 2DEG level for a concentration $N_D = 1.1 \times 10^{14}\,\mathrm{m}^{-2}$ of impurities located 40 nm above the 2DEG, multiplied by a disorder strength scale factor $K = 39$.

Once the potential at the 2DEG level has been obtained, the transmission matrix of the device has been evaluated using the recursive Green's function technique,

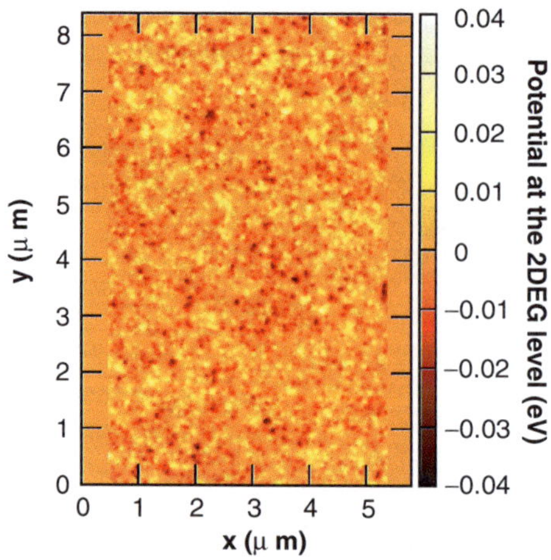


Fig. 1. Map of the potential obtained at the 2DEG level for a concentration $N_D = 1.1 \times 10^{14}\,\mathrm{m}^{-2}$ of impurities located 40 nm above the 2DEG, multiplied by a scale factor $K = 39$.

with a representation over the transverse modes in the confined direction and in real space in the transport direction [24, 25]. From the transmission matrix $t$, the conductance $G$, the shot noise power spectral density $S_I$ and the Fano factor $\eta$ have been obtained using the relations [26]:

$$G = \frac{2\,e^2}{h}\sum_i w_i, \quad S_I = 4\,\frac{e^3}{h}\,|V|\sum_i w_i(1-w_i), \quad \eta = \frac{\sum_i w_i\,(1-w_i)}{\sum_i w_i}, \tag{1}$$

where $V$ is the mean value of the externally applied voltage, $h$ is Planck's constant and the $w_i$'s are the eigenvalues of the matrix $t^\dagger t$.

In these calculations we have separately averaged the conductance and noise results (and thus the numerator and denominator of $\eta$ in Eq. (1)) over 41 energy values uniformly spaced in an energy range of 80 $\mu$eV around $E_F = 9\,\mathrm{meV}$.

In Fig. 2 we report the Fano factor that we have obtained for seven values of the dopant concentration $N_D$, as a function of the disorder strength scale factor $K$. We see that for elevated concentrations the interval of disorder strength within which the curves approach the value 1/3 is very narrow, while it gets wider for low concentrations. However, in this latter case the diffusive behavior is obtained only for very large scale factors $K$. If we focus our attention on the potential deriving, at the 2DEG level, from each single charged impurity and in particular on the portion that affects transport the most, i.e., above the Fermi energy, we see that for these values of $K$ its spatial extent is of the order of hundreds of nanometers. Since this extension is unrealistic for semiconductor devices (while it could be reasonable for metallic conductors, characterized by the presence of large grains), we conclude that in semiconductor structures it is quite unlikely to obtain a 1/3 shot noise suppression factor within a reasonably wide parameter range. Finally, for the lowest concentration ($N_D = 1.1 \times 10^{12}\,\mathrm{m}^{-2}$) the Fano factor remains well

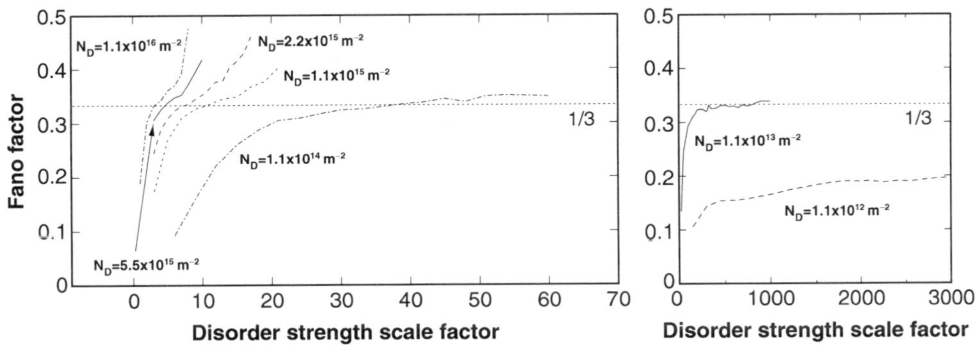


Fig. 2. Fano factor as a function of the disorder scale factor $K$ for seven values of the dopant concentration $N_D$ and for $E_F = 9\,\text{meV}$.

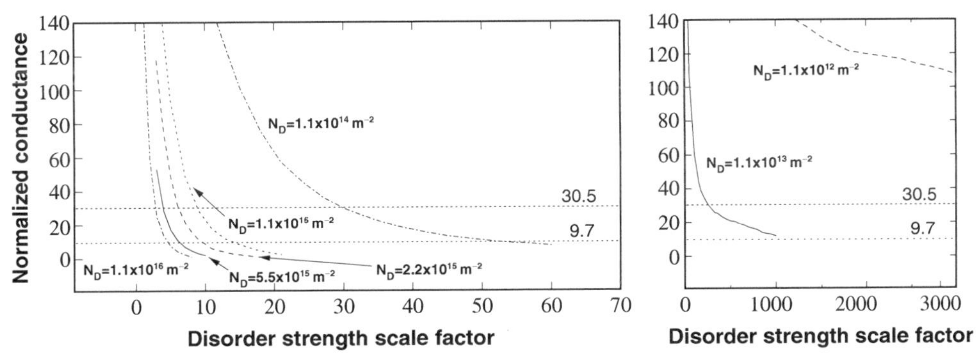


Fig. 3. Conductance (normalized with respect to the conductance quantum $G_0$) as a function of the disorder scale factor $K$ for seven values of the dopant concentration $N_D$ and for $E_F = 9\,\text{meV}$.

below 1/3 and thus we have a substantially ballistic regime for all the considered disorder strengths.

In Fig. 3 we show the normalized conductance $G/G_0$ (where $G_0$ is the conductance quantum $2e^2/h$) as a function of the disorder strength scale factor $K$ for the same impurity concentrations $N_D$. Since we have that $G/G_0 \approx Nl/(L+l)$ [4], in this case, being $N = 336$, the condition for diffusive transport $l \ll L \ll L_l$ (with $L_l \approx Nl$) is satisfied by a factor of 10 for both inequalities if $9.7 < G/G_0 < 30.5$. We again observe that the interval in which the condition for diffusive transport is satisfied is very narrow for elevated concentrations, while it widens for low concentrations, for which, however, large disorder strengths are needed. Finally, in the case of $N_D = 1.1 \times 10^{12}\,\text{m}^{-2}$, the conductance never satisfies such a condition.

The comparison between Figs. 2 and 3 confirms that the Fano factor takes on values near 1/3 in the same parameter range in which the condition $l \ll L \ll L_l$ is satisfied. This range can clearly be widened by increasing the number $N$ of propagating modes, but this can be obtained only by considering wider conductors, with macroscopic dimensions.

Instead, in the case of metallic conductors, due to the higher number of propagating modes and to the possibility of extended scatterers, diffusive transport can actually be reached and experimentally measured, although with a peculiar choice of size and geometries.

Finally, in order to compare these conclusions with some previously obtained results [27], we have performed some simulations using a more simplified model for the disordered potential. In detail, in [27] a discussion on the exact conditions needed to obtain the diffusive transport regime was presented, considering wires where the disordered potential was simulated with a random distribution of square obstacles. Here we have repeated the calculation corresponding to the upper curve of Fig. 6 of [27], obtaining the results reported in the upper panel of Fig. 4. The simulation has been performed considering a wire with a width $W = 5\,\mu$m and a length $L = 8\,\mu$m, containing 300 hard-wall square obstacles, with a 100 nm edge. In the figure we show the Fano factor as a function of the Fermi energy of the impinging electrons. We see that, as soon as the Fermi energy has reached the value corresponding to a number of propagating modes for which the condition $l \ll L \ll Nl$ is satisfied, the Fano factor reaches the value 1/3.

In order to verify whether the diffusive regime is preserved over a large range of scatterer concentrations, we have made some simulations varying the number of

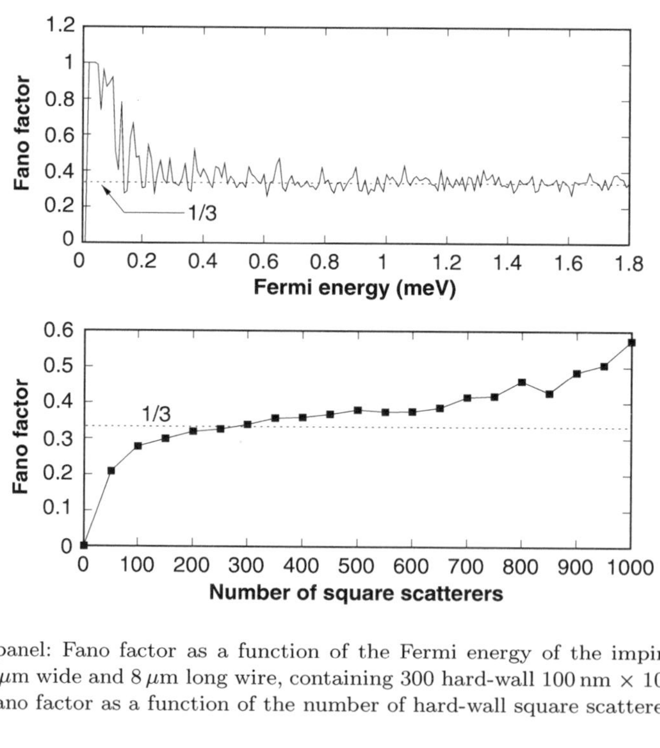


Fig. 4. Upper panel: Fano factor as a function of the Fermi energy of the impinging electrons, obtained in a $5\,\mu$m wide and $8\,\mu$m long wire, containing 300 hard-wall 100 nm × 100 nm obstacles. Lower panel: Fano factor as a function of the number of hard-wall square scatterers in the wire.

square scatterers inside the wire and computing the Fano factor around a value of the Fermi energy, $E_F = 1.25\,\text{meV}$, for which the diffusive regime has been reached in the case of 300 scatterers. In particular, we have separately averaged the conductance and noise results over a set of 41 uniformly spaced energy values in a range of $80\,\mu\text{eV}$ around 1.25 meV. In the lower panel of Fig. 4 we show the behavior of the Fano factor as a function of the number of square scatterers inside the conductor. We see that also in this case the Fano factor does not settle around 1/3 for a wide range of scatterer numbers, but just crosses the 1/3 value in correspondence of about 300 scatterers, which is indeed the situation for which the conditions for diffusive regime have been investigated in [27]. Thus, this was a particular case, not representative of the general behavior.

### 3. One-Dimensional Disorder

The case of strictly one-dimensional disorder [28, 29], i.e., of unevenly spaced tunnel barriers located in an otherwise purely ballistic device with any dimensionality, is a bit different from the cases of 2D or 3D disorder, even though, according to semiclassical studies of this structure [17, 30], also in this case the shot noise suppression factor should approach 1/3 as the number of cascaded barriers is let go to infinity. The same result can be obtained also with a semiclassical approach such as the one in [31], which is based on an equivalent circuit description, an approach often useful for the treatment of a variety of physical phenomena [32].

A series of barriers can be defined, for example, in a heterostructure-based device defining a series of gates on the surface of the device. Such gates, when negatively biased, locally deplete the 2DEG, each generating a tunnel barrier for the electrons traveling in the device.

We consider, for simplicity, idealized rectangular barriers, although we have verified that the main results hold also for realistic barriers [33]. Since a wire with a series of such rectangular transversal barriers (each extending across the whole cross-section) can be seen as made up of a series of sections differing only for the value of their constant potential, the wave functions associated with the transverse modes are the same in all of the sections and the tunnel barriers do not introduce any mode-mixing. Therefore, the transport calculation can be subdivided into many purely 1D problems. Using a scattering matrix approach, we have found the transmission coefficient $t_i$ for each mode $i$ through the device, from which the conductance, shot noise power spectral density and Fano factor can then be obtained.

In Fig. 5(a) we report, for a series of identical barriers, the Fano factor as a function of the number of the unevenly spaced barriers for three values of the barrier transparency $\Gamma$. In detail, we have considered an $8\,\mu\text{m}$ wide structure and we have averaged our conduction and noise results over 500 energies uniformly distributed in a range of $40\,\mu\text{eV}$ around 9.03 meV. We have considered 0.425 nm thick barriers, with heights equal to 0.8, 0.25 and 0.07 eV. We define the transparency $\Gamma$ as the squared modulus of the transmission coefficient through each barrier, averaged

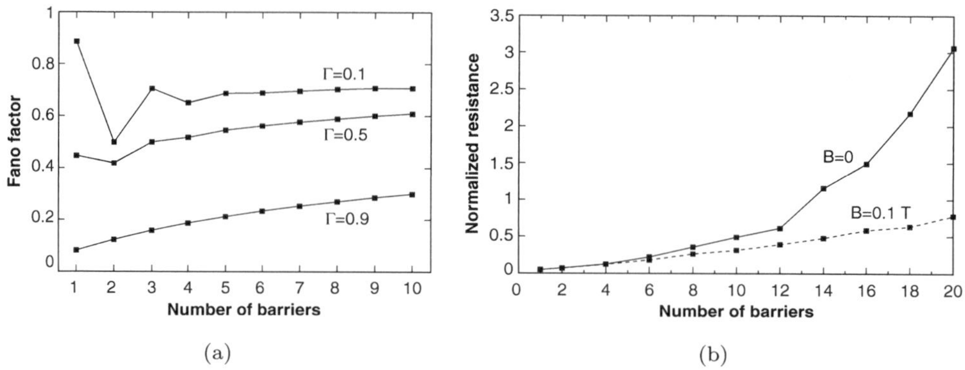


Fig. 5. (a) Fano factor as a function of the number of barriers for three values of the barrier transparency Γ, averaged over 50 different sets of interbarrier distances. (b) Resistance (normalized with respect to the resistance quantum $h/(2e^2)$) as a function of the number of cascaded tunnel barriers in a $1\,\mu$m wide conductor with 66 meV high and 1.56 nm thick barriers, with $E_F = 9.03$ meV; the exponential behavior has been obtained without the magnetic field, while the nearly linear one has been found applying an orthogonal magnetic field $B = 0.1$ T.

over all the propagating modes. For the three considered values of barrier height the transparency Γ is about equal to 0.1, 0.5 and 0.9, respectively. In order to obtain a general behavior, we have averaged the results over 50 different sets of interbarrier distances, which is equivalent to introducing dephasing with a simple phase randomization model preserving localization effects [34]. These averages show that, contrary to what was expected from semiclassical analyses [17], no common asymptotic 1/3 value for the Fano factor is reached increasing the number of barriers.

The reason is that, in the absence of mode-mixing, the overall transport problem is just a collection of intrinsically one-dimensional problems; therefore we can consider $N = 1$, and thus $L_l = 2l$ and it is practically impossible to satisfy the condition for diffusive transport $l \ll L \ll L_l$. Since semiclassical descriptions do not take into account the effect of phase coherence on transport and thus do not include localization, this effect cannot be predicted using semiclassical arguments, but results only from a quantum-mechanical analysis.

In [33] we have also shown that the presence of a realistic amount of edge-roughness in the depletion gates defining the barriers, introducing only a small degree of mode-mixing, does not alter significantly the described results. On the other hand, the amount of 2D disorder that we should add to the device in order to create mode-mixing and reach the diffusive regime would be such that it would lead to diffusive transport even in the absence of the barriers, and therefore of the 1D disorder. Therefore, in order to reach the diffusive regime while preserving the 1D nature of the disorder, we have to introduce a different source of mode-mixing, for example a magnetic field threading the device.

It has been shown in the case of diffusive samples that the presence of a magnetic field does not alter the shot noise suppression factor in the diffusive regime [10, 13], but, breaking the time-reversal symmetry and thus changing the symmetry index $\beta$ from 1 to 2, it can affect the value of the localization length: being $L_l = (\beta N+2-\beta)l$, $L_l$ increases in a multi-mode system, while it does not change if $N = 1$ [5–7, 35–42].

However, in the particular case of a series of tunnel barriers, we have shown elsewhere [28] that, especially in the case of barriers with high transparency, the presence of a magnetic field makes it possible to reach a 1/3 value for the Fano factor. The reason is that the magnetic field, introducing mode-mixing, changes the effective dimensionality of the system from 1D to 2D; therefore a greater $N$ has to be considered in the previous expression for $L_l$ and the localization length increases with respect to the mean free path.

Here we show the results obtained for a 1 $\mu$m wide conductor containing a series of 66 meV high and 1.56 nm thick barriers, with an average transparency at the considered Fermi energy (9.03 meV) $\Gamma = 0.5$. The transport calculation has been carried out using the recursive Green's function technique, and adopting, for the representation of the vector potential, a Landau gauge with nonzero component only along the transverse direction [43]. The conductance and noise results have been averaged over a set of 25 energy values uniformly distributed over a range of 40 $\mu$eV around 9.03 meV. The final results have been averaged over 20 different sets of interbarrier distances. We see in Fig. 5(b) that, while in the absence of magnetic field we observe an exponential behavior of the resistance as a function of the number of the barriers (characteristic of the strong localization regime), applying an orthogonal magnetic field $B = 0.1$ T the resistance behavior becomes approximately linear, i.e., we approach the diffusive regime. In particular, we have found that in our structure $L_l$ increases by about a factor of 7 when the magnetic field is applied.

However, in order to obtain a diffusive regime over a really wide range of parameters, the mode-mixing introduced by the magnetic field has to be combined with the presence of a large number of propagating modes, which requires, also in this case, macroscopic dimensions, where shot noise is generally not observable.

## 4. Conclusion

We have investigated the possibility of diffusive transport, and thus of suppression of shot noise down to 1/3, in mesoscopic semiconductor devices with 2D and 1D disorder. From our numerical simulations, in which different representations for the potential disorder have been adopted, we have concluded that it should be very difficult and rather uncommon to obtain fully diffusive transport in mesoscopic semiconductor devices, due to the insufficient number of propagating modes. In addition, in the case of 1D disorder the absence of mode-mixing makes it impossible to reach the diffusive regime, unless a source of mode-mixing, such as a magnetic field, is present. Our results seem to be supported by existing experimental evidence, which has not shown a clear 1/3 suppression of shot noise in semiconductor devices.

In order to achieve a 1/3 suppression factor, more propagating modes would be needed, but, to properly satisfy the inequality $l \ll L \ll L_l$, this would imply an increased length, thus most likely leading to a resistive behavior without shot noise if the Debye length is exceeded [21] and a macroscopic size is reached.

## References


[1] Ya. M. Blanter and M. Büttiker, Shot noise in mesoscopic conductors, *Phys. Rep.* **336** (2000) 1.

[2] M. J. M. de Jong and C. W. J. Beenakker, Shot noise in mesoscopic systems, in *Mesoscopic Electron Transport*, eds. L. L. Sohn, L. P. Kouwenhoven and G. Schön, NATO ASI Series E345 (Kluwer Academic Publishers, Dordrecht, 1997), p. 225.

[3] W. Schottky, Über spontane Stromschwankungen in verschiedenen Elektrizitätsleitern, *Ann. Phys. (Leipzig, Ger.)* **57** (1918) 541.

[4] S. Datta, *Electronic Transport in Mesoscopic Systems* (Cambridge University Press, Cambridge, 1995).

[5] C. W. Beenakker, Random-matrix theory of quantum transport, *Rev. Mod. Phys.* **69** (1997) 731.

[6] J.-L. Pichard, Random transfer matrix theory and conductance fluctuations, in *Quantum Coherence in Mesoscopic Systems*, ed. B. Kramer, NATO ASI Series B254 (Plenum, New York, 1991), p. 369.

[7] S. Kettemann and R. Mazzarello, Magnetolocalization in disordered quantum wires, *Phys. Rev. B* **65** (2002) 085318.

[8] A. Kolek, A. W. Stadler and G. Haładaś, Numerical simulations of shot noise in degenerate disordered conductors in reduced dimensions, *Phys. Rev. B* **64** (2001) 075202.

[9] R. A. Jalabert, J.-L. Pichard and C. W. J. Beenakker, Universal quantum signatures of chaos in ballistic transport, *Europhys. Lett.* **27** (1994) 255.

[10] P. A. Mello and J.-L. Pichard, Maximum-entropy approaches to quantum electronic transport, *Phys. Rev. B* **40** (1989) 5276.

[11] M. J. M. de Jong and C. W. J. Beenakker, Sub-Poissonian shot noise in a diffusive conductor, in *Coulomb and Interference Effects in Small Electronic Structures*, eds. D. C. Glattli and M. Sanquer (Editions Frontières, France, 1994), p. 427.

[12] Yu. V. Nazarov, Limits of universality in disordered conductors, *Phys. Rev. Lett.* **73** (1994) 134.

[13] C. W. J. Beenakker and M. Büttiker, Suppression of shot noise in metallic diffusive conductors, *Phys. Rev. B* **46** (1992) 1889.

[14] B. L. Altshuler, L. S. Levitov and A. Yu. Yakovets, Nonequilibrium noise in a mesoscopic conductor: A microscopic analysis, *JETP Lett.* **59** (1994) 857.

[15] Ya. M. Blanter and M. Büttiker, Shot-noise current-current correlations in multiterminal diffusive conductors, *Phys. Rev. B* **56** (1997) 2127.

[16] K. E. Nagaev, On the shot noise in dirty metal contacts, *Phys. Lett. A* **169** (1992) 103.

[17] M. J. M. de Jong and C. W. J. Beenakker, Semiclassical theory of shot-noise suppression, *Phys. Rev. B* **51** (1995) 16867.

[18] M. Henny, S. Oberholzer, C. Strunk and C. Schönenberger, 1/3-shot-noise suppression in diffusive nanowires, *Phys. Rev. B* **59** (1999) 2871.

[19] F. Liefrink, J. I. Dijkhuis, M. J. M. de Jong, L. W. Molenkamp and H. van Houten, Experimental study of reduced shot noise in a diffusive mesoscopic conductor, *Phys. Rev. B* **49** (1994) 14066.

[20] W. Song, A. K. M. Newaz, J. K. Son and E. E. Mendez, Drastic reduction of shot noise in semiconductor superlattices, *Phys. Rev. Lett.* **96** (2006) 126803.

[21] G. Gomila, T. Gonzalez and L. Reggiani, Shot-noise anomalies in nondegenerate elastic diffusive conductors, *Phys. Rev. B* **66** (2002) 245423.

[22] L. Bonci, G. Fiori, M. Macucci, G. Iannaccone, S. Roddaro, P. Pingue, V. Piazza, M. Cecchini and F. Beltram, Analysis of shot-noise suppression in disordered quantum wires, *Physica E* **19** (2003) 107.

[23] F. Stern and W. E. Howard, Properties of semiconductor surface inversion layers in the electric quantum limit, *Phys. Rev.* **163** (1967) 816.

[24] F. Sols, M. Macucci, U. Ravaioli and K. Hess, Theory for a quantum modulated transistor, *J. Appl. Phys.* **66** (1989) 3892.

[25] M. Macucci, A. Galick and U. Ravaioli, Quasi-three-dimensional Green's-function simulation of coupled electron waveguides, *Phys. Rev. B* **52** (1995) 5210.

[26] M. Büttiker, Scattering theory of thermal and excess noise in open conductors, *Phys. Rev. Lett.* **65** (1990) 2901.

[27] M. Macucci, G. Iannaccone, G. Basso and B. Pellegrini, Numerical investigation of shot-noise suppression in diffusive conductors, *Phys. Rev. B* **67** (2003) 115339.

[28] P. Marconcini, M. Macucci, G. Iannaccone and B. Pellegrini, Quantum analysis of shot noise suppression in a series of tunnel barriers, *Phys. Rev. B* **79** (2009) 241307(R).

[29] P. Marconcini, M. Macucci, G. Iannaccone and B. Pellegrini, Effect of localization on the Fano factor of cascaded tunnel barriers, *Proc. 20th Int. Conf. Noise and Fluctuations (ICNF-2009). AIP Conf. Proc.*, Pisa, Italy, Vol. 1129 (2009), p. 423.

[30] R. C. Liu, P. Eastman and Y. Yamamoto, Inhibition of elastic and inelastic scattering by the Pauli exclusion principle: Suppression mechanism for mesoscopic partition noise, *Solid State Commun.* **102** (1997) 785.

[31] M. Macucci, P. Marconcini and G. Iannaccone, Equivalent resistance and noise of cascaded mesoscopic cavities, *Int. J. Circ. Theor. Appl.* **35** (2007) 295.

[32] A. Musolino, M. Raugi and B. Tellini, 3-D field analysis in tubular induction launchers with armature transverse motion, *IEEE Trans. Magn.* **35** (1999) 154.

[33] M. Totaro, P. Marconcini, D. Logoteta and M. Macucci, Transport and noise behavior of cascaded realistic tunnel barriers, *Proc. 9th IEEE Conf. Nanotechnology (IEEE Nano 2009). IEEE Conf. Proc.*, Genoa, Italy (2009), p. 611.

[34] M. G. Pala and G. Iannaccone, Statistical model of dephasing in mesoscopic devices introduced in the scattering matrix formalism, *Phys. Rev. B* **69** (2004) 235304.

[35] P. A. Lee and D. S. Fisher, Anderson localization in two dimensions, *Phys. Rev. Lett.* **47** (1981) 882.

[36] K. B. Efetov and A. I. Larkin, Kinetics of a quantum particle in a long metallic wire, *Sov. Phys. JETP* **58** (1983) 444.

[37] O. N. Dorokhov, Electron localization in a multichannel conductor, *Sov. Phys. JETP* **58** (1983) 606.

[38] J.-L. Pichard, M. Sanquer, K. Slevin and P. Debray, Broken symmetries and localization lengths in Anderson insulators: Theory and experiment, *Phys. Rev. Lett.* **65** (1990) 1812.

[39] H. Tamura and T. Ando, Conductance fluctuations in quantum wires, *Phys. Rev. B* **44** (1991) 1792.

[40] R. Blümel and U. Smilansky, Symmetry breaking and localization in quantum chaotic systems, *Phys. Rev. Lett.* **69** (1992) 217.

[41] H. Schomerus and C. W. J. Beenakker, Search for two-scale localization in disordered wires in a magnetic field, *Phys. Rev. Lett.* **84** (2000) 3927.

[42] M. Weiss, T. Kottos and T. Geisel, Spreading and localization of wavepackets in disordered wires in a magnetic field, *Phys. Rev. B* **63** (2001) 081306(R).
[43] P. Marconcini and M. Macucci, Numerical techniques for the evaluation of conductance and noise in the presence of a perpendicular magnetic field, *J. Comput. Electron.* **2** (2003) 387.

# SHOT NOISE SUPPRESSION IN INDIVIDUAL AND SERIES ARRAYS OF MAGNETIC TUNNEL JUNCTIONS

AISHA GOKCE, RYAN STEARRETT* and E. R. NOWAK

*Department of Physics and Astronomy*
*University of Delaware, Newark, DE 19716, USA*
**stearrer@udel.edu*

C. NORDMAN

*Nonvolatile Electronics*
*11409 Valley View Road, Eden Prairie*
*MN 55344, USA*

Received 6 July 2011
Accepted 28 July 2011

Communicated by Arkady Victorovich Yakimov

Charge-current shot noise is investigated in single magnetic tunnel junctions and devices having multiple junctions that are connected in series. The ratio of the measured shot noise in single junctions to the expected Poisson value, namely the Fano factor, $F$, is observed to vary from 1 to well below 0.5. Deviations from $F = 1$ are attributed to localized states (defects) located in the tunnel barrier or at the interfaces with the magnetic electrodes. For series arrays of junctions, the Fano factor scales inversely with the number ($1 \leq N \leq 30$) of junctions in series, even for junctions exhibiting sub-Poissonian ($F < 1$) shot noise. The $1/N$ scaling is consistent with the incoherent tunneling of electrons across junctions and indicates that each junction behaves as an individual noise source. The advantages of incorporating series arrays of magnetic tunnel junctions into devices for magnetic field sensing are discussed.

*Keywords*: Shot noise suppression; series array; annealing; magnetic tunnel junctions.

## 1. Introduction

Charge-current shot noise arises from the time-dependent fluctuations in the electrical current due to the discreteness of charge. It was first discovered in vacuum tubes and manifests itself in many modern solid state devices such as metal-insulator-metal tunnel junctions, *p-n* junctions, and mesoscopic conductors [1, 2]. Shot noise is typically more sensitive to the microscopic mechanisms of transport than the average conductance. Particularly, it has been a major tool for gathering information about the charge and statistics of quasiparticles relevant for transport and the effects of interactions between charge carriers [1, 3]. In addition to its fundamental importance, shot noise has applications in thermometry [4] and device reliability evaluation, for example, to reveal the presence of stress-induced leakage currents in metal-oxide-semiconductor capacitors [5].

Electron shot noise is characterized by a frequency-independent power spectrum $S_I$ of current fluctuations such that $S_I \propto eI$, where $I$ is the average current flowing through the device. If the transport of electrons through the device is random and uncorrelated, then the shot noise takes on its full (Poisson) value $S_I = 2eI \equiv S_I^{\text{Poisson}}$, as is well-known from electron tubes and tunnel junctions. Shot noise can be suppressed below or enhanced above the full value, depending on the nature of correlations and electron–electron interactions during electron transmission through the device. The Fano factor, $F = S_I/2eI$, is often used to characterize sub-Poissonian ($F < 1$) and super-Poissonian ($F > 1$) shot noise. The Pauli exclusion principle and Coulomb interactions lead to negative correlations in electron transport and result in shot noise suppression, $F < 1$ [1]. A well-known example of shot noise suppression is in mesoscopic diffusive conductors, where a universal $F = 1/3$ suppression factor was predicted [6, 7] and experimentally confirmed [8].

In the particular case of a tunnel junction device of resistance $R$, the low-frequency current noise spectral density is $S_I = (2eV/R)\coth(eV/2k_BT)$ [9], where $V$ and $T$ are the voltage across and the temperature of the junction, respectively. This expression gives thermal noise $S_I = 4k_BT/R$ at low voltage biases, $eV \ll k_BT$, and Poisson shot noise in the opposite limit. Direct tunneling yields $F = 1$, but shot noise suppression can occur if transport involves sequential tunneling via defects (e.g., traps, impurities) in the barrier which can introduce negative correlations in a number of ways, including Pauli blocking (Fermi statistics) acting on each trap, Coulomb repulsion limiting the occupancy of a trap, and possibly the inhibition of tunneling in a surrounding region by means of Coulomb repulsion [5]. A universal $F = 3/4$ suppression factor is predicted for resonant tunneling through strongly localized states that are randomly distributed inside the tunnel barrier [10]. This physical situation has similarities to tunneling through double-barrier tunnel junctions [11–13] and double-barrier resonant tunneling (DBRT) conductors [14–17], for which the corresponding Fano factor depends on the asymmetry in the (voltage bias-dependent) tunneling rates onto and off of the defect or quantum well. The Fano factor expression for one-dimensional (1D) double-barrier systems is independent of whether the tunneling process preserves full quantum coherence or if it is fully incoherent (sequential tunneling model) [1] and it yields values which range between 1/2 (symmetric barrier) and 1 (very asymmetric barrier). A recent comparative analysis of coherent and incoherent tunneling models by Aleshkin *et al.* [18] found that shot noise with $F < 1/2$ is possible for coherent tunneling in DBRT structures when the transparency of the barriers is sufficiently high. Super-Poissonian shot noise has also been observed in DBRT structures and is attributed to interacting localized states [19] and electron–electron interactions [20, 21].

Suppression of shot noise also occurs for incoherent transmission through a 1D series array of $N$ tunnel junctions. For an array of identical junctions, standard circuit theory [22, 23] shows that the Fano factor can be very small, $F \propto 1/N \ll 1$ with respect to a single junction of equivalent resistance. Heuristically, the shot noise is suppressed from its full Poisson value because the current noise from each junction is effectively shunted by the junction resistance, which is much less than the total resistance of the array. Rosini *et al.* [24] have argued that in the absence of correlations among the electrons, the Fano factor must be one at vanishing frequencies, and reaches the value $1/N$ at frequencies higher than the inverse transit time through the array. The presence of other excess noise sources (e.g., $1/f$ noise), however, makes it difficult to detect $F = 1$ at sufficiently low frequencies.

More recently, shot noise is being used to study spin-polarized charge transport in magnetic systems [25–34] and pure spin currents in semiconductor nanostructures [35]. Theoretical investigations include predictions for the dependence of shot noise on parameters such as the relative alignment of ferromagnetic electrodes [33, 34] and spin blockade effects in quantum dots [36]. By comparison, there are few experimental studies of shot noise in magnetic nanostructures such as MTJs. Early studies in MTJs with $AlO_x$ tunnel barriers reported both full [37, 38] and suppressed shot noise [39]. Later, Guerrero *et al.* [40] found shot noise suppression that was dependent on the magnetization alignment. The suppression was attributed to sequential tunneling through nonmagnetic and paramagnetic impurity levels, some of which were deliberately introduced into the tunnel barrier. Full shot noise has been observed in epitaxial [41] and sputtered-deposited [42] MgO-based MTJs, irrespective of the relative magnetic configuration of the electrodes. However, a recent investigation [43] has revealed sub-Poissonian shot noise in the parallel (P) configuration due to spin-dependent coherent electron tunneling in crystalline MgO-based MTJs. In all these works, the Fano factor was close to unity or below. And in one case [39], it is as low as 0.45. Enhanced, spin-dependent shot noise has also been observed in Co-$AlO_x$-Co MTJs. The super-Poissonian shot noise was attributed to tunneling through two interacting localized states with spin-dependent tunneling rates [44].

In this letter, we report on shot noise in single element and one-dimensional arrays of MTJs having $AlO_x$ and MgO tunnel barriers. Sub-Poissonian shot noise is found in MgO-based MTJs that were annealed at low temperatures, whereas shot noise in optimally annealed devices shows the Poisson value with $F$ close to unity. Strong suppression ($F < 1/2$) was observed in some of the former devices. In arrays having $N$ MTJs connected in series, we find a $1/N$ suppression of the shot noise which has advantages for certain applications such as magnetic field sensors.

## 2. Methods and Results

In total, six different types of MTJ devices were investigated. Here, we report on shot noise in three types (A, B, and C). Table 1 summarizes the structures and common electrical properties of our devices, including their area, resistance-area product (RAP), and the tunneling magnetoresistance (TMR). Types A and C have

Table 1. MTJ devices are characterized by their materials stacks. Devices of the same type have undergone similar processing steps and differ only by their areas or the thicknesses of their barrier. RAP and TMR values are at room temperature. The Fano factor is reported for single MTJ elements in types A and C devices and it was measured at 10 K and at a voltage bias of approximately +100 mV. Type B devices could be probed only at the ends of the array. Their reported Fano factor represents the average value for a single MTJ element assuming a $1/N$ scaling for the shot noise in the series array.

| Type | Materials Stack (thicknesses in nm) |
|---|---|
| A | Ru(4)/IrMn(11)/CoFeB(5)/MgO($x$)/CoFeB(10)/Ta(5)/Ru(4)/Ta(50) |
| B | Ru(2.4)/IrMn(11)/CoFeB(4)/$AlO_x$ ($x$)/ CoFeB(12)/Ru(1.5)/Ta(2)/Ru(1.5)/CoFeB(2.5)/Ru(1.5)/Ta(50) |
| C | Ta(10)/PtMn(15)/CoFeB(2.5)/Ru(0.8)/CoFeB(3)/MgO($x$)/ CoFeB(3)/Ta(1) |

| Type | Device | Barrier | Thickness, $x$ (nm) | Area ($\mu m^2$) | RAP ($k\Omega\,\mu m^2$) | TMR (%) | $F \pm 0.09$ |
|---|---|---|---|---|---|---|---|
| A | A1 | MgO | 2.0 | 48 | 30 | 65 | 0.79 |
| | A2 | MgO | 2.0 | 240 | 48 | 35 | 0.27 |
| | A3 | MgO | 2.0 | 114 | 106 | 50 | 0.15 |
| | A4 | MgO | 2.5 | 48 | 190 | 94 | 0.98 |
| | A5 | MgO | 2.5 | 246 | 246 | 80 | 0.86 |
| | A6 | MgO | 2.5 | 164 | 125 | 80 | 0.95 |
| | A7 | MgO | 2.5 | 240 | 137 | 70 | 0.72 |
| B | B | $AlO_x$ | 2.0 | 534 | 2937 | 60 | 0.98 |
| C | C | MgO | 2.5 | 12.25 | 49 | 233 | 1.0 |

MgO tunnel barriers whereas type B have alumina barriers. The materials stacks for type A and B devices were deposited at NVE Corp. using dc/rf magnetron sputtering in a six-target Shamrock sputtering system. A magnetic field was applied during the deposition to induce the magnetic easy axis and the pinning direction. Annealing was done in forming gas under a 400 mT applied magnetic field. Several photolithographic and ion beam milling steps were used to pattern and fabricate the devices. Similar methods were used to fabricate type C devices. Type C stacks, however, were deposited using an Anelva C7100 sputtering system and then patterned at the MINT Center at the University of Alabama.

Type A devices consisted of a series array of 20 junctions. The devices had leads that allowed junctions to be probed individually or across a number, $N$, of junctions connected in series. Figure 1 illustrates the cross-sectional and top views of a segment of the array. The junctions were connected using the bottom electrode (the IrMn and Ru layers) and a 500 nm thick Al metal interconnect top layer. Type B devices consisted of a series array of 30 junctions that could only be probed at the ends of the array. Type C devices consisted of individual junctions having a cross geometry. Devices having MgO barriers require a high temperature annealing treatment to improve the degree of crystallization in the CoFeB electrodes and to achieve high TMR. Type A devices were annealed for 1 hour at 300°C, a relatively low temperature. The resulting TMR values were rather low, ranging from 30 to 90%. Typically, devices are annealed at lower temperatures if there is a need to emphasize good pinning, high blocking temperatures, and low coercive fields of free

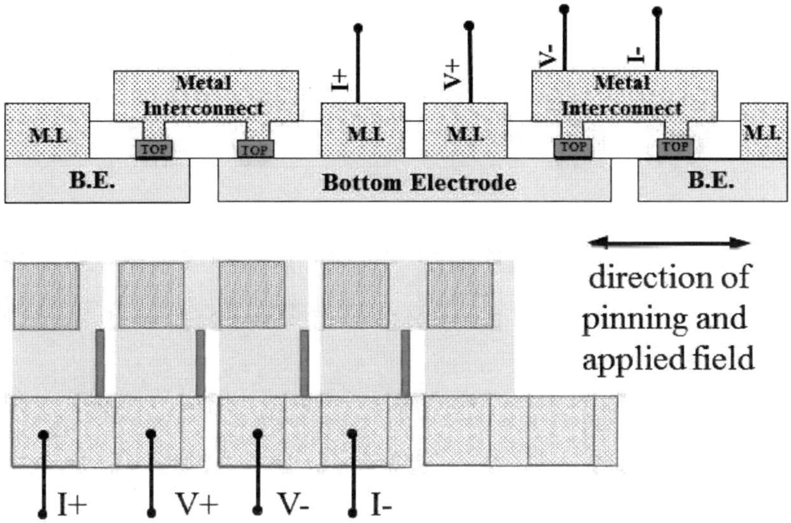


Fig. 1. (Color online.) Drawings illustrating the cross-sectional and top views of an array of MTJs connected in series. Adjacent MTJs are connected using their common bottom electrode or through aluminum metal interconnects (M.I.). The upper drawing shows an example of a four-probe wiring configuration on a single MTJ, whereas the bottom drawing shows the corresponding case for a series combination of two MTJs. An external magnetic field is applied in-plane and perpendicular to the long axis of the MTJ.

layer. Type C devices were annealed at a higher temperature (360°C for 2 hours) and exhibited much higher TMR, up to 250% at room temperature.

Devices were fabricated on silicon wafers which were diced into small die that were varnished onto ceramic leaded chip carriers. Al wire-bonding was used to make connections between Al metal contact pads on the die and gold contact pads on the chip carrier. The chip carriers were then mounted onto a custom fixture that was inserted into a commercial Physical Properties Measurement System made by Quantum Design. Transport measurements were made at low temperatures, typically at 10 K, in order to minimize thermal Johnson noise and $1/f$ barrier-resistance noise. A constant current source (typically a lead-acid battery and large ballast resistor) was used to bias the junctions. Four-probe measurements were conducted whenever the device geometry allowed for it. A sufficiently large external magnetic field could be applied in the plane of the device in order to switch the free magnetic layer so that it could be oriented either P or antiparallel (AP) to the direction of the magnetization in the pinned magnetic layer.

The tunneling resistances of all junction types were independent of temperature below roughly 100 K and exhibited negative temperature coefficients, which is consistent with good quality tunnel barriers that are free of metallic-like pinholes. Figure 2 plots the differential resistance, $R_d(V) = dV/dI$, as a function of voltage bias. Data were obtained using a differential conductance measurement

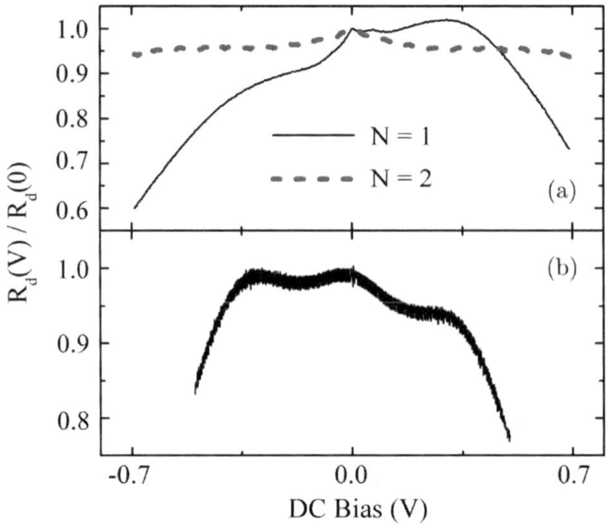


Fig. 2. (Color online.) Differential resistance as a function of the DC voltage bias. The resistance values are normalized with respect to the zero-bias value. Panel (a) shows data for type A4 devices measured at 10 K. The solid line is for a single MTJ and the dashed line is the differential resistance obtained by passing current through two adjacent MTJs. Panel (b) shows the corresponding data for a single type C device measured at 300 K.

setup consisting of a model 6221 current source and 2182A nanovoltmeter from Keithley Instruments. The measurement involved superimposing a small AC current on an incrementally varying DC current and measuring the AC and DC voltages. Figure 2(a) plots $R_d(V)$ across an individual junction and a series combination of two adjacent junctions in a type A4 device. $R_d(V)$ is bias-dependent and asymmetric for a single junction. When measured across two adjacent junctions, $R_d(V)$ is nearly symmetric because they are oppositely biased. At low temperatures, we routinely observed a resistance peak near zero bias. The size of the peak varies among different types of devices. Such zero-bias anomalies have been attributed to defects or impurities in the tunnel barrier [40, 45] and disorder in the magnetic electrodes [46]. For comparison, Fig. 2(b) shows a representative $R_d(V)$ curve for type C junctions which are fully annealed and exhibit high TMR.

For noise measurements, the voltage fluctuations across the junctions were first amplified, anti-alias filtered, and digitized. The power spectral density, $S_V(f)$, was then calculated as a function of frequency, $f$. A typical spectrum is plotted in the main panel of Fig. 3. At low frequencies, the spectrum is dominated by $1/f$-like noise due to resistance fluctuations [37, 47]. Above a characteristic frequency, $f_c$, the spectrum shows a roll-off which is due mainly to parasitic capacitance associated with the electrical wiring. After correcting for the electrical circuit's frequency response which scales as $[1 + (f/f_c)^2]^{-1}$, and subtracting the $1/f$ noise (dashed line in Fig. 3) and amplifier noise contributions, the resulting spectrum is white. That

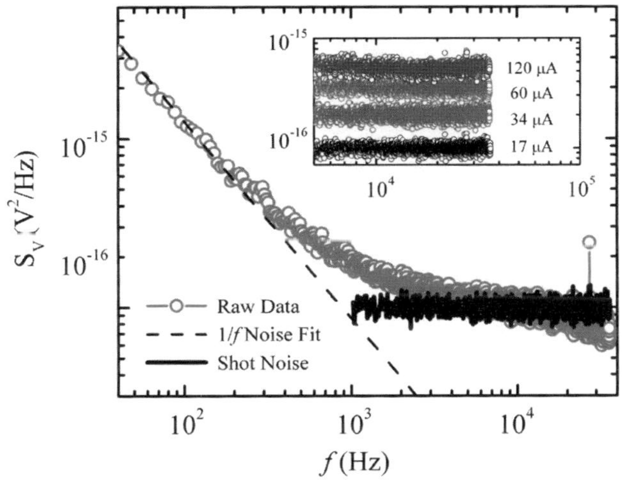


Fig. 3. (Color online.) Noise power spectral density obtained at 10 K for a type A4 device. The solid black line shows the white noise spectrum after the raw data is corrected for capacitive roll-off, $1/f$ noise (dashed line) and amplifier noise. The inset shows that the resulting white noise power scales approximately linearly with current bias, which is consistent with current shot noise.

is, it is frequency-independent. The inset to Fig. 3 plots the corrected noise spectra at several current biases. At low temperatures, thermal noise from the device resistance is negligible and the white noise is entirely due to shot noise which exhibits a nearly linear dependence on current bias. Unless noted otherwise, shot noise is reported in the P state.

Figure 4 plots the measured shot voltage noise, $S_V^{\text{shot}}(I)$, as a function of the Poisson prediction, $2eIR_d^2$, which is calculated using a value for $R_d$ corresponding to the measurement current. $S_V^{\text{shot}}$ is taken as the average white noise power over a bandwidth of several kHz (4000 points) centered around 10 kHz. The fractional standard deviation of $S_V^{\text{shot}}$ was less than 9%. The solid diagonal line in Fig. 4 depicts a Fano factor, $F \equiv S_V^{\text{shot}}/2eIR_d^2$, of unity. The Fano factor is independent of biasing conditions such as the mean voltage or current, if the data points for a given device fall on a straight line parallel to the Poisson reference line.

The Fano factors for the devices shown in Fig. 4 range from unity for type C devices to 0.15 for device A10. Strikingly, data for most type A devices is below the diagonal line, indicating sub-Poissonian shot noise. Most of the data points for a given device lie on a straight line indicating that, to a large extent, the Fano factor is independent of bias. Deviations were evident in some type A devices near zero bias and at the highest biases (>200 mV) where an onset of sub-linear behavior was observed. Please see, devices A3 and A7 in Fig. 4 for examples. We also note that within our measurement error, $F$ did not depend on the magnetic state (P or AP) of the device.

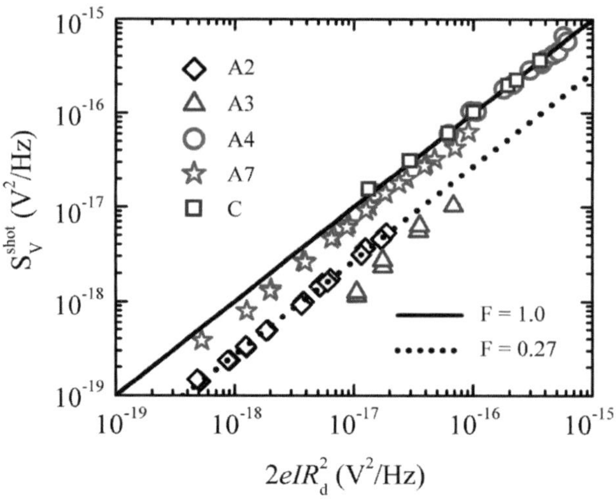


Fig. 4. (Color online.) Shot noise values at 10 K for various samples are plotted against the expected Poisson noise values. Shot noise levels at several different current biases are shown for each sample. The solid line indicates full shot noise, having Fano factor $F = 1$, and the dotted line depicts $F = 0.27$.

The Fano factors for our devices are summarized in Table 1. Types B and C devices exhibited full shot noise, $F = 1.0 \pm 0.09$, whereas considerable variation in $F$, ranging from 0.98 to 0.15, was found among different type A devices. Although data is limited, examination of Table 1 suggests some trends. Firstly, the lowest Fano factor values are found in type A devices which have thinner (2 nm) MgO barriers and exhibit lower TMR (<70%). Secondly, higher Fano factors (close to or equal to unity) are observed in devices having thicker (2.5 nm) MgO barriers and higher TMR. Optimally annealed type C devices show TMR > 225% and $F \approx 1$, whereas type A devices, which were annealed at a lower temperature, show sub-Poissonian shot noise.

Although Fano factors can vary considerably among different type A devices, the MTJs within an array were found to have very similar values and bias dependences. The inset to Fig. 5 shows an example where we plot $F(V)$ for three MTJs in array A6. Consistency amongst MTJs within an array also exists in cases for which the Fano factors are much less than 1, as discussed below. Exceptions include arrays in which an MTJ has been irreversibly damaged, for example, by a deliberately applied high voltage pulse or by electrostatic discharge. Such MTJs exhibit very small Fano factors ($F < 0.1$) and significantly lower resistance. And TMR values than their neighbors.

Next, we examine how the white noise across an array scales with the number, $N$, of MTJs connected in series. Both the differential resistance and noise were measured across a varying number (from 1 to 20) of MTJs in type A device arrays. The measured white noise was compared to the voltage noise, $2eIR_{eq}^2$, that would

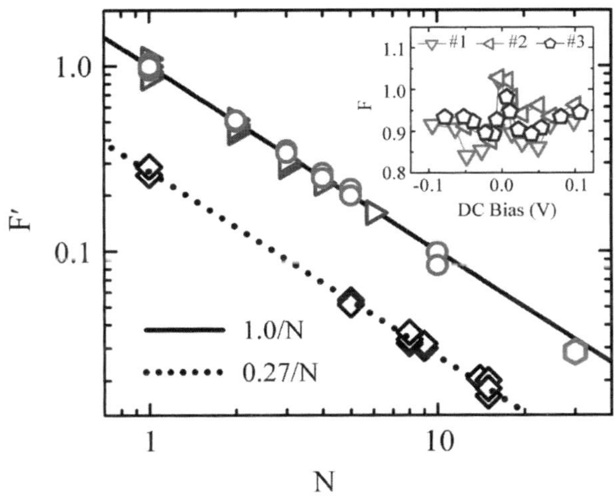


Fig. 5. (Color online) The main panel plots the Fano factor, $F'$, measured across an array of $N$ series connected MTJs. The lines depict the $1/N$ scaling prediction for incoherent tunneling across an array of identical MTJs. Solid line corresponds to each MTJ showing Poissonian ($F = 1$) shot noise and the dotted line depicts sub-Poissonian ($F = 0.27$) shot noise. Data is shown for type A2 (diamond), A4 (circle), A6 (triangle), and and B (hexagon) device arrays. The inset shows $F(V)$ for three type A6 devices that are part of a series array. All data is for 10 K.

arise from full shot noise due to the same current bias passing through a single MTJ having the equivalent resistance of $N$ MTJs, $R_{eq} = NR_d$. We define a corresponding Fano factor, $F' \equiv S_V^{\text{shot}}/2eIR_{eq}^2$, where $F' = 1$ indicates the voltage noise across the series of $N$ MTJs is the same as that across a single MTJ having the equivalent resistance of the array. If we assume that the MTJs in the array are identical with each one exhibiting full shot noise and that the tunneling processes are not correlated among the MTJs, then $S_V^{\text{shot}} \propto N$ and so $F' = 1/N$.

Figure 5 plots $F'$ as a function of $N$ for several different arrays. The data points for the A4 array correspond to a 50 mV across each MTJ. For this bias, each MTJ has a Fano factor approximately equal to unity. The data scales inversely with $N$ and the points lie very close to the solid line which depicts $F' = 1/N$. A data point for a type B device having $N = 30$ is also shown and it, too, is in good agreement with the $1/N$ scaling. Also shown in Fig. 5 is data for a type A6 array which also closely follows solid line. On the other hand, type A2 devices showed sub-Poissonian shot noise with $F = 0.27$. The dotted line in Fig. 5 indicates that the inverse scaling with $N$ also applies to an array of type A2 devices but with a proportionality constant less than unity, namely $F' = 0.27/N$.

## 3. Discussion

Sub-Poissonian shot noise was found only in type A devices. Type C devices also have MgO tunnel barriers, but they consistently showed Poisson shot noise. One difference between the two types of devices is that type A samples were annealed at lower temperatures of 300°C versus 360°C for type C devices. A 300°C anneal for 1 h is not sufficient to attain optimal crystallization of the CoFeB electrodes, which is necessary for coherent tunneling and high TMR [48]. And, in fact, type A devices show low TMR (<100%), consistent with a poor degree of crystallization. In contrast, type C junctions have TMR >225% and exhibit a double maximum in the differential resistance near 300 mV. Please see Fig. 2(b). These properties are characteristic of clean and highly crystalline CoFeB/MgO interfaces [49]. In addition to promoting crystallization of the magnetic electrodes, a 375°C annealing treatment was shown to reduce disorder and defects in thin MgO films [50]. Randomly distributed point-like defects, such as oxygen vacancies, can give rise to strongly localized states within the barrier. If the tunneling process is mediated by such defects, shot noise suppression occurs with $F = 3/4$ [10]. This value is consistent with the Fano factors found for devices A1, A5, and A7. On the other hand, devices A4 and A6 show $F$ close to unity indicating direct tunneling and the absence of electron interactions and/or sequential tunneling phenomena.

It is tempting to attribute the very low Fano factors ($F < 1/2$) in devices A2 and A3 to damaged tunnel barriers. However, these devices did show nonlinear $I$–$V$ curves and a negative temperature coefficient for the tunneling resistance. Such properties are considered opposite to those usually exhibited by damaged MTJs or metallic-like pinholes bridging the tunnel barrier. Moreover, series arrays of these devices showed shot noise suppression scaling as $1/N$. It seems unlikely that all the devices in the array would be damaged in a manner that would yield identically low Fano factors. A possible explanation involves disorder and/or defect density in the barrier which, if sufficiently high, has been found to form band tails in the band gap of MgO that extend near the Fermi levels of the magnetic electrodes [50]. We speculate that the resulting defect bands lead to electron correlations that strongly suppress the shot noise. A similar idea was invoked to explain the decrease in barrier-resistance noise with increasing voltage bias that was observed in as-prepared (not annealed) CoFeB/MgO/CoFeB MTJs [47]. Upon increasing the annealing time at a fixed annealing temperature, the quality of the barrier improved and the noise as well as its bias dependence were found to decrease. Such defect bands may also give rise to non-trivial voltage/current bias dependences in the shot noise, such as sub-linear behavior observed at higher biases for devices A3 and A7 in Fig. 4.

The $1/N$ scaling of shot noise indicates that electron transmission occurs by incoherent tunneling through our one-dimensional series arrays of MTJs. The $1/N$ scaling applies both to arrays in which the tunneling process across a single array element is direct ($F = 1$) or defected-mediated ($F < 1$). This is expected since coherent tunneling across the entire array is suppressed by the macroscopic scale of the interconnecting Al electrodes and inelastic scattering processes. Although we report on MTJs, the $1/N$ scaling also pertains to tunnel junctions with nonmagnetic electrodes.

The $1/N$ scaling has practical implications for improving the high-frequency performance of MTJ-based magnetic sensors. The low-frequency noise contributions in MTJ-based magnetic sensors were recently summarized in [51]. A typical sensor uses a Wheatstone bridge architecture with a MTJ in each leg of the bridge. By magnetically shielding two legs of the bridge, the imbalance voltage becomes proportional to the external applied field. Large output signals ($\sim$1 V) are desirable which would require that the MTJs be biased at high voltages. Higher biases result in lower TMR and higher shot noise. The shot noise contribution to the overall intrinsic noise of the sensor can be reduced by replacing the single MTJ in each leg with $N$ identical, series-connected MTJs each having $1/N$ of the bridge's equivalent resistance. Not only does this reduce the shot voltage noise power by a factor of $N$, but it also keeps the voltage bias across each MTJ lower, thereby maintaining a higher TMR. It also makes the device less susceptible to electrostatic damage. Since the magnetic and barrier-resistance fluctuations in each MTJ element are uncorrelated, a series array can be used to increase the effective volume of the sensor, thereby reducing the fractional resistance $1/f$ noise.

In conclusion, we have measured the voltage noise arising from charge-current shot noise in both single MTJ devices and one-dimensional arrays of $N$ MTJs connected in series. Incoherent tunneling across individual MTJ elements leads to a Fano factor for the array that is a factor of $N$ smaller than that of a single MTJ having the equivalent resistance of the array. This property can be used to reduce high-frequency white noise in certain device applications, such as magnetic field sensors. Individual MTJs having $AlO_x$ barriers or MgO barriers and high TMR (>225%) show Poisson shot noise indicative of direct tunneling. MgO-based MTJs annealed below 350°C, however, exhibit low TMR and sub-Poissonian shot noise. Suppressed shot noise with Fano factors ranging from 0.7 to 0.9 is attributed to electron interactions and/or sequential tunneling processes involving localized states in the barrier. The localized state may be due to oxygen vacancy defects that were not removed during the lower temperature anneal. Strongly suppressed shot noise ($F < 1/2$) in MgO-based MTJs may be related to correlated transport involving defect bands in the band gap of MgO.

## Acknowledgments

We thank A. Gupta and coworkers at the MINT Center at the University of Alabama for providing the type C samples. Types A and B device fabrication was supported by ONR under STTR award N00014-07-C-0355. Measurements and graduate stipend for A. Gokce were supported equally by ONR and DOE under award DE-FG02-07ER46374. R. S. was supported by a NASA Space Grant Fellowship.

## References

[1] Ya. M. Blanter and M. Büttiker, Shot noise in mesoscopic conductors, *Phys. Rep.* **336** (2000) 2–166.

[2] A. Van der Ziel, *Noise in Solid State Devices and Circuits* (Wiley, New York, 1986).

[3] Ya. M. Blanter, Recent advances in studies of current noise, arXiv:cond-mat/0511478 (2005).
[4] L. Spietz, K. W. Lehnert, I. Siddiqi and R. J. Schoelkopf, Primary electronic thermometry using the shot noise of a tunnel junction, *Science* **300** (2003) 1929–1932.
[5] G. Iannaccone, F. Crupi, B. Neri and S. Lombardo, Theory and experiments of suppressed shot noise in stress-induced leakage currents, *IEEE Trans. Electr. Dev.* **50** (2003) 1363–1369.
[6] C. W. J. Beenakker and M. Büttiker, Suppression of shot noise in metallic diffusive conductors, *Phys. Rev. B* **46** (1992) 1889–1892.
[7] K. E. Nagaev, On the shot noise in dirty metal contacts, *Phys. Lett. A* **169** (1992) 103–107.
[8] M. Henny, S. Oberholzer, C. Strunk and C. Schönenberger, 1/3-shot-noise suppression in diffusive nanowires, *Phys. Rev. B* **59** (1999) 2871–2880.
[9] G. Lecoy and L. Gouskov, Noise in thin film metal-oxide-metal Al-$Al_2O_3$-Al, *Phys. Status Solidi* **30** (1968) 9–17.
[10] Y. V. Nazarov and J. J. R. Struben, Universal excess noise in resonant tunneling via strongly localized states, *Phys. Rev. B* **53** (1996) 15466–15468.
[11] H. Birk, M. J. M. de Jong and C. Schönenberger, Shot-noise suppression in the single-electron tunneling regime, *Phys. Rev. Lett.* **75** (1995) 1610–1613.
[12] S. Hershfield, J. H. Davies, P. Hyldgaard, C. J. Stanton and J. W. Wilkins, Zero-frequency current noise for the double-tunnel-junction Coulomb blockade, *Phys. Rev. B* **47** (1993) 1967–1979.
[13] A. N. Korotkov, D. V. Averin, K. K. Likharev and S. A. Vasenko, *Single-Electron Tunneling and Mesoscopic Devices*, eds. H. Koch and H. Lübbig (Springer Series in Electronics and Photonics, Berlin, 1992).
[14] L. Y. Chen and C. S. Ting, Theoretical investigation of noise characteristics of double-barrier resonant-tunneling systems, *Phys. Rev. B* **43** (1991) 4534–4537.
[15] L. Y. Chen and C. S. Ting, Noise characteristics of sequential tunneling through double-barrier junctions, *Phys. Rev. B* **46** (1992) 4714–4717.
[16] J. H. Davies, P. Hyldgaard, S. Hershfield and J. W. Wilkins, Classical theory of shot noise in resonant tunneling, *Phys. Rev. B* **46** (1992) 9620–9633.
[17] Y. P. Li, A. Zaslavsky, D. C. Tsui, M. Santos and M. Shayegan, Noise characteristics of double-barrier resonant-tunneling structures below 10 kHz, *Phys. Rev. B* **41** (1990) 8388–8391.
[18] V. Ya Aleshkin, L. Reggiani and M. Rosini, Comparative analysis of sequential and coherent tunneling models in resonant diodes, *Phys. Rev. B* **73** (2006) 165320.
[19] S. S. Safonov, A. K. Savchenko, D. A. Bagrets, O. N. Jouravlev, Y. V. Nazarov, E. H. Linfield and D. A. Ritchie, Enhanced shot noise in resonant tunneling via interacting localized states, *Phys. Rev. Lett.* **91** (2003) 136801.
[20] V. Ya Aleshkin and L. Reggiani, Electron transport and shot noise in double-barrier resonant diodes: The role of Pauli and Coulomb correlations, *Phys. Rev. B* **64** (2001) 245333.
[21] G. Iannaccone, G. Lombardi, M. Macucci and B. Pellegrini, Enhanced shot noise in resonant tunneling: Theory and experiment, *Phys. Rev. Lett.* **80** (1998) 1054–1057.
[22] A. N. Korotkov and K. K. Likharev, Shot noise suppression at one-dimensional hopping, *Phys. Rev. B* **61** (2000) 15975–15987.
[23] R. Landauer, Mesoscopic noise: Common sense view, *Physica B.* **227** (1996) 156–160.
[24] M. Rosini, V. Ya Aleshkin, A. Reklaitis and L. Reggiani, Noise and diffusion of particles obeying asymmetric exclusion processes, *Proc. Fourth Int. Conf. on Unsolved*

*Problems of Noise and Fluctuations in Physics, Biology, and High Technology* **800** (2005) 183–187.

[25] W. Belzig and M. Zareyan, Spin-flip noise in a multiterminal spin valve, *Phys. Rev. B* **69** (2004) 140407.

[26] J. Carlos Egues, G. Burkard, D. S. Saraga, J. Schliemann and D. Loss, Shot noise and spin-orbit coherent control of entangled and spin-polarized electrons, *Phys. Rev. B* **72** (2005) 235326.

[27] F. Elste and C. Timm, Transport through anisotropic magnetic molecules with partially ferromagnetic leads: Spin-charge conversion and negative differential conductance, *Phys. Rev. B* **73** (2006) 235305.

[28] A. Lamacraft, Shot noise of spin-polarized electrons, *Phys. Rev. B* **69** (2004) 081301.

[29] E. G. Mishchenko, Shot noise in a diffusive ferromagnetic-paramagnetic-ferromagnetic spin valve, *Phys. Rev. B* **68** (2003) 100409.

[30] K. E. Nagaev and L. I. Glazman, Current fluctuations in a spin filter with paramagnetic impurities, *Phys. Rev. B* **73** (2006) 054423.

[31] O. Sauret and D. Feinberg, Spin-current shot noise as a probe of interactions in mesoscopic systems, *Phys. Rev. Lett.* **92** (2004) 106601.

[32] M. Zareyan and W. Belzig, Shot noise of spin current in ferromagnet-normal-metal systems, *Europhys. Lett.* **70** (2005) 817.

[33] B. R. Bulka, J. Martinek, G. Michalek and J. Barnas, Shot noise in ferromagnetic single-electron tunneling devices, *Phys. Rev. B* **60** (1999) 12246.

[34] Y. Tserkovnyak and A. Brataas, Shot noise in ferromagnet–normal metal systems, *Phys. Rev. B* **64** (2001) 214402.

[35] B. K. Nikolic and R. L. Dragomirova, What can we learn about the dynamics of transported spins by measuring shot noise in spin–orbit-coupled nanostructures? *Semicond. Sci. Technol.* **24** (2009) 064006.

[36] A. Cottet, W. Belzig and C. Bruder, Positive cross correlations in a three-terminal quantum dot with ferromagnetic contacts, *Phys. Rev. Lett.* **92** (2004) 206801.

[37] L. Jiang, E. R. Nowak, P. E. Scott, J. Johnson, J. M. Slaughter, J. J. Sun and R. W. Dave, Low-frequency magnetic and resistance noise in magnetic tunnel junctions, *Phys. Rev. B* **69** (2004) 054407.

[38] E. R. Nowak, M. B. Weissman and S. S. P. Parkin, Electrical noise in hysteretic ferromagnet–insulator–ferromagnet tunnel junctions, *Appl. Phys. Lett.* **74** (1999) 600–602.

[39] L. Jiang, J. F. Skovholt, E. R. Nowak and J. M. Slaughter, *Proc. SPIE-Int. Soc. Opt. Eng.* **5469** (2004) 13.

[40] R. Guerrero, F. G. Aliev, Y. Tserkovnyak, T. S. Santos and J. S. Moodera, Shot noise in magnetic tunnel junctions: Evidence for sequential tunneling, *Phys. Rev. Lett.* **97** (2006) 266602.

[41] R. Guerrero, D. Herranz, F. G. Aliev, F. Greullet, C. Tiusan, M. Hehn and F. Montaigne, High bias voltage effect on spin-dependent conductivity and shot noise in carbon-doped Fe(001)/MgO(001)/Fe(001) magnetic tunnel junctions, *Appl. Phys. Lett.* **91** (2007) 132504.

[42] K. Sekiguchi, T. Arakawa, Y. Yamauchi, K. Chida, M. Yamada, H. Takahashi, D. Chiba, K. Kobayashi and T. Ono, Observation of full shot noise in CoFeB/MgO/CoFeB-based magnetic tunneling junctions, *Appl. Phys. Lett.* **96** (2010) 252504.

[43] T. Arakawa, K. Sekiguchi, S. Nakamura, K. Chida, Y. Nishihara, D. Chiba, K. Kobayashi, A. Fukushima, S. Yuasa and T. Ono, Sub-Poissonian shot noise in CoFeB/MgO/CoFeB-based magnetic tunneling junctions, *Appl. Phys. Lett.* **98** (2011) 202103.

[44] S. Garzon, Y. Chen and R. A. Webb, Enhanced spin-dependent shot noise in magnetic tunnel barriers, *Physica E* **40** (2007) 133–140.

[45] E. R. Nowak, P. Spradling, M. B. Weissman and S. S. P. Parkin, Electron tunneling and noise studies in ferromagnetic junctions, *Thin Solid Films* **377–378** (2000) 699–704.

[46] J. Bernos, M. Hehn, F. Montaigne, C. Tiusan, D. Lacour, M. Alnot, B. Negulescu, G. Lengaigne, E. Snoeck and F. G. Aliev, Impact of electron-electron interactions induced by disorder at interfaces on spin-dependent tunneling in Co-Fe-B/MgO/Co-Fe-B magnetic tunnel junctions, *Phys. Rev. B* **82** (2010) 060405.

[47] R. Stearrett, W. G. Wang, L. R. Shah, A. Gokce, J. Q. Xiao and E. R. Nowak, Evolution of barrier-resistance noise in CoFeB/MgO/CoFeB tunnel junctions during annealing, *J. Appl. Phys.* **107** (2010) 064502.

[48] W. G. Wang, C. Ni, G. X. Miao, C. Weiland, L. R. Shah, X. Fan, P. Parson, J. Jordan-Sweet, X. M. Kou, Y. P. Zhang, R. Stearrett, E. R. Nowak, R. Opila, J. S. Moodera and J. Q. Xiao, Understanding tunneling magnetoresistance during thermal annealing in MgO-based junctions with CoFeB electrodes, *Phys. Rev. B* **81** (2010) 144406.

[49] C. Tiusan, F. Greullet, M. Hehn, F. Montaigne, S. Andrieu and A. Schuhl, Spin tunnelling phenomena in single-crystal magnetic tunnel junction systems, *J. Phys.: Condens. Matter* **19** (2007) 165201.

[50] P. G. Mather, J. C. Read and R. A. Buhrman, Disorder, defects, and band gaps in ultrathin (001) MgO tunnel barrier layers, *Phys. Rev. B* **73** (2006) 205412.

[51] W. F. Egelhoff Jr, P. W. T. Pong, J. Unguris, R. D. McMichael, E. R. Nowak, A. S. Edelstein, J. E. Burnette and G. A. Fischer, Critical challenges for picoTesla magnetic-tunnel-junction sensors, *Sens. Actuators A* **155** (2009) 217.

# ON THE ORIGIN OF 1/F NOISE IN MOSFETS

L. K. J. VANDAMME
*Eindhoven University of Technology (PT 9.13), 5600 MB Eindhoven, The Netherlands*
*L.K.J.Vandamme@tue.nl*

Received 8 August 2007
Revised 16 August 2007
Accepted 16 August 2007
Communicated by Jamal Deen

Often the 1/f noise in MOSFETs is stated to be an ensemble of many RTS with different time constants. The majority of literature on 1/f noise is overlooking the contribution due to mobility fluctuations that are uncorrelated with number fluctuations. Here, we demonstrate that the so-called proofs for ΔN can also be obtained from the empirical relation. The following misunderstandings and controversial topics on the surface and bulk contributions to the low-frequency noise will be addressed: 1) 1/f and RTS noise can have different physical origins. An analysis in time domain shows that the low-frequency noise with RTS is nothing else than a superposition of a two level noise with a Lorentzian spectrum and a Gaussian noise with a pure 1/f spectrum and different bias dependency. 2) It is very unlikely that in a spectrum consisting of one strong two level RTS and a pure 1/f noise, the 1/f noise is a superposition of many RTS with different time constants. 3) The spreading in $WLS_I/I^2$ below a critical WL is not a proof for the ΔN origin. 4) The typical shape in the double log plot from sub threshold to strong inversion of $S_I/I^2$ versus I, is also not a proof for the ΔN origin.

*Keywords*: MOSFET; CMOS; 1/f-noise; RTS-noise; number-fluctuation; mobility-fluctuation; low-frequency noise; noise-source.

## 1. Introduction

The origin of 1/f noise in MOSFETs has been discussed for half a century. There are two conflicting schools of thought:
1: $\Delta\mu$ noise. The observed 1/f noise in the conductance is caused by fluctuations in the mobility of the free carriers in the conducting channel of the MOSFET. The noise of Ohmic homogenous layers can be expressed by Hooge's empirical relation as

$$\frac{S_R}{R^2} = \frac{\alpha}{Nf} \tag{1}$$

where $S_R$ is the spectral power density of resistance, $R$ the resistance of the sample, $f$ the frequency at which the noise is measured, $N$ the total number of free electrons, and $\alpha$ is an empirical dimensionless parameter often indicated by $\alpha_H$ and with values usually between $10^{-6}$ and $10^{-4}$. There is no simple theory behind (1).

2: *ΔN* noise. The 1/f noise is due to generation-recombination noise in the electron transitions between the conduction band of the channel material and the traps in the oxide layer of the MOSFET. This is the so-called McWhorter model.

In about thousand publications one can find ample support for any theory, and many counter arguments against the alternative. The majority of literature on 1/f noise in MOSFETs is overlooking the 1/f noise contribution due to mobility fluctuations that are uncorrelated with number fluctuations, for example [1–7]. The very useful characterization of the 1/f noise in terms of Hooge's empirical relation and α- value is often not done, except for example in [8–11].

Here, we demonstrate that the so-called proofs for ΔN are no proofs because it can also be obtained from the empirical relation. The following experimentally observed trends are often considered in favor of the ΔN origin and will be discussed here:
i) Spectra consisting of a strong two level RTS and a pure 1/f noise (so-called proof that the 1/f noise is a superposition of many RTS with different time constants)
ii) The spreading in $WLS_I/I^2$ below a critical WL (so-called proof for the ΔN origin)
iii) The often observed shape in the double log plot of $S_I/I^2$ versus I considered as a proof for ΔN will be explained without invoking number fluctuations.
Attempts to reduce the 1/f noise such as: 1) bias switching to inactivate the traps and 2) an inversion channel in a SiGe MOSFET away from the notorious interface, are based on the ΔN-hypothesis and are not so successful in a strong 1/f noise reduction. The RTS contribution can be reduced or just appear by switching bias but not the 1/f noise.

An important point here is that the dimensionless 1/f noise parameter α between $10^{-6}$ and $2\text{x}10^{-6}$ as observed in Si bulk resistors is also found in the best MOSFETs (with the lowest 1/f noise). Higher noise can be due to lattice damage by implantation or by a 1/f-like noise due to traps on top of the 1/f noise stemming from a bulk effect. The so-called low-frequency noise is often an addition of 1/f, generation recombination and RTS with different physical origins.

### 1.1. *1/f noise in a MOSFET calculated with a Δμ source*

The conducting channel is considered as parallel layers. The empirical relation is applied to each of them. The noise of a MOSFET is found by integration over all layers. The result is that the empirical relation can be written down for the complete channel with N electrons in total with an average mobility <μ> and with $\mu_i$ the mobility of electron i.

$$G = cN\bar{\mu} = cN\frac{1}{N}\sum_{i=1}^{N}\mu_i = c\sum_{i=1}^{N}\mu_i \quad and \quad \Delta G = c\sum_{i=1}^{N}\Delta\mu_i \tag{2}$$

$$\overline{(\Delta G)^2} = c^2\sum_{i=1}^{N}\overline{(\Delta\mu_i)^2} = c^2 N\overline{(\Delta\mu)^2} \Rightarrow because\ \overline{\Delta\mu_i\Delta\mu_j} = o \quad for\ i \neq j \tag{3}$$

$$\alpha_\mu = fN\frac{S_G}{G^2} \propto N\frac{\overline{(\Delta G)^2}}{G^2} = N\frac{c^2\sum_{i=1}^{N}\overline{(\Delta\mu_i)^2}}{(c\sum_{i=1}^{N}\mu_i)^2} = N\frac{N\overline{(\Delta\mu_i)^2}}{N^2<\mu>^2} = \frac{\overline{(\Delta\mu)^2}}{<\mu>^2} \tag{4}$$

$\alpha_\mu$ is independent of *N*. This is a typical result for p-channel MOSFETs.

If the inversion layer has a noisy parallel path with negligible conductance $G_p$ and non negligible noise $< (\Delta G_p)^2>$, then the calculated effective $\alpha_{eff}$ -value is inversely proportional to N and to the effective gate voltage $V_G^* = V_G - V_T$, because above threshold $N = V_G^* C_{ox} WL/q$. The proportionality $\alpha \propto 1/N \propto 1/V_G^*$ is typical for n-channel MOSFETs and is considered as a strong proof for the ΔN origin of 1/f noise that results from the flat band potential fluctuations, induced by the charge capture at the dielectric-Si interface [1–9]. For the parallel connection of G and $G_p$ holds, without evoking number fluctuations:

$$G_t = G + G_p \cong G = cN < \mu > \quad and \quad < (\Delta G_t)^2 > = c^2 N \overline{(\Delta\mu)^2} + < (\Delta G_p)^2 > \tag{5}$$

$$\alpha_{eff} = fN \frac{S_{G_t}}{G_t^2} \cong \alpha_\mu + \frac{K_p}{Nc^2 < \mu >^2} \quad with \quad S_{G_p} = \frac{K_p}{f}. \tag{6}$$

At threshold there is a balance between the total charge in the inversion layer and the depletion layer. The lowest value for N in sub threshold, to calculate α from experimental results using the empirical relation is $N_0$ [10, 11]:

$$N_o = \frac{kT}{q} \frac{C_{ox} WL}{q}. \tag{7}$$

## 1.2. *The McWhorter model for ΔN noise in MOSFETs*

Much experimental work proves that the 1/f noise in homogeneous samples is mobility noise, whereas there hardly is any theoretical treatment of mobility noise. With noise in MOSFETs the situation is reverse: a generally accepted model, but little experimental evidence independent from the noise that the model applies. The observed noise can be calculated from a reasonable assumption on the number of traps, but the trap distribution and the number of slow traps are not verified independently. A model for 1/f noise must explain two things: i) The shape of the spectrum. The power spectral density S is inversely proportional to the frequency *f* over a wide range of frequencies. ii) The magnitude of the noise. At a very early stage McWhorter [12] presented a simple model for the 1/f shape of the spectrum, due to an oxide layer on a semiconductor crystal. He used a mathematical result of Surdin [13], who showed that a wide range of Lorentzian spectra, with a statistical weight, g (τ), proportional to 1/τ produces a 1/f spectrum

$$S(\omega,\tau) \propto \frac{\tau}{1+\omega^2\tau^2} \quad and \quad g(\tau) \propto \frac{1}{\tau} \Rightarrow S = \int_0^\infty g(\tau)\, S(\omega,\tau) d\tau \propto \int_0^\infty \frac{d\tau}{1+\omega^2\tau^2} \propto \frac{1}{\omega} \arctan\Big|_0^\infty \propto \frac{1}{f} \tag{8}$$

McWhorter proposed a physical model of the statistical weight, required in Surdins mathematical model. The semiconductor is covered by an oxide, in which traps, *x*, are homogeneously distributed along the distance, *s*, from the interface: *dx/ds,* is constant. The electrons in the conducting channel reach the traps by tunneling, which gives a trapping probability exponentially depending on the trap distance, *s*.

$$\tau(s) = \tau(0)\, e^{s/\lambda} \Rightarrow g(\tau) = \frac{dx}{d\tau} = \frac{dx}{ds} \cdot \frac{ds}{d\tau} \propto \lambda \frac{1}{\tau} \quad with\ \lambda = 0.1 nm \text{ "McWhorter constant"} \tag{9}$$

The generation-recombination noise –GR noise– from each trap is a Lorentzian with τ(s). The summation of the GR spectra results in a 1/f spectrum, because of Surdins integral in Eq. (8). GR noise is ΔN noise. McWhorters model in Eq. (9) is the physical basis of the ΔN model and λ is the attenuation depth of the electron wave function into the oxide.

## 2. Problems and Misconceptions

### 2.1. *1/f noise and RTS noise have different origins*

The observed low-frequency noise versus time often looks like Fig. 1a, which is a summation of real 1/f and RTS noise. The bias dependence and size dependence of the two noise components can be studied separately. From different bias dependencies it can be concluded that the 1/f and RTS component have different physical origins [14].

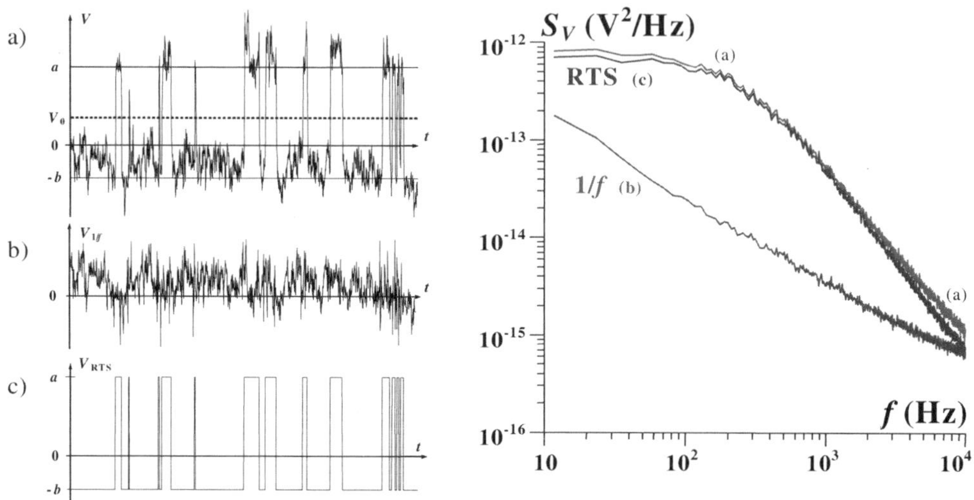


d) Observed noise and its decomposition in a 1/*f* and RTS spectrum

Fig. 1. The noise in a) equals the sum of the contributions in b) and c); V= $V_{1/f}$ + $V_{RTS}$. a) Observed low-frequency noise V (t) versus time. There are two peaks in the amplitude probability density function (pdf) at the levels "a" and "-b". b) The $V_{1/f}$ (t) component has a Gaussian pdf. c) The $V_{RTS}$ (t) component. d) The spectra of: a), b), c) [14].

Neglecting the different origins of 1/f and RTS noise has led to some misunderstanding on how to reduce low-frequency noise. In for example, gate oxide nitridation studies a frequency index $\gamma > 1.2$ or $\gamma < 0.8$ is often found. We consider this 1/f-like noise as the result of trapping of carriers on the top of (bulk) 1/f noise. In non mature technologies with new oxides, high 1/f or 1/f like noise is observed [8, 15–18]. The so-called 1/f noise reduction by additional cleaning steps before oxidation or the switching bias technique [19] often turned out to be a reduction in the contribution of Lorentzians (RTS) but not often in the 1/f noise part of the spectrum.

The relative 1/f noise has 1/N dependence, while the relative RTS noise has $1/N^2$ dependence. This explains why for modern sub-micron devices hold [20–22]: i) Smaller devices are faster but have more 1/f noise. ii) A Lorentzian from one trap can appear on top of 1/f noise if $N = C_{gate}V_G^*/q < 1/(4\pi\alpha) \equiv N_{crit}$. This results in a 1/f-like spectrum. This criterion is normally fulfilled in sub-micron devices with an area $LW < 1\mu m^2$. We successfully verified this criterion on spectra in [20, 21]. All spectra in Ref. [7] from submicron MOSTs that fulfill the number criterion show Lorentzians on top of the 1/f noise. The calculated α-values from the 1/f part of their spectra are $\alpha = 5x10^{-5}$. The $1/N^2$

dependence for the RTS explains the increase in the low-frequency noise, normalized for area and current, $WLS_I/I^2$ in sub micron devices [2–4, 7, 20–22].

### 2.2. *Is 1/f noise an addition of Lorentzians?*

Surdin [13] showed that 1/f noise could be obtained from the addition of Lorentzians stemming from traps with a $1/\tau$ distribution. McWhorter [12] gave a physical explanation for the $1/\tau$ distribution. If all traps in the distribution have *the same variance* $<(\Delta N)^2>$ and have a distribution $g(\tau) = dN_t(\tau)/d\tau = K/\tau$ in the range $\tau_2 < \tau < \tau_3$ then:

$$\frac{S_I}{I^2} = \frac{S_N}{N^2} = \frac{1}{N^2}\int_{\tau_2}^{\tau_3} \frac{4\overline{(\Delta N)^2}\tau\, g(\tau)d\tau}{1+(2\pi f\tau)^2} = \frac{4\overline{(\Delta N)^2}}{N^2}\int_{\tau_2}^{\tau_3} \frac{Kd\tau}{1+(2\pi f\tau)^2}. \tag{10}$$

K is defined by the total number of traps to create a 1/f spectrum. We assume as lowest value K=1. Then we can still use a distribution function and the integral approach instead of the addition of Lorentzians of discrete traps to calculate the spectrum. For K=1 holds one trap per factor e in $\tau$, or 27 traps to cover a 1/f noise from $10^{-6}$ Hz to $10^6$ Hz.

$$N_t = \int_{\tau_2}^{\tau_3} g(\tau)\,d\tau \Rightarrow N_t = K\ln(\tau_3/\tau_2) \to K = \frac{N_t}{\ln(\tau_3/\tau_2)}. \tag{11}$$

The solution of the integral in Eq. (10) is:

$$\int_{\tau_2}^{\tau_3} \frac{Kd\tau}{1+(2\pi f\tau)^2} = \frac{K}{2\pi f}\left[arctg(2\pi f\tau_3) - arctg(2\pi f\tau_2)\right]. \tag{12}$$

We now consider the case of: i) one single trap with $\tau_1$, which generates RTS noise and ii) the usual distribution ($g(\tau) \sim 1/\tau$) of traps with Lorentzian spectra, which together generate 1/f noise. The spectrum of the RTS noise from the single trap with $\tau_1$ is:

$$\frac{S_I}{I^2} = \frac{1}{N^2}\frac{4\tau_1\overline{\Delta N^2_{RTS}}}{1+(2\pi\tau_1 f)^2} = \frac{1}{N^2}\frac{\tau_1}{1+(2\pi\tau_1 f)^2} \quad if \quad \overline{\Delta N^2_{RTS}} = 1/4. \tag{13}$$

The variance and the characteristic time $\tau_1$ of *one single trap* characterized by average emission- and capture times, $\tau_e$ and $\tau_c$ are given by:

$$\overline{\Delta N^2_{RTS}} = \frac{1}{2+\tau_c/\tau_e+\tau_e/\tau_c} \quad and \quad \frac{1}{\tau_1} = \frac{1}{\tau_e} + \frac{1}{\tau_c}. \tag{14}$$

For a symmetric trap ($\tau_e = \tau_c$) the variance equals ¼. For strong asymmetric traps ($\tau_e << \tau_c$ or $\tau_c << \tau_e$) the variance becomes zero. If the variance of the RTS trap has its maximum value ¼, then the total spectrum of the RTS trap with $\tau_1$ and the distribution is:

$$\frac{S_I}{I^2} = \frac{1}{N^2}\left\{\frac{4\overline{(\Delta N)^2}N_t}{\ln(\tau_3/\tau_2)}\frac{\left[arctg\,(2\pi f\tau_3) - arctg\,(2\pi f\tau_2)\right]}{2\pi f} + \frac{\tau_1}{1+(2\pi f\tau_1)^2}\right\}. \tag{15}$$

Equation (15) can be approximated for $f < 1/2\pi\tau_3$, $f > 1/2\pi\tau_2$ and 1/f region by, respectively:

$$\frac{S_I}{I^2}=\frac{1}{N^2}\frac{4\overline{(\Delta N)^2}N_t}{\ln(\tau_3/\tau_2)}\tau_3 \; and \; \frac{S_I}{I^2}=\frac{1}{N^2}\frac{4\overline{(\Delta N)^2}N_t}{\ln(\tau_3/\tau_2)}\frac{1}{(2\pi f)^2}\frac{1}{\tau_2} \qquad (16a, 16b)$$

$$for\ 1/\tau_3 << 2\pi f << 1/\tau_2 \Rightarrow \frac{S_I}{I^2}=\frac{1}{N^2}\left\{\frac{\overline{(\Delta N)^2}N_t}{\ln(\tau_3/\tau_2)}\frac{1}{f}+\frac{\tau_1}{1+(2\pi f\tau_1)^2}\right\}. \qquad (16c)$$

The α-value characterizing the 1/f noise in this ΔN model is then proportional to the *variance per trap*, $<(\Delta N)^2>$ belonging to the 1/τ-set and from Eq. (16c) follows:

$$\alpha=\frac{\overline{(\Delta N)^2}N_t}{\ln(\tau_3/\tau_2)}\frac{1}{N}\propto\frac{1}{C_{ox}V_G^*}\propto t_{ox}. \qquad (17)$$

The α-value is independent of the device size, because $N_t$ and N at a fixed $V_G^*$ are proportional to WL. We try to compare the RTS and 1/f noise at conditions where both components appear. The symmetric RTS noise can best be observed at $f=1/2\pi\tau_1$ on top of the 1/f noise if $N<1/(4\pi\alpha)\equiv N_{crit}$ [20, 21]. For a 1/f noise generated by traps, then we find with Eq. (17) and the RTS number criterion $N < N_{crit}$:

$$N<\frac{N\ln(\tau_3/\tau_2)}{4\pi\overline{(\Delta N)^2}N_t}\Rightarrow\overline{(\Delta N)^2}<\frac{1}{4\pi} \quad if\ N_t=\ln\frac{\tau_3}{\tau_2}\ or\ K=1. \qquad (18)$$

However, to fulfill the condition to observe an RTS at $f=1/2\pi\tau_1$ on top of the 1/f we need a very low *variance per trap*, $<(\Delta N)^2>$ for all gate voltages compared to the variance of the single RTS trap, $<(\Delta N_{RTS})^2>=1/4$. We assume a trap distribution in the dielectric given by $D_{ox}$ [$cm^{-3}\ eV^{-1}$] $\approx N_t/kTWLs_3$ with $s_3$ the largest tunnel distance to the interface from the trap with a characteristic time $\tau_3$. Assuming a small device, where we can expect RTS on top of 1/f noise with: WL = 0.1μm$^2$, kT = 1/40 eV, $s_3$ = 2.7 nm and the lowest number of active traps in the distribution $N_t$ = 27 and $t_{ox}$ =2.7 nm; we find $D_{ox}=4\times10^{17}$ [$cm^{-3}\ eV^{-1}$] and $N=8\times10^3$ for $V_G^*$ = 1V. The value $D_{ox}$ is in the range of published results in [3, 17].

The ratio of the total noise over the 1/f noise at $f=1/2\pi\tau_1$ is calculated from (16c) as:

$$\frac{RTS+1/f}{1/f}=1+\frac{N_{crit}}{N}=1+\frac{1}{4\pi\overline{(\Delta N)^2}} \quad if\ N_t=\ln\frac{\tau_3}{\tau_2}. \qquad (19)$$

In the calculation used in Fig. 2 and Table 1 we used: $t_{ox}$ = 2.7 nm; $V_G^*$ = 1V; $\tau_1=10^{-2}$ s; $\tau_2=10^{-6}$ s; $\tau_3=10^6$ s; $<(\Delta N_{RTS})^2>$ = ¼ and WL = 0.1μm$^2$ for the devices on row 1, 2, 3 and WL = 1μm$^2$ for the device on row 4 in Table 1.

On the first row a high noise device is characterized with $D_{ox}=4\times10^{18}\ cm^{-3}\ eV^{-1}$ and a variance per trap $<(\Delta N)^2>$ =1/4. The RTS noise is not visible at all see black diamonds in Fig. 2.

On the second and third row a normal and low noise device are shown with $<(\Delta N)^2>$ =1/4 (triangles in Fig. 2) and 1/40, respectively. The RTS component is visible only if $N<1/4\pi\alpha\equiv N_{crit}$ and a small variance per trap $<(\Delta N)^2> << 1/4\pi$ as is shown by open squares in Fig. 2.

On the fourth row the results of a low noise "large" device with WL = 1μm$^2$ with $D_{ox} = 4x10^{16}$ cm$^{-3}$ eV$^{-1}$ are shown and presented by dots in Fig. 2.

Table 1. The parameters used in the simulation presented in Fig. 2 and results.

| N | $N_t$ | $<(\Delta N)^2>$ | α-value | $N_{crit} \equiv 1/4\pi\alpha$ | (RTS+1/f)/(1/f) | Symbol |
|---|---|---|---|---|---|---|
| $10^4$ | 300 | 1/4 | $2.7x10^{-4}$ | $2.9x10^2$ | 1.03 | Diamonds |
| $10^4$ | 30 | 1/4 | $2.7x10^{-5}$ | $2.9x10^3$ | 1.3 | Triangles |
| $10^4$ | 30 | **1/40** | $\mathbf{2.7x10^{-6}}$ | $2.9x10^4$ | 3.9 | squares (RTS) |
| $10^5$ | 30 | 1/4 | $2.7x10^{-6}$ | $2.9x10^4$ | 1.3 | Dots |

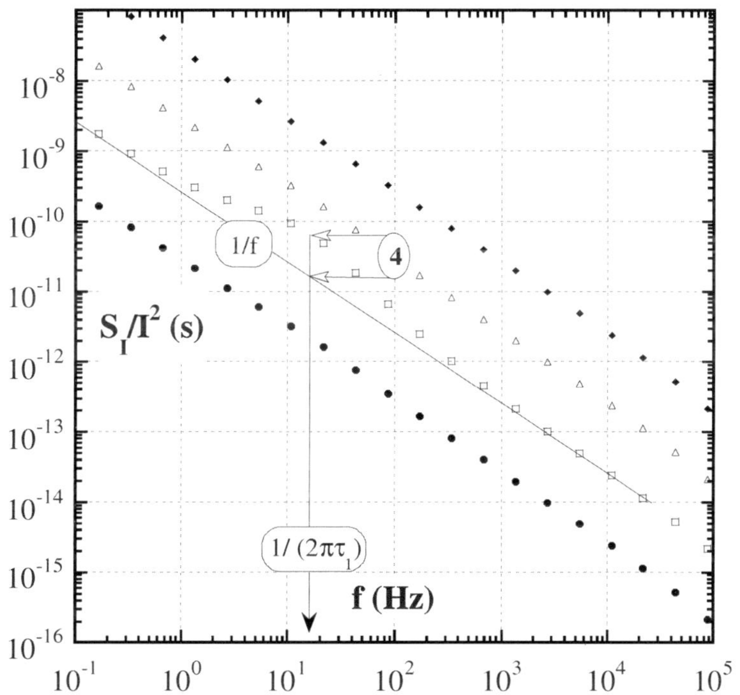


Fig. 2. The calculated 1/f spectra from a distribution of Lorentzians and one RTS using Eq. (15, 16c) with the parameters in Table 1.

The RTS noise is visible only under the very unlikely condition that at an arbitrary gate voltage, the variance per trap in the distribution is at least 10 times smaller then the maximum variance = ¼ of the single RTS trap.

Hence, it is very unlikely that the observed 1/f noise contribution in a low-frequency noise, consisting of a pure 1/f with a $2x10^{-6} < \alpha < 3x10^{-4}$ and an RTS, stems from the addition of distributed traps as given in the (16c).

On the contrary, a 1/f noise with a strong RTS contribution for N < 1/4πα is a clear indication that the 1/f noise does not result from an addition of Lorentzians. This is in sharp contrast with the statement by Kirton and Uren and also for example, Scholten et al. that the 1/f noise in large area MOSFETs is generated by a superposition of many Lorentzians with distributed time constants [5, 6].

### 2.3. *Spreading in normalized noise is used as improper argument for ΔN fluctuations*

The spreading in $WLS_I / I^2$ versus size below a critical L or WL value is used improperly as proof of 1/f noise being a ΔN fluctuation [1–4]. The noise normalized for size may decrease or increase below a critical size. Reduction was observed at a constant normalized current level [1–4]. The so-called ΔN proof is based on the idea that in sub-micron MOSFETs we can have a number of traps quite different from the average expected for that size. Some devices will have a number of traps well below the average and some will have a non 1/τ-ditribution, others will have a larger number of traps than the average. Hence, too little or no 1/f noise or too much low-frequency noise could be the result. Here, a model for noise reduction and noise enhancement is proposed. We demonstrate that such results can be explained without using the hypothesis that 1/f noise is an addition of RTS.

#### 2.3.1. *Model to explain the reduction in $WLS_I/I^2$ with decreasing size*

The model is based on the assumption that channel resistance and noise consists of two contributions. Our calculation takes into account the special bias condition that was used in Ref. [1–4]: "all measurements at the same normalized current *IL/W* and all results at a fixed drain source voltage to compare the normalized noise from MOSFETs with different length and size". The channel and series resistance contributions are taken into account as follows: i) channel resistance, $R_{ch}$ with its 1/f noise, $S_{Rch}$. ii) drain series resistance inversely proportional to channel width and with negligible noise contribution. The specific series resistance, $R_{do}$ (Ω μm) = $R_d$ (Ω) W is technology dependent with a value of 60 kΩ μm for digital circuits and of 2 kΩ μm for analogue MOSFETs due to higher implantation energy [23, 24].

This implies that, due to the increasing effect of drain series resistance, shorter channel MOSTs need a higher effective gate voltage to keep the same normalized current level equal to IL/W. The channel resistance R, current I and the normalized current $K_I$ are:

$$R = R_d + R_{ch} = R_d + \frac{L}{W\mu C_{ox}V_G^*} = \frac{L}{W}\left(\frac{R_{do}}{L} + \frac{1}{\mu C_{ox}V_G^*}\right) \Rightarrow R_d \equiv \frac{R_{do}}{W} \tag{20a}$$

$$I = \frac{V_D}{R} = \frac{W}{L}\frac{V_D}{\left(R_{do}/L + 1/\mu C_{ox}V_G^*\right)} \Rightarrow K_1 \equiv \frac{IL}{W} = \frac{V_D}{\left(R_{do}/L + 1/\mu C_{ox}V_G^*\right)}. \tag{20b}$$

All devices are compared at the same drain source voltage $V_D$. The specific series resistance $R_{do}$ is the same for all devices with the same technology. The effective gate voltage $V_G^*$ must increase slightly for shorter L, in order to keep the normalized current $K_I$ constant (see Eq. (20b)). From Eq. (20b) follows $V_G^*$, $V_{G0}^*$ and $L_{crit}$:

$$V_G^* = \frac{1}{R_{do}C_{ox}\mu}\left[\frac{L_{crit}L}{L - L_{crit}}\right] = V_{GO}^*\frac{L}{L - L_{crit}} \tag{21a}$$

$$V_{GO}^* \equiv \frac{K_1}{V_D C_{ox}\mu};\ L_{crit} \equiv \frac{R_{do}K_1}{V_D} = \frac{ILR_{do}}{WV_D} = R_{do}V_{GO}^*C_{ox}\mu. \tag{21b}$$

For a comparison at fixed $V_D$ (is the applied voltage across the device with its series resistance) and the same technology, the values $K_l$ and $L_{crit}$ are constant. The effective gate voltage $V_{GO}^*$ is the gate voltage applied on a long channel device to obtain the normalized current ($IR_d << V_D$). The critical length $L_{crit}$, depends on the specific series resistance and bias $V_D$ and the ratio $L_{crit}/L = IR_d/V_D = R_d/R < 1$.

The relative noise normalized for area WL, frequency and current is for long channel devices ($L >> L_{crit}$) denoted by the subscript "∞" and is given by:

$$\left[\frac{f\,WL\,S_I}{I^2}\right]_\infty = \frac{\alpha q}{C_{ox}V_{GO}^*}. \tag{22}$$

The normalized relative 1/f noise for short channels (without RTS contribution in the frequency range of investigation) is given by:

$$\left[\frac{f\,WL\,S_I}{I^2}\right]_L = \frac{1}{(1+R_d/R_{ch})^2}\frac{\alpha q}{C_{ox}V_G^*} = \frac{\alpha q}{C_{ox}V_{GO}^*}\left[1-\frac{L_{crit}}{L}\right]^3. \tag{23}$$

We calculate the reduction factor with Eqs. (22, 23). The ratio between the normalized noise for short channel denoted by index L and a very long channel denoted by index ∞ ($R_d$ <<Rch) is

$$\frac{\left[f\,WL\,S_I/I^2\right]_L}{\left[f\,WL\,S_I/I^2\right]_\infty} = \left[1-\frac{L_{crit}}{L}\right]^3 \quad with \quad \frac{L_{crit}}{L} = \frac{R_d}{R} = \frac{IR_d}{V_D} < 1. \tag{24}$$

The power 3 in Eq. (24) becomes 2 in a model with comparison under constant $V_G^*$ instead of constant IL/W. Hence, a reduction in normalized noise can be expected with decreasing lengths mainly due to a series resistance effect.

### 2.3.2. *Model to explain the enhancement in $WLS_I/I^2$ with decreasing size*

The noise enhancement can be explained as the consequence of the fallacy regarding the different physical origins for RTS and 1/f noise. For simplicity, we compare devices with different length biased close to the threshold or in sub threshold. Then $N=N_0$ as given by Eq. (7) and the 1/f noise and RTS noise at the characteristic frequency $f = 1/2\pi\tau_1$ is:

$$\frac{S_{I\,1/f}}{I^2} = \frac{\alpha}{N_0 f} \Rightarrow at\ f = \frac{1}{2\pi\tau_1}\ at\,threshold\ \frac{S_{I\,1/f}}{I^2} = \frac{2\pi\tau_1\alpha q}{WLC_{ox}(kT/q)} \tag{25}$$

$$\frac{S_{I\,RTS}}{I^2} = \frac{1}{N_0^2}\frac{\tau_1}{2}\quad at\ f = \frac{1}{2\pi\tau_1} \tag{26}$$

$$\frac{S_{I\,1/f}+S_{I\,RTS}}{S_{I\,1/f}} = 1+\frac{1}{4\pi\alpha N_0} \equiv 1+\frac{N_{crit}}{N_0} \equiv 1+\frac{L_{RTS}}{L} \tag{27a}$$

$$L_{RTS} \equiv \frac{q^2}{4\pi\alpha kTC_{ox}W} \,. \tag{27b}$$

The parameter $L_{RTS}$ is proportional to the oxide thickness and inversely proportional to the channel width and α. In the calculation presented in Fig. 3 we used for the noise enhancement part Eq. (27a) with: W = 1μm, the gate oxide thickness $t_{ox}$ = 3nm, T = 300K and $\alpha = 4\text{x}10^{-5}$ which results with Eq. (27b) in $L_{RTS}$ = 1μm.

For the noise reduction part we calculated with Eq. (21b) $L_{crit}$ = 0.1 μm from the following reasonable values: $t_{ox}$ = 3nm, $V_{GO}^{*}$ = 0.1 V; μ = 200 $cm^2$/Vs; and a specific series resistance $R_{do}$ = 4.3 kΩ μm.

The scattering in noise normalized for area at low L is not an argument for the ΔN origin of the 1/f noise. The scattering shows that submicron MOSFETs can suffer from series resistance effects and an increased low-frequency noise due to Lorentzians stemming from RTS on top of 1/f noise. We have shown that the abundant literature on spreading with shrinking dimensions, for example [2–4] is explained without using the hypothesis that 1/f noise is an addition of RTS.

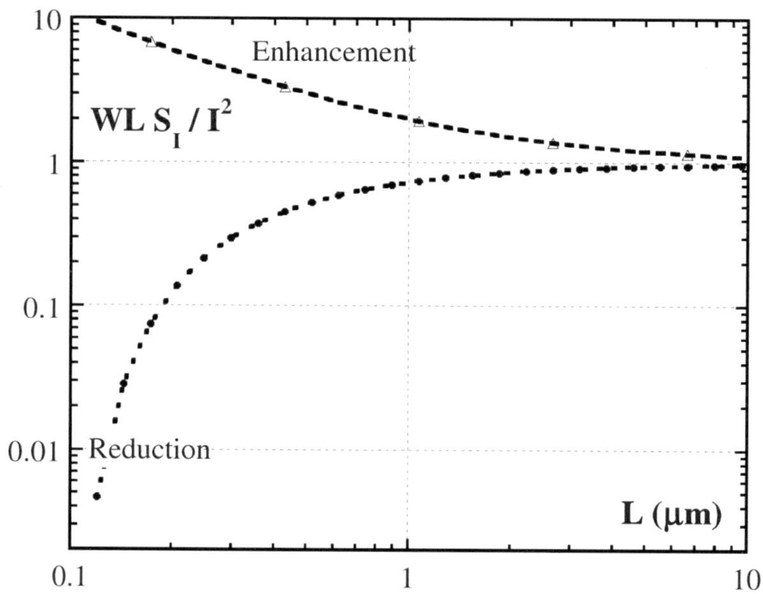


Fig. 3. The spreading in normalized relative noise versus channel length L calculated with Eqs. (24, 27) and $L_{crit}$ = 0.1 μm and $L_{RTS}$ = 1 μm. The enhancement is explained by the 1/N and $1/N^2$ dependences of the 1/f and RTS contributions. Reduction is explained by the series resistance effect.

### 2.4. *Is the typical $S_I / I^2$ - I plot at low fixed $V_{DS}$ an indisputable prove of the ΔN origin of 1/f noise?*

A typical plot of $S_I/ I^2$ versus I is shown in the Fig. 4 for two p-channel MOSFETs without series resistance complications (channel width W = 15μm and channel lengths L = 15 μm and 0.35 μm). The leveling off is due to sub threshold, where the $S_I / I^2 = \alpha/fN_0$ with $N_0 = (kT/q)\, C_{gate}/q$. The α-value over the whole range of currents is for both devices about $2\text{x}10^{-6}$, typical for bulk silicon of high quality. The same devices were used successfully to check the RTS number criterion: $N < N_{crit} = 1/(4\pi\alpha)$ in [20]. If the criterion $N < N_{crit}$ holds then it often leads to an expected RTS on top of 1/f noise.

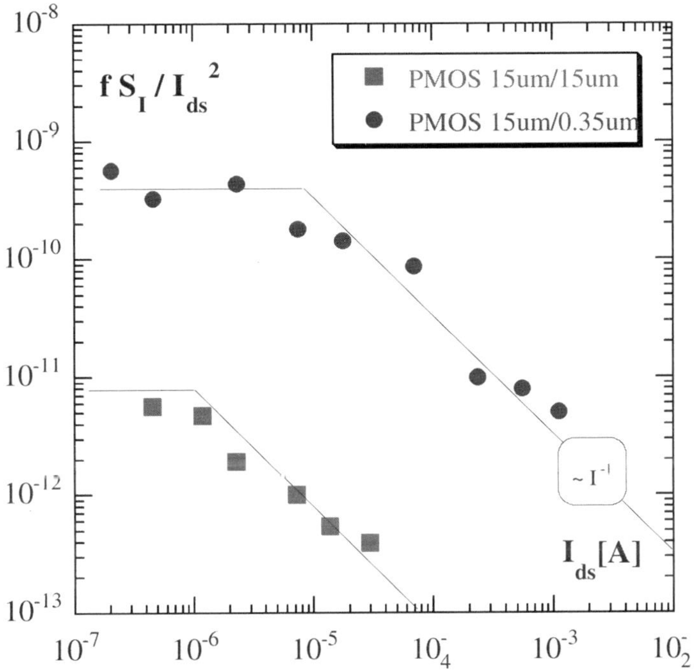


Fig. 4. Experimental results showing the relative saturation current noise versus current for $V_G = V_d$ and $V_T = -0.83\text{V}$. The high-quality p-channel MOSFETs with: $t_{ox} = 6.5$ nm, W = 15μm and channel lengths L = 15 μm and 0.35 μm are biased in the range of $-0.872 > V_G\,(\text{V}) > -1.96$ and $-0.65 > V_G\,(\text{V}) > -1.95$, respectively. The leveling off is typical for the sub threshold region where the 1/f noise is given by $S_I/I^2 = \alpha/(N_0 f)$ with $N_0 = (kT/q)\,(C_{gate}/q)$ and the 1/f noise parameter is $\alpha = 2\text{x}10^{-6}$. The proportionality: $fS_I/I^2 \sim I^{-1}$ in the ohmic region is typical for a constant α-value and negligible series resistance effect.

Submicron devices often show series resistance complications [23]. The current-voltage relation for the ohmic region, taking into account the series resistor $R_d$ is:

$$I = \frac{W\mu C_{ox} V_G^*(V_D - IR_d)}{L} \Rightarrow I = \frac{\beta V_G^* V_D}{1+\beta V_G^* R_d} \; with \; \beta = \frac{W\mu C_{ox}}{L}. \tag{28}$$

Equation (28) is equivalent with Eq. (20b). Now we consider two cases: i) a noise-free $R_d$ and ii) a noisy $R_d$ in series with the channel resistance $R_{ch}$. The resistance, $R = R_{ch}+R_d = V_D/I$, $R_{ch}$, and $R_{ch}/R$ at low $V_D$ are

$$\frac{1}{R} = \frac{I}{V_D} = \frac{\beta V_G^*}{1+\beta V_G^* R_d} \Rightarrow R_{ch} = \frac{1}{\beta V_G^*} \Rightarrow \frac{R_{ch}}{R} = \frac{1}{1+\beta V_G^* R_d} = 1 - \frac{L_{crit}}{L}. \tag{29}$$

For case i) a noise-free $R_d$ with $R_{d0} = R_d W$, $S_R$ and relative noise at low $V_D$ are

$$S_R = S_{R_{ch}} \Rightarrow \frac{fS_I}{I^2} = \frac{fS_R}{R^2} = \frac{fS_{Rch}}{R_{ch}^2}\left[\frac{R_{ch}}{R}\right]^2 = \frac{\alpha}{N+N_0}\left[\frac{R_{ch}}{R}\right]^2 . \tag{30}$$

The relative current noise as function of the effective gate voltage $V_G^*$ is

$$\frac{fS_I}{I^2} = \frac{\alpha q}{WLC_{ox}(V_G^* + kT/q)(1+R_d \beta V_G^*)^2} \text{ with } V_G^* = 0 \text{ in sub threshold} . \tag{31}$$

For case ii) with a noisy drain series resistance with $S_{R_d} = C_{R_d} R_d^2 / f$ we find:

$$S_R = S_{R_{ch}} + C_{R_d} R_d^2 / f \Rightarrow \frac{fS_R}{R^2} = \frac{fS_I}{I^2} = \frac{\alpha}{N+N_0}\left[\frac{R_{ch}}{R}\right]^2 + C_{R_d}\left[\frac{R_d}{R}\right]^2 . \tag{32}$$

The relative current noise calculated from the empirical relation with Eqs. (31, 32) taken into account sub threshold and series resistance will now level off at low and at high gate voltages as often observed [1–4, 9, 17].

The simulated results with series resistance and without are shown by 3 curves in Fig. 5 for the parameters mentioned in Table 2. The full line shows the naïve application of the empirical relation without sub threshold and series resistance corrections. The discrepancy between experimental results that are often given by curve 2 or 3 and the naïve application of the empirical relation are often used to demonstrate erroneously the non validity of the empirical relation.

The solution of the Poisson equation for a MOSFET biased at constant gate voltage below threshold explains the change of N into $N_0$ as in Eqs. (7, 30, 31, and 32). From charge continuity follows that a perturbation with a charge $\Delta Q$ at the dielectric interface is compensated by the same charge $-\Delta Q$, but shared over the inversion and depletion layer. The reduced fluctuation in the inversion layer explains the leveling off in RTS noise and in $S_I/I^2$ versus I in sub threshold. The solution of the Poisson equation for MOS structure at constant gate voltage is insensitive on how the perturbation was created (either by trapping or by a mobility fluctuation) [10, 11].

In saturation we find a reduced number of 2/3N with N the number of carriers at zero drain-source voltage. The relative current noise in saturation, $S_{Is}/I_s^2$ is twice the value of $S_I/I^2$ in the ohmic region at the same effective gate voltage for the $\Delta\mu$ and $\Delta N$ model; at least for a healthy MOSFET without series resistance complications [24]. Therefore, the noise is often characterized in the ohmic region in a relative way as function of the over-drive gate voltage $V_G^* = V_G - V_T$ on double log scales in order to verify the proportionality between $S_I/I^2$ and $1/V_G^{*2}$ or $1/V_G^*$ or $\alpha$ constant with $V_G^*$ or $\alpha$ proportional to $1/V_G^*$. Designers are interested in the saturation region (above threshold) and can read from such diagram the relative current noise in saturation for a given $V_G^*$.

Table 2. Simulation parameter values for the calculation of current and relative current noise for a MOSFET biased at $V_{DS} = 0.1$ V and $0 < V_G^*$ [V] $< 4$.

| W (μm) | L(μm) | $C_{ox}$ (F/cm²) $t_{ox}$=5nm | μ (cm²/Vs) | α | Specific series resistance $R_{do}$ (Ω μm) and $C_{RD}$ |
|---|---|---|---|---|---|
| 5 | 0.25 | $6.9\times10^{-7}$ | 200 | $6.2 \times 10^{-5}$ | $2 \times 10^3$ (Ω μm) in 2, 3 and $C_{RD} = 10^{-9}$ in 3 |

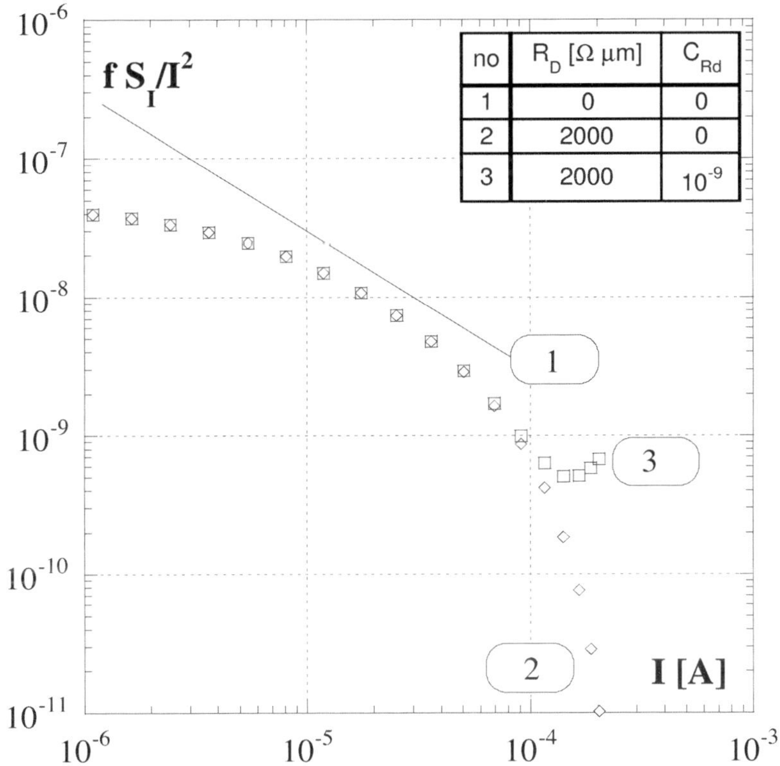


Fig. 5. The relative current noise calculated for MOSFETs with the parameter values as indicated in Table 2. Line 1 indicates the over simplification without taking into account series resistance effects and the typical sub threshold behavior of 1/f noise in MOSFETs. Line 2 is calculated with Eq. (31) and shows the effect of sub threshold an a noise free series resistance resulting in a steep decrease at high current. Line 3 is calculated from Eq. (32) and takes into account the effect of sub threshold and a noisy series resistance. The curl up in line 3 and the steep decrease in line 2 at high currents was often observed in experimental results [1, 3, and 4].

## 3. α A Figure of Merit for 1/f Noise Comparison

The 1/f noise parameter α is a good figure of merit to compare the 1/f noise of different technologies. Without suggesting a ΔN or Δμ origin for the 1/f part of the spectrum it can be used to compare 1/f noise reduction strategies in a quantitative way. For example, it is misleading to present a noise reduction by a factor 4 in a SiGe MOSFET compared to a rather noisy Si reference device if the reference device has a not mentioned $\alpha = 2.4\text{x}10^{-4}$ and the SiGe MOSFET a not mentioned α-value of $6\text{x}10^{-5}$, which is still a factor 60 higher than the best bulk p-channel MOSFETs [24].

Therefore, we need absolute values to compare noise and in Fig. 6 we present calculated α-values versus oxide thickness from experimental results. For MOSTs where α depends on gate voltage, we choose $V_G^*=1\text{V}$ to calculate α. For the discussion some references are mentioned in Fig. 6 and for clarity some are not mentioned. It is very helpful to calculate α-values from experimental results with the equations in [24, 22, 9–11]. Some remarks on typical results in Fig. 6 from n- and p-channel MOSFETs [23, 25–33]:

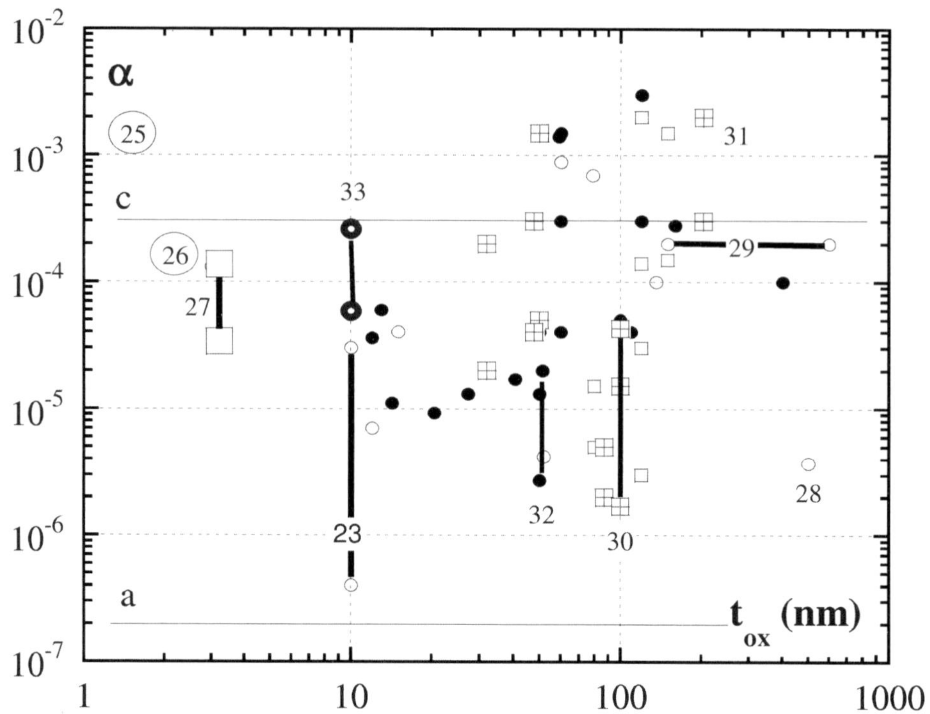


Fig. 6. Calculated α-values from data collected from literature over 40 years, just to compare results, not to suggest an origin.

Reference [23] in Fig. 6 shows the very low 1/f noise in bulk and higher noise for surface p-channels.

The circles [25] and [26] stem from high-K dielectric p-channels with an equivalent oxide thickness (EOT of 1.5 and 2.3 nm). It cannot be excluded that the high α-values are due to ΔN contributions in [25, 26].

The results [27] show the impact of post oxidation annealing on the 1/f noise of p-channels.

Reference [28] shows a modern device with $t_{ox}$ = 500 nm (for irradiation dosimeter) compared to old devices [29] with oxide thickness independent α ($t_{ox}$ = 150 and 600 nm). The international technology roadmap for semiconductors (ITRS) is based on ΔN and predicts α proportional to $t_{ox}$ as can be expected from Eq. (17).

Ref. [30] indicates results on n-channels with the same oxide thickness but a noise level proportional to the surface state density.

Ref. [31] shows the result of a commercially available (40 years ago) n-channel MOST. There is no difference in α with a modern MOSFET [25].

Ref. [32] shows the n-channel in the temperature range 4.2K-295K.

Ref. [33] shows the calculated α-value from the sub threshold for the noisy Si reference device with $\alpha = 2.4\text{x}10^{-4}$ and the SiGe p-channel with $\alpha = 6\text{x}10^{-5}$.

The observed $t_{0x}$ dependence of α cannot be used as argument in favor for a ΔN or a Δμ origin (see Eqs. (4, 6, and 17)). The calculated α do not rule out the McWhorter or the Δμ-model. The "reconciliation" between ΔN and Δμ by adding to the main noise source ΔN a correlated Δμ(ΔN) source is not more than curve fitting [10, 37].

The horizontal lines in Fig. 6 indicated by "a" and "c" show the range of observed α-values in bulk silicon resistors [34–36]. Experimental results on IC-graded silicon bulk resistors at 77K and 300K show $10^{-6} < \alpha < 2\text{x}10^{-6}$ [35]. The role of the $SiO_2$-Si interface in these experiments could be ignored; because constriction dominated contact resistors were used. Poorly annealed p-type buried resistors show $\alpha = 3\text{x}10^{-4}$ as given by line "c" and buried resistors annealed at 900°C result in $\alpha < 5\text{x}10^{-6}$ [34]. Accumulation of electrons at the Si-silicon oxide interface of a gated n-type resistor resulted in $\alpha = 10^{-5}$ and pushing the current path away from the interface with a negative gate voltage resulted in depletion at the interface and $\alpha \approx 3\text{x}10^{-7}$ indicated by line "a" [36].

A nice demonstration of the implantation effect on the 1/f noise in MOSFETs expressed in α-values is given in Fig. 7 below. Deeper implantations (more lattice damage and traps) can result in more lattice damage below the inversion layer and hence, in lower α-values. Shallow implantations with a lower projected range, $R_p$ can have the peak of less lattice damage in the inversion layer and result in higher α-values [38]. The peak of the lattice damage is at about ½ $R_p$.

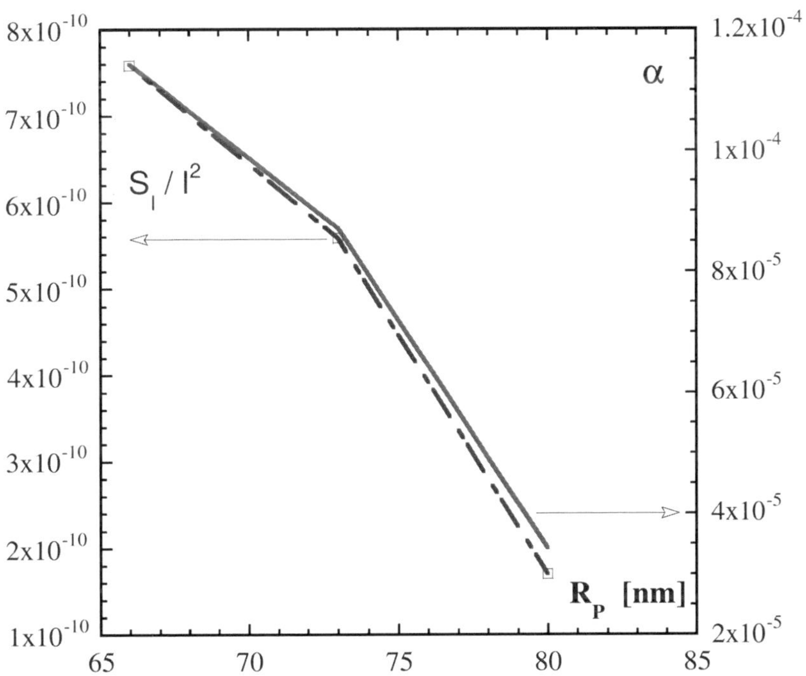


Fig. 7. The impact of implantation expressed in projected range on the 1/f noise expressed on the right hand scale in α-values and on the left hand scale in relative noise (at 1Hz). The lattice damage in the Si is mainly concentrated at a distance $1/2R_p$. The α-values are calculated from data in [38].

## 4. Discussion and Conclusion

The bulk contribution to the 1/f noise in MOSFETs is strongly under estimated in literature. Some misunderstandings on the low-frequency noise can be avoided by decomposing spectra in pure 1/f and Lorentzians and by characterizing the 1/f noise in an α-value as a general figure of merit. So we can compare the low-frequency noise in a MOSFET with what is found in bulk silicon resistor with α-values as low as $10^{-6}$ [35, 39]. So we can learn more about the origin of different noise sources, for example:

i) α-values higher than $10^{-6}$ can be expected due to additional crystal damage in the inversion layer. This can be caused by implantation or due to strain in the inversion layer as a result of high-K gate stacks on top of the inversion layer. A 1/f noise (or 1/f-like) with a ΔN origin and $\alpha = 4\text{x}10^{-5}$ cannot be excluded [40]. Such α-values with the $1/V_G^*$ proportionality fit very well with the original ΔN models for MOSFETs with reasonable values for the oxide trap density [41, 42]. Christensson et al. and Hsu applied Eq. (9) and obtained an expression for the 1/f noise, here written as a relative current noise [41, 42]:

$$\frac{S_N}{N^2} = \frac{S_I}{I^2} = \frac{kTWL\lambda}{fN^2} D_{ox}(eV^{-1}cm^{-3}) = \frac{q^2 kT\lambda D_{ox}(eV^{-1}cm^{=3})}{fWLC_{ox}^2 V_G^{*2}}. \quad (33)$$

From Eqs. (1, 33) follows a relation between α and trap density in the oxide $D_{ox}$ shown in Fig. 8 and given by:

$$\alpha \equiv \frac{fS_I N}{I^2} = \frac{kTWL\lambda}{N} D_{ox} = \frac{qkT\lambda D_{ox}}{C_{ox} V_G^*}. \quad (34)$$

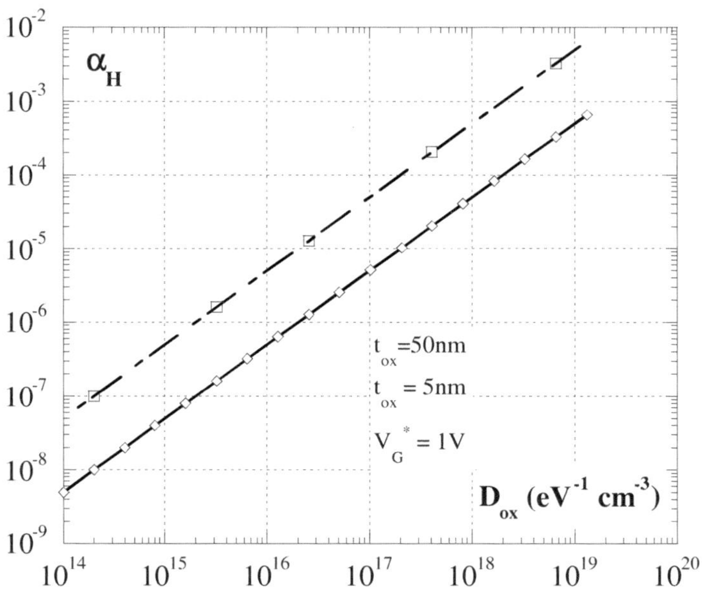


Fig. 8. The α parameter versus the oxide trap density $D_{ox}$ calculated from Eq. (34) for a $V_G^* = 1V$ with the upper curve an oxide thickness $t_{ox}$ = 50 nm and the lower curve $t_{ox}$ = 5 nm. Trap densities measured by techniques independent from the noise are in the range of $10^{17} - 10^{18}$ ($cm^{-3}$ $eV^{-1}$ ). Hence, $\alpha < 10^{-6}$ is hardly to explain from ΔN.

ii) Additional ΔN noise can be expected with non mature technologies, leading to $1/f^\gamma$ spectra with $0.5 < \gamma < 1.7$. We often see that the ΔN contribution is very sensitive for technology improvements, irradiation damage or anneal [43]. The 1/f bulk component that can have $\alpha \leq 10^{-6}$ is more sensitive for lattice damage in the inversion layer.

iii) The comparison in α-values facilitates a quantitative judgment on "low" or "high" noise. A quantitative comparison makes more sense between a Si reference MOSFET, high-K dielectric MOSFET, SiGe MOSFETs, different types of gate electrode material and the 1/f noise in a buried Si resistor.

iv) At higher bias current often a reduction in current drive (different from mobility degradation) is observed. It is very useful to distinguish channel and series resistance contributions to the noise of the MOSFET.

The following experimental trends often claimed as proof for ΔN are no proofs at all:
1) α proportional to $1/V_G^*$, see Eq. (6). A gate voltage independent α is easily explained with the Δμ model (Eq. (4)) but also with complications in the ΔN model. A proportionality $\alpha \sim 1/V_G^*$ is easily explained with ΔN (Eq. (17)) but also with Δμ (Eq. (6)).
2) The spreading in normalized noise WLf $S_I/I^2$ for submicron devices (enhancement and reduction compared to the average value for large devices) is explained by Eqs. (24, 27) and is not a proof for ΔN origin of 1/f noise in MOSFETs.
3) The typical shape of the plot $S_I/I^2$ versus I can be calculated from the empirical relation taking into account the sub threshold and series resistance effects in Eqs. (31, 32).
4) RTS on top 1/f noise is due to trapping (ΔN). This is rather normal for areas WL < 1 $\mu m^2$ and more precisely if $N < 1/(4\pi\alpha) = N_{crit}$. But this is not a proof that in larger devices the 1/f part is an addition of different RTS, see Fig. 2 and Eqs. (16c, 18, and 19).
5) Annealing can result in a reduction of interface traps and reduce the 1/f noise in MOSFETs as shown in Fig. 6 of Ref. [27]. But annealing can also result in lower α-values in buried Si resistor [34]. Implantation damage that is shifted below the inversion layer can result in a reduction in α, see Fig. 7. Such results are no proof for the bulk or surface origin of 1/f noise.

The question on: "Surface effect due to ΔN or bulk effect due to Δμ?" is well discussed in [44]. The assumption of mobility fluctuations (uncorrelated with number fluctuations) as origin of 1/f noise lead to better fits between experimental results on the 1/f noise in thermal voltage and Hall voltage than the number fluctuation assumption. Nothing is proven on the origin of 1/f noise in MOSFETs. The low-frequency noise in MOSFETs seems to be an addition of a bulk and surface contribution with different origins.

Most experimental results indicate that avoiding traps and improving the lattice at the inversion layer position can reduce the low-frequency noise to a level of $\alpha \approx 10^{-6}$ as observed in bulk Si resistors where trapping at an interface could be ignored [35]. The low frequency noise in MOSFETs with α-values of $10^{-6}$ and lower stems from a bulk origin (Δμ). MOSFETs with α-values between $10^{-5}$ and $10^{-3}$ can have quite different and uncorrelated bulk and surface contributions. MOSFETs with α-values beyond $10^{-3}$ as often observed for nitrided oxides or high-K dielectrics have a ΔN origin. All devices with $\alpha > 10^{-6}$ are "too noisy".

## References

[1] I. M. Hafez, G. Ghibaudo and F. Balestra, Flicker noise in metal-oxide-semiconductor transistors from liquid helium to room temperature, *J. Appl. Phys.* **66**(5) (1989) 2211–2213.

[2] G. Ghibaudo, O Roux-dit -Buisson and J. Brini, Impact of scaling down on low frequency noise in silicon MOS transistors, *Physica Status Solidi A* **132** (1992) 501–507.

[3] G. Ghibaudo and T. Boutchacha, Electrical noise and RTS fluctuations in advanced CMOS devices, *Microelectronics and Reliability* **42**(4-5) (2002) 573–582.

[4] M. Fadlallah, G. Ghibaudo, J. Jomaah and G. Guegan, Static and low frequency noise characterization in surface-and burried-mode 0.1mm PMOSFETS, *Solid Sate Electr.* **47** (2003) 155–1160.

[5] M. J. Kirton and M. J. Uren, Noise in solid-state microstructures: A new perspective on individual defects, interface states and low-frequency (1/f) noise, *Advances in Physics* **38**(4) (1989) 367–468.

[6] A. J. Scholten, L. F. Tiemeijer, R. van Langevelde, R. J. Havens, A. T. A. Zegers-Van Duijnhoven and V. C. Venezia, Noise modeling for RF CMOS circuit simulation, *IEEE Trans. Electr. Dev.* **50**(3) (2003) 618–632.

[7] J.-H. Lee, S.-Y. Kim, I. Cho, S. Hwang and J.-H. Lee, 1/f noise characteristics of sub-100nm MOS transistors, *J. Semiconductor Technology and Science JSTS* **6**(1) (2006) 38–42.

[8] F. Crupi, P. Srinivasan, P. Magnone, E. Simoen, C. Pace, D. Misra and C. Claeys, Impact of the interfacial layer on the low-frequency noise (1/f) behavior of MOSFETs with advanced gate stacks, *IEEE Electron DeviceLetters* **27**(8) (2006) 688–691.

[9] M. Valenza, A. Hoffmann, A. Laigle, D. Rigaud and M. Marin, Impact of scaling down on 1/f noise in MOSFETs, in *Noise in Dev. and Circuits*, (ed.) M. J. Deen *et al.*, Bellingham: SPIE-Int. Soc. Opt. Eng (2003) pp. 29–43.

[10] E. P. Vandamme and L. K. J. Vandamme, Unsolved problems on 1/f noise in MOSFETs and possible solutions, UPoN'99, AIP Conference Proceedings **511** (1999) 395–400.

[11] J. Rhayem, D. Rigaud, A. Eya'a, M. Valenza and A. Hoffmann, 1/f noise in metal-oxide-semiconductor transistors biased in weak inversion, *J. Appl. Phys.* **89**(7) (2001) 4192–4194.

[12] A. L. McWhorter. 1/f Noise and germanium surface properties, R. H. Kingston. Philadelphia: Univ. Pennsylvania Press. (Semiconductor surface physics, *Proc. Conf. of Physics of Semiconductor Surfaces*) (1957) pp. 207–228.

[13] M. Surdin, Fluctuations de courant thermionique et le "flicker effect", *J. de Physique & Radium.* (Paris.) **10** (1939) 188–189.

[14] A. E. Belyakov, L. K. J. Vandamme, M. Y. Perov and A. I. Yakimov, The different physical origins of 1/f noise and superimposed RTS noise in light-emitting quantum dot diodes, *Fluctuation and Noise Letters* **3**(3) (2003) L325–L339.

[15] N. Lukyanchikova, M. Petrichuk, N. Garbar, E. Simoen, A. Mercha, C. Claeys, H. van Meer and K. De Meyer, The 1/f(1.7) noise in submicron SOI MOSFETs with 2.5 nm nitrided gate oxide, *IEEE Trans. Electr. Dev.* **49**(12) (2002) 2367–2370.

[16] E. Simoen and C. Claeys, On the flicker noise in submicron silicon MOSFETS, *Solid-State Electr.* **43**(5) (1999) 865–882.

[17] R. Jayaraman and C. G. Sodini, 1/f noise interpretation of gate oxide nitridation and reoxidation of dielectric traps, *IEEE Trans. Electr. Dev.* (1) (1990) 305–309.

[18] H. Wong and Y. C. Cheng, Gate dielectric-dependent flicker noise in metal-oxide-semiconductor transistors, *J. Appl. Phys.* **67**(2) (1990) 863–867.

[19] E. A. M. Klumperink, S. L. J. Gierkink, A. P. van der Wel and B. Nauta, Reducing MOSFET 1/f noise and power-consumption by switched biasing, *IEEE J. Solid-State Circuits* **35**(7) (2000) 994–1001.

[20] M. Toita, L. K. J. Vandamme, S. Sugawa, A. Teramoto and T. Ohmi, Geometry and bias dependence of low-frequency random telegraph signal and 1/f noise levels in MOSFETs, *Fluctuation and Noise Letters* **5**(4) (2005) L539–L548.

[21] L. K. J. Vandamme and M. Macucci, 1/f and RTS noise in submicron devices: Faster is noisier, Unsolved Problems of Noise and Fluctuations UPoN, (eds.) L. Reggiani *et al.*, AIP 0-7354-0289-2/05. (CP800) (2005) 436–443.

[22] L. K. J. Vandamme, 1/f Noise in MOSTs: Faster is Noisier, in *Adv. Exp. Methods for Noise Research in Nanoscale Electronic Devices*, Kluwer Acad. Publ. (2004) 109–120.

[23] X. Li, C. Barros, E. P. Vandamme and L. K. J. Vandamme, Parameter extraction and 1/f noise in a surface and a bulk-type, p-channel LDD MOSFET, *Solid-State Elect.* **37**(11) (1994) 1853–1862.

[24] L. K. J. Vandamme, X. Li and D. Rigaud, 1/f noise in MOS devices, mobility or number fluctuations? *IEEE Trans. Elect. Dev.* **41**(11) (1994) 1936–1945.

[25] G. Giusi, E. Simoen, G. Eneman, P. Verheyen, F. Crupi, K. De Meyer, C. Claeys and C. Ciofi, Low-frequency (1/f) noise behavior of locally stressed $HfO_2$/TiN gate-stack pMOSFETs, *IEEE Electr. Dev. Lett.* **27**(6) (2006) 508–510.

[26] P. Srinivasan, E. Simoen, R. Singanamalla, H. Y. Yu, C. Claeys and D. Misra, Gate electrode effects on low-frequency (1/f) noise in p-MOSFETs with high-k dielectrics, *Solid-State Elect.* **50** (2006) 992–998.

[27] A. K. M. Ahsan and D. K. Schroder, Impact of post-oxidation annealing on low-frequency noise, threshold voltage, and subthreshold swing of p-channel MOSFETs, *IEEE Electr. Dev. Lett.* **25**(4) (2004) 211–213.

[28] N. G. Tarr, C. Plett, A. Yeaton, G. F. Mackay and I. Thomson, Limitations on MOSFET dosimeter resolution imposed by 1/f noise, *IEEE Trans. Nuclear Sc.* **43**(5) (1996) 2492–2495.

[29] F. M. Klaassen, Characterization of low 1/f noise in MOS transistors, *IEEE Trans. Electr. Dev.* ED-**18**(10) (1971) 887–891.

[30] L. K. J. Vandamme and R. G. M. Penning de Vries, Correlation between MOST 1/f noise and CCD transfer inefficiency, *Solid-State Electr.* **28**(10) (1985) 1049–1056.

[31] M. Decker, L. Gouskov and D. Rigaud, Noise measurements in Metal-Oxide-Semiconductor transistors below saturation, *Electr. Lett.* **3**(12) (1967) 565–566.

[32] E. A. Hendriks and R. J. J. Zijlstra, 1/f noise in (100) n-channel Si MOSFETS from T = 4.2 K to T = 295 K, *Solid-State Electr.* **31**(6) (1988) 1105–1111.

[33] M. J. Prest, D. J. F. Fulgoni, A. R. Bacon, T. J. Grasby, E. H. C. Parker and T. E. Whall, Low-frequency noise mechanisms inSi and pseudomorphic SiGe p-channel field effect transistors, *Appl. Phys. Lett.* **85**(24) (2004) 6019–6021.

[34] L. K. J. Vandamme and S. Oosterhoff, Annealing of ion-implanted resistors reduces the 1/f noise, *J. Appl. Phys.* **59**(9) (1986) 3169–3174.

[35] R. H. M. Clevers, Volume and temperature dependence of the 1/f noise parameter α in Si, *Physica B* **154** (1989) 214–224.

[36] L. K. J. Vandamme, Bulk and Surface 1/f Noise, *IEEE Trans. Electr. Dev.* **36**(5) (1989) 987–992.

[37] E. P. Vandamme and L. K. J. Vandamme, Critical discussion on unified 1/f noise models for MOSFETs, *IEEE Trans. Electr. Dev.* **47**(11) (2000) 2146–2152.

[38] P. Fantini, D. Riccardi and G. Croce, Buried CMOS structure with a straddle-gate architecture for low-noise analog applications, *Fluctuation and Noise Letters* **3**(3) (2003) 349–355.

[39] R. D. Black, M. B. Weissman and P. J. Restle, 1/f noise in silicon wafers, *J. Appl. Phys.* **53**(9) (1982) 6280–6284.

[40] G. A. Garfunkel and M. B. Weissman, 1/f noise and the field effect in gated resistors, *J. Appl. Phys.* **57**(2) (1985) 634–636.

[41] S. Christensson, I. Lundstrom and C. Svensson, Low frequency noise in MOS transistors-I, *Solid-State Electronics* **11** (1968) 797–812.

[42] S. T. Hsu, Surface state related 1/f noise in MOS transistors, *Solid-State Electronics* **13** (1970) 1451–1459.

[43] H. D. Xiong, D. M. Fleetwood, B. K. Choi and A. L. Sternberg, Temperature dependence and irradiation response of 1/f-noise in MOSFETs, *IEEE Trans Nuclear Sc.* **49**(6) (2002) 2718–2723.

[44] F. N. Hooge, T. G. M. Kleinpenning and L. K. J. Vandamme, Experimental studies on 1/f noise, *Rep. Prog. Phys.* **44**(5) (1981) 479–532.

# LOW-FREQUENCY NOISE IN AlGaN/GaN HETEROSTRUCTURE FIELD EFFECT TRANSISTORS AND METAL OXIDE SEMICONDUCTOR HETEROSTRUCTURE FIELD EFFECT TRANSISTORS

S. L. RUMYANTSEV*, N. PALA† and M. S. SHUR,

*Department of Electrical, Computer, and Systems Engineering and Center for Integrated Electronics and Electronics Manufacturing, CII 9017, Rensselaer Polytechnic Institute, Troy NY 12180-3590, USA*

†*palan@rpi.edu*

M. E. LEVINSHTEIN and P. A. IVANOV

*Solid State Electronics Division, The Ioffe Physical-Technical Institute of Russian Academy of Sciences, 194021, St. Petersburg, Russia*

M. ASIF KHAN, G. SIMIN and J. YANG

*Department of Electrical and Computer Engineering, University of South Carolina, Columbia, SC, 29208, USA*

X. HU, A. TARAKJI and R. GASKA

*Sensor Electronic Technology, Inc., 21 Cavalier Way, Latham, NY 12110, USA*

Received 15 November 2001
Revised 6 December 2001
Accepted 7 December 2001

The dependence of the *1/f* noise on 2D electron concentration in the channel $n_{Ch}$ of AlGaN/GaN Heterostructure Field Effect Transistors and Metal Oxide Semiconductor Heterostructure Field Effect Transistors has been studied and compared. The dependencies of Hooge parameter $\alpha_{Ch}$ for the noise sources located in the channel of the transistors on sheet electron concentration are found identical for both types of devices. The increase of the Hooge parameter $\alpha_{Ch}$ with the decrease of the channel concentration observed in both types of devices confirms that the noise sources are located in the region under the gate in the AlGaN/GaN heterostructure and that electron tunneling from the 2D electron gas into the traps in GaN or AlGaN layers is a probable noise mechanism.

*Keywords:* 1/f noise; AlGaN/GaN; HFETs; MOSHFETs; 2DEG concentration.

## 1. Introduction

GaN-based Heterojunction Field Effect Transistors (HFETs) are expected to find applications in high power, high frequency, and low noise electronics [1–3]. Recently, a novel $SiO_2$/AlGaN/GaN Metal-Oxide-Semiconductor Heterostructure Field Effect Transistor

*On leave from the Ioffe Institute of Russian Academy of Sciences, 194021 St-Petersburg, Russia

(MOSHFET) device has been demonstrated [4,5]. The gate leakage current of this devices is 4–6 orders of magnitude lower than that of a conventional AlGaN/GaN HFET. The general noise properties of AlGaN/GaN MOSHFETs have been studied in Refs. [6,7].

The investigation of the gate voltage dependence of the drain current noise is an effective tool to analyze the physical nature and localization of the noise sources (see, for example, [8–11]). Our recent studies of HFETs using this method showed that the tunneling of the electrons from the two dimensional electron gas in the device channel into traps in the bulk GaN or AlGaN might cause the 1/f noise [11].

In this paper, we report on the comparative study of the AlGaN/GaN HFETs and MOSHFETs that confirms this conclusion.

## 2. Experimental details

The AlGaN/GaN heterostructures used in this study were grown by metalorganic chemical vapor deposition (MOCVD) on a semi-insulating 4H-SiC substrate. They consisted of a 50-nm-thick AlN buffer layer, 0.4-μm-thick undoped GaN layer, followed by $Al_{0.2}Ga_{0.8}N$ barrier layer, which was doped with silicon to approximately $2\times10^{18}$ $cm^{-3}$. As always in our growth, we added trace amounts of indium to all nitride layers, which considerably improves the materials quality. HFETs and MOSHFETs were fabricated on the same wafer. In order to fabricate MOSHFET and HFET devices on the same wafer, prior to the gate metallization a 7 nm $SiO_2$ layer was deposited on a part of the heterostructures using Plasma Enhanced Chemical Vapor Deposition.

The transistors had the source-drain spacing of 4–5 μm, the gate length, $L$, of 1–1.5 μm, and a gate width, $W$, in the range of 50–150 μm. The Transmission Line Model (TLM) structures and transistors were fabricated on the same wafers.

All dc and noise measurements were performed in linear (Ohmic) regime at a small drain-source dc bias.

## 3. Results and Discussion

For both HFETs and MOSHFETs, the capacitance–voltage measurements were performed at frequency of 1 kHz on the test transistors with the large gate area. For the entire range of the gate biases, $V_g$, from zero to the threshold voltage, $V_T$, the gate capacitance was practically constant manifesting the linear dependence of the electron channel sheet concentration $n_{Ch}$ on the gate bias. The measured gate threshold voltages were equal to –6 V and –2 V for MOSHFETs and HFETs, respectively. A more negative threshold voltage of the MOSHFET is partially linked to an additional voltage drop across the silicon dioxide layer. The thickness of GaAlN and $SiO_2$ layers extracted from the capacitance measurements was in good agreement with the data obtained from technological estimations.

The noise spectra of the drain current fluctuations had the form of $1/f^{\Gamma}$ noise with $\Gamma$ close to unity ($\Gamma$= 1.0–1.17). At low drain biases, the spectral noise density was proportional to the square of the drain voltage $S_{Id} \sim V_d^2$. Using the techniques described in details in Ref. [12], we checked that neither contact noise nor noise from gate leakage current contributed much to the total output noise of HFETs.

Figure 1 shows the dependence of the relative spectral density of short circuit current fluctuations $S_{Id}/I_d^2$ multiplied to the gate area $A_{Ch}$ on normalized gate voltage $V_{gn} = (V_g - V_T)/V_T$ for HFETs and MOSHFETs.

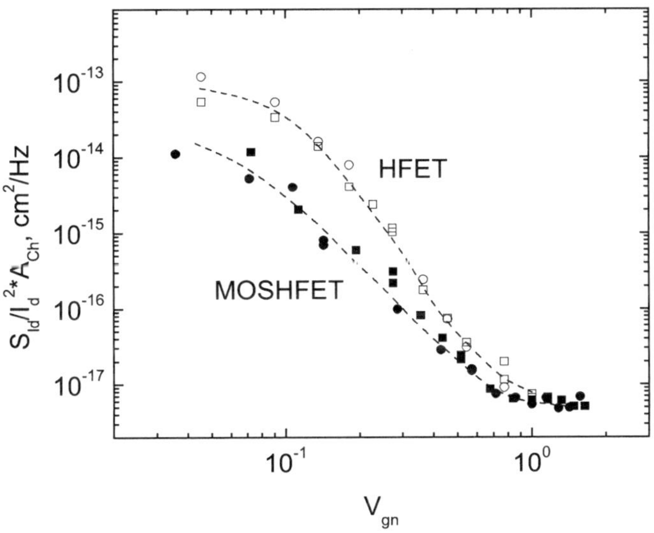


Fig. 1. The dependencies of the relative spectral noise density of the drain current fluctuations on normalized gate voltage $V_{gn} = (V_g - V_t)/V_t$ for HFETs and MOSHFETs. Frequency of analysis $f = 1$Hz.

As seen in Fig.1, the gate voltage dependencies of noise for both types of the devices are identical in spite of different threshold voltages for the HFETs and MOSHFETs. At low values of $V_{gn}$, the noise in MOSHFETs is somewhat smaller than that in HFETs. One of the reasons for this difference might be the different electron concentration even at the same values of $V_{gn}$. Another possible reason for the different noise level can be the different ratio of the channel resistance to the contact resistance for HFETS and MOSHFETs. In order to take into account these two factors, we estimated the Hooge parameter $\alpha_{Ch}$ in the channel under the gate [13]:

$$\alpha_{Ch} = \frac{S_{RCh}}{R_{Ch}^2} fN \tag{1}$$

where $N$ is the total number of the conduction electrons in the channel under the gate, $f$ is the frequency of the analysis, $R_{Ch}$ is the channel resistance, $S_{RCh}/R_{Ch}^2$ is the relative spectral density of the channel resistance fluctuations.

Assuming that the noise sources are not correlated and located in the channel, and in the source-gate, and gate-drain regions, the spectral noise density of the drain current fluctuations can be presented as follows:

$$\frac{S_{Id}}{I_d^2} = \frac{S_{RCh} + S_{ac}}{R_{ds}^2} \tag{2}$$

where $R_{ds} = R_{ch}+R_s+R_d+R_c$ is the drain-source resistance at low drain bias, $R_s$ and $R_d$ are the series resistance of source–gate and gate–drain regions, respectively $R_c$ is the contact resistance, $S_{Rch}$ and $S_{ac}$ are spectral densities of the $R_{ch}$ and $(R_s+R_d)$ fluctuations, respectively. The spectral noise density $S_{ac}$ does not depend on gate voltage, as the contribution from the channel resistance $S_{Rch}$ varies with the gate voltage. Assuming that both $S_{Rch}$ and $S_{ac}$ comply with the Hooge formula (1), the dependence of $\alpha_{Ch}$ on the channel concentration $n_{Ch}$ can be found as:

$$\alpha_{Ch} = \left[ \frac{S_I}{I_d^2} - \frac{\alpha_0 (R_s + R_d)^2}{A_{ac} n_{ac} R_{ds}^2} \right] \frac{A_{Ch} n_{Ch} R_{ds}^2}{R_{Ch}^2} \tag{3}$$

where $A_{ac}$ is the area of the drain-gate, gate-source openings, $n_{ac}$ is the electron concentration in these regions,

$$n_{Ch} = \frac{1}{q} \int_{Vg1}^{V_g} C dV_g \tag{4}$$

$C$ is the gate capacitance per unit area, $V_{g1}$ is the gate voltage which was taken to be smaller than the threshold voltage $V_T$. Since at $V_g < V_T$ the gate capacitance is very small, the chosen value of $V_{g1}$ does not significantly affect the dependence $n_{Ch}(V_g)$ given by Eq. (4). The channel resistance $R_{Ch}$ was found as $R_{Ch} = R_{ds} - R_c - R_d - R_s$. The values $R_c$, $R_s$, and $R_d$, were found from the TLM measurements [14] and from the known transistor dimensions. In Eq. (3), $\alpha_0$ is Hooge parameter for ungated regions of the transistor. We estimated the value of $\alpha_0$ assuming that at zero gate voltage, the value of the Hooge parameter is the same for the channel and ungated regions.

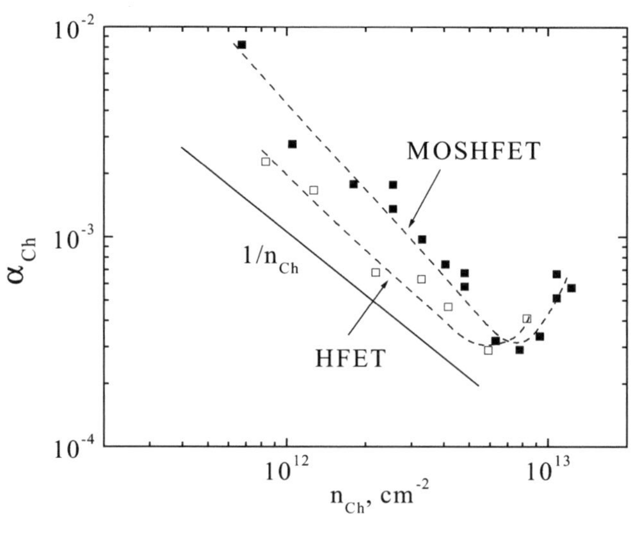


Fig. 2. The dependencies of the Hooge parameter $\alpha_{Ch}$ on 2D sheet channel concentration $n_{Ch}$ for HFETs and MOSHFETs.

Figure 2 shows the dependencies of the Hooge parameter $\alpha_{Ch}$ on 2D sheet channel concentration $n_{Ch}$ calculated for the HFETs and MOSHFETs. As seen in Fig. 2, both the absolute values of $\alpha_{Ch}$ and their dependencies on the channel concentration are identical for both types of the devices. This clearly shows that $SiO_2$ layer does not affect the noise properties.

## 4. Conclusions

At low channel concentrations, $\alpha_{Ch}$ decreases with the increase of $n_{Ch}$ approximately as $\alpha_{Ch} \sim 1/n_{Ch}$, reaches a minimum and then increases with a further increase of $n_{Ch}$. The dependence $\alpha_{Ch} \sim 1/n_{Ch}$ indicates that the spectral density of the channel resistance fluctuations $S_{Rch}$ does not depend on channel concentration. This is a very typical situation for Si MOSFETs where the noise arises from tunneling of electrons from semiconductor to traps in the oxide. The amplitude of that surface noise depends only on the trap density at the Fermi level [15,16]. Similar noise mechanism consisting of tunneling of electrons from 2D gas to GaN or AlGaN might be dominant in GaAlN/GaN-based transistors as well.

The increase of $\alpha_{Ch}$ at high concentrations, $n_{Ch}$, can be attributed to the electron spillover from the 2D-channel to a parallel low electron-mobility "parasitic conduction" channel (see also [11]). The noise level in such "parasitic channel" should be very large and corresponds to $\alpha \sim (10^{-2} - 1)$ for the bulk GaN [17–19].

### Acknowledgment

The work at RPI was supported by the Office of Naval Research. Project monitor Dr. J. Zolper. The work at USC was supported by the Ballistic Missile Defense Organization (BMDO) under Army SSDC contract DASG60-98-1-0004, monitored by Dr. Brian Strickland and Dr. Kepi Wu. The work at SET, Inc. was supported by BMDO under SBIR Phase II contract F33615-01-C-1911 and monitored by Dr. J. Gillespie of WPAFB. The authors would like to thanks Dr. E. Borovitskaya for useful discussions.

### References

[1] A. Tarakji, G. Simin, N. Ilinskaya, X. Hu, A. Kumar, A. Koudymov, J. Zhang, M. Asif Khan, M.S. Shur and R. Gaska, *Appl. Phys. Lett.* **78**(15) (2001) 2169–2171.

[2] S. T. Sheppard, K. Doverspike, W. L. Pribble, S. T. Allen, J. W. Palmour, L. T. Kehias, and T. J. Jenkins, *IEEE Elect. Dev. Lett.* **20** (1999) 161.

[3] M. E. Levinshtein, S. L. Rumyantsev, M. S. Shur R. Gaska, and M. Asif Khan, *Low frequency and 1/f noise in wide gap semiconductors: Silicon Carbide and Gallium Nitride, Review, Special Issue of 'IEE Proc. - Circuits, Devices, and Systems*, to be published, (Oct. 2001).

[4] M. Asif Khan, X. Hu, G. Simin, A. Lunev, J. Yang, R. Gaska, and M. S. Shur, *IEEE Elect. Dev. Lett.* **21** (2000) 63.

[5] M. Asif Khan, X. Hu, A. Tarakji, G. Simin, J. Yang, R. Gaska, and M. S. Shur, *Appl. Phys. Lett.* **77**(9) (2000) 1339–1341.

[6] N. Pala, R. Gaska, S. L. Rumyantsev, M. S. Shur, Asif Khan, X. Hu, G. Simin, and J. Yang, *Low frequency noise in AlGaN/GaN MOS-HFETs, Elect. Lett,* **36**(3) (2000) 268–270.

[7] S. L. Rumyantsev, N. Pala, M. S. Shur, M. E. Levinshtein, R. Gaska, X. Hu, J. Yang, G. Simin, and M. Asif Khan, *Low frequency noise in GaN-based transistors, Seventeenth Biennial IEEE/Cornell University Conference on Advanced Concepts in High Performance Devices,* (August 7–9, 2000) 257–264.

[8] J. A. Garrido, F. Calle, E. Munoz, I. Izpura, J. L. Sanches-Rojas, R. Li, and K. L. Wang, *Elect. Lett.* **34** (1998) 2357.
[9] J. A. Garrido, B. E. Foutz, J. A. Smart, J. R. Shealy, M. J. Murphy, W. J. Schaff, and L. F. Eastman, *Low-frequency noise and mobility fluctuations in AlGaN/GaN heterostructure field-effect transistors, Appl. Phys. Lett.* **76**(23) (2000) 3442–3444.
[10] A. Balandin, *Gate-voltage dependence of low-frequency noise in GaN/AlGaN heterostructure field-effect transistors, Elect. Lett.* **36**(10) (2000) 912–913.
[11] S. L. Rumyantsev, N. Pala, M. S. Shur, R. Gaska, M. E. Levinshtein, P. A. Ivanov, M. Asif Khan, G. Simin, X. Hu, and J. Yang, *Concentration dependence of the 1/f noise in AlGaN/GaN Heterostructure Field Effect Transistors, J. of Appl. Phys.,* submitted for publication.
[12] S. L. Rumyantsev, N. Pala, M. S. Shur, R. Gaska, M. E. Levinshtein, M. A. Khan, G. Simin, X. Hu, and J. Yang, *Effect of gate leakage current on noise properties of AlGaN/GaN field effect transistors, J. of Appl. Phys.* **88**(11) (2000) 6726–6730.
[13] F. N. Hooge, T. G. M. Kleinpenning, and L. K. J. Vandamme, *Experimental studies on 1/f noise, Rep. Progr. Phys.* **44**(5) (1981) 479–532.
[14] M. S. Shur, *GaAs devices and circuits*, Plenum Press, New York (1987).
[15] L. K. J. Vandamme, X. Li, and D. Rigaud, *1/f noise in MOS devices, mobility or number fluctuations?, IEEE Transaction Electron Devices* **41** (1994) 1936–1945.
[16] M. J. Kirton and M. J. Uhren, *Noise in solid-state microstructures: A new perspective on individual defects, interface states and low-frequency (1/f) noise, Adv. in Phys.* **38** (1989) 367–468.
[17] M. E. Levinshtein, F. Pascal, S. Contreras, W. Knap, S. L. Rumyantsev, R. Gaska, J. W. Yang, and M. S. Shur, *Low-frequency noise in GaN/GaAlN heterojunctions, Appl. Phys. Lett* **72** (1998) 3053–3055.
[18] M. E. Levinshtein, S. L. Rumyantsev, R. Gaska, J. W. Yang, and M. S. Shur, *AlGaN/GaN high electron mobility field effect transistors with low 1/f noise, Appl. Phys. Lett.* **73**(8) (1998) 1089–1091.
[19] N. V. Dyakonova, M. E. Levinshtein, and S. L. Rumyantsev, *Nature of the Bulk 1/f Noise in GaAs and Si, Sov. Phys. Semicond,* **25**(12) (1991) 1241–1265.

# LOW FREQUENCY NOISE CONVERSION IN FETs UNDER NONLINEAR OPERATION

F. DANNEVILLE, B. TAMEN and A. CAPPY
*IEMN, UMR CNRS 8520, Avenue Poincaré, BP 69, 59652 Villeneuve d'Ascq, France*
*E-mail: francois.danneville@iemn.univ-lille1.fr*

J-B JURAVER, O. LLOPIS and J. GRAFFEUIL
*LAAS-CNRS, 7 Av. du Colonel Roche, 31077 Toulouse, France*

Received 31 August 2001
Revised 20 September 2001
Accepted 26 September 2001

The conversion mechanisms of microscopic low frequency noise sources (e.g. generation-recombination noise sources) located in the channel of a FET (Field Effect Transistor), in the presence of a large RF signal, are investigated. It is shown that the base-band (low frequency) input gate noise voltage spectral density is strongly dependent on the magnitude of the input RF power applied to the FET. Moreover, the microscopic generation-recombination noise sources distributed along the channel are also responsible of up-converted input gate noise voltage spectral density around the RF frequency.

*Keywords*: FETs; large signal operation; microscopic noise modeling; low frequency noise conversion mechanisms.

## 1. Introduction

One of the most difficult problem encountered in FET noise modeling concerns the conversion of base-band low frequency *(LF)* noise when the FET is driven by a large microwave signal. Such operation corresponds to a case of non-autonomous circuits requiring an external large signal as amplifiers (device operating around the compression point) or mixers, it can also be the case of autonomous circuits (oscillators with no external large signal required). In the present work the study has been carried out by considering only the case of a device operating in a non-autonomous circuit; obviously, the results can be formally extrapolated to the autonomous one. Let us go back to a FET operating in linear operation; basically, part of the LF noise components of a FET is located at the surface between source and gate, or can be located in the substrate; this leads generally to noise sources associated with extrinsic resistances such as the gate resistance or source resistance which are simply added to the intrinsic part of the device; most often, they are identified to be flicker noise components (1/f) which are slightly bias dependent. However, part of the noise is also located in the channel of the FET. The corresponding microscopic noise sources are related to carriers number fluctuations (generation-recombination noise, *GR* noise) and are distributed along the channel. The overall contribution of *GR* noise is extracted most often experimentally and is presented as a macroscopic equivalent input noise voltage. If one considers now a device operating under non linear operation, these microscopic *GR* noise sources will produce at the external electrodes not only low frequency noise sources (base band), but also noise sources which have been converted around the RF output signal frequency and its harmonics [1]. The treatment of such problem requires basically the development of a device physics based noise modeling [1,2] that is not easy to carry out. To avoid such a complexity, the concept of noisy nonlinear uniform line model has been used since it has already provided in both linear and nonlinear case results in agreement with experiments [3,4].

## 2. Theoretical analysis

### 2.1. *Large signal analysis*

In order to conduct the large signal analysis, let us consider the intrinsic part of the FET described by a uniform active line, and that the device operates as the active part of a FET amplifier, under large signal operation (Fig. 1).

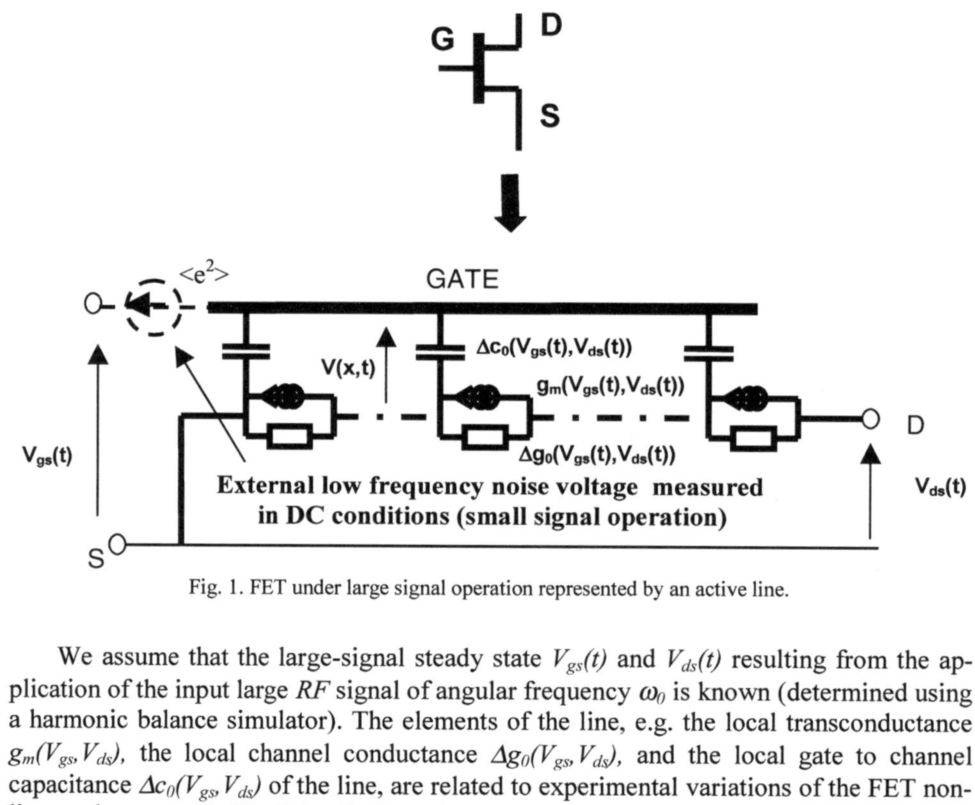


Fig. 1. FET under large signal operation represented by an active line.

We assume that the large-signal steady state $V_{gs}(t)$ and $V_{ds}(t)$ resulting from the application of the input large *RF* signal of angular frequency $\omega_0$ is known (determined using a harmonic balance simulator). The elements of the line, e.g. the local transconductance $g_m(V_{gs}, V_{ds})$, the local channel conductance $\Delta g_0(V_{gs}, V_{ds})$, and the local gate to channel capacitance $\Delta c_0(V_{gs}, V_{ds})$ of the line, are related to experimental variations of the FET nonlinear elements $G_m(V_{gs}, V_{ds})$, $G_d(V_{gs}, V_{ds})$, $C_{gs}(V_{gs}, V_{ds})$, and $C_{gd}(V_{gs}, V_{ds})$, which can be determined quite easily. Since time-varying voltages $V_{gs}(t)$ and $V_{ds}(t)$ are applied at the input and output of the FET, the line elements are also time dependent and can be developed in Fourier Series.

In the following, we are going to describe the different steps leading to the noise correlation matrix (NCM) calculation of the FET.

## 2.2. *Noise analysis*

If we consider small perturbations around the large signal steady state, each element of the line can be replaced by its conversion matrix [4]. The elements of the conversion matrix *[Gm], [ΔG0]* and *[ΔC0]* are determined in a straightforward way by the Fourier components of $g_m(t)$, $\Delta g_0(t)$ and $\Delta c_0(t)$. To perform the noise analysis, we have chosen a representation with an open circuit between gate and source and a short circuit between drain and source (Fig. 2). A key point is to define the spectral density $S_{in}$ of the local channel noise sources $<i_n^2>$ (Fig. 2), which does not change along the channel (uniform line model).

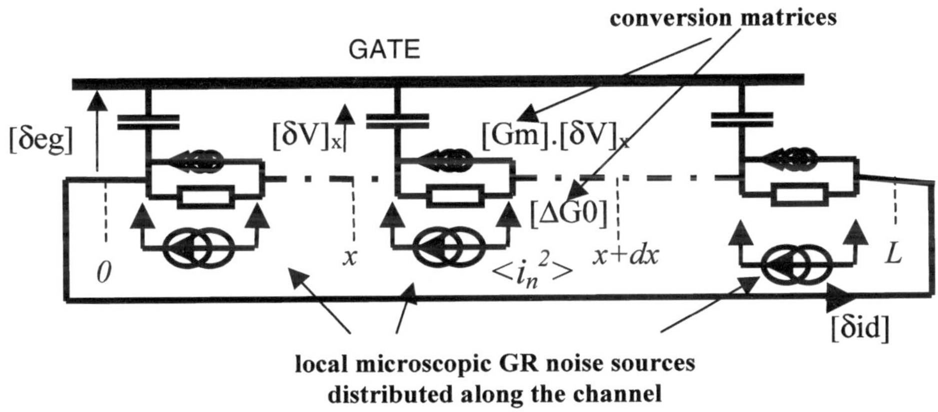


Fig. 2. Circuit used for noise calculation.

A *GR* microscopic noise source has characteristics time $\tau_c$ ranged typically between $1\mu s$ and $1ms$. This means that the period of the microwave RF signal will be much lower than $\tau_c$, consequently, we will consider that $<i_n^2>$ is not modulated by the large signal, and is defined by only a low frequency spectral density $S_{in}(\omega)$ [1]. A problem encountered is to relate $S_{in}(\omega)$ to the spectral density $Se(\omega)$ of the external input gate noise voltage $<e^2>$ (Fig. 1), determined experimentally using a low frequency noise measurement system. The following relationship can be derived from a linear noise analysis of the circuit shown in Fig. 2 [3], when the FET is biased in saturation (active operation):

$$S_{in}(\omega) = \left(\frac{L}{dx}\right) * g_{0\,DC} * g_{m\,DC} * 2 * Se_{GR}(\omega) \tag{1}$$

L is the gate length, $dx$ the discrete calculation step, $Se_{GR}(\omega)$ is the Generation-Recombination noise component of $Se(\omega)$, $g_{0DC}$ and $g_{mDC}$ correspond respectively to the DC values of the channel conductance and transconductance. Once the microscopic noise sources are defined, the objective lies in the determination of the noise correlation matrix of the circuit shown in Fig. 3. The problem is to calculate the influence of any microscopic noise source located between x and x+dx at the gate and drain electrodes. This calculation won't be developed for it is similar to those we have used previously in the case of FET mixer noise analysis [4]. By the end of the calculation, the FET can be described by its noise conversion and noise correlation matrices (Fig. 3). The input ports and output ports are defined for each angular frequency $\omega + K.\omega_0$, with $\omega_0$ corresponding to the RF angular frequency and $\omega$ to the angular frequency deviation from $\omega_0$. Note that the 1/f component of $<e^2>$, considered to be external of the intrinsic circuit (Fig. 1), is simply added to the base band noise source $<eg_0^2>$ (Fig. 3) in order to allow some comparisons with experiments.

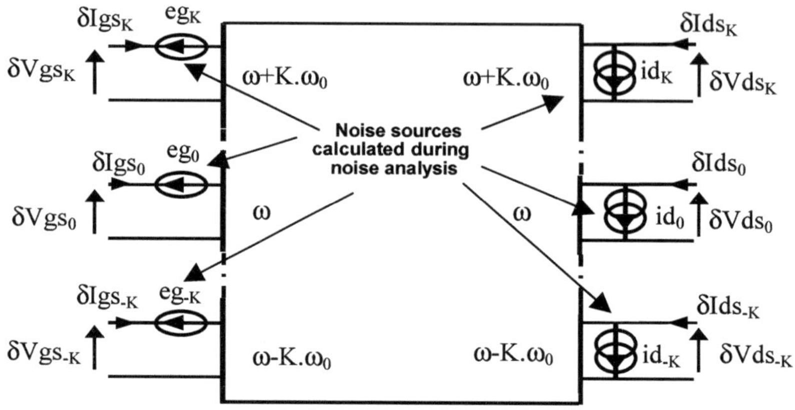


Fig. 3. FET representated by a linear noisy multifrequency network.

## 3. Results

We have performed a complete simulation of a *FET* amplifier driven by a input *RF* large signal at 3.5 GHz (Fig. 4).

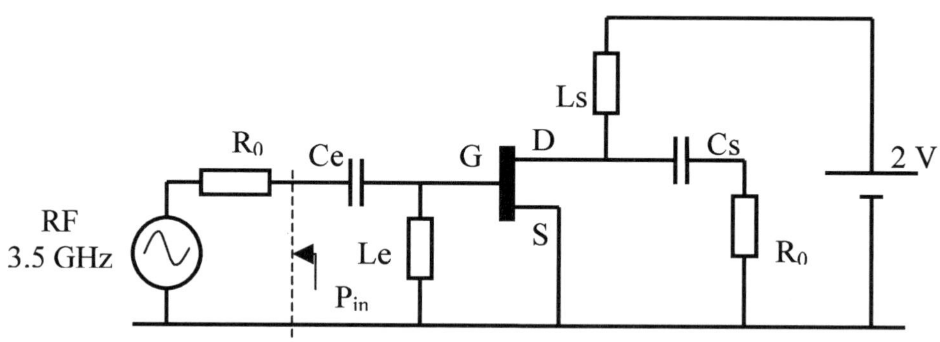


Fig. 4. Circuit used for the simulation.

The input and output RF terminations $R_0$ of the amplifier are equal to $50\ \Omega$, the device is a pseudomorphic HEMT $(0.25*200\ \mu m^2)$ biased at $V_{gs} = 0V$ and $V_{ds} = 2V$. It was found that the correlation between any input gate noise voltage and any output drain noise current (Fig. 3) is zero, which is consistent with what has been observed in *FET* mixer noise analysis [4]. Due to the uniform character of the line, the drain noise current source presents only a spectral density in base band ($\omega$) and to describe how the input gate noise voltage behaves, we have performed simulations of the amplifier for a large variation of the input *RF* power magnitude $P_{in}$. The spectral densities of the theoretical base-band voltage noise source at the input $<eg_0^2(\omega)>$ (including flicker and *GR* noise components) are represented on Fig. 5. We observe that the spectral density $Seg_0(\omega)$ of the base-band voltage noise $<eg_0^2(\omega)>$ depends strongly on the magnitude of the input power magnitude $P_{in}$. For the lower power level, $Seg_0(\omega)$ is slightly dependent on the *RF* power $P_{in}$ and close to the spectral density measured in linear operation. However, there is a strong effect on its magnitude for higher level of $P_{in}$, especially for frequency deviations where the *GR* noise components are of primary importance.

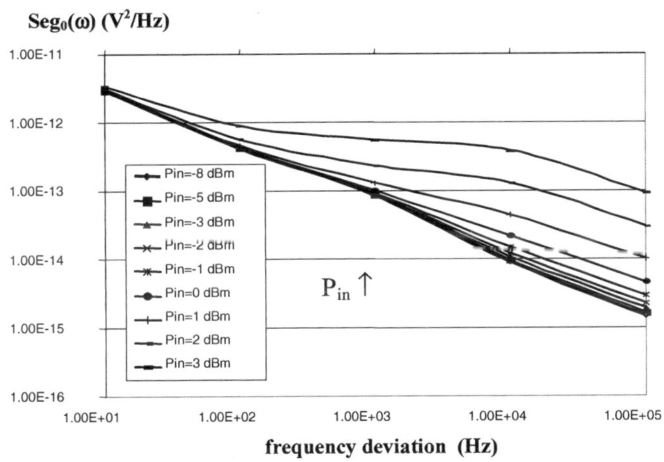


Fig. 5. Modeled Equivalent input LF noise spectral density (-8 dBm < $P_{in}$ < 3 dBm) versus different input RF power at 3.5 GHz.

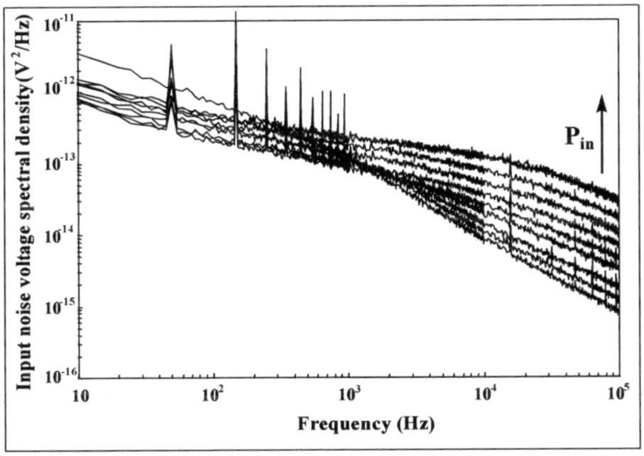


Fig. 6. Measured Equivalent input LF noise spectral density (-8 dBm < $P_{in}$ < 4 dBm) versus different input RF power at 3.5 GHz.

We can compare the results with experimental data shown in Fig. 6. We observe a significant difference for the lower frequency deviations (10 Hz); indeed, we use the 1/f noise spectral density measured in DC conditions (small signal operation) in Fig. 5 (modeled results) while it varies experimentally (Fig. 6). However, both the shape and the magnitude are in quite good agreement with the experimental data shown in Fig. 6 [5]; the differences are certainly explained by the use of a nonlinear uniform line model to describe the FET.

It is interesting as well to see how the modeled up-converted low frequency Generation-Recombination noise sources behave around the RF frequency (Fig. 7). We observe that their spectral densities (for $\omega = 2.\pi.10^3\ rd/s$) are reaching values in the same order of magnitude to those of $Seg_0(\omega)$ for the highest level of $P_{in}$.

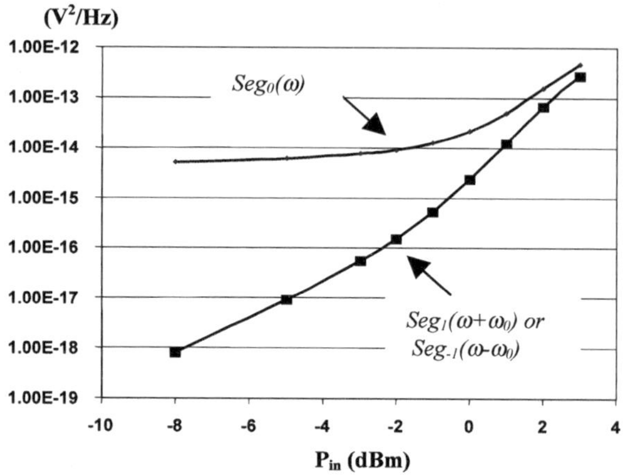


Fig. 7. Modeled Base-band ($Seg_0(\omega)$) and Up-converted ($Seg_1(\omega+\omega_0)$ or $Seg\text{-}1(\omega-\omega_0)$) spectral densities of the input gate noise voltage as a function of the input microwave Power $P_{in}$.

## 4. Conclusion

Low frequency noise conversion mechanisms in a *FET* have been studied theoretically using a microscopic approach and compared with experimental data. It was shown that the spectral density of the base-band voltage noise at the input of the device is strongly dependent of the input power level. Moreover, the noise modeling carried out has clearly made evidence of an important effect : the up-conversion of base-band *GR* microscopic noise sources located in the channel around the *RF* frequency at the device electrodes. This noise modeling constitutes a theoretical support for the use of new FET compact noise model [5] either those of a distributed approach [6] to model accurately the phase noise of FET's amplifier or oscillator.

## References

[1] A. Cappy, F. Danneville, G. Dambrine and B. Tamen, *Noise Analysis in devices under nonlinear operation, Solid-State Electronics* **43** (1999) 21–26.

[2] F. Bonani, S. D. Donati Guerrieri, G. Ghione and M. Pirola, *A TCAD Approach to the Physics-Based Modeling of Frequency Conversion and Noise in Semiconductor Devices Under Large-Signal Forced Operation, IEEE Trans. Electr. Dev.* **48** (2001) 966–977.

[3] F. Danneville, H. Happy, G. Dambrine, J-M Belquin and A. Cappy, *Microscopic Noise Modeling and Macroscopic Noise Models: How Good a Connection?, IEEE Trans. Electr. Dev.* **41** (1994) 779–786.

[4] F. Danneville, G. Dambrine and A. Cappy, *Noise Modeling in MESFET and HEMT Mixers Using a Uniform Noisy Line Model, IEEE Trans. Electr. Dev.* **45** (1998) 2207–2212.

[5] O. Llopis, J.-B. Juraver, B. Tamen, F. Danneville, M. Chaubet, A. Cappy and J. Graffeuil, *Nonlinear noise modeling of a PHEMT device through residual phase noise and low frequency noise measurements, Proc. IEEE MTT-S Symposium, Phoenix* (2001) 831–834.

[6] J. Obregon, M. Prigent, J.C. Nallatamby, M. Camiade, D. Rigaud and R. Quere, *Key aspects of modeling and design of low phase noise microwave transistor-oscillators, IEEE MTT-S International Microwave Symposium, Workshop on Microwave Oscillators, Boston* (2000).

# PHYSICAL MODEL FOR LOW FREQUENCY NOISE IN AVALANCHE BREAKDOWN OF PN JUNCTIONS

O. MARINOV and M. JAMAL DEEN
*Electrical and Computer Engineering Department (CRL)*
*McMaster University, Hamilton, Ontario, Canada L8S 4K1*

Received 21 October 2003
Accepted 31 March 2004

A physically-based transient model for low frequency noise of both microplasma and impact ionization in PN diodes is discussed and implemented in the SPICE simulator. The simulation indicates that the model correctly describes the non-monotonic behavior of both the DC and the noise characteristics of diode at the onset of avalanche breakdown. The model is based on a new microplasma switching theory, and the results of simulation confirm the findings of this theory. The microplasma switching threshold is the condition of equality of free- to space charge concentration in the depletion layer. The microplasma turn-on is initialized by the charge generation due to few recombination centers in the microplasma region at high avalanche multiplication due to impact ionization. The microplasma on-current is approximately twice the threshold current and the on-current sustains until the low, but larger than 1, avalanche multiplication compensates for the carrier diffusion from microplasma region into the depletion layer. When the multiplication becomes lower than the diffusion, the microplasma switches off.

*Keywords*: Microplasma noise; multiplication low-frequency noise; avalanche breakdown.

## 1. Introduction

The investigation on fluctuation of the reverse current in semiconductor diode's breakdown began in 1950's. To date, several theories and models have been proposed: shot noise of avalanche multiplication [1,2]; burst noise of trap occupation [3,4]; RTS noise of microplasma fluctuation [5,6,7,8,9,10]; thermal noise; and 1/f noise [11,12,13,14,15]. Despite the complications arising due to change of the charge transport with breakdown and the non-monotonic DC and noise characteristics, rules are provided for distinguishing the trap occupation noise from the microplasma noise [4], and for the evolution of the noise with breakdown [16,17,18].

Two noise mechanisms dominate in avalanche breakdown — avalanche multiplication noise [1,2] and bistable microplasma noise [6,10]. A new theory of microplasma switching [8] explains the non-monotonic diode characteristics. It also relates the microplasma to the semiconductor parameters [18]. A physical model is proposed in [16] for description of both the microplasma and avalanche multiplication noise. The elaboration and implementation of this model is goal of current work. Therefore, an introduction to basic models for low-frequency noise in junction breakdown is presented in next section. Then, the physical model for microplasma switching is explained, and values for model parameters are estimated. The results from implementation of the model in SPICE simulator are discussed and compared to experimental data.

One important issue is that the bistable microplasma fluctuation is a large signal non-linear process, and only the transient analysis in time domain is reasonable for simulation. However, the microplasma switching process is fast, in the range of the transit time $\tau_t$~(10-100)ps of free carriers in junction depletion layer. On the other hand, the microplasma fluctuation has a time constant in the range of 100μs. The large ratio (~100μs/10ps=$10^7$) results in difficulties in computer simulation, and appropriate "scaling-up" of microplasma switching is mandatory for acceleration of the simulation.

A preliminary version [19] of this paper was presented at the Fluctuation and Noise Symposium in Santa Fe, New Mexico, USA in June 2003.

## 2. Basic Theories for Noise in Avalanche Breakdown

One can separate the theories for the noise in reverse biased diodes in two branches — "current fluctuation in avalanche charge multiplication" [1,2] and "microplasma fluctuation" [5,6,7]. Recent developments of these theories considered the charge acceleration before ionizing collisions in submicron charge multiplication layers [2], and explained the physics of microplasma switching [8,18].

The most popular "current fluctuation" theory [1] explains the noise in breakdown onset in terms of the charge multiplication due to impact ionization. In this theory, an assumption is made, that the electron and hole ionization coefficients α and β vary in the same manner with electric field, and the approximation $\beta(x)=k\cdot\alpha(x)$ is reasonable one, where k is a constant. With this approximation, the power spectral density $S_{IR}$ of the noise current at a high value of the multiplication coefficient $M=I_R/I_{IN}\rightarrow\infty$ is

$$S_{I_R} = 2\cdot q\cdot I_R\cdot M_\alpha\cdot M_\beta,\ \text{with}\ M_\beta = k\cdot M_\alpha \tag{1}$$

where $M_\alpha$ and $M_\beta$ are the multiplication coefficients for electrons and holes, $I_R$ is the reverse current, $I_{IN}$ is the initial current, and $q=1.6\times10^{-19}$ C is the electron charge. $I_{IN}$ is due to the generation-recombination process in depletion layer for packaged Si diodes at room temperature [20,21]. $I_{IN}$ is higher for photodiodes due to photogeneration. By introducing a parameter $\gamma = M_\beta\cdot M_\alpha/M^2$, it is shown [17,18] that Eq. (1) can be rewritten as

$$S_{I_R} = 2\cdot k\cdot\gamma\cdot q\cdot I_{IN}\cdot M^3 \propto {I_R}^3,\ \text{with}\ k\cdot\gamma\approx\text{constant} \tag{2}$$

and the fluctuation $\Delta_{IR}$ of the reverse current around its mean value $I_R$ becomes

$$\Delta_{I_R} = \sqrt{S_{I_R}\Delta f} = \Delta_{I_{IN}}\cdot M^{3/2}\sqrt{k\cdot\gamma} \tag{3}$$

where $\Delta_{IIN}$ is the fluctuation of the initial current around its mean value $I_{IN}$.

The "microplasma fluctuation" branch of noise modeling has been established in [10] with a stochastic model, based on observed bistable microplasma current. The random telegraph signal (RTS) noise was used, with turn-on ($t_{on}$) and turn-off ($t_{off}$) mean times of the microplasma states. For these times, the Poisson distribution of the charge fluctuation is shown in [5,22] to be the most reasonable distribution. A characterization procedure for the microplasma model has been developed, including the parameters of the bias circuit [6].

Ellouze *et al.* [7] provided some insight into the explanation of microplasma switching, assuming a branching process of the carrier multiplication. The results of this stochastic modeling [7] are good, but are less popular than McIntyre's theory [1]. McIntyre's theory has the advantage of a physical relation between the junction noise and the semiconductor properties at high electric fields, but it does not take into account the microplasma noise. Recently, the mechanism of microplasma switching has been explained physically in [8], and some basic relations between the microplasma noise and the junction properties have also been found [18]. These are now briefly reviewed.

- The microplasma switches on, if the free charge concentration n exceeds the impurity concentration N in the lower doped drift region of junction. The threshold condition $n = N$ can be rewritten in terms of current density J, or current I as

$$n = N \Leftrightarrow J = q \cdot v_s \cdot N \Leftrightarrow I = A_\mu \cdot J = A_\mu \cdot q \cdot v_s \cdot N \Leftrightarrow \text{microplasma switches} \quad (4)$$

  where $A_\mu$ is the microplasma cross-section area, and $v_s$ is the saturation velocity of the free charges ($v_s \approx 10^7$ cm/s in Si). The condition n=N follows from the continuity equation $\partial n/\partial t \propto -\partial J/\partial x$. The sign of the gradient of the current $\partial J/\partial x$ depends on the sign of the gradient of the electric field $\partial E/\partial x \propto (N\ n)$, because the direction of J follows the vector of E. Therefore, the sign of $\partial n/\partial t$ is the same as the sign of (n-N). Thus, if n>N, then n increases with time, and microplasma switches on, whereas it switches off when n<N, because n decreases with time.

- The microplasma conducting steady state is described by equal concentration of electrons, holes and space charge $n = p = N$. Thus, the concentration of the free carriers is $n_\mu = 2N$. This leads to current density $J = 2 \cdot q \cdot v_s \cdot N$, or current

$$n_\mu = 2 \cdot N \Leftrightarrow I_\mu = A_\mu \cdot J_\mu = 2 \cdot A_\mu \cdot q \cdot v_s \cdot N \quad \Leftrightarrow \quad \text{microplasma current.} \quad (5)$$

  The equality $n = p$ arises from the fact that the potentials along the microplasma have to be equal to the potentials of the surrounding regions of the junction, that is the net charge is $p - n + N = N$, in order that no charge drifts from the microplasma to the depletion layer. Also, the majority carriers in the semiconductor cannot exceed the ionized impurities in drift regions, because the excess charge will decay exponentially with a very short relaxation time constant. Therefore, $n = N$ (or $p = N$, if the diode base is of p–type). Equation (5) is useful for an estimation of the microplasma's cross-section $A_\mu$, because $I_\mu$ is measurable and N can be easily found from the breakdown voltage. Also, a current generator $I_\mu$ can model the microplasma turn-on state.

- The microplasma turns on at a high value of the multiplication coefficient

$$M(U) = I/I_{IN} \geq N/2 \cdot n_g \quad \Leftrightarrow \quad \text{high multiplication for microplasma turn-on} \quad (6)$$

  where U is the turn-on reverse voltage across the junction for M(U), and $n_g$ is the initial charge concentration, which is thermally or optically generated.

- The microplasma conducting state requires a low multiplication of $M_\mu \approx (2 \text{ to } 5) > 1$, in order to compensate for the carrier diffusion from the microplasma into the depletion layer of the junction. Therefore, the microplasma turn-off voltage $U_B$ is lower than the turn-on voltage, and it is related to low multiplication coefficient $M_B$ by

$$M < M_B(U_B) = \left(1 + 2\sqrt{\frac{D_{eff} \cdot \tau_t}{A_\mu}}\right)^2 \sim (2 \text{ to } 5) \Leftrightarrow \text{microplasma turns off.} \tag{7}$$

Here, $D_{eff}$ is the effective diffusion coefficient of electrons and holes in depletion layer of diode junction. Note that $U_B$ is not zero since $M_B > 1$.

On the basis of these relations between the properties of the microplasma and the junction, a model was introduced in [16]. This model is developed in next section.

The microplasma switching theory also explains the trap-assisted noise in avalanche breakdown of the diode. The trap-aided fluctuation is very strong, and it can be observed as instability during the DC measurements since the time constant of the charge trap can be in a range of seconds. Fortunately, this phenomenon is rare, because the volume of the microplasma is very small, then the charge trap concentration is low and the probability of having a charge trap in microplasma region is also very low. Therefore, charge trap-aided microplasma fluctuation could be excluded from the models.

## 3. Noise Modeling in Avalanche Breakdown

The avalanche multiplication noise is relatively well-expressed with Eqs. (1), (2) and (3) for frequency and time domains in large and small devices, unless the device is very small of submicron-scale size [2]. The use of the shot noise approximation for the fluctuation of the initial current $I_{IN}$ with Poisson distribution usually gives satisfactory accuracy. $I_{IN}$ can be easily obtained from the DC measurements of diode's leakage current.

However, in submicron devices, or when analyzing small volumes in the depletion layer of the diode, the discrete nature of $I_{IN}$ becomes an important issue. The estimated values for the physical quantities in an abrupt Si junction with a breakdown knee voltage of BV=27V gives [18,20,21].

Space charge concentration $N = 2.5 \times 10^{16}\ \text{cm}^{-3}$ in lower-doped region (Ref. 20, p. 228, Fig. 8.12). (8)

Junction width $w = \sqrt{\dfrac{2 \cdot \varepsilon \cdot (V_i + BV)}{q \cdot N}} = 1.2 \times 10^{-4}\ \text{cm}$ at a built-in voltage $V_i$=0.94V and for $\varepsilon = 1\times10^{-12}$ F/cm. (9)

Transit time $\tau_t = w/v_s = 12 \times 10^{-12}$ s at a saturation velocity of $v_s = 1\times10^{7}$ cm/s. (10)

Microplasma cross-section $A_\mu = I_\mu/2 \cdot q \cdot v_s \cdot N = 1.25\times10^{-9}\ \text{cm}^2$ for an on-current $I_\mu$=0.1mA. (11)

Number of impurity atoms in the microplasma $N_\mu = N \cdot w \cdot A_\mu = 3760$. (12)

Number of recombination centers in microplasma $N_t \cdot w \cdot A_\mu \sim N \cdot w \cdot A_\mu/2000 = 1.5$, at $N/N_t \sim 2000$. (13)

Equation (13) indicates that the initial current $I_{IN}$ in the microplasma region is due to a single recombination center and $I_{IN}$ is discrete. The calculation for carrier generation rate from a single recombination center of gold showed a mean time of 400μs of occurrence of initial carrier in the microplasma region at room temperature. But the temperature in the microplasma region is high, and it is reasonable to assume higher generation rate. Therefore, the mean time of 40μs is used in simulation. Thus, the initial impulse current can be estimated as

$$I_{IN\_SH} = \frac{q}{\tau_t} = 13.3\text{nA with mean time of occurance } \tau_g = 40\mu\text{s}. \tag{14}$$

Comparing $I_{IN\_SH}$ to the microplasma switching threshold $I_\mu/2$ (see Eqs. (4) and (5)), one can see that multiplication of 5000 is required for activation of the microplasma. Since the multiplication is approximately an exponential function of breakdown over-voltage $\log_{10}(M)$ = (V-BV)/b, with b ≈ 50mV/dec for diodes with a breakdown knee voltage BV ~ 30V [18], (V-BV) ~ 200mV is sufficient for the microplasma to turn on by a single initial carrier. On the other hand, $\tau_g$ is larger than the electrical time constant $C_b \cdot R_L$ of the junction barrier capacitance $C_b$ < 2nF and the load resistance $R_L$ < 10kΩ of the bias circuit. Therefore, the microplasma switching follows the random occurrence of the free carrier generated by a recombination center localized in the microplasma region. The result is random microplasma switching, and this is the basis for stochastic microplasma models.

The RTS model [10] of microplasma noise uses constant amplitude $I_\mu$ of the microplasma current pulses. The DC current $I_R$ is the average of the pulses, and it is

$$I_R = I_\mu \cdot t_{on}/(t_{on}+t_{off}). \tag{15}$$

As a RTS, the microplasma current noise has spectral density, given by

$$S_I = 4 \cdot I_\mu{}^2 \frac{t_{on} \cdot t_{off}}{(t_{on}+t_{off})^2} \frac{\tau}{1+(\omega \cdot \tau)^2}, \text{ where } 1/\tau = 1/t_{on} + 1/t_{off}. \tag{16}$$

Several investigations [5,7,22] proved that Poisson distribution of the charge fluctuation is the most reasonable one. Therefore, this distribution is used for modeling of the time constant $\tau_g$ for the initial carrier generation from a single recombination center. That is, the time step is selected low (~ $\tau_g/40$) in order that the random number generation with a Poisson distribution will produce values of 0 and 1, and a maximum of 2, but not higher.

In contrast, less randomness is observed for the microplasma turn-off [6]. This is because the number of free carriers is high when the microplasma is conducting, and it is

$$\text{Number of carriers in microplasma conducting state } 2 \cdot N_\mu = 2 \cdot N \cdot w \cdot A_\mu = 7520. \tag{17}$$

Since the standard deviation of a Poisson distribution is the square root of the average, then the random distribution of carrier number can be used for a description of the fluctuation in the initial current for the microplasma conducting state, with a constant time step for the noise pattern. Then, the fluctuation in the microplasma current can be simulated using Eq. (3).

Since the multiplication is approximately an exponential function of the breakdown voltage, the difference ΔU between the microplasma turn-on and turn-off voltages is proportional to $\log(M/M_B)$ of the multiplication coefficients for the microplasma turn-on and turn-off. Subtracting the voltage drop across the microplasma access resistance $R_S$, the amplitude $\Delta V$ of the voltage fluctuation in the diode terminals is

$$\Delta V = \Delta U - I_\mu \cdot R_S \approx b \cdot \log_{10}(M/M_B) - I_\mu \cdot R_S. \tag{18}$$

Note that $\Delta U > 0$, because $M > M_B$. Since microplasma area $A_\mu$ is small, the access resistance $R_S$ is not negligible [5,23]. For an abrupt junction, $R_S$ is [23]

$$R_S \approx 400 \frac{W}{\sqrt{A_\mu}}, \text{ohm}. \tag{19}$$

According to above discussion, the equivalent circuit of microplasma is presented in Fig. 1. The multiplication M is obtained by monitoring of the voltage across the microplasma region. M multiplies the amplitude of the initial shot current $I_{IN\_SH}$ and the pattern of occurrence of this current with Poisson distribution $1\mu s/\tau_g$. Once the product reaches the microplasma switching threshold $I_\mu/2$, the microplasma current $I_\mu$ turns on.

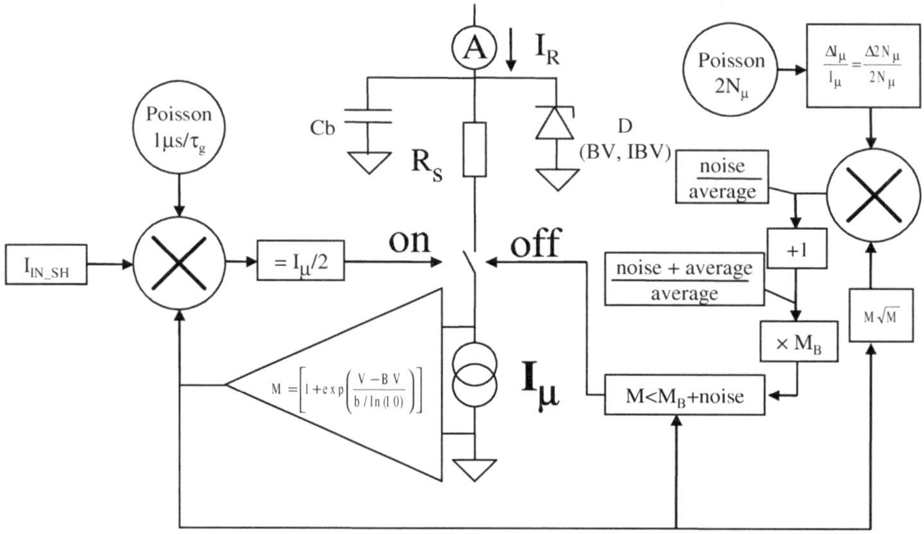


Fig. 1. Structure of the model. The left side represents the microplasma turn-on. The right side of the diagram represents microplasma turn-off.

The noise of microplasma current $I_\mu$ is converted into a random turn-off threshold ($M_B$ + noise), as follows. In the equivalent circuit of Fig. 1, the fluctuation $\Delta I_\mu$ in the microplasma current is introduced with the deviation from the average of the microplasma carrier number $2N_\mu$ which has a Poisson distribution. The deviation is normalized to the average for convenience, since $\Delta I_\mu/I_\mu = \Delta 2N_\mu/2N_\mu$. The normalized fluctuation $\Delta I_\mu/I_\mu$ is then multiplied by $M^{3/2}$, according to Eq. (3) for impact ionization in microplasma. Adding 1 to the product, the average is reconstructed back and the avalanche noise is converted into randomized off-threshold ($M_B$ + noise), multiplying with required multiplication $M_B$ for the sustaining microplasma. Once M becomes lower than the randomized threshold ($M_B$ + noise), the microplasma switches-off.

The equivalent circuit in Fig. 1 is completed with the access resistance $R_S$ of microplasma to the junction, the barrier capacitance $C_b$ and the junction D with voltage BV and current IBV of the breakdown knee. The total breakdown current of the diode is $I_R$ and is monitored by an ammeter A.

## 4. Simulation of Microplasma Switching in Avalanche Breakdown

The details for implementation of microplasma model in SPICE simulator are given in [19]. Several adjustments are made in order to scale-up the fast microplasma transients in range of microseconds, and to maintain the convergence of the simulator at microplasma switching thresholds. This reduced the computation time to several minutes for transient analysis. The simulation data for interval from 1ms to 11ms are analyzed for DC and frequency components in the frequency band from 100Hz to 1kHz. The model parameters are set to the values estimated above, and correspond to surge suppressors MR2535. The DC and noise characteristics of these diodes are previously measured [17,18]. These settings provide reliable verification of the model by comparison between simulation and experiment.

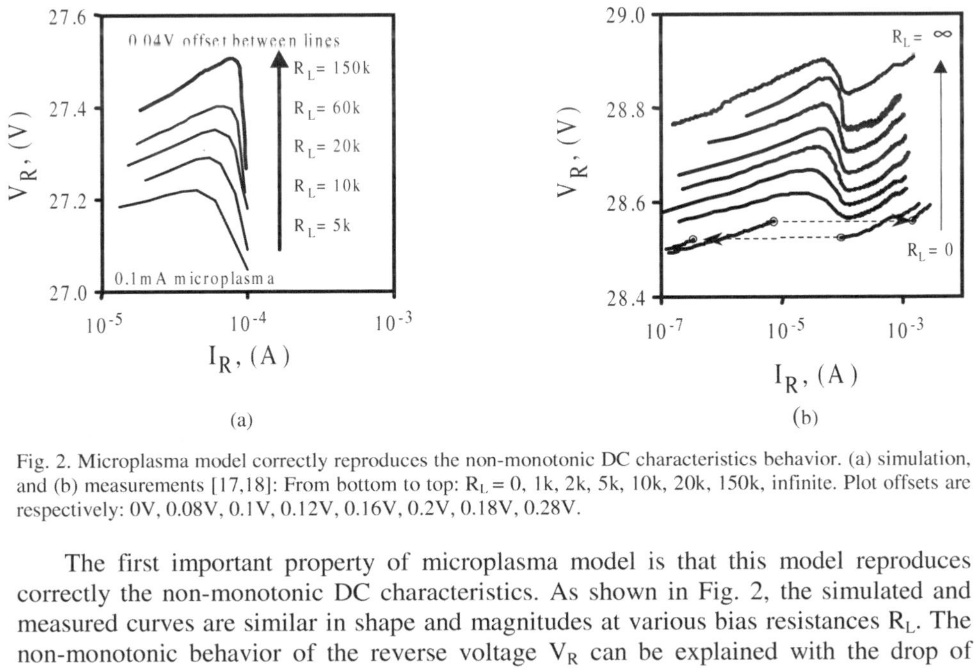


Fig. 2. Microplasma model correctly reproduces the non-monotonic DC characteristics behavior. (a) simulation, and (b) measurements [17,18]: From bottom to top: $R_L$ = 0, 1k, 2k, 5k, 10k, 20k, 150k, infinite. Plot offsets are respectively: 0V, 0.08V, 0.1V, 0.12V, 0.16V, 0.2V, 0.18V, 0.28V.

The first important property of microplasma model is that this model reproduces correctly the non-monotonic DC characteristics. As shown in Fig. 2, the simulated and measured curves are similar in shape and magnitudes at various bias resistances $R_L$. The non-monotonic behavior of the reverse voltage $V_R$ can be explained with the drop of multiplication in the microplasma on-state. The multiplication and the voltage $V_\mu$ are lower during the on-state in comparison to those in the off-state. Increasing $I_R$, $V_R$ decreases because the average duty cycle $t_{on}/(t_{on}+t_{off})$ of the microplasma fluctuation increases and $V_\mu$ contributes more to the average $V_R$. Obviously, $I_R \approx I_\mu/2$ for maximum of $V_R$.

The second important property of the microplasma model is that this model reproduces correctly the non-monotonic behavior of noise as function of the reverse current. As seen from Fig. 3, the maximum of the noise level is at a reverse current higher than the current for maximum of reverse voltage. This is predicted in [10] and elaborated in [16] for the bistable microplasma fluctuation, because the spectral density of RTS with constant amplitude $I_\mu$ and average value $I_R$ at low frequencies has a maximum at $I_R = 2 \cdot I_\mu/3$.

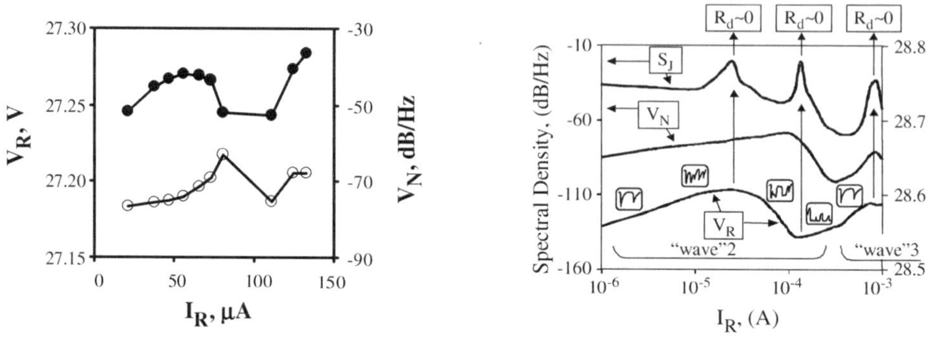


(a) (b)

Fig. 3. Microplasma model successfully reproduces the non-monotonic behavior of noise characteristics with shift between maximums of noise ($V_N$) and DC voltage ($V_R$) in respect to reverse current $I_R$. (a) simulation: solid symbols are DC voltage $V_R$; open symbols: noise spectral density $V_N$, and (b) measurements [17,18].

Note that the values of model parameters are set only by physical assumptions and fitting parameters are not used. Therefore, the similarity in the shape and values for the DC and noise characteristic presented in Fig. 2 and Fig. 3 demonstrates the excellent ability of the model to predict the microplasma behavior in avalanche breakdown.

## 5. Conclusion

This work presents a physically-based model for microplasma fluctuation in avalanche breakdown of PN junctions. The model parameters are set in accordance with the properties of the semiconductor junction such as: doping concentration, concentration of free carriers and recombination centers; recombination time constant; saturation velocity of carriers in high electric fields; breakdown voltage, current and slope. This model is based on a new theory for microplasma switching, incorporating also the Poisson distribution for fluctuations in the generation time constant and in the carrier concentration, and avalanche multiplication for obtaining the properties of the stochastic transient model. The excellent ability of the model to predict the noise and the DC characteristics of the junction at the onset of avalanche breakdown is demonstrated by comparison of the simulated data to the results from measurements of diodes.

Since the model is based on a new theory for microplasma switching, the results of simulation confirm the findings of this theory. These are, as follows.

- First, the microplasma switching threshold is the condition of equality of free- to space charge concentration in depletion layer.
- Second, when conducting, the microplasma current is approximately twice the turn-on threshold current.

- Third, the microplasma turn-on is initialized by the charge generation due to few recombination centers in microplasma region at high avalanche multiplication, while the microplasma turn-off is due to carrier diffusion from microplasma region into the depletion layer at low, but larger than 1, avalanche multiplication.

Since no fitting parameters are used in microplasma model, one can conclude that the bridge between semiconductor properties and microplasma properties is established successfully, and the empirically introduced amplitudes and time constants of bistable microplasma fluctuation have been physically explained.

### Acknowledgments

The authors would like to acknowledge the Natural Sciences and Engineering Research Council (NSERC) of Canada for partial financial support of this work through a research grant. We are also grateful to some of our colleagues in our research group for their useful input and advice.

### References

[1] R. McIntyre, *Multiplication noise in uniform avalanche photodiodes, IEEE Transactions on Electron Devices* **13**(1) (1966) 164–168.

[2] R. McIntyre, *A new look at impact ionization — Part I: A theory of gain, noise, breakdown probability, and frequency response, IEEE Transactions on Electron Devices* **46**(8) (1999) 1623–1631.

[3] S. Hsu, R. Whitter and C. Mead, *Physical model for burst noise in semiconductor devices, Solid - State Electronics* **13**(7) (1970) 1055–1071.

[4] S. Hsu, *Bistable noise in p-n junctions, Solid - State Electronics* **14**(6) (1971) 487–497.

[5] R. J. McIntyre, *Theory of microplasma instability in silicon, Journal of Applied Physics* **32**(6) (1961) 983–995.

[6] R. Haitz, *Model for electrical behavior of a microplasma, Journal of Applied Physics* **35**(5) (1964) 1370–1376.

[7] N. Ellouze, J. C. Hoffmann and B. Lacaze, *A stochastic model for microplasma noise, Journal of Applied Physics* **49**(1) (1978) 297–300.

[8] O. Marinov and M. J. Deen, *Theory and physical explanation of the microplasma fluctuation in the avalanche breakdown of silicon diodes, unpublished* (2000). Email: omarinov@yahoo.com or jamal@mcmaster.ca

[9] K. McKay, *Avalanche breakdown in silicon, Physical Revie, Series II* **94**(4) (1954) 877–884.

[10] K. Champlain, *Microplasma fluctuations in silicon, Journal of Applied Physics* **30**(7) (1959) 1039–1050.

[11] X. Zhao, M. Deen and L. Tarof, *Low frequency noise in separate absorption, grading, charge, and multiplication (SAGCM) avalanche photodiodes, Electronics Letters* **32**(3) (1996) 250–252.

[12] S. An, M. Deen, A. Vetter, W. Clark, J.-P. Noel and F. Shepherd, *Effect of mesa overgrowth on low-frequency noise in planar separate absorption, grading, charge, and multiplication avalanche photodiodes, IEEE Journal of Quantum Electronics* **35**(8) (1999) 1196–1202.

[13] S. An and M. J. Deen, *Low-frequency noise in single growth planar separate absorption, grading, charge, and multiplication avalanche photodiodes, IEEE Transactions on Electron Devices* **47**(3) (2000) 537–543.

[14] A. Zaklikiewiez, *1/f Noise of avalanche noise, Solid - State Electronics* **43**(1) (1999) 11–15.

[15] O. Marinov, *The low frequency noise in HFETs estimates the effect of electrical stress, Microelectronics Reliability* **40**(11) (2000) 1959–1963.

[16] O. Marinov and M. Deen, *Low frequency noise in avalanche breakdown of PN junction diodes, Proc. 16$^{th}$ International Conference on Noise in Physical Systems and 1/f Fluctuations*

*(ICNF 2001)*, ed. G. Bosman, World Scientific, New Jersey, 187-192, (Gainesville, Florida, October 22–25), (2001).

[17] O. Marinov, M. Deen, V. Loukanov and V. Velikov, *The low frequency noise in reverse biased rectifier diodes, IEEE Transactions on Electron Devices* **49**(1) (2002) 184–187.

[18] O. Marinov, M. Deen, V. Loukanov and V. Velikov, *Low frequency noise of reverse biased rectifier diodes in the avalanche breakdown regime, Journal of Applied Physics* **91**(11) (2002) 9232–9240.

[19] O. Marinov and M. Deen, *Physical model for low frequency noise in avalanche breakdown of PN junctions, Proc. FaN: Noise in Devices and Circuits (FaN 2003)*, eds. M. Deen, Z. Celik-Butler, M. Levinshtein, Proc. of SPIE **Vol. 5113**, 484–493, (Santa Fe, New Mexico, June 2–4), (2003).

[20] M. Tyagi, *Introduction to Semiconductor Materials and Devices*, John Wiley & Sons, New York, USA, (1991).

[21] C.-T. Sah, *Fundamentals of Solid - State Electronics*, World Scientific Publishing (1991).

[22] P. Koktavy and J. Sikula, *Reverse biased P-N junction noise in GaAsP diodes with avalanche breakdown induced microplasmas, Fluctuations and Noise Letters* **2**(2) (2002) L65–L70.

[23] W. Shockley, *Problems related to p-n Junctions in silicon, Solid - State Electronics* **2**(1) (1961) 35–67.

# DEDICATED INSTRUMENTATION FOR HIGH SENSITIVITY, LOW FREQUENCY NOISE MEASUREMENT SYSTEMS

C. CIOFI, G. GIUSI and G. SCANDURRA
*Dipartimento di Fisica della Materia e TFA and INFM, Salita Sperone,31 I-98166, Messina, Italy*

B. NERI
*Dipartimento di Ingegneria dell'Informazione, Via Diotisalvi, 2 I-56100, Pisa, Italy*

Received 8 August 2003
Accepted 21 March 2004

Low Frequency Noise Measurements (LFNM) can be used as very sensitive tool for the characterization of the quality and the reliability of electron devices. However, especially in those cases in which the frequency range of interest extends below 1 Hz, instrumentation with an acceptable low level of background noise is not easily found on the market. In fact, at very low frequencies, the flicker noise introduced by the electronic components which make up the instrumentation becomes predominant and several interesting phenomena which could be detected by means of LFNM may result completely hidden in the background noise. This consideration is not limited to the case of input preamplifiers but does extend to any piece of instrumentation that contributes to the LFNM systems, and in particular to the power supplies used for biasing the Device Under Test. During the last few years, our research groups have been strongly involved in the design of very low noise instrumentation for application in the field of LFNM. In this work we report the main results which we have obtained together with a discussion of the design guidelines that have allowed us, in a few cases, to reach noise levels not to be equalled by any instrumentation available on the market.

*Keywords*: Noise measurements; spectral analysis; low noise instrumentation; electron device.

## 1. Introduction

Since the late 1960s, Low Frequency Noise Measurement (LFNM) technique has been used as a tool for the analysis of conduction mechanisms in solid state devices and for the characterization of the quality of electron devices[1]. The analysis of the Low Frequency Noise (LFN) has become more and more important with time as we have moved towards the present VLSI era and it is bound to become even more important as we approach device dimensions in the nanometric range. In fact, the increasing miniaturization leads to a significant increase in the noise to signal ratio, thus making noise one of the main limiting factors for pursuing the ultimate miniaturization limits. Moreover, as LFN can be used as a probe capable of investigating the phenomena occurring at a microscopic scale, it can be useful for characterizing the quality and the reliability of microelectronic materials and devices. The most important peculiarities of LFNM technique with respect to this specific application field can be summarized as follows: (a) it is not destructive; (b) it is highly sensitive to localized phenomena; (c) it does not require complicated procedures for the preparation of the samples and expensive instrumentation.

Therefore the LFNM technique can be advantageously used for a not invasive and not expensive reliability analysis of many failure mechanisms occurring in electron devices. Several evidences of the capability of LFNM to serve as a tool for the characterization of the reliability of electron devices are reported in the literature since '70 [2,3] and a comprehensive review of the application of LFMN to the characterization of different failure mechanisms can be found in [4] and in the references therein.

Unfortunately, notwithstanding the potential advantages of LFNMs based characterization techniques, their use is not yet as widespread as they would deserve. Sensible noise measurements, especially at very low frequencies ($f <$ 1Hz), are not easily performed as several spurious effects often superimpose to the noise signal to be detected. Besides the external interferences which could be, at least in principle, completely removed by proper electrical, mechanical and thermal shielding, the instrumentation itself may become the limiting factor because it contributes with its own internal noise to set the minimum level of detectable signal. At very low frequencies, thermal noise does not usually play an important role. It is the flicker noise ($1/f$ noise), particularly the one introduced by the active devices used in electronic instrumentation, which plays the most important role. Commercial instrumentation, with rare exceptions, is characterized by a quite high level of flicker noise thus considerably limiting the field of application of LFNMs technique.

The feeling of our research groups, founded on more than 15 years of experience in the field, is that a skill in the design and realization of low noise instrumentation can represent a key factor for obtaining the full advantages out of the application of LFNM technique to the characterisation of electron devices and materials. The main purpose of this paper is to provide the reader with a review of the several issues connected with the realization of specific instrumentation for LFNMs and with a discussion of the solutions which we have adopted for solving several specific problems.

## 2. Dedicated Instrumentation for LFNM

A simplified block diagram of the several pieces of instrumentation which make up almost any conceivable LFNM system is reported in Fig. 1.

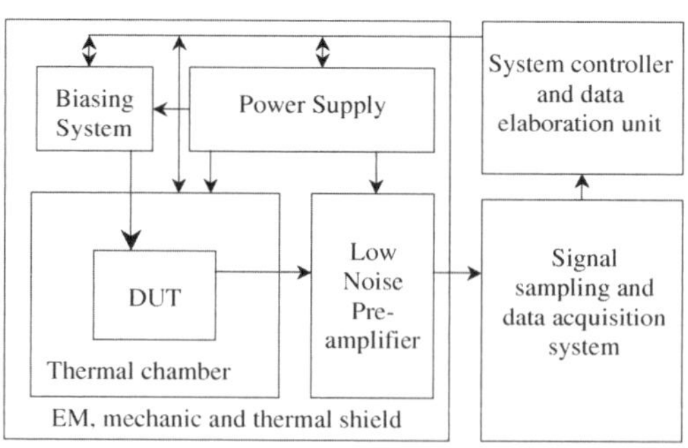


Fig.1. Block diagram of an automated LFNM system.

It refers to the ideal and desirable case in which all the measurement parameters, especially the temperature of the thermal chamber and the bias of the DUT can be set by means of an external controller, usually a Personal Computer (PC), in order to make automated noise measurements possible. However, this is seldom the case, especially because although a few examples of low noise programmable biasing sources have been demonstrated [5,6], their use is still limited either because of their size and high cost, or because of the complication involved in their integration as a standard piece of instrumentation in LFNM systems. In the following subsections we will address a few issues relative to each specific component of a LFNM system.

## 2.1. *Input preamplifiers*

By means of a careful design, the Equivalent Input Voltage Noise (EIVN) of a voltage amplifier can be reduced to that of the active device used in the first stage of the amplifying chain. In order to avoid disturbances coming from the power supply and other common mode interferences (temperature fluctuations, EM interferences and so on) the use of differential configurations is mandatory. As the best performances in terms of EIVN are to be found in discrete devices, we usually resort to discrete components input stages. We have selected the matched BJT pair SSM2220 by PMI as the devices with the lowest EIVN (1.5, 3, 10 nV/[Hz]$^{1/2}$ at 1, 0.1, and 0.01 Hz, respectively). The performances of an ultra low noise amplifier based on the SSM2220 can be found in [7]. With such a design, we obtain the best performances in term of EIVN (curve labelled BJT in Fig. 3). However, when the equivalent impedance of the source is above a few tens of ohms, the effect of the Equivalent Input Current Noise (EICN) of the BJTs becomes predominant. In these cases a JFET input stage is preferable, although the best discrete devices we have been able to select (IF3601 discrete JFET by InterFet) have a significantly higher EIVN. In the case of the JFETs, however, the EICN is enormously lower thus making them the ideal choice for high impedance DUTs. Moreover, as the required biasing current is in the order of a few pA, AC coupling down to the mHz range is readily obtained. Figure 2 shows an example of a JFET input stage low noise preamplifier.

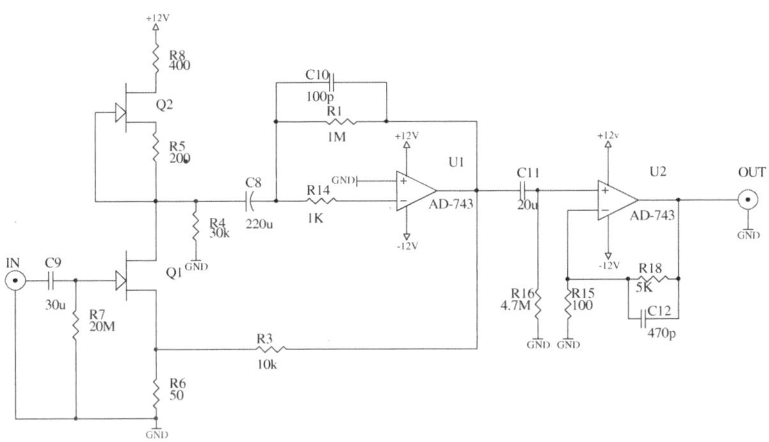


Fig. 2. Schematic of a JFET input stage ultra low noise preamplifier.

The first stage of the amplifier can be regarded as a cascade of a JFET transconductance amplifier coupled to a transresistance amplifier, which realize a two stage high gain voltage amplifier in a shunt series feedback configuration with a gain of about 200 (set by the ratio R3/R6). The transistor Q2 acts as a current source for providing high gain and high PSSR notwithstanding that we do not use a differential configuration. The second stage, which is AC coupled to avoid saturation due to the output offset of the first stage, provides for an overall 80 dB voltage gain. The input and intermediate high pass filters provide for a low corner frequency of about 2 mHz.

When the performances of our amplifiers are compared to those of commercially available instrumentation, the improvement which can be obtained at very low frequencies is remarkable. As an example, the EIVN spectra of our BJT and JFET preamplifiers are compared to that of the popular Brookdeal 5003 in Fig. 3.

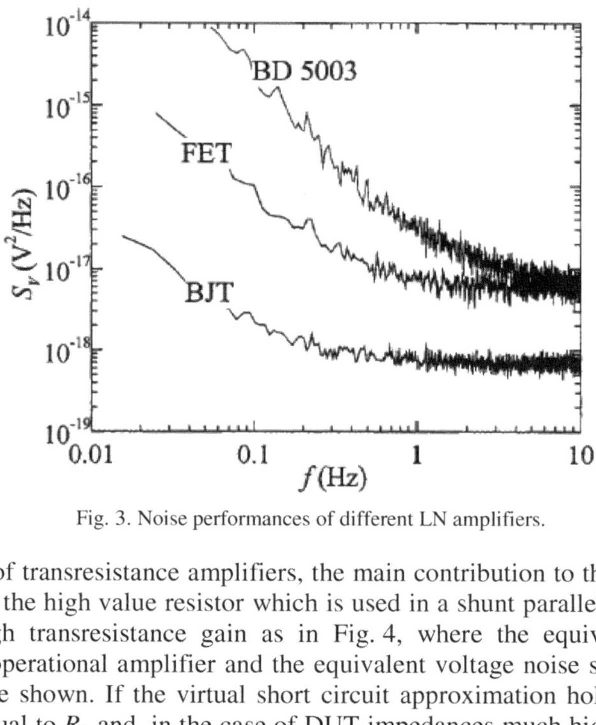


Fig. 3. Noise performances of different LN amplifiers.

In the case of transresistance amplifiers, the main contribution to the output noise often comes from the high value resistor which is used in a shunt parallel configuration for obtaining a high transresistance gain as in Fig. 4, where the equivalent input noise sources of the operational amplifier and the equivalent voltage noise source of the feedback resistor are shown. If the virtual short circuit approximation holds, the transresistance gain is equal to $R_R$ and, in the case of DUT impedances much higher than the input impedance of the transresistance amplifier, the spectral density $S_{ineq}$ $(f)$ of the equivalent input current noise source of the entire system can be estimated as follows:

$$S_{ineq}(f) = S_{in} + \frac{S_{vn}}{R_R^2} + \frac{4kT}{R_R}. \tag{1}$$

In the case of a transresistance gain of 100 MΩ, the noise due to $R_R$ (the third term in Eq. 1) amounts to about $166\times10^{-30}$ A$^2$/Hz. In order for the second term (noise due to the equivalent voltage input source of the operational amplifier) to be comparable with such a value, $S_{vn}$ should be in the order of $1.6\times10^{-12}$ V$^2$/Hz (that is 1.3 μV/[Hz]$^{1/2}$ ). This clearly demonstrates that for the realization of low noise transresistance amplifiers we must resort to operational amplifiers with a very low EICN, while quite high values of EIVN can be tolerated. In our recent designs we resort to the MOSFET input operational amplifier TLC070 which is characterized by an EICN as low as $3.6\times10^{-31}$ A$^2$/Hz and by and EIVN of about 90 nV/[Hz]$^{1/2}$@1 Hz. Eq. 1 also shows that increasing $R_R$ is beneficial as far as the equivalent input noise is concerned. However, there are limitations to the value that $R_R$ can assume. In fact, it is sometimes difficult to AC couple the device under test to the input of the amplifier and therefore the DC component of the current must flow across $R_R$ without saturating the operational amplifier. Moreover, the bandwidth of the amplifier is often limited by the parasitic capacitance in parallel to $R_R$. Since for a given resistor technology and package such a capacitance is almost independent of the resistance value, the bandwidth decreases proportionally to $R_R$. The observation that the BN of a transresistance amplifier is mostly due to the feedback resistor, has recently lead us to explore the possibility of designing transresistance amplifiers which use a noiseless impedance in the feedback loop rather than a resistance. This can be done by employing a feedback capacitor instead of the resistor $R_R$ in Fig. 4. This idea has lead us to the design of a new topology for the realization of a high gain, low noise, high bandwidth transresistance amplifiers. The schematic of the new topology is reported in Fig. 5.

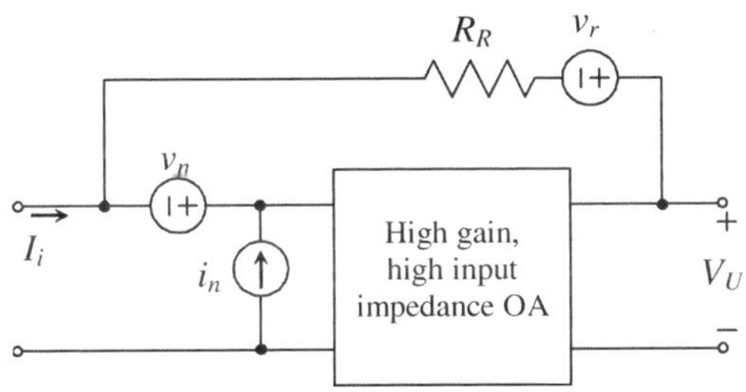


Fig. 4. Simplified schematic of a transresitance amplifier.

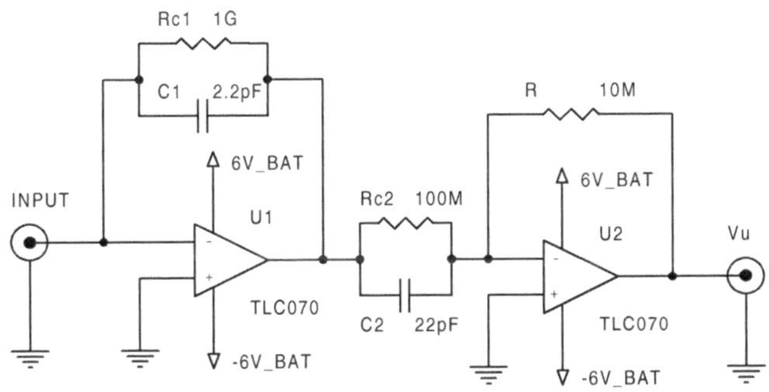


Fig. 5. New topology for a low noise, high bandwidth transresistance amplifier configuration.

The resistor $Rc_1$ provides for a DC feedback path for the first operational amplifier. The condition $Rc_1 \times C_1 = Rc_2 \times C_2$ must be satisfied in order to have a flat response down to DC. In the virtual short circuit approximation, the transresistance gain $A_{R0}$ results:

$$A_{R0} = \frac{C_2}{C_1} R = 100\ \text{M}\Omega\,. \tag{2}$$

It can be demonstrated that the new approach allows, for the same transresistance gain, to obtain an increase of the bandwidth by a factor $C_2/C_1$ and a reduction of the EICN by the same factor[8]. This last advantage can be obtained in the case in which the resistance $R_{C1}$ can be made large enough not to significantly contribute to the output noise. However, the larger $R_{C1}$, the lower is the DC input current that can be tolerated by the amplifier without causing the saturation of the first stage. A comparison between the performances which can be obtained by using the new approach with respect to the conventional one (for the same transresistance gain and using the very same operational amplifiers and resistance types) is given in Fig. 6.

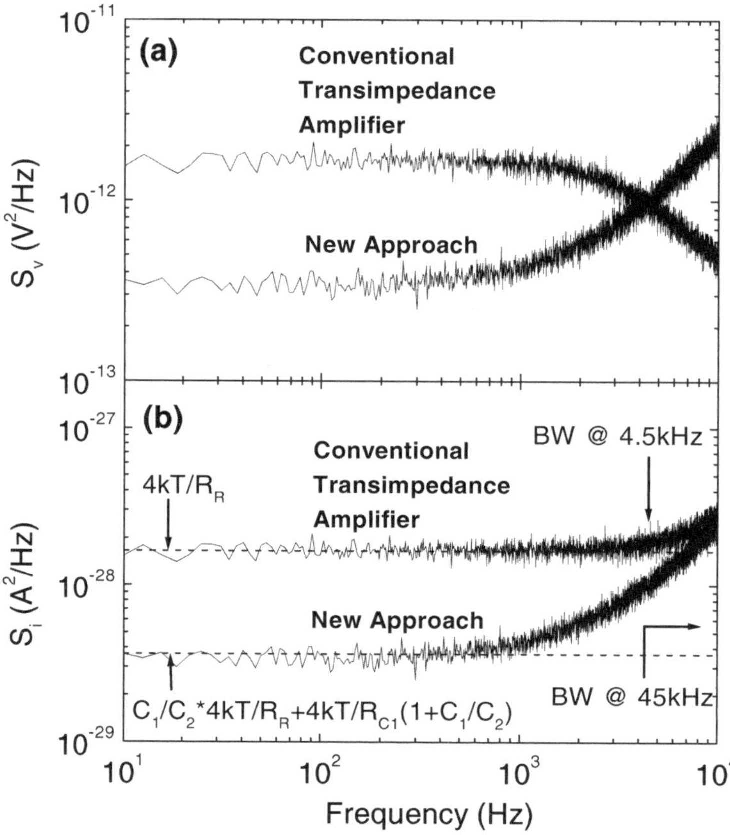


Fig. 6. Comparison of the noise performances between the new and the conventional transresistance amplifiers (open input). In (a), the power spectra of the voltage fluctuations at the output are reported. In (b) the spectra of the equivalent input current noise source corresponding to the curves in (a) are reported.

The increase of the output noise above 1 kHz is due to the combined effect of the EIVN and of the common mode parasitic input capacitance of the operational amplifier. The same effect is present in the conventional design, but it is hidden because of the limited bandwidth and the much higher level of white noise. This is clearly shown in Fig. 6b, where the calculated EICN power spectral densities are reported. If we compare our new design with commercial instrumentation (the popular SR570 by Stanford Research, and two among the best amplifiers available on the market by FEMTO), we obtain better performances both when we compare the bandwidth for the same equivalent input noise and when we compare the equivalent input noise for the same bandwidth (Table 1).

Table 1. Comparison of the performances of different transresistance amplifiers.

| Same equivalent input noise ($\approx 4\times10^{-29}$ $A^2/Hz$) | | | |
|---|---|---|---|
| | New design ($4\times10^{-29}$ $A^2/Hz$) | SR570 ($2.5\times10^{-29}$ $A^2/Hz$) | LCA-4K-1G ($4.2\times10^{-29}$ $A^2/Hz$) |
| Bandwidth: | 45 kHz | 10 Hz | 4 kHz |
| Same bandwidth ($\approx$45 kHz) | | | |
| | New design (45 kHz) | SR570 (20 kHz) | LCA-40K-100M (40 kHz) |
| Eq. Inp. Noise: | $4\times10^{-29}$ $A^2/Hz$ | $3.6\times10^{-25}$ $A^2/Hz$ | $3.6\times10^{-28}$ $A^2/Hz$ |

### 2.2. *Bias sources*

Solid state bias sources are normally useless in the case of LFNM because of the high level of $1/f$ noise produced by the zener which is used as a reference in all these systems. Therefore, batteries are normally used for providing voltage bias to the DUT. Current bias is usually obtained by connecting a few batteries in series and using a conveniently large series resistor. Batteries behave as low frequency noise voltage or current sources provided that they do not supply too much current. As a rule of thumb (for measurements down below 100 mHz) a battery should not supply a current higher than that needed to fully discharge it in less than 100 hours. In any case, the voltage or current supplied by a battery drops with time and therefore it is never the same in repeated measurements. Moreover, only a discrete set of voltages are available. One can resort to resistive voltage dividers, but in this case, for the system to behave as a low internal resistance voltage source, the current supplied by the battery must be quite large thus making the problem of the discharge more and more important. Remote control of the voltage value is not possible and therefore a lot of time is wasted in changing the configuration of the battery packs or for just monitoring its actual value. The problem of realizing reliable low noise voltage or current sources (either programmable or not) is never an easy one because the level of noise introduced by such devices directly appears at the ends of the DUT. In the course of the last few years we have addressed the problem of the realization of fixed and programmable bias sources and we believe that some of our results are remarkable and deserve to be mentioned here.

#### 2.2.1. *Fixed value low noise voltage and current sources*

Almost any battery, when supplying a very low current, may behave as a very low noise, high stability voltage reference. In fact, while the exact value of the generated voltage does depend on the charge state of the batteries (for rechargeable batteries) or on their age (for non rechargeable ones), if the battery does not supply current, its value may be assumed almost constant. Accurate measurements performed in a highly stabilized temperature chamber (30 ±0.005 °C) have shown that the voltage drop (mainly due to self discharge) is in the order of 100 ppm/h in the case of lead acid batteries (Sonnenschein Dryfit A300, 6.8 V, 1.1 Ah), while in the case of lithium batteries (Duracell type 223, 6 V) the voltage drift can be as low as 3 ppm/h.

Even in the case of the lead acid battery, the voltage drift results 14 mV in 24 hours which is largely acceptable for many applications. Therefore, in most applications we use small capacity lead acid battery as references for low noise solid state voltage sources. A simple circuit which behaves as a low noise voltage source is illustrated in Fig. 7(a).

The high value resistor $R$ in series with the battery is necessary in order to avoid that the battery discharges through the Gate-Source junction of the JFET when $V_{DD}$ is turned off. The capacitor $C$ is then required for filtering out the thermal noise generated by the resistor. With a careful design of the $RC$ time constant, the power spectrum of the voltage fluctuations at the output of the system reduces to that introduced by the EIVN of the JFET. The resistance $R_D$ is needed in order to maintain the drain-source voltage below the value for which impact ionization in the channel becomes significant. When employing the large area JFET IF3601 by Interfet, voltage noise as low as $5\times10^{-15}$, $10^{-16}$, $10^{-17}$ $V^2/Hz$ at 0.01, 0.1 and 1 Hz, respectively, can be obtained.

Using a battery as a reference, we can also realize a low noise current source using the configuration reported in Fig. 7(b). Here the reference battery together with the source resistance $R_S$ sets the current value according to the following equation:

$$I_{OUT} = \frac{V_B - V_{GS}}{R_S} . \tag{3}$$

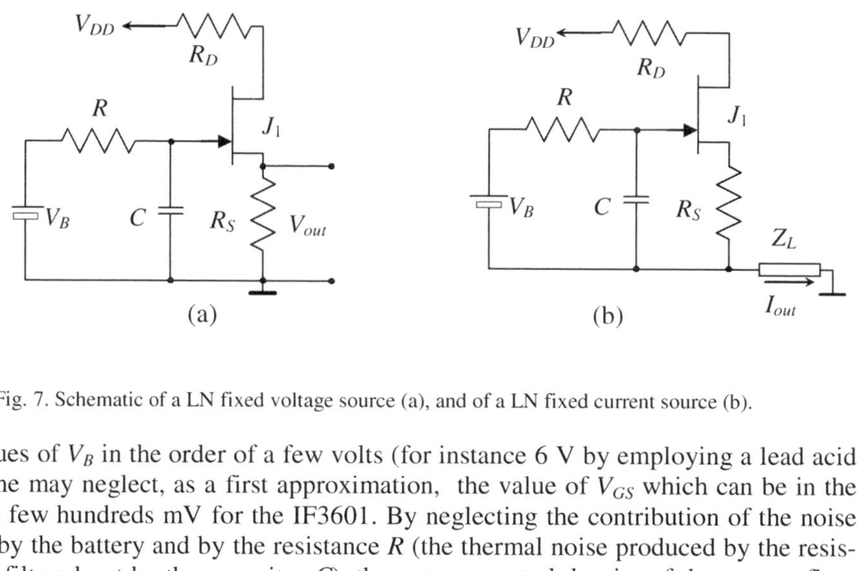


Fig. 7. Schematic of a LN fixed voltage source (a), and of a LN fixed current source (b).

Using values of $V_B$ in the order of a few volts (for instance 6 V by employing a lead acid battery) one may neglect, as a first approximation, the value of $V_{GS}$ which can be in the range of a few hundreds mV for the IF3601. By neglecting the contribution of the noise produced by the battery and by the resistance $R$ (the thermal noise produced by the resistance $R$ is filtered out by the capacitor $C$), the power spectral density of the current fluctuations delivered to the load can be estimated as follows:

$$S_{Iout} = (S_{enFET} + 4kTR_S) \times \left( \frac{g_m}{1 + g_m R_S} \right)^2 \tag{4}$$

where $S_{enFET}$ is the power spectral density of the EIVN of the JFET and $g_m$ is the transconductance of the JFET. When employing the large area JFET IF3601, values of $g_m$ in the order of 200 mA/V are easily obtained. The actual circuit used for realizing a practical low noise current noise source is more complex with respect to the one shown in Fig. 7, because, as in the case of the voltage sources, we must avoid too large values of $V_{DS}$ independently of the load impedance $Z_L$. The circuit we use is described in [9]. As an example of the performances which can be obtained, the power spectra at the output of our current source are reported in Fig. 8 for different values of the current ($I_G$ in the figure) supplied to a 200 Ω load. Such a current source has been successfully used to demonstrate the feasibility of low noise measurements at wafer level as described in [10].

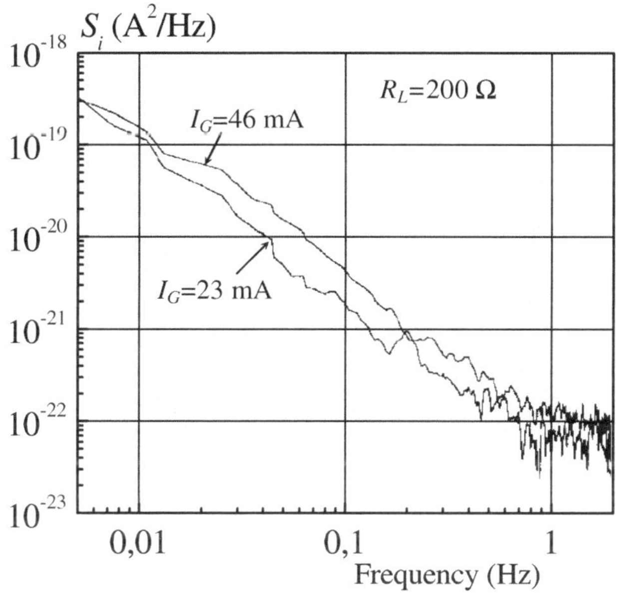


Fig. 8. Current noise at the output of an ultra low noise current source.

### 2.2.2. *Programmable low noise voltage sources*

Designing programmable low noise voltage or current sources is not an easy task since standard DA converters produce a high level of low frequency noise. Moreover, the digital circuitry needed for the operation of a programmable source is likely to introduce interferences which degrade the noise performances. Up to now we have however successfully addressed the issue of designing and realizing programmable, high accuracy, ultra low noise voltage sources. In one case, we used a low noise fixed value voltage source as a reference for a low noise resistive voltage divider ladder network. The output of such a ladder network was used as the input of a low noise power voltage follower to provide low output impedance and high current driving capability. Such a piece of instrumentation behaves as a very low noise 12 bit DA converter capable of generating voltages in the range from 0 to 8 V with an accuracy better than ±1.5 mV and delivering to the load currents in the range of a few hundred mA. Typical values of the spectral density of the voltage fluctuations at the output are $10^{-12}$, $10^{-15}$, $10^{-16}$ V$^2$/Hz at 0.01, 0.1 and 1 Hz, respectively [5].

In many applications, however, the ability to supply currents in the range of a few mA may be sufficient. This prompted us to verify the possibility of realizing a programmable low noise voltage reference by directly filtering the output noise of a standard solid state DA converter. With reference to Fig. 9(a), if the time constant of the RC filter in the figure is sufficiently high, the high level of noise produced by a standard solid state DA converter (more than 50 dB above that corresponding to the EIVN of the IF3601 in the frequency range below 1 Hz) could be filtered out starting from a conveniently low frequency. However, it may be calculated that at $f$=100 mHz, in order to reduce the noise generated by a "low noise" commercial DA converter (in the order of $10^{-10}$ $V^2$/Hz at 100 mHz) to a level at least 10 dB below that of the EIVN of the JFET ($10^{-16}V^2$/Hz at 100 mHz), a time constant of about 30 minutes would be required, thus making the use of the device impractical because one would have to wait a few hours before having a stable output upon any change of the voltage to be generated. In order to overcome this limitation, we resorted to the circuit shown in Fig. 9(b) which is part of a microcontroller based control system as described in [6]. The optically controlled analog switch (H11F1) behaves as a very high resistance with the driving LED switched off ($R_{off}$>300 MΩ), while it behaves as a small value resistance when the LED is turned on. The LED may be automatically operated by the microcontroller, being switched ON in order to speed up the voltage change transient. When the new steady state condition is reached, the LED is switched off in such a way as to allow the low pass filter to develop its full action.

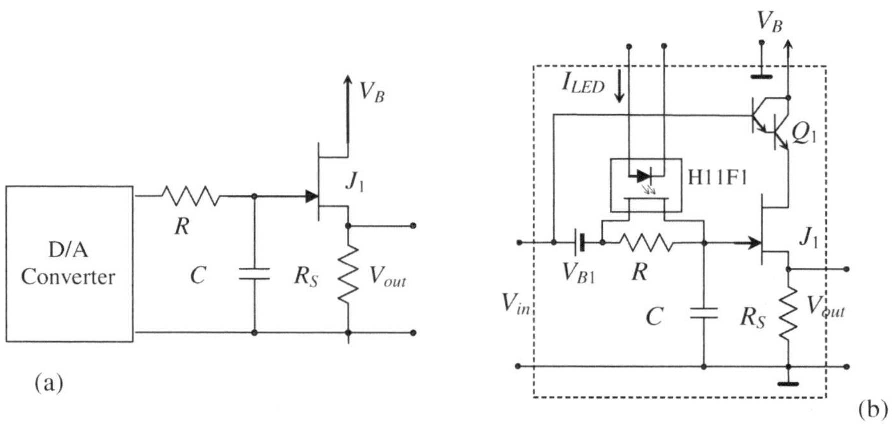


Fig. 9. A simple implementation of a low noise voltage reference (a); in (b) the actual implementation of the output stage, that allows to obtain reduced set-up times, is shown.

In fact, while the output voltage is changing, no noise measurement would be meaningful and by means of the procedure described above, the new steady state is reached in a matter of a few minutes. We decided to employ two high capacity 12 V, 6Ah lead acid batteries, one for the control board and one for the output buffer, in order to reduce the size of the entire system. For the output voltage to reach 0 V we had therefore to insert the battery $V_{B1}$ which insures that the JFET is pinched off when the input voltage is close to 0 V. At the same time the battery allows, together with the darlington Q1, to maintain the JFET in saturation with the lowest possible value of $V_{DS}$. The battery $V_{B1}$, through which only the negligible gate current flows, is obtained as the series of 3 tiny 100 mAh NiCd cells. The voltage at the output of the circuit in Fig. 9(b) is continuously monitored by means of a high resolution AD converter and a digital control loop is used in order to drive the DA converter section for obtaining high accuracy and stability. Proper countermeasures were taken in order to reduce the interferences generated by the action of the control loop [6]. The noise performances which can be obtained are impressive as can be noted in Fig. 10, where, together with the power spectrum of the voltage fluctuations $S_{out}$ measured at the output of the voltage source, we have reported the power spectrum at the output of the solid state voltage source (REF192) which provides the reference for the DA converter used in the design. Accuracy and stability of the supplied voltage are within ±250 μV in the entire output range from 0 to 5 V.

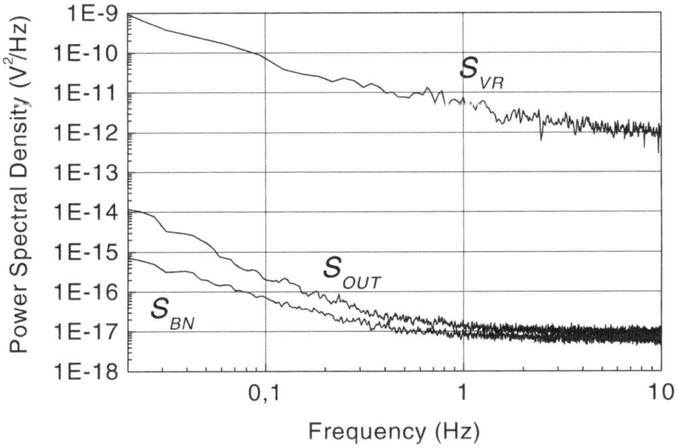


Fig. 10. Noise performances of the digitally programmable low noise voltage source. $S_{BN}$ is the BN noise of the measurement system, $S_{VR}$ is the noise of the voltage reference of the DA employed in the design.

### 2.3. *Low noise thermal chamber*

Often, LFNMs must be performed on samples maintained at a fixed temperature [11]. Standard thermal chambers for reliability tests cannot be used because of the high level of temperature fluctuations and insufficient shielding. In the past we have been using miniaturized heater that were battery supplied in order to allow to put them in the shielded measurement environment. More recently, because of the need for a thermal chamber large enough for hosting several samples and with the capability of allowing long term noise measurements [12], we had to face the problem of the design of a high stability thermal chamber capable of meeting these requirements. In a thermostatic chamber, the residual thermal fluctuations are primarily due to the environmental perturbations and to the fluctuation of the power source which supplies the heaters. In the case of our design [13], the choice has been made to use a two stage thermal structure in which two coaxial aluminum vessels are separated by a thermal insulating layer (PTFE). This structure, shown in Fig. 11, acts as a second-order low-pass filter, effectively reducing the effects of environmental perturbations and heaters power fluctuations. This solution has allowed a reduction of more than 90 dB of the temperature fluctuations with respect to an oven chamber of equivalent dimensions and using the same mass of aluminum and the same external insulation structure. The sample chamber can reach a maximum temperature of 250$^0$C with a stability better than 0.1$^0$C. The effectiveness of the design was such that we were not able, using the best instrumentation we had available, to evaluate the residual thermal fluctuations within the sample chamber [13]. The internal heater directly in contact with the inner chamber (Fig. 11) is supplied only during start up in order to reduce the time required for reaching a steady state temperature in the sample chamber. The aluminum bottom cap, together with a metallic pipe running without any interruption trough the outer insulation layer and across the entire shielded box into which the thermal chamber lays, allows an effective insulation from the electromagnetic interferences originating from the cables which connect the heaters to the external solid state power supply.

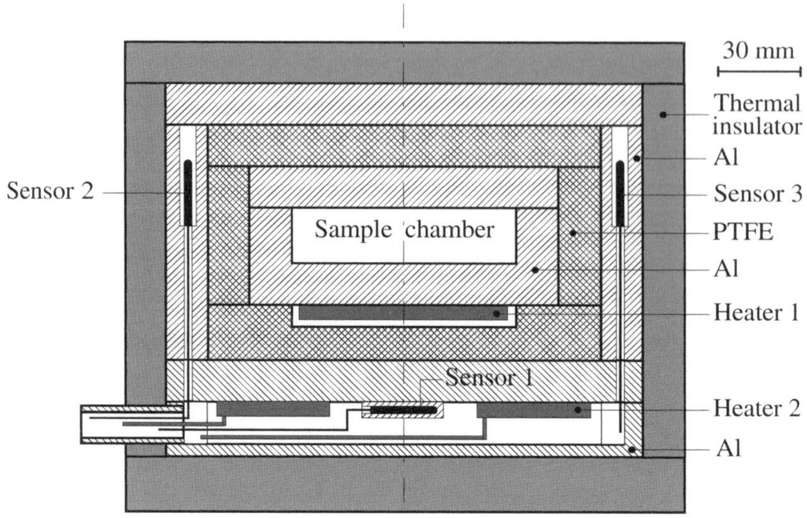


Fig.11. Double stage, low temperature noise thermal chamber for LFNM.

### 2.4. *Wafer level low frequency noise measurements*

Low frequency noise measurements on electron devices are more easily performed on packaged samples rather than at wafer level. There are at least two reasons for this. The first one is the high level of low frequency noise which is generated at the point contacts of the probe tips in the case of wafer level noise measurements [14]. The probe contact noise is mainly $1/f$ noise whose amplitude is proportional to the square of the DC current that crosses the contact and therefore its effect is particularly important in the case of measurements performed with relatively high currents (a few tens of mA) and at very low frequencies ($f < 1$ Hz) as in the case of the electromigration experiments [11]. The second reason which makes wafer level LFNM difficult is the need for an accurate mechanical and electromagnetic shielding of the large volume occupied by the probe station. Probe stations with excellent mechanical stability and high immunity to external interferences may be certainly found on the market, but at a quite high cost. We have recently addressed this specific issue and we have devised a micro probe system dedicated to LFNM on devices available at wafer level. Notwithstanding its low cost, the performances of the system have resulted fully compatible with the requirements for high sensitivity noise measurements [15]. A schematic diagram of the system is reported in Fig. 12. We renounced to the possibility of realizing a probe system capable of hosting an entire 8-inch wafer. The sample holder (7) top plate has a diameter of 3 cm and can therefore host quite large dies. The entire structure is obtained starting from two 4 mm thick aluminum boxes, the cover plates of both being screwed together to form the base (1), which is 14 cm × 22 cm in size.

One of the boxes is used to form the bottom section (2), which hosts the biasing system and the noise preamplifiers (3) together with the battery pack (4). The other box forms the top cover (5). A ferromagnetic flat base (6) allows the sample holder (7), which has four magnetic disks inserted in its base (8), to slide for a coarse positioning of the DUT. A thin (0.1 mm) PTFE foil (9) allows for electrical insulation. Up to four probes can be hosted onto the base. The probes consist of standard tungsten tips (10) embedded in an insulating holder (11) each of which is mounted on an XYZ precision translator (12). The inner conductor of a RG174 coaxial cable (13) is soldered to a miniature screw wire holder, which is secured to the tungsten tip.

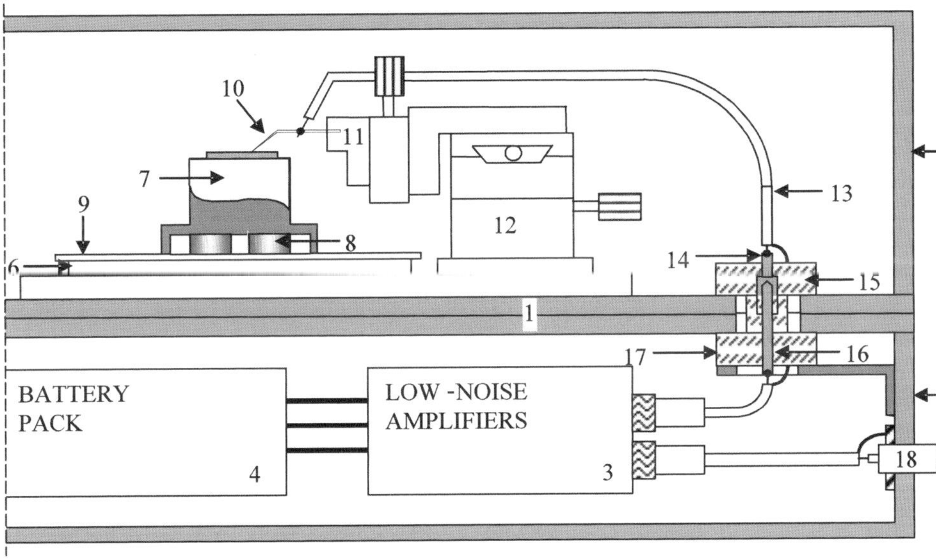


Fig.12. Schematic view of the low noise probe system used for LFNM on wafer level samples.

The other end of the cable is connected to a 4 mm female connector (14), which is inserted in a highly insulating PTFE holder (15). Each female connector has a corresponding male connector (16), which is inserted in a second PTFE holder (17) secured to the bottom section box. Once the base is inserted into the bottom section, electric continuity from the tips to the electronic section is automatically obtained and the electronic section is completely shielded from the environment. At this point one may position the sample holder and contact the DUT pads with the aid of a microscope (not shown in Fig. 12), and put the top cover in place thus obtaining a structure in which the sample is shielded from the electronic section and both are completely shielded from the environment. The results of the measurements of the shot noise of the gate leakage current on a MOS test structure is reported in Fig. 13. Note that the lowest measured shot noise level is as low as 5.65 fA/√Hz, thus confirming the very high sensitivity that can be obtained with our dedicated probe system.

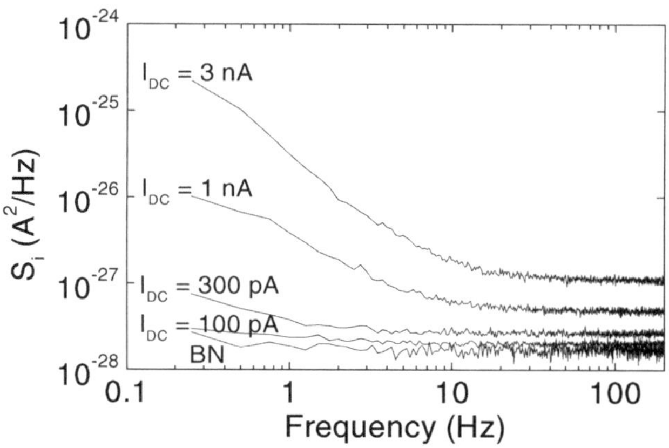


Fig.13. Power spectral density of the current noise measured in a polycrystalline silicon gate MOS structure with a 7 nm thick oxide at different DC current levels. The measurements were performed using the new probe system. The BN of the transresistance amplifier is also shown.

## 3. Signal Acqusition and Elaboration

Noise spectra estimation in the case of LFNMs is usually performed by means of dynamic signal analyzers. Commercial FFT spectrum analyzers have, however, a few limitations. Usually, only one or two input channels are available thus making the possibility of using LFNMs for performing reliability tests on a large number of samples [12] either impossible or time consuming. Moreover, after performing the calculation of the FFT, the acquired time record is discarded to make room for another one. This fact, when we remember that in order to perform power spectra estimation of down to a few mHz a time record of several thousands of seconds need to be elaborated, not to talk of the case of long term noise measurements [12], leads to a few drawbacks. In the first place, it is seldom possible for an operator to constantly monitor the measurement for detecting possible interferences which may cause spurious increase in the measured spectra. Moreover, one has to make the choice, at the beginning of the measurement, of either privileging frequency resolution at the cost of bandwidth or of privileging bandwidth at the cost of resolution. If one realizes that a different bandwidth-frequency resolution combination was necessary, one has to repeat the measurement with the new parameters. All these problems would be solved if it were possible to store the entire duration of the time record in such a way as to be able to defer the spectral estimation at the conclusion of the measurements. In this way, one could play back the acquired data at a faster speed and make sure that no anomalous behavior occurred during the measurements while selecting the proper combination of bandwidth and frequency resolution. We have in fact designed and realized such a system which, in its early version, is described in [7]. The system we use at present is based on a DSP board that we have purposely developed for LFNMs and uses a PC for controlling the acquisition, storing the data and performing spectral estimation. The system has 8 input channels and the maximum acquisition frequency is 512 Hz, which is largely sufficient for LFNMs. For the connection from the front end and the sampling boards to the PC, we used a fiber optical link in order to avoid the electromagnetic interferences generated by the PC to reach the low noise preamplifiers.

Signal elaboration, either before or after sampling, may be used in order to reach equivalent BN which can be below that of the input preamplifiers. The possibility of performing reliable noise measurements when the noise produced by the DUT is comparable to, or lower than, the BN of the input preamplifier has been demonstrated using different methods [16–18]. We have recently proposed another method for performing reliable very low noise measurements which, although in principle equivalent to the one described in [17], is based on a rather different approach [19]. The circuit configuration of the system is reported in Fig. 14 for the case of voltage noise measurements. The DUT voltage noise $V_s$ is simultaneously amplified by two different channels with identical gains and the two amplified signals are fed to the analog addition and subtraction blocks. The output voltages $x_1(t)$ and $x_2(t)$ can be written as:

$$\begin{cases} x_1(t) = c(t) + u_1(t) \\ x_2(t) = c(t) + u_2(t) \end{cases} \tag{5}$$

where $c(t)$ is due to the input signal and to the EICN of the two amplifiers, $u_1(t)$ is due to the EIVN of the first amplifier and $u_2(t)$ is due to the EIVN of the second amplifier. Therefore, at the output of the addition and subtraction blocks we have:

$$\begin{cases} a(t) = 2c(t) + u_1(t) + u_2(t) \\ s(t) = u_1(t) - u_2(t). \end{cases} \tag{6}$$

At least in the case of LFNM, one can assume that $u_1(t)$, $u_2(t)$ and $c(t)$ are uncorrelated and, therefore, at the output of each channel of the spectrum analyzer in Fig. 14 we have:

$$\begin{cases} S_a(f) = 4S_c(f) + S_{u1}(f) + S_{u2}(f) \\ S_s(f) = S_{u1}(f) + S_{u2}(f) \end{cases} \tag{7}$$

where $S_x(f)$ is the Power Spectral Density (PSD) of the signal $x(t)$. By subtracting $S_s(f)$ from $S_a(f)$ we obtain an estimate of $S_c(f)$ alone. It is clear that, in the frequency range in which the contribution of the EICN of the amplifiers is negligible, we obtain a direct estimate of the PSD of the noise signal we are interested in. Since the contributions of the EIVN of the amplifiers cancel out, the sensitivity of the measurement method is limited by the contributions of the EICN, whose effect depends on the equivalent impedance of the DUT. This fact implies that in order to measure very low voltage noise, the EICN of the amplifier represents the limiting factor rather than the EIVN. Therefore, using the measurement configuration we propose, one may obtain better noise performances from MOSFET input stage amplifiers (high EIVN and low EICN) rather than from those employing BJTs (low EIVN and high EICN), even in the case of low DUT impedances.

In order to verify the effectiveness of our approach, we used two voltage noise amplifier based on the TLC2201 by Texas Instruments which is characterized by a relatively high EIVN (10 nV/√Hz, $f$>100 Hz) and by an excellent level of EICN (0.6 fA/√Hz). Two INA141 instrumentation amplifier were used for realizing the adding and subtracting blocks. A resistance $R_{10}$ of 10 Ω was used as DUT. At room temperature the value of the PSD of the voltage fluctuation to be measured is about 407 pV/√Hz. As it can be verified in Fig. 15, the estimation of DUT noise as obtained by using the proposed method is rather good, with an average value of 409 pV/√Hz and a standard deviation of 270 pV/√Hz after $2^{16}$ averages with a frequency resolution of 25 Hz. It must be noted that the noise produced by the DUT is more than 25 dB below the EIVN of the input. The same method can be used also in the case of current noise measurements [19].

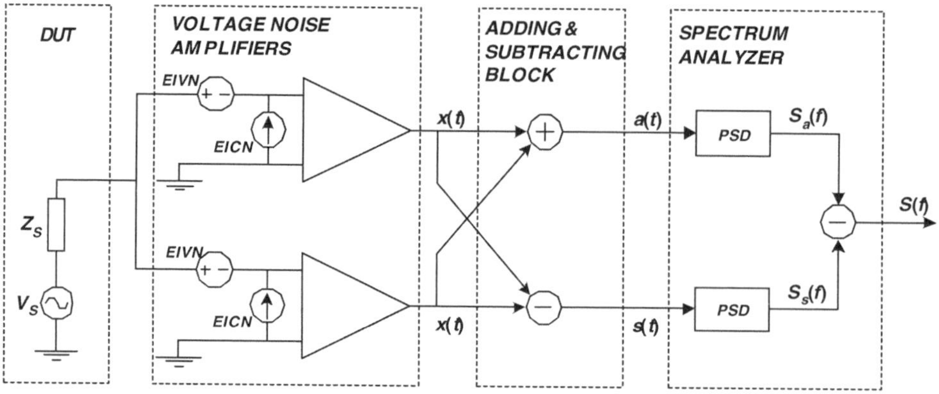


Fig. 14. Block diagram of the new voltage noise measurement system.

Another possible method for reducing the influence of the low frequency noise introduced by the input preamplifiers consists in modulating the bias in the DUT in such a way as to translate the noise to be measured at a frequency where the $1/f$ noise of the amplifier can be neglected. The original noise signal is detected, after amplification, by resorting to synchronous detection schemes [20]. While this method may result effective in a few cases, it must be noted that its application results in a quite complex measurement configuration. It is for this reason that we strongly feel that its use should be avoided whenever possible and that avoiding such complications fully justify the efforts toward the design of very low noise amplifiers for direct coupling to the DUT as those discussed in Sec. 2.2.

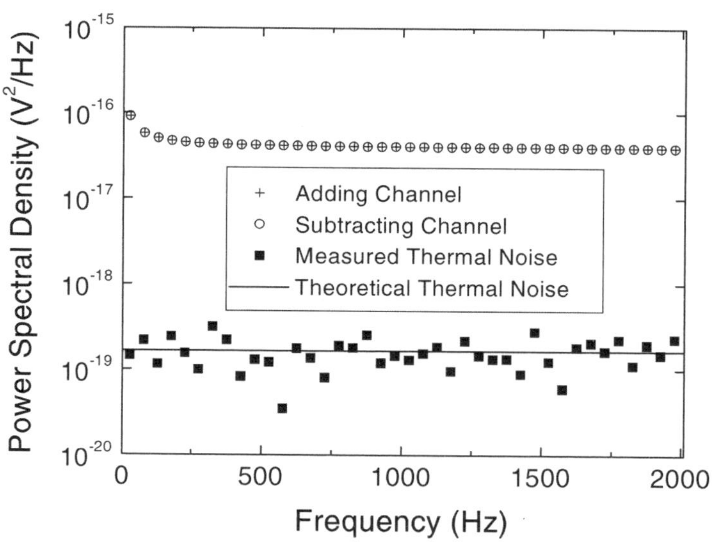


Fig. 15. Power spectra at the output of the adding and subtracting channel and estimated spectrum of the input voltage noise source (a 10 Ω resistor).

## 4. Conclusions

Low frequency noise measurements may represent a powerful tool for the characterization of the quality and reliability of electron devices. The sensitivity which can be obtained by means of this characterization technique may be very high, provided that the BN produced by the instrumentation which is used to build the measurement chain is minimized. Unfortunately, especially in the case in which one is interested in the very low frequency range of the spectrum ($f$<1 Hz), commercially available instrumentation does not allow to reach sufficient sensitivity for many applications. Therefore one must often resort to purposely designed instrumentation in order to reduce the BN to an acceptable level. In this paper we have reviewed the work done by our research groups in this field over the past few years. In particular we have discussed the design and the performances of very low noise voltage and transresistance amplifiers, of very low noise programmable voltage and current sources, of a high stability thermal chamber, of a dedicated probe systems for LFNM on wafer level devices and of a dedicated multichannel acquisition system. We have also discussed data elaboration methods for increasing LFNM sensitivity. While in most cases we have obtained excellent results as far as the BN at very low frequencies is concerned, there are still several important issues that need to be addressed, and among them the design of compact size, very low noise, programmable voltage and current sources is one of the most challenging ones.

## References

[1] F. N. Hooge and A. M. Hoppenbrouwers, *1/f noise in continuous thin gold films*, *Physica* **45** (1969) 386–392.

[2] J. L. Vossen, *Screening of metal film defects by current noise measurements*, *Appl. Phys. Lett* **23** (1973) 287–289.

[3] G. Bertotti, M. Celasco, F. Fiorillo and P. Mazzetti, *Thermal equilibrium properties of vacancies in metal through current noise measurements*, *J. Appl. Phys.* **50** (1979) 6948.

[4] C. Ciofi and B. Neri, *Low-frequency noise measurements as a characterization tool for degradation phenomena in solid-state devices*, *Journal of Physics D* **33** (2000) R199–R216.

[5] L. Baracchino, G. Basso, C. Ciofi and B. Neri, *Ultralow-noise, programmable, voltage source*, *IEEE Trans. Instr. Meas.* **46** (1997) 1256–1261.

[6] C. Pace, C. Ciofi and F. Crupi, *Very low noise, high accuracy, programmable voltage reference*, *IEEE Trans. Instr. Meas.* **52** (2003) 1251–1254.

[7] C. Ciofi, M. De Marinis and B. Neri, *Ultralow-noise PC-based measurement system for the characterization of the metallizations of integrated circuits*, *IEEE Trans. Instr. Meas.* **46** (1997) 798–793.

[8] C. Ciofi, F. Crupi, C. Pace and G. Scandurra, *Improved trade-off between noise and bandwidth in op-amp based transimpedance amplifier*, Proceedings of IMTC 2004, Como, Italy.

[9] C. Ciofi, R. Giannetti, V. Dattilo and B. Neri, *Ultra low-noise current sources*, *IEEE Trans. Instr. Meas.* **47** (1998) 78–81.

[10] C. Ciofi, M. De Marinis and B. Neri, *Wafer level measurement system for SARF characterization of metal lines*, *Microelectron. Reliab.* **36** (1996) 1851–1855.

[11] B. Neri, C. Ciofi and V. Dattilo, *Noise and fluctuations in submicrometric Al-Si interconnect lines*, *IEEE Trans. Electr. Dev.* **44** (1997) 1454–1459.

[12] V. Dattilo, B. Neri and C. Ciofi, *Low frequency noise evolution during lifetime tests of lines and vias subjected to electromigration*, *Microelectron. Reliab.* **40** (2000) 1323–1327.

[13] C. Ciofi, I. Ciofi, S. Di Pascoli and B. Neri, *Temperature controlled oven for low noise measurement systems*, *IEEE Trans. Instr. Meas.* **49** (1999) 546–549.

[14] A. M. Yassine, T. M. Chen and B. A. Beitman, *Characterization of probe contact noise for probes used in wafer-level testing*, *IEEE Elect. Dev. Lett.* **12** (1991) 200–203.

[15] C. Ciofi, F. Crupi, C. Pace and G. Scandurra, *Micro-prober for wafer-level low-noise measurements in MOS devices*, *IEEE Trans. Instr. Meas.* **52** (2003) 1533–1536.

[16] M. Macucci and B. Pellegrini, *Very sensitive measurement method of electron device current noise*, *IEEE Trans. Instrum. Meas.* **40** (1991) 7–12.

[17] M. Sampietro, L. Fasoli and G. Ferrari, *Spectrum analyzer with noise reduction by cross correlation technique on two channels*, *Rev. Sci. Instrum.* **70** (1999) 2520–2525.

[18] E. Rubiola and V. Giordano, *A correlation-based noise measurement scheme showing sensitivity below the thermal floor*, Proceedings of International Conference on Noise in Physical Systems and 1/f Fluctuations, Hong Kong (1999) 483–486.

[19] C. Ciofi, F. Crupi and C. Pace, *A new method for high sensitivity noise measurements*, *IEEE Trans. Instr. Meas.* **51** (2002) 656–659.

[20] A. M. Yassine and C. T. Chen, *Electromigration noise measurements using a novel AC/DC wafer-level noise measurement system*, *IEEE Trans. Elect. Dev.* **44** (1997) 180–184.

# On the Theory and Design of Cold Resistors

Jiaao Song* and Laszlo B. Kish†
*Department of Electrical & Computer Engineering*
*Texas A&M University*
*Mailstop 3128, College Station, TX 77841-3128, USA*
**jiaaosong@tamu.edu; jiaao.song1@gmail.com*
*†Laszlokish@tamu.edu*

Received 20 May 2020
Accepted 26 May 2020
Published 29 September 2020

Communicated by Boris M. Grafov

The concept of using special electrical circuit design to realize "cold resistors", i.e., an active resistor circuitry with lower effective noise temperature, was first introduced about 80 years ago. Later on, various kinds of artificial resistors were applied in different research areas, such as gravitational wave detection, photo-amplifiers and quartz oscillators. Their proofs of concepts were experimentally proved. Unfortunately, the complete theory was not found even though several attempts had been published, sometimes with errors. In this paper, we describe a correct and complete circuit theoretical model of a cold resistor system. The results are confirmed by computer simulations. A design tool for this circuit is also shown.

*Keywords*: Thermal noise; negative feedback; low-noise resistors; theory; design.

## 1. Introduction

The exploration of *cold resistors* (*CRs*) has a long history but no flawless theory. The goal of this paper is to present and demonstrate a correct and complete theory for the related circuit designs. In this section, we briefly survey the foundation and some of the history of these efforts.

### 1.1. *The Second Law of Thermodynamics*

The thermal noise voltage (Johnson noise) in two parallel resistors $R_h$ and $R_c$ with temperatures $T_h > T_c$ generates a power flow $P_{hc}$ (Fig. 1).

*Corresponding author.

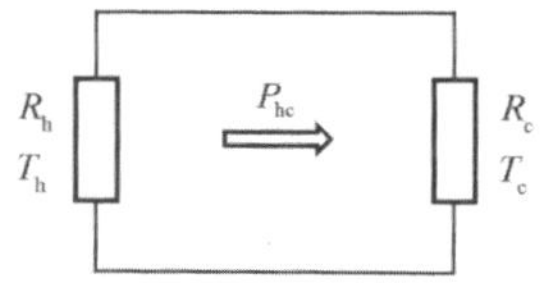


Fig. 1. $T_h$ and $T_c$ are the temperatures of resistors $R_h$ and $R_c$, respectively, where $T_h > T_c$. $P_{hc}$ is the thermal power flow between these two parallel resistors.

The power flow $P_{hc}$, from $R_h$ to $R_c$, in the frequency bandwidth $\Delta f$ can be expressed as [1]

$$P_{hc} = 4k(T_h - T_c)\frac{R_h R_c}{(R_h + R_c)^2}\Delta f, \tag{1}$$

where $k$ is the Boltzmann constant.

In the case of homogeneous temperature, $T_h = T_c$, the power flow is zero in accordance with the Second Law of Thermodynamics. Conversely, if we want to reduce the thermal noise of a resistor below the theoretical level, we need an active electronic device.

### 1.2. *Definition of cold resistors*

A *CR* is a device emulating a resistor which has a lower effective noise temperature $T_{\text{eff}}$ than the temperature of its environment. Here, $T_{\text{eff}}$ is defined by the Johnson–Nyquist noise formula [2]

$$S_u(f) = 4kT_{\text{eff}}R, \tag{2}$$

where $S_u(f)$ is the power density spectrum of the thermal noise voltage, $k$ is the Boltzmann constant and $R$ is the resistance of the resistor.

The CR system can also be seen as a black box, see Fig. 2. The ambient temperature is $T$. At proper conditions, the CR behaves as a resistor with effective resistance $R_{\text{eff}}$ and effective noise temperature $T_{\text{eff}}$, where $T_{\text{eff}} < T$.

### 1.3. *Approaches for the realization of cold resistors*

Cold resistors have a long history beginning in the 1940s [3–11]. Yet, this research is quite isolated and relatively unknown in the field of electronic noise. This is the

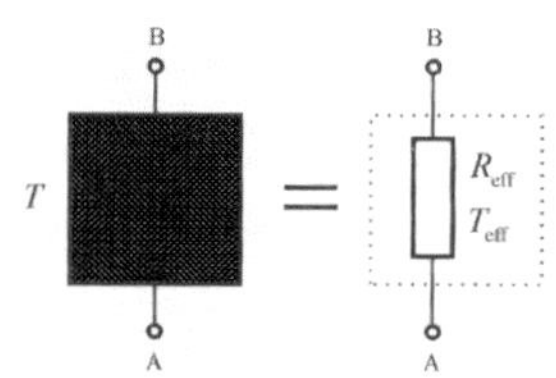


Fig. 2. The CR system as a black box. $T$ is the ambient temperature. $R_{\text{eff}}$ is the effective (dynamical) resistance and $T_{\text{eff}}$ is the effective noise temperature of the CR, where $T_{\text{eff}} < T$.

probable reason that the possibility of producing an active device with Ohmic behavior and reduced noise temperature, i.e., a cold resistor, is not mentioned in the textbooks, and this fact was a surprise to most noise researchers even at the *2018 International Conference on Unsolved Problems of Noise* [12].

### 1.3.1. *The cold input resistance approach for CR design*

One type of CRs [3–6] are based on the effective resistance of the input (virtual ground) during a specific type of negative feedback, see Fig. 3. The dynamical resistance between the inverting input $B$ and the ground and the related noise temperature are reduced. The amplifier can be as simple as a transistor.

The effective noise temperature $T_{\mathrm{eff}}$ is given as

$$T_{\mathrm{eff}} = \frac{T}{1 + |A_1|}, \tag{3}$$

where $A_1$ is the closed-loop gain of the amplifier.

The equivalent circuit, see Fig. 3, is a grounded resistor with reduced resistance $R_{\mathrm{eff}}$,

$$R_{\mathrm{eff}} = \frac{R_0}{1 + |A_1|}, \tag{4}$$

where $R_0$ is the resistance of the feedback resistor.

### 1.3.2. *The cold output resistance approach for CR design*

Another type of CRs [7, 8] are designed by the cold output resistance approach. This approach is based on the fact that the output resistance of an amplifier and a spurious voltage at its output are reduced by different scaling during the negative feedback.

The proposed circuit realization first increases the output resistance of an amplifier of gain $A_1$ by a serial resistor placed at its output. Then the negative feedback is established by a second amplifier of gain $A_2$, see Fig. 4.

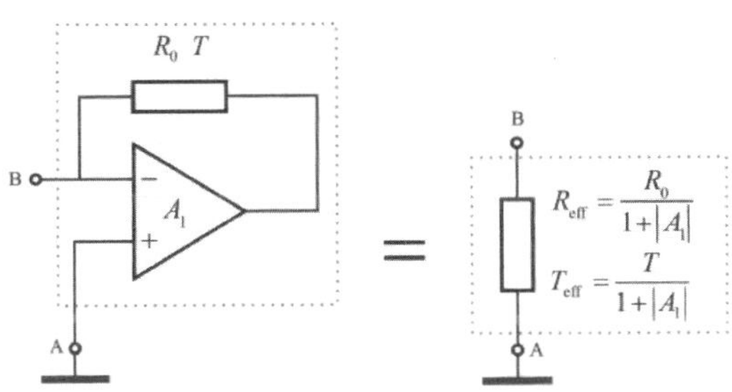


Fig. 3. CR system with cold input resistance approach. In the system, $A_1$ is the closed-loop gain of the amplifier and $R_0$ is the resistance of the feedback resistor.

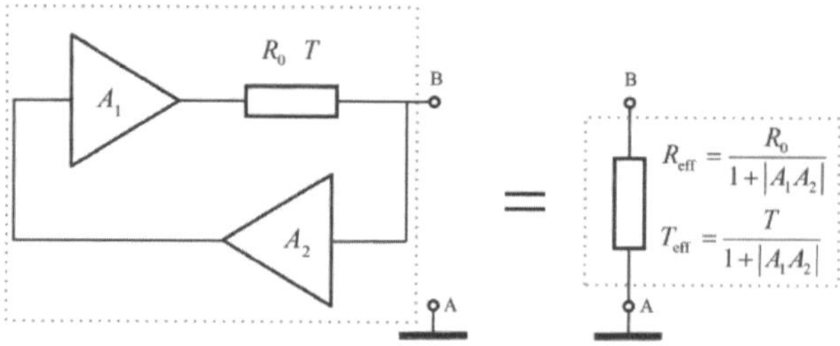


Fig. 4. CR system with cold output resistance approach. In the system, $A_1$ and $A_2$ are the closed-loop gains of the first amplifier and second amplifier, respectively. $R_0$ is the resistance of the resistor at the output.

The effective resistance $R_{\text{eff}}$ of the CR is [8]

$$R_{\text{eff}} = \frac{R_0}{1 + |A_1 A_2|}, \tag{5}$$

where $R_0$ is the resistance of the resistor at the output, and $A_1$ and $A_2$ are the closed-loop gains of the amplifiers.

The effective noise temperature $T_{\text{eff}}$ is [8]

$$T_{\text{eff}} = \frac{T}{1 + |A_1 A_2|}. \tag{6}$$

The cold resistor appears between the ground and the output of the first amplifier — resistor system.

### 1.3.3. *The AC-coupled diode approach for CR design*

A recent patent [9] by Linear Technology Corporation proposes a completely different method to create a CR. In the circuit, a diode is forward-biased by a low-noise DC current source, and it is connected to the external circuit by a coupling capacitor, see Fig. 5. The circuit performs as a non-linear dynamical resistor with reduced noise temperature.

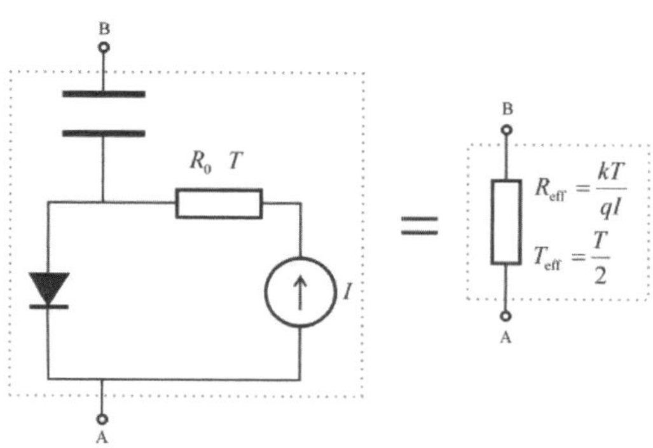


Fig. 5. Forward-biased, AC-coupled diode as a CR [9]. $I$ is the current of the DC source and $R_0$ is the resistance of the resistor in the circuit.

If the AC voltage between point $A$ and point $B$ is much less than $kT/q$, where $q$ is the elementary charge, the dynamical resistance is fairly linear and the effective noise temperature $T_{\text{eff}}$ is approximately half of the environment temperature $T$, provided the diode's current–voltage characteristic is near to ideal.

Specifically [9], the effective resistance of the equivalent resistor $R_{\text{eff}}$ is

$$R_{\text{eff}} = \frac{kT}{qI}, \tag{7}$$

where $k$ is the Boltzmann constant, $I$ is the DC bias current and the effective noise temperature $T_{\text{eff}}$ is

$$T_{\text{eff}} = \frac{T}{2}. \tag{8}$$

### 1.4. *Former CR theories and claims*

In this subsection, we briefly survey some of the most important former theoretical claims about CRs via negative feedback.

#### 1.4.1. *The CR design proposed for transductors*

CR was proposed for transductors in gravitational wave detection [4]. The generic scheme is the same as the one outlined in Sec. 1.3.1. The CR was created by connecting a resistor $R_0$ between the output and the inverting input of a low-noise operational amplifier, see Fig. 6, where $I_n$ and $U_n$ are the input current and voltage noise sources of the amplifier, respectively, and $U_0$ is the thermal noise source of the resistor.

The CR's effective resistance is given as

$$R_{\text{eff}} = \frac{R_0}{1 + A_{\text{opt}}}, \tag{9}$$

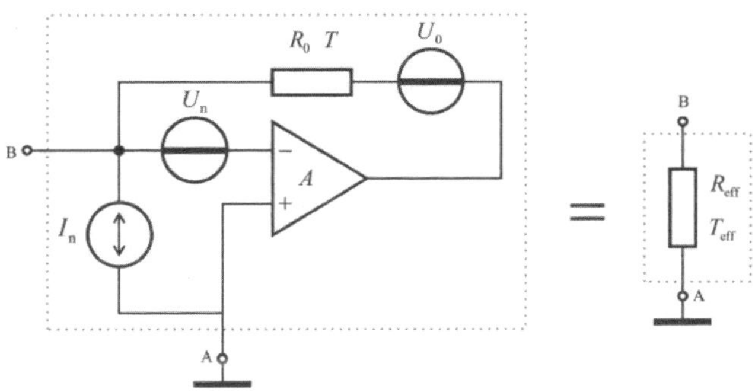


Fig. 6. Noise analysis circuit of the CR for gravitational wave detection. $I_n$ and $U_n$ are the input current and voltage noise sources of the amplifier, respectively, and $U_0$ is the thermal noise source of the resistor $R_0$ with temperature $T$.

where $A_{\text{opt}}$ is the so-called optimum voltage gain of the operational amplifier [4],

$$A_{\text{opt}} = \sqrt{\frac{4kTR_0 + R_0^2 I_n}{U_n}}. \quad (10)$$

The resulting minimum effective noise temperature between point $A$ and point $B$ is [4]

$$T_{\text{eff,min}} = \frac{\sqrt{(4kTR_0 + R_0^2 I_n)U_n}}{2kR_0}. \quad (11)$$

This theory [4] reaches the minimum effective noise temperature at the so-called optimum voltage gain. The derivation is straightforward and correct.

However, the theory is incomplete. Particularly, the variable $R_0$ contributes not only to the value of the effective resistance in Eq. (9) but also to the minimum effective noise temperature in Eq. (11). The following are some open questions that we will answer in this paper:

- Will there also be an optimum effective resistance?
- What is the lowest effective temperature for a given effective resistance?
- Is there an absolute lowest effective temperature that can be reached?

### 1.4.2. *The CR design proposed for quartz oscillators*

Huang [5] proposed a CR approach to achieve low-temperature mode in mechanical resonators. His theoretical and experimental studies indicated that a CR scheme can be used to cool the mechanical mode of resonators as effectively as by liquid nitrogen.

In his system, the negative feedback between the output and inverting input takes place via a PID $RC$ filter for a frequency-compensation purpose, see Fig. 7. $U_t$ is the total thermal noise source of the filter.

The effective impedance between point $A$ and point $B$ is [5]

$$Z_{\text{eff}}(f) = \frac{R_1R_2 + R_2R_3 + R_1R_3}{R_2 + [1 + A(f)]R_3}, \quad (12)$$

where $R_1$, $R_2$ and $R_3$ are the resistances of the resistors in the circuit, respectively, and $A(f)$ is the closed-loop gain of the operational amplifier.

The effective temperature of $Z_{\text{eff}}(f)$ is [5]

$$T_{\text{eff}} = \frac{4kT\Re[Z_f(f)] + U_n|A(f)|^2 + I_n|Z_f(f)|^2}{4k\Re\{Z_f(f)[1 + A(f)]\}}, \quad (13)$$

where $Z_f(f)$ is the generalized feedback impedance of the CR system,

$$Z_f(f) = [1 + A(f)]Z_{\text{eff}}(f) = \frac{[1 + A(f)](R_1R_2 + R_2R_3 + R_1R_3)}{R_2 + [1 + A(f)]R_3}. \quad (14)$$

Note that there is a fundamental error (perhaps typo) in the above result [5] — missing $T$ in the numerator in the equation giving the effective temperature. In Eq. (13), we already made this necessary correction.

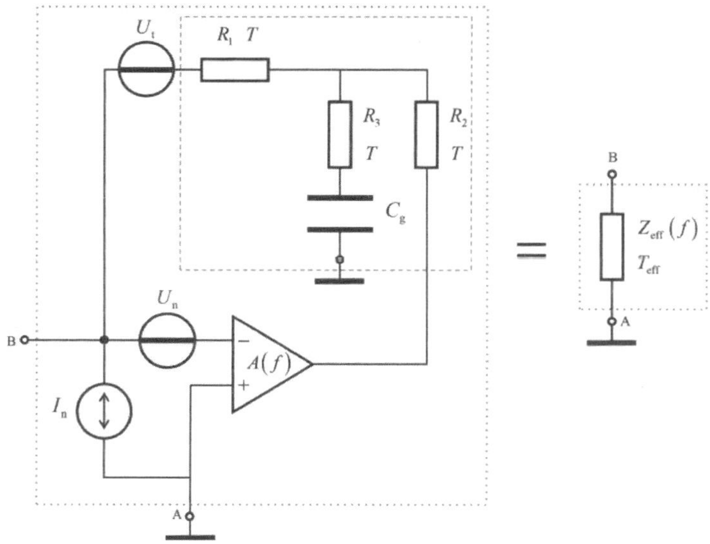


Fig. 7. Noise analysis circuit of the CR applied to quartz oscillators for cooling the loss component of the mechanical resonance. $R_1$, $R_2$, $R_3$ and $C_g$ form a PID filter. $U_t$ is the total thermal noise source of this filter. $I_n$ and $U_n$ are the input current and voltage noise sources of the amplifier, respectively, and $Z_{\text{eff}}\ (f)$ is the effective (cold) impedance between point $A$ and point $B$.

Finally, similarly to the incompleteness of the theory, which we mentioned in Sec. 1.4.1, this CR theory [5] also misses specifying the lowest possible effective temperature versus the effective resistance, for a given device.

In the next section, we deduce the correct and complete theory of CR systems with negative feedback.

## 2. Cold Resistor System Analysis

To deduce the correct theory of cold resistors, our formerly published [7, 8] system will be used for the analysis. It is important to note that the difference between the output resistance-based and input resistance-based schemes is only a question of viewing the role of the resistor $R_0$ because point $B$ in Fig. 4 can be viewed not only as an output (of $A_1$) but also as an input (of $A_2$).

### 2.1. *Circuit analysis of the CR system*

The practical CR circuitry is shown in Fig. 8. A passive resistor with resistance $R_0$ is connected to the output of a linear amplifier $LA_1$ with effective amplification $A_1 > 0$. The "new" output is point $B$. Another two linear amplifiers, $LA_2$ (in follower arrangement) and $LA_3$ with total effective amplification $A_2 < 0$ provide the negative feedback for $LA_1$ [7, 8]. This arrangement serves a high input impedance for $LA_2$ at point $B$. Thus the current load at point $B$ by $LA_2$ is negligible.

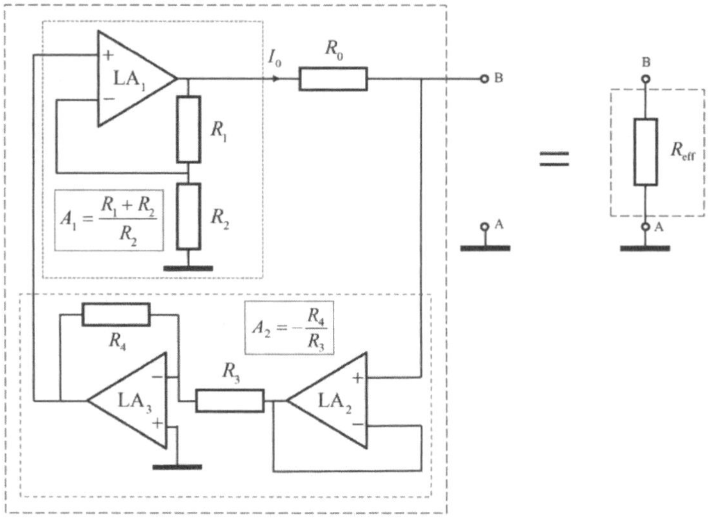


Fig. 8. In the CR system, $LA_1$, $LA_2$ and $LA_3$ are the linear operational amplifiers. $A_1$ is the effective amplification of $LA_1$. $A_2$ is the total effective amplification of $LA_2$ and $LA_3$. Resistors $R_1$, $R_2$, $R_3$ and $R_4$ are chosen so small that their thermal noise contributions are negligible. $R_0$ is the resistance that is getting transformed to become a dynamical resistor with reduced resistance and temperature. $I_0$ is the current through $R_0$.

In the high open-loop amplification limit,

$$A_1 = \frac{R_1 + R_2}{R_2}, \tag{15}$$

$$A_2 = -\frac{R_4}{R_3}. \tag{16}$$

The (absolute) loop amplification for this system is

$$A_L = |A_1 A_2| = -A_1 A_2 = \frac{(R_1 + R_2)R_4}{R_2 R_3}. \tag{17}$$

If the external measurement current $I_0$ is injected into point $B$, the voltage $U_B$ there satisfies

$$U_B = A_1 A_2 U_B + I_0 R_0. \tag{18}$$

Thus

$$U_B = \frac{I_0 R_0}{1 - A_1 A_2}. \tag{19}$$

Therefore, the effective (output) resistance of the CR is reduced by the loop amplification as

$$R_{\text{eff}} \equiv \frac{U_B}{I_0} = \frac{R_0}{1 + |A_1 A_2|}. \tag{20}$$

### 2.2. *Noise analysis of the CR system*

For the noise analysis, see Fig. 9, the thermal noise voltage source $U_0(t)$ of the resistor $R_0$, the voltage noise source $U_{n1}(t)$ and current noise source $I_{n1}(t)$ of $\mathrm{LA}_1$, the noise voltage source $U_{n2}(t)$ and current noise source $I_{n2}(t)$ of $\mathrm{LA}_2$ and the voltage noise source $U_{n3}(t)$ and current noise source $I_{n3}(t)$ of $\mathrm{LA}_3$ all contribute to the total noise. We assume that since the other resistance values are chosen so small their thermal noise contributions are negligible. So we explore only their current-induced voltage noise contributions for the proper choice of the amplifiers, see Eq. (21).

$R_0$ is chosen much larger than the other resistors thus only its thermal noise is the observable Johnson noise contribution $U_0(t)$. The amplifier noise voltages and currents result in

$$A_1\{A_2[U_B(t) + U_{n2}(t) + U_{n3}(t) + I_{n3}(t)R_4] + U_{n1}(t) - I_{n1}(t)R_1\} \\ + I_{n2}(t)R_0 + U_0(t) = U_B(t), \tag{21}$$

where $U_B(t)$ is the instantaneous noise voltage between point $A$ and point $B$.

At this point, we select the values of $R_1$, $R_2$, $R_3$ and $R_4$ and input current noise of the amplifiers so that the current-noise-induced voltage noise terms are negligible in Eq. (21). Thus, we get a system with practical relevance:

$$A_1\{A_2[U_B(t) + U_{n2}(t) + U_{n3}(t)] + U_{n1}(t)\} + I_{n2}(t)R_0 + U_0(t) = U_B(t). \tag{22}$$

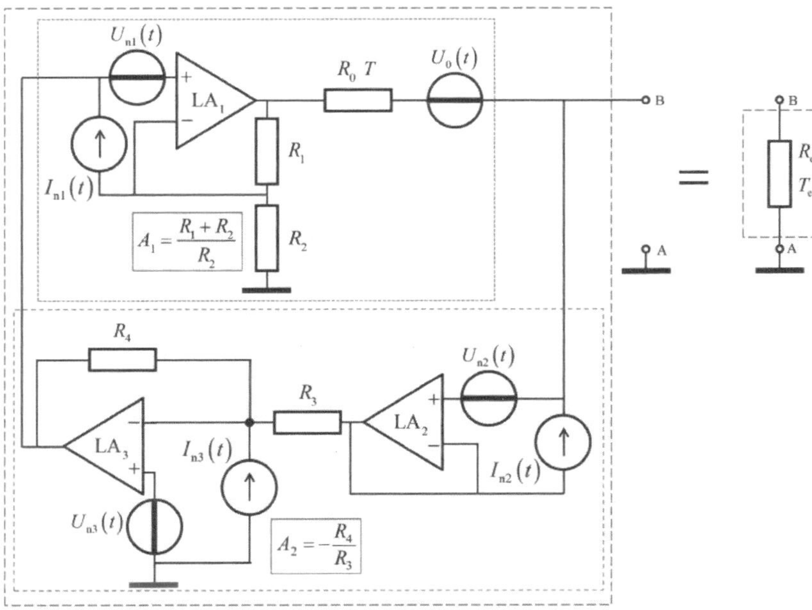


Fig. 9. Noise analysis of the CR system. $U_0(t)$ is the thermal noise voltage source of the resistor $R_0$. $U_{n1}(t)$, $U_{n2}(t)$ and $U_{n3}(t)$ are the voltage noise sources of $\mathrm{LA}_1$, $\mathrm{LA}_2$ and $\mathrm{LA}_3$, respectively. $I_{n1}(t)$, $I_{n2}(t)$ and $I_{n3}(t)$ are the current noise sources of $\mathrm{LA}_1$, $\mathrm{LA}_2$ and $\mathrm{LA}_3$, respectively.

From Eq. (22), we get the noise voltage at the output:

$$U_B(t) = \frac{A_1 A_2}{1+A_L}[U_{n2}(t) + U_{n3}(t)] + \frac{A_1}{1+A_L}U_{n1}(t) + \frac{1}{1+A_L}R_0 I_{n2}(t) + \frac{1}{1+A_L}U_0(t). \tag{23}$$

The total noise in Eq. (23) originates from three fundamental components:

(i) The voltage noise sources of the linear amplifiers (first and second terms).
(ii) The current noise source of the linear amplifier $LA_2$.
(iii) The thermal noise source of the resistor $R_0$.

These noise sources are independent, thus the noise spectra of these terms are additive. The spectrum of the resultant noise voltage at point $B$ is

$$\begin{aligned} S_{\text{eff}}(f) &= \frac{A_L^2}{(1+A_L)^2}[S_{u2}(f) + S_{u3}(f)] + \frac{A_1^2}{(1+A_L)^2}S_{u1}(f) \\ &\quad + \frac{1}{(1+A_L)^2}R_0^2 S_{i2}(f) + \frac{1}{(1+A_L)^2}S_0(f), \end{aligned} \tag{24}$$

where $S_{i2}(f)$ is the noise current spectrum of $LA_2$, $S_{u1}(f)$, $S_{u2}(f)$ and $S_{u3}(f)$ are the input noise voltage spectra of the amplifiers and $S_0(f)$ is the thermal noise spectrum of the resistor $R_0$.

The effective thermal noise spectrum of CR is

$$S_{\text{eff}}(f) = 4kT_{\text{eff}}R_{\text{eff}}, \tag{25}$$

where $R_{\text{eff}}$ and $S_{\text{eff}}(f)$ are given by Eqs. (20) and (24), respectively, and $T_{\text{eff}}$ is the effective noise temperature of the CR. From Eqs. (20), (24) and (25), we obtain

$$T_{\text{eff}} = \frac{A_L^2[S_{u2}(f) + S_{u3}(f)]}{(1+A_L)4kR_0} + \frac{A_1^2 S_{u1}(f)}{(1+A_L)4kR_0} + \frac{R_0 S_{i2}(f)}{(1+A_L)4k} + \frac{T}{1+A_L}, \tag{26}$$

where $T$ is the temperature of the resistor $R_0$.

In the large loop amplification limit, $A_L \gg 1$, the expression of $T_{\text{eff}}$ can be simplified as

$$\begin{aligned} T_{\text{eff}} &\approx \frac{A_L[S_{u2}(f) + S_{u3}(f)]}{4kR_0} + \frac{R_0 S_{i2}(f)}{4kA_L} + \frac{T}{A_L} \\ &= \frac{A_L[S_{u2}(f) + S_{u3}(f)]}{4kR_0} + \frac{R_0[S_{i2}(f) + S_i(f)]}{4kA_L}, \end{aligned} \tag{27}$$

where $S_i(f)$ is the thermal noise current spectrum of the resistor $R_0$. From Eq. (27),

$$\frac{dT_{\text{eff}}}{dR_0} = \frac{-A_L[S_{u2}(f) + S_{u3}(f)]}{4kR_0^2} + \frac{S_{i2}(f) + S_i(f)}{4kA_L} = 0. \tag{28}$$

By solving Eq. (28), we can obtain the optimal resistance that makes $T_{\text{eff}}$ minimal:

$$R_{0,\text{opt}} = A_L \sqrt{\frac{S_{u2}(f) + S_{u3}(f)}{S_{i2}(f) + S_i(f)}}. \tag{29}$$

Then the minimal effective noise temperature $T_{\text{eff,min}}$ can be calculated from $R_{0,\text{opt}}$,

$$T_{\text{eff,min}} = \frac{\sqrt{[S_{i2}(f) + S_i(f)][S_{u2}(f) + S_{u3}(f)]}}{2k}. \tag{30}$$

The effective resistance of CR is

$$R_{\text{eff,opt}} = \sqrt{\frac{S_{u2}(f) + S_{u3}(f)}{S_{i2}(f) + S_i(f)}}. \tag{31}$$

Note that the same results can be obtained by optimizing versus $A_L$. From Eq. (27), we get

$$\frac{dT_{\text{eff}}}{dA_L} = \frac{S_{u2}(f) + S_{u3}(f)}{4kR_0} - \frac{R_0[S_{i2}(f) + S_i(f)]}{4kA_L^2} = 0. \tag{32}$$

From Eq. (32), the optimal loop amplification $A_{L,\text{opt}}$ minimizing $T_{\text{eff}}$ is

$$A_{L,\text{opt}} = \sqrt{\frac{[S_{i2}(f) + S_i(f)]R_0^2}{S_{u2}(f) + S_{u3}(f)}}. \tag{33}$$

When the loop amplifier gain is set to $A_{L,\text{opt}}$, the minimal effective noise temperature is

$$T_{\text{eff,min}} = \frac{\sqrt{[S_{i2}(f) + S_i(f)][S_{u2}(f) + S_{u3}(f)]}}{2k}, \tag{34}$$

which is the same result as in Eq. (30) and the optimal effective resistance of the CR system is

$$R_{\text{eff,opt}} = \sqrt{\frac{S_{u2}(f) + S_{u3}(f)}{S_{i2}(f) + S_i(f)}}, \tag{35}$$

which is the same result as in Eq. (31).

From Eqs. (30) and (31), or Eqs. (34) and (35), we can deduce a practically useful design rule between $T_{\text{eff,min}}$ and $R_{\text{eff,opt}}$:

$$T_{\text{eff,min}} = \frac{S_{u2}(f) + S_{u3}(f)}{2kR_{\text{eff,opt}}}. \tag{36}$$

With Eq. (36), the designer can determine the lowest possible effective temperature for the aimed effective temperature, in the high loop amplification limit with the given amplifiers, provided $A_L$ is $\gg 1$.

## 3. Test Results of the Cold Resistor System

For the CR circuitry in Fig. 8, we chose $R_1 = 1\,\text{k}\Omega$, $R_2 = 34\,\Omega$, $R_3 = 100\,\Omega$ and $R_4 = 1\,\text{k}\Omega$. $LA_1$, $LA_2$ and $LA_3$ are ADA4625-1 operational amplifiers from Analog Devices Corporation. $R_0$ was varied. The simulations were carried out by the LTspice XVII industrial circuit tester/simulator system.

In Fig. 10, the blue solid line shows the effective resistance $R_{\text{eff}}$ versus $R_0$ at the ideal loop amplification by Eq. (33). Red dots, green, orange and purple dashed lines are the SPICE simulation results under 1, 10, 20 and 30 kHz, respectively. The practical results are limited by non-ideal features such as parasitic capacitances, phase shift, frequency characteristics, etc.

We also verified Eq. (36), the relationship between CR's lowest effective noise temperature and its effective resistance. For the circuitry in Fig. 9, we chose $R_1 = 1\,\text{k}\Omega$, $R_3 = 100\,\Omega$ and $R_4 = 1\,\text{k}\Omega$. Both $R_0$ and $R_2$ were varied to find the optimum. Figure 11 shows the practical relationship between the lowest effective noise temperature $T_{\text{eff,min}}$ and its effective noise temperature $R_{\text{eff,opt}}$ for different frequency domains.

The results in Fig. 11 show that at 1 kHz, the simulation results match the theory well. There is some derivation at 100 kHz for $R_{\text{eff,opt}} > 100\,\text{k}\Omega$ which can be predicted from the conclusion of Fig. 10. In the much lower frequency domain, the optimal temperature increases because of the $1/f$ noise which we could not take into the account in our equations due to the lack of information.

For the higher frequency domain (1 MHz), we see that the valid range of the theory versus $R_{\text{eff,opt}}$ is sharply limited by non-ideal features in the circuitry — large phase shift and reduced open-loop amplification of the operational amplifiers.

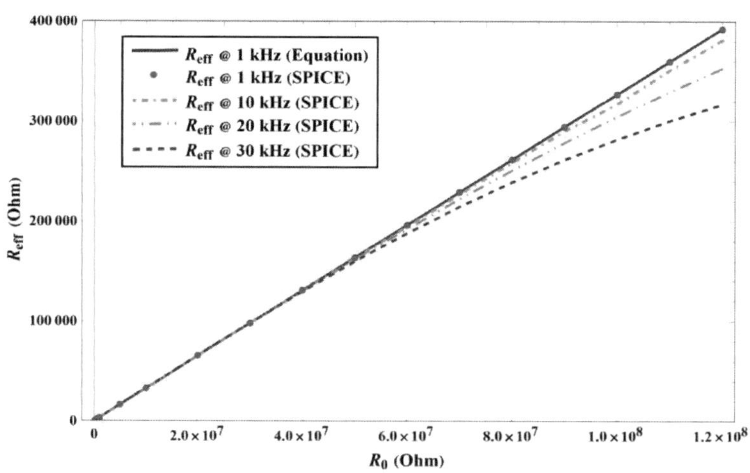


Fig. 10. The blue solid line shows the effective resistance $R_{\text{eff}}$ versus $R_0$ at the ideal loop amplification by Eq. (33). Red dots, green, orange and purple dashed lines are the SPICE simulation results under 1, 10, 20 and 30 kHz, respectively.

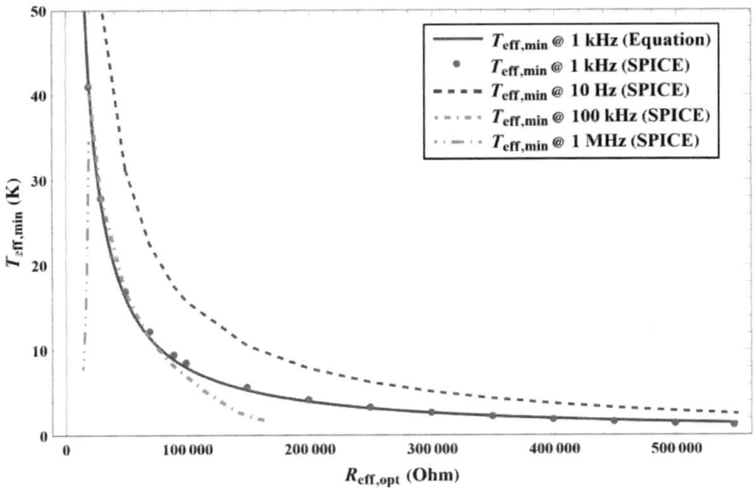


Fig. 11. The blue solid curve shows the theoretical relationship [Eq. (36)] between the lowest effective noise temperature $T_{\text{eff,min}}$ and the effective resistance $R_{\text{eff,opt}}$. The red dots and the purple, green and orange dashed curves represent the SPICE simulation results under 1 kHz, 10 Hz, 100 kHz and 1 MHz, respectively.

## 4. Conclusions

We gave the correct/complete circuit theoretical model of CR systems. The results are confirmed by computer simulations. We also deduced a practical "design tool" that shows the lowest effective temperature versus the effective resistance. With our test amplifiers, the lowest effective noise temperature that we could reach was 1.16 K.

## References


[1] L. B. Kish and T. Horvath, Notes on recent approaches concerning the Kirchhoff-law–Johnson-noise-based secure key exchange, *Phys. Lett. A* **373** (2009) 2858–2868.

[2] C. D. Motchenbacher and J. A. Connelly, *Low-Noise Electronic System Design* (Wiley-Interscience, New York, 1993).

[3] M. J. O. Strutt and A. van der Ziel, Suppression of spontaneous fluctuations in amplifiers and receivers for electrical communication and for measuring devices, *Physica* **9** (1942) 513–527.

[4] K. Oide, Y. Ogawa and H. Hirakawa, Artificial cold resistors, *Jpn. J. Appl. Phys.* **17** (1978) 429–432.

[5] M. Huang, Cooling a mechanical resonator with an artificial cold resistor, Master's thesis, Leiden University (2015).

[6] R. L. Forward, Reducing noise in photoamplifier circuits with low noise artificial resistors, *J. Appl. Phys.* **53** (1982) 3365–3371.

[7] J. Song and L. B. Kish, Linear resistor plugin with ultra-low noise temperature without cooling, Provisional Patent Application, Texas A&M University, No. 62/720, 786, August (2018).

[8] J. Song and L. B. Kish, Does a standalone, "cold" (low-thermal-noise), linear resistor exist without cooling?, *Fluct. Noise Lett.* **17** (2018) 1850030.

[9] M. T. Engelhardt, Active differential resistors with reduced noise, US Patent, Application No. US9866245B2, Linear Technology LLC, January 9 (2018).

[10] R. L. Forward and G. D. Thurmond, Network for simulating low-noise temperature resistors, US Patent, Application No. 4156859, Hughes Aircraft Company, May 29 (1979).

[11] R. L. Forward and R. W. Peterson, Cold damping of mechanical structures, US Patent, Application No. 4199989, Hughes Aircraft Company, April 29 (1980).

[12] L. B. Kish, Standalone cold virtual resistor? Can we cool by a linear amplifier without non-linearity, switches, or phase transition?, in *Proc. 8th Int. Conf. Unsolved Problems on Noise (UPoN2018)* (2018).

# ON THE STATISTICAL ANALYSIS OF NOISE IN CHEMICAL SENSORS AND ITS APPLICATION FOR SENSING

JANUSZ SMULKO

*Faculty of Electronics, Telecommunication and Informatics, Technical University of Gdansk, 80-952 Gdansk, ul. G. Narutowicza 11/12, Poland*

CLAES-GÖRAN GRANQVIST

*Department of Materials Science, The Ångström Laboratory, Uppsala University, P.O. Box 534, SE-75121 Uppsala, Sweden*

LASZLO B. KISH

*Department of Electronic Engineering, Texas A&M University, College Station, TX 77843-3128, USA*

Received 16 July 2001
Revised 14 August 2001
Accepted 23 August 2001

Resistance noise data from a single gas sensor can be utilized to identify gas mixtures. We calculated the power spectral density, higher order probability densities and the bispectrum function of the recorded noise samples; these functions are sensitive to different natural vapors and can be employed to select a proper detection criterion for gas composites and odors.

*Keywords:* Chemical sensing; gas sensing; electronic nose; electronic tongue; noise analysis.

## 1. Introduction

Much research and development are presently carried out on electronic sensors for detecting gases and vapors [1–3]. The usual instrument, based on a sensor array, is not convenient for several practical reasons. However, it was recently shown that resistance noise data of a single sensor could be used for gas recognition [4,5]. Noise, i.e. a broadband signal, is able to provide more information than dc resistance changes of the sensor arrangement. The resistance noise has been analyzed by evaluating its power spectral density $S(f)$. The average slope of the $f{\cdot}S(f)$ product, fixed for different decades of frequency $f$, has been proposed as one of the possible constitutive parameters for gas recognition [5]. Other noise data features have not been investigated up to now, though. However, it is physically plausible to assume that the microscopic processes, which result in the measured noise, are nonlinear and non-Gaussian. Additional information on the origin of resistance noise may then be extractable by a thorough statistical analysis of the measured noise and its linearity. Such additional information can be very useful in building criteria for the recognition of different gas or vapor mixtures. In this paper, we explore some possible ways of such an analysis and elucidate their effectiveness.

## 2. Statistical Analysis of Gaussianity and Linearity of Noise

A stationary random signal can be characterized separately by its power spectral density and probability density. Usually, it is assumed that the noise is a Gaussian process, which implies that nonlinear cross-products do not exist in the spectrum for linear combinations of frequencies. Such an assumption implies that all statistical information about the process is contained in second order distribution functions, such as second two-time probability density, autocorrelation function, or power spectrum. However, some published data on noise, generated in semiconductor devices, display easily detectable deviations from Gaussian behavior [6]. The presence of non-Gaussian noise components was recognized in commercial diodes and bipolar transistors, in GaAs epitaxial films and hydrogenated amorphous silicon [6–8]. The parameters of the observed deviations from Gaussian distribution, in a case of random telegraph noise, can be even used for reliability prediction of semiconductor devices [9]. These facts indicate that, at certain cases involving small sensor size, non-Gaussianity can be a possible measure for gas sensing.

The distribution of the recorded noise data can be investigated by parameters such as skewness $\gamma_3$ or kurtosis $\gamma_4$, which are computed for a discrete–time random process $x(n)$ having standard deviation $\sigma_x$, given as:

$$\gamma_3 = \frac{1}{\sigma_x^3} E\left[(x(n) - E[x(n)])^3\right] \tag{1}$$

$$\gamma_4 = \frac{1}{\sigma_x^4} E\left[(x(n) - E[x(n)])^4\right] - 3 \tag{2}$$

where $E[\ ]$ denotes average. These measures are equal to zero for a Gaussian distribution. The estimates of skewness and kurtosis can be used for construction of a statistical test that indicates whether the investigated data have Gaussian (normal) distribution. For example, the Jarque-Bera procedure constitutes such a statistical test [10]. Under the null hypothesis, that assumes a normal distribution of $N$ random data samples, the Jarque-Bera test statistic $\chi_{JB}^2$ defined by

$$\chi_{JB}^2 = N\left[\frac{\gamma_3^2}{6} + \frac{\gamma_4^2}{24}\right] \tag{3}$$

has a Chi-square distribution with two degrees of freedom. The null hypothesis applies if $\chi_{JB}^2$ is smaller than or equal to the value $\chi_{2;\alpha}^2$ at the $\alpha$ level of significance (i.e. the value $\chi_{2;\alpha}^2$ is such that the probability $P[\chi_2^2 > \chi_{2;\alpha}^2]$ is equal to $\alpha$, where $\chi_2^2$ is a statistical Chi-square variable with two degrees of freedom). Otherwise, the alternative hypothesis, that the investigated data do not have a normal distribution, is assumed.

Hinich has developed algorithms to test non-skewness (Gaussianity) and linearity [11]. The underlying idea is that a bispectrum function is calculated, and it is then checked whether a mean square of $S_{3x}(f_1, f_2)$ is zero or not. The bispectrum function, being the function of two frequencies $f_l$ and $f_2$, is defined by

$$S_{3x}(f_1, f_2) = \sum_{k=-\infty}^{\infty} \sum_{l=-\infty}^{\infty} C_{3x}(k, l) e^{-j2\pi f_1 k} e^{-j2\pi f_2 l} \tag{4}$$

where $C_{3x}(k,l) = E[x(n)x(n+k)x(n+l)]$ is the third-order autocorrelation function of the process $x(n)$. The bispectrum function is equal to zero for processes with zero skewness, i.e. for Gaussian processes. This implies that Gaussian components in the recorded noise will be eliminated when the bispectrum function is used for analysis.

The linearity, investigated by Hinich's algorithm, can be evaluated provided that $S_{3x}(f_1, f_2)$ is nonzero. Then, it is possible to test whether the bicoherence function $P_{3x}(f_1, f_2)$ of the process $x(n)$ is constant or not. This function can be calculated from

$$|P_{3x}(f_1, f_2)|^2 = \frac{|S_{3x}(f_1, f_2)|^2}{S_x(f_1 + f_2) S_x(f_1) S_x(f_2)} \tag{5}$$

where $S_x(f)$ is the power spectral density of the process $x(n)$. If this function is constant for any $f_1$ and $f_2$, the process $x(n)$ is called linear, since there is no apparent indication of cross products at linear combination of frequencies $f_1$ and $f_2$.

Noise in a device can be generated by various mechanisms, and the linearity and intensity of the physical cause of the noise can affect the statistical properties of the recorded noise samples. These properties can be investigated by a special time function that represents an average behavior before and after a momentary fluctuation [6]. When $x(n)$ is a discrete-time signal, this time function is defined as

$$R(k, x_0) = E[x(n) \cdot x(n+k)]_{x(n)=x0} \tag{6}$$

for a given noise sample value $x_0$. For linear systems, the function $R(k, x_0)$ does not depend on the choice of $x_0$. For nonlinear systems, $R(k, x_0)$ depends on $x_0$, but no distinction can be made between different nonlinear mechanisms.

In this paper, as an illustration of the above considerations, we measure and compute some of the statistical quantities mentioned above for random signals using a limited number of samples.

### 3. Data from Noise Measurements

A commercial carbon monoxide gas sensor of type *RS 286-636*, manufactured by Nemoto & Co., Ltd., was applied in the prepared measurement setup. The sensor was heated by a battery and was inserted in a grounded steel box with a volume of 12 liter. Different sources of natural vapors were placed in the box a few centimeters away from the sensor, and the box was then closed. The noise was recorded for at least 10 minutes after closing the box, when the sensor resistance had stabilized and the noise had become stationary. The noise data were recorded using a commercial data acquisition board, controlled by a PC computer, at a sampling frequency of 4kHz. A measuring current of

$100\,\mu A$ was run through the sensor, and a nominal heating voltage of 1.2 V was applied at the heater electrodes. We recorded voltage fluctuations in the sensor film, which are proportional to its resistance fluctuations.

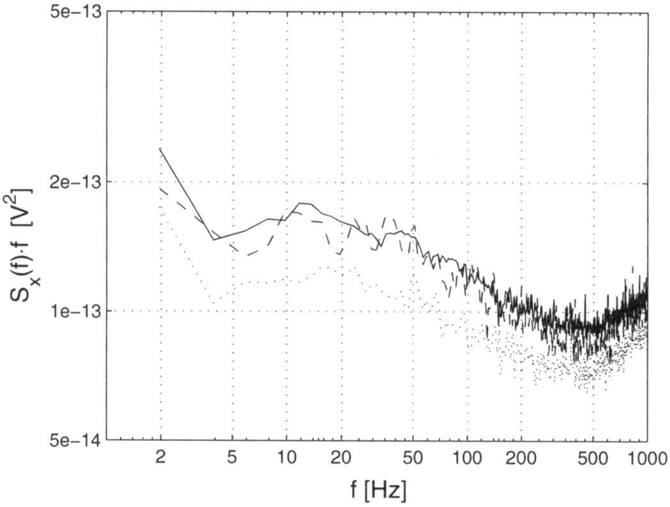


Fig. 1. Voltage noise spectrum multiplied by frequency, for the sensor exposed to different natural vapors. Solid line: laboratory air; rarely scattered line: instant coffee; tightly scattered line: cherry flavored tee.

Figure 1 shows noise spectra of three natural vapors. The presented data on $f{\cdot}S_x(f)$ were derived by averaging over 400 spectra, so the random error does not exceed 5%.

Figure 2 illustrates the relative probability of voltage noise occurrence on a semilog scale. There are visible but small differences between the measured amplitude distribution $p(x)$ of noise for different vapors.

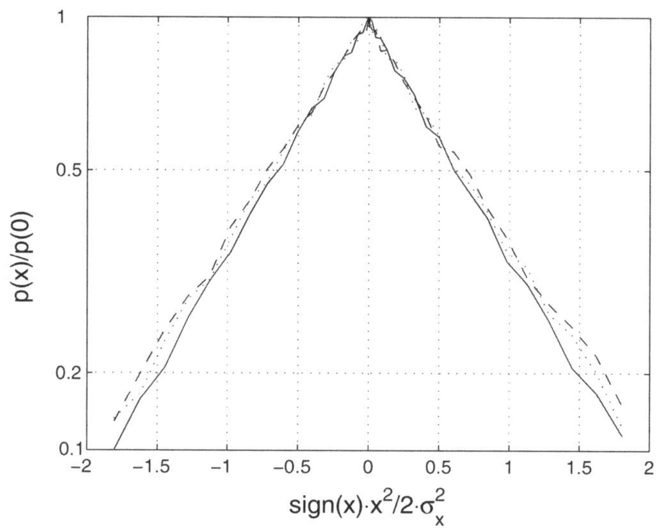


Fig. 2. The measured amplitude distribution $p(x)$ of voltage noise samples during exposure to different natural vapors. Solid line: laboratory air; rarely scattered line: instant coffee; tightly scattered line: cherry flavored tee.

The number of noise samples exceeded $10^6$ in order to assure that the random error of the estimated probability density remained less than 2% even in the worst case. The observed small differences between $p(x)/p(0)$ values, obtained for different flavors, are up to a three times greater than the mentioned random error of $p(x)/p(0)$ function but only for relatively high noise amplitude. The estimated parameters skewness $\gamma_3$ and kurtosis $\gamma_4$, that measure deviations from Gaussian distribution of the estimated $p(x)$ function for the whole range of noise amplitudes, confirmed the presence of the observed differences. The calculated values of $\gamma_3$ and $\gamma_4$ are gathered in Table 1.

Table 1. $\gamma_3$ and $\gamma_4$ parameters of voltage noise samples for the sensor exposed to different natural vapors.

| VAPOR | $\gamma_3$ | $\gamma_4$ |
|---|---|---|
| laboratory air | 0.071 | 0.48 |
| cherry flavored tea | 0.128 | 0.18 |
| instant coffee | 0.054 | –0.04 |

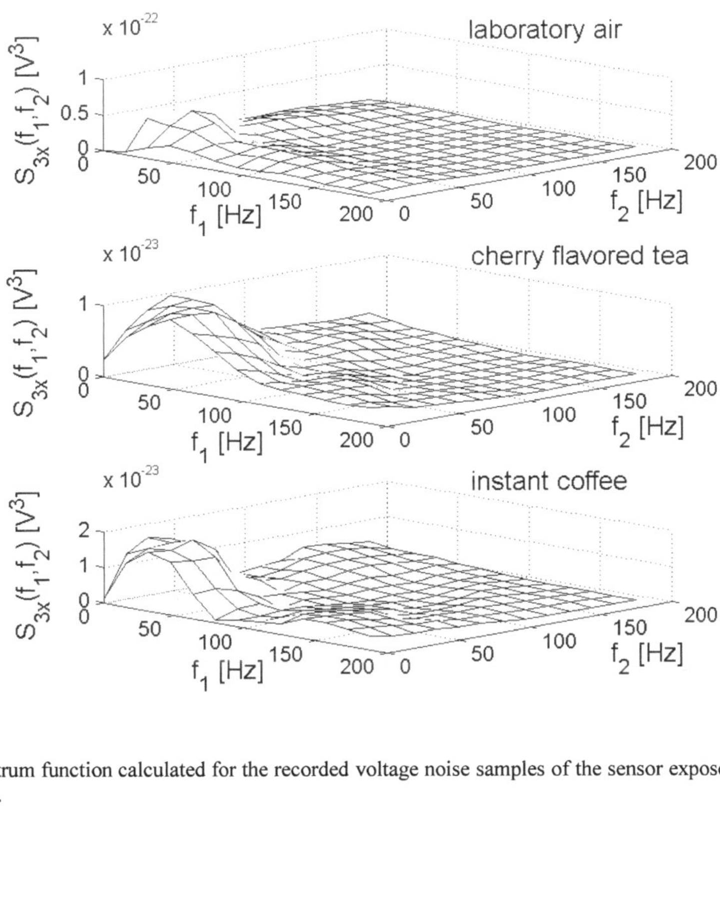


Fig. 3. Bispectrum function calculated for the recorded voltage noise samples of the sensor exposed to different natural vapors.

One can observe changes of $|\gamma_4|$ by as much as one order of magnitude for different vapors. The variances of the estimated $\gamma_3$ and $\gamma_4$ values are inversely proportional to the number $N$ of noise samples for large $N$ [12]. The standard deviation of the $\gamma_3$ and $\gamma_4$ parameters does not exceed 0.005 for the assumed N• $10^6$ samples, that is much smaller than the values given in Table 1.

The results of a Jarque-Bera test confirm the mentioned conclusion. The values of $\chi^2_{JB}$, computed for all the investigated vapors, enabled us to reject the null hypothesis, that assumed normal distribution, at the level of significance being a = 5%.

Figure 3 presents bispectral functions, which are nonzero but relatively small for a limited range of low frequencies. These small values of the bispectral function imply that the tests of Gaussianity and linearity using Hinich algorithms do not reject a Gaussian hypothesis, that is different than the results of using the Jarque-Bera test. Very different shapes of the bispectrum function are observed for the different vapors. It is evident that the small, but characteristic, values of the bispectrum functions for different vapors give useful information for identifying odors. The values of the $S_{3x}(f_1, f_2)$ function were derived with the same number of voltage noise samples as for the previously presented parameters, so the random error of the estimated function can be neglected.

The function $R(k,x_0)/R(0,x_0)$ was calculated for two values of $x_0$. Results, for the resistance noise of the sensor recorded in the presence of three natural vapors, are given in Fig. 4. Similar results were obtained for $k < 0$. The function $R(k,x_0)/R(0,x_0)$ depends on $x_0$ but no clear relationship between the form of $R(k,x_0)/R(0,x_0)$ and a specific vapor could be identified.

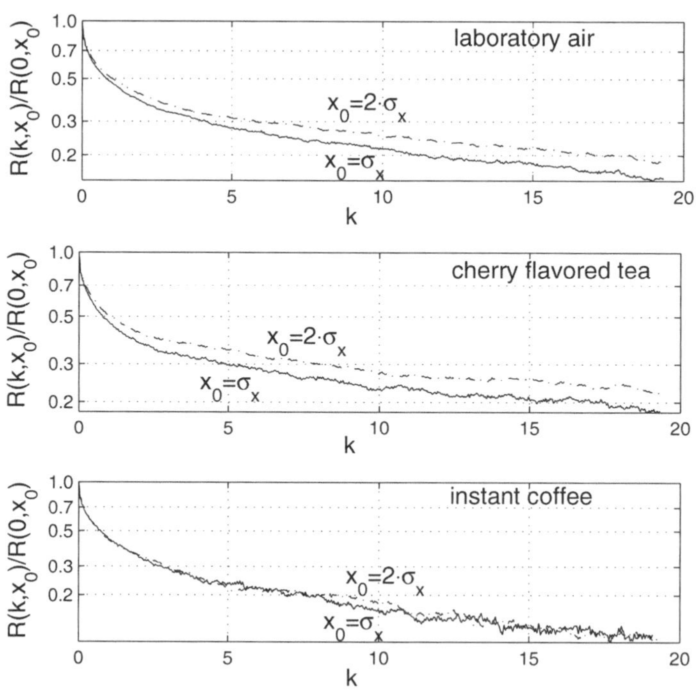


Fig. 4. Normalized $R(k,x_0)$ function calculated for two chosen values of $x_0$ for the noise samples of the sensor exposed to different natural vapors.

## 4. Summary

The results described in this paper confirm the presence of nonlinear components in the observed resistance noise of a carbon monoxide gas sensor. These components appear characteristic for different vapors, so a connection between a specific vapor and the nonlinearities of the registered noise can be assumed. Therefore, the additional information derived by a sophisticated statistical analysis can be useful in selecting a detection criterion for different vapors exposed to a single sensor electronic nose. The most valuable input for such a criterion appears to be the bispectrum function at small frequencies and the parameters such as skewness and kurtosis.

## References

[1] J. W. Gardner, W. S. Hyun and E. L. Hines, *An electronic nose system to diagnose illness, Sensors & Actuators B* **70** (2000) 19–24.

[2] D. Dumitrescu, B. Lazzerini, and F. Marcelloni, *A fuzzy hierarchical classification system for olfactory signals, Pattern Analysis & Applications* **3** (2000) 325–334.

[3] S. Capone, P. Siciliano, F. Quaranta, R. Rella, M. Epifani and L. Vasanelli, *Analysis of vapours and foods by means of an electronic nose based on a sol-gel metal oxide sensors array, Sensors & Actuators B* **69** (2000) 230–235.

[4] L. B. Kish, R. Vajtai and C. G. Granqvist, *Extracting information from the noise spectra of chemical sensors: Electronic nose and tongue by one sensor, Sensors & Actuators B* **71** (2000) 55–59.

[5] L. B. Kish, C. G. Granqvist and J. Söderlund, *Detection of chemicals based on resistance fluctuation-spectroscopy, Swedish patent*, No: 9803019-0, Publ. No.: 513148.

[6] R. F. Voss, *Linearity of 1/f noise mechanisms, Phys. Rev. Lett.* **40** (1978) 913–916.

[7] D. Quicker, P. W. West and J. Kakalios, *Long-range disorder and metastability in amorphous silicon, J. Non-Crystalline Solids* **266-269** (2000) 397–400.

[8] A. V. Yakimov and F. N. Hooge, *A simple test of the Gaussian character of noise, Physica B* **291** (2000) 97–104.

[9] R. Crook and B. K. Jones, *Noise and DC characteristics of power silicon diodes, Microelectronics and Reliability* **37** (1997) 1635–1638.

[10] A. Spanos, *Statistical foundations of econometric modeling*, Cambridge University Press, (1993).

[11] M.J. Hinich, *Testing for Gaussianity and linearity of a stationary time series, J. Time Series Analysis.* **3** (1982) 169–176.

[12] H. Cramer, *Mathematical methods of statistics*, Princeton University Press, Seventh edition, (1957).

[13] J. S. Bendat and A. G. Piersol, *Random data analysis and measurement procedures*, Wiley, (2000).

# METHODS OF ELECTROCHEMICAL NOISE ANALYSIS FOR INVESTIGATION OF CORROSION PROCESSES

JANUSZ SMULKO

*Faculty of Electronics, Telecommunications and Informatics, Technical University of Gdansk, 80-952 Gdansk, ul. G. Narutowicza 11/12, Poland*

*jsmulko@sunrise.pg.gda.pl*

Received 3 February 2006
Revised 16 February 2006
Accepted 20 February 2006
Communicated by Boris M. Grafov

Electrochemical corrosion processes can be investigated by observation of charge flows between the electrolyte and the corroding metal. Usually, the charge flows are observed as spontaneous current and voltage fluctuations (electrochemical noise) in a three-electrode setup. Different types of corrosion processes can be recognized by electrochemical noise analysis. Uniform corrosion rate can be evaluated by estimation of polarization resistance between the metal and electrolyte. Local corrosion events (breakdowns of the passive layer) that produce characteristic transients observed in noise can be detected as well. Different methods of electrochemical noise analysis are presented in a brief review. The limitations and advantages of the proposed methods for corrosion monitoring and research are underlined. The experimental results are also discussed.

*Keywords:* Noise; corrosion; polarization resistance; power spectrum; bispectrum; stationarity.

## 1. Introduction

A flow of charge between electrolyte and metal can be observed when an electrochemical corrosion process occurs. Investigation of these charge flows can be applied to monitor and recognize corrosion processes [1–3]. The three-electrode setup is usually applied for the research [4]. The metal electrodes can be treated as stochastic sensors that recognize the susceptibility of this metal to corrosion processes in the given chemical environment.

The three-electrode setup consists of two nominally identical working electrodes and one reference electrode. All the three electrodes are immersed into the investigated chemical environment. The current fluctuations are observed between the short circuited leads of the working electrodes. The voltage fluctuations are observed between the potential of the reference electrode and the potential of the working electrodes. In practice, the reference electrode is often prepared in the same way as the working electrodes.

The method of corrosion recognition by electrochemical noise measurement and analysis does not involve any exterior polarization of the electrodes. The observed current and voltage noise is generated by spontaneous charge flows during occurrence of the electrochemical corrosion reactions. It means that this method does not disturb interactions between metal and electrolyte by outside polarization that is contrary to other electrochemical methods applied for corrosion investigation (linear polarization resistance, electrochemical impedance spectroscopy) [5]. Moreover, electrochemical noise observation can be used to recognize the type of corrosion processes, which is more difficult by applying other electrochemical methods. The method is applied presently in industry even though it sometimes leads only to rough conclusions about corrosion intensity and it needs further improvement.

The main disadvantage of the presented method is a huge amount of data that have to be analyzed. Moreover, conditions at the metal surface can change significantly within the time of noise measurement due to changes of a corroding environment and development of corrosion (width of corrosion products that cover the metal surface). Special precautions have to be undertaken for testing the stationarity of the analyzed noise and the removal of the detected non-stationary components.

This paper discusses the presently proposed methods of electrochemical noise analysis in a brief review and their use for recognition of corrosion type and estimation of its rate. The worked out methods are proposed for uniform and local corrosion separately (Fig. 1). These methods are verified by different experiments conducted in a laboratory using real corroding systems.

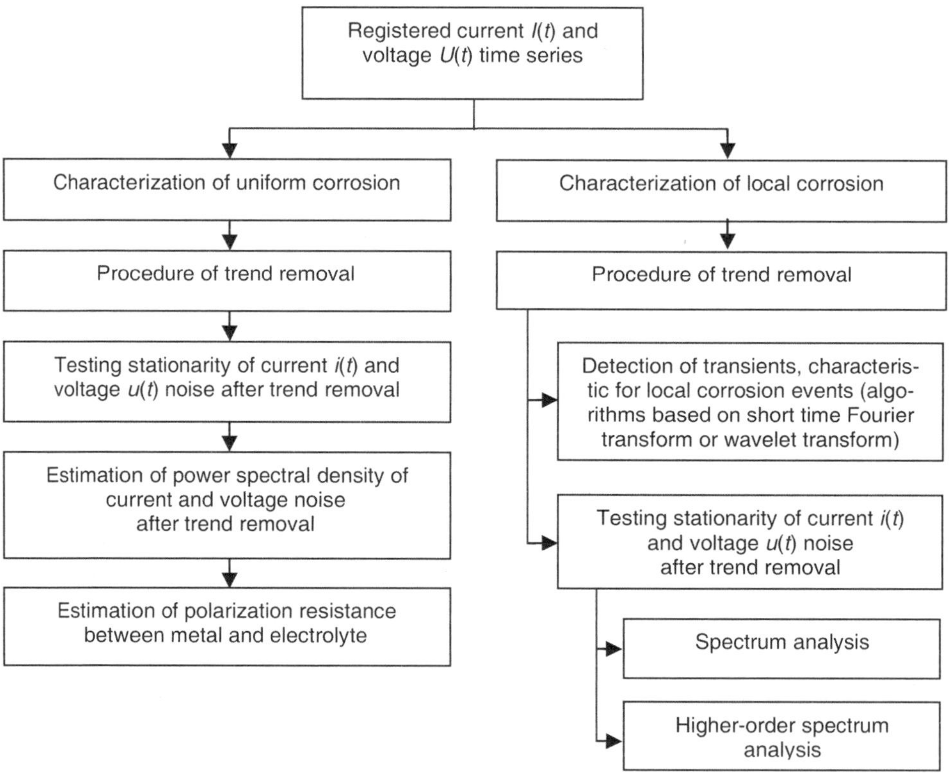


Fig. 1. Methods of electrochemical noise analysis for the investigation of corrosion processes.

## 2. Recognition and Evaluation of Uniform Corrosion

Uniform corrosion processes are distributed independently and uniformly in time and space on the whole corroding metal surface. The charge flows between metal and electrolyte can be represented by a noise source that has Gaussian distribution and $1/f-$ like power spectral density [6]. The observed voltage $U(t)$ and current $I(t)$ fluctuations have visible trends that are induced by unavoidable differences between nominally identical working electrodes. The trends have to be removed from the registered noise records before a further analysis.

The rate of uniform corrosion can be evaluated by estimation of a polarization resistance $R_p$ that exists between the metal and electrolyte [7]. Its value can be estimated by electrochemical noise measurements. A rough estimation of $R_p$ can be obtained by dividing the adequate standard deviations $\sigma_u, \sigma_i$ of the voltage $u(t)$ and current $i(t)$ noise after trends removal. The calculated value $R_n = \sigma_u / \sigma_i$ is called noise resistance and is equal to $R_p$ only in some cases [8]. The other proposed method of $R_p$ calculation is based on the estimation of the impedance module $|Z(f)|$ using the adequate power spectral density $S_i(f), S_u(f)$ of the voltage $u(t)$ and current $i(t)$ noise records separately

$$|Z(f)| = k \cdot \sqrt{S_u(f)/S_i(f)} \,. \tag{1}$$

The $R_p$ value is estimated by averaging $|Z(f)|$ over the low frequency range when it does not depend on frequency. The coefficient $k$ is established for the applied type of a reference electrode. When the reference electrode is noiseless the coefficient $k=1$ or when it is the same as the working electrodes $k = 1/\sqrt{3}$ . These results are derived by assuming that a metal-electrolyte layer is represented by a parallel connection of the border layer capacity $C_p$, polarization resistance $R_p$ and a current noise source $S_{iRp}(f)$ that represents the intensity of corrosion processes (Fig. 2) [8]. The thermal noise $S_{iT\,\mathrm{Re}}(f)$ of the electrolyte resistance $R_e$ can be neglected when the observed noise is a $1/f-$ like noise. It suggests that the white noise source $S_{iT\,\mathrm{Re}}(f)$ is overwhelmed by the source $S_{iRp}(f)$ or inner noise of the applied measurement system.

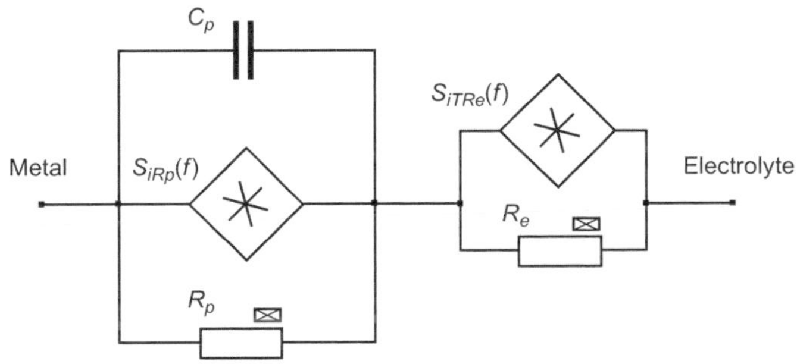


Fig. 2. The equivalent circuit of a metal-electrolyte boundary layer.

The recorded current $I(t)$ and voltage $U(t)$ for carbon steel electrodes and its random components $i(t), u(t)$ after trend removal are presented in Fig. 3. Different methods of trend removal can be applied [10]. Good results are obtained by the Donoho algorithm that is based on wavelet transform and therefore preserves uniform distribution of energy between the frequency sub-bands (details) when $1/f$ noise is analyzed [11]. The algorithm is available in the wavelet toolbox of MatLab software. The estimated $R_p \approx 99\Omega$ by averaging $|Z(f)|$ (Fig. 4) was in good agreement with $R_p$ values estimated by the electrochemical method in Ref. [9]. The relative random error of $R_p$ does not exceed 4% [12, 13].

The attached measurement results were obtained for a setup of three carbon steel electrodes ST3S, each of 1 cm$^2$ and placed into a pipe of the municipal water supply system. The measurement system enables electrochemical noise registration within long time intervals. Thus a corrosion rate (polarization resistance) can be repeatedly calculated.

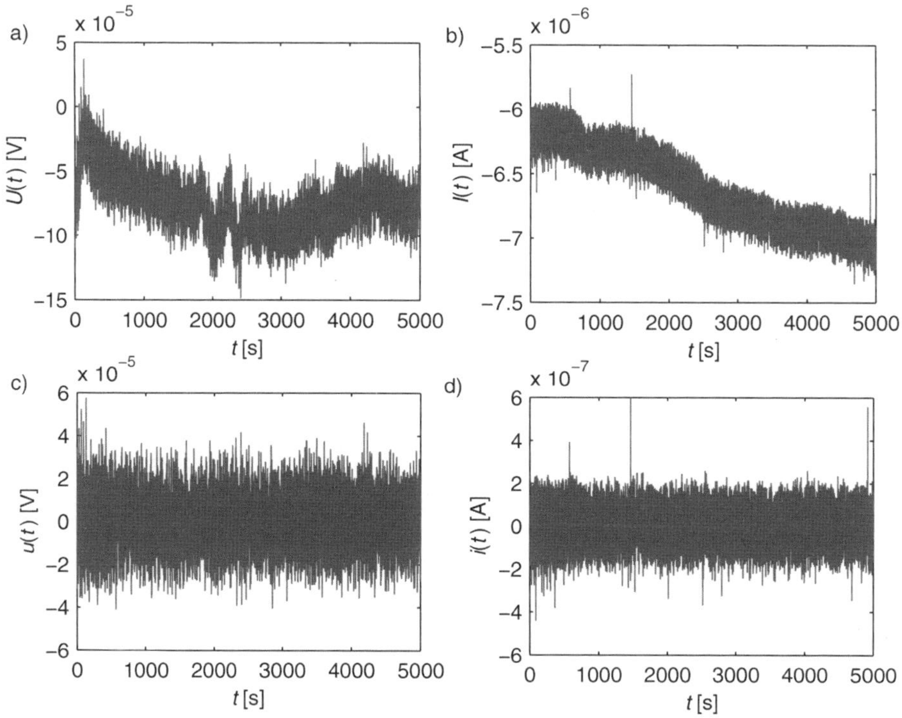


Fig. 3. Electrochemical noise in a carbon steel ST3S three-electrode setup immersed in municipal water: (a) voltage $U(t)$, (b) current $I(t)$ and their components (c) $u(t)$, (d) $i(t)$ after trend removal using the Donoho algorithm.

The presented method of polarization resistance estimation is equivalent to the impedance measurement when external polarization is replaced by noise that is generated by randomly occurring electrochemical corrosion processes. It means that any outside polarization of metal-electrolyte layers is not needed and does not disturb corrosion processes there. The metal electrodes can be treated as a set of sensors that investigate how the electrochemical environment interacts with the applied metal. This property is important in practice only when a relatively high uniform corrosion rate might be changed by the applied external polarization within the time of measurement. Another advantage of the method based on electrochemical noise measurements is a plausible identification of any corrosion rate change when the polarization resistance is repeatedly estimated. This feature can be desirable in practice in different cases (e.g., monitoring of reinforcement corrosion rate when concrete is soaked by a solution of NaCl during the winter season).

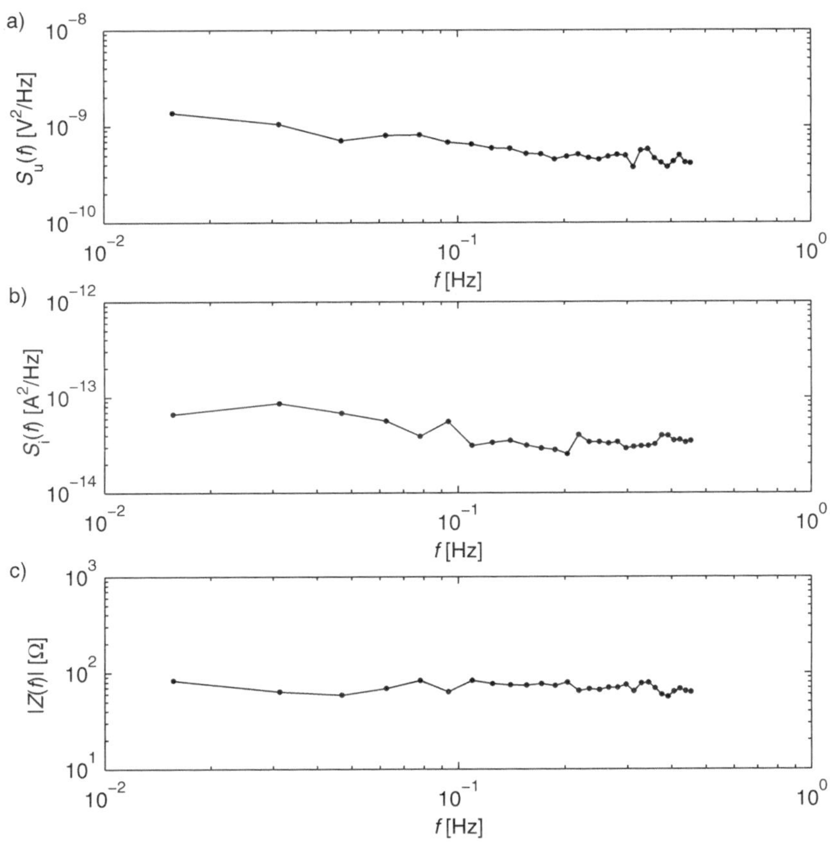


Fig. 4. Power spectra $S_u(f), S_i(f)$ of the observed: (a) voltage $u(t)$, (b) current $i(t)$ and the derived (c) impedance $|Z(f)|$.

## 3. Local Corrosion Recognition and Evaluation

Local corrosion events can cause local damage of a corroding metal when the rest of the metal surface is not corroding or is corroding very slowly. Pitting corrosion, that can produce deep local pits in metal at surface, has an important practical meaning since even a single pit can cause catastrophic effects. The events are mainly initiated by local changes in acidity on an area of the passive layer that exists between the metal and electrolyte [14]. The passive layer can locally break down and initiate a slow process of its recovery.

The whole process involves an intensive charge flow between the metal and electrolyte and can be observed as a characteristic transient in the electrochemical noise time series. The shape of the transients depends on the metal type. A sudden drop of instantaneous current and voltage values followed by its exponential return is observed in carbon steel (Fig. 5). This process is called metastable pitting. Only some of the observed local events can transform into stable pits that cause local damage of the metal surface.

The process of a stable pit formation is preceded by abundant metastable pitting that can be identified in electrochemical noise records [15]. The probability of a stable pitting formation is proportional to the intensity of the metastable pitting. Different methods of the observed transient detection are proposed [15, 16]. The presence of transients can be recognized by a change of a slope of the power spectral density versus frequency characteristics. The problem is that the intensity of metastable pitting can be non-stationary and its power spectrum density cannot be estimated. Another, more universal solution is transient detection by applying joined time-frequency analysis or wavelet transform. An efficient algorithm is described in detail in [17]. The algorithm can be applied for detection of transients of given shape and random amplitudes. Moreover, the process of transient detection can be self-calibrated by recognition of the most intensive transient in the time series and can be performed by the prepared software (e.g., virtual instrument in LabView software). This method does not require stationarity of the registered noise but presumes that the transient shape is the same at least within the analyzed noise record.

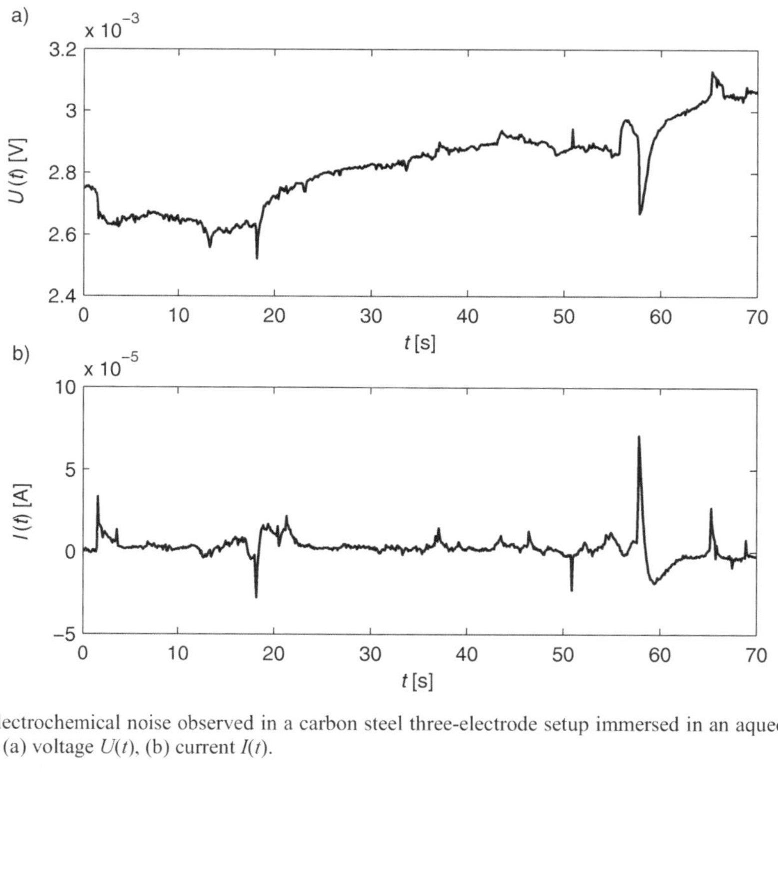


Fig. 5. Electrochemical noise observed in a carbon steel three-electrode setup immersed in an aqueous solution of $FeCl_3$: (a) voltage $U(t)$, (b) current $I(t)$.

The other proposed method of transient characterization is higher-order statistic (e.g., bispectrum function) [17, 18]. The bispectrum of a discrete signal $i(n)$ is its two dimensional Fourier transform

$$S_{3i}(f_1, f_2) = \sum_{k=-\infty}^{\infty} \sum_{l=-\infty}^{\infty} E[i(n)i(n+k)i(n+l)]e^{-j2\pi f_1 k} e^{-j2\pi f_2 l} \tag{2}$$

where the operator $E[..]$ denotes averaging. This function detects an asymmetry of the probability distribution of the analyzed signal. The bispectrum uncovers transients that have non-Gaussian distribution and dumps all the remaining Gaussian components (e.g., generated by uniform corrosion or measurement setup). Moreover, it provides an average time constant that is characteristic of the analyzed transients. Thus, the bispectrum can characterize processes of the passive layer breakdowns but can be applied only when

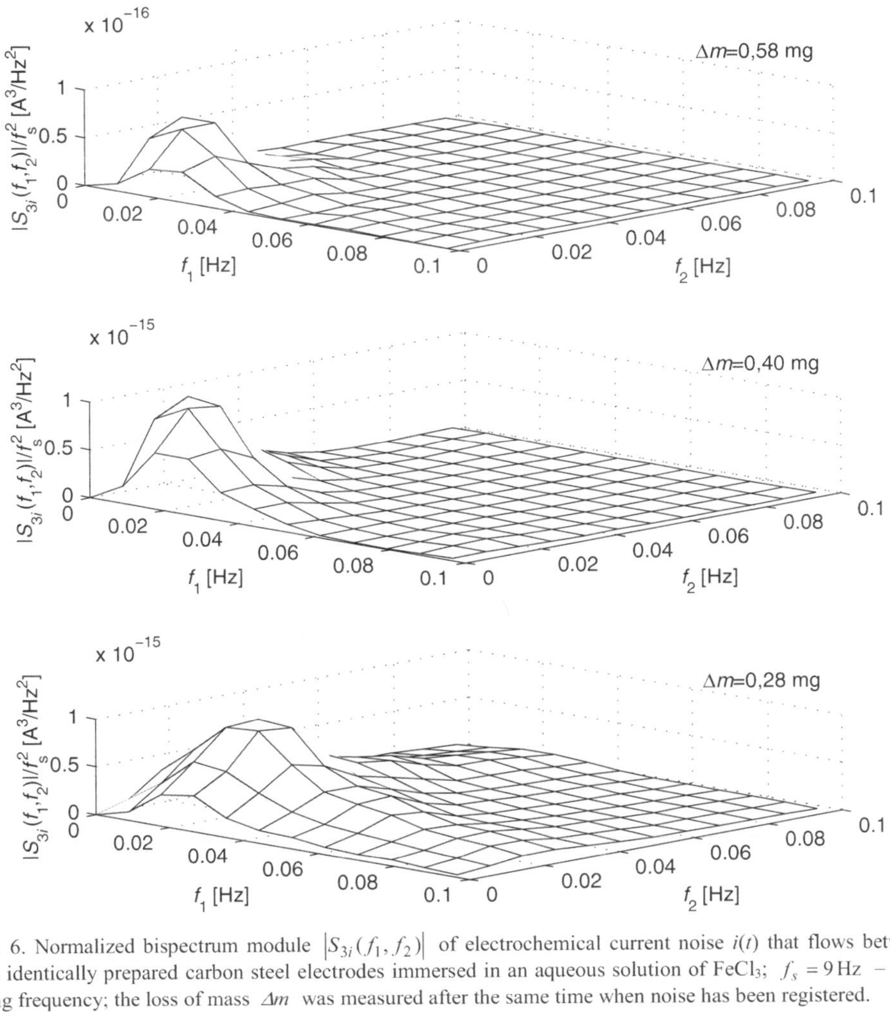


Fig. 6. Normalized bispectrum module $|S_{3i}(f_1, f_2)|$ of electrochemical current noise $i(t)$ that flows between two identically prepared carbon steel electrodes immersed in an aqueous solution of $FeCl_3$; $f_s = 9\,\text{Hz}$ – sampling frequency; the loss of mass $\Delta m$ was measured after the same time when noise has been registered.

their occurrence is stationary within the time of measurement. The presented preliminary results suggest that slower processes of passive layer recovery (that are adequate to the lower frequencies where the bispectrum module reaches a maximum) are responsible for greater loss of the electrode mass (Fig. 6).

The attached results are provided for current noise records only because the current represents charge flows plausibly responsible for loss of mass. Similar results may be expected for voltage noise records due to the time correlation of transients observed simultaneously in current and voltage noise records (Fig. 5). Therefore, the bispectrum function can be utilized for the characterization of the pitting corrosion events.

## 4. Conclusions

The presented methods of electrochemical noise analysis can be applied to determine the type of corrosion and for evaluation of its rate. It is very important in practice since the method can be applied for long-term corrosion monitoring in industry or laboratory as well. Different methods for different types of corrosion processes have to be applied. The described methods enable estimation of the presented noise parameters that can reduce huge number of noise samples into a single parameter or function that recognize the type of corrosion and estimate its rate.

When uniform corrosion processes prevail, noise measurements are equivalent to impedance measurements when the internal noise source replaces the external polarization that would influence on the recognized corrosion rate. The method that is based on electrochemical noise measurements is important when the corrosion rate is relatively high.

When local corrosion events are present, the electrochemical noise observation is much more interesting from a practical point of view for the others electrochemical methods apply the external polarization that can drastically change the probability of occurrence of corrosion events. The analysis of electrochemical noise can detect the presence of transients typical for local corrosion events and estimates its rate that is proportional to the probability of transient occurrence. The attached results of the bispectrum function expose the processes responsible for the passive layer breakdowns that provide additional information about the local corrosion events resulting in metal damage.

## Acknowledgements

A preliminary version of the results was presented at the SPIE Conference "Noise and Information in Nanoelectronics, Sensors and Standards", Santa Fe 2003 [19].

## References

[1] A. M. Brennenstuhl, G. Palumbo, F. S. Gonzalez and G. P. Quirk, The use of electrochemical noise to investigate the corrosion resistance of UNS Alloy N04400 nuclear heat exchanger tubes, in *Proc. ASTM STP 1277*, eds. J. R. Kearns, J. R. Scully, P. R. Roberge, D. L. Reichert and J. L. Dawson, Philadelphia, PA (1996), pp. 266–287.

[2] E. Almeida, L. Mariaca, A. Rodriguez and J. U. Chavarín, Characterization of prerusted steels in some Ibero-American atmospheres by electrochemical potential noise measurement, in *Proc. ASTM STP 1277*, eds. J. R. Kearns, J. R. Scully, P. R. Roberge, D. L. Reichert and J. L. Dawson, Philadelphia, PA (1996), pp. 411–426.

[3] B. M. Grafov and I. B. J. Grafova, Discrete Laplace transformation as a wavelet tool to analyze the electrochemical random time series, *J. Electroan. Chem.* **502** (2001) 47–50.

[4] J. Smulko, K. Darowicki and P. Wysocki, Digital measurement system for electrochemical noise, *Polish J. Chem.* **72** (1998) 1237–1241.

[5] H. J. De Bruyn, Current corrosion monitoring trends in the petrochemical industry, *Int. J. Pres. Ves. & Piping* **66** (1996) 293–303.

[6] R. A. Cottis, Interpretation of electrochemical noise data, *Corrosion* **57** (2000) 265–285.

[7] A. Lowe, H. Eren, Y. J. Tan, B. Kinsella and S. Bailey, Continuous corrosion rate measurement by noise resistance calculation, *IEEE Trans. Instrum. Meas.* **50** (2001) 1059–1963.

[8] F. Huet, A. Bautista and U. Bertocci, Listening to corrosion, in the *Electrochemical Society Interface*, Winter (2001) 40–43.

[9] J. Smulko, K. Darowicki and A. Zielinski, Electrochemical noise analysis of uniform corrosion in water supply system, in *Proc. XVII IMECO World Congress*, Croatia-Dubrovnik (June 2003), pp. 659 662.

[10] U. Bertocci, F. Huet, R. P. Nuguerra and P. Rousseau, Drift removal procedures in the analysis of electrochemical noise, *Corrosion*, **58** (2002) 337–347.

[11] D. L. Donoho, De-noising by soft-thresholding, *IEEE Trans. on Inf. Theory* **41** (1995) 613–627.

[12] J. S. Bendat and A. G. Piersol, *Random Data Analysis and Measurement Procedures* (Wiley, New York, 2001).

[13] J. Smulko, On accuracy of polarization resistance estimation by electrochemical noise measurements, in *Proc. 17th Int. Conf. Noise and Fluctuations*, Prague, Czech Republic (August 2003) ,707–710.

[14] T. Shibata, Statistical and stochastical approaches to localized corrosion, *Corrosion* **52** (1996) 813–830.

[15] A. Aballe, M. Bethencourt, F. J. Botana, M. Marcosy and J. M. Sánchez-Amaya, Use of wavelets to study electrochemical noise transients, *Electrochimica Acta* **46** (2001) 2353–2361.

[16] J. Smulko, K. Darowicki and A. Zielinski, Detection of random transients caused by pitting corrosion, *Electrochimica Acta* **47** (2002) 1297–1303.

[17] R. A. Cottis, Interpretation of electrochemical noise data, *Corrosion* **57** (2001) 265–285.

[18] J. Smulko and K. Darowicki, Nonlinearity of electrochemical noise caused by pitting corrosion, *J. Electroanalytical Chemistry* **545** (2003) 59–63.

[19] J. Smulko, Electrochemical noise analysis methods for the investigation of corrosion processes, in *Proc. SPIE, Noise and Information in Nanoelectronics, Sensors and Standards*, Santa Fe, USA **5115** (June 2003), pp. 86–91.

# LOCATING CRITICAL LINES IN HIGH-VOLTAGE ELECTRICAL POWER GRIDS

PAOLO CRUCITTI
*Scuola Superiore di Catania*
*Via S. Paolo 73, 95123 Catania, Italy*

VITO LATORA
*Dipartimento di Fisica e Astronomia, Università di Catania*
*and INFN sezione di Catania, Corso Italia 57, 95129 Catania, Italy*

MASSIMO MARCHIORI
*W3C and Lab. for Computer Science*
*Massachusetts Institute of Technology, USA*

Received 7 December 2004
Revised 28 April 2005
Accepted 13 May 2005
Communicated by Werner Ebeling and Bernardo Spagnolo

Electrical power grids are among the infrastructures that are attracting a great deal of attention because of their intrinsic criticality. Here we analyze the topological vulnerability and improvability of the spanish 400 kV, the french 400 kV and the italian 380 kV power transmission grids. For each network we detect the most critical lines and suggest how to improve the connectivity.

*Keywords*: Complex networks; electric power grids; vulnerability; improvements.

## 1. Introduction

Nowadays the issue of vulnerability and protection of critical infrastructure is attracting a great deal of attention among scientific researchers. One of the most used approaches is that of complex networks. A critical infrastructure (e.g. a power grid) is represented as a graph in which nodes represent the main components of the network (e.g. power plants or trasmission substations) and edges are the physical connections among them (i.e. the electric lines in the above example). Following this approach, several scientific works have been produced with the purpose of analyzing error and attack resilience of both artificially generated topologies and real world networks.

In literature, failures and attacks have been simulated as the removal of a certain percentage either of nodes [1–5] or of edges [2, 6] of the network.

In particular, a general method to find the critical components of a network has been introduced in Ref. [7]. There, the authors assume that the performance of a network $G$ can be measured by a single variable $\Phi(G) > 0$, and propose to use the variation in the performance to spot the critical components of $G$.

The performance of $G$ is evaluated from the network structure as: $\Phi(G) \equiv E(G) = \frac{1}{N(N-1)} \sum_{i \neq j \in G} \frac{1}{d_{ij}}$, where $E(G)$ is the global efficiency of $G$ defined from the shortest path lengths $\{d_{ij}\}$ (number of steps) from node $i$ to node $j$ [8].

Different from the characteristic path length [1], the efficiency $E$ is a well-defined quantity also for non-connected graphs [3, 8]. Indicating with $D$ the set of possible damages, and with $DAMAGE(G, d)$ a map that gives the network resulting from $G$ after a damage $d \in D$ (for instance after the removal of single nodes, or single edges, or couples of edges), the importance of the damage $d$ is measured by the relative drop in performance $\Delta\Phi^-/\Phi$, with $\Delta\Phi^- = \Phi[G] - \Phi[DAMAGE(G, d)]$.

The critical damages are those that maximize $\Delta\Phi^-/\Phi$, and the larger $\Delta\Phi^-/\Phi$ the higher is the vulnerability of $G$ under the class of damages $D$. Locating the critical components in an infrastructure is a very important task. However, also finding the most useful possible improvements is fundamental in real world networks. In an analogous way, given a set of improvements I (e.g. the addition of edges), and defined for any improvement $i \in I$ the map $IMPROVE(G, i)$ that gives the network obtained after the improvement $i$, the importance of the improvement $i$ can be defined as $\Delta\Phi^+/\Phi$, with $\Delta\Phi^+ = \Phi[IMPROVE(G, d)] - \Phi[G]$. The best improvements are those that maximize $\Delta\Phi^+/\Phi$, and the larger $\Delta\Phi^+/\Phi$ the higher is the improvability of $G$ under the class of improvements $I$.

In this paper we apply the method of Ref. [7] to study the topological vulnerability and improvability of three different power grids: (1) the spanish 400 kV electric network, with $N_S = 98$ substations and $K_S = 175$ lines; (2) the french 400 kV electric network, with $N_F = 146$ and $K_F = 223$; (3) the italian 380 kV electric network, with $N_I = 127$ and $K_I = 171$. The power grids are the same studied by Rosato *et al.* in Ref. [9].

We consider three different classes of damages $D$, namely the removal of single edges, couples of edges and triples of edges, and one class of improvements, namely the addition of single edges. An alternative possibility, not explored in this paper, is the removal of nodes.

The results of the analysis are shown in Sec. 2. In Sec. 3 we study the damage probability distribution obtained by removing a higher number of edges (from 5 to 25) and, in Sec. 4, we draw some conclusions.

## 2. Critical Damages and Improvements

The three power grids studied are shown in Figs. 1–3 [10].

In Table 1 we report the four most critical edges, obtained by considering single-edge removals. The italian network shows to be the most vulnerable one: the removal of a single edge is sufficient to generate a drop in the efficiency close to 5%, while in the french and in the spanish power grids the drop in the efficiency is limited to about 3%. The reason for this difference has to be found in the stretched shape of the italian peninsula (and consequently of the related power grid), in which, especially in the extreme South, nodes are poorly linked (See Fig. 3). For instance

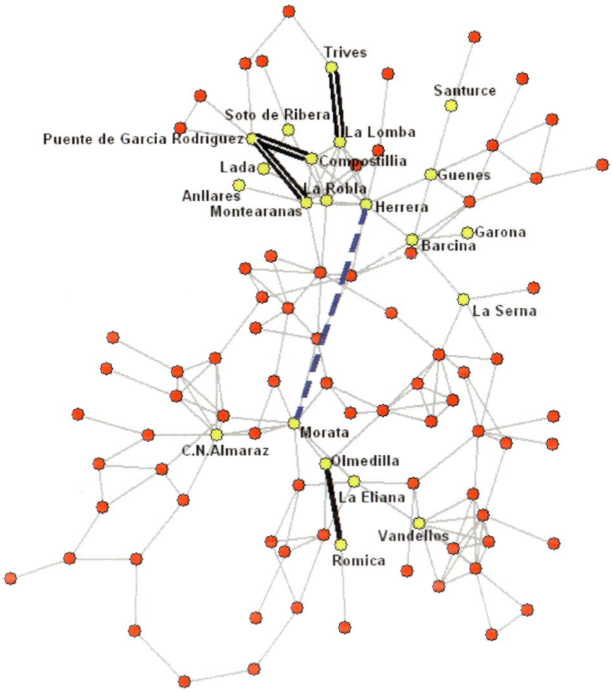


Fig. 1. The spanish 400kV power grid with the most vulnerable edge (black thick line), the most vulnerable triple of edges (black double lines), and the edge that, if added to the network, improves the performance of the system best (dashed blue line). Nodes reported in tables are marked with a lighter color and with a label indicating the name of their respective substations.

the removal of a single edge, as the line connecting Laino and Rossano, is sufficient to isolate seven nodes from the rest of the network.

Analogous observations are valid when the class of possible damages is restricted to couples of edges (Table 2). By removing the line connecting Laino and Rossano, and that connecting Vignole B. and La Spezia, the efficiency of the italian power grid experiences a drop of 8.32%. It is worthy noting that not always the joint removal of the two most important edge of Table 1 is among the most damaging strategies for removals of couples of edges. In fact, if we remove both the Laino–Rossano and the Rossano–Scandale lines from the italian grid, we found a drop in efficiency only equal to 5.2%, and such a removal ranks only 220th in the list.

The differences between the italian and the other two networks is even more evident when three edges are removed jointly (Table 3). In fact, with removals of triples of nodes, the italian power grid can be broken into two different subgraphs of approximately the same size, thus decreasing the efficiency by 31%. This result is in agreement with those obtained in Ref. [9] by means of a spectral bisection method.

Bisection methods allow to find the division of the vertices of a graph into two nearly equally-sized vertex subsets $G_1$ and $G_2$, with the minimum number of edges running between them. The edges connecting the two subgraphs found by the

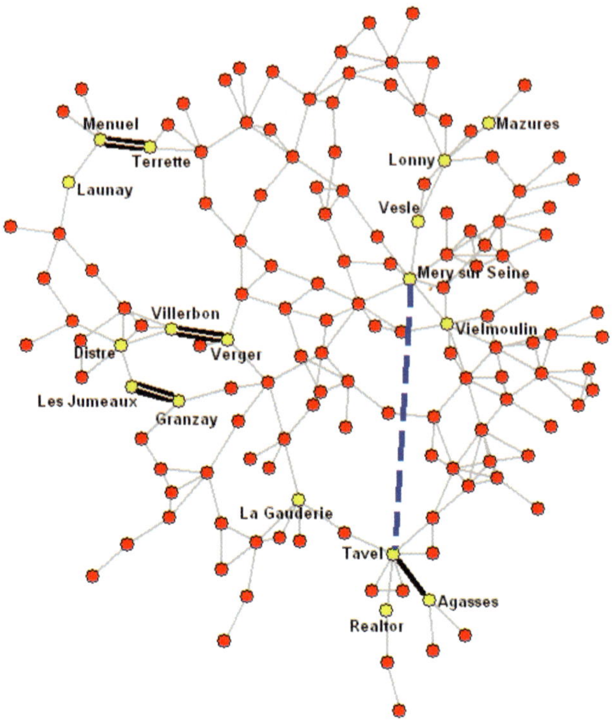


Fig. 2. The french 400kV power grid with the most vulnerable edge (black thick line), the most vulnerable triple of edges (black double lines), and the edge that, if added to the network, improves the performance of the system best (dashed blue line). Nodes reported in tables are marked with a lighter color and with a label indicating the name of their respective substations.

authors of Ref. [9] for the italian grid, are three and are those at the 3rd rank in Table 3.

Finally, the removal of the three most important edges in Table 1 is not equivalent to perform an important three-edges removal: the joint removal of the lines Mery sur Seine–Vesle, Tavel–Agasses, and Tavel–Realtor is at the 20th place in the rank for the french grid and the joint removal of the lines Laino–Rossano, Rizziconi–Scandale, and Rossano–Scandale ranks 33302nd for the italian grid. The most important couple of edges of Table 2 is not either meaningful for the three-edges removals: the most important three-edges removal that involves the couple of edges Laino–Rossano and Vignole B.-La Spezia ranks only 65th.

Turning now to the improvements, we report in Table 4 the best addition of single edges. The italian power grid not only is the most vulnerable network, but is also the most improvable one. For instance, the addition of a single edge between Piacenza and Benevento causes an improvement in efficiency of almost 8%. Obviously, it is not easy to realize a physical connection between two distant substations: a careful trade-off between realization cost and efficiency improvement should be taken into account. However, in this paper we ignore it, dealing only with topological connections.

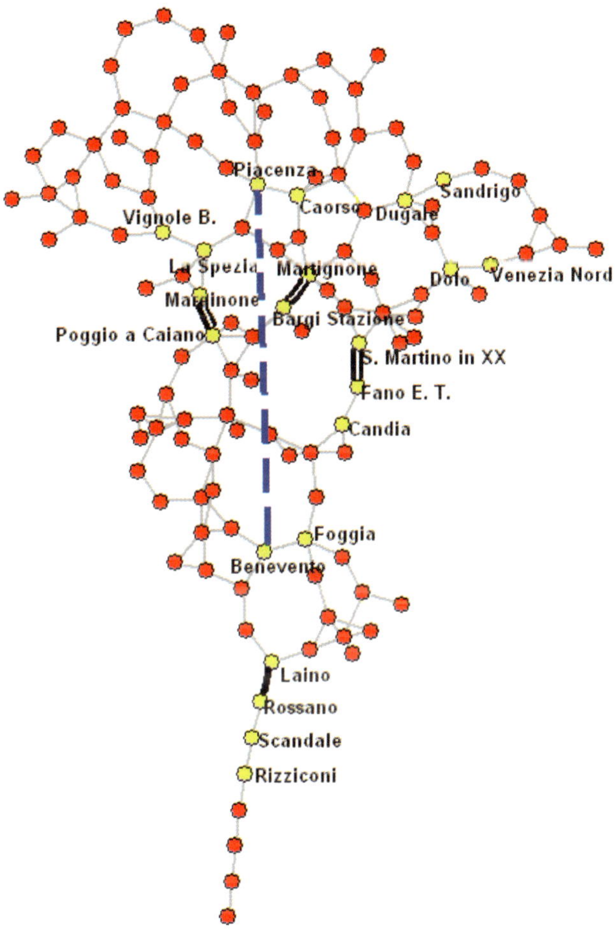


Fig. 3. The italian 380kV power grid with the most vulnerable edge (black thick line), the most vulnerable triple of edges (black double lines), and the edge that, if added to the network, improves the performance of the system best (dashed blue line). See tables. Nodes reported in tables are marked with a lighter color and with a label indicating the name of their respective substations.

## 3. A Glimpse to Distributions

An evaluation of all the possible multiple removals is too much time-consuming for a number of edges larger than 2 or 3. In this section we consider five different classes of damage, namely the removal of $m = 5, 10, 15, 20, 25$ edges. For each class of damage we calculate the drop in the efficiency for 10000 randomly selected different cases, out of the total number of possibilities which is of the order of $K^m$ (with $K$ being the number of lines in the network).

In Figs. 4 and 5 we report the damage distributions, i.e. the probability of causing a certain damage. The distribution is gaussian both for the spanish and the french networks. Obviously, when the number of removed edges increases, the probability of causing high damages grows, i.e. the distribution curves move to

Table 1. The four most critical lines and the relative damage (in percentage).

| | S-400kV | $\frac{\Delta\Phi^-}{\Phi}$ | F-400kV | $\frac{\Delta\Phi^-}{\Phi}$ | I-380kV | $\frac{\Delta\Phi^-}{\Phi}$ |
|---|---|---|---|---|---|---|
| 1. | Olmed.-Romica | 3.22 | Tavel-Agasses | 2.96 | Laino-Ross. | 5.01 |
| 2. | Gueñes-Santurce | 3.03 | Tavel-Realtor | 2.84 | Rossano-Scand. | 4.18 |
| 3. | Barcina-La Serna | 2.00 | Mery sur S.-Vesle | 2.50 | Rizz.-Scand. | 3.42 |
| 4. | Barcina-Garoña | 1.90 | Mazures-Lonny | 2.22 | Vign.B.-La Sp. | 3.39 |

Table 2. The four most critical couples of lines and the relative damage (in percentage).

| | S-400kV | $\frac{\Delta\Phi^-}{\Phi}$ | F-400kV | $\frac{\Delta\Phi^-}{\Phi}$ | I-380kV | $\frac{\Delta\Phi^-}{\Phi}$ |
|---|---|---|---|---|---|---|
| 1. | Olmed.-Romica<br>Gueñes-Santurce | 6.22 | Tavel-Agasses<br>Tavel-Realtor | 5.67 | Laino-Ross.<br>Vign. B.-La Sp. | 8.32 |
| 2. | Olmed.-Romica<br>Barcina-La Serna | 5.22 | Mery sur S.-Vesle<br>Tavel-Agasses | 5.43 | Dugale-Sandrigo<br>Dolo-Ven. N. | 8.20 |
| 3. | Barcina-Garoña<br>Olmed.-Romica | 5.10 | Mery sur S.-Vesle<br>Tavel-Realtor | 5.31 | Mart.-Bargi St.<br>Magin.-Pogg. a C. | 7.72 |
| 4. | Mont.-Anllares<br>Olmed.-Romica | 5.04 | Mazures-Lonny<br>Tavel-Agasses | 5.16 | Laino-Ross.<br>Magin.-Pogg. a C. | 7.71 |

Table 3. Critical triples of lines and relative damage (in percentage).

| | S-400kV | $\frac{\Delta\Phi^-}{\Phi}$ | F-400kV | $\frac{\Delta\Phi^-}{\Phi}$ | I-380kV | $\frac{\Delta\Phi^-}{\Phi}$ |
|---|---|---|---|---|---|---|
| 1. | P.de G.R.-Compost.<br>P.de G.R.-Mont.<br>Trives-La Lomba | 8.72 | Menuel-Terrette<br>Vill.-Verger<br>Granzay-Les Jum. | 14.23 | Mart.-Bargi St.<br>S.M. in XX-Fano<br>Marg.-Pogg. a C. | 30.96 |
| 2. | Olmed.-Romica<br>Soto de R.-La Robla<br>Lada-La Robla | 8.09 | Menuel-Terrette<br>Distre-Les Jum.<br>Marg.-Pogg. a C. | 13.57 | Mart.-Bargi St.<br>Vill.-Verger<br>Fano-Candia | 30.73 |
| 3. | Olmed.-Romica<br>Barcina-La Serna<br>Gueñes-Sant. | 8.06 | Menuel-Launay<br>Vill.-Verger<br>Granzay-Les Jum. | 12.43 | S.M. in XX-Fano<br>Marg.-Pogg. a C.<br>Bargi St.-Calenz. | 30.21 |
| 4. | Olmed.-Romica<br>Barcina-Garoña<br>Gueñes-Santurce | 8.05 | Launay-Domloup<br>Vill.-Verger<br>Granzay-Les Jum. | 11.84 | Marg.-Pogg. a C.<br>Bargi St.-Calenz.<br>Fano-Candia | 29.93 |

Table 4. The four best improvement and the relative increase in performance (in percentage).

| | S-400kV | $\frac{\Delta\Phi^+}{\Phi}$ | F-400kV | $\frac{\Delta\Phi^+}{\Phi}$ | I-380kV | $\frac{\Delta\Phi^+}{\Phi}$ |
|---|---|---|---|---|---|---|
| 1. | Herr.-Morata | 4.39 | Mery s. S. -Tavel | 4.41 | Piac.-Benev. | 7.93 |
| 2. | Herr.-Vandellos | 4.15 | Distre-Mery s. S. | 4.32 | Piac.-Foggia | 7.89 |
| 3. | Herr.-La Eliana | 4.14 | Distre-Vielmoulin | 4.24 | Caorso-Benev. | 7.87 |
| 4. | Herr.-C. N. Alm. | 4.05 | Mery s. S. -La Gaud. | 4.21 | Caorso-Foggia | 7.77 |

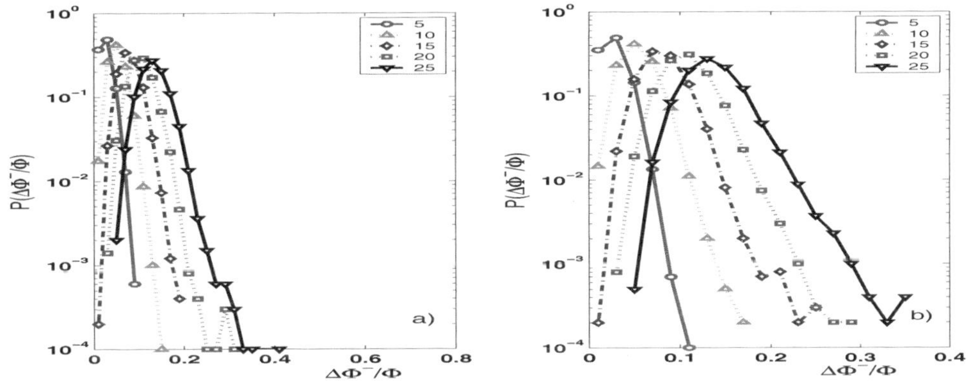


Fig. 4. Damage distribution in log-linear scale for a) the spanish and b) the french 400 kV power grid, and for a number of removed edges equal to 5,10,15,20,25.

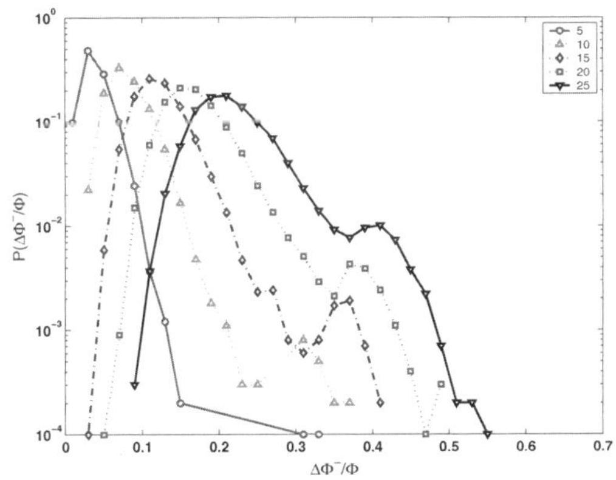


Fig. 5. Damage distribution in log-linear scale for the italian 380kV power grid, and for a number of removed edges equal to 5,10,15,20,25.

the right side of figures. An interesting difference can be noticed for the italian power grid, where the damage distribution is bimodal, especially for a number of removed edges in the range from 10 to 20, and can be fitted by the sum of two gaussians. Again, the reason for this particular behavior has to be found in the particular shape of the italian power grid. In fact when a certain number of edges are removed, the damage with the highest probability is low, but there is also a high probability to break the entire network at least into two big subgraphs (as in the case of the removal of the most important triples of edges in Table 3). It is not accidentally that the peaks of the secondary gaussian curves are close to the value that the drop in efficiency has when the removal concerns only the minimum number of nodes (three) for breaking the network into two big subgraphs. Obviously, when a huge number of edges is removed, also other meaningful lines can happen to be removed and this explains why the peak moves to the right for increasing number of removals.

## 4. Conclusions

In this paper we have studied and compared the topological properties of three electric power grids. We have used a general method that allows to find the critical components of a graph, i.e. the nodes and edges that when removed affect the most the structure of the graph. The method also allows to find the edges that, if added to the network, generate the best improvements. Moreover, we have analyzed the damage distributions and related them to the shape and the connectivity of the networks. Results have shown that the stretched shape of the italian power grid makes it structurally very different from the other two systems. In fact, on one hand, it shows to be so vulnerable that the joint removal of only three edges is sufficient to break dramatically the network and to cause a drop in the efficiency close to 30%. On the other hand, it is also the network that can be improved most with the addition of a single edge.

Obviously, in order to design an improvement to real electrical networks, further aspects concerning both the modeling of dynamical cascading failures (see [5, 11, 12] and references therein), and the inclusion of euclidean distances between graph nodes, should be considered.

### Acknowledgments

We thank V. Rosato for kindly providing us with the networks studied in Ref. [9].

### References

[1] R. Albert, H. Jeong and A.-L. Barabási, *Attack and error tolerance in complex networks*, *Nature* **406** (2000) 387–482.

[2] P. Holme, B. J. Kim, C. N. Yoon and S. K. Han, *Attack vulnerability of complex networks*, *Phys. Rev. E* **65** (2002) 056109.

[3] P. Crucitti, V. Latora, M. Marchiori and A. Rapisarda, *Efficiency of scale-free networks: error and attack tolerance*, *Physica A* **320** (2003) 622.

[4] R. Albert, I. Albert and G. L. Nakarado, *Structural vulnerability of the North American power grid*, *Phys. Rev. E* **69** (2004) 025103(R).

[5] P. Crucitti, V. Latora and M. Marchiori, *Model for cascading failures in complex networks*, *Phys. Rev. E* **69** (2004) 045104(R).

[6] A. E. Motter, T. Nishikawa and Y. Lai, *Range-based attack on links in scale-free networks*, *Phys. Rev. E* **66** (2002) 065103.

[7] V. Latora and M. Marchiori, *Vulnerability and protection of infrastructure networks*, *Phys. Rev. E* **71** (2005) 015103(R).

[8] V. Latora and M. Marchiori, *Efficient behavior of small-world networks*, *Phys. Rev. Lett.* **87** (2001) 198701.

[9] V. Rosato, S. Bologna and F. Tiriticco, *Topological properties of high-voltage electrical transmission networks*, submitted to *Electric Power Systems Research.*

[10] The figures have been generated with the Kamada-Kawai free energy of the software Pajek (from http://vlado.fmf.uni-lj.si/pub/networks) and, thus, the position of the nodes does not necessarily correspond to the geographical one.

[11] P. Crucitti, V. Latora and M. Marchiori, *A topological analysis of the Italian electric power grid*, *Physica A* **338** (2004) 92.

[12] R. Kinney, P. Crucitti, R. Albert and V. Latora, *Modeling Cascading Failures in the North American Power Grid*, *cond-mat/0410318*(2004).

# 6

# Noise, Computation and Energy Dissipation

# Comments on "Sub-$k_BT$ Micro-Electromechanical Irreversible Logic Gate"

Laszlo B. Kish

*Department of Electrical and Computer Engineering*
*Texas A&M University*
*College Station, TX 77843-3128, USA*
*laszlo.kish@ece.tamu.edu*

Received 29 June 2016
Accepted 30 June 2016
Published 10 August 2016

Communicated by Andras Der

In a recent paper, [M. López-Suárez, I. Neri and L. Gammaitoni, Sub-$k_BT$ micro-electromechanical irreversible logic gate, *Nat. Commun.* **7** (2016) 12068] the authors claimed that they demonstrated sub-$k_BT$ energy dissipation at elementary logic operations. However, the argumentation is invalid because it neglects the dominant source of energy dissipation, namely, the charging energy of the capacitance of the input electrode, which totally dissipates during the full (0-1-0) cycle of logic values. The neglected dissipation phenomenon is identical with the mechanism that leads to the lower physical limit of dissipation (70–100 $k_BT$) in today's microprocessors (CMOS logic) and in any other system with thermally activated errors thus the same limit holds for the new scheme, too.

*Keywords*: Energy dissipation of control; dissipation limits in logic gates; error probability.

In a recent paper [1], the authors claim that they demonstrate sub-$k_BT$ energy dissipation at elementary logic operations. In the experiments, they use micro-electromechanical cantilevers (MEC) where the attractive electrostatic forces towards the input electrode(s) control the position of the cantilever tip.

The authors discuss an OR gate however, for the sake of simplicity but without the loss of generality, we investigate a follower logic gate in this paper. Two separate tip positions define the two logic values, 0 and 1, respectively. These logic values and tip positions correspond to driving voltages 0 and $U_1$ on the input electrode. The energy dissipation (due to friction) in the cantilever is measured during changing the logic value and the authors correctly conclude that the energy dissipation due to friction can be made smaller than $k_BT$.

*Does this approach account for all the major energy loss phenomena determining the lower limit of energy dissipation or is there any dominant but neglected component?*

The answer is straightforward:

The considerations in Ref. [1] neglected the energy dissipation due to charging and discharging the input capacitance, which are the same dissipation phenomena [2–6] that determine the lower limits of dissipation (70–10)$k_B T$ [3] not only in today's microprocessors (CMOS logic) but also in any other system with thermally activated errors. Thus, the same dissipation limit (70–100 $k_B T$) holds for the new scheme in Ref. [1], too.

While the flaw in Ref. [1] is now obvious due to the above arguments and no further considerations are needed to deny the validity of the sub-$k_B T$ energy dissipation claim, for the readers who are less familiar with the topic, we provide more details below. In voltage-controlled logic, see Fig. 1, when the switch $S_1$ is closed, the voltage on the capacitor is changing from the value 0 to $U_1$ while the logic (bit) value is switching between 0 and 1.

Thus, in the case of the 0 ==> 1 bit value change, the energy of the capacitor is changing from 0 value to

$$E_1 = 0.5CU_1^2, \tag{1}$$

while exactly the same amount of energy ($E_1$) is dissipated in the resistor $R$ due to its heating by the charging current. Here, $C$ represents the input capacitance (the gate capacitance in CMOS logic) and $R$ is the resistance of the closed switch (the source-drain resistance in CMOS). During erasure by resetting, $S_1$ is off and $S_2$ is closed, and again energy $E_1$ is dissipated in the other resistor. However, the resistors create Johnson noise (thermal noise) leading to thermally activated bit errors depending on the magnitude of $U_1$, which separates the logic states 0 and 1. Considerations based on these effects result in the following formula for the lower limit of energy dissipation in arbitrary systems with thermally activated errors [2–4]:

$$Q_T \approx k_B T \ln \frac{1}{\varepsilon}, \tag{2}$$

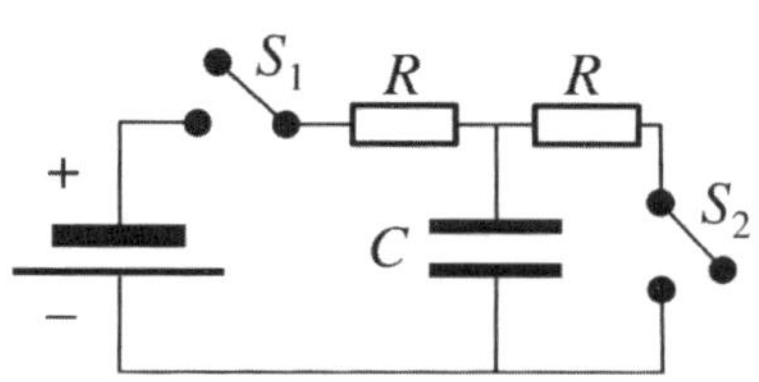


Fig. 1. Elementary circuit model to show the major energy dissipation in voltage-controlled logic [3].

where $\varepsilon(< 0.5)$ is the error probability during the observation time, and the formula is valid in the short observation time limit, that is, when the observation time of error events is less than the correlation time $\tau$ of thermal fluctuations activating the errors (the Johnson noise on the capacitor). The value of $Q_\tau$ is around 70 $k_\mathrm{B}T$ at today's error probability expectations [3].

In the long observation time, $t_o$, limit, Eq. (2) is slightly modified [5, 6]:

$$Q_t \approx k_\mathrm{B}T\left[\ln\left(\frac{1}{\varepsilon}\right) + \ln\left(\frac{t_o}{\tau}\right)\right], \tag{3}$$

which results in a slightly higher energy dissipation, $Q_t \approx 100\,k_\mathrm{B}T$ with today's typical bandwidth and $t_o = 1$ year.

The logic gate described in Ref. [1] is also a voltage-controlled gate with the unavoidable Johnson noise during charging and discharging the input capacitance [6], thus, Eqs. (2) and (3) and the corresponding dissipation of 70–100 $k_\mathrm{B}T$ holds there, too.

However, the argumentation is invalid because it neglects the dominant source of energy dissipation, the 50% loss of the charging energy of the capacitance of the control gate and the total loss of the remaining part during discharging. The neglected dissipation phenomenon is identical with the mechanism that leads to the ultimate dissipation limit in today's microprocessors (CMOS logic), thus the same limit holds for both.

To pre-empt the question, if there is a way to charge and discharge a capacitor without the loss of energy shown above we show the resonant circuit solution used in switching power supplies, see Fig. 2. Suppose $C_1$ is charged, $C_2$ is not. Closing $S_1$ causes a sinusoidal current flow in the coil $L$. At the peak value of the current, all the energy is magnetic and $C_1$ has discharged state. Then $S_2$ closes and $S_1$ opens. When the current reaches the zero value, all the energy is transferred into $C_2$. However, such energy saving scheme is efficient only at large ($\gg 70\,k_\mathrm{B}T$) charging energies. The situation is much worse at the logic gate energy range because then two new switches must also operate, each one with energy dissipation similar to the energy we want to save!

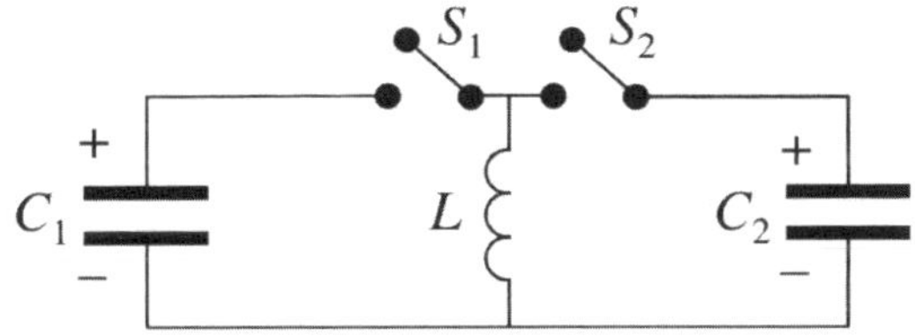


Fig. 2. "Tank" LC circuit example to bounce the charging energy in a lossless way between two capacitors with timed switches and LC resonator. Due to the energy dissipation of controlling the switches, it fails at the charging energy range where logic gates operate. All similar efforts require to increase the number of switching events thus lead to increased energy dissipation at the energy levels of logic gates.

Finally, without arguments or details, we summarize our general results [7, 8] in the related matter of energy dissipation in memories and logic gates:

(i) *Information entropy cannot be interrelated with dissipation, in general* [8].
(ii) *Writing of data (running of logic gates) is dissipation costly.*
(iii) *Erasure, on the other hand, can be completely dissipation-free, in special cases* [7].

There are interesting alternative considerations with related conclusions, see e.g., [9, 10].

## References

[1] M. López-Suárez, I. Neri and L. Gammaitoni, Sub-$k_{\rm B}T$ micro-electromechanical irreversible logic gate, *Nat. Commun.* **7** (2016) 12068.
[2] L. Brillouin, *Science and Information Theory* (Academic, New York, 1962).
[3] L. B. Kish, Moore's Law and the energy requirement of computing versus performance, *IEE Proc. — Circ. Dev. Syst.* **151** (2004) 190–194.
[4] R. Alicki, Stability versus reversibility in information processing, *Int. J. Mod. Phys. Conf. Ser.* **33** (2014) 1460353.
[5] L. B. Kish and C. G. Granqvist, Energy requirement of control: Comments on Szilard's engine and Maxwell's demon, *Europhys. Lett.* **98** (2012) 68001.
[6] L. B. Kish and C. G. Granqvist, Electrical Maxwell demon and Szilard engine utilizing Johnson noise, measurement, logic and control, *PLoS ONE* **7** (2012) e46800.
[7] L. B. Kish, C. G. Granqvist, S. P. Khatri and F. Pepper, Zero and negative energy dissipation at information-theoretic erasure, *J. Comput. Electron.* **15** (2015) 335–339.
[8] L. B. Kish, C. G. Granqvist, S. P. Khatri and F. Peper, Response to Comment on 'Zero and negative energy dissipation at information-theoretic erasure', *J. Comput. Electron.* **15** (2015) 343–346.
[9] J. D. Norton, All shook up: Fluctuations, Maxwell's demon and the thermodynamics of computation, *Entropy* **15** (2013) 4432–4483.
[10] W. Porod, R. O. Grondin and D. K. Ferry, Dissipation in computation, *Phys. Rev. Lett.* **52** (1984) 232–235.

# NOISE-BASED LOGIC: WHY NOISE? A COMPARATIVE STUDY OF THE NECESSITY OF RANDOMNESS OUT OF ORTHOGONALITY

HE WEN*,† and LASZLO B. KISH*

*Department of Electrical and Computer Engineering
Texas A&M University, College Station
TX 77843-3128, USA

†College of Electrical and Information Engineering
Hunan University, Changsha 410082, China

Received 11 April 2012
Accepted 2 June 2012
Published 11 October 2012

Communicated by Zoltan Giugl

Although noise-based logic shows potential advantages of reduced power dissipation and the ability of large parallel operations with low hardware and time complexity the question still persist: Is randomness really needed out of orthogonality? In this Letter, after some general thermodynamical considerations, we show relevant examples where we compare the computational complexity of logic systems based on orthogonal noise and sinusoidal signals, respectively. The conclusion is that in certain special-purpose applications noise-based logic is exponentially better than its sinusoidal version: Its computational complexity can be exponentially smaller to perform the same task.

*Keywords*: Noise-based logic; random telegraph waves; sinusoidal signals; orthogonality; randomness.

## 1. Introduction

Recently, new, nonconventional ways of deterministic (nonprobabilistic) multi-valued *noise based logic* (*NBL*) system [1], stealth communications [2], and unconditionally secure communications [3] that are inspired by the fact that brain uses noise and its statistical properties for information processing, has been introduced for lower energy consumption and higher complexity parallel operations in post-Moore-law-chips.

The *NBL* uses electronic noise as information carrier, as shown in Fig. 1, where the logic 0 and 1 signals are represented by independent stochastic noise sources. In Fig. 1, an orthogonal system of random noise processes forms the reference signal system (orthogonal base) of logic values.

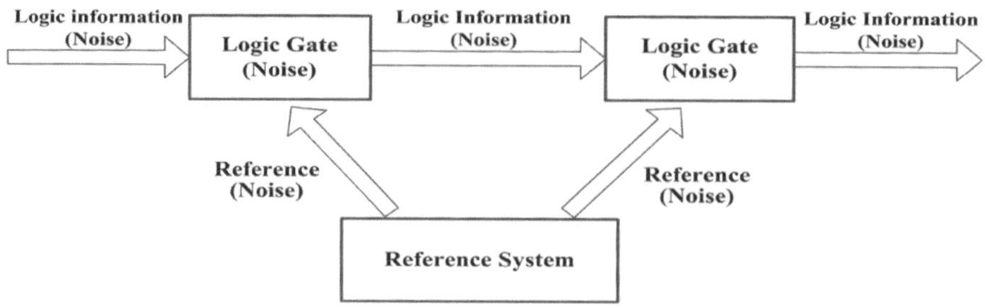


Fig. 1. Basic structure of noise-based logic system.

Within the *NBL* framework, *instantaneous noise-based logic* (*INBL*) [4, 5], has been introduced where the logic values are encoded into nonzero, bipolar, independent *random telegraph waves* (*RTW*). The *RTW*s used in the *INBL* are random square waves, which take the value of $+1$ or $-1$ with probability 0.5 at the beginning of each clock period and stay with this value during the rest of the clock duration. In the nonsqueezed *INBL*, for the $r$th noise bit, there are two reference *RTW*s, $H_r(t)$ and $L_r(t)$, representing its logic values, respectively.

Moreover, logic schemes based on other orthogonal processes, e.g., sinusoidal signals [6], have also been proposed to identify a signal without ambiguity. Thus, although the *NBL* has several potential advantages, such as reduced error propagation and power dissipation, it is still interesting to answer the question: Why noise? Is randomness really needed out of orthogonality? In this paper, we compare the *RTW*s and sinusoidal signals to illustrate the necessity and importance of randomness when representing logical values with orthogonal signals. For the sake of simplicity, this paper will concentrate on the instantaneous logic scheme and special-purpose applications.

In the rest of the paper, we will compare *NBL* with an alternative orthogonal deterministic realizations using sinusoidal signals. It was already pointed out for universal, correlator-based *NBL* that sinusoidal representations are inferior [1] and a subsequent circuit-analysis-based study confirmed the negative expectations [7]. However no analysis or comparison exists between instantaneous *NBL* and its sinusoidal alternatives.

## 2. General Arguments: The Case of Entropy

Preliminary considerations of the energy dissipation issue in biological informatics have been given in [8–10]. Noise-based logic was inspired by the stochastic neural signals in the brain [1] and a particular brain logic scheme utilizing stochastic signals was also proposed [9,10]. The human brain consumes about 10–20 Watt power while it operates with noise (stochastic neural spike sequences) as information carrier. The natural questions occurs: Is there a physical principle that favors stochastic processes for energy-friendly computation?

In this section, we consider the related aspects for noise-based logic by incorporating Brillouin's negentropy principle [11, 12]. When energy dissipation is an issue, the problem of entropy can provide a generic answer why certain types of noises may serve better than deterministic signals. Let us imagine an isolated system in thermal equilibrium.

According to Brillouin's negentropy principle [11, 12] if we set up an oscillator to generate a deterministic signal in a subvolume of an isolated thermodynamical system, that will unavoidably lead to entropy production, heat, emitted to the rest of the system. According to Brillouin's negentropy principle, the deterministic signal has negative entropy (negentropy) $-dS$ due to its reduced relative uncertainty. According to the second law of thermodynamics, the total entropy of the whole system cannot decrease, which means that the entropy in the rest of the system (out of the signal channel) must increase by $dS$. That means generating heat. Moreover, oscillators have loss, which means, to keep the oscillation going, a continuous energy injection is needed, which results in a steady heating power. Therefore, if we set up a system of $2N$ sinusoidal oscillators with different frequencies to provide orthogonal signals for the $N$-bits reference system, itself having the signal system will require an upfront energy dissipation and a subsequent steady power dissipation.

On the other hand, spontaneous fluctuations of thermal origin (thermal noise) are present in the system. As a relevant example for noise-based logic, we can utilize these thermal fluctuations as reference signals. For example, we can use the thermal noise of $2N$ resistors as the orthogonal signals for the reference signal system of $N$ bits noise-based logic. Having this signal system requires zero energy dissipation, it is granted by the existence of thermal fluctuations with zero energy requirement.

Even though, the above considerations are only about the reference signal system, and computations would require energy dissipation even with noise, this simple example indicates that noise can have a key role when energy-friendly computation is necessary.

## 3. Hardware Complexity

A general problem with today's binary logic is that, the DC voltage levels representing the different bits and bit values can be considered one-dimensional vectors [1]. These vectors are not orthogonal to each other thus, the classic binary logic cannot perform high parallelism with low hardware complexity.

However, the *NBL* can represent a multidimensional vector system utilizing the orthogonality of independent noises [1]. Generally, $2N$ independent (orthogonal) noises form $N$ noise-bits, where in a single noise bit $X$, one noise represents the $L$ and another the $H$ value of the bit, respectively. Products of the base vectors lead out from the base and form a *hyperspace* with $2^N$ dimensions [1, 13]. A logic signal that propagates in a single wire can be constructed by the binary superposition (on or off) of hyperspace vectors, which results in $2^{2^N}$ different logic values [2]. Consequently, $N$ noise-bits correspond to $2^N$ classical bits in a single wire [13].

For applications mimicking quantum computing, the $N$-bit long hyperspace vectors (product states), $W = X_1 X_2 \cdots X_N$ and their superpositions have key importance [13], where these product states represent the basis vectors of the $2^N$ dimensional Hilbert space of quantum informatics. Then, there are $2^N$ different products in an $N$-bit logic system. Thus, in a classical computer or utilizing the sinusoidal signals, the corresponding $2^N$ dimensional *universe*, which is the uniform superposition of all the $N$-bit long product states has complexity of $O(N^* 2^N)$ [13].

For applications, it is essential that the universe can be synthesized with low-complexity operations. With the Achilles ankle method, it can be generated [13] with the following operation that has only $O(2N)$ complexity

$$Y(2^N) = \prod_{r=1}^{N} (L_r + H_r). \tag{1}$$

The easy synthesis of the universe and the resulting high parallelism when manipulating it with simple, low-complexity operations, is very attractive. For example, when the number of noise-bits $N = 200$, a proper algebraic operation acting on this superposition in a single wire can potentially achieve a (special-purpose) parallel operation that would require a $200^* 2^{200}$ bits operation in classical computing [14].

## 4. Bandwidth (Time Complexity) of Product Strings

As described above, to set up a *hyperspace* with $2^N$ dimensions, $2N$ independent (orthogonal) noises are required to form $N$ noise-bits. Similarly, sinusoidal signals can also be used to represent the noise-bit values [1,7]. In this section, we compare the time complexity of *RTW* hyperspace based logic with the corresponding sinusoidal one. It will be shown that the required bandwidth of the sinus-based product string (hyperspace vector) will scale with $2^N$ to avoid the *degeneracy* of the system.

Time complexity in these special-purpose applications has two sides:

(a) The time complexity needed to set up the hyperspace vectors.
(b) The time complexity needed to analyze/decode the result which, in the practically important cases, is a single product string because, while large-scale parallel calculations are being executed on the superposition of hyperspace vectors; the final result, on which the actual measurement is done, is typically a single vector.

### 4.1. *Time complexity needed to set up the hyperspace vectors*

Using *RTWs* to represent the bit values results in no time complexity increase when setting up a hyperspace vector, because the *RTW* representing the hyperspace vector has the same clock frequency as its components.

Concerning the sinusoidal signal system, first let us suppose that we fill up the "harmonics space" linearly with the bit values; for example, the subsequent

Table 1. The logic values and corresponding frequencies for both the linear and exponential frequency representation.

| Bit | Logic value | Frequency | |
|---|---|---|---|
| | | Linear Representation | Exponential Representation |
| 1st | $L_1$ | $f_0$ | $f_0$ |
| | $H_1$ | $2f_0$ | $2f_0$ |
| 2nd | $L_2$ | $3f_0$ | $4f_0$ |
| | $H_2$ | $4f_0$ | $8f_0$ |
| — | — | — | — |
| $N$th | $L_N$ | $(2N-1)f_0$ | $2^{2N-2}f_0$ |
| | $H_N$ | $2Nf_0$ | $2^{2N-1}f_0$ |

odd harmonics represent the subsequent Low bit values and the subsequent even harmonics the subsequent High values (*linear frequency representation*, see Table 1). Thus to represent $N$ "sinus-based" bits, $2N$ different harmonics are needed. In this case, we can define the logic values of the $r$th bit as

$$L_r(t) = \mathbf{e}^{j2\pi(2r-1)f_0t} \quad \text{and} \quad H_r(t) = \mathbf{e}^{j2\pi 2rf_0t}. \tag{2}$$

In this case, the highest frequency in the system will be the sum of all the utilized harmonics from 1 to $2N$, that is, $N(2N+1)f_0$, which is scaling with $N^2$. However, this system will be *degenerated* because of the frequency-overlapping when forming the hyperspace vector. For example, the product $L_1H_2$ and $H_1L_2$ will produce the same harmonics. This kind of degeneracy will cause significant information loss and make the system virtually useless.

Second, let us suppose that we fill up the "harmonics space" exponentially (*exponential frequency representation*) where the frequencies of harmonics form a 2-based geometric sequence. Then the subsequent bit values are represented by harmonics that scale with the power of 2 (see also Table 1):

$$L_r(t) = \mathbf{e}^{j2\pi 2^{2r-2}f_0t} \quad \text{and} \quad H_r(t) = \mathbf{e}^{j2\pi 2^{2r-1}f_0t}. \tag{3}$$

As a result, the highest frequency in the system will be the sum of all the utilized harmonics which is $(2^{2N}-1)f_0$ that is scaling with $2^{2N}$. In conclusion to set up a single hyperspace vector of a sinus-based system requires exponential time complexity $O(2^{2N})$.

### 4.2. *Time complexity needed to analyze/decode the result (hyperspace vector)*

To read out the bit values of this hyperspace vector requires a Fourier analysis and a time window of $1/f_0$. However, this time window contains frequency components as high as $(2^{2N}-1)f_0$, which means that the time complexity of the read-out operation is also $O(2^{2N})$.

Concerning the reading out the noise-bit values in a hyperspace vector Stacho [15] has developed an algorithm that provides an exponential speedup compared to a "brute force" determination. Stacho's algorithm provides that, when utilizing less than $t_\epsilon = N(\log_2 N)^{(1+\epsilon)}$ clock periods, the probability $P$ that the measurement algorithm fails is scaling as $P \propto 2^{-N}$ in the limit of large $N$, where for $N \to \infty$, $\epsilon \to 0$.

In time-shifted *NBL*, there is a polynomial (linear in $2N$) time complexity in setting up the hyperspace vector and also the same $O(N)$ complexity to read out the values, specifically, $2N * \log_4(N/P)$ of time steps are needed, which in this system means $\log_4(N/P)$ clock periods [14].

## 5. Conclusions

Our goal was to answer the following question: *Does noise-based logic really need "noise" (stochasticity) or an orthogonal system of sinusoidal signals is enough with similar computational complexity*?

Our answer is "noise" must be an essential component of these special-purpose, orthogonal-signal-based logic systems for various reasons including low power dissipation (low entropy production) and low computational complexity.

## Acknowledgments

Discussions with Sergey Bezrukov, Andreas Klappenecker and Laszlo Stacho are appreciated. This work has partially been supported by the National Natural Science Foundation of China under grant 61002035.

## References

[1] L. B. Kish, Noise-based logic: Binary, multi-valued, or fuzzy, with optional superposition of logic states, *Phys. Lett. A* **373** (2009) 911–918.
[2] L. B. Kish, Stealth communication: Zero-power classical communication, zero-quantum quantum communication and environmental-noise communication, *Appl. Phys. Lett.* **87** (2005) 234109.
[3] L. B. Kish, Totally secure classical communication utilizing Johnson (-like) noise and Kirchoff's law, *Phys. Lett. A* **352** (2006) 178–182.
[4] L. B. Kish, S. Khatri and F. Peper, Instantaneous noise-based logic, *Fluct. Noise Lett.* **9** (2010) 323–330.
[5] F. Peper and L. B. Kish, Instantaneous, non-squeezed, noise-based logic, *Fluct. Noise Lett.* **10** (2011) 231–237.
[6] M. Ueda, M. Ueda, H. Takagi, M. J. Sato, T. Yanagida, I. Yamashita and K. Setsune, Biologically-inspired stochastic vector matching for noise-robust information processing, *Physica A: Stat. Mech. Appl.* **387** (2008) 4475–4481.
[7] K. C. Bollapalli, S. P. Khatri and L. B. Kish, Implementing digital logic with sinusoidal supplies, *Proc. Automation & Test in Europe Conference & Exhibition (DATE)*, (2010), pp. 315–318.
[8] S. M. Bezrukov and L. B. Kish, How much power does neural signal propagation need?, *Smart Mater. Struct.* **11** (2002) 800.

[9] S. M. Bezrukov and L. B. Kish, Deterministic multivalued logic scheme for information processing and routing in the brain, *Phys. Lett. A* **373** (2009) 2338–2342.

[10] Z. Gingl, S. Khatri and L. B. Kish, Towards brain-inspired computing, *Fluct. Noise Lett.* **9** (2010) 403–412.

[11] L. Brillouin, *Scientific Uncertainty and Information* (Academic Press, New York, 1964).

[12] L. Brillouin, *Science and Information Theory* (Academic Press, New York, 1962).

[13] L. B. Kish, S. Khatri and S. Sethuraman, Noise-based logic hyperspace with the superposition of 2(N) states in a single wire, *Phys. Lett. A* **373** (2009) 1928–1934.

[14] H. Wen, L. B. Kish, A. Klappenecker and F. Peper, New noise-based logic representations to avoid some problems with time complexity, *Fluct. Noise Lett.* **11**(2) (2012) 1250003 (8 pages); http://arXiv.org/abs/1111.3859.

[15] L. Stacho, Fast measurements of hyperspace vectors in noise-based logic, *Fluct. Noise Lett.*, in press.

# INSTANTANEOUS, NON-SQUEEZED, NOISE-BASED LOGIC

FERDINAND PEPER* and LASZLO B. KISH†

*National Institute of Information and Communications Technology
Kobe, 651-2492 Japan

†Texas A&M University
Department of Electrical and Computer Engineering
College Station
TX 77843-3128, USA

Received 16 December 2010
Accepted 2 February 2011

Communicated by Igor Goychuk

Noise-based logic, by utilizing its multidimensional logic hyperspace, has significant potential for parallel operations in beyond-Moore-chips. However universal gates for Boolean logic thus far had to rely on either time averaging to distinguish signals from each other or, alternatively, on squeezed logic signals, where the logic-high was represented by a random process and the logic-low was a zero signal. A major setback is that squeezed logic variables are unable to work in the hyperspace, because the logic-low zero value sets the hyperspace product vector to zero. This paper proposes Boolean universal logic gates that alleviate such shortcomings. They are able to work with non-squeezed logic values where both the high and low values are encoded into nonzero, bipolar, independent random telegraph waves. Non-squeezed universal Boolean logic gates for spike-based brain logic are also shown. The advantages versus disadvantages of the two logic types are compared.

*Keywords*: Noise-based logic; noise-driven informatics; brain logic; multi-valued logic; deterministic logic.

## 1. Introduction

Noise-based informatics has received increasing attention in recent years. In noise-based logic [1–6], the information is carried by the noise, while in noise-driven information systems [7–10], noise is used to drive the system. In noise-based communication, the randomness of noise is utilized for secure key exchange [11–17].

*Instantaneous noise based logic* (*INBL*) [1–3] is a noise-based logic scheme [1–6] in which the logic 0 and 1 signals are represented by independent noise sources, whereby operations on the signals do not involve time-averaging. Unlike the noise-based schemes in [4, 5], where the noise sources are continuously valued and time-averaging is intensively employed in the input correlators of gates, *INBL* relies on so-called Random-Telegraph Waves (*RTW*) [1, 6] or random neural spike sequences [2, 3]. *RTWs* are random square waves, valued $-1$ or $+1$, both with probabilities 0.5 at discrete points in time [1, 6]. *INBL* schemes like in [1] thus allow fast logic operations without time averaging. Similarly to the continuum noise based logic schemes [4, 5], signals represented by *RTWs* can be combined into hyperspace vectors (see [6] for an application) by multiplication. In the brain logic scheme [2, 3], which is also *INBL* [1], the hyperspace is combined by the addition of spike sequences [2, 3].

The Boolean *INBL* schemes proposed in [1] have an important problem in the formation of hyperspace: only the logic *High* ($H$) value is represented by *RTWs*, while the logic *Low* ($L$) value is squeezed to zero, out of necessity to have a useful AND gate logical operation. This squeezing of waves causes zero vectors to appear in the product hyperspace of *RTWs*, and that fact prohibits the direct use of combinations of such $L$ logic values in a hyperspace product containing several bits because it would reset the whole product vector to zero. A similar situation occurs in the Boolean brain logic scheme with non-overlapping spike sequences in [1], where the hyperspace is additive.

This paper presents *RTW* and spike based Boolean *INBL* schemes that represent both their $H$ and $L$ values by non-zero noise voltages. We call them non-squeezed *INBL* schemes. They have a better potential to utilize the multidimensional logic hyperspace.

## 2. Random Telegraph Wave Based Non-Squeezed Logic

Let us suppose that the noise functions $H(t)$ and $L(t)$ representing the Boolean logic values are independent (orthogonal) *RTWs* with independent random $+1$ or $-1$ values. The $H$ and $L$ RTWs must be distributed in the whole processor as reference signals (see [4]). These two independent RTWs can for example be generated by amplifying/thresholding the noise of two separate CMOS transistors. If we are able to realize the NOT and the AND Boolean logic functions with this system then this logic scheme is universal [1].

The NOT gate operation on an input value $X(t)$ of this type non-squeezed *INBL* logic can be realized in two different ways. The first way is additive:

$$Y = \text{NOT } X = U(t) - X(t), \tag{1}$$

where the universe $U(t)$ is defined as

$$U(t) = H(t) + L(t). \tag{2}$$

The universe is not an *RTW* because its value can be $-2, -1, 0, 1,$ and 2. It follows immediately from this definition that if $X(t) = L(t)$ then $Y(t) = H(t)$ and if $X(t) = H(t)$, then $Y(t) = L(t)$.

The above NOT gate can be realized in hardware by a linear differential amplifier, or by a proper logic circuitry that can input multiple voltage levels ($-2, -1, 0, 1,$ and 2) and output the proper $+1/-1$ levels for *RTW* generation. The last circuit would be quite involved due to the need of comparators to detect the five different levels of the actual universe value at the input.

The other, more advantageous way of realizing the NOT gate is via the relation:

$$Y = \text{NOT } X = X(t)H(t)L(t). \tag{3}$$

This function is a NOT function because

$$L^2(t) = H^2(t) = 1 \tag{4}$$

and because multiplication of *RTWs* is commutative. The advantage of the NOT function defined by Eq. (3) is that it does not require the use of the additive universe in Eq. (2), so the amplitude values remain +1 and −1 during the computation. Such a circuit can be realized by switches and binary logic operations because only binary signals exist in Eq. (3).

The AND gate operation on inputs $X_1$ and $X_2$ is defined by

$$Y = X_1 \text{ AND } X_2 = \frac{1}{4}[H(t) - L(t)][X_1(t) - L(t)][X_2(t) - L(t)] + L(t). \tag{5}$$

When one or more of the inputs is $L(t)$, the first term of (5) becomes 0, which makes the output $Y(t)$ equal to the remaining additive term $L(t)$. When both inputs are $H(t)$, we obtain a more complicated expression, which can be simplified by using the relation:

$$\left\{\frac{[H(t) - L(t)]}{2}\right\}^3 = \frac{H(t) - L(t)}{2}. \tag{6}$$

This relation is easily verified by expanding the left-hand-side of Eq. (6) and using Eq. (4) to simplify the expression.

The AND gate can be realized in hardware either by analog circuitry (three differential amplifiers, two multipliers, and one adder) or by proper binary logic combined with analog comparator circuitry, which can receive and analyze multiple discrete voltage levels and output the proper $+1/-1$ levels for *RTW* generation. The last circuit would be quite involved.

Since both the AND and NOT gates can be realized, we have proven that the RTW-based non-squeezed instantaneous logic scheme defined above is universal.

## 3. Random Spike Sequence (Brain) Based Non-Squeezed Logic

In [2], a model for instantaneous, deterministic, multi-valued logic was presented, as a possible brain logic scheme. In this logic, set-theoretical operations on randomly occurring, non-overlapping (orthogonal) uniform and unipolar neural spike trains were used to construct logic superpositions and identify components in them. The key neural circuit element is the *orthon*, see Fig. 1, which consists of two neurons. The idealized neurons (see Fig. 1, left hand side) have two inputs, an excitatory (+) and an inhibitory (−) one. Pulses arriving at (+) will propagate to the output of the neuron, unless at the same time a pulse arrives also at the (−) input. These idealistic

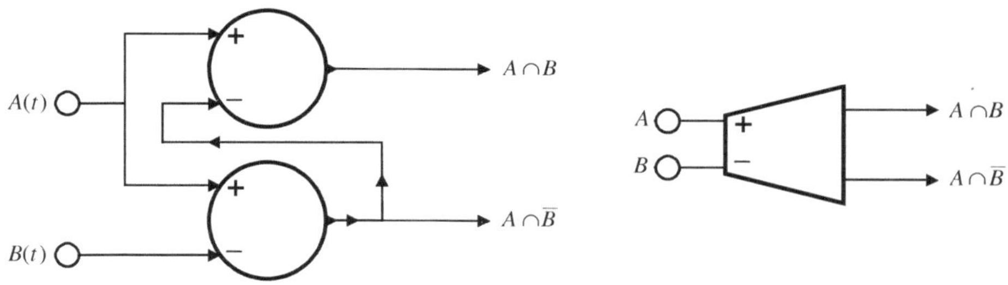


Fig. 1. Left: the neural circuit of the orthon [2]. Right: its symbol.

neurons are assumed to be free of delays [2, 3]. The orthon [2] has two inputs and the $A$ and $B$ spike trains are treated as sets of spikes. The upper output provides the set-theoretical $AB$ ($A \cap B$) operation, where the overlapping spikes of $A$ and $B$ are kept and the rest are discarded. The lower output, on the other hand, provides the set theoretical $A\bar{B}$ ($A \cap \bar{B}$) operation, where the spikes of $A$ not overlapping with $B$ are kept and the rest are discarded.

In [1], universal Boolean logic gates for the squeezed $H(t) \neq 0$, $L(t) = 0$, version of spike-based logic were shown. Here we introduce the universal gates for the non-squeezed logic, where the logic values $H$ and $L$ are represented by the orthogonal (non-overlapping) spike trains $H(t)$ and $L(t)$, respectively, which are non-empty sequences. The orthogonality condition is

$$H(t) \cap L(t) = 0, \tag{7}$$

that is, the spikes in the two sequences do not overlap. We assume that the amplitude of a spike is 1, and that the inverse signal $\bar{X}(t)$ of $X(t)$ at time $t$ is defined by $\bar{X}(t) = 1 - X(t)$.

The NOT gate is then defined as:

$$Y(t) = \text{NOT } X = \bar{X}(t) \cap U(t) = \bar{X}(t) \cap [L(t) \cup H(t)], \tag{8}$$

where the universe $U(t)$ is the union (sum) of $H(t)$ and $L(t)$, i.e.,

$$U(t) = L(t) \cup H(t). \tag{9}$$

This immediately implies that, for $X(t) = H(t)$, the output becomes $Y(t) = L(t)$ and, for $X(t) = L(t)$, the output becomes $Y(t) = H(t)$.

The AND gate is defined as:

$$Y = X_1 \text{ AND } X_2 = [X_1(t) \cap X_2(t) \cap H(t)] \cup [X_1(t) \cap L(t)] \cup [X_2(t) \cap L(t)]. \tag{10}$$

The existence of both a NOT gate and an AND gate implies that the non-squeezed spike-based instantaneous logic scheme defined above is universal as well.

The neural hardware of the orthon-based representation of the NOT gate is shown in Fig. 2.

The neural circuitry of the AND gate is shown in Fig. 3. It consists of four orthons and an "adder" neuron with three excitatory inputs.

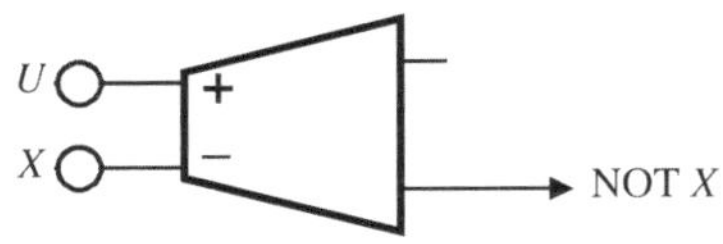


Fig. 2. The random spike sequence based binary NOT gate utilizes the universe and the orthon element of the brain logic scheme. The orthon [2] is defined in Fig. 1. The upper output is not used in the circuit.

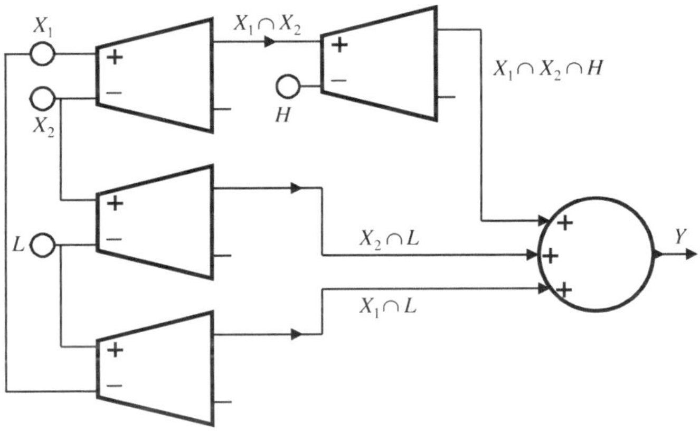


Fig. 3. The AND gate of the brain logic. The orthon [2] is defined in Fig. 1. The output neuron is summing the input pulses received at its excitatory inputs in an additive fashion (set theoretical union function).

The spike based logic scheme is clearly an instantaneous logic scheme. This is because the output is determined, without uncertainty, as soon as the first spike of the given logic value ($H$ or $L$) emerges.

## 4. Conclusions and Discussion

Two schemes of non-squeezed noise-based logic are described that are instantaneous. The lack of a need to average signals over time translates in a potentially fast on-line operation, in which inputs are operated upon as they flow in. The first scheme, which is based on Random Telegraph Waves, delivers a result of which the reliability increases as the bits arrive. The probability of ambiguity i.e., when the $H(t)$ and $L(t)$ are identical for $n$ clock steps, is $0.5^n$ [6], which translates to high reliability with relatively few bits. To achieve a reliability of $10^{-25}$ (the reliability of an idealistic CMOS gate), for example, only 83 bits are required [6].

The spike-based brain logic does even better: after only one spike, the result is established, with the remaining spikes serving as a verification of the result. This high speed of the spike-based logic takes advantage of the property that spikes in different orthogonal trains do not overlap. Naturally, the high speed relates to the question of how long one should continue to receive spikes before cutting off a signal and switching to the next spike train. This time appears to be limited by the required reliability and, possibly, by the time period over which a certain spike train is needed to perform a particular function.

The hyperspace, in which signals can be combined before operations are conducted on them collectively, is described in more detail in [1, 6]. The logic schemes presented here are more suitable for use in hyperspace, because they do not suffer from zero-multiplications in the squeezed instantaneous noise logic [1].

## Acknowledgments

We are grateful for discussions with Sergey Bezrukov (National Institute of Health) and Sunil Khatri (Texas A&M University). LBK's visit to NICT, Japan, was supported by the NetSci subproject of NICT.

## References

[1] L. B. Kish, S. Khatri and F. Peper, Instantaneous noise-based logic, *Fluctuation and Noise Letters* **9** (2010) 323–330.
[2] S. M. Bezrukov and L. B. Kish, Deterministic multivalued logic scheme for information processing and routing in the brain, *Phys. Lett. A* **373** (2009) 2338–2342.
[3] Z. Gingl, S. Khatri and L. B. Kish, Towards brain-inspired computing, *Fluctuation and Noise Letters* **9** (2010) 403–412.
[4] L. B. Kish, Noise-based logic: Binary, multi-valued, or fuzzy, with optional superposition of logic states, *Phys. Lett. A* **373** (2009) 911–918.
[5] L. B. Kish, S. Khatri and S. Sethuraman, Noise-based logic hyperspace with the superposition of $2^N$ states in a single wire, *Phys. Lett. A* **373** (2009) 1928–1934.
[6] L. B. Kish, S. Khatri and T. Horvath, Computation using noise-based logic: Efficient string verification over a slow communication channel, *European Journal of Phys. B*, (2010). http://arXiv.org/abs/1005.1560.
[7] A. K. Kikombo, T. Asai, T. Oya, Y. Leblebici and Y. Amemiya, A neuromorphic single-electron circuit for noise-shaping pulse-density modulation, *Int. J. Nanotechnology and Molecular Computation* **1** (2009) 80–92.
[8] K. Leibnitz, N. Wakamiya and M. Murata, Biologically inspired self-adaptive multipath routing in overlay networks, *Commun. ACM* **49** (2006) 63–67.
[9] F. Peper, J. Lee and T. Isokawa, Brownian cellular automata, *J Cell. Automat.* **5** (2010) 185–206.
[10] J. Lee and F. Peper, On Brownian cellular automata, *Proc. Automata 2008*, pp. 278–291.
[11] L. B. Kish, Totally secure classical communication utilizing Johnson (-like) noise and Kirchoff's law, *Physics Lett. A* **352** (2006) 178–182.
[12] R. Mingesz, Z. Gingl and L. B. Kish, Johnson(-like)–Noise–Kirchhoff-loop based secure classical communicator characteristics, for ranges of two to two thousand kilometers, via model-line, *Phys. Lett. A* **372** (2008) 978–984.
[13] L. B. Kish, Protection against the man-in-the-middle-attack for the Kirchhoff-loop-Johnson(-like)-noise cipher and expansion by voltage-based security, *Fluctuation and Noise Letters* **6** (2006) L57–L63.
[14] L. B. Kish and P. Mingesz, Totally secure classical networks with multipoint telecloning (teleportation) of classical bits through loops with Johnson-like noise, *Fluctuation and Noise Letters* **6** (2006) C9–C21.

[15] D. Bar-Lev and J. Scheuer, Enhanced key-establishing rates and efficiencies in fiber laser key distribution systems, *Phys. Lett. A* **373** (2009) 4287–4296.
[16] P.-L. Liu, Prediction accuracy of band-restricted random signals and security risk in statistical key exchange, *Fluctuation and Noise Letters* **9** (2010) 413–422.
[17] L. L. Kish, M. Zhang and L. B. Kish, Cracking the Liu key exchange protocol in its most secure state with Lorentzian spectra, *Fluctuation and Noise Letters* **9** (2010) 37–45.

# FAST MEASUREMENT OF HYPERSPACE VECTORS IN NOISE-BASED LOGIC

L. L. STACHÓ

*Bolyai Institute, Szeged University*
*Aradi Vértanúk tere 1, Szeged, H-6720, Hungary*
*stacho@math.u-szeged.hu*

Received 26 March 2012
Revised 26 March 2012
Published 9 July 2012

Communicated by Zoltan Gingl

We propose a method to determine single hyperspace vectors (product strings of noise-bits) by classical means with the same effectiveness as the results using time shifted noise-based logic. A system of binary linear equations based on the amplitudes of the hyperspace vector and the reference noise-bits is set up and solved after enough independent information is collected. The resulting error probability (the chance of getting no answer) has approximately an exponential decay with the time of measurement.

*Keywords*: Noise-based logic (NBL); hyperspace vector; random telegraph wave (RTW).

## 1. Introduction

Noise-based logic (NBL) [1–9], on the contrary of its name, is a deterministic logic scheme. The logic information is carried by independent stochastic processes, their products and their superpositions [1]. In nonsqueezed NBL, $2N$ independent reference noises are needed to form $N$ noise-bits: one for each $L$ and another one for each $H$ value and $N$-bit long products (product-strings) represent hyperspace vectors.

The identification of a single product-string (hyperspace vector) is necessary after a quantum-computing-type calculation with a simple output (such as the binary representation of a prime number) and it is one of the challenging questions. Recently, "time shifted noise-based logic" has been introduced [9], which is constructed by shifting each reference signal with a small time delay. This modification implies an exponential speedup of single hyperspace vector identification compared to the former case and it requires the same, $O(N)$ complexity as in quantum computing.

The goal of the present paper is to show that the identification of such a single hyperspace vector can be done with similar complexity without the need of time shifted noise-based logic. Thus this task can be carried out within the original noise-based logic framework, which uses the same clock timing for each reference component.

## 2. Problem Setting

Let $N$ denote the number of bits and for $k = 1, 2, \ldots, N$ let

$$A_k, B_k : \{1, 2, \ldots\} \to \{-1, 1\} \tag{1}$$

be random telegraph waves (RTW). As for the physical realization, each function $A_k$ or $B_k$ represents a sequence of electric impulses with intensity $A_k(t)$ (or $B_k(t)$) in every time interval $(t-1, t)$ between consecutive clock signals for $t = 1, 2, \ldots$. We are given a binary string $x_0, x_1, \ldots, x_N$ consisting of terms 0 or 1 as usually which will be transformed into the *product wave*

$$W := X_1 X_2 \cdots X_N \quad \text{where} \quad \begin{aligned} X_k &= A_k \quad \text{if } x_k = 1 \quad \text{and} \\ X_k &= B_k \quad \text{if } x_k = 0. \end{aligned} \tag{2}$$

**Problem 1.** (Measurement problem). *Knowing only the product wave $W$, determine the* 0-1 *string $x_1, \ldots, x_N$ satisfying* (2).

Our purpose will be to develop an algorithm based on an infinite system of binary linear algebraic equations describing the relationships encoded in (2). Given the waves $W, A_1, B_1, A_2, B_2, \ldots, A_N, B_N$, for $T = 1, 2, \ldots$, our algorithm reads the impulses $W(T), A_1(T), B_1(T), \ldots, A_N(T), B_N(T)$ and it determines if there is only a unique string $x_1, \ldots, x_N$ suitable for solving Problem 1 on the basis of the gathered data as far. Having found the first time

$$S = S(W, A_1, B_1, \ldots, A_N, B_N) \tag{3}$$

with this uniqueness property, the algorithm provides the precise guess for the sequence $x_0, x_1, \ldots, x_N$ and then stops. We shall call $S$ the *time requirement for complete measurement.* In Algorithm 11 below we formulate a precise algebraic construction for $S$. Notice that the possibility $S = \infty$ cannot be excluded as e.g., if $W(t) = A_k(t) = B_k(t) \equiv 1$. However, such algebraically extreme cases are practically improbable due to the random construction of the waves $A_k$ and $B_k$. Actually, throughout the whole paper we regard the terms $A_k(t), B_k(t)$ ($k = 1, \ldots, N$; $t \in \mathbf{Z}_+$) to be *completely independent random variables* such that

$$\mathbf{Prb}(A_k(t) = (-1)^s) = \mathbf{Prb}(B_k(t) = (-1)^s) = 1/2 \quad (s = 0, 1). \tag{4}$$

**Definition 2.** Introduce the below *transformed variables* respectively *product wave*:

$$C_k := A_k B_k \ (k = 1, \dots, N), \quad U := W B_1 B_2 \cdots B_N. \tag{5}$$

**Observation 3.** Due to the identities $A_k^2 = B_k^2 = 1$, we have $X_k = C_k^{x_k} B_k$, and hence $X_1 X_2 \cdots X_N = W \iff (X_1 B_1)(X_2 B_2) \cdots (X_N B_N) = U$ that is

$$C_1^{x_1} C_2^{x_2} \cdots C_N^{x_N} = U. \tag{6}$$

**Definition 4.** Introduce the matrix $\mathbf{C} := \big[c_{tk}\big]_{t=1}^{\infty}{}_{k=1}^{N}$ respectively the column vector $\mathbf{u} := \big[u_t\big]_{t=1}^{\infty}$ as follows: we let

$$c_{tk} := [1 \text{ if } C_k(t) = -1, \ 0 \text{ if } C_k(t) = 1], \tag{7}$$

$$u_t := \big[1 \text{ if } U(t) = -1, \ 0 \text{ if } U(t) = 1\big]. \tag{8}$$

**Observation 5.** For any time $t$ we have

$$(-1)^{u(t)} = U(t) = C_1(t)^{x_1} \cdots C_N(t)^{x_N} = (-1)^{c_{t1}x_1 + \cdots + c_{tN}x_N}.$$

Hence the relationship $U = C_1^{x_1} \cdots C_N^{x_N}$ is equivalent to the following (infinite) system of linear equations **over the field** $\mathbf{Z}_2 := \{0, 1\}$ of two elements

$$\begin{array}{c} c_{11}x_1 + c_{12}x_2 + \cdots + c_{1N}x_N = u_1 \\ c_{21}x_1 + c_{22}x_2 + \cdots + c_{2N}x_N = u_2 \\ \vdots \qquad \vdots \qquad \vdots \qquad \vdots \\ c_{t1}x_1 + c_{t2}x_2 + \cdots + c_{tN}x_N = u_t \\ \vdots \qquad \vdots \qquad \vdots \qquad \vdots \end{array} \tag{9}$$

**Remark 6.** The answer to Problem 1 is nothing else than to find the solution $x_1, \dots, x_N \in \mathbf{Z}_2$ of the above infinite system of linear equations. Notice that this system has a solution by assumption. On the other hand, for $T \to \infty$ the columns of the submatrix $\mathbf{C}^{(T)} := [c_{tk}]_{t=1}^{\infty}{}_{k=1}^{N}$ become linearly independent (over $\mathbf{Z}_2$) with probability 1 since the variables $C_k(t)$ $(1 \le t \le T, \ 1 \le k \le N)$ are completely independent in stochastic sense.

**Definition 7.** Let $S$ denote the minimal index for which the columns of $\mathbf{C}^{(S)}$ are linearly independent (over $\mathbf{Z}_2$). Notice that $S$ is also a random variable with $\lim_{s \to \infty} \mathbf{Prb}(S > s) = 0$. Henceforth let us write

$$\pi_s := \mathbf{Prb}(S = s), \quad \overline{\pi}_t := \mathbf{Prb}(S \ge t). \tag{10}$$

Notice that $\pi_t$ and $\overline{\pi}_t$ are the probabilities for our measurement algorithm to *stop exactly at time* $t$ and to *fail before time* $t$, respectively. Actually we have $\overline{\pi}_t = \sum_{s=t}^{\infty} \pi_s$.

**Problem 8.** *Find sharp asymptotic estimates for* $\overline{\pi}_t$.

**Problem 9.** *Considering large values $N$ of bits, is it possible to find a function $N \mapsto t_N$ such that $t_N \ll 2^N$ and $\overline{\pi}_{t_N}$ tends to zero with exponential rate as $N \to \infty$.*

**Remark 10.** $2^N$ is the lowest time limit $T$ for the complete orthogonality of the product functions $C_1^{x_1} \cdots C_N^{x_N}$ when restricted to the interval $\{1, \ldots, T\}$.

## 3. Results

To answer Problems 8 and 9, we are going to investigate the Gaussian elimination algorithm applied to the matrix $\mathbf{C}$ establishing row indices $t_1 < t_2 < \cdots < t_N$ such that row $t_1$ is the first nonvanishing row in $\mathbf{C}$ and, for each $k = 2, \ldots, N$, row $t_k$ is the least index with row $t_k$ being linearly independent of rows $t_1, \ldots, t_{k-1}$ in $\mathbf{C}$. To do so, we can proceed to find entries with indices $(t_1, k_1), (t_2, k_2), \ldots, (t_N, k_N)$ in the following manner.

**Algorithm 11.** Let $\mathbf{G}_0 := \mathbf{C}$ the matrix introduced in Definition 2, $K_0 := \{1, \ldots, N\}$.

- Step (1) Let $t_1$ be the row index of the first nonzero row in $\mathbf{G}_0$ and let $k_1$ be the column index of the first nonzero entry in this row. Then we set $K_1 := K_0 \backslash \{k_1\}$ and we let $\mathbf{G}_1$ to be the matrix obtained with elimination killing the subcolumn below the entry $t_1, k_1)$.

  ⋮

- Step (d) Let $t_d$ be the first row index $> t_{d-1}$ such that row $t_d$ in $\mathbf{G}_{d-1}$ does not vanish but all the rows of $\mathbf{G}_{d-1}$ vanish between row $t_{d-1}$ and $t_d$. Then let $k_d$ be the minimal index in $K_{d-1}$ such that the entry of $\mathbf{G}_{d-1}$ with index $(t_d, k_d)$ does not vanish. We define $K_d := K_{d-1} \backslash \{k_d\}$ and the matrix $\mathbf{G}_d$ is obtained with elimination from $\mathbf{G}_{d-1}$ by killing the subcolumn below the entry $(t_d, k_d)$.

**Theorem 12.** *Let $0 = t_0 < t_1 < t_2 < \cdots < t_N$ be arbitrarily given integers and let $n_d := t_d - t_{d-1} - 1$ $(d = 1, \ldots, N)$. Then we have*

$$\mathbf{Prb}(\text{the Algorithm produces the row indices } t_1, \ldots, t_N)$$
$$= \prod_{d=0}^{N-1} [2^{n_d(N-d)}(1 - 2^{d-N})].$$

**Proof.**

$$\mathbf{Prb}(\text{the Algorithm produces the row indices } t_1, \ldots, t_N)$$
$$= \Bigg[\mathbf{Prb}(\text{the } Nn_0 \text{ entries of the first } n_0 \text{ rows vanish})$$
$$\times \sum_{r=0}^{N-1} \mathbf{Prb}(r \text{ zeros stand before a term 1 in row } t_1)\Bigg]$$
$$\times [\mathbf{Prb}(\text{the } (N-1)n_1 \text{ entries of the first } n_1 \text{ rows after the } t_1\text{th vanish})$$

$$\times \sum_{r=0}^{N-2} \mathbf{Prb}(r \text{ zeros before a 1 in the part with column indices in } K_1 \text{ of row } t_2)]$$

$$\vdots$$

$$\times \Bigg[\mathbf{Prb}(\text{the } (N-1)n_1 \text{ entries of the first } n_1 \text{ rows after the } t_1\text{th vanish})$$

$$\times \sum_{r=0}^{N-2} \mathbf{Prb}(r \text{ zeros before a 1 in the part with column indices in } K_1 \text{ of row } t_2)\Bigg].$$

Since the entries of $\mathbf{C}$ assume the values 0-1 with probability 1/2 independently, hence

$$\begin{aligned}
&\mathbf{Prb}(\text{the Algorithm produces the row indices } t_1, \ldots, t_N) \\
&= \prod_{d=0}^{N-1} [2^{-n_d(N-d)}(2^{-1} + 2^{-2} + \cdots + 2^{d-N})] \\
&= \prod_{d=0}^{N-1} [2^{-n_d(N-d)}(1 - 2^{d-N})].
\end{aligned}$$

□

**Corollary 13.** *For any $s \geq N$ we have $\pi_s \leq \binom{s}{N} 2^{N-s}$.*

**Proof.** We know that $\pi_s = \sum_{0=t_0<\cdots<t_N=s} \mathbf{Prb}(\text{the Algorithm produces the row indices } t_1, \ldots, t_N)$. Therefore

$$\begin{aligned}
\pi_s &= \sum_{n_0+\cdots+n_{N-1}=s-N} \prod_{d=0}^{N-1} [2^{-n_d(N-d)}(1 - 2^{d-N})] \\
&\leq \sum_{n_0+\cdots+n_{N-1}=s-N} \exp_2\left(-\sum_{d=0}^{N-1} n_d(N-d)\right) \\
&\leq \sum_{n_0+\cdots+n_{N-1}=s-N} \exp_2\left(-\sum_{d=0}^{N-1} n_d\right) = \binom{s}{N} 2^{N-s},
\end{aligned}$$

due to the combinatorial fact that $\#\{(n_0, \ldots, n_{N-1}) : n_0 + \cdots + n_{N-1} = s - N\}$ coincides with the number of repeated combination of dividing $s - N$ elements into $N$ parts. □

**Theorem 14.** *For $t \geq N$ we have $\overline{\pi}_t \leq 2^{N+1-t} \sum_{k=0}^{N} \binom{t}{k}$.*

**Proof.** Observe that, in general, $\binom{s}{N} x^{N-s} = \frac{1}{N!} \frac{d^N}{dx^N} x^s$. Thus, by Corollary 13,

$$\overline{\pi}_t = \sum_{s=t}^{\infty} \pi_s \leq \sum_{s=t}^{\infty} \binom{s}{N} \left(\frac{1}{2}\right)^{N-s} = \frac{1}{N!} \frac{d^N}{dx^N} x^t (1-x)^{-1}\Bigg|_{x=1/2}. \tag{11}$$

According to Leibniz' rule,

$$\begin{aligned}\frac{d^N}{dx^N}x^t(1-x)^{-1} &= \sum_{k=0}^{N}\binom{N}{k}\frac{d^k}{dx^k}x^t\frac{d^{N-k}}{dx^{N-k}}(1-x)^{-1}\\ &= \sum_{k=0}^{N}\binom{N}{k}t(t-1)\cdots(t-k+1)x^{t-k}(-1)^{N-k}\\ &\quad\times(N-k)!(1-x)^{-(N-k+1)}.\end{aligned}$$

Taking into account that $x^{t-k}(1-x)^{-(N-k+1)}|_{x=1/2} = 2^{k-t+N-k+1} = 2^{N+1-t}$, from (11) we get

$$\overline{\pi}_t \leq \frac{1}{N!}\sum_{k=0}^{N}\binom{N}{k}\frac{t!}{(t-k)!}(N-k)!\,2^{N+1-t} = \sum_{k=0}^{N}\binom{t}{k}2^{N+1-t}. \qquad \square$$

**Corollary 15.** *For $t \geq N$ we have $\overline{\pi}_t \leq 2et^N/2^{t-N}$.*

**Proof.** Putting the inequalities $\binom{t}{k} = t(t-1)\cdots(t-k+1)/k! \leq t^N/k!$ $(k \leq N)$ in Theorem 14, we get $\overline{\pi}_t \leq 2t^N 2^{N-t}\sum_k k!^{-1}$.

We are now ready to answer Problem 9. Notice that as far we used an arbitrarily fixed number $N$ of bits (and we suppressed the parameter $N$ in the notation in the notation $\overline{\pi}_t = \overline{\pi}_t^{(N)}$). We are now interested in finding reasonably small function $N \mapsto t(N)$ such that $\overline{\pi}_t^{(N)}$ decreases exponentially to zero as $N \to \infty$, or which is the same, we have $\limsup_{N\to\infty} N^{-1}\log_2 \overline{\pi}_{t(N)}^{(N)} < 0$ or even $= -\infty$. $\square$

**Proposition 16.** *Given any small constant $\varepsilon > 0$, with the choice $t(N) := N(\log_2 N)^{1+\varepsilon}$ we have $\lim_{N\to\infty} N^{-1}\log_2 \overline{\pi}_{t(N)}^{(N)} = -\infty$.*

**Proof.** In this case

$$\begin{aligned}\frac{1}{N}\log_2 \overline{\pi}_{t(N)}^{(N)} &\leq \frac{1}{N}\log_2(2e[N(\log_2 N)^{1+\varepsilon}]^N 2^{N(\log_2 N)^{1+\varepsilon}-N}]\\ &= \frac{1+\log_2 e}{N} + 1 + (1+\varepsilon)\log_2\log_2 N + \log_2 N\\ &\quad - (\log_2 N)^{1+\varepsilon} \to -\infty. \qquad \square\end{aligned}$$

**Remark 17.** In conclusion, however small $\varepsilon > 0$ we choose, the probability that the measurement algorithm fails in less than $t_\varepsilon = N(\log_2 N)^{1+\varepsilon}$ time steps is scaling as $2^{-N}$ in the limit of large $N$.


## Acknowledgments

I would like to express my sincere gratitude to Prof. L. Kish for his valuable suggestions and the stimulating discussions on the topics. The author was supported partially by the project TAMOP-4.2.1/B-09/-1/KONV-2010-0005.

## References

[1] L. B. Kish, Noise-based logic: Binary, multi-valued, or fuzzy, with optional superposition of logic states, *Phys. Lett. A* **373** (2009) 911–918.

[2] L. B. Kish, S. Khatri and S. Sethuraman, Noise-based logic hyperspace with the superposition of 2N states in a single wire, *Phys. Lett. A* **373** (2009) 1928–1934.

[3] S. M. Bezrukov and L. B. Kish, Deterministic multivalued logic scheme for information processing and routing in the brain, *Phys. Lett. A* **373** (2009) 2338–2342.

[4] L. B. Kish, S. Khatri and F. Peper, Instantaneous noise-based logic, *Fluct. Noise Lett.* **9** (2010) 323–330.

[5] F. Peper and L. B. Kish, Instantaneous, non-squeezed, noise-based logic, *Fluct. Noise Lett.* **10** (2011) 231–237.

[6] L. B. Kish, S. Khatri and T. Horvath, Computation using noise-based logic: Efficient string verification over a slow communication channel, *Eur. Phys. J. B* **79** (2011) 85–90.

[7] Z. Gingl, S. Khatri and L. B. Kish, Towards brain-inspired computing, *Fluct. Noise Lett.* **9** (2010) 403–412.

[8] L. B. Kish, S. M. Bezrukov, S. P. Khatri, Z. Gingl and S. Sethuraman, Noise-based logic and computing: From boolean logic gates to brain circuitry and its possible hardware realization, *Nat. Comput.* **2** (2010) 13–22.

[9] H. Wen, L. B. Kish, A. Klappenecker and F. Peper, New noise-based logic representations to avoid some problems with complexity, Submitted to *Fluct. Noise Lett.* http://arcXiv.org/abs/1111.3859.

# Noise-Based Logic Gates by Operations on the Reference System

Laszlo B. Kish*,‡ and Walter C. Daugherity†

*Department of Electrical and Computer Engineering
Texas A&M University
College Station, TX 778433128, USA

†Department of Computer Science & Engineering
Texas A&M University
College Station, TX 778433112, USA
‡Laszlo.Kish@ece.tamu.edu

Received 26 March 2018
Accepted 5 June 2018
Published 20 July 2018

Communicated by Laszlo Stacho

We propose a new, low-complexity solution to realize multi-input-bit gates acting on exponentially large superpositions in noise-based logic processors. Two examples are shown, the NOT gate and the CNOT gate. The operations can be executed and repeated with polynomial time and hardware complexity. The lack of a solution for this problem had been one of the major issues prohibiting the efficient realization of Shor's algorithm by Instantaneous Noise-Based Logic, which runs on a classical Turing computer with a true random number generator. With the method described in this paper, we are one step closer to this goal.

*Keywords*: Quantum algorithms; noise-based logic; exponential speedup; low complexity; parallel operations.

## 1. Introduction

### 1.1. *On noise-based logic*

Noise-based logic (NBL) [1–14], including the brain [5–7], utilizes a reference system of uncorrelated (orthogonal) stochastic processes (noises), their superpositions, and their products to carry the logic information. There are many different realizations and types of NBL. Among them, the instantaneous NBL (INBL) schemes [3, 4, 8–14] including the brain [5–7] do not require time averaging to identify the logic state and they produce their output superposition immediately. Practical examples are, see Fig. 1, the random-telegraph-wave (RTW) based INBL schemes [3, 4, 7–14] described in Secs. 1.2 and 1.3 below, that by utilizing *Achilles' heel logic functions* [2] are able to synthesize an exponentially large superposition of bit strings with polynomial hardware and time complexity, similarly to the original quantum computing schemes. Moreover, the single-bit quantum gate operations acting on exponentially large superpositions can also be implemented/run with polynomial complexity [9], indicating an exponential speedup due to the large parallelism represented by the exponentially large superposition. Thus there is hope that Shor's and other quantum algorithms can be run in an INBL-based processor, that is, in a classical binary computer with polynomial resolution of degree $2N$ and a physical random number generator [12]. Here, $N$ is the required quantum bit resolution of the equivalent quantum computer. If such a goal is achieved, quantum algorithms could be realized in classical desktop and laptop computers, with the same computational complexity class as on a quantum computer, thus making today's computation-based encryption schemes insecure.

However, the 2-bit gate operations, such as the CNOT, had been problematic on superpositions. In the most successful case to date [9], exponential speedup with polynomial complexity was attained by a single CNOT operation on an exponentially large, $O(2^N)$, superposition. However, each additional CNOT operation doubled the required hardware complexity, so cascaded CNOT operations led to exponentially large and thus infeasible hardware [9].

In the present paper, we introduce a new philosophy for the INBL gate operations on superpositions by implementing active circuitry manipulating the reference system that supplies the INBL processor with the reference noises. For the sake of simplicity of exposition, but without restricting generality, we present the simplest RTW-based INBL, even though many of its superpositions produce zero amplitudes most of the time. This characteristic makes the simplest scheme impractical and only academically/educationally interesting, as a toy model. However, it is straightforward to implement the new logic gates introduced in this paper using the more advanced/practical types of RTW-based schemes that avoid the zero-problem, namely, asymmetric [8], time-shifted [8, 13] and complex RTWs [9], because the utilized features of RTWs are similar in each of these systems.

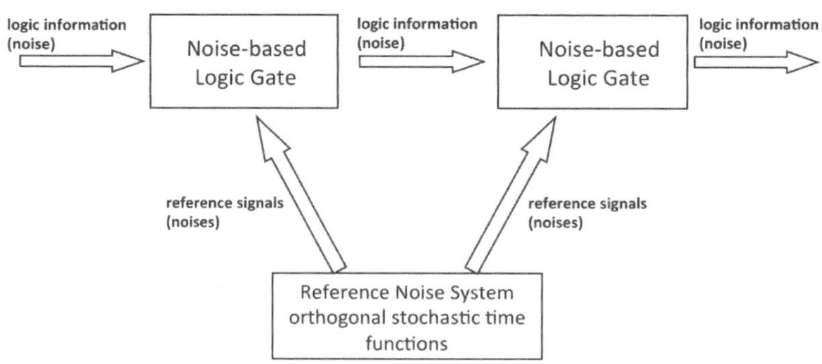


Fig. 1. Essential elements in a noise-based logic processor. The reference system supplies the $2N$ statistically independent noises (stochastic AC signals). The gates form the superpositions representing the logic states and act upon them.

For the convenience of the reader, in Sec. 1.2, we briefly outline the noise-based product strings used to represent integer numbers and their superpositions in INBL. Then in Sec. 1.3, we briefly describe the simplest scheme where the noises belong to the simplest type of RTWs. That system will be used to illustrate the new type of logic gate operations on superpositions that will be introduced in Sec. 2.

### 1.2. *The superposition of integer numbers in instantaneous noise-based logic*

In a system with $N$ noise-bits [1], the reference base consists of $2N$ independent, uncorrelated noise sources, $W_{i,j}(t)$, each representing the fingerprint of a bit value $j \in \{0, 1\}$, in the $N$ noise-bit system, $i \in \{0, \ldots, N-1\}$. This system, through the superpositions of its bit strings (see Eq. (1) below), can represent $2^{2^N}$ different logic values [1, 2], that is, its logic depth corresponds to $2^N$ classical bits. Any integer number $R$ in the range of $[0, 2^{N-1}]$ is represented by its corresponding bit string, which is a new RTW, $X_R(t)$, formed by the product of the relevant $W_{i,j}(t)$ functions [2, 7, 11, 13]:

$$X_R(t) = \prod_{i=0}^{2N-1} W_{i,j(i,R)}(t), \tag{1}$$

where the $j(i, R)$ function provides the value (0 or 1) of the $i$-th bit in the product string. Note that, for continuum noises, the bandwidth must be limited by filters or the natural bandwidth of the system, to avoid increasing the time complexity. This scheme, similarly to quantum computing schemes, is able to represent the *superposition* of states (signals) representing integer numbers in the form of a sum $Y(t)$ of amplitudes of the relevant product strings:

$$Y(t) = X_{R1} + X_{R2} + \cdots . \tag{2}$$

A powerful feature of INBL that is the superposition of all integer numbers, that is, the universe $U(t)$, and many of its sub-superpositions, can be generated by utilizing Achilles' heel functions with polynomial complexity [2], (for concrete examples, see [13]), even though the sum representing the universe consists of an exponentially large ($2^N$) number of elements. For example, the universe is

$$U(t) = \prod_{i=0}^{2N-1} [W_{i,0}(t) + W_{i,1}(t)]. \tag{3}$$

### 1.3. *Random telegraph waves as noise base*

In the rest of this paper, we deal with the simplest random telegraph waves. For more advanced RTW types [8, 9, 13], proper modifications of the new gates will be necessary. The situation also depends on the given application and may need additional steps.

The simplest RTW-based INBL system has a deterministic and identical clock for all the logic values. At the beginning of the clock period, the value of the RTW, $-1$ or $+1$, is randomly chosen with identical (0.5) probability and it remains static during the whole clock period. In this way, the time and amplitude complexity does not increase during multiplication. The essential properties are zero mean and orthogonality:

$$\langle W_{i,j}(t)\rangle = 0, \tag{4}$$

$$\langle W_{i,j}(t)W_{m,n}(t)\rangle = \delta(i,m)\delta(j,n), \tag{5}$$

where $\delta(r,s)$ is the Kronecker symbol. The product of two orthogonal RTWs is a new RTW, $[W_{i,j}(t)W_{m,n}(t)] = W_{p,q}(t)$, that is orthogonal to the original RTWs. With $i \neq m$ and/or $j \neq n$ this is shown by

$$\langle W_{p,q}(t)W_{i,j}(t)\rangle = \langle W_{m,n}(t)\rangle = 0 \quad \text{and} \quad \langle W_{p,q}(t)W_{m,n}(t)\rangle = \langle W_{i,j}(t)\rangle = 0, \tag{6}$$

so $W_{p,q}(t)$ is a new RTW that is uncorrelated with $W_{i,j}(t)$ and $W_{m,n}(t)$.

The NOT gate is a simple product operation

$$\mathrm{NOT}_i = W_{i,j}(t)W_{i,k}(t)^*, \tag{7}$$

where $\mathrm{NOT}_i[W_{i,j}(t)] = W_{i,k}(t)$ with $j \neq k$, $j \in \{0,1\}$.

The NOT gate also works on any selected bit in a product strings or in a superposition, for example,

$$\begin{aligned} &\mathrm{NOT}_i[W_{1,a}(t)W_{2,b}(t),\ldots,W_{i,j}(t),\ldots,W_{N,z}(t)] \\ &\quad = W_{1,a}(t)W_{2,b}(t),\ldots,W_{i,k}(t),\ldots,W_{N,z}(t), \end{aligned} \tag{8}$$

where $j \neq k$, $j \in \{0,1\}$.

We will show in Sec. 2 that the CNOT operations acting on arbitrary bits in a superposition can be built up by a combination of NOT gates acting on proper reference wires in a noise-based logic processor.

### 1.4. *Earlier attempts for achieving CNOT operations*

Many attempts were carried out, with only the version based on "ghost bits" being published [9], yet they all had a common problem. A single CNOT operation yielded the expected exponential speedup of $O(2^N)$ compared to classical binary logic gates because the gate simultaneously acted on all the product strings in the superposition. However, when the CNOT operation was repeated on various bit pairs, the hardware and/or time complexity grew exponentially with the number of the CNOT operations. Thus, after the CNOT gate was applied to each subsequent bit pair in the product string the complexity approached that of the classical computers.

The new scheme described in this paper avoids this problem.

## 2. The New NOT and CNOT Gates

Figure 2 shows the generic noise-based logic processor scheme where the "Hyperspace Synthesizer/Processor" unit generates the given logic superposition state $Y$ on which

Fig. 2. The generic noise-based logic processor scheme generating the given logic superposition state $Y$ on which the logic gates will act.

the subsequent logic gates will act. Examples are Eq. (3) or its truncated versions to generate the universe or its fractions.

In such systems, we can do operations not only directly on $Y$ but also on the reference wires. We show two examples: the realization of the NOT and the CNOT gates.

The new NOT gate is simple; it acts on the reference wires of the chosen bit by applying Eq. (7) on both wires (which represent the 1 and 0 bit values). See Fig. 3, in which the realization of the $\text{NOT}_3$ gate acts on the 3rd bit of each product string in the superposition $Y$ by inverting the value of its 3rd bit. The operation represents a $O(2^N)$ speedup compared to classical computers because it acts on $O(2^N)$ number of orthogonal strings.

The realization of the CNOT gates requires similar techniques but is more complex when multiple gates are involved because the operations may be non-commutative. The CNOT gate inverts the target bit if the control bit has high value. The method is as follows:

(a) First, the NOT operator of the target bit is inserted into the reference wire carrying the value 1 of the control bit. This circuitry will invert the target bit in all the product strings that contain value 1 of the control bit.
(b) Then, whenever it is necessary (such as with a sequence of CNOT operations which is not commutative), add the proper NOT operators applied to other reference wires to correct for excess NOTs and to supply missing NOTs. The examples below demonstrate this straightforward procedure.

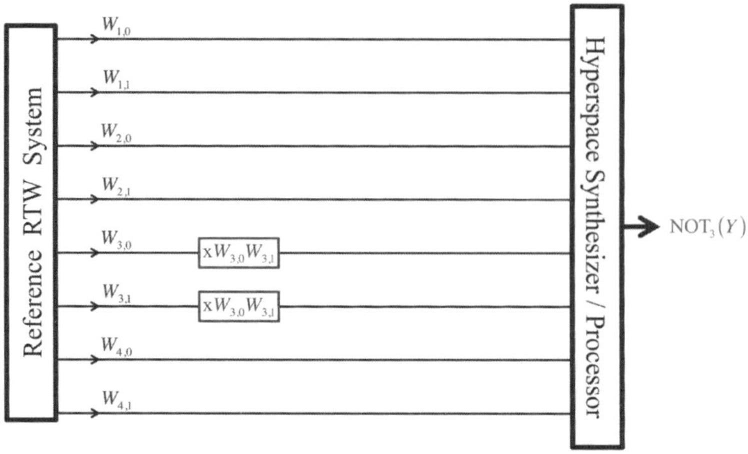


Fig. 3. Implementation of the $NOT_3(Y)$ gate; see Eq. (7). It acts simultaneously and instantaneously on each product string in the $O(2^N)$ superposition that contains any value of bit 3 by inverting that value.

### 2.1. *Single CNOT operation*

Implementation of the $CNOT_{2,3}(Y)$ gate is displayed in Fig. 4. It acts simultaneously and instantaneously on each product string in the superposition that contains value 1 of the control bit (bit 2) by inverting the value of the target bit (bit 3).

### 2.2. *Two or more cascaded CNOT operations*

In this case, we must consider whether or not the sequence of CNOT operations is independent. When the order of CNOT operations is such that they do not act on the

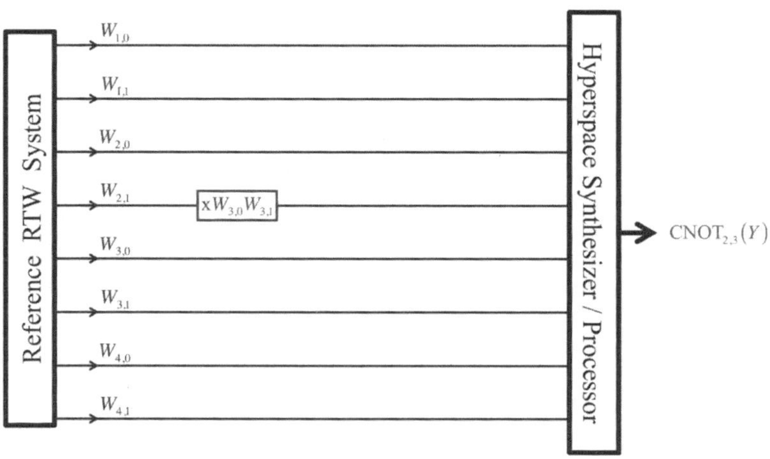


Fig. 4. Implementation of the $CNOT_{2,3}(Y)$ gate. It acts simultaneously and instantaneously on each product string in the superposition that contains value 1 of the control bit (bit 2) by inverting the value of the target bit (bit 3).

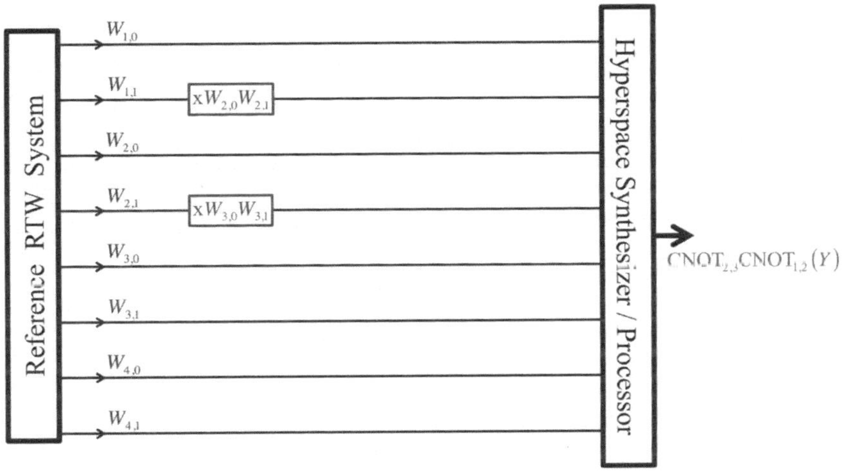


Fig. 5. The non-interacting cascade $\text{CNOT}_{2,3}\text{CNOT}_{1,2}$ gate.

control bit of the next CNOT, the NBL gate is simply the combination of the single CNOT operations shown above; see Fig. 5, which shows the example of the $\text{CNOT}_{2,3}$ $\text{CNOT}_{1,2}(Y)$ cascaded gates.

However, when the order of CNOT operations is such that a CNOT gate in the cascade acts on the control bit of a subsequent CNOT, that is, the gates "interact," then extra NOT operations must be added to achieve the correct result. See Fig. 6, which shows the example of the $\text{CNOT}_{1,2}\text{CNOT}_{2,3}(Y)$ gate. We start with the circuitry of the non-interacting $\text{CNOT}_{2,3}\text{CNOT}_{1,2}(Y)$ gate (Fig. 5). That will not yield the correct result for the target bit (bit 3) if the value of the first control bit (bit 1) is

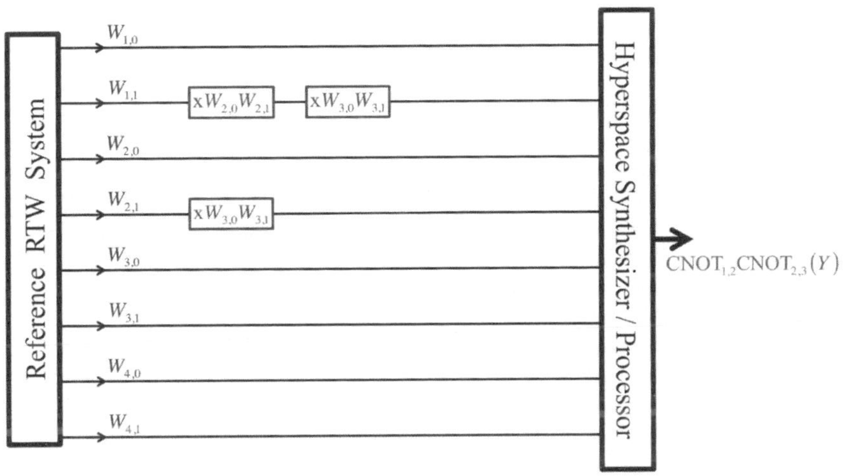


Fig. 6. The interacting cascade $\text{CNOT}_{1,2}\text{CNOT}_{2,3}$ gate.

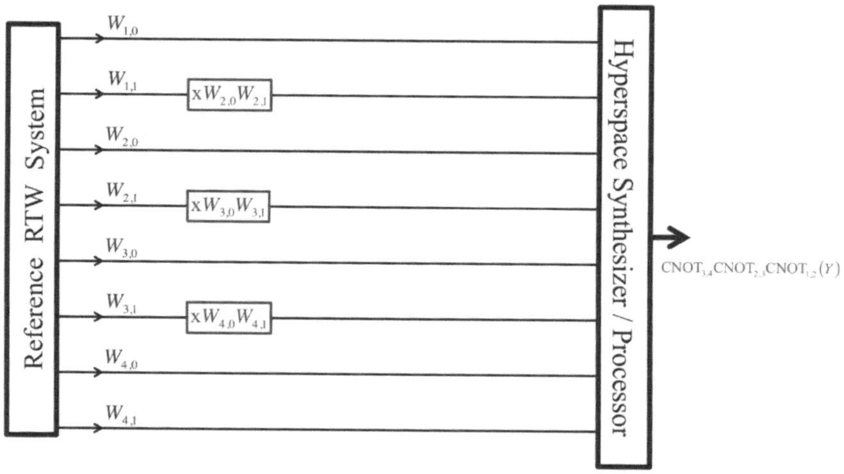


Fig. 7. The non-interacting cascade $CNOT_{3,4}CNOT_{2,3}CNOT_{1,2}$ gate.

1, so this problem is corrected by the $NOT_3$ operation inserted in reference wire $W_{11}$; see Fig. 6.

Finally, Fig. 7 presents the $CNOT_{3,4}CNOT_{2,3}CNOT_{1,2}(Y)$ as an example for another non-interacting CNOT cascade and, in Fig. 8, the $CNOT_{1,2}CNOT_{2,3}$ $CNOT_{3,4}(Y)$ is shown as another example for non-commutative interacting gates.

### 2.3. *Hardware and time complexity*

These examples indicate that cascaded CNOT systems have only polynomial hardware complexity versus $N$ while they are acting on exponentially large, $O(2^N)$,

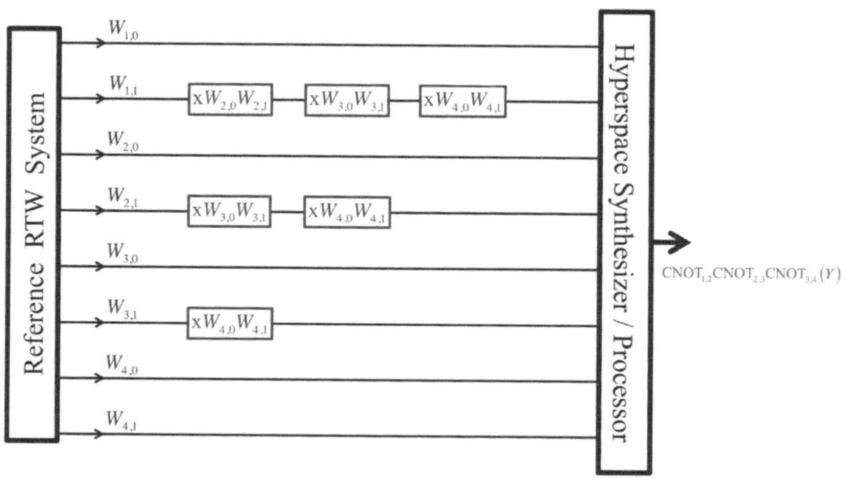


Fig. 8. The interacting cascade $CNOT_{1,2}CNOT_{2,3}CNOT_{3,4}$ gate.

superpositions of strings (that is, integer numbers) with time complexity $O(1)$ because the result is instantaneously produced. These facts indicate exponential speedup of these operations compared to the case of classical digital computers. Our conjecture is that the number $M$ of NOT hardware elements required to construct an $N$ long cascade of CNOT gates is bounded as:

$$N \leq M \leq 0.5N(N+1). \tag{9}$$

Note that though our conjecture is plausible and satisfied by the simple examples we have shown, a rigorous mathematical proof is missing at the moment.

## 3. Conclusions

A successful effort has been carried out to keep polynomial hardware and time complexity with cascaded CNOT parallel operations on exponentially large superpositions in Instantaneous Noise-Based Logic operating with classical physical elements and physical noise, that is, true random numbers. The lack of a solution of this problem had been one of the major issues prohibiting the efficient realization of Shor's algorithm by Instantaneous Noise-Based Logic, which runs on a classical Turing computer with a true random number generator. With the method described in this paper, we are one step closer to this goal.

We also note that the results may help to understand the wiring and signal structure of the brain [5–7] and its neural code.

## Acknowledgment

Valuable discussions with Andreas Klappenecker are appreciated.

## References

[1] L. B. Kish, Noise-based logic: binary, multi-valued, or fuzzy, with optional superposition of logic states, *Phys. Lett. A* **373** (2009) 911–918 http://arxiv.org/abs/0808.3162.

[2] L. B. Kish, S. Khatri and S. Sethuraman, Noise-based logic hyperspace with the superposition of $2^N$ states in a single wire, *Phys. Lett. A* **373** (2009) 1928–1934 http://arxiv.org/abs/0901.3947.

[3] L. B. Kish, S. Khatri and F. Peper, Instantaneous noise-based logic, *Fluct. Noise Lett.* **9** (2010) 323–330 http://arxiv.org/abs/1004.2652.

[4] F. Peper and L. B. Kish, Instantaneous, non-squeezed, noise-based logic, *Fluct. Noise Lett.* **10** (2011) 231–237. Open access.

[5] F. Grueneis, An interpretation of 1/f Fluctuations in Neuronal Spike trains, *Biol. Cybern.* **60** (1989) 161–169.

[6] S. M. Bezrukov and L. B. Kish, Deterministic multivalued logic scheme for information processing and routing in the brain, *Phys. Lett. A* **373** (2009) 2338–2342.

[7] L. B. Kish, C. G. Granqvist, S. M. Bezrukov and T. Horvath, Brain: Biological noise-based logic, in *4th Int. Conf. Cognitive Neurodynamics*, June 2013, Sigtuna, Sweden; *Proc. Adv. Cognitive Neurodynamics* (IV), ed. H. Liljenström (Springer, 2014),

pp. 319–322, doi:10.1007/978-94-017-9548-7_45. http://link.springer.com/chapter/10.1007%2F978-94-017-9548-7_45.

[8] H. Wen and L. B. Kish, Noise-based logic: Why noise? A comparative study of the necessity of randomness out of orthogonality, *Fluct. Noise Lett.* **11** (2012) 1250021 http://arxiv.org/abs/1204.2545.

[9] H. Wen, L. B. Kish, A. Klappenecker and F. Peper, New noise-based logic representations to avoid some problems with time complexity, *Fluct. Noise Lett.* **11** (2012) 1250003 http://arxiv.org/abs/1111.3859.

[10] H. Wen, L. B. Kish and A. Klappenecker, Complex noise-bits and large-scale instantaneous parallel operations with low complexity, *Fluct. Noise Lett.* **12** (2013) 1350002 http://vixra.org/abs/1208.0226.

[11] L. L. Stachó, Fast measurement of hyperspace vectors in noise-based logic, *Fluct. Noise Lett.* **11** (2012) 1250001.

[12] L. B. Kish, S. Khatri and T. Horvath, Computation using noise-based logic: Efficient string verification over a slow communication channel, *Euro. J. Phys. B* **79** (2011) 85–90 http://arxiv.org/abs/1005.1560.

[13] L. B. Kish, C. G. Granqvist, T. Horvath, A. Klappenecker, H. Wen and S. M. Bezrukov, Bird's-eye view on noise-based logic, *Int. J. Mod. Phys. Conf. Ser.* **33** (2014) 1460363. Open access. http://www.worldscientific.com/doi/pdfplus/10.1142/S2010194514603639.

[14] B. Zhang, L. B. Kish and C. G. Granqvist, Drawing from hats by noise-based logic, *Int. J. Parallel, Emerg. Distrib. Syst.* **32** (2017) 244–251, http://dx.doi.org/10.1080/17445760.2016.1140168; http://arxiv.org/abs/1511.03552.

# 7

# Noise in Finance

# THE KEY ROLE OF LIQUIDITY FLUCTUATIONS IN DETERMINING LARGE PRICE CHANGES

FABRIZIO LILLO
*INFM Palermo Unit and Dipartimento di Fisica e Tecnologie Relative, Università di Palermo viale delle Scienze, Edificio 18, I-90128, Palermo, Italy*

J. DOYNE FARMER
*Santa Fe Institute*
*1399 Hyde Park Road, Santa Fe, NM 87501, USA*

Received 7 December 2004
Revised 25 February 2005
Accepted 13 May 2005
Communicated by Werner Ebeling and Bernardo Spagnolo

Recent empirical analyses have shown that liquidity fluctuations are important for understanding large price changes of financial assets. These liquidity fluctuations are quantified by gaps in the order book, corresponding to blocks of adjacent price levels containing no quotes. Here we study the statistical properties of the state of the limit order book for 16 stocks traded at the London Stock Exchange (LSE). We show that the time series of the first three gaps are characterized by fat tails in the probability distribution and are described by long memory processes.

*Keywords*: Financial markets; long-memory processes; power-law distributions.

## 1. Introduction

One of the best known statistical regularities of financial time series is the fact that the empirical distribution of asset price changes is fat tailed, i.e. there is a higher probability of extreme events than in a Gaussian distribution [1–3]. Moreover, there are strong indications that the part of the distribution describing large price changes is a power-law [4]. Many mechanisms explaining this empirical fact have been suggested, including subordinated stochastic processes [5] and volume fluctuations [6]. By investigating the fine details of the process of price formation in the London Stock Exchange, we have recently shown that large price fluctuations are driven by liquidity fluctuations, i.e. variations in the markets ability to absorb new orders [7]. We have shown that even for the most liquid stocks there can be substantial gaps in the limit order book, corresponding to a block of adjacent price levels containing no quotes (see below for a definition). When such a gap exists next to the best price, a new order can remove the best quote, triggering a large price change. Thus,

Fig. 1. A schematic illustration of the continuous double auction mechanism. We adopt the arbitrary convention that buy orders are negative and sell orders are positive.

the distribution of large price changes merely reflects the distribution of gaps in the limit order book. This is a finite size effect, caused by the granularity of order flow: in a market where participants place many small orders uniformly across prices, such large price fluctuations would not happen.

In this paper we investigate the statistical properties of the first three gaps of 16 stocks traded at the LSE. Specifically we investigate whether the fat tail properties of the first gap discovered in [7] are shared by other gaps. We then investigate the time correlation properties of gaps and the synchronous cross correlation between the size of different gaps. The analysis shows that the liquidity of the book as described by the multidimensional stochastic process of the gap sizes has interesting properties, such as power-law distribution, long memory and synchronous behavior.

In order to have a representative sample of high volume stocks we select 16 companies traded on the London Stock Exchange (LSE) in the 4-year period 1999-2002. The stocks we analyzed are Astrazeneca (AZN), Baa (BAA), BHP Billiton (BLT), Boots Group (BOOT), British Sky Broadcasting Group (BSY), Diageo (DGE), Gus (GUS), Hilton Group (HG.), Lloyds Tsb Group (LLOY), Prudential (PRU), Pearson (PSON), Rio Tinto (RIO), Rentokil Initial (RTO), Reuters Group (RTR), Sainsbury (SBRY), Shell Transport & Trading Co. (SHEL). These stocks were selected because they have high volume and they are all continuously traded during the full period. The number of transactions vary from $226,000$ for BAA to $723,000$ for LLOY. The total number of considered transactions is 6.5 million.

## 2. Market Mechanism

To understand our results it is essential that the reader understand the double continuous auction, which is the standard mechanism for price formation in most modern financial markets. Agents can place different types of orders, which can be grouped into two categories: Impatient traders submit *market orders*, which are requests to buy or sell a given number of shares immediately at the best available price. More patient traders submit *limit orders*, or *quotes* which also state a limit price $\pi$, corresponding to the worst allowable price for the transaction. Limit orders often fail to result in an immediate transaction, and are stored in a queue called the *limit order book*. Buy limit orders are called *bids*, and sell limit orders are called *offers*. At any given time there is a best (lowest) offer to sell with price $a(t)$, and a best (highest) bid to buy with price $b(t)$. These are also called the *best prices*. The price gap between them is called the *spread* $s(t) = a(t) - b(t)$. Prices are not continuous, but rather change in discrete quanta called *ticks*. The number of shares in an order is called either its *size* or its *volume*. As market orders arrive they are matched against limit orders of the opposite sign in order of first price and then arrival time, as shown in Fig. 1. A high density of limit orders per price results in high *liquidity* for market orders, i.e., it implies a small movement in the best price when a market order is placed. When a market order arrives it can cause changes in the best prices. This is called *market impact* or *price impact*. Buy market orders can increase the midprice $m(t) = (a(t) + b(t))/2$, and sell orders can decrease it. Price changes are typically characterized as *returns* $\Delta p_\tau(t) = \log m(t) - \log m(t-\tau)$.

## 3. Importance of the Granularity of Limit Order Book

We investigate the unconditional probability density function of price changes at the level of individual transactions. To be more precise, for each transaction at time $t$ we compute the price return, defined as $\Delta p(t) = \log(m(t+\epsilon)/m(t-\epsilon))$, where $m$ is the midprice and $\epsilon$ is an infinitesimal time. The quantity $\Delta p$ is termed the *price impact* and its average and conditional properties have been extensively investigated recently (see for example [8,9]). We have investigated the unconditional statistical properties of the price impact [7]. In Fig. 2 we plot the unconditional cumulative distribution of individual price changes for the stock AZN. The empirical distribution is consistent with a power-law distribution, i.e a probability distribution decaying asymptotically as $P(x > z) \sim z^{-\alpha}$, where $\alpha$ is called the tail index. A least squares fit on the tail of the distributions gives a value of the tail index equal to $\alpha = 2.19$ for the positive tail and $\alpha = 2.17$ for the negative tail. These results indicates that the stochastic process of the price impact has finite variance. At first sight the large fluctuations of price impact are puzzling. How can price changes due to individual transactions fluctuate so much?

A possible explanation is the role of the volume. Large individual price changes could be triggered by large market orders that cause transactions with many limit orders present in the book at different price levels. The effect of such a large market order would be a large price change. We have shown [7] that this is not the most frequent scenario. Specifically, the volume of the market order is important in determining whether the midprice changes, but when the midprice changes, the price return induced by the transaction is almost independent of the market order size. Thus the role of the volume of the market order in determining the individual price change is marginal. In other words, at the level of individual transactions, the impact function is not deterministic and the fluctuations of price impact are very large. These results show that the fluctuations in the state of the book have a key role in determining price changes. But how can small volume transactions create large price changes?

We have investigated this problem by studying the state of the limit order book before a individual market order was able to trigger a large price change. We have

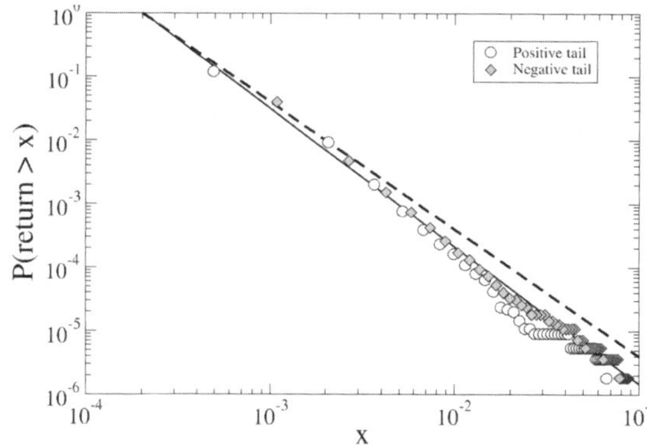


Fig. 2. Distribution of large price returns triggered by individual transactions (price impact). The continuous line is the best fit for the negative tail, whereas for illustrative purposes the dashed line indicates the power law behavior separating the regime with finite variance from the Lévy regime (infinite variance). The data refer to the stock Astrazeneca (AZN).

found [7] that an important way of describing the state of the book at a given time is through the properties of the *gaps*. To understand what we mean by a gap, let us consider the limit orders on the sell side of the book. By definition, the lowest price where shares are offered for sale is the best ask $\pi_0(t) \equiv a(t)$. The next price level occupied by (at least) one sell limit order will be denoted by $\pi_1(t) > \pi_0(t)$. In general the $n$th price level inside the book occupied by at least one limit order is $\pi_n(t)$, $n = 1, 2, ... N_s(t)$, where $N_s(t) + 1$ is the number of occupied offer price levels at time $t$. The first gap is defined as $g_1(t) = \log \pi_1(t) - \log a(t)$, and the $n$th order gap is defined as $g_n(t) = \log \pi_n(t) - \log \pi_{n-1}(t)$. Likewise the gaps for the side of the book occupied by buy limit orders is $g_{-n}(t) = \log \pi_{n-1}(t) - \log \pi_n(t)$ and $\pi_0(t) = b(t)$. Thus $g_i$ is the $i$th gap starting from the best prices on the bid side for $i < 0$ and on the offer side for $i > 0$.

The size of the gaps (and the number of occupied price levels) present in the limit order book at a given time are a measure of the liquidity of the asset at that time. In fact a book with few occupied price levels and large gaps, i.e. a *sparse book*, is an indication of a lack of liquidity and reluctance to trade the asset by the investors. On the other hand when many orders are present and the gaps are small (*a dense book*) the amount of liquidity of the asset is high. The empirical analysis shows that the first gap size is a highly fluctuating quantity. Specifically it has been shown that the first gap size distribution $p(g_1)$ and $p(g_{-1})$ are well approximated in the tail by a power-law distribution with a small exponent. This implies that it is not uncommon that the best available limit price beside the best is very far from the best price. Although this is a clear indication of liquidity crisis, we still have not determined what causes these crises. We have shown that the first gap distribution is a key quantity to understand large individual price changes [7]. In fact we have shown that the first gap distribution is very similar to the distribution of nonvanishing price changes. The reason for that is that investors tend to place market orders of size smaller or equal to the size of the order at the best available

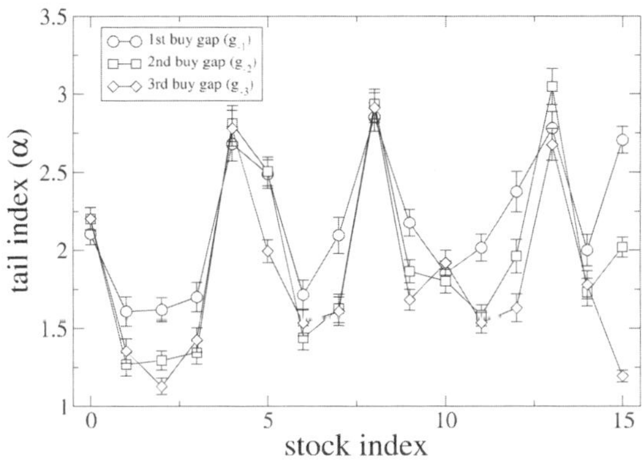


Fig. 3. Hill estimator of the tail index $\alpha$ of the tail of the distribution of the size of the first three gaps on the bid size of the limit order book ($g_{-1}$, $g_{-2}$, and $g_{-3}$). On the $x$ axis we sort alphabetically the 16 stocks under investigation.

price (best bid or best ask). Thus when the size of the market order is smaller than the volume at the corresponding best, the midprice does not change and the price impact is zero. When the volume of the market order is equal to the volume at the best the midprice changes of a quantity $m(t+\epsilon) - m(t-\epsilon) = g_{\pm 1}(t)/2$, where the plus (minus) sign refers to buy (sell) market orders. This argument shows why the distribution of first gap and individual price change are closely related.

In conclusion, our previous analysis has shown that the most important factor determining large price changes is not the transaction volume. We have shown instead that a liquidity crisis, characterized by a sparse state of the book, causes a situation in which a small market order can trigger a large price change. Even if we still do not have an explanation for what creates the conditions for the sparseness of the book, we believe that this is an important factor determining the dynamics of asset prices. In this sense it is important to investigate in more detail the statistical properties of the granularity of the limit order book. In this paper we provide a first empirical analysis of these properties.

A first remark to the above discussed explanation of the large price changes in terms of the fat tail of the first gap distribution concerns whether the first gap is in some sense special or whether the whole limit order book (i.e. also the second, third, etc. gaps) is characterized by granularity and large fluctuations. To this end we analyzed the probability distribution of the first three gaps both of sell and of buy limit orders. Specifically we compute the empirical pdf of gap size and we characterize the tail of the distribution using the Hill estimator [10]. This is a semi-parametric statistical estimator of the tail index $\alpha$ of a power-law distribution. Figure 3 shows the Hill estimator of the tail index $\alpha$ of the first three gaps of the buy side of the limit order book of the 16 stocks investigated (similar results are observed for gaps on the sell side). Figure 3 shows that the asymptotic statistical properties of the distribution of the first three gaps of a given stock are quite similar. This result supports the idea that the whole book follows a time dynamic in which the balance between the flows of limit orders, market orders and cancellations makes

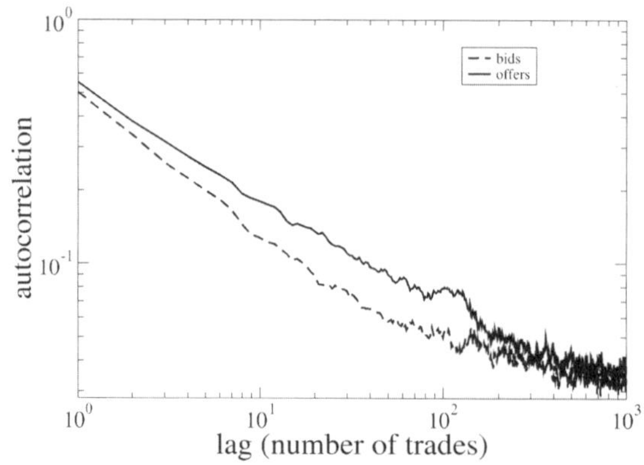


Fig. 4. Autocorrelation function of the first gap size for bids ($g_{-1}(t)$) and offers ($g_1(t)$) in a log-log plot. The data refer to Astrazeneca.

the book more or less granular (or sparse). The error bars given in Fig. 3 are 95% confidence intervals of the estimator, under the assumption that the data are IID. Unfortunately, these data are far from IID. The real error bars are certainly larger than these, due to the long-memory property of the data discussed below. It is difficult to assign proper error bars for long-memory processes, and doing so is beyond the scope of this paper. However, it is worth noting that the variation of the tail index across different stocks is much larger than the variation of the tail index of different gaps in the same stock. It is our belief that the variations between the tail exponents of gaps are not statistically significant.

The idea that the sparseness of the limit order book is not concentrated in "space", i.e. it is a property shared by the whole book, suggests that it is worth investigating whether the sparseness of the book is also persistent in time. More specifically, if the book is sparse at a given instant of time, will the sparseness persist in time, or is a sparse book an event that is strongly localized in time? To answer this question we have studied the time correlation properties of gap size. Figure 4 shows the autocorrelation function of the first buy and sell gap size of AZN in double logarithmic scale. The figure indicates that the first gap size is a persistent quantity. The autocorrelation function is well fit by a power-law $\rho(\tau) \sim \tau^{-\beta}$ with $\beta = 0.24$ for the first gap of buy limit orders (bids) and $\beta = 0.34$ for the first gap of sell limit orders (offers). These values are consistent with long memory processes characterized by an Hurst exponent $H = 1 - \beta/2$ larger than $1/2$. Long memory processes [11] are stochastic processes characterized by the lack of a typical time scale and they seems to provide a good description of several financial quantities such as the volatility [14] and the sign of orders [12,13]. It is known that the power-law fit of the autocorrelation function is not a good statistical estimator of the Hurst exponent. We have thus measured the Hurst exponent of the first three gaps for bids and offers by making use of the Detrended Fluctuation Analysis (DFA) [15] which is known to be a good estimator of long memory dependence. The result for AZN is $H(g_{-1}) = 0.75$, $H(g_{-2}) = 0.76$, $H(g_{-3}) = 0.76$ for gaps in bids and $H(g_1) = 0.78$,

Table 1. Correlation coefficient of the first three gap size $g_i$ on the buy side ($i = -3, -2, -1$) and on the sell side ($i = 1, 2, 3$) of the limit order book. The data shown refer to the stock Astrazeneca.

| $\rho$ | $g_{-3}$ | $g_{-2}$ | $g_{-1}$ | $g_1$ | $g_2$ | $g_3$ |
|---|---|---|---|---|---|---|
| $g_{-3}$ | 1.00 | 0.35 | 0.24 | 0.10 | 0.08 | 0.08 |
| $g_{-2}$ | 0.35 | 1.00 | 0.27 | 0.11 | 0.08 | 0.08 |
| $g_{-1}$ | 0.24 | 0.27 | 1.00 | 0.15 | 0.15 | 0.13 |
| $g_1$ | 0.10 | 0.11 | 0.15 | 1.00 | 0.33 | 0.30 |
| $g_2$ | 0.08 | 0.08 | 0.15 | 0.33 | 1.00 | 0,41 |
| $g_3$ | 0.08 | 0.08 | 0.13 | 0.30 | 0.41 | 1.00 |

$H(g_2) = 0.77$, $H(g_3) = 0.79$ for gaps in offers. We have not attempted to compute error bars for these values, but we would be very surprised if these differences are statistically significant. All these values demonstrate that the gap size is described by a long memory process, i.e. a stochastic process without a typical time scale. The fact that the granularity of the limit order book is well described by a long memory process indicates that the lack of liquidity in the market is not a temporary state, but rather one that persists in time.

The slow dynamics of the book as described by the persistence of the sparseness of the book and thus of the lack of liquidity suggests that when the first gap is large also the other gaps (on the same or on different side) are large. In other words the sparseness of the book is a property shared by the whole book *at the same instant of time*. To test this idea quantitatively we have computed the cross-correlation coefficient of the size of the first three gaps on both sides of the book. The result of this analysis is the symmetric correlation matrix shown in Table 1. First of all we note that all the correlation coefficients are positive, indicating a synchronous behavior of all the gaps. The correlation coefficient is significantly larger (of the order of $0.24 - 0.41$) when it is computed between gaps on the same side of the book. This is reasonable because it indicates that the sparseness of the book needs not to be shared simultaneously by both sides of the book. In fact at a given time the market can experience a liquidity crisis on the buy side, for example, but not on the sell side. Moreover it is worth noting that the correlation coefficient is stronger between the second and the third gap on both sides of the book. The data suggest that there may be a statistically significant correlation between gaps on different sides of the book. The $3\sigma$ level for the investigated sample is of the order of 0.004. However, these are long-memory processes, implying much larger error bars. For a long-memory process the error scale as $N^{-(1-H)}$. For a Hurst exponent of 0.75, for example, the errors scale as $N^{-0.25}$. Thus the standard error of 0.004 becomes roughly 0.065. Thus if these values are significant they are only just barely so. If real, this correlation indicates that a lack of liquidity on one side of the book makes a lack of liquidity on the other side more likely.

In conclusion we have shown that the multidimensional stochastic process which describes the state of the limit order book of a stock traded in a financial market has remarkable statistical properties. We have shown that the gaps closer to the best available prices are characterized by large fluctuations which are important to explain large price fluctuations in terms of liquidity fluctuations. We have also shown that the liquidity availability described by the gaps size is a long memory process, indicating that the book dynamics lacks a typical time scale. Finally we have investigated the synchronous behavior of the size of different gaps showing that significant cross correlations between different gaps are present. These findings have relevant practical importance because they give some understanding of what determines financial risks, and gives some clues about how to reduce them.

### Acknowledgments

FL thanks partial funding support from research projects MIUR 449/97 "Dinamica di altissima frequenza nei mercati finanziari" and MIUR-FIRB RBNE01CW3M.

### References

[1] B. B. Mandelbrot, *The variation of certain speculative prices*, *J. Business* **36** (1963) 394–419.

[2] R. N. Mantegna and H. E. Stanley, *Scaling behavior in the dynamics of an economic index Nature* **376** (1995) 46–49.

[3] R. N. Mantegna and H. E. Stanley, *Introduction to econophysics: correlations and complexity in finance*, Cambridge University Press, Cambridge (1999).

[4] P. Gopikrishnan, M. Meyer, L. A. N. Amaral and H. E. Stanley, *Inverse cubic law for the distribution of stock price variations*, *Eur. Phys. J. B* **3** (1998) 139–140.

[5] P. K. Clark, *A subordinated stochastic process model with finite variance for speculative prices*, *Econometrica* **41** (1973) 135–256.

[6] X. Gabaix, P. Gopikrishnan, V. Plerou and H. E. Stanley, *A theory for power-law distributions in financial market fluctuations*, *Nature* **423** (2003) 267–270.

[7] J. D. Farmer, L. Gillemot, F. Lillo, S. Mike and A. Sen, *What really causes large price changes?*, *Quantitative Finance* **4** (2004) 383–397. Preprint available at xxx.lanl.gov/cond-mat/0312703.

[8] F. Lillo, J. D. Farmer and R. N. Mantegna, *Master curve for price-impact function*, *Nature* **421** (2003) 129–130.

[9] M. Potters and J.-P. Bouchaud, *More statistical properties of order books and price impact*, *Physica A* **324** (2002) 133–140.

[10] B. M. Hill, *A simple general approach to inference about the tail of a distribution*, *Annals Statistics* **3** (1975) 1163–1173.

[11] J. Beran, *Statistics for Long-Memory Processes*, Chapman & Hall (1994).

[12] J.-P. Bouchaud, Y. Gefen, M. Potters and M. Wyart, *Fluctuations and response in financial markets: the subtle nature of 'random' price changes*, *Quantitative Finance* **4** (2004) 176–190.

[13] F. Lillo and J. D. Farmer, *The long memory of efficient market*, *Studies in Nonlinear Dynamics and Econometrics* **8** (2004) 1.

[14] Z. Ding, C. W. J. Granger and R. Engle, *A long-memory property of stock market returns and a new model*, *J. Empirical Finance* **1** (1993) 83–106.

[15] C.-K. Peng, S. V. Buldyrev, S. Havlin, M. Simons, H. E. Stanley and A. L. Goldberger, *Mosaic organization of DNA nucleotides*, *Physical Review E* **49** (1994) 1685–1689.

# Market Correlation Structure Changes Around the Great Crash: A Random Matrix Theory Analysis of the Chinese Stock Market

Rui-Qi Han

*Department of Mathematics, School of Science*
*Research Center for Econophysics*
*East China University of Science and Technology*
*Shanghai 200237, P. R. China*

Wen-Jie Xie*

*Department of Finance, School of Business*
*Postdoctoral Research Station*
*Research Center for Econophysics*
*East China University of Science and Technology*
*Shanghai 200237, P. R. China*
*wjxie@ecust.edu.cn*

Xiong Xiong and Wei Zhang

*College of Management and Economics*
*China Center for Social Computing and Analytics*
*Tianjin University, Tianjin 300072, P. R. China*

Wei-Xing Zhou†

*Department of Finance, School of Business*
*Department of Mathematics, School of Science*
*Research Center for Econophysics*
*East China University of Science and Technology*
*Shanghai 200237, P. R. China*
*wxzhou@ecust.edu.cn*

Received 23 July 2016
Accepted 11 February 2017
Published 2 May 2017

Communicated by Fabrizio Lillo

The correlation structure of a stock market contains important financial contents, which may change remarkably due to the occurrence of financial crisis. We perform a comparative analysis of the Chinese stock market around the occurrence of the 2008 crisis based on the random matrix analysis of high-frequency stock returns of 1228 Chinese

*Corresponding author.
†Corresponding author.

stocks. Both raw correlation matrix and partial correlation matrix with respect to the market index in two time periods of one year are investigated. We find that the Chinese stocks have stronger average correlation and partial correlation in 2008 than in 2007 and the average partial correlation is significantly weaker than the average correlation in each period. Accordingly, the largest eigenvalue of the correlation matrix is remarkably greater than that of the partial correlation matrix in each period. Moreover, each largest eigenvalue and its eigenvector reflect an evident market effect, while other deviating eigenvalues do not. We find no evidence that deviating eigenvalues contain industrial sectorial information. Surprisingly, the eigenvectors of the second largest eigenvalues in 2007 and of the third largest eigenvalues in 2008 are able to distinguish the stocks from the two exchanges. We also find that the component magnitudes of the some largest eigenvectors are proportional to the stocks' capitalizations.

*Keywords*: Econophysics; random matrix theory; Partial correlation; stock market; financial crisis; eigenvalue.

## 1. Introduction

Financial markets evolve in a self-organized manner with the interacting elements forming complex networks at different levels [1], including international markets [2–5], individual markets [6–11], and security trading networks [12–21]. There are well-documented stylized facts of stock return time series within individual markets unveiled by the random matrix theory (RMT) analysis [7, 22]: (1) The largest eigenvalue reflects the market effect such that its eigenportfolio returns are strongly correlated with the market returns; (2) Other largest eigenvalues contain information of industrial sectors; and (3) The smallest eigenvalues embed stock pairs with large correlations. However, for stock exchange index returns [2] and housing markets [23, 24], the largest eigenvalues can be used to extract geographic traits. Moreover, the signs of eigenvector components contain information of local interactions [23, 25, 26].

Financial crises occur more frequently than people usually expect, during which financial markets experience abrupt regime changes like phase transitions [27, 28]. The market correlation structure changes around financial crashes. The average correlation after the critical point of crash is higher than that before the crash [4, 22, 29]. Also, the correlation network becomes more connected after the crash [30], and the network measures fluctuate accordingly [31]. It is natural that the absorption ratio of the largest eigenvalue serves as a measure of systemic risk [32]. It is well documented that market correlation structure is non-stationary [33], which can be used to identify evolving quasi-stationary market states [34–37]. The fluctuations in market correlation structure also have financial implications in risk management [38].

On the other hand, the partial correlation analysis, which is a powerful tool for investigating the intrinsic correlation between two time series affected by common factors [39], has been applied to financial markets [10, 22–24, 40–43]. An intriguing feature is that the partial correlation analysis is able to identify influences among different time series [42].

Applying the random matrix theory analysis to the 1-min high-frequency returns of Chinese stocks, we investigate in this work the correlation structure changes around the breakout of the Great Crash at the end of 2007 [44]. We focus on unveiling the information contents embedded in the deviating eigenvalues and their associated eigenvectors of the raw and partial correlation matrices. Novel results are found.

## 2. Data Sets

We investigate the 1-min return time series of 489 A-share stocks traded on the Shenzhen Stock Exchange (SZSE) and 739 A-share stocks traded on the Shanghai Stock Exchange (SHSE) in 2007 and 2008, totally 1228 stocks, which were kindly provided by RESSET (http://resset.cn/). These stocks are chosen such that they have been listed before 2007 and had at least 180 trading days in each year.

The SZSE stocks belong to 16 industrial sectors including manufacturing (C, 306, 62.58%), real estate (K, 39, 7.98%), wholesale and retail industry (F, 39, 7.98%), electric power, heat, gas and water production and supply (D, 26, 5.32%), transport, storage and postal service (G, 14, 2.86%), mining (B, 11, 2.25%), information transmission, software and information technology services (I, 10, 2.04%), and 9 other industries. The SHSE stocks belong to 18 industrial sectors including manufacturing (C, 397, 53.72%), wholesale and retail industry (F, 74, 10.01%), real estate (K, 54, 7.31%), transport, storage and postal service (G, 47, 6.36%), electric power, heat, gas and water production and supply (D, 41, 5.55%), mining (B, 23, 3.11%), information transmission, software and information technology services (I, 21, 2.84%), and 11 other industries.

The 1-min logarithmic returns of stock $i$ are calculated as follows

$$r_i(t) = \ln P_i(t) - \ln P_i(t-1), \tag{1}$$

where $P_i(t)$ denotes the price of stock $i$ at time $t$ and $t = 1, 2, \ldots, T$. The returns are calculated at the intraday manner and no overnight returns are considered.

## 3. Results

### 3.1. *Distributions of correlation coefficients and partial correlation coefficients*

For each year, 2007 or 2008, we calculate the correlation coefficient $c_{ij}$ between the returns of stock $i$ and stock $j$, which form the correlation matrix $\mathbf{C}$, as follows:

$$c_{ij} = \frac{\langle [r_i - \langle r_i \rangle][r_j - \langle r_j \rangle] \rangle}{\sigma_{r_i} \sigma_{r_j}}, \tag{2}$$

where $\sigma_{r_i}$ and $\sigma_{r_j}$ are the standard deviations of $r_i(t)$ and $r_j(t)$. It is common that the 1-min trading time sequences of stock $i$ and stock $j$ do not overlap. Under such circumstance, we discard those times appeared in only one stock. For two arbitrary return time series $r_i(t)$ and $r_j(t)$, we can extract their idiosyncratic components $\varepsilon_i(t)$ by removing a common collective component $r_m(t)$ and calibrating the following simple univariate factor model:

$$r_i(t) = \alpha_i + \beta_i r_m(t) + \varepsilon_i(t), \tag{3}$$

where $r_m(t)$ is the 1-min return time series of the Shanghai Stock Exchange Composite Index (SSCI). The partial correlation coefficient $\rho_{ij}$ between $r_i(t)$ and $r_j(t)$ with respect to $r_m(t)$ is defined as the correlation coefficient between the two residuals $\varepsilon_i(t)$ and $\varepsilon_j(t)$:

$$\rho_{ij} = \frac{\langle [\varepsilon_i - \langle \varepsilon_i \rangle][\varepsilon_j - \langle \varepsilon_j \rangle] \rangle}{\sigma_{\varepsilon_i} \sigma_{\varepsilon_j}}, \tag{4}$$

where $\sigma_{\varepsilon_i}$ and $\sigma_{\varepsilon_j}$ are the standard deviations of $\varepsilon_i(t)$ and $\varepsilon_j(t)$. We denote $\mathbf{P} = [\rho_{ij}]$ the partial correlation matrix. Simple algebraic manipulations result in the following equation [39, 41]

$$\rho_{ij} = \frac{c_{ij} - c_{im} c_{jm}}{\sqrt{(1 - c_{im}^2)(1 - c_{jm}^2)}}, \tag{5}$$

where $c_{im}$ ($c_{jm}$) is the correlation coefficient between $r_i$ ($r_j$) and $r_m$.

Figure 1 illustrates the distributions $f(c_{ij})$ and $f(\rho_{ij})$ of correlation coefficients $c_{ij}$ and partial correlation coefficients $\rho_{ij}$ in 2007 and 2008. The overwhelming majority of $c_{ij}$ are positive and the distributions are rightly skewed. The average correlation coefficient in 2007 ($\langle c_{ij,2007} \rangle = 0.098$) is slightly smaller than that in

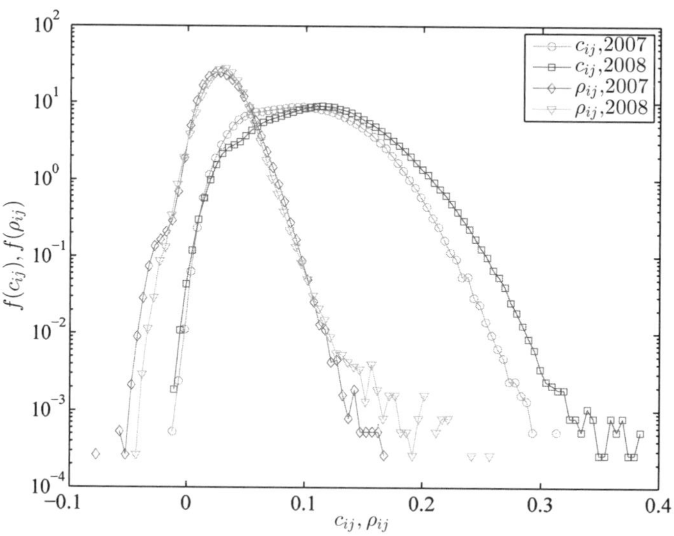


Fig. 1. (Color online.) Probability distributions $f(c_{ij})$ and $f(\rho_{ij})$ of correlation coefficients $c_{ij}$ and partial correlation coefficients $\rho_{ij}$ of stock return time series in 2007 and 2008.

2008 ($\langle c_{ij,2008}\rangle = 0.112$), implying that the Chinese stock market was in high risk and this systemic risk is higher in the panic bearish period than in the mania bullish period. The average correlation coefficients of 1-min returns are significantly smaller than those of daily returns (about 0.35) [29, 45], which is due to the fact that high-frequency returns are more noisy. The maximum correlation coefficient is 0.314 in 2007 and 0.386 in 2008.

After removing the influence of SSCI, the distributions $f(\rho_{ij,2007})$ and $f(\rho_{ij,2008})$ become narrower with the average partial correlation coefficients closer to 0 when compared with $f(c_{ij,2007})$ and $f(c_{ij,2008})$, the bulk parts with $-0.02 < \rho < 0.12$ of the two distributions almost overlap, and the discrepancy between the two distributions $f(\rho_{ij,2007})$ and $f(\rho_{ij,2008})$ becomes much smaller. It suggests that the discrepancy between $f(c_{ij,2007})$ and $f(c_{ij,2008})$ is mainly caused by a market effect. We further observe that there are more negative partial correlation coefficients in 2007 and more positive partial correlation coefficients in 2008. However, the proportions are low.

### 3.2. *Distributions of eigenvalues*

For the correlation matrix and the partial correlation matrix, we can determine their eigenvalues and the associated eigenvectors. When the observed time series are i.i.d. and have zero mean and unit variance, in the limit $N \to \infty$, $T \to \infty$ where $Q = T/N \geq 1$ is fixed, the random matrix theory predicts that the distribution $f_{\rm rmt}(\lambda)$ of eigenvalues $\lambda$ can be expressed as [46–48]

$$f_{\rm rmt}(\lambda) = \frac{Q}{2\pi}\frac{\sqrt{(\lambda_{\max}-\lambda)(\lambda-\lambda_{\min})}}{\lambda} \tag{6}$$

for $\lambda \in [\lambda_{\min}, \lambda_{\max}]$, where $\lambda_{\min}$ and $\lambda_{\max}$ are respectively given by

$$\lambda_{\min}^{\max} = (1 \pm \sqrt{1/Q})^2, \tag{7}$$

which predicts a finite range of eigenvalues determined by the ratio $Q = T/N$. Here, Eq. (6) applies to the uncorrelated Wishart model, not for the correlated Wishart model [49–54].

There are $N = 1228$ stocks in our sample. For the raw and partial correlation matrices in 2007, $T = 56{,}882$ and thus $Q = 46.32$. We obtain that $\lambda_{\min} = 0.728$ and $\lambda_{\max} = 1.315$. For the raw and partial correlation matrices in 2008, $T = 58{,}310$ and thus $Q = 47.48$. We obtain that $\lambda_{\min} = 0.731$ and $\lambda_{\max} = 1.311$. These characteristic values are summarized in Table 1. Note that the results are essentially the same if we construct the random matrix from shuffled return time series [23, 55].

To identify that the estimated cross-correlations between stock returns are not a result of randomness, we compare in Fig. 2 the empirical eigenvalue distributions $f(\lambda)$ of the raw and partial correlation matrices **C** and **P** with the RMT prediction $f_{\rm rmt}(\lambda)$ given by Eq. (6). It is somewhat "trivial" to observe that all the empirical eigenvalue distributions deviate significantly from the RMT prediction.

Table 1. Characteristic values of the eigenvalue distributions of the raw and partial correlation matrices in 2007 and 2008. $p < \lambda_{\min}$ and $p > \lambda_{\max}$ are respectively the percentages of empirical eigenvalues which are less than $\lambda_{\min}$ and greater than $\lambda_{\max}$.

| | $\lambda_{\min}$ | $\lambda_{\max}$ | $\lambda_1$ | $p < \lambda_{\min}$ (%) | $p > \lambda_{\max}$ (%) |
|---|---|---|---|---|---|
| $c_{2007}$ | 0.728 | 1.315 | 130.8 | 27.04 | 3.75 |
| $c_{2008}$ | 0.731 | 1.311 | 148.9 | 28.34 | 2.20 |
| $\rho_{2007}$ | 0.728 | 1.315 | 41.89 | 15.39 | 7.00 |
| $\rho_{2008}$ | 0.731 | 1.311 | 41.90 | 14.01 | 5.94 |

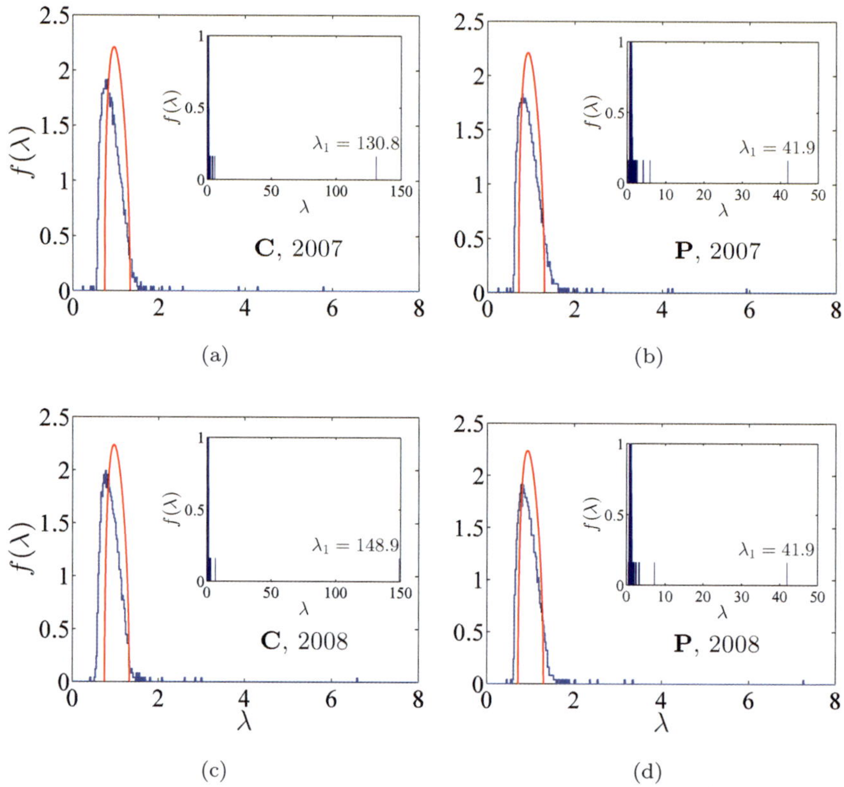


Fig. 2. (Color online.) Probability distributions $f(\lambda)$ of the eigenvalues obtained from the raw and the partial correlation matrices of stock return time series in 2007 and 2008. The red smooth curve in each plot is the eigenvalue distribution predicted by the random matrix theory. The insets show the largest eigenvalues $\lambda_1$.

For the raw correlation matrices **C**, there are 46 eigenvalues (3.75%) exceeding $\lambda_{\max} = 1.315$ in 2007 and 27 eigenvalues (2.20%) exceeding $\lambda_{\max} = 1.311$ in 2008. However, the largest eigenvalue $\lambda_1 = 148.9$ in 2008 is greater than $\lambda_1 = 130.8$ in 2007. Since $\lambda_1/N$ is a measure of systemic risk [32], we argue that the Chinese stock market has a higher systemic risk in 2008 than in 2007. Moreover, the largest eigenvalue captures 11.6% of the variations in the return time series in 2007 and 12.1% of the variations in 2008. We also find that 332 eigenvalues (27.04%) are less than $\lambda_{\min} = 0.728$ in 2007 and 348 eigenvalues (28.34%) are less than $\lambda_{\min} = 0.731$ in 2008. All these deviating eigenvalues and the associated eigenvectors might contain significant economic information.

For the partial correlation matrices **P**, there are 86 eigenvalues (7.00%) exceeding $\lambda_{\max} = 1.315$ in 2007 and 73 eigenvalues (5.94%) exceeding $\lambda_{\max} = 1.311$ in 2008. Surprisingly, the largest eigenvalue $\lambda_1 = 41.90$ in 2008 is almost equal to $\lambda_1 = 41.89$ in 2007, indicating that the higher systemic risk in 2008 was mainly introduced by the co-movements of stocks. In addition, the largest eigenvalue accounts for about 3.4% of the variations in the return residual time series in 2007 and 2008. We also find that 189 eigenvalues (15.39%) are less than $\lambda_{\min} = 0.728$ in 2007 and 172 eigenvalues (14.01%) are less than $\lambda_{\min} = 0.731$ in 2008. Overall, after removing the influence of SSCI, the eigenvalue distribution becomes much closer to the RMT prediction.

### 3.3. *Eigenvectors of the largest eigenvalues*

We now turn to unveil the economic information embedded in the first few largest eigenvalues of each matrix. Figure 3 shows the associated eigenvectors of the five

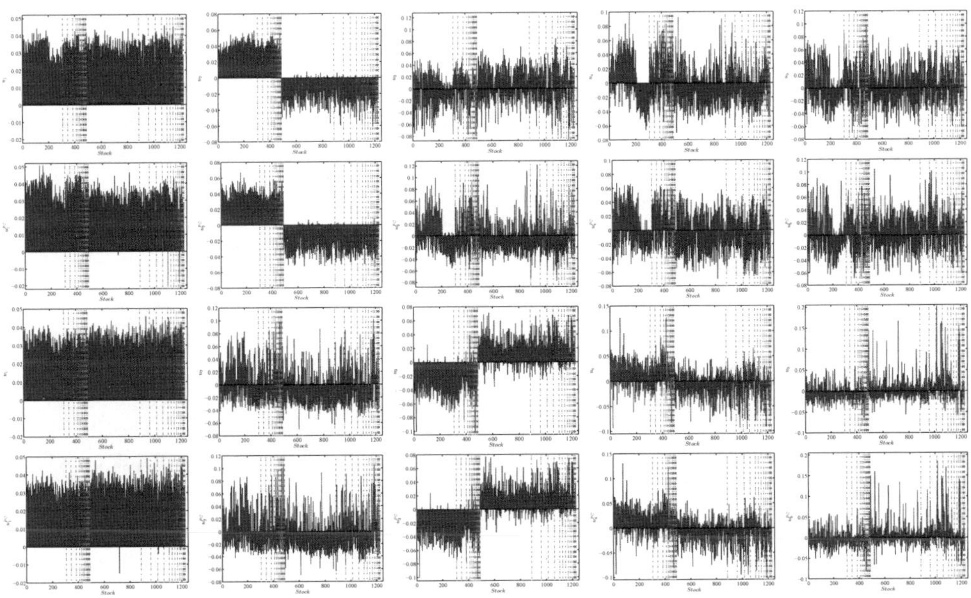

Fig. 3. Components of the eigenvectors $u_1$ to $u_5$ corresponding to the five largest eigenvalues of the raw and partial correlation matrices in 2007 and 2008. The vertical dashed lines are used to separate in turn 489 SZSE stocks (39.8%) and 739 SHSE stocks (60.2%) according to their industries.

Table 2. Percentages of stocks traded on the SZSE and the SHSE whose components are positive (or negative) in the eigenvector $u_2$ or $u_3$ of the raw correlation matrix $\mathbf{C}$ in 2007 or 2008.

| | | 2007 | | 2008 | |
|---|---|---|---|---|---|
| | | + (%) | − | + (%) | − (%) |
| $u_2$ | SZSE | 93.86 | 0 | 49.64 | 34.77 |
| | SHSE | 6.14 | 100% | 50.36 | 65.23 |
| $u_3$ | SZSE | 35 | 45% | 3.79 | 78.28 |
| | SHSE | 65 | 55% | 96.21 | 21.72 |

largest eigenvalues of the raw and partial correlation matrices in 2007 and 2008. Strikingly, there are no significant differences between the two eigenvectors associated with the $k$th largest eigenvalue of the raw and partial correlation matrices in a given year. However, we observe significant differences for the eigenvectors in different years. We thus focus on the discussions of the raw correlation matrices below.

The most intriguing pattern is observed in the eigenvector $u_2$ of the second largest eigenvalue in 2007 and in the eigenvector $u_3$ of the third largest eigenvalue in 2008. In the eigenvector $u_2$ of 2007, 93.86% of the positive components correspond to SZSE stocks and 6.14% to SHSE stocks, while all the negative components are associated with SHSE stocks, as shown in Table 2. In the eigenvector $u_3$ of 2008, 3.79% of the positive components correspond to SZSE stocks and 96.21% to SHSE stocks, while 78.28% of the negative components are associated with SZSE stocks. Therefore, the component signs of $u_2$ in 2007 and $u_3$ in 2008 are able to distinguish SZSE stocks from SHSE stocks. This feature is not clearly observed for other eigenvectors. For instance, in the eigenvector $u_2$ of 2008, half of the positive components come from SZSE stocks and 1/3 of the negative components from SZSE stocks.

### 3.4. *Market effect*

The deviating eigenvalues capture the collective behaviors of different groups of stocks and in particular the largest eigenvalue usually reflects the market effect [8]. The characteristic of a market effect is the eigenvector $u_1$ of the largest eigenvalue $\lambda_1$ has roughly equal components on all of the $N$ stocks, showing a nice linear relationship between the returns of the eigenportfolio constructed from $u_1$ and of the market index [22]. Usually, other deviating eigenvalues do not reflect a market effect but the comovement of stocks in the same industrial sector [22], the same traits shared by stocks [45], or geographic localization [56]. However, it is also possible that other deviating eigenvalues also reflect a market effect, such as the USA housing market [23].

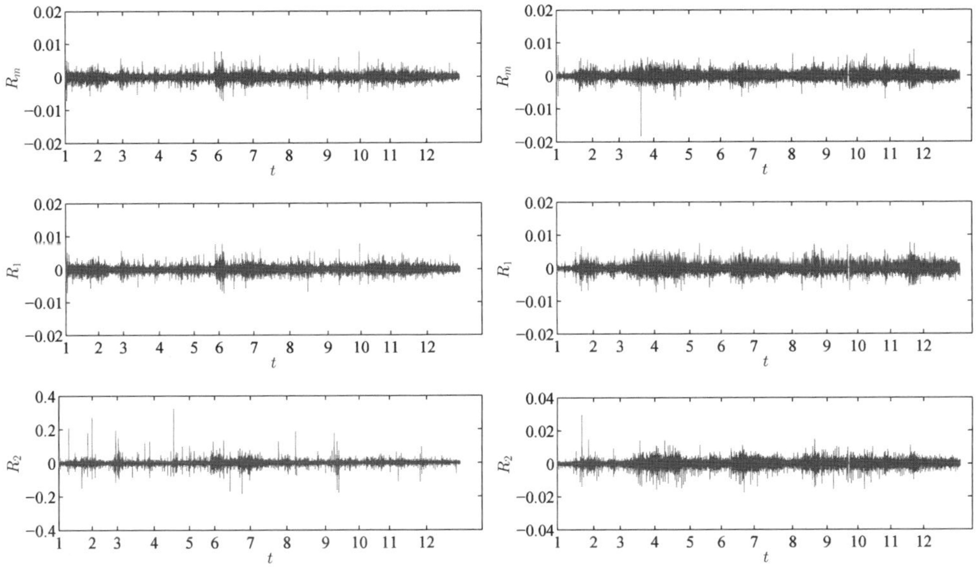

Fig. 4. Return time series of the SSCI (top row), the $\lambda_1$ eigenportfolio (middle row) and the $\lambda_2$ eigenportfolio (bottom row) constructed from the partial correlation matrices $\mathbf{P}$ in 2007 (left) and 2008 (right).

For each eigenvector $\mathbf{u}_k = [u_{k1}, \ldots, u_{ki}, \ldots, u_{kN}]^{\mathrm{T}}$ associated with eigenvalue $\lambda_k$, we construct its eigenportfolio, whose returns are calculated by

$$R_k(t) = \frac{\mathbf{u}_k^{\mathrm{T}}\mathbf{r}(t)}{\sum_{i=1}^{N} u_{ki}} \tag{8}$$

where $\mathbf{r}(t) = [r_1(t), \ldots, r_i(t), \ldots, r_N(t)]^{\mathrm{T}}$, and $\mathbf{u}_k^{\mathrm{T}}\mathbf{r}$ denotes the projection of the return time series on the eigenvector $\mathbf{u}_k$. The use of the normalization term is to ensure that the return $R_k$ constructed is the return of a portfolio. Using other normalization terms will not change the conclusion. The return time series $R_m(t)$ of the SSCI, $R_1(t)$ of the first eigenportfolio and $R_2(t)$ of the second eigenportfolio of the partial correlation matrices are plotted in left column of Fig. 4 for 2007 and in the right row of Fig. 4 for 2008. The results are almost the same for the raw correlation matrices, which is natural due to the similarity of corresponding eigenvectors of $\mathbf{P}$ and $\mathbf{C}$ shown in Fig. 3. It is found that $R_1(t)$ is quite similar to $R_m(t)$ in each year, while $R_2(t)$ is very different from $R_m(t)$.

Figure 5 presents the scatter plots of market returns $R_m(t)$ against the eigenportfolio returns $R_k(t)$ $(k = 1, 2)$ constructed from $\mathbf{P}$. We find that there is no linear dependence between $R_m(t)$ and $R_2(t)$. No linear dependence is observed either for $R_k(t)$ with $k > 2$ (not shown in Fig. 5). In contrast, there is a nice linear dependence between $R_m(t)$ and $R_1(t)$ in each year. The slope is 0.815 in 2007 and 0.733 in 2008. For $\mathbf{C}$, the slope is $0.829 \pm 0.004$ in 2007 and $0.741 \pm 0.004$in 2008. These

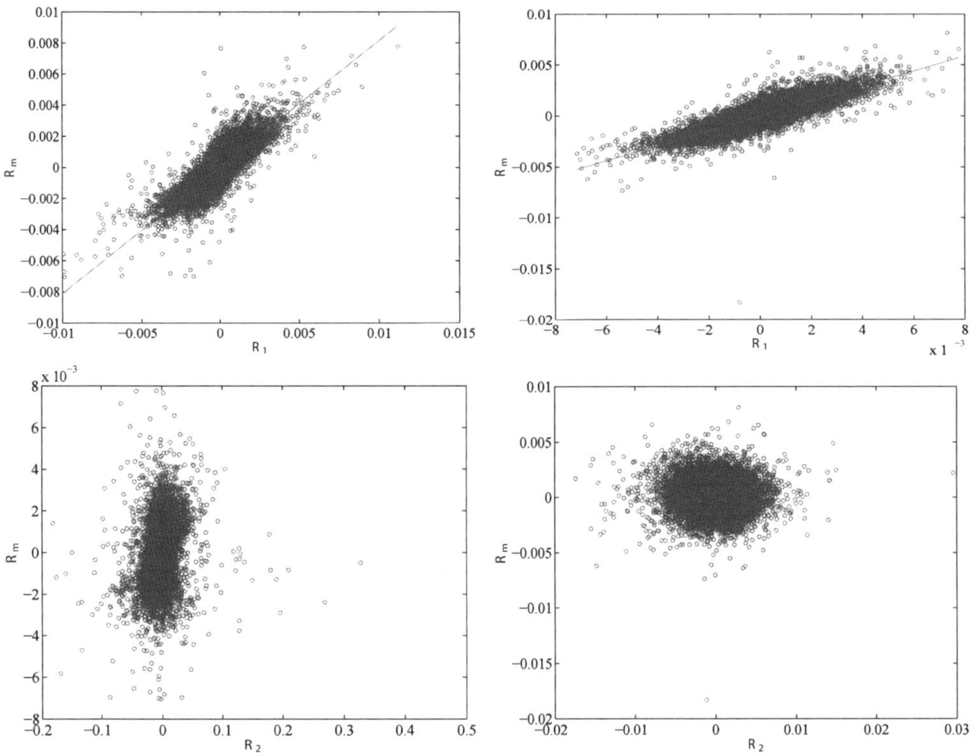

Fig. 5. Market effect of the largest eigenvalues $\lambda_1$ of the partial correlation matrices $\mathbf{P}$ in 2007 (left) and 2008 (right). The results for the raw correlation matrices $\mathbf{C}$ are almost the same. The outlier point in right column is caused by an abrupt price change at 10:30 on 20 March 2008 shown in the top-right plot of Fig. 4.

observations suggest that the largest eigenvalue quantifies a common influence on all stocks, while the rest of deviating eigenvalues contain no information about such a market effect.

The results for the raw correlation matrices $\mathbf{C}$ are well established for diverse stock markets, especially on the daily level [22, 45]. However, the results for the partial correlation matrices $\mathbf{P}$ are somewhat surprising. It was expected that the largest eigenvalue is no longer associated with the market, since the effect of the index has been removed [40]. According to our results, this natural conjecture is surprisingly not true. On the contrary, similar phenomena were observed for daily stock returns [57].

### 3.5. *Eigenvector components and stocks' market capitalizations*

We checked the components of eigenvectors associated with the smallest eigenvalues. There are a few components whose magnitudes are significantly greater than the

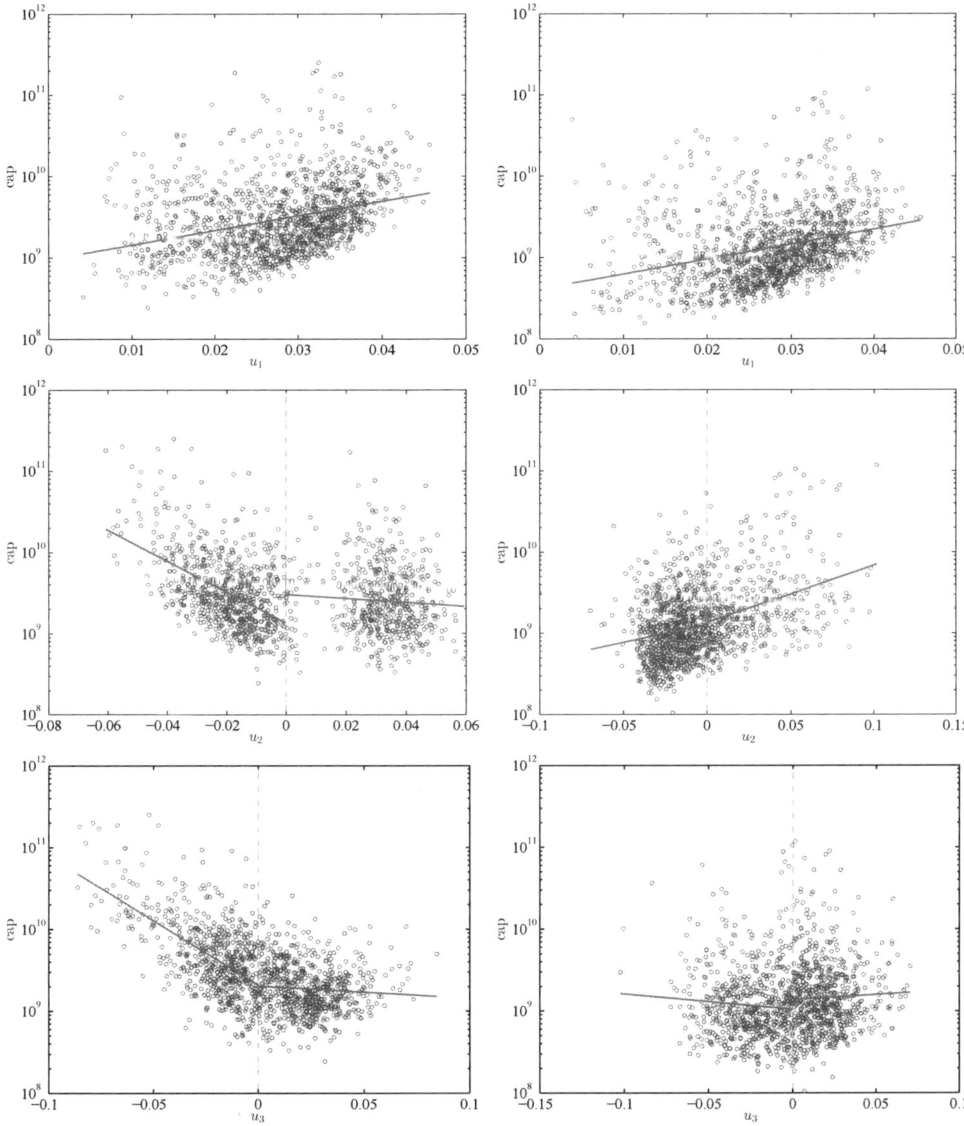


Fig. 6. Relationship between the eigenvector components of the raw correlation matrix and stocks' market capitalizations in 2007 (left) and 2008 (right).

averages. However, we did not observe solid evidence that the correlations of the corresponding return time series of these components are among the largest, which is different from the U.S. stock market [22] and the global crude oil market [56].

We investigated the relationship between the eigenvector components and stocks' market capitalizations. Figure 6 illustrates the results for $u_1$, $u_2$ and $u_3$

of the raw correlation matrix. The results for the partial correlation matrices are similar, as elaborated by the eigenvectors in Fig. 3. The eigenvector components of $u_1$ are positively correlated with the corresponding capitalizations, as shown in the first row of Fig. 6. For other largest eigenvalues, such positive correlations between the component magnitudes and capitalizations are observed for positive components or negative components. These findings suggest that stocks with low volatility and high liquidity exhibit high magnitudes of the eigenvector components.

## 4. Conclusion

We have conducted a comparative analysis of the information contents embedded in the largest eigenvalues of the raw and partial correlation matrices constructed from the 1-min high-frequency returns of Chinese stocks in 2007 and 2008. We identified market correlation structure changes around the Great crash in several aspects. In addition, although the correlation coefficient distributions and the largest eigenvalues of the raw and partial correlation matrix are significantly different in each period, the eigenvectors of the raw and partial correlation matrices in each period are strikingly similar. We found that the largest eigenvalue of each matrix reflects the whole market mode. It is found that the eigenvectors of the second largest eigenvalues in 2007 and of the third largest eigenvalues in 2008 are able to distinguish the stocks from the two exchanges, which are different from the cases of the U.S.A. stock market [22] and the Chinese stock market when daily returns are analyzed [29, 45]. We also found that the component magnitudes of the some largest eigenvectors are proportional to the stocks' capitalizations.

Based on the current work, there are possible extensions for future researches. One possibility is to consider the effect of different timescales on the results. We conjecture that the main results will be qualitatively similar with quantitative differences. For instance, the distribution of partial correlation coefficients will be narrower than that of the conventional correlation coefficients, while the average correlation coefficient will be larger for large timescales since $\langle c \rangle \approx 0.1$ for the 1-min returns in this work and $\langle c \rangle \approx 0.37$ for daily returns reported in Ref. [45]. This is indeed the Epps effect [58]. Another possibility is to investigate other financial markets around the Great Crash using high-frequency data.

The results reported in this work have potential applications, which is common for RMT analysis of financial markets. The calculated eigenvalues can be used to construct a measure of systemic risk and the eigenportfolio of the largest eigenvalue can be employed to construct a new stock market index. In this vein, a strategy of moving windows should be adopted. These topics deserve further separate studies.

## Acknowledgments

This work was supported by National Natural Science Foundation of China (71532009 and 71131007) and Fundamental Research Funds for the Central Universities (222201718006).

## References


[1] M. Tumminello, F. Lillo and R. N. Mantegna, Correlation, hierarchies, and networks in financial markets, *J. Econ. Behav. Org.* **75** (2010) 40–58.

[2] D.-M. Song, M. Tumminello, W.-X. Zhou and R. N. Mantegna, Evolution of worldwide stock markets, correlation structure, and correlation based graphs, *Phys. Rev. E* **84** (2011) 026108.

[3] S. Kumar and N. Deo, Correlation and network analysis of global financial indices, *Phys. Rev. E* **86** (2012) 026101.

[4] A. Nobi, S. E. Maeng, G. G. Ha and J. W. Lee, Random matrix theory and cross-correlations in global financial indices and local stock market indices, *J. Korean Phys. Soc.* **62** (2013) 569–574.

[5] A. Nobi, S. M. Lee, D. H. Kim and J. W. Lee, Correlation and network topologies in global and local stock indices, *Phys. Lett. A* **278** (2014) 2482–2489.

[6] R. N. Mantegna, Hierarchical structure in financial markets, *Eur. Phys. J. B* **11** (1999) 193–197.

[7] V. Plerou, P. Gopikrishnan, B. Rosenow, L. A. N. Amaral and H. E. Stanley, Universal and nonuniversal properties of cross correlations in financial time series, *Phys. Rev. Lett.* **83** (1999) 1471–1474.

[8] L. Laloux, P. Cizeau, J.-P. Bouchaud and M. Potters, Noise dressing of financial correlation matrices, *Phys. Rev. Lett.* **83** (1999) 1467–1470.

[9] J. Kwapień and S. Drożdż, Physical approach to complex systems, *Phys. Rep.* **515** (2012) 115–226.

[10] L. Uechi, T. Akutsu, H. E. Stanley, A. J. Marcus and D. Y. Kenett, Sector dominance ratio analysis of financial markets, *Physica A* **421** (2015) 488–509.

[11] C. Curme, M. Tumminello, R. N. Mantegna, H. E. Stanley and D. Y. Kenett, Emergence of statistically validated financial intraday lead-lag relationships, *Quant. Financ.* **15** (2015) 1375–1386.

[12] Z.-Q. Jiang and W.-X. Zhou, Complex stock trading network among investors, *Physica A* **389** (2010) 4929–4941.

[13] J.-J. Wang, S.-G. Zhou and J.-H. Guan, Characteristics of real futures trading networks, *Physica A* **390** (2011) 398–409.

[14] X.-Q. Sun, X.-Q. Cheng, H.-W. Shen and Z.-Y. Wang, Distinguishing manipulated stocks via trading network analysis, *Physica A* **390** (2011) 3427–3434.

[15] X.-Q. Sun, H.-W. Shen, X.-Q. Cheng and Z.-Y. Wang, Degree-strength correlation reveals anomalous trading behavior, *PLoS One* **7** (2012) e45598.

[16] J.-J. Wang, S.-G. Zhou and J.-H. Guan, Detecting potential collusive cliques in futures markets based on trading behaviors from real data, *Neurocomputing* **92** (2012) 44–53.

[17] M. Tumminello, F. Lillo, J. Piilo and R. N. Mantegna, Identification of clusters of investors from their real trading activity in a financial market, *New J. Phys.* **14** (2012) 013041.

[18] Z.-Q. Jiang, W.-J. Xie, X. Xiong, W. Zhang, Y.-J. Zhang and W.-X. Zhou, Trading networks, abnormal motifs and stock manipulation, *Quant. Financ. Lett.* **1** (2013) 1–8.

[19] X.-Q. Sun, H.-W. Shen and X.-Q. Cheng, Trading network predicts stock price, *Sci. Rep.* **4** (2014) 3711.

[20] M.-X. Li, Z.-Q. Jiang, W.-J. Xie, X. Xiong, W. Zhang and W.-X. Zhou, Unveiling correlations between financial variables and topological metrics of trading networks: Evidence from a stock and its warrant, *Physica A* **419** (2015) 575–584.

[21] F. Musciotto, L. Marotta, S. Miccichè, J. Piilo and R. N. Mantegna, Patterns of trading profiles at the Nordic Stock Exchange. A correlation-based approach, *Chaos, Solitons Fractals* **88** (2016) 267–278.

[22] V. Plerou, P. Gopikrishnan, B. Rosenow, L. A. N. Amaral, T. Guhr and H. E. Stanley, Random matrix approach to cross correlations in financial data, *Phys. Rev. E* **65** (2002) 066126.

[23] H. Meng, W.-J. Xie, Z.-Q. Jiang, B. Podobnik, W.-X. Zhou and H. E. Stanley, Systemic risk and spatiotemporal dynamics of the US housing market, *Sci. Rep.* **4** (2014) 3655.

[24] H. Meng, W.-J. Xie and W.-X. Zhou, Club convergence of house prices: Evidence from China's ten key cities, *Int. J. Mod. Phys. B* **29** (2015) 1550181.

[25] X.-F. Jiang and B. Zheng, Anti-correlation and subsector structure in financial systems, *EPL* (*Europhys. Lett.*) **97** (2012) 48006.

[26] X.-F. Jiang, T.-T. Chen and B. Zheng, Structure of local interactions in complex financial dynamics, *Sci. Rep.* **4** (2014) 5321.

[27] D. Sornette, *Why Stock Markets Crash* (Princeton University Press, Princeton, 2003).

[28] D. Sornette, Critical market crashes, *Phys. Rep.* **378** (2003) 1–98.

[29] F. Ren and W.-X. Zhou, Dynamic evolution of cross-correlations in the Chinese stock market, *PLoS ONE* **9** (2014) e97711.

[30] A. Nobi, S. E. Maeng, G. G. Ha and J. W. Lee, Effects of global financial crisis on network structure in a local stock market, *Physica A* **407** (2014) 135–143.

[31] L.-F. Zhao, W. Li and X. Cai, Structure and dynamics of stock market in times of crisis, *Phys. Lett. A* **380** (2016) 654–666.

[32] M. Billio, M. Getmansky, A. W. Lo and L. Pelizzon, Econometric measures of connectedness and systemic risk in the finance and insurance sectors, *J. Financ. Econ.* **104** (2012) 535–559.

[33] F. Meudt, M. Theissen, R. Schäfer and T. Guhr, Constructing analytically tractable ensembles of stochastic covariances with an application to financial data, *J. Stat. Mech. Theor. Exp.* (2015) P11025.

[34] M. C. Münnix, T. Shimada, R. Schäfer, F. Leyvraz, T. H. Seligman, T. Guhr and H. E. Stanley, Identifying states of a financial market, *Sci. Rep.* **2** (2012) 644.

[35] P. Rinn, Y. Stepanov, J. Peinke, T. Guhr and R. Schäfer, Dynamics of quasi-stationary systems: Finance as an example, *EPL* (*Europhys. Lett.*) **110** (2015) 68003.

[36] D. Chetalova, R. Schäfer and T. Guhr, Zooming into market states, *J. Stat. Mech. Theor. Exp.* (2015) P01029.

[37] D. Chetalova, M. Wollschläger and R. Schäfer, Dependence structure of market states, *J. Stat. Mech. Theor. Exp.* (2015) P08012.

[38] T. A. Schmitt, R. Schäfer and T. Guhr, Credit risk: Taking fluctuating asset correlations into account, *J. Credit Risk* **11** (2015) 73–94.

[39] K. Baba, R. Shibata and M. Sibuya, Partial correlation and conditional correlation as measures of conditional independence, *Aust. N. Z. J. Stat.* **46** (2004) 657–664.

[40] D. Y. Kenett, Y. Shapira and E. Ben-Jacob, RMT assessments of the market latent information embedded in the stocks' raw, normalized, and partial correlations, *J. Prob. Stat.* **2009** (2009) 249370.

[41] D. Y. Kenett, M. Tumminello, A. Madi, G. Gur-Gershgoren, R. N. Mantegna and E. Ben-Jacob, Dominating clasp of the financial sector revealed by partial correlation analysis of the stock market, *PLoS One* **5** (2010) e15032.

[42] D. Y. Kenett, X.-Q. Huang, I. Vodenska, S. Havlin and H. E. Stanley, Partial correlation analysis: Applications for financial markets, *Quant. Financ.* **15** (2015) 569–578.

[43] X.-Y. Qian, Y.-M. Liu, Z.-Q. Jiang, B. Podobnik, W.-X. Zhou and H. E. Stanley, Detrended partial cross-correlation analysis of two nonstationary time series influenced by common external forces, *Phys. Rev. E* **91** (2015) 062816.
[44] Z.-Q. Jiang, W.-X. Zhou, D. Sornette, R. Woodard, K. Bastiaensen and P. Cauwels, Bubble diagnosis and prediction of the 2005–2007 and 2008–2009 Chinese stock market bubbles, *J. Econ. Behav. Org.* **74** (2010) 149–162.
[45] J. Shen and B. Zheng, Cross-correlation in financial dynamics, *EPL* (*Europhys. Lett.*) **86** (2009) 48005.
[46] V. A. Marčhenko and L. A. Pastur, Distribution of eigenvalues for some sets of random matrices, *Math. USSR-Sbornik* **1** (1967) 457–483.
[47] A. Edelman, Eigenvalues and condition numbers of random matrices, *SIAM. J. Matrix Anal. Appl.* **9** (1988) 543–560.
[48] A. M. Sengupta and P. P. Mitra, Distributions of singular values for some random matrices, *Phys. Rev. E* **60** (1999) 3389–3392.
[49] T. W. Anderson, *An Introduction to Multivariate Statistical Analysis*, 3rd edn. (Wiley, New Jersey, 2003).
[50] R. J. Muirhead, *Aspects of Multivariate Statistical Theory*, 2nd edn. (Wiley, New Jersey, 2005).
[51] C. Recher, M. Kieburg and T. Guhr, Eigenvalue densities of real and complex Wishart correlation matrices, *Phys. Rev. Lett.* **105** (2010) 244101.
[52] Vinayak and A. Pandey, Correlated Wishart ensembles and chaotic time series, *Phys. Rev. E* **81** (2010) 036202.
[53] G. Akemann, T. Guhr, M. Kieburg, R. Wegner and T. Wirtz, Completing the picture for the smallest eigenvalue of real Wishart matrices, *Phys. Rev. Lett.* **113** (2014) 250201.
[54] T. Wirtz, M. Kieburg and T. Guhr, Limiting statistics of the largest and smallest eigenvalues in the correlated Wishart model, *EPL* (*Europhys. Lett.*) **109** (2015) 20005.
[55] W.-X. Zhou, G.-H. Mu and J. Kertész, Random matrix approach to the dynamics of stock inventory variations, *New J. Phys.* **14** (2012) 093025.
[56] Y.-H. Dai, W.-J. Xie, Z.-Q. Jiang, G. J. Jiang and W.-X. Zhou, Correlation structure and principal components in global crude oil market, *Empir. Econ.* **51** (2016) 1501–1519.
[57] D.-M. Song, Complexity Researches in Financial Markets, Ph.D. thesis, East China University of Science and Technology, Shanghai, June, 2013.
[58] B. Tóth and J. Kertész, The Epps effect revisited, *Quant. Financ.* **9** (2009) 793–802.

# 8

# Noise and Security in Communications

# PROTOCOLS FOR DISTRIBUTING KEY STRINGS BASED ON PHASE AND INTENSITY FLUCTUATION

AJUNG KIM
*Dept. of Opt. Eng., Sejong University*
*Seoul 143-747, Korea*

Communicated by Laszlo Kish

A cryptology protocol for key distribution using phase and intensity fluctuation is proposed. It provides ways to detect the extent of eavesdropping. Detecting test factors and system design rules are suggested and the performances are analyzed for various signal keyings and cryptofunctions. It has a wide range of implementations, including the use of additive white Gaussian noise, background noise, and polarization modification in wire/wireless and optical communications.

*Keywords*: Fluctuation; key distribution; cryptology.

## 1. Introduction

As communication systems and networks have been widely installed, system security and reliability have become issues of great importance. The advent of the quantum computer [1, 2] with the ability of massive parallelism and storing superposition of numbers has been a threat to conventional cryptosystems. This has brought a strong demand for establishing security systems not only based on computational processing but also on physical signal processing.

In cryptosystems, key is a sequence of bits that ciphers a plain text, as an input to encryption functions. In shared key cryptographic systems, the secret private key is distributed and agreed between legitimate users and the secure distribution and management of keys play an important role in the total security of message transmission.

In this paper, a key agreement protocol using fluctuation is proposed. A drawback of the classical cryptosystem is that it can always be monitored without the legitimate users being aware of any eavesdropping taking place. The proposed cryptographic system, however, provides ways to detect the extent of eavesdropping. The performance is evaluated in terms of mutual information between legitimate users and an eavesdropper, and their error rates. Detecting test factors, as criteria to estimate the degree of eavesdropping, and system design rules are suggested as well. The proposed system is based on fluctuations and has a wide range of implementations as the sources of employed noise are ubiquitous. In addition, unlike quantum cryptography [3, 4], this cryptosystem enables signal amplification and implementation over a long distance.

## 2. Protocol

One of its simple implementations is the use of additive white Gaussian noise (AWGN). Suppose Alice sends to Bob a sequence of signal bits which are susceptible to additive noise. For example, in a binary signal system with phase reverse keying (PRK) [5] and a coherent measurement, Alice sends signals $s_0(t)$ and $s_1(t)$ represented as

$$s_1(t) = \sqrt{\frac{2E}{T}}\cos(2\pi f_c t) = \sqrt{E}\varphi(t)$$
$$s_0(t) = -\sqrt{E}\varphi(t) \tag{1}$$

where $E$, $T$ and $f_c$ are bit energy, pulse duration and carrier frequency, respectively, and $\int_T \varphi^2(t)dt = 1$. The signal constellation with the PRK is depicted in Fig. 1.

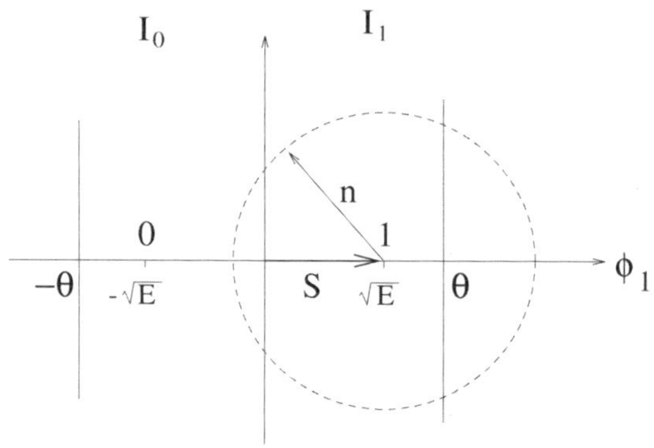


Fig. 1. Signal constellation with PRK.

The received signal is a vector sum of the transmitted signal and noise in a signal state space. The optimal detector measures $r = \int_T r(t)\phi(t)dt$ where $r(t) = s(t) + n(t)$ and $n(t)$ is a white Gaussian noise with power spectral density $N_0$.

The estimation with optimal decision region divisions[5] identifies signals $r(t)$ falling in the region $I_1$ as the signal '1' and $r(t)$ falling in $I_0$ as the signal '0'. An eavesdropper, Eve, might try to detect each bit of signals according to the maximum a posteriori (MAP) decision rule; the signal is '1' if $r>0$, and '0' if $r<0$. On the other hand, Bob makes detection decisions at a higher threshold ( $\theta > 0$) and discards the unreliable bits falling in the erroneous region below the threshold. Bob then tells Alice via a public channel which bits were accepted as a key string without telling the values of his bits. After the public announcement, Alice and Bob take the sequence of the accepted bits as a private key string shared between the two of them for ciphering a text. In this protocol, even though Eve measured every bit transmitted to Bob and heard which bits were accepted from the public announcement, she cannot tell the exact values of the key bits accepted between Alice and Bob. Eve rather ends up with a higher error rate in her key string than Bob. Moreover, Eve's measured bit stream is uncorrelated with Bob's, because the device noise of Eve's detector is uncorrelated with that of Bob's. Thus Eve can obtain neither the correct key string, nor the same key string that Bob accepts.

## 3. Performance

The performance can be analyzed for various signal keyings and cryptofunctions. In such systems where Eve makes a decision with the MAP detection rule, the error probability of Eve is obtained as

$$P_e^E = \frac{1}{2} erfc\left(\sqrt{\frac{E}{N_0^E}}\right) \tag{2}$$

where

$$erfc(u) = \frac{2}{\sqrt{\pi}} \int_u^{\infty} e^{-r^2} dr.$$

On the other hand, Bob measures each bit at a threshold $\theta$ with the following decision rule; the signal is1 if $r > \theta$, and 0 if $r < -\theta$. Then, Bob's error probability without eavesdropping is shown below with SNR(signal-to-noise ratio).

$$P_e^B = \frac{1}{2} erfc\left(\frac{\theta + \sqrt{E}}{\sqrt{N_0^B}}\right). \tag{3}$$

Figure 2 shows error rates for Eve and Bob with this scheme as a function of SNR for various threshold values. Bob takes a value of $\theta$ that makes $P_e{}^B$ considerably small.

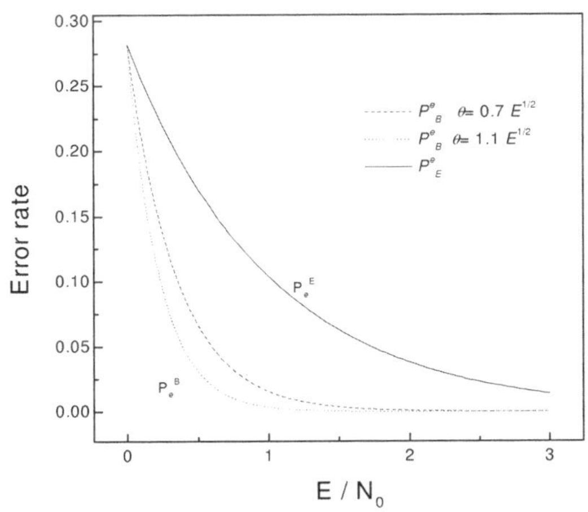


Fig. 2. Error rate for Eve and Bob in an AWGN-based cryptosystem as a function of SNR. The dashed line refers to Bob's error rate with a threshold $\theta = 0.7\sqrt{E}$, and the dotted line refers to $\theta = 1.1\sqrt{E}$.

According to the protocol, Bob discards the data that fall below the threshold value. Then, the data rate per use is

$$\begin{aligned} R &= \frac{1}{2} erfc\left(\frac{\theta - \sqrt{E}}{\sqrt{N_0^B}}\right) & [\theta > \sqrt{E}] \\ R &= 1 - \frac{1}{2} erfc\left(\frac{\sqrt{E} - \theta}{\sqrt{N_0^B}}\right) & [\theta < \sqrt{E}] \end{aligned} \tag{4}$$

A signal with a low SNR introduces a high error rate of Eve and a high threshold value induces a low error rate of Bob. The penalty of a high threshold value which confirms a high security level is a reduced data rate.

Eve cannot effectively eavesdrop with an intercept/resend scheme [6] without increasing $P_e^B$ by a significant amount, if the cryptosystem is designed to make $P_e^E$ sufficiently large. Moreover, as device noise $n_B$ and $n_E$ are uncorrelated with each other, the measurement results are uncorrelated as well. Therefore, Eve cannot perform an effective eavesdropping with a correlator [7]. In addition, Eve's attempt to split the signal into several samples or to amplify the signal is futile, as splitting reduces the signal to the level where the noise effects are more significant, and amplifying disturbs the states and introduces noise large enough to nullify the advantage of eavesdropping on strong signals.

With the presence of Eve, Eve's erroneous decisions and retransmissions introduce errors in Bob's bit string. In the eavesdropping detection, Bob sets up a tolerable value $Q$ for his error rate, in the region below where error correction is available and thus the transmission is considered safe. To suppress $P_e^B$ below $Q$, Eve attempts to eavesdrop only on a fraction $\varepsilon$ of Alice's string. Then, the error probability for Bob when the eavesdropping exists, $\overline{P_e^B}$, is increased by Eve's erroneous measurements and retransmissions of wrongly fabricated bits. Thus Bob can tell the presence of Eve from the increased error rate in his bits. It is expressed as

$$\begin{aligned}\overline{P_e^B} &= \varepsilon\left[\frac{1}{4}erfc\left(\sqrt{\frac{E}{N_0^E}}\right)erfc\left(\frac{\theta-\sqrt{E}}{\sqrt{N_0^B}}\right)+\frac{1}{2}\left\{1-\frac{1}{2}erfc\left(\sqrt{\frac{E}{N_0^E}}\right)\right\}erfc\left(\frac{\theta+\sqrt{E}}{\sqrt{N_0^B}}\right)\right] \\ &= \varepsilon P \\ &= Q. \end{aligned} \tag{5}$$

Partial information between Eve and Bob, which are to be hashed out at private amplification, can be obtained as

$$I_{EB} = \varepsilon[1 + P_e^B \log_2 P_e^B + (1-P_e^B)\log_2(1-P_e^B). \tag{6}$$

where $\varepsilon = Q/P$ from Eq. (5). For partial information between Eve and Alice, $I_{AE}$, is

$$I_{AE} = \varepsilon[1 + P_e^E \log_2 P_e^E + (1-P_e^E)\log_2(1-P_e^E)]. \tag{7}$$

Thus Eve's mutual information with Alice and Bob is significantly reduced to the level not high enough to succeed in obtaining the right key string with a plain-text attack.

Implementation of this concept is not limited to the source in the thermal noise dominant region. It can be applied to other noise sources, for example, phase noise-induced intensity fluctuation. In interferometric measurements, phase noise is converted to intensity noise in the interfering stage, which induces intensity fluctuation. Other available noise includes polarization modification induced by couplers, and background noise in wireless communications.

## 4. Practical Implementation

The system is designed with various parameters in practice. One of criteria in cryptosystems is mutual information, which will be hashed out at private amplification. However, since mutual information has only asymptotic significance for noisy systems, we would not be concerned with possible privacy amplification and would design our cryptographic system according to the design procedure based on constraints for countermeasurement against various possible attacks. We will consider three kinds of possible attacks that Eve can perform: a key trial attack, an intercept/resend attack and a partial intercept/resend attack.

With a key trial scheme, Eve may make a copy of the transmitted signals, measure them and attempt to guess a correct key string by changing arbitrarily the value of $n$ bits out of her $|k|$ bit key string in a possible known plain text attack. Suppose Eve has the capability of trying $N_E$ key strings, which yields a probability of success $P_c^E$. For a given noise power spectral density $N_0^E$, this determines a maximum signal energy $E_1$ allowed between Alice and Bob. From the constraints, it turns out that $P_c^E$ is smaller than $10^{-37}$ with a key length $|k|\sim10^3$ for optical stream ciphers. For $|k|\sim50$ and $P_e^E\sim0.1$, some hashing for privacy amplification is necessary to make $P_c^E$ small. Instead, we propose using a simple mod 2 sum of 10 different bits each to get a key bit from a 500-bit string. With such a key string of $|k|\sim50$, we have $P_e^E\sim0.4463$ and $P_c^E\sim5.63\text{x}10^{-8}$ for $N_E\sim2^{20}$.

Countermeasurement for another kind of possible attack, an intercept/resend attack, yields another constraint on $E$, which will be denoted by $E_2$. With the intercept/resend attack, Eve might correlate Bob's measurement results with hers, but her erroneous decisions introduce a great deal of errors in Bob's bit string. The error rate $\overline{P_e^B}$ with this attack is obtained as

$$\overline{P_e^B} = P_e^E(1-P_e^B)+(1-P_e^E)P_e^B\ . \tag{8}$$

For $E\sim N_0$, $\theta\sim2N_0$, we have $P_e^B\sim10^{-3}$, $P_e^E\sim0.1$ and $\overline{P_e^B}\sim10^{-1}$ so the users can easily detect Eve's presence. Provided a probability of intercept $P_I^E$ is defined as a probability of Eve evading the error check test, its value depends on $E$ in relations to $P_e^E$ and determines an upper bound $E_2$. With the two possible attacks in mind, Alice and Bob would take the smaller value of $E_1$ and $E_2$ as the signal level.

In order for Eve to evade the error check test, she can devise another type of attack, a partially intercept/resend attack, where Eve resends a fraction of the bits tapped from Alice's string. If the users set a safety region of, say, three standard deviations, Eve's resulting correlations with Bob would apply only to one tenth of the tapped bit string in order to evade the test with a high probability. Eve may use the resulting 10% deterministic "hard bit" [8] information to shorten the length of the trial string in a key trial attack. Nevertheless, the success probability $P_c^E$ with this attack is less than $10^{-36}$ for $P_e^E\sim0.1$, which is still very low. Thus the partial intercept/resend attack is not quite effective, although it helps to evade the test with a high probability. These results are obtained from the binomial distribution and valid for any binary scheme. In the Gaussian case with $\theta=mE$ and $|k|=10^3$, the resulting data rate $R\sim0.043$ bit/pulse for $m\sim2.72$, $E_2\sim N_0$ and $P_e^E\sim0.16$, which is comparable to the four-state cryptographic system [8], whereas $R\sim0.13$ for $m\sim1.90$, $E_2\sim1.3N_0$ and $P_e^E\sim0.1$. The data rates are large compared to the usual values in the quantum case. These results are due to the fact that large SNR signals can be used in our case to take more precise advantage of Eve's limitations.

Alice and Bob also determine the threshold value $\theta$ with consideration of the tradeoff between the error rate $P_e^B$ and the data rate $R$. A larger $\theta$ will decrease the data rate as well as the error rate. In practice, since a reconciliation procedure [8,9], such as parity checks, is required to eliminate possible errors in the key string agreed between Alice and Bob, $P_e^B$ is to be set with this overhead cost in mind.

With the analysis above, the system parameter values are quite encouraging for practical applications. At microwave frequencies including mobile cellular transmission [10], the independent AWGN model for both Bob and Eve are often valid, and a signal data rate $R_0$ can be in the gigabit range so that the key distribution data rate $R$ ~10 *Mbits*. Even after overhead costs such as reconciliation and the above opaque eavesdropping check by a different bit string are taken into account, the resulting actual rate $R$ may still compare favorably to the well-known public-key methods based on computational complexity. The situation is even better in optical fiber systems where $R_0$ can be made 10 *Gbps* or higher. Currently, *Gbps* high-speed direct detection systems are thermal noise limited [11]. Thus, the uncorrelated noise model above is essentially valid because Bob and Eve each see the thermal noise in their respective detectors. The total photon number in the above estimates would be high enough to make the standard Gaussian approximation [11] valid, and thus the model and estimates are directly applicable. The on-off signal case is just a scaled version of antipodal signals [5]. The $P_e^E$ value is also valid for the coherent unipolar signal with amplitude shift keying (ASK) with the amplitude level of $E_0$ and $E_1$ satisfying $\sqrt{E_0} - \sqrt{E_1} = 2\sqrt{E}$ so that the level spacing remains unchanged. However, PRK that needs less signal power for a specified level spacing is favorable in terms of the power budget and security. The model and estimates here are applicable in a broad area of communication systems and can establish a robust physical security layer in practice.

## 5. Conclusion

We suggested a key agreement protocol using phase and intensity fluctuation. Detecting test factors and system design rules were analyzed. The two basic goals in realizing this protocol for secret key agreement are i) to make it difficult for Eve to distinguish between transmitted states and ii) to make Eve unable to obtain closely correlated to the legitimate users' results. The former can be achieved by using signals susceptible to additive noise and the latter can be fulfilled by exploiting the fact that the eavesdropper's device noise is uncorrelated to the legitimate receiver's. The proposed protocol provides ways to detect the extent of eavesdropping by checking the increase in the legitimate receiver's key string error rate.

This cryptosystem based on any physical noise has a broad range of implementations by using uncorrelated noise such as AWGN(Additive White Gausian Noise), background noise, and polarization modification in wire/wireless and optical communications. The implementation is easier and requires simpler processing at error correction and private amplification. It constitutes a new approach to establishment of secure communications.

## Acknowledgements

This work was supported by KRF D00392, the FRF of Sejong.

## References

[1] P. W. Shor, *Proc. 35 Ann. Symp. Found. of Computer Science*, (ed.) S. Goldwasser, IEEE Computer Society Press, Los Alamitos, Ca., (1994) 124.
[2] P. K. Amiri, Quantum computers, *IEEE Potentials* **21**(5) (2003) 6–9.
[3] S. Wiesner, Conjugate coding, *SIGACT News* **15** (1983) 78.
[4] P. D. Townsend, Experimental investigation of the performance limits for first telecommunications-window. Quantum cryptography systems, *IEEE Phot. Tech. Lett.* **10**(7) (1998) 1048–1050.

[5] H. L. Van Trees, *Detection, Estimation, and Modulation Theory*, Part 1, Wiley (1968).
[6] A. K. Ekert, B. Huttner, G. M. Palma and A. Peres, Eavesdropping on quantum-cryptographical systems, *Phys. Rev. A* **50** (1994) 4298.
[7] H. P. Yuen, Communication and measurement with squeezed *states in Quantum Squeezing*, (eds.) P. D. Drummond and Z. Ficek, Springer-Verlag, New York, Sec. 3 (2003).
[8] C. H. Bennett, F. Bessette, G. Brassard, L. Salvail and J. Smolin, Experimental quantum cryptography, *J. of Cryptology* **5** (1992) 3.
[9] C. H Bennett, G. Brassard, C. Crepeau and U. M. Maurer, Generalized privacy amplification, *IEEE Trans. on Info. Theory* **41**(6) (1995) 1915–1923.
[10] A. Mishra, M. Shin, N. L. Petroni, C. Clancy and W. A. Arbaugh, Proactive key distribution using neighbor graphs, *IEEE Wireless Comm.* (2004) 26–36.
[11] P. E. Green Jr., *Fiber Optics Networks*, Wiley, Ch. 8 (1993).

# What Kind of Noise Guarantees Security for the Kirchhoff-Law–Johnson-Noise Key Exchange?

Robert Mingesz*, Gergely Vadai and Zoltan Gingl
*Department of Technical Informatics, University of Szeged*
*Árpád tér 2, 6720 Szeged, Hungary*
**mingesz@inf.u-szeged.hu*

Received 5 May 2014
Accepted 26 May 2014
Published 27 June 2014

Communicated by Laszlo B. Kish

This paper is a supplement to our recent one about the analysis of the noise properties in the Kirchhoff-law–Johnson-noise (KLJN) secure key exchange system [Gingl and Mingesz, *PLOS ONE* **9** (2014) e96109, doi: 10.1371/journal.pone.0096109]. Here, we use purely mathematical statistical derivations to prove that only normal distribution with special scaling can guarantee security. Our results are in agreement with earlier physical assumptions [Kish, *Phys. Lett. A* **352** (2006) 178–182, doi: 10.1016/j.physleta.2005.11.062]. Furthermore, we have carried out numerical simulations to show that the communication is clearly unsecure for improper selection of the noise properties. Protection against attacks using time and correlation analysis is not considered in this paper. Related simulations are available at www.noise.inf.u-szeged.hu/Research/kljn/.

*Keywords*: KLJN; secure key exchange; unconditionally secure communication; secure key distribution; noise.

## 1. Introduction

At present, the security of the communication is mostly provided by software-based cryptographic solutions. Since the security is ensured only by the assumption that the eavesdropper does not have enough processing capability to break the code, considerable efforts have been made to develop unconditionally secure communication protocols. One promising research area is the quantum encryption, where security is based on the laws of quantum mechanics. However, recently an alternative communication scheme has been proposed, the Kirchhoff-law–Johnson-noise (KLJN) protocol, which is based only on the laws of classical physics [2]. One of the main advantages of the KLJN protocol is that it can provide at least the same security as quantum systems at orders of magnitude lower cost. Although until now there are only a few real implementations of the system [3, 4], many potential applications,

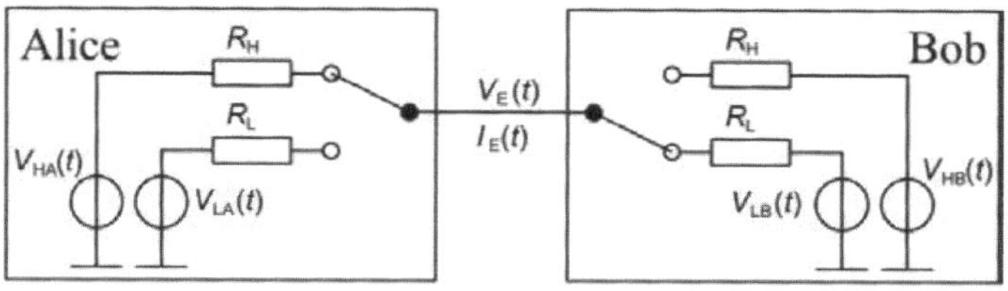


Fig. 1. Simplified diagram of the KLJN system (HL state is shown).

such as key distribution over Smart Grid [5], uncloneable hardware keys [6] or securing computer hardware [7] have been proposed. While several attack methods have been discussed [8–13], the debate is still going on concerning the security of the system [14, 15]. Furthermore, new, extended protocols have been proposed for enhance the security of non-ideal devices [16].

The simplified diagram of the communication system is shown in Fig. 1. During the key exchange, both Alice and Bob randomly select an L or H bit value. Then, they select the corresponding resistor ($R_{\mathrm{L}}$ and $R_{\mathrm{H}}$) and connect it to the wire. The noise sources, $V_{\mathrm{L}}(t)$ and $V_{\mathrm{H}}(t)$ represent the thermal noise of the resistors. During the communication, the voltage and current noise measured in the wire ($V_{\mathrm{E}}(t)$ and $I_{\mathrm{E}}(t)$) are determined by the selected resistors and can be measured not only by Alice and Bob, but also by the eavesdropper, Eve. The security of the system is based on the assumption that even if Eve can measure these signals, she cannot differentiate between the LH state and HL state.

In real applications, the thermal noise of resistors is too low therefore voltage noise generators are typically used to emulate high enough temperature [13]. It has already been stated that the security requires the use of Johnson-like noise, namely, the noise must have normal distribution and the standard deviance must be scaled as the root of the resistance [2]. We have proven this statement using purely mathematical statistical tools [1], and in the present paper we will show that these noise properties are not only needed, but also guarantee absolute security against statistical attacks. Note that in this paper we do not address protection against attacks based on the analysis of the time dependence of the signals.

## 2. Results

Equations (1) and (2) show the voltage and current values that can be measured by the eavesdropper during the two secure states, LH and HL, respectively. The notation used in the equations is introduced in Fig. 1. The communication is secure if Eve cannot distinguish between these two states. The voltage and current measured by Eve in the LH state can be expressed as:

$$V_{\mathrm{E,LH}}(t) = \frac{V_{\mathrm{LA}}(t) \cdot R_{\mathrm{H}} + V_{\mathrm{HB}}(t) \cdot R_{\mathrm{L}}}{R_{\mathrm{L}} + R_{\mathrm{H}}} \quad \text{and} \quad I_{\mathrm{E,LH}}(t) = \frac{V_{\mathrm{HB}}(t) - V_{\mathrm{LA}}(t)}{R_{\mathrm{L}} + R_{\mathrm{H}}}. \tag{1}$$

The voltage and current signal measured by Eve in the HL state (corresponding to Fig. 1) are given by the following equations:

$$V_{\mathrm{E,HL}}(t) = \frac{V_{\mathrm{HA}}(t) \cdot R_{\mathrm{L}} + V_{\mathrm{LB}}(t) \cdot R_{\mathrm{H}}}{R_{\mathrm{L}} + R_{\mathrm{H}}} \quad \text{and} \quad I_{\mathrm{E,HL}}(t) = \frac{V_{\mathrm{LB}}(t) - V_{\mathrm{HA}}(t)}{R_{\mathrm{L}} + R_{\mathrm{H}}}. \tag{2}$$

For secure communication, the joint probability density function $p_{\mathrm{LH}}(I_{\mathrm{E}}, V_{\mathrm{E}})$ and $p_{\mathrm{HL}}(I_{\mathrm{E}}, V_{\mathrm{E}})$ must be the same. If $I_{\mathrm{E}}$ and $V_{\mathrm{E}}$ are independent, this is satisfied.

As it has been proven [17], linear combinations $Y_A$ and $Y_B$ of two independent random variables $X_1$ and $X_2$ in Eq. (3) will be statistically independent if and only if each random variable is normally distributed and Eq. (4) is satisfied:

$$Y_A = A_1 \cdot X_1 + A_2 \cdot X_2 \quad \text{and} \quad Y_B = B_1 \cdot X_1 + B_2 \cdot X_2, \tag{3}$$

$$A_1 \cdot B_1 \cdot \sigma_1^2 + A_2 \cdot B_2 \cdot \sigma_2^2 = 0, \tag{4}$$

where $\sigma_1$ and $\sigma_2$ are the standard deviations of $X_1$ and $X_2$, respectively. In our case, we obtain:

$$V_{\mathrm{E,HL}}(t) = \frac{R_{\mathrm{L}}}{R_{\mathrm{L}} + R_{\mathrm{H}}} V_{\mathrm{HA}}(t) + \frac{R_{\mathrm{H}}}{R_{\mathrm{L}} + R_{\mathrm{H}}} V_{\mathrm{LB}}(t), \tag{5}$$

$$I_{\mathrm{E,HL}}(t) = -\frac{1}{R_{\mathrm{L}} + R_{\mathrm{H}}} V_{\mathrm{HA}}(t) + \frac{1}{R_{\mathrm{L}} + R_{\mathrm{H}}} V_{\mathrm{LB}}(t), \tag{6}$$

$$\frac{R_{\mathrm{L}}}{R_{\mathrm{L}} + R_{\mathrm{H}}} \cdot \left(-\frac{1}{R_{\mathrm{L}} + R_{\mathrm{H}}}\right) \cdot \sigma_{\mathrm{VH}}^2 + \frac{R_{\mathrm{H}}}{R_{\mathrm{L}} + R_{\mathrm{H}}} \cdot \frac{1}{R_{\mathrm{L}} + R_{\mathrm{H}}} \cdot \sigma_{\mathrm{VL}}^2 = 0, \tag{7}$$

therefore

$$-R_{\mathrm{L}} \cdot \sigma_{\mathrm{VH}}^2 + R_{\mathrm{H}} \cdot \sigma_{\mathrm{VL}}^2 = 0, \tag{8}$$

where $\sigma_{\mathrm{VH}}$ and $\sigma_{\mathrm{VL}}$ are the standard deviations of $V_{\mathrm{HA}}$ and $V_{\mathrm{LB}}$, respectively. Note that we get a similar equation for the LH case. According to this, the distribution

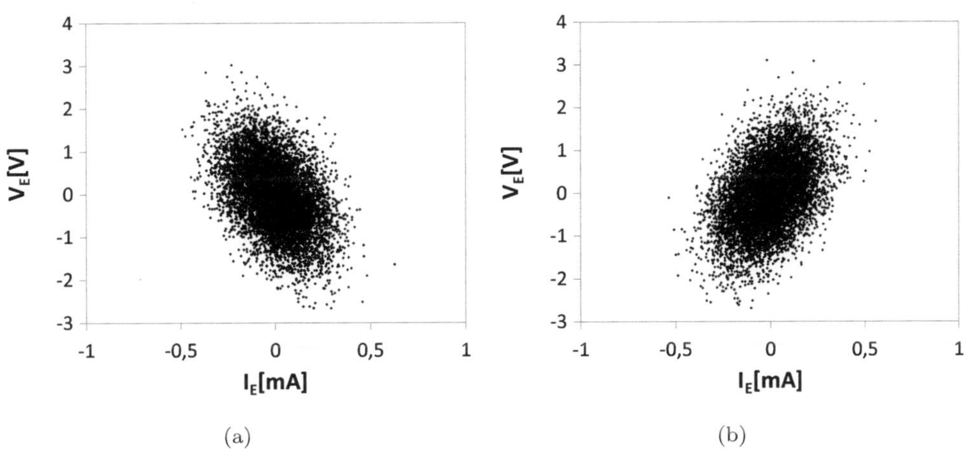


Fig. 2. Scatter plot for cases LH (a) and HL (b) using noise with normal distribution if Eq. (9) is not satisfied. $R_{\mathrm{L}} = 1\,\mathrm{k\Omega}$, $R_{\mathrm{H}} = 10\,\mathrm{k\Omega}$, $\sigma_{\mathrm{VH}}/\sigma_{\mathrm{VL}} = 1,5$.

of $V_{\mathrm{HA}}$ and $V_{\mathrm{LB}}$ must be normal, and the scaling of the standard deviation must follow the rule:

$$\frac{\sigma_{\mathrm{VH}}}{\sigma_{\mathrm{VL}}} = \sqrt{\frac{R_{\mathrm{H}}}{R_{\mathrm{L}}}}, \tag{9}$$

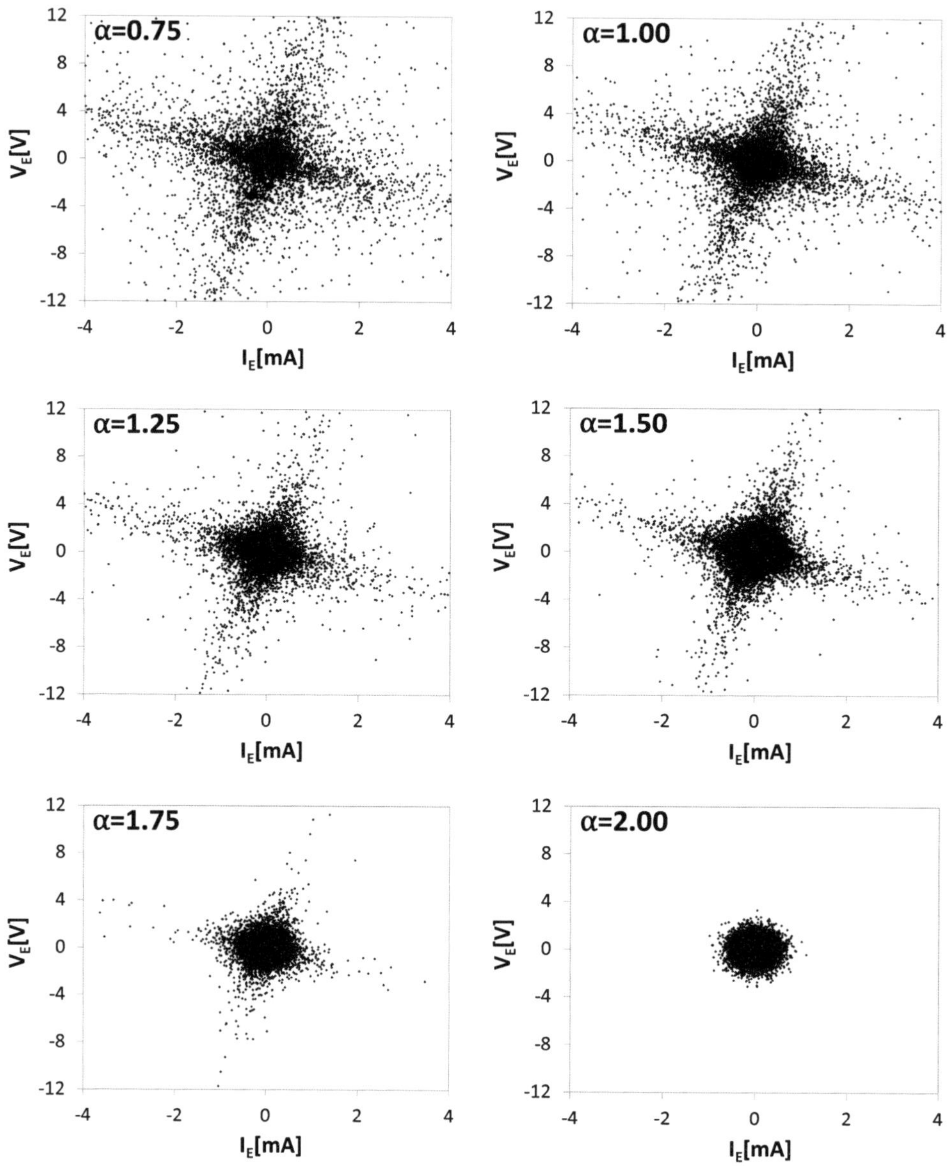


Fig. 3. Scatter plot for case HL using distributions with different values of $\alpha$. Note that $\alpha = 1$ and $\alpha = 2$ correspond to Cauchy and normal distribution, respectively. $R_{\mathrm{L}} = 1\,\mathrm{k}\Omega$, $R_{\mathrm{H}} = 10\,\mathrm{k}\Omega$, $w_{\mathrm{VH}}/w_{\mathrm{VL}} = (R_{\mathrm{H}}/R_{\mathrm{L}})^{1/2}$.

in agreement with the results presented in [1]. We have carried out numerical simulations[a] to obtain the joint statistics of $I_{\mathrm{E}}$ and $V_{\mathrm{E}}$. We have generated $2^{13}$ samples both for the current and voltage and made scatter plots for several cases. Figure 2 demonstrates what happens with the joint distribution of $I_{\mathrm{E}}$ and $V_{\mathrm{E}}$ if Eq. (9) is not satisfied: there is an asymmetry in the distribution that depends on the actual state, LH or HL.

In order to achieve a secure communication, the linear combination of noises must give the same type of probability distribution as the original one [1]. Such distributions are called stable distributions; here, we consider symmetric $\alpha$-stable distributions that include normal distribution as a special case. Assuming distributions symmetric around zero their characteristic function is defined by the following equation:

$$\varphi(t) = e^{-w^{\alpha}|t|^{\alpha}}, \tag{10}$$

where $\alpha$ is the stability parameter in the range from 0 to 2 and $w$ is the scaling factor of the probability density function. Note that $\alpha = 2$ corresponds to normal distribution and $\alpha = 1$ corresponds to Cauchy distribution. However, according to [17], $I_{\mathrm{E}}$ and $V_{\mathrm{E}}$ are not independent except in the case of normal distribution ($\alpha = 2$) as can be seen in Fig. 3. Note that not all of such distributions have finite variance, therefore the scaling of the noise voltages was based on the scaling factor $w$ which can be associated with the voltage noise magnitude and is defined in Eq. (10). Thus, the higher and lower noise voltages have scaling factors $w_{\mathrm{VH}}$ and $w_{\mathrm{VL}}$, respectively.

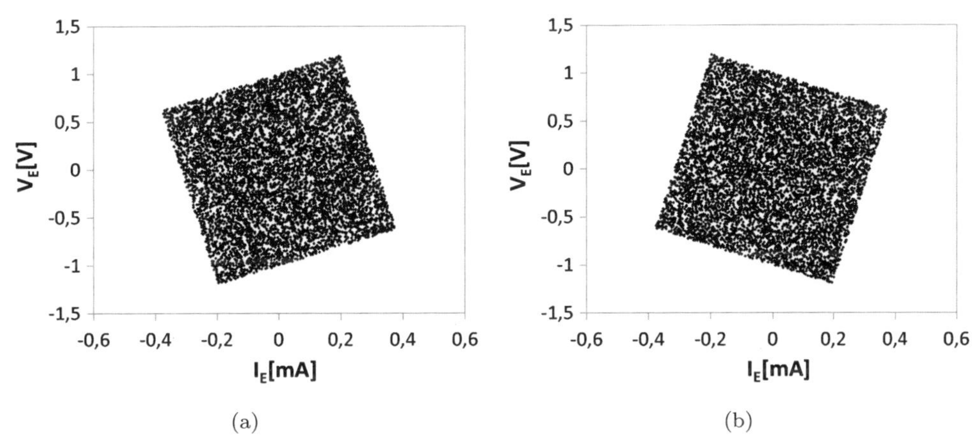


Fig. 4. Scatter plot for cases LH (a) and HL (b) using noise with uniform distribution. $R_{\mathrm{L}} = 1\,\mathrm{k}\Omega$, $R_{\mathrm{H}} = 10\,\mathrm{k}\Omega$, $\sigma_{\mathrm{VH}}/\sigma_{\mathrm{VL}} = (R_{\mathrm{H}}/R_{\mathrm{L}})^{1/2}$.

[a]Related simulations are available at www.noise.inf.u-szeged.hu/Research/kljn/.

Digital implementations of KLJN [3] require numerically generated random numbers. Pseudo-random number generators typically provide uniform distribution and generating normally distributed numbers is more complicated. However, in agreement with our results, Fig. 4. clearly shows that the scatter plots for LH and HL cases are different for uniformly distributed signals, therefore they cannot be used for secure communication.

## 3. Conclusion

We have shown that communication using the KLJN protocol is secure if and only if noise voltages with normal distribution are used and the variance of the noise voltages follow the scaling defined by Eq. (9). This result is based on mathematical statistical derivation and it is in agreement with previous results [1, 2]. Note that protection against attacks using time and correlation analysis is not considered and can be addressed in subsequent publications. Further analysis can clarify how the time domain properties of the noise influence the security of the system.

## Acknowledgments

We thank the anonymous reviewer of our previous paper [1] for drawing our attention to the problem of statistical dependence of voltage and current fluctuations in the communication wire related to absolute security.

This research was supported by the European Union and the State of Hungary, co-financed by the European Social Fund in the framework of TÁMOP 4.2.4. A/2-11-1-2012-0001 "National Excellence Program".

## References

[1] Z. Gingl and R. Mingesz, Noise properties in the ideal Kirchhoff-law–Johnson-noise secure communication system, *PLoS ONE* **9** (2014) e96109, doi: 10.1371/journal.pone.0096109.

[2] L. B. Kish, Totally secure classical communication utilizing Johnson(-like) noise and Kirchhoff's law, *Phys. Lett. A* **352** (2006) 178–182, doi: 10.1016/j.physleta.2005.11.062.

[3] R. Mingesz, Z. Gingl and L. B. Kish, Johnson(-like)-noise–Kirchhoff-loop based secure classical communicator characteristics, for ranges of two to two thousand kilometers, via model-line, *Phys. Lett. A* **372** (2008) 978–984, doi: 134 10.1016/j.physleta.2007.07.086.

[4] R. Mingesz, L. B. Kish, Z. Gingl, C. G. Granqvist, H. Wen, *et al.*, Unconditional security by the laws of classical physics, *Metrol. Measurement Syst.* **XX** (2013) 3–16, doi: 10.2478/mms-2013-0001.

[5] E. Gonzalez, L. B. Kish, R. S. Balog and P. Enjeti, Information theoretically secure, enhanced Johnson noise based key distribution over the smart grid with switched filters, *PLoS ONE* **8** (2013) e70206, doi: 10.1371/journal.pone.0070206.

[6] L. B. Kish and C. Kwan, Physical uncloneable function hardware keys utilizing Kirchhoff-law–Johnson-noise secure key exchange and noise-based logic, *Fluct. Noise Lett.* **12** (2013) 1350018, doi: 10.1142/S0219477513500181.

[7] L. B. Kish and O. Saidi, Unconditionally secure computers, algorithms and hardware, *Fluct. Noise Lett.* **8** (2008) L95–L98, doi: 10.1142/s0219477508004362.

[8] F. Hao, Kish's key exchange scheme is insecure, *IEE Proc. Inform. Soc.* **153** (2006) 141–142, doi: 10.1049/ip ifs:20060068.

[9] L. B. Kish, Response to Feng Hao's paper "Kish's key exchange scheme is insecure, *Fluct. Noise Lett.* **6** (2006) C37–C41, doi: 10.1142/s021947750600363x.

[10] J. Scheuer and A. Yariv, A classical key-distribution system based on Johnson (like) noise — How secure? *Phys. Lett. A* **359** (2006) 737–740, doi: 10.1016/j.physleta.2006.07.013.

[11] L. B. Kish, Response to schuer-Yariv, A classical key-distribution system based on Johnson (like) noise — How secure? *Phys. Lett. A* **359** (2006) 741–744, doi: 10.1016/j.physleta.2006.07.037.

[12] L. B. Kish and T. Horvath, Notes on recent approaches concerning the Kirchhoff-law–Johnson-noise-based secure key exchange, *Phys. Lett. A* **373** (2009) 901–904, doi: 10.1016/j.physleta.2009.05.077.

[13] L. B. Kish and J. Scheuer, Noise in the wire: The real impact of wire resistance for the Johnson(-like) noise based secure communicator, *Phys. Lett. A* **374** (2010) 2140–2142, doi: 10.1016/j.physleta.2010.03.021.

[14] C. H. Bennett and C. J. Riedel, On the security of key distribution based on Johnson-Nyquist noise (2013), http://arxiv.org/abs/1303.7435.

[15] L. B. Kish, D. Abbott and C. G. Granqvist, Critical analysis of the Bennett–Riedel attack on secure cryptographic key distributions via the Kirchhoff-law–Johnson-noise scheme, *PLoS ONE* **8** (2013) e81810, doi: 10.1371/journal.pone.0081810.

[16] L. B. Kish, Enhanced secure key exchange systems based on the Johnson-noise scheme, *Metrol. Measurement Syst.* **XX** (2013) 191–204, doi: 10.2478/mms-2013-0017.

[17] E. Lukacs and E. P. King, A property of normal distribution, *The Annals Math. Statist.* **25** (1954) 389–394.

# PROTECTION AGAINST THE MAN-IN-THE-MIDDLE-ATTACK FOR THE KIRCHHOFF-LOOP-JOHNSON(-LIKE)-NOISE CIPHER AND EXPANSION BY VOLTAGE-BASED SECURITY [1]

L. B. KISH

*Department of Electrical Engineering, Texas A&M University, College Station, TX 77843-3128, USA*
*Laszlo.Kish@ece.tamu.edu*

Received 19 December 2005
Revised 13 January 2006
Accepted 16 January 2006
Communicated by Robert Vajtai

It is shown that the original Kirchhoff-loop-Johnson(-like)-noise (KLJN) cipher is naturally protected against the man-in-the-middle (MITM) attack, if the eavesdropper is using resistors and noise voltage generators just like the sender and the receiver. The eavesdropper can extract zero bit of information before she is discovered. However, when the eavesdropper is using noise current generators, though the cipher is protected, the eavesdropper may still be able to extract one bit of information while she is discovered. For enhanced security, we expand the KLJN cipher with the comparison of the instantaneous voltages via the public channel. In this way, the sender and receiver has a full control over the security of measurable physical quantities in the Kirchhoff-loop. We show that when the sender and receiver compare not only their instantaneous current data but also their instantaneous voltage data then the zero-bit security holds even for the noise current generator case. We show that the original KLJN scheme is also zero-bit protected against that type of MITM attack when the eavesdropper uses voltage noise generators, only. In conclusion, within the idealized model scheme, the man-in-the-middle-attack does not provide any advantage compared to the regular attack considered earlier. The remaining possibility is the attack by a short, large current pulse, which described in the original paper as the only efficient type of regular attacks, and that yields the one bit security. In conclusion, the KLJN cipher is superior to known quantum communication schemes in every respect, including speed, robustness, maintenance need, price and its natural immunity against the man-in-the-middle attack.

*Keywords*: Totally secure communication without quantum; man-in-the-middle attack; stealth communication; noise.

## 1. Introduction: Totally Secure Communication Without Quantum Information with Kirchhoff-Loop and Johnson(-like) Noise

Recently, a totally secure classical communication scheme was introduced [1, 2] utilizing two pairs of resistors and noise voltage generators, the physical properties of an idealized

[1] This preprint was uploaded at arxiv.org on December 19, 2005; http://arxiv.org/abs/physics/0512177

Kirchhoff-loop and the statistical physical properties thermal noise. In the idealized scheme of the Kirchhoff-loop-Johnson-(like)-noise (KLJN) cipher, the passively observing eavesdropper can extract zero bit of information. The *intrusive eavesdropper*, who emits a large and short current pulse in the channel, can extract *only one bit* of information while she is getting discovered [1] because the sender and the receiver are measuring their instantaneous current amplitudes and compare them via a public channel (like non-jamable radio transmission). The issue of the man-in-the-middle (MITM) attack was not studied in the original paper because of the author's earlier belief that no core physical secure layer is protected against such an attack and any security against the MITM attack has to be provided by additional (mainly software) tools, signatures, trusted third party, etc.

In this paper, it is shown that the originally proposed KLJN scheme [1] is naturally protected against the MITM attack. Moreover, we propose an easy enhancement of the original security by the exchange and comparison of not only the instantaneous current amplitudes but also the instantaneous voltage amplitudes. This step further enhances the security against the MITM attack and it can be beneficial against other types of attacks of any practical realizations.

## 2. General Clarifications about the Issue of Security of Physical Secure Layers

In secure communication, any one of the following cases implies absolute security, however the relevant cases for physical secure layers are points 3 and 4:

1. The eavesdropper cannot physically access the information channel.
2. The sender and the receiver have a shared secret key for the communication.
3. *The eavesdropper has access and can execute measurements on the channel but the laws of physics do not allow extracting the communicated information from the measurement data.*
4. *The eavesdropper can extract the communicated information however, when that happens, it disturbs the channel so that the sender and receiver discover the eavesdropping activity.*

Keeping points 3 and 4 in mind, we can classify the focus topics of research of physical secure layers as follows:

i) *Absolute security of the idealized situation.* This is the most fundamental scientific part of the research and the mathematical model of the idealized physical system is developed and tested. The basic question is that how much information can be extracted from the data by the physical measurements allowed in the idealized situation? The original paper and the present study aim the investigation of this question.

ii) *Absolute security of the practical situation.* This part of the research requires an interdisciplinary effort including the fields of physics, engineering and data security. Because no real system can totally match the physical properties of the ideal mathematical model system, this kind of absolute security *does not exist in reality*; it is rather approached by an *only practically absolute* security. For example, in quantum communication, we have no ideal single photon source, no noise-free channel, and no noise-free detectors, and any of these deficiencies compromise absolute security.

Similarly, with the KLJN cipher, cable resistance [3] and cable capacitance can cause information leak because the eavesdropper can execute measurements along the cable. This effect can be controlled and minimized by the particular design and choosing proper driving resistances and noise bandwidth that prohibit to make an acceptable statistics about the deviation of the noise strength along the cable within the clock period. Similarly, fast switching of the resistors can violate the *no-wave bandwidth rule* described in [1] (Eq. (9)) but this problem can be avoided by using slow switches and/or filters at the line input.

iii) *If the code is broken, how many bits can be extracted by the eavesdropper before she is discovered due to the disturbance of the channel?* This question can also be treated at both the idealized-fundamental level and at the practical one. Some answers for the idealized case: RSA: infinite number of bits; Quantum: 20 – 10000 bits; KLJN cipher: 1 bit.

In the rest of the paper, we strictly focus on question i) while we are discussing the MITM attack in terms of point 4 above.

## 3. Security of the KLJN Cipher Against the Man-In-The-Middle Attack and Expanding the System with Voltage Based Security

In this section we show that the original KLJN cipher arrangement [1] is secure against the MITM attack. The goal is to study the idealized model and show that the eavesdropper is discovered when she executes the attack. Finally, we propose a simple expansion of the scheme by comparing the voltage data, too, for enhanced security.

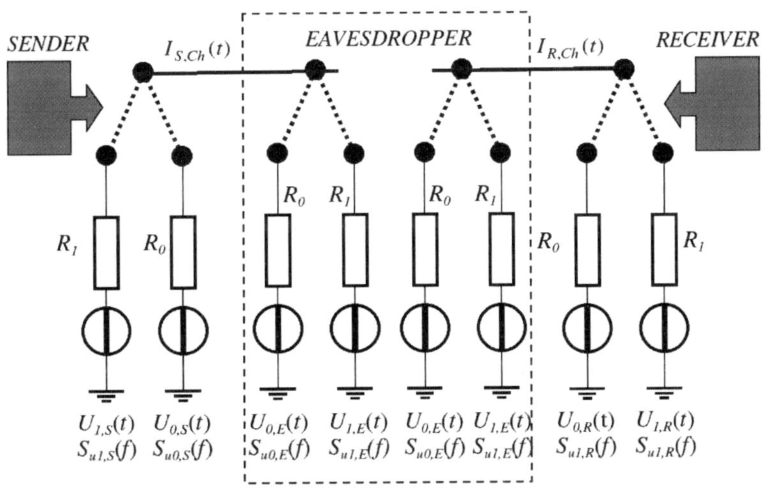


Fig. 1. Man-in-the-middle-attack by using resistors with the same values and noise voltage generators with the same parameters as those of the sender and the receiver.

Figure 1 shows the MITM attack by using resistors with the same values and noise voltage generators with the same parameters as those of the sender and the receiver. The eavesdropper breaks the line at the middle and installs two KLJN communicators, one for the sender and another one for the receiver. According to the original scheme, the instantaneous current amplitudes, at the sender's end and at the receiver's end, are compared via a public channel. Because the eavesdropper's noise generators are different representations of the corresponding stochastic processes, their instantaneous amplitudes are different from those of the sender's and the receiver's noise generators. Therefore, the current amplitudes are different in the two loops and the eavesdropper is discovered within the reciprocal of the bandwidth, within the time resolution of the communicated noise, before extracting a single bit of information. We remind the reader [1] that the extraction of information from noise needs the making of a (short range) statistics through an averaging time window, which is the clock period, thus a single time (noise) sample provides zero information.

Mathematically, we can write as follows. Let $i \in 0,1$; $m \in 0,1$; $k \in 0,1$; $p \in 0,1$. Then all the following voltages $U_{i,S}(t)$, $U_{m,E}(t)$, $U_{k,R}(t)$ and $U_{p,E}(t)$, see Fig. 1, are statistically independent Gaussian stochastic processes with zero mean. The current at the sender's end and the receiver's end can be written as

$$I_{S,Ch}(t) = \frac{U_{i,S}(t) - U_{m,E}(t)}{R_i + R_m} \quad \text{and} \quad I_{R,Ch}(t) = \frac{U_{k,R}(t) - U_{p,E}(t)}{R_k + R_p}, \tag{1}$$

respectively. As the most pessimistic case, let us suppose that the denominators are equal and the nominators also contain the same type of noise generators, for example, $i = k$ and $m = p$. Then the RMS value of $I_{S,Ch}(t)$ and that of $I_{R,Ch}(t)$ are equal. Then the probability $P_0$ that $I_{S,Ch}(t_0) = I_{R,Ch}(t_0)$ at a given time moment $t_0$ is roughly equal to the ratio of the amplitude resolution D of the measurement system and the RMS value of these currents:

$$P_0 \approx \frac{\Delta}{I_{RMS}} \tag{2}$$

Let us suppose 7 bits resolution of the measurement (a pessimistic value), then $P_0 = 1/128$, which is less than 1% chance of staying hidden. On the other hand, $P_0$ is the probability that the eavesdropper can stay hidden during the correlation time t of the noise, where t is roughly the inverse of the noise bandwidth. Because the KLJN cipher works with statistics made on noise, the actual clock period $T$ is $N >> 1$ times longer than the correlation time of the noise used [1]. Thus, during the clock period, the probability of staying hidden is:

$$P_{\text{clock}} = P_0^N \tag{3}$$

Supposing a practical $T = 10\tau$ (see [1]) the probability at the other example $P < 10^{-20}$.

This is the estimated probability that, in the given system the eavesdropper can extract a single bit without getting discovered. The probability that she can stay hidden while extracting 2 bits is $P < 10^{-40}$, for 3 bits it is $P < 10^{-60}$, etc. In conclusion, we can safely say that the eavesdropper is discovered immediately before she can extract a single bit of information. The probability of staying hidden can be estimated in the same manner in the rest of this paper, however we skip these calculations because of their trivial nature, and the straightforward considerations below.

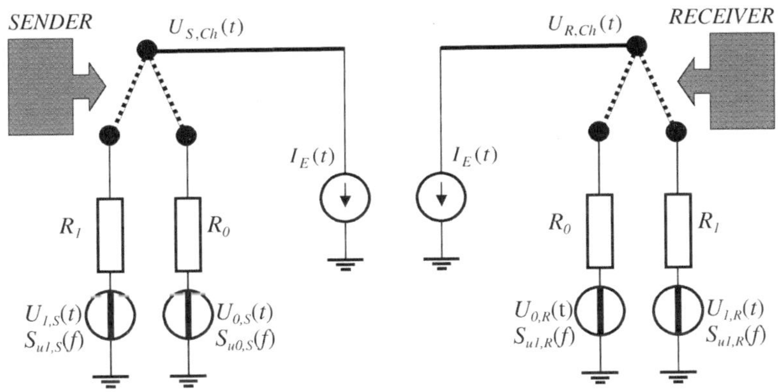


Fig. 2. Man-in-the-middle-attack by using twin noise current generators with the same instantaneous current amplitudes.

Because the defense described above is based on the difference of the instantaneous current amplitudes at the two sides, a natural question arises. Can we provide security when the eavesdropper executes the MITM attack by using *two noise current generators* with the same instantaneous current amplitudes and RMS values imitating a reasonable noise current in the channel? Figure 2 shows the MITM attack by using twin noise current generators with the same instantaneous current amplitudes. In this case, the equality of the current amplitudes at the two sides is guaranteed by the twin noise current generators, therefore comparing the current amplitudes cannot be used for protection. For the best protection, we can expand the original KLJN cipher [1] by a voltage-based security enhancement. The sender and the receiver can compare the instantaneous voltage amplitudes at their end, via a public channel. Because the sender's noise generator and receiver's one have either different parameters or they are different representations of the same stochastic processes, their instantaneous amplitudes are different at most of the time. Therefore, the voltage amplitudes are different at the two ends and the eavesdropper is discovered practically immediately, before extracting a single bit of information.

However, it is important to note that the original scheme is also secure against this last kind of attack though not at the zero-bit security level but at the one-bit level. Normally the power density spectrum $S_{u,Ch}(f)$ of the voltage noise in the channel is *smaller* than that of the actual noise voltage generator $S_u(f)$ of the sender and the receiver, respectively, because it is proportional to the parallel resultant of those resistances:

$$S_{u,Ch}(f) < S_u(f) \tag{4}$$

However, during the MITM attack shown in Fig. 2, the following relations hold:

$$S_{u,Ch}(f) = S_u(f) + R^2 S_{i,E}(f) \quad \text{therefore} \quad S_{u,Ch}(f) > S_u(f) \; . \tag{5}$$

That means, the MITM attack will be discovered after the sufficient statistics is made on the spectrum during the clock period. During the communication of the secure bit, the side using the larger resistor will discover the attack faster. At the same time, the eavesdropper, while getting discovered, can extract one bit of information. Thus, the expansion by the voltage-based security measure described above provides an enhanced security.

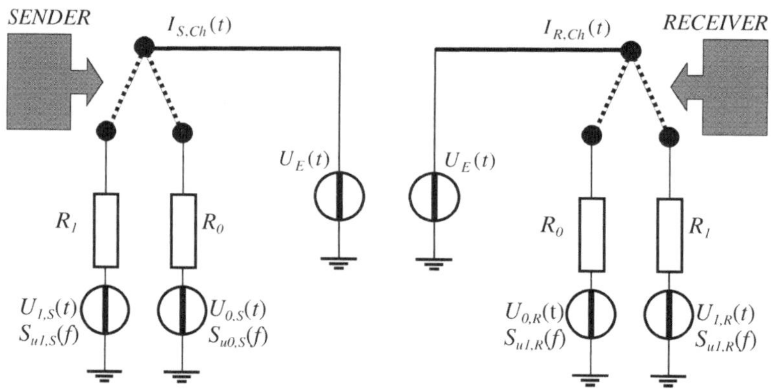


Fig. 3. Man-in-the-middle-attack by using twin noise voltage generators with the same instantaneous current amplitudes.

Because the defense described in the context of Fig. 2 is based on the difference of the voltages at the two sides, a natural question arises. Can we provide security when the eavesdropper executes the MITM attack by using *two noise voltage generators* with the same instantaneous voltage amplitudes and RMS values imitating a reasonable noise voltage in the channel? Figure 3 shows the man-in-the-middle-attack by using twin noise voltage generators with the same instantaneous current amplitudes. In this case, the twin noise voltage generators guarantee the equality of the instantaneous voltage amplitudes at the two sides. Therefore comparing the voltage amplitudes cannot be used for protection. For the protection, the sender and the receiver can compare the instantaneous current amplitudes, via a public channel. Because the sender's noise generator and receiver's one have either different parameters or they are different representations of the same stochastic processes, their instantaneous amplitudes are different. Therefore the current amplitudes are different in the two loops and the eavesdropper is discovered immediately, before extracting a single bit of information.

## 4. Conclusion

The KLJN cipher with public channel for comparing currents [1] is naturally protected against the MITM attack and the eavesdropper is discovered with a very high probability while or before she can extract a single bit. Enhanced security can be reached by comparing the voltages and then the eavesdropper is discovered with a very high probability before she can extract a single bit of information. Thus within the idealized model scheme, the man-in-the-middle-attack does not provide any advantage compared to the regular attack considered earlier. The remaining possibility is the attack by a short, large current pulse, which described in the original paper [1] as the only efficient type of regular attacks, and that yields the one bit security.

Therefore, the KLJN cipher is superior to known quantum communication schemes in every respect, including speed, robustness, maintenance need, price and its natural immunity against the man-in-the-middle attack.

**Acknowledgments**

Pre-published on the arxiv.org server on December 19, 2005 as http://arxiv.org/abs/physics/0512177 . Discussions with a number of security experts and scientists/engineers challenging the security of the KLJN scheme and emphasizing the importance of the MITM attack are appreciated. Without the ability to provide a complete list, here are some of the most significant commenters: Jonathan Wolfe, Bruce Schneier, Steven M. Bellovin, David Wagner, Charlie Strauss, Adrian Cho, Janos Bergou, Julio Gea-Banacloche, Matthew Skala, Greg Kochanski, Christian Vandenbroeck, Frank Moss, Derek Abbott, Bob Biard.

## References

[1] L. B. Kish, Totally secure classical communication utilizing Johnson (-like) noise and Kirchoff's law, *Physics Letters A*, in press; preprint at http://dx.doi.org with code doi:10.1016/j.physletb.2003.10.071; also at http://arxiv.org/physics/0509136.

[2] A. Cho, Simple noise may stymie spies without quantum weirdness, *Science* **309** (2005) 2148.

[3] Janos Bergou, statement in [2].

# On KLJN-based Secure Key Distribution in Vehicular Communication Networks

X. Cao[*,‡,§], Y. Saez[*,¶], G. Pesti[†,||] and L. B. Kish[*,**]

[*]*Department of Electrical and Computer Engineering*
*Texas A&M University, College Station*
*TX 77843-3128, USA*

[†]*Texas A&M Transportation Institute*
*Texas A&M University, College Station*
*TX 77843-3135, USA*

[‡]*College of Automotive Engineering*
*Jilin University, Changchun*
*Jilin 130025, P. R. China*
[§]*caoxiaolin@jlu.edu.cn*
[¶]*yessica.saez@tamu.edu*
[||]*g-pesti@tamu.edu*
[**]*laszlo.kish@ece.tamu.edu*

Received 20 September 2014
Accepted 23 October 2014
Published 8 December 2014

Communicated by Zoltan Gingl

In a former paper [*Fluct. Noise Lett.* **13** (2014) 1450020] we introduced a vehicular communication system with unconditionally secure key exchange based on the Kirchhoff-Law–Johnson-Noise (KLJN) key distribution scheme. In this paper, we address the secure KLJN key donation to vehicles. This KLJN key donation solution is performed lane-by-lane by using roadside key provider equipment embedded in the pavement. A method to compute the lifetime of the KLJN key is also given. This key lifetime depends on the car density and gives an upper limit of the lifetime of the KLJN key for vehicular communication networks.

*Keywords*: Security; vehicular communication networks; Kirchhoff-Law–Johnson-Noise; unconditional security.

## 1. Introduction

After more than 100 years of development on modern vehicle technology, we are on our way to smarter cars and much more intelligent transportation systems.

¶Corresponding author.

Nowadays, people pay more attention to safety and comfort in vehicles, rather than traditional traction ability, fuel cost, handling, and stability. Therefore, it is believed that safety and mobility information such as road and traffic information (e.g., emergency braking, vehicle collision, congestion, toll collection, etc.), weather forecast warnings, (e.g., water or ice on the pavement), and local services, (e.g., route maps, gas or restaurant locations, etc.) [1–4] should be collected and provided to drivers and passengers in the vehicle.

### 1.1. *Vehicular communication networks*

Vehicular communication networks have become a reasonable intelligent transportation solution that can satisfy these demands effectively. A typical vehicular communication network is shown in Fig. 1 [1, 5–9].

As summarized in previous publications [1, 5–9], Vehicles, Roadside Devices (RSDs), and Certification Authorities (CAs) are the three basic nodes in most of vehicular communication networks. Vehicles are mobile terminal nodes that are in charge of collecting road and traffic information, reporting events to the CAs through the RSDs, and exchanging warnings with nearby vehicles. The RSDs are intermediate nodes in charge of transferring messages between vehicles and CAs in two-ways. The CAs are the host nodes that manage information related to vehicles. These nodes also generate secure keys and provide certifications for all vehicles in the network, control message exchanges of the whole network, and distribute

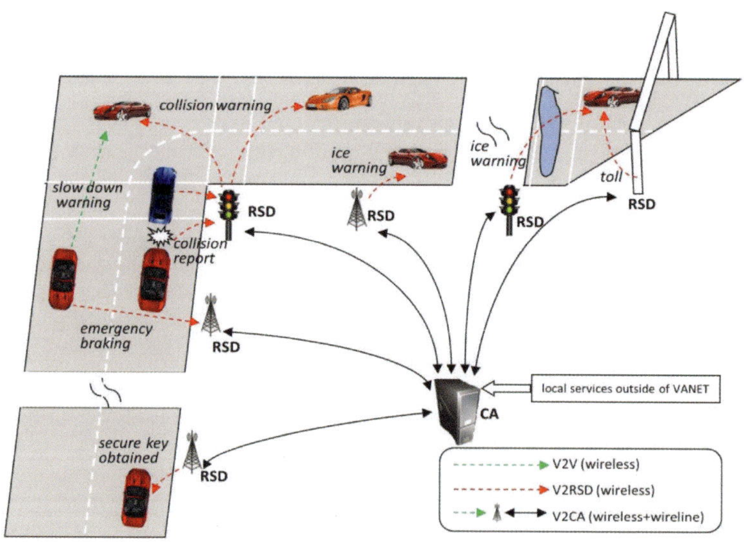


Fig. 1. A typical vehicular communication network. Three basic nodes are encountered in this type of network: Vehicles, Roadside Devices (RSDs) and Certification Authorities (CAs). The types of communication within vehicular communication networks include [1, 5–9]: Vehicle-to-Vehicle (V2V), Vehicle-to-Roadside-Device (V2RSD), and Vehicle-to-Certification-Authority (V2CA).

local information obtained outside the local vehicular communication network. Accordingly, the types of communication within vehicular communication networks include [1, 5–9]: Vehicle-to-Vehicle (V2V), Vehicle-to-Roadside-Device (V2RSD), and Vehicle-to-Certification-Authority (V2CA). Communications within vehicular communication networks raise concerns for security and privacy. For example, the identity of vehicles, emergency braking, and vehicle collision warnings among vehicles must be transmitted securely to avoid malicious activities. The private financial information used in toll collection when cars pass by RSDs also needs to be protected.

In order to solve these fundamental security-related issues for promising vehicular communication network applications, several security protocols have been proposed by different researchers. In [10–11], the authors proposed a security infrastructure that is based on public key infrastructure (PKI). Later, other solutions based on PKI were proposed [2, 4, 12, 13]. The authors of [2] provided a "lightweight" authenticated key scheme that integrates blind signature techniques for V2V and V2RSD communications. In [4], the authors presented an approach that combines the traditional PKI and identity-based public key cryptography for vehicular communication networks. In [12], a secure scheme with session keys (pairwise and group keys) used in non-safety-related applications (e.g., "chatting in platoon") was designed. In [13], temporary anonymous certified keys (TACKs) were constructed, and a key management scheme based on TACKs was proposed for vehicular communication networks. Besides PKI, group signatures are another important category of proposed security methods. Based on the strong Diffie–Hellmanand linear assumption (DHM), the authors of [14] introduced the under-200 bytes group signature scheme that has a similar security level to the Rivest, Shamir and Adleman (RSA public-key cryptosystem) signature of the same length. A group signature-based protocol using tamper-resistance devices and a probabilistic signature verification scheme was proposed in [3]. In [15], the authors constructed an identity-based batch verification scheme for V2RSD communication in vehicular communication networks. In [16], a software-based roadside unit-aided messages authentication protocol for V2V communications was proposed. In addition, a software-based solution that uses secure and privacy enhancing communication schemes for vehicular sensor networks was provided in [17].

Most of the above security schemes or protocols are constructed based on software encryption mechanisms. The security on these software-based methods is based on the premise that the eavesdroppers have *limited computational power*. Thus, these security schemes offer just a *computationally conditional security* [18–22]. Moreover, these architectures focus their attention on V2V or V2RSD communications and although there is significant information transmitted in the Roadside-Device-to-Certification-Authority (RSD2CA) communication [9], it is very rare to find works related to securing this particular communication channel.

In [9], a novel unconditionally secure vehicular communication architecture that utilizes the Kirchhoff-law–Johnson-noise (KLJN) key distribution scheme was proposed. In this architecture, a new node called the Roadside-Key-Provider (RSKP) was introduced to provide the cars with KLJN keys. Based on this work, we discuss the KLJN-based secure key generation, donation, and lifetime in vehicular communication networks. The remainder of this paper is organized as follows. In Sec. 2.1, we discuss the key generation process in vehicular networks by describing the message exchange and specifying the different network lines used during the process. In Sec. 2.2, we propose a lane-by-lane KLJN key donation solution for the vehicular communication architecture with unconditionally secure key exchange proposed in [9]. An upper limit for the KLJN key lifetime in vehicular communication networks is computed in Sec. 2.3. The results are demonstrated with practical considerations. Section 3 concludes the paper.

### 1.2. *On the KLJN key exchange*

The illustration of the ideal KLJN key exchange scheme is shown in Fig. 2 [19, 21–24].

The KLJN secure key exchange scheme was proposed in 2005 [19] as a statistical/physical competitor to Quantum key distribution (QKD) [20]. The KLJN scheme provides unconditional security based on the Kirchhoff's loop law of quasi-static electrodynamics and the fluctuation–dissipation theorem [19, 21–24]; for a general security proof, see [25]. Several potential applications have been proposed such as: classical networks [26], smart power grids [27], and secure computers, algorithms, and hardware [28].

In this ideal KLJN scheme, the two communicating parties, Alice and Bob, communicate via a wireline channel. There is a switch and an identical pair of resistors ($R_L$ and $R_H, R_L \neq R_H$) on each communicator side, where $R_L$ represents

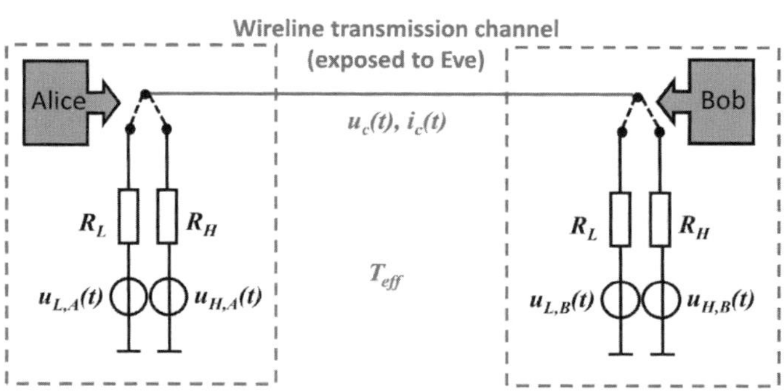


Fig. 2. Illustration of the ideal KLJN key distribution scheme. There is a switch and an identical pair of resistors ($R_L$ and $R_H, R_L \neq R_H$) on each communicator (referred to as Alice and Bob) side, where $R_L$ represents the low, $L$ bits, $R_H$ represents the high, $H$ bits; $u_{L,A}(t), u_{H,A}(t)$ and $u_{L,B}(t), u_{H,B}(t)$ are the thermal (Johnson) noise voltages (Gaussian noise voltage generators) at temperature $T_{\rm eff}$ of $R_L$ and $R_H$ of Alice and Bob, respectively, $u_c(t)$ is the channel noise voltage, and $i_c(t)$ is the noise current in the wire.

the low, $L$ bit and $R_H$ represents the high, $H$ bit. The Gaussian noise voltage generators $u_{L,A}(t), u_{H,A}(t)$ and $u_{L,B}(t), u_{H,B}(t)$ represent the enhanced thermal (Johnson) noise voltages at temperature $T_{\text{eff}}$ of $R_L$ and $R_H$ of Alice and Bob, respectively, while the $u_c(t)$ is the channel noise voltage, and the $i_c(t)$ is the noise current in the wire. At the beginning of the bit sharing period, both Alice and Bob *randomly* choose one of the resistors ($R_L$ or $R_H$) and the corresponding Gaussian noise voltage generator ($u_{L,A}(t)$ or $u_{H,A}(t), u_{L,B}(t)$ or $u_{H,B}(t)$, respectively) and connect them to the wireline. The possible permutations of the resistors connected to the channel will be: $LL$, $LH$, $HL$, and $HH$. In the cases of $LL$ or $HH$, the location of the resistors and the exchanged bits are publicly known, thus these bits are discarded [19]. On the other hand, $LH$ and $HL$ are secure bit exchange situations, because Eve cannot differentiate between the situations $LH$ and $HL$.

The security of the ideal KLJN scheme is based on the second law of thermodynamics [19, 21–24], that is the difficulty to crack the ideal KLJN system is similar to that of to build a perpetual-motion machine of the second kind. In addition, the KLJN system is robust and not sensitive to vibrations [24] and is easy to be integrated on chips [28]. Based on the above core scheme, some advanced schemes were proposed to enhance the speed of the KLJN system [23].

In order to protect against active (invasive) attacks (and also against passive attacks on non-ideal systems), the KLJN system continuously monitors or measures the instantaneous current and voltage at the two ends of the line [22]. These measurements are compared via an authenticated public channel. Therefore, any intruder causing changes in the circuitry, and thus affecting the instantaneous measurements, will cause an alarm to go off and Alice and Bob will discover the intrusion. This defense mechanism is illustrated in Fig. 3.

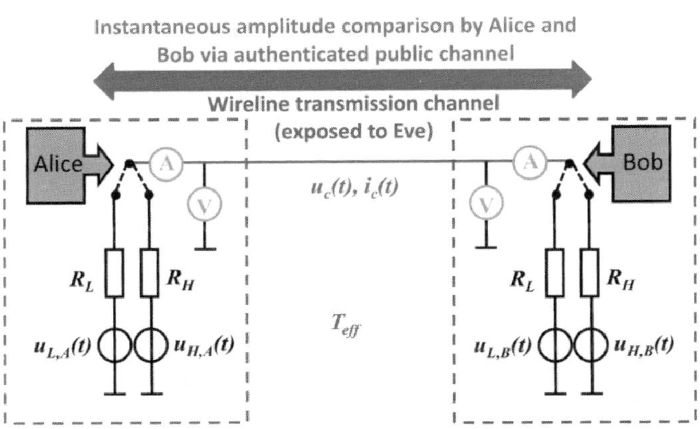


Fig. 3. KLJN system protection against invasive (active) attacks. Alice and Bob measure the instantaneous channel voltage and current amplitudes and compare them via an authenticated public channel. Alice and Bob learn all the information Eve can have. Additional elements to prevent hacking — such as line filters, line capacitance killer arrangement, etc. — are not shown. The notation is the same as in Fig. 2.

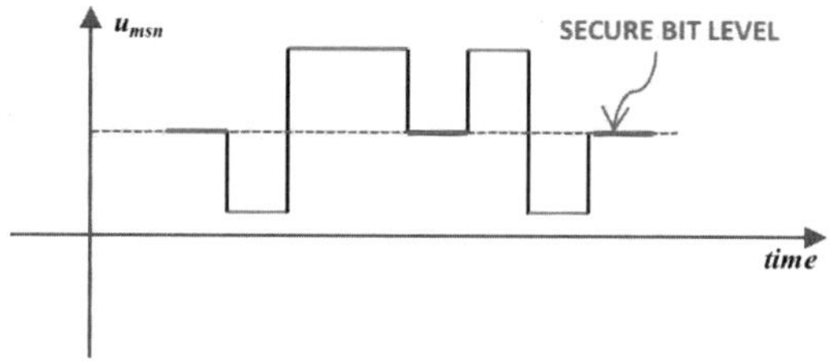


Fig. 4. The secure bit generation in the KLJN scheme. The intermediate mean-square noise level represent the bit situations *LH* or *HL*, that is when a secure bit exchange takes place.

It is important to note that the instantaneous current and voltage data contain all the information related to the key that Eve could have. Thus, it is impossible for Eve to extract key information without letting the system know of such activity. In consequence, Alice and Bob can decide whether to discard the compromised bits according to a previously agreed maximum allowed level of information leak toward Eve [22].

According to the working principle of the KLJN scheme, the secure bit exchange takes place when the resistor states of the two communicators (i.e., the CA and RSD and/or RSKP in vehicular communication networks) are different, i.e., *LH* or *HL*. This is indicated by an intermediate level of the mean-square noise voltage ($u_{\text{msn}}$) on the line, or that of the noise current in the wire [19]. This concept is shown in Fig. 4.

It is important to mention that the two communication parties must previously and publicly agree on which one of them will invert the exchanged bit to have identical keys at the two ends.

## 2. KLJN Secure Key in Vehicular Communication Networks

Based on the working principle of the KLJN scheme [19] and the vehicular communication network model with unconditional secure key exchange proposed in [9], we discuss the generation, donation, and lifetime of the KLJN secure key in vehicular communication networks.

### 2.1. *KLJN key generation in vehicular communication networks*

According to the vehicular communication network model with unconditional secure key distribution proposed in [9], there is a KLJN line connecting the CA to the RSDs and RSKP (see Fig. 5).

The KLJN key generation process is performed as follows:

(i) When a vehicle needs a secure key, it sends a message (via wireless communication) to the closest RSKP with the key request.
(ii) The RSKP will use the extra wire (i.e., the high speed communication line) to inform the CA in charge about the key request.

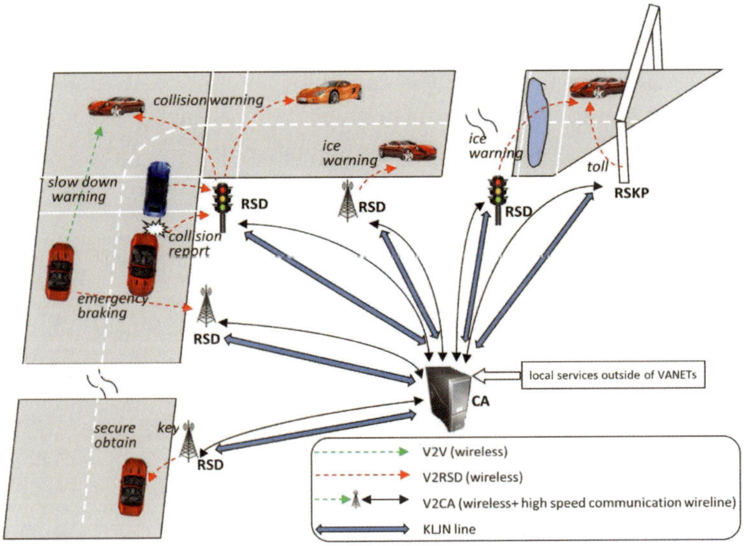


Fig. 5. Vehicular communication networks with unconditional secure key exchange. The network nodes remain the same except for a new node: the roadside key provider (RSKP) and extra wires for KLJN key exchange between the CA and RSD and/or RSKP. The existing wirelines between the RSDs/RSKPs and the CA are kept for high speed communication purposes.

(iii) A key generation process will take place between the RSKP and the CA.
(iv) The RSKP will then provide the cars with the unconditional secure keys by using a near field communication wireless technology [9].
(v) The RSDs also use their KLJN lines that connect them to the CA to generate KLJN keys that are used to secure the communication between RSDs and the CA.

Note that the KLJN line is used only to secretly generate and share the KLJN keys that are going to be used to secure the communication between two nodes. The rest of the communication is done either via wireless communication or using a high speed communication wireline.

## 2.2. *KLJN key donation in vehicular communication networks*

It is important to mention that the RSKP key donation that was proposed in [9], where RSKPs were visualized as gates, might not be as efficient as expected. This is because vehicles would have to slow down in order to get sufficiently close to the RSKPs (as proximity is needed for secure key donation). Therefore, we also propose a lane-by-lane key donation using RSKP equipment embedded in the pavement. In this way, vehicles will not have to slow down to obtain their keys. To detect vehicles in each lane, either loop detectors [29] or high-definition digital wave radars [30] deployed on the side of the roadway can be used. Both the RSKPs and the radar units can be connected to RSDs through a high speed wireline connection. Thus, the KLJN key generation is performed between RSDs and the CA only, while the RSKP will be only in charge of providing the cars with the unconditionally secure KLJN keys. Moreover, this key donation process would be encrypted with the former key, therefore, even if an eavesdropper is listening, he/she would not be able to extract the key information unless he/she has the former key. Figure 6 illustrates this solution.

### 2.3. *Upper limit of the KLJN key lifetime*

The lifetime of the KLJN key in vehicular communication networks is a very important technical parameter that needs to be discussed. This is because the longer the KLJN key is used, the more susceptible it is to attacks. In order to find out the lifetime of the KLJN key in vehicular communication networks, we proceed as follows.

First of all, the noise bandwidth $B_{\text{KLJN}}$ is determined by the distance $L$ between the two communicating parties, which in the case of vehicular communication networks depends on the length of the KLJN line segment between RSDs and the CA. Thus, the following relationship must be satisfied [19] $B_{\text{KLJN}} \ll \frac{c}{L}$, where $c$ is the speed of electromagnetic waves in the wireline. Suppose that $0 < \Theta \ll 1$ and the noise bandwidth is:

$$B_{\text{KLJN}} = \Theta \frac{c}{L}. \tag{1}$$

Also, the duration of the bit sharing period $\tau$ must be long enough compared to the correlation time of the noise $\tau_{\text{KLJN}}$, i.e., $\tau_{\text{KLJN}} \approx \frac{1}{B_{\text{KLJN}}}$, in order to correctly distinguish between the different resistors situations [31, 32]. The frequency of *secure* bit exchange is:

$$f_{\text{sec}} = \frac{1}{2} \frac{B_{\text{KLJN}}}{\gamma}, \tag{2}$$

where $\gamma \gg 1$, see [31, 32] and the factor $\frac{1}{2}$ is due to the fact that a secure bit exchange occurs (on average) 50% of the time.

The lifetime of the KLJN key $\tau_k$ in vehicular communication networks depends on the vehicle density. For the sake of simplicity, first we assume homogenous car

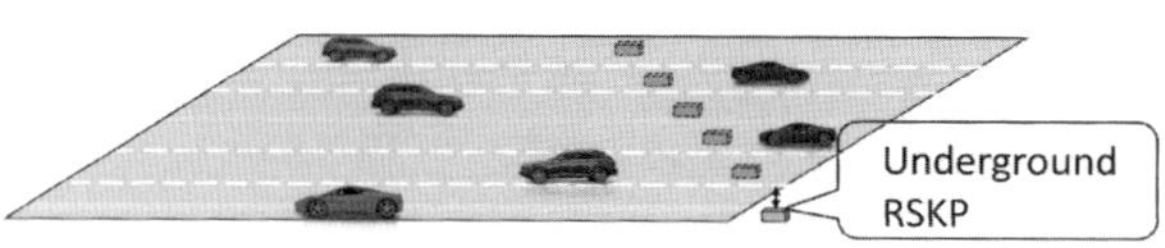


Fig. 6. Key donation with RSKP equipment embedded in the pavement. RSKPs are located underground of each lane.

density:

$$n_c = \frac{N_c}{N_{\mathrm{KLJN}}}, \tag{3}$$

where $N_c$ is the number of cars and $N_{\mathrm{KLJN}}$ is the number of RSDs with KLJN units. Thus, a KLJN unit serves $n_c$ cars. Consequently, the frequency of *secure* bit donation to a single car is:

$$f_c = \frac{f_{\square\mathrm{oc}}}{n_c}. \tag{4}$$

If the length of the KLJN key is defined as $N_k$, then by combining Eqs. (1)–(4), we find that the lifetime of the KLJN key in vehicular communication networks is:

$$\tau_k = \frac{N_k}{f_c} = \frac{2N_k n_c \gamma L}{\Theta c}. \tag{5}$$

Note that this result represents a pessimistic estimation for inhomogeneous vehicular communication networks when $n_c$ is the upper limit of the number of cars any RSD is handling. Thus, Eq. (5) gives an upper limit of the lifetime of the KLJN key in vehicular communication networks. To demonstrate the results, we assign possible practical values to the parameters. Let $L = 1000\,\mathrm{m}, c = 2*10^8\,\mathrm{m/s}, \gamma = 100$ (since $\gamma = \frac{B_{\mathrm{KLJN}}}{f_B}$, where $f_B \approx \frac{1}{\tau}$ should be low enough compared to $B_{\mathrm{KLJN}}$, [30, 31]), $N_k = 100\,\mathrm{bits}$, $n_c = 1000$ vehicles, and $\Theta = 0.1$ (in order to satisfy $B_{\mathrm{KLJN}} \ll \frac{c}{L}$, that is the no-wave limit condition [22]). Then the lifetime of KLJN key is $\tau_k = 10^3\,\mathrm{s}$.

Techniques such as building parallel channels by using chip and multi-wire cables can be used to enhance the speed of the KLJN scheme and to decrease $\tau_k$ [19]. There is also a possibility to increase the security of physically exchanged keys in the case of repeated usage [33].

## 3. Conclusion

In this paper, we reviewed the communication infrastructure and discussed some security-related aspects of vehicular communication networks. We have proposed a KLJN key donation solution for vehicular communication networks. The KLJN key generation in vehicular communication networks has also been discussed and an upper limit for the lifetime of this KLJN key was computed.

## Acknowledgments

X. Cao's contribution is supported by China Scholarship Council. Y. Saez is grateful to IFARHU/SENACYT for supporting her PhD studies at Texas A&M.

## References

[1] P. Papadimitratos, L. Buttyan, T. Holczer, E. Schoch, J. Freudiger, M. Raya, Z. Ma, F. Kargl, A. Kung and J. Hubaux, Secure vehicular communication system: Design and architecture, *IEEE Commun.* **46** (2008) 100–109.

[2] C. T. Li, M. S. Hwang and Y. P. Chu, A secure and efficient communication scheme with authenticated key establishment and privacy preserving for vehicular ad hoc networks, *Comput. Commun.* **31** (2008) 2803–2814.

[3] J. Guo, J. P. Baugh and S. Wang, A group signature based secure and privacy-preserving vehicular communication framework, *Proc. Conf. on Mob. Network. for Vehic. Environ.*, Anchorage, Alaska, United States of America, May (2007), pp. 103–108.

[4] K. D. Kim and P. R. Kumar, An MPC-based approach to provable system-wide safety and liveness of autonomous ground traffic, accepted for publication on, *IEEE Trans. Autom. Control* **59** (2014) 3341–3356, http://ieeexplore.ieee.org/stamp/stamp.jsp?tp=&arnumber=6882808.

[5] M. Raya, P. Papadimitratos and J. P. Hubaux, Securing vehicular communications, *IEEE Wireless Commun.* **13** (2006) 8–15.

[6] M. Raya and J. P. Hubaux, The security of vehicular ad hoc networks, *Proc. 3rd ACM Workshop on Security of Ad Hoc Sensor Networks*, Alexandria, VA, USA, November (2005), pp. 11–21.

[7] P. Papadimitratos, F. La, K. Evenssen, R. Bringnolo and S. Cosenza, Vehicular communication systems: Enabling technologies, applications and future outlook on intelligent transportation, *IEEE Commun.* **47** (2009) 84–95.

[8] P. Ardelean and P. Papadimitratos, Secure and privacy-enhancing vehicular communication: Demonstration of implementation and operation, *Proc. 68th IEEE Conf. on Vehicular Technology*, Calgary, CA, September (2008), pp. 1–2.

[9] Y. Saez, X. Cao, L. B. Kish and G. Pesti, Securing vehicle communication systems by the KLJN key exchange protocol, *Fluct. Noise Lett.* **13** (2014) 1450020.

[10] J. Blum and A. Eskandarian, The threat of intelligent collisions, *IT Professional* **6** (2004) 24–29.

[11] M. Raya and J. P. Hubaux, Securing vehicular ad hoc networks, *J. Comput. Sec.* **15** (2007) 39–68.

[12] N. Wang, Y. Huang and W. Chen, A novel secure communication scheme in vehicular ad hoc networks, *Comput. Commun.* **31** (2008) 2827–2837.

[13] A. Studeret, E. Shi, F. Bai and A. Perrig, TACKing together efficient authentication revocation, and privacy in VANETs, *Proc. 6th Annual Conf. on Sensor, Mesh and Ad Hoc Communications and Networks*, Rome, Italy, June (2009), pp. 1–9.

[14] D. Boneh, X. Boyen and H. Shacham, Short group signatures, *Proc. 24th Annual Int. Cryptology Conference*, Santa Barbara, California, United States of America, August (2004), pp. 41–55.

[15] C. Zhang, R. Lu, X. Lin, P. Ho and X. Shen, An efficient identity-based batch verification scheme for vehicular sensor networks, *Proc. 27th Conf. on Computer Communications*, Phoenix, Arizona, United States of America, April (2008), pp. 13–18.

[16] C. Zhang, X. Lin, R. Lu and P. Ho, RAISE: An efficient RSU-aided message authentication scheme in vehicular communication networks, *Proc. IEEE Int. Conf. on Communications*, Beijing, China, May (2008), pp. 1451–1457.

[17] T. W. Chim, S. M. Yiu, L. C. K. Hui and V. O. K. Li, SPECS: Secure and privacy enhancing communications schemes for VANETs, *Ad Hoc Networks* **9** (2011) 189–203.

[18] Y. Liang, H. V. Poor and S. Shamai, Information theoretic security, *Foundations Trends Commun. Inform. Theory* **5** (2008) 355–580.

[19] L. B. Kish, Totally secure classical communication utilizing Johnson (-like) noise and Kirchhoff's law, *Phys. Lett. A* **352** (2006) 178–182.

[20] C. H. Bennett, G. Brassard, S. Breidbart and S. Wiesner, Quantum cryptography, or unforgeable subway tokens, *Adv. in Cryptol.: Proc. Crypto '82*, Plenum Press, Santa Barbara (1982) 267–275.

[21] R. Mingesz, L. B. Kish, Z. Gingl, C. G. Granqvist, H. Wen, F. Peper, T. Eubanks and G. Schmera, Unconditional security by the laws of classical physics, *Metrol. Meas. Syst.* **20** (2013) 3–16.

[22] L. B. Kish, D. Abbott and C. G. Granqvist, Critical analysis of the Bennett–Riedel attack on secure cryptographic key distributions via the Kirchhoff-law–Johnson-noise scheme, *PLoS ONE* **8** (2013) e81810. Open access: http://www.plosone.org/article/info%3Adoi%2F10.1371%2Fjournal.pone.0081810.

[23] L. B. Kish, Enhanced secure key exchange systems based on the Johnson-Noise scheme, *Metrol. Meas. Syst.* **20** (2013) 191–204.

[24] R. Mingesz, Z. Gingl and L. B. Kish, Johnson (-like) -noise–Kirchhoff-loop based secure classical communicator characteristics, for ranges of two to two thousand kilometers, via model-line, *Phys. Lett. A* **372** (2008) 978–984.

[25] L. B. Kish and C. G. Granqvist, On the security of the Kirchhoff-law–Johnson-noise (KLJN) communicator, *Quantum Inf. Process.* **13** (2014) 2213–2219.

[26] L. B. Kish and R. Mingesz, Totally secure classical networks with multipoint telecloning (teleportation) of classical bits through loops with Johnson-like noise, *Fluct. Noise Lett.* **6** (2006) C9–C21.

[27] E. Gonzalez, L. B. Kish and R. S. Balog, Information theoretically secure, enhanced Johnson noise based key distribution over the smart grid with switched filters, *PLoS ONE* 8 (2013) e70206. Open access: http://www.plosone.org/article/info%3Adoi%2F10.1371%2Fjournal.pone.0070206.

[28] L. B. Kish and O. Saidi, Unconditionally secure computers, algorithms and hardware, such as memories, processors, keyboards, flash and hard drives, *Fluct. Noise Lett.* **8** (2008) L95–L98.

[29] P. Lingenfelser and P. Thilo, Loop detectors for measuring road traffic, *Siemens Rev.* **37** (1970) 332–337.

[30] D. Middleton, H. Charara and R. Longmire, Alternative vehicle detection technologies for traffic signal systems: Technical report, *Research Report FHWA/TX-09/0-5845-1*, Texas Transportation Institute, The Texas A&M University System, College Station, Texas, United States of America, February (2009).

[31] Y. Saez and L. B. Kish, Errors and their mitigation at the Kirchhoff-Law–Johnson-Noise secure key exchange, *PLoS ONE* **8** (2013) e81103. Open access: http://www.plosone.org/article/info%3Adoi%2F10.1371%2Fjournal.pone.0081103.

[32] Y. Saez, L. B. Kish, R. Mingesz, Z. Gingl and C. G. Granqvist, Current and voltage based bit errors and their combined mitigation for the Kirchhoff-law–Johnson-noise secure key exchange, *J. Comput. Electron.* **13** (2014) 271–277.

[33] L. B. Kish, Enhanced usage of keys obtained by physical, unconditionally secure distributions, manuscript: http://arxiv.org/abs/1408.5800 (2014), version 1.

# Author Index